A Specialist Periodical Report

Organic Compounds of Sulphur, Selenium, and Tellurium
Volume 4

A Review of the Literature Published between
April 1974 and March 1976

Senior Reporter
D. R. Hogg, *Department of Chemistry, University of Aberdeen*

Reporters
G. C. Barrett, *Oxford Polytechnic*
R. J. S. Beer, *University of Liverpool*
E. Block, *University of Missouri, U.S.A.*
D. C. Dittmer, *Syracuse University, New York, U.S.A.*
G. L. Dunn, *Smith Kline and French Laboratories, Pennsylvania, U.S.A.*
T. Durst, *University of Ottawa, Ontario, Canada*
U. Eisner, *Trent Polytechnic, Nottingham*
J. G. Gleason, *Smith Kline and French Laboratories, Pennsylvania, U.S.A.*
S. Gronowitz, *University of Lund, Sweden*
M. Haake, *Institut für Pharmazeutische Chemie, Marburg, West Germany*
B. Iddon, *University of Salford*
F. Kurzer, *Royal Free Hospital School of Medicine, London*
P. A. Lowe, *University of Salford*
P. Metzner, *Université de Caen, France*
G. Prota, *Università di Napoli, Italy*
J. Voss, *Universität Hamburg, West Germany*
W. Walter, *Universität Hamburg, West Germany*

The Chemical Society
Burlington House, London, W1V 0BN

ISBN: 0 85186 289 6

ISSN: 0305-9812

Library of Congress Catalog Card No. 77-23818

Organic formulae composed by Wright's Symbolset method

Printed in Great Britain by John Wright and Sons Ltd. at The Stonebridge Press, Bristol BS4 5NU

5011/8/78NF

Preface

The general aims of this fourth Volume of 'Organic Compounds of Sulphur, Selenium, and Tellurium' remain those set out in the Preface to Volume 1.

Change of personnel has led to the division of the wide-ranging chapters on thiocarbonyl compounds and on ylides into more manageable parts. In keeping with more recent work, the short sections on sulphines and sulphenes have been confined to the chapter on thiocarbonyl compounds. To avoid possible inconsistencies, the chemistry of sulphonyl, sulphinyl, and sulphenyl carbanions has been largely included in the chapter on ylides. The order of the chapters has been revised so that the chemistry of acyclic sulphur compounds and sulphur-containing functional groups is covered before cyclic systems.

In general, these Reports cover the period from April 1974 to March 1976 unless otherwise stated. Due to necessary limitations of space, the topics have, in parts, been treated more summarily than in previous Reports. A review of the chemistry of thiepins and dithiins, omitted from Volume 3, has been included in this volume, but accounts of the chemistry of thiazepines and of heterocyclic compounds of quadricovalent sulphur have been held over to Volume 5.

As was stated in earlier volumes, constructive criticism, comments, and advice will be welcomed, particularly if they enable subsequent Reports to be orientated more exactly with the requirements of research workers.

D. R. H.

Contents

Contents   vii

Chapter 14 Thiadiazoles and Selenadiazoles   417
By F. Kurzer

Abbreviations

The following abbreviations have been used:

u.v.	ultraviolet
i.r.	infrared
n.m.r.	nuclear magnetic resonance
Eu(fod)$_3$	tris[1,1,1,2,2,3,3]heptafluoro-7,7-dimethyl-4,6-octanedionato-europium(III)
Eu(tfn)$_3$	tris[1,1,1,2,2,3,3,7,7,8,8,9,9,9]tetradecafluoro-4,6-nonanedionato-europium(III)
o.r.d.	optical rotatory dispersion
c.d.	circular dichroism
e.s.r.	electron spin resonance
dopa	3,4-dihydroxy-L-phenylalanine
g.l.c.	gas–liquid chromatography
NBS	N-bromosuccinimide
CNDO	complete neglect of differential overlap
DABCO	diazabicyclo-octane
DCCI	dicyclohexylcarbodi-imide
DME	dimethoxyethane
DMF	NN-dimethylformamide
DMSO	dimethyl sulphoxide
TMEDA	tetramethylethylenediamine
DMAD	dimethyl acetylenedicarboxylate
TFA	trifluoroacetic acid
HMPA	hexamethylphosphoric triamide
THF	tetrahydrofuran
py	pyridine
Vol. 3, Vol. 2, Vol. 1	Reference back to a preceding Volume of this series of 'Specialist Periodical Reports'

1
Aliphatic Organo-sulphur Compounds, Compounds with Exocyclic Sulphur Functional Groups, and their Selenium and Tellurium Analogues

BY G. C. BARRETT

The literature covering compounds containing organosulphur functional groups is reviewed in this chapter, which is split, as in previous Volumes, into sections dealing with the different classes of organosulphur compounds. Preliminary sections list textbooks and reviews, and spectroscopic and other physical properties, dealt with in an integrated manner. Later chapters in this volume review the literature of organosulphur compounds containing thiocarbonyl functional groups, and compounds enclosing organosulphur functional groups as integral parts of cyclic systems.

1 Textbooks and Reviews

Specialist textbook coverage is becoming less sparse in this area of organic chemistry; the plenary lectures and abstracts of papers presented at the VIth International Symposium on Organic Sulphur Chemistry [1] have been published, and a number of specific topics have been reviewed in monograph form.[2] Thorough coverage of organoselenium [3] and organotellurium chemistry [4] has been published.

Recent developments in organosulphur chemistry have been broadly reviewed,[5] and other more general reviews cover applications of electron spectroscopy [6] and dynamic conformational analysis of organosulphur compounds.[7]

General comments on the problems of handling organotellurium compounds, which are sensitive to oxygen in solution, are given in ref. 138.

[1] 'Organic Sulphur Chemistry', ed. C. J. M. Stirling, Butterworths, London, 1975.
[2] (a) 'Topics in Sulphur Chemistry', ed. A. Senning, Vol. 1, Thieme Verlag, Stuttgart, 1975; (b) 'The Chemistry of the Thiol Group', ed. S. Patai, Wiley–Interscience, London and New York, 1974; (c) Y. M. Torchinskii, 'Sulphydryl and Disulphide Groups of Proteins', Plenum Press, London, 1974; (d) 'Metabolic Pathways. Vol. 7. Metabolism of Sulphur Compounds', ed. D. M. Greenberg, Academic Press, New York, 1975; (e) 'Chemistry and Biochemistry of Thiocyanic Acid and its Derivatives', ed. A. A. Newman, Academic Press, New York, 1975; (f) E. Kühle, 'The Chemistry of the Sulphenic Acids', Thieme Verlag, Stuttgart, 1973; (g) D. Martin and H. G. Hauthal, 'Dimethyl Sulphoxide', Van Nostrand, London, 1975.
[3] K. J. Irgolic and M. V. Kudchadker, in 'Selenium', ed. R. A. Zingaro and W. C. Cooper, Van Nostrand–Reinhold, New York, 1974, p. 408.
[4] K. J. Irgolic, 'Organic Chemistry of Tellurium', Gordon & Breach, New York, 1974; K. J. Irgolic, J. Organometallic Chem., 1975, 103, 91.
[5] P. Neumann and F. Voegtle, Chem.-Ztg., 1974, 98, 138; 1975, 99, 308.
[6] B. J. Lindberg, Internat J. Sulfur Chem., Part C, 1972, 7, 33.
[7] C. H. Bushweller, Internat. J. Sulfur Chem., 1973, 8, 103.

1

Reviews of the following topics have appeared: alkynethiols and thioketens,[8a] addition of SCl_2 to unsaturated systems,[8b] pyridyl sulphides and selenides,[8c] 1-lithiocyclopropyl phenyl sulphide and diphenylsulphonium cyclopropylide as alkylating agents,[8d] sulphuranes,[8e, 8f] α-halogenosulphoxides and related compounds,[8g] allylic sulphoxides and sulphides in organic synthesis,[8h] stereochemistry of sulphoxides,[8i] α-sulphinylcarbanions,[8j] aliphatic diazosulphones,[8k] base-induced rearrangements of sulphones,[8l, 8m] sulphoximides,[8n] stereochemistry of substitution reactions at sulphur,[8o] insertion of SO_2 into $Si-C$[8p] and transition-metal–carbon bonds,[676] 1,3-dipolar cycloaddition reactions of isothiocyanates,[8q] aromatic sulphenic acid derivatives,[8r] and sulphenamides.[8s]

2 Spectroscopic and Other Physical Properties

Like the preceding section, relatively little beyond a statement of the content of each paper is provided for most of the citations dealing with spectroscopic and other physical studies of organosulphur compounds in the following paragraphs. However, all the references include interpretation of the data in terms of conformation, bonding, or insight into reactivity; mere data compilations are excluded.

Spectroscopic methods for the study of thiols, sulphides, and disulphides have been reviewed.[9]

General Conformational Studies and Molecular Orbital Calculations.—Factors favouring the increasing proportion of the *trans,trans*-rotamer through the series divinyl sulphide, sulphoxide, and sulphone have been considered.[10] The molecular mechanics (or force-field) method has been extended to thiols and sulphides,[11] leading to reasonable agreement between predicted conformations and those established by physical studies. Conformational characteristics of sulphoxides suggest repulsion, rather than non-bonded attraction, between sulphoxide oxygen and nearby hydrogen atoms.[12] Attractive forces between sulphur and carbon atoms which have a 1,4-relationship in aliphatic sulphides and disulphides are implied by M.O. calculations of dihedral angle–energy relationships.[13]

[8] (a) R. Mayer and H. Kroeber, *Z. Chem.*, 1975, **15**, 91; (b) M. Mühlstädt and D. Martinetz, *Z. Chem.*, 1974, **14**, 297; (c) H. L. Yale, in 'The Chemistry of Heterocyclic Compounds', Vol. 14, Part 4, 1975, 189; (d) B. M. Trost, *Accounts Chem. Res.*, 1974, **7**, 85; (e) N. E. Hester, *Internat. J. Sulfur Chem.*, 1973, **8**, 119; (f) J. C. Martin and E. F. Perozzi, *Science*, 1976, **191**, 154; (g) C. G. Venier and H. J. Barager, *Org. Prep. Proced. Internat.*, 1974, **6**, 79, *Prepr., Div. Pet. Chem.*, *Amer. Chem. Soc.*, 1974, **19**, 191; (h) D. A. Evans and G. C. Andrews, *Accounts Chem. Res.*, 1974, **7**, 147; (i) L. Van Acker, *Ind. chim. belge*, 1974, **39**, 125; (j) T. Durst and R. Viau, *Intra-Sci. Chem. Reports*, 1973, **7**, 63; (k) A. L. Fridman, Y. S. Andreichikov, V. L. Gein, and S. S. Novikov, *Uspekhi. Khim.*, 1975, **44**, 2284; (l) J. Skarzewski and Z. Skrowaczewska, *Wiad. Chem.*, 1974, **28**, 155; (m) V. N. Drozd, *Internat. J. Sulfur Chem.*, 1973, **8**, 443; (n) P. D. Kennewell and J. B. Taylor, *Chem. Soc. Rev.*, 1975, **4**, 189; (o) D. J. Cram, J. Day, D. C. Garwood, D. R. Rayner, D. M. von Schrilz, T. R. Williams, A. Nudelman, F. G. Yamagishi, R. E. Booms, and M. R. Jones, *Internat. J. Sulfur Chem., Part C*, 1972, **7**, 103; (p) U. Kunze and J. D. Koola, *J. Organometallic Chem.*, 1974, **80**, 281; (q) E. van Loock, *Ind. chim. belge*, 1974, **39**, 661; (r) E. Vinkler and F. Klivenyi, *Internat. J. Sulfur Chem.*, 1973, **8**, 111; (s) F. A. Davis, *Internat. J. Sulfur Chem.*, 1973, **8**, 71.
[9] G. Jung and M. Ottnad, *Chem.-Ztg.*, 1974, **98**, 147.
[10] A. B. Remizov, T. G. Mannafov, and F. R. Tantasheva, *Zhur. obshchei Khim.*, 1975, **45**, 1402.
[11] N. L. Allinger and M. J. Hickey, *J. Amer. Chem. Soc.*, 1975, **97**, 5167.
[12] N. L. Allinger and J. Kao, *Tetrahedron*, 1976, **32**, 529.
[13] H. E. van Wart, L. L. Shipman, and H. A. Scheraga, *J. Phys. Chem.*, 1975, **79**, 1428, 1436.

Stabilization of an α-carbanion by divalent sulphur is the result of the polarizability of the sulphur atom rather than p_π–d_π conjugation;[14, 15] and the same basis is suggested for the enhanced acidity of an α-proton in aliphatic sulphides.[15] π-Donating abilities for various heteroatoms towards an adjacent carbonium ion centre follow the series $P > S \geqslant N > O > Cl > F$,[16] implying (in contrast to current ideas) that a Period 3 element in the system $X-CH_2^+$ is a better π-electron donor than the corresponding Period 2 element.

Ultraviolet Spectra.—Interaction between the chromophores in phenylsulphonyl-guanidines through an empty d-orbital on S, and between the sulphone π-system and the guanidine chromophore, is indicated by u.v. spectra,[17] while related studies show the lack of homoconjugation (through-space conjugation) in aryl benzyl sulphides.[18] U.v. fluorescence and phosphorescence data indicate exciton interaction between the phenyl chromophores in diphenyl sulphide.[19]

Infrared, Raman, and Microwave Spectra.—Interpretation of i.r. data in terms of hydrogen-bonding equilibria has been reported for thiols,[20, 21] sulphides,[22] and $MeSO_2NMe_2$.[23] A diminished H-bonding capacity is demonstrated for a sulphonamide compared with the corresponding carboxamide.[23]

Aliphatic thiols have been subjected to i.r., Raman,[24] and microwave spectroscopic [25] studies. Methyl and ethynyl groups are *gauche* in ethyl vinyl sulphide, as judged from microwave data.[26] I.r./Raman studies of aliphatic sulphides [27] and disulphides [28] similarly include interpretation in terms of conformation or bonding. Comparative i.r./Raman studies of dimethyl sulphide, sulphoxide, and sulphone [29] and the corresponding diphenyl series [30] have been described. I.r. spectra of selenuranes [31] and sulphones [32] and sulphimides [33] have been inter-

[14] A. Streitwieser and J. E. Williams, *J. Amer. Chem. Soc.*, 1975, **97**, 191.
[15] F. Bernardi, I. G. Csizmadia, A. Mangini, H. B. Schlegel, M.-H. Whangbo, and S. Wolfe, *J. Amer. Chem. Soc.*, 1975, **97**, 2209.
[16] F. Bernardi, I. G. Csizmadia, and N. D. Epiotis, *Tetrahedron*, 1975, **31**, 3085.
[17] A. Rastelli, P. G. de Benedetti, A. Albasini, and P. G. Pecorari, *J.C.S. Perkin II*, 1975, 522.
[18] V. Mancini, O. Piovesana, and S. Santini, *Z. Naturforsch.*, 1974, **29b**, 815.
[19] P. G. Russell, *J. Phys. Chem.*, 1975, **79**, 1475.
[20] B. Wladislaw, P. R. Olivato, and R. Rittner, *Rev. Latinoamer. Quim.*, 1974, **5**, 206 (*Chem. Abs.*, 1975, **82**, 85 596).
[21] F. D. Mamedov and M. A. Salimov, *Izvest. Akad. Nauk Azerb. S.S.R., Ser. Fiz. Tekh. Mat. Nauk*, 1973, 104 (*Chem. Abs.*, 1974, **80**, 119 753).
[22] B. A. Trofimov, N. I. Shergina, E. I. Kositsyna, E. P. Vyalykh, S. V. Amosova, N. K. Gusarova, and M. G. Voronkov, *Reakts. spos. org. Soedinenii*, 1973, **10**, 747.
[23] M. Jarra, M. Saastamoinen, and P. O. I. Virtanen, *Finn. Chem. Letters*, 1974, 169.
[24] C. S. Hsu, *Spectroscopy Letters*, 1974, **7**, 439; L. D. Neff and S. C. Kitching, *J. Phys. Chem.*, 1974, **78**, 1648; D. Bhaumik, *Indian J. Phys.*, 1974, **48**, 324; J. N. Som and D. K. Mukherjee, *J. Mol. Structure*, 1975, **26**, 120.
[25] M. Hayashi, H. Imaishi, and K. Kuwada, *Bull. Chem. Soc. Japan*, 1974, **47**, 2382; M. Hayashi, J. Nakagawa, and K. Kuwada, *Chem. Letters*, 1975, 1267; A. M. Mirri, F. Scappini, and H. Maeder, *J. Mol. Spectroscopy*, 1975, **57**, 264; R. E. Schmidt and C. R. Quade, *J. Chem. Phys.*, 1975, **62**, 3864; J. H. Griffiths and J. E. Boggs, *J. Mol. Spectroscopy*, 1975, **56**, 257.
[26] A. Bjorseth, *J. Mol. Structure*, 1974, **23**, 1.
[27] M. Ohsaku, *Bull. Chem. Soc. Japan*, 1975, **48**, 707; *Spectrochim. Acta*, 1975, **31A**, 1271; N. Nogami, H. Sugeta, and T. Miyazawa, *Bull. Chem. Soc. Japan*, 1975, **48**, 2417, 3573.
[28] H. Sugeta, *Spectrochim. Acta*, 1975, **31A**, 1729; H. E. van Wart, F. Cardinaux, and H. A. Scheraga, *J. Phys. Chem.*, 1976, **80**, 625.
[29] E. I. Gritsaev, S. V. Dozmorov, and Y. G. Slizhov, *Zhur. fiz. Khim.*, 1975, **49**, 3010.
[30] B. Nagel, T. Steiger, J. Fruwert, and G. Geiseler, *Spectrochim. Acta*, 1975, **31A**, 255.
[31] V. Horn and R. Paetzold, *Spectrochim. Acta*, 1974, **30A**, 1489.
[32] K. Dathe and K. Doerffet, *J. prakt. Chem.*, 1974, **316**, 621; **317**, 757.
[33] J. T. Shah, *Canad. J. Chem.*, 1975, **53**, 2381.

preted in terms of deformation modes of these functional groups. I.r. spectroscopic studies with a similar objective have been reported for isothiocyanato-esters,[34] ethyl thiocyanate,[35] selenocyanates,[36] methanesulphenyl chloride,[37] diaryltellurium dihalides,[38] *N*-alkyl sulphinamides,[39] anilinesulphonic acids, their sodium salts, and deuteriated analogues,[40] bis(aryl) sulphonimides,[41] methanesulphonyl chloride,[42] and methanesulphonamide.[43]

Nuclear Magnetic Resonance Spectra.—A broad study of ^{77}Se n.m.r. chemical shifts in selenides, diselenides, and selenocyanates has been published.[44] ^{13}C N.m.r. data have been collected for 2-adamantanethiol,[45] alkyl vinyl sulphides,[46] and ring-substituted methyl phenyl sulphides, sulphoxides, and sulphones.[47]

Eu(fod)$_3$ forms weak complexes with 2-adamantanethiol,[45] and with sulphides and disulphides.[48] Larger shifts for sulphides and disulphides are induced by Eu(tfn)$_3$, in which the Lewis acidity of the lanthanide atom is enhanced by the larger degree of fluorination of the organic moiety.[48] The chemical shift non-equivalence observed in isopropyl methanesulphinate is larger than that of the analogous sulphoxide, and is enhanced further by complex formation with benzene.[49]

2-Substituted 2-fluoroethyl phenyl sulphones, RCHF·CH$_2$·SO$_2$Ph (R = SPh, SOPh, SO$_2$Ph, or Br), exist predominantly in the conformation in which the most bulky groups are *anti*-periplanar.[50] N.m.r. spectra give more reliable conformational information for di-4-pyridyl disulphide than that deduced from Kerr-effect data.[51] Full details have now been published [52] on the empirical relationship between ΔJ_{gem} and the relative orientation of a —CH$_2$— group attached to O, S, SO, or SO$_2$. The chemical shifts of the CH$_2$ protons in *para*-substituted phenyl alkyl and phenylsulphonyl alkyl sulphones correlate well with Hammett substituent constants, indicating the transmission of electronic effects through the sulphone grouping.[53] Analogously, the MeSO$_2$ substituent is shown [54] to possess a lower mesomeric electron-accepting power than a NO$_2$ group when substituted in the

[34] L. Floch and S. Kovac, *Coll. Czech. Chem. Comm.*, 1975, **40**, 2845.
[35] A. Bjorseth, *Acta Chem. Scand.*, 1974, **28**, 113.
[36] W. J. Franklin, R. L. Werner, and R. A. Ashby, *Spectrochim. Acta*, 1974, **30A**, 387.
[37] F. Winther, A. Guarnieri, and O. F. Nielsen, *Spectrochim. Acta*, 1975, **31A**, 689.
[38] N. S. Dance and W. R. McWhinnie, *J.C.S. Dalton*, 1975, 43.
[39] A. Kolbe and E. Wenschuh, *J. Mol. Structure*, 1975, **28**, 359.
[40] W. H. Evans, *Spectrochim. Acta*, 1974, **30A**, 543.
[41] Y. Tanaka and Y. Tanaka, *Chem. and Pharm. Bull. (Japan)*, 1974, **22**, 2546.
[42] K. Hanai, T. Okuda, and K. Machida, *Spectrochim. Acta*, 1975, **31A**, 1227.
[43] K. Hanai, T. Okuda, T. Uno, and K. Machida, *Spectrochim. Acta*, 1975, **31A**, 1217.
[44] A. Fredga, S. Gronowitz, and A. B. Hornfeldt, *Chem. Scripta*, 1975, **8**, 15.
[45] H. Duddeck and W. Dietrich, *Tetrahedron Letters*, 1975, 2925.
[46] B. A. Trofimov, G. A. Kalabin, V. M. Bzhesovskii, N. K. Gusarova, D. F. Kushnarev, and S. V. Amosova, *Reakts. spos. org. Soedinenii*, 1974, **11**, 367; G. A. Kalabin, B. A. Trofimov, V. M. Bzhezovskii, D. F. Kushnasev, S. V. Amosova, N. K. Gusarova, and M. L. Alpert, *Izvest. Akad. Nauk. S.S.S.R.*, *Ser. khim.*, 1975, 576.
[47] G. W. Buchanan, C. Reyes-Zamora, and D. E. Clarke, *Canad. J. Chem.*, 1974, **52**, 3895.
[48] T. C. Morrill, R. A. Clark, D. Bilobran, and D. S. Youngs, *Tetrahedron Letters*, 1975, 397.
[49] R. V. Norton and I. B. Douglass, *Org. Magn. Resonance*, 1974, **6**, 89.
[50] G. Marchese, F. Naso, D. Santo, and O. Sciacovelli, *J.C.S. Perkin II*, 1975, 1100.
[51] A. Forchioni and G. C. Pappalardo, *Spectrochim. Acta*, 1975, **31A**, 1367.
[52] R. Davies and J. Hudec, *J.C.S. Perkin II*, 1975, 1395.
[53] E. V. Konovalov, R. G. Dubenko, T. Y. Lavrenyuk, V. M. Neplynev, and P. S. Pelkis, *Zhur. org. Khim.*, 1974, **10**, 427.
[54] D. D. MacNicol, *Tetrahedron Letters*, 1975, 2599.

4-position of *NN*-dimethyl-2-nitro-aniline. N.m.r. provides a particularly direct technique for the study of the kinetics of proton exchange, exemplified in studies of arenethiols in AcOH.[55]

Both ^{1}H and ^{19}F n.m.r. data have been employed in the assignment of 2,5-difluorophenyl methyl sulphide as the structure of the product of reaction between MeS^{-} Na^{+} and 1,2,5-trifluorobenzene.[56]

Mass Spectra.—Organosulphur compounds are well represented in mass spectrometric studies, and the various sulphur functional groups promote individualistic rearrangement behaviour in certain molecular ions. While mercaptopyridines,[57] dialkyl sulphoxides,[58] and dithioacetals and xanthates [59] undergo unspectacular fragmentation (although the formation of sulphenium ions in the latter classes [59] is notable), rearrangements of (arylsulphonyl)methyl arenesulphonates to α-disulphones with extrusion of HCHO,[60] of aryl propargyl sulphones to the SO$_2$-extrusion product,[61] of methyl- or benzyl-sulphonylmethyl methyl sulphones to analogous Ramberg–Bäcklund rearrangement products,[62] and O to S migration in *N*-substituted sulphoximides under electron impact [63] illustrate more involved processes. A further example,[64] the rearrangement of thiolsulphonates to sulphenylsulphinates after electron loss, proceeds through an unprecedented [2,2,1]-bicyclic transition state.

Detailed studies of mass spectrometric behaviour of aryl methyl and aryl chloromethyl sulphones,[65] chlorosulphonylmethyl methyl sulphones,[66] and α-diazosulphones [67] have been reported. Broad studies of mass spectrometric behaviour of di- and tri-sulphides [64] and α-disulphones [64] are described in a key reference for this area. A limited study of di-organotellurium dicarboxylates has been reported.[68] The α-cleavage fragmentation path is followed by the molecular ion from methyl alkanesulphinates, resulting in parent ions of low abundance and the hydrocarbon ion as the base peak.[69]

Modified mass spectrometric techniques are illustrated in chemical ionization studies of methanethiol [70] and in field desorption studies of sulphonic acids and esters.[71]

[55] V. K. Pogorelyi, V. P. Prokopenko, and I. P. Gragerov, *Teor. i eksp. Khim.*, 1974, **10**, 399.
[56] G. Haegele, J. Richter, and M. E. Peach, *Z. Naturforsch.*, 1974, **29b**, 619.
[57] A. Maquestiau, Y. van Haverbeke, C. de Meyer, A. R. Katritzky, and J. Frank, *Bull. Soc. chim. belges*, 1975, **84**, 465.
[58] P. Potzinger, H. U. Stracke, W. Kuepper, and K. Gollnick, *Z. Naturforsch.*, 1975, **30a**, 340.
[59] D. Schumann, E. Frese, K. Praefcke, and W. Knoefel, *Org. Mass. Spectrometry*, 1975, **10**, 527.
[60] T. Graafland, J. B. F. N. Engberts, and W. D. Weringa, *Org. Mass Spectrometry*, 1975, **10**, 33.
[61] D. K. Bates and B. S. Thyagarajan, *Internat. J. Sulfur Chem.*, 1973, **8**, 57.
[62] R. F. Langler, W. S. Mantle, and M. J. Newman, *Org. Mass Spectrometry*, 1975, **10**, 1135; W. R. Hardstaff and R. F. Langler, *ibid.*, p. 215.
[63] C. P. Whittle, C. G. MacDonald, and G. F. Katekar, *Org. Mass. Spectrometry*, 1974, **9**, 422.
[64] E. Block, M. D. Bentley, F. A. Davis, I. B. Douglass, and J. A. Lacadie, *J. Org. Chem.*, 1975, **40**, 2770.
[65] I. Pratanata, L. R. Williams, and R. N. Williams, *Org. Mass Spectrometry*, 1974, **8**, 147; **9**, 418.
[66] W. R. Hardstaff, R. F. Langler, and M. J. Newman, *Org. Mass Spectrometry*, 1974, **9**, 1156.
[67] O. Luinenberg, J. B. F. N. Engberts, and W. D. Weringa, *Org. Mass Spectrometry*, 1974, **9**, 837.
[68] T. J. Adley, B. C. Pant, and R. T. Rye, *Canad. J. Spectroscopy*, 1975, **20**, 71.
[69] W. G. Filby, R. D. Penzhorn, and L. Stieglitz, *Org. Mass Spectrometry*, 1974, **8**, 409.
[70] B. O. Jonsson and J. Lind, *J.C.S. Faraday II*, 1974, **70**, 1399.
[71] H. R. Schulten and D. Kuemmler, *Z. analyt. Chem.*, 1976, **278**, 13.

Photoelectron Spectra.—The practitioners of this technique are now moving from the simplest representative compounds to bifunctional series, providing useful information on preferred conformations. Studies of methanethiol [70] and dimercaptomethane [72] have been described, and suggest that the latter compound preferentially adopts a non-planar conformation in the gas phase. 2,4-Dithiapentane, MeSCH$_2$SMe, has been shown to be planar.[72]

Fundamental information on bonding is provided by photoelectron spectroscopy of thioanisoles [73] and other alkyl aryl sulphides.[74] Two conformers are predominant for these compounds, RSPh, and the proportion of the conformer with maximum p–π overlap decreases through the series R = H, Me, Et, Pr^i, or Bu^t.[74] Related studies for sulphoxides provide correlation diagrams for the effects of substituents R in RS(O)R, as well as information on preferred conformations.[75] An orienting survey of hexavalent sulphur functional groups, *viz.* alkyl, vinyl, and aryl sulphones, SS-dialkyl sulphimides, sulphuryl halides, sulphoximides, and sulphurdi-imides, is available.[76]

Electron Diffraction.—Bond-length and bond-angle data have become available for trifluoromethanethiol,[77] phenyl vinyl sulphide,[78] dimethyl disulphide,[79] ethyl methyl disulphide,[79] dimethyl sulphurdi-imide,[80] and benzenesulphonyl chloride.[81]

Dipole-moment and Kerr-effect Studies.—Information on preferred conformations is provided by these techniques, although only a relatively limited precision is possible in estimates of molecular geometry.

Aromatic disulphides and diselenides adopt a conformation in which the torsion angle between the C—S—S— and —S—S—C'— planes lies between 70 and 90°.[82–84] Related studies of aliphatic disulphides have also been undertaken.[83] Comparisons of S, Se, and O analogues represented by ArXMe,[85] (Ar = Ph or *meta*-substituted phenyl) and aryl diarylmethyl sulphides [86] have been made, in terms of their conformational behaviour. Studies of diaryl ditellurides ArTeTeAr and phenyl tellurobenzoates [87] PhCOTeR provide dipole-moment data consistent with a non-planar C_2 conformation and planar Z conformation,

[72] C. Guimon, M. F. Guimon, and G. Pfister-Guillouzo, *Tetrahedron Letters*, 1975, 1413.

[73] F. Bernardi, G. Distefano, A. Mangini, S. Pignataro, and G. Spunta, *J. Electron Spectroscopy Related Phenomena*, 1975, **7**, 457.

[74] P. S. Dewar, E. Ernstbrunner, J. R. Gilmore, M. Godfrey, and J. M. Mellor, *Tetrahedron*, 1974, **30**, 2455.

[75] H. Bock and B. Solouki, *Chem. Ber.*, 1974, **107**, 2299.

[76] B. Solouki, H. Bock, and R. Appel, *Chem. Ber.*, 1975, **108**, 897.

[77] C. J. Marsden, *J. Mol. Structure*, 1974, **21**, 168.

[78] N. M. Zaripov and T. G. Mannafov, *Zhur. strukt. Khim.*, 1974, **21**, 168.

[79] A. Yokozeki and S. H. Bauer, *J. Phys. Chem.*, 1976, **80**, 618.

[80] J. Kuyper, P. H. Isselmann, F. C. Mijlhoff, A. Speltos, and G. Renes, *J. Mol. Structure*, 1975, **29**, 247.

[81] J. Brunvoll and J. Hargittai, *J. Mol. Structure*, 1976, **30**, 361.

[82] L. A. Sorokina, L. M. Kataeva, and A. B. Remizov, *Zhur. fiz. Khim.*, 1974, **48**, 1559; S. G. Vulfson, L. A. Sorokina, L. M. Kataeva, A. N. Vereshchagin, and E. G. Kataev, *Doklady Akad. Nauk S.S.S.R.*, 1974, **219**, 1363; S. G. Vulfson, L. A. Sorokina, L. M. Kataeva, and A. N. Vereshchagin, *ibid.*, **216**, 837.

[83] M. J. Aroney, S. W. Filipczuk, and D. V. Radford, *J.C.S. Perkin II*, 1975, 695.

[84] G. C. Pappalardo and S. Gruttadauria, *Gazzetta*, 1975, **105**, 427.

[85] G. Y. Cheryukanova, G. A. Chmutova, and A. N. Vereshchagin, *Zhur. fiz. Khim.*, 1975, **49**, 234.

[86] S. Sorriso, G. Reichenbach, S. Santini, and A. Ceccon, *J.C.S. Perkin II*, 1974, 1588.

[87] V. Jehlicka, J. L. Piette, and O. Exner, *Coll. Czech. Chem. Comm.*, 1974, **39**, 1577.

respectively. The series RXXR and RCOXR (X = O, S, Se, or Te) adopt the non-planar and planar conformations respectively, and in this connection the Group VI atom exerts an insignificant effect. Routine collection of data for diphenyl sulphone,[83] dialkyl sulphoxides,[88] substituted benzenesulphonamides,[89] and aryl selenocyanates [90] has been undertaken. The additional information available from dipole-moment data is illustrated in the last-mentioned study, which confirms the electron-withdrawing property of the —SeCN grouping as a substituent on an aromatic ring system.[90]

Optical Rotatory Dispersion and Circular Dichroism Spectra.—The chirospectroscopic properties of derivatives of optically active thiols have been surveyed.[91] This provides a valuable background for the assignment of absolute configuration to chiral thiols by o.r.d. and c.d. methods, requiring the thiol to be converted into a sulphide, disulphide, Δ^2-thiazoline, or *S*-acyl derivative, and exploiting the Cotton-effect behaviour of these classes of organosulphur compound.[91] Continuing interest in the o.r.d. behaviour of chiral disulphides of natural origin, particularly L-cystine and its derivatives, is confirmed by detailed calculations of the Cotton-effect parameters associated with various conformations of the disulphide chromophore in L-cystine.[92]

Other Physical Properties.—Conformational information for thiols and sulphides can be obtained from enthalpy of formation data.[93] Proton-transfer of benzylmercaptan (protonation at sulphur with CF_3SO_3H, and de-protonation by imidazole) has been studied.[94] Thermodynamic parameters for acidity constants of substituted benzenethiols,[95] and for hydrogen-bonding interactions between alkanols and di-n-octyl sulphide and the analogous ether and *N*-methylamine,[96] reveal a dominant electronic influence of the *ortho*-substituent,[95] and substantially higher hydrogen-bond acceptor ability for the aliphatic sulphide than is generally assumed.[96]

3 Thiols

Preparation.—Standard procedures developed in recent years have been used for the synthesis of 2-mercaptophenol from catechol,[97] and 3,5-dimethoxybenzenethiol from 3,5-dimethoxyphenol [98] by way of the thionocarbamate–thiolcarbamate rearrangement (ArOH → ArOCSNMe$_2$ → ArSCONMe$_2$ → ArSH). Further development of 'sulphurated sodium borohydride' (NeBH$_2$S$_3$) as a reagent for the preparation of thiols from benzyl halides, giving high yields of both α-toluenethiols and benzyl polysulphides,[99] has been reported.

[88] D. M. Petkovic and M. S. Pavlovic, *Z. phys. Chem. (Frankfurt)*, 1974, **88**, 54.
[89] P. Ruostesuo, *Finn. Chem. Letters*, 1975, 191.
[90] G. A. Chmutova, L. A. Sorokina, L. M. Kataeva, and N. S. Podkovyrina, *Zhur. fiz. Khim.*, 1974, **48**, 282.
[91] C. Toniolo, *Internat. J. Sulfur Chem.*, 1973, **8**, 89.
[92] R. W. Strickland, J. Webb, and F. S. Richardson, *Biopolymers*, 1974, **13**, 1269.
[93] J. W. H. Kao and A. Chung-Phillips, *J. Chem. Phys.*, 1975, **63**, 4152.
[94] J. J. Delpuech and D. Nicole, *J.C.S. Perkin II*, 1974, 1025.
[95] P. de Maria, A. Fini, and F. M. Hall, *J.C.S. Perkin II*, 1974, 1443.
[96] H. L. Liao and D. E. Martire, *J. Amer. Chem. Soc.*, 1974, **96**, 2058.
[97] R. M. Dodson and J. B. Hanson, *J.C.S. Chem. Comm.*, 1975, 926.
[98] H. Wolfers, U. Kraatz, and F. Korte, *Synthesis*, 1975, 43.
[99] J.-R. Brindle and J.-L. Liard, *Canad. J. Chem.*, 1975, **53**, 1480.

While thiones give sulphides through thiophilic addition of alkyl- or aryl-lithiums (see also Chapter 3, p. 131), di-t-butyl thioketone reacts through an alternative reaction path [100] (Scheme 1), leading to hindered thiols. β-Phenylethyl

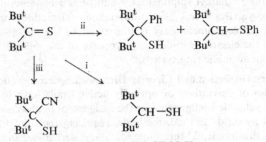

Reagents: i, Bu^nLi or NaOEt; ii, PhLi; iii, $NaCN–CF_3SO_3H$

Scheme 1

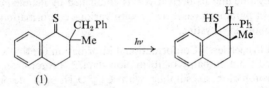

(1)

thiones, *e.g.* (1), give cyclopropanethiols on irradiation with light of wavelength longer than *ca.* 460 nm.[101]

Long-established methods for synthesis of thiols for which modified procedures have been worked out include the use of $Na_2^{35}SSO_3$ for the synthesis of ^{35}S-labelled thiols from alkyl bromides,[102] the use of $Na_2S + S$ in DMF for synthesis of dithiols from dihalides or disulphonates,[103] and the analogous use of xanthates for the synthesis of optically active thiols from toluene-*p*-sulphonates under mild conditions.[104] Modifications of known methods were also used in the syntheses of glucopyranoside-6-thiols [105a] and 1,3,4,6-tetrathio-D-iditol.[105b] Related substitution processes have been illustrated in the recent literature, leading to 2,3-dimercaptopropanoic acid and the corresponding alcohol,[106] and

(2)   (3)

[100] A. Ohno, K. Nakamura, and S. Oka, *Chem. Letters*, 1975, 983; A. Ohno, K. Nakamura, M. Uohama, S. Oka, T. Yamabe, and S. Nagata, *Bull. Chem. Soc. Japan*, 1975, **48**, 3718.

[101] A. Couture, M. Hoshino, and P. de Mayo, *J.C.S. Chem. Comm.*, 1976, 131.

[102] G. A. Bagiyan, S. A. Grachev, I. K. Koroleva, and N. V. Soroka, *Zhur. org. Khim.*, 1975, **11**, 900.

[103] E. L. Eliel, V. S. Rao, S. Smith, and R. O. Hutchins, *J. Org. Chem.*, 1975, **40**, 524.

[104] E. Beretta, M. Cinquini, S. Colonna, and R. Fornasier, *Synthesis*, 1974, 425.

[105] (a) D. Trimnell, E. I. Stout, W. M. Doane, C. R. Russell, V. Beringer, M. Saul, and G. van Gessel, *J. Org. Chem.*, 1975, **40**, 1337; (b) G. E. McCasland, A. B. Zanlungo, and L. J. Durham, *ibid.*, 1976, **41**, 1125.

[106] L. N. Owen and M. B. Rahman, *J.C.S. Perkin I*, 1974, 2413.

to mono- and di-thiosquarate dianions (2) and (3) from diethoxy- or bis(alkyl-amino)-precursors,[107] and to the dithiocyclobutenedithione dianion analogue.[108] A novel synthesis of thiols from α-alkylamino-nitriles [109] by reaction with H_2S probably proceeds by way of the *gem*-dithiol:

$$R^1R^2CH(NHR^3)CN \rightarrow R^1R^2CHSH$$

Ring-opening reactions leading to thiols involve 6-nitrobenzothiazole with MeO^-,[110, 810a] 2-nitrothiophen with secondary amines,[810b] 2-alkylthiothiazolium salts with OH^-,[111] 2-aminothiazole and 2-aminothiazoline with $[PtCl_4]^{2-}$ or $[Pd_2Cl_6]^{2-}$,[112] and mixtures of thiols (4)—(6) from penicillin derivatives by

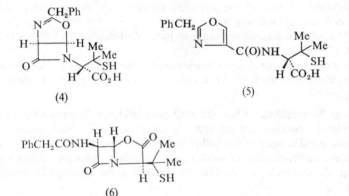

(4) (5)

(6)

reaction with mercury(II) acetate followed by treatment with H_2S.[113] More conventional ring-opening procedures involve the reactions of oxirans with xanthates [114] and of thiirans with RSH.[115] A route to 1-amino-2-methylpropane-2-thiol has been re-worked, so as to provide authentic material.[116]

Ethane-1,2-diselenol, $HSeCH_2CH_2SeH$, is accessible through reduction of the corresponding di-selenocyanate with H_3PO_2.[117]

Heterocyclic Thiols.—Representative synthetic methods described in the recent literature cover the synthesis of tetrachloropyridine-2-thiol from pentachloro-pyridine *N*-oxide and thiourea,[118] the synthesis of 2-phenylthiazole-5-thiols from the *S*-acetyl compounds,[119] and syntheses of 4-phenyl-1,2-dithiolan-3-

[107] D. Coucouvanis, F. J. Hollander, R. West, and D. Eggerding, *J. Amer. Chem. Soc.*, 1974, **96**, 3006.
[108] G. Seitz, K. Mann, R. Schmiedel, and R. Matusch, *Chem.-Ztg.*, 1975, **99**, 90.
[109] R. Crossley and A. C. W. Curran, *J.C.S. Perkin I*, 1974, 2327.
[110] G. Bartoli, F. Ciminale, and P. E. Tedesco, *J.C.S. Perkin II*, 1975, 1472.
[111] Y. Gelernt and P. Sykes, *J.C.S. Perkin I*, 1974, 2610.
[112] J. Dehand and J. Jordanov, *J.C.S. Chem. Comm.*, 1975, 743.
[113] R. J. Stoodley and N. S. Watson, *J.C.S. Perkin I*, 1975, 883.
[114] M. E. Ali, N. G. Kardouche, and L. N. Owen, *J.C.S. Perkin I*, 1975, 748.
[115] A. V. Fokin, A. F. Kolomiets, L. S. Rudnitskaya, and V. I. Shevchenko, *Izvest. Akad. Nauk S.S.S.R., Ser. khim.*, 1975, 660.
[116] J. L. Corbin and D. E. Work, *J. Org. Chem.*, 1976, **41**, 489.
[117] M. V. Lakshmikantham, M. P. Cava, and A. F. Garito, *J.C.S. Chem. Comm.*, 1975, 383.
[118] B. Iddon, H. Suschitzky, A. W. Thompson, and E. Ager, *J.C.S. Perkin I*, 1974, 2300.
[119] G. C. Barrett and R. Walker, *Tetrahedron*, 1976, **32**, 583.

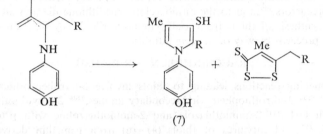

(7)

thione-5-thiol,[120] and of the pyrrole-3-thiols [121] (7), totally unexpectedly, by thiation reactions involving S_8.

(6*H*)-1,3,4-Thiadiazines rearrange to pyrazole-4-thiols with $LiPr^i_2N$ in THF at $-110\ °C$.[122]

The thiol tautomeric form predominates for pyrazole-5-thiols,[123] and for 2-phenylthiazole-5-thiols,[119] in contrast to the behaviour of their oxygen analogues.

Thiols as Nucleophiles.—Citations describing nucleophilic substitution reactions and related processes are grouped together here, and nucleophilic addition reactions are discussed in the following section. Some of the routes leading to sulphides (see Preparation of Sulphides, p. 16) will be relevant reading for those seeking all references in this Chapter describing the nucleophilic reactions of thiols.

Studies of aromatic substitution reactions with thiolates have been described for halogenobenzofurans,[124] polyfluorobenzenes (an interesting generalization is that all but two fluorine substituents can be replaced by reaction with MeS^-),[56, 125] benzenediazonium salts (with $AgSCF_3$),[126] aryl iodides (with $CuSCF_3$),[127] and 3,4,5-trichloropyridine-2,6-dicarbonitrile (all three Cl substituents are replaced by alkyl- or aryl-thiolates).[128] Kinetic studies of substitution reactions of thiolates with chlorobenzenes,[129] halogenonitrobenzenes,[130] and 2-fluoropyridines [131] have been described. An example of reversible addition (8) $\rightleftharpoons$ (9) (X = H or D), characterized by n.m.r., provides a near analogy to the formation of Meisenheimer complexes.[132]

Photostimulated reactions of thiolates with iodobenzenes in liquid NH_3, leading to sulphides in good yields, proceed through $RS^- + ArI \rightarrow R\overset{\cdot}{S}Ar \rightarrow R\cdot +$

[120] J. P. Brown and M. Thompson, *J.C.S. Perkin I*, 1974, 863.
[121] I. R. Gelling and M. Porter, *Tetrahedron Letters*, 1975, 3089.
[122] R. R. Schmidt and H. Huth, *Tetrahedron Letters*, 1975, 33.
[123] J. J. Bergman and B. M. Lynch, *J. Heterocyclic Chem.*, 1974, **11**, 135.
[124] L. Di Nunno, S. Florio, and P. E. Todesco, *Tetrahedron*, 1974, **30**, 863.
[125] M. E. Peach and A. M. Smith, *J. Fluorine Chem.*, 1974, **4**, 341, 399.
[126] N. V. Kondratenko and V. P. Sambur, *Ukrain. khim. Zhur.*, 1975, **41**, 516 (*Chem. Abs.*, 1975, **83**, 58 321).
[127] L. M. Yagupolskii, N. V. Kondratenko, and V. P. Sambur, *Synthesis*, 1975, 721.
[128] C. T. Goralski and T. E. Evans, *J. Org. Chem.*, 1975, **40**, 799.
[129] K. V. Solodova, N. V. Kozhevnikova, and S. M. Shein, *Zhur. org. Khim.*, 1974, **10**, 214.
[130] G. Guanti, C. Dell'Erba, S. Thea, and G. Leandri, *J.C.S. Perkin II*, 1975, 389.
[131] R. A. Abramovitch and A. J. Newman, *J. Org. Chem.*, 1974, **39**, 3692.
[132] C. A. Veracini and F. Pietra, *J.C.S. Chem. Comm.*, 1974, 623.

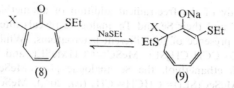

ArS⁻ → mixed sulphides, indicating the operation of an aromatic $S_{RN}1$ reaction.[133]

Conditions have been worked out for the synthesis of methyl polyfluoroalkyl sulphides[134] and 2-bromo-1,1-difluoroethyl phenyl sulphide[135] by halogen-substitution reactions involving sodium thiolates. *p*-Nitrothiophenol reacts with 1,2-dichlorohexafluorocyclopentene to give the 1,2-bis(*p*-nitrophenylthio)-analogue.[136] Aryl alkyl tellurides PhTeR[137] and bis(phenyltelluro)methane PhTeCH₂TePh[138] can be obtained from PhTeLi and an alkyl halide, or di-iodomethane, respectively.

The nucleophilicity of thiols towards an sp^2 ester carbon atom has previously been little studied.[139] A detailed report has appeared showing that rates of reaction of substituted arenethiols with *p*-nitrophenyl acetate can be correlated with pK_a values of the thiols, after substantial solvent effects have been taken into account; this is a key reference for this area.[139]

S_N2-Type processes are involved in the ring-opening reactions of oxirans, under mild conditions (by thiols, adsorbed on alumina),[140] and in the ring-closure of a penicillin-derived enethiolate.[141]

Addition Reactions of Thiols.—Base-catalysed (*i.e.* ionic) addition of thiols R¹SH to acetals R²CH=CXCH(OEt)₂ gives R¹SCR²=CHCH(OEt)₂,[142] while addition of PhCH₂SH to isopropenyl ketones gives PhCH₂SCH₂CMeCOR in partly resolved form when Ca or Ba (−)-amyl oxides are used as bases.[143] 2-Phenyl-thiazole-5-thiols behave as typical heteroaromatic thiols in addition reactions,[119] while their oxygen analogues undergo cycloaddition reactions with the same unsaturated systems. This is a result of the respective dominance of the keto-tautomeric and meso-ionic forms.

Radical-addition processes have been studied using vinylphosphines (radical addition at either C or P),[144] allylhydrazine,[145] α-methylstyrenes,[146] and methyl oleate (C-9 and C-10 adducts are formed in nearly equal amounts).[147]

[133] P. J. Bunnett and X. Creary, *J. Org. Chem.*, 1974, **39**, 3173; 1975, **40**, 3740.
[134] B. Haley, R. N. Haszeldine, B. Hewitson, and A. E. Tipping, *J.C.S. Perkin I*, 1976, 525.
[135] V. A. Korinko, Y. A. Serguchev, and L. M. Yagupolskii, *Zhur. org. Khim.*, 1975, **11**, 1268.
[136] R. F. Stockel, *Canad. J. Chem.*, 1975, **53**, 2302.
[137] K. J. Irgolic, P. J. Busse, R. A. Grigsby, and M. R. Smith, *J. Organometallic Chem.*, 1975, **88**, 175.
[138] D. Seebach and A. K. Beck, *Chem. Ber.*, 1975, **108**, 314.
[139] G. Guanti, C. Dell'Erba, F. Pero, and G. Leandri, *J.C.S. Perkin II*, 1975, 212.
[140] G. H. Posner, D. Z. Rogers, C. M. Kinzig, and G. M. Gurria, *Tetrahedron Letters*, 1975, 3597.
[141] A. G. W. Baxter and R. J. Stoodley, *J.C.S. Chem. Comm.*, 1975, 251.
[142] N. A. Keiko, L. G. Stepanova, G. I. Kirillova, I. D. Kalikhman, and M. G. Voronkov, *Izvest. Akad. Nauk S.S.S.R., Ser. khim.*, 1975, 2274.
[143] K. Yamaguchi and Y. Minoura, *Chem. and Ind.*, 1975, 478.
[144] D. H. Brown, R. J. Cross, and D. Millington, *J.C.S. Dalton*, 1976, 334.
[145] G. Y. Gadzhiev and V. A. Budagov, *Azerb. khim. Zhur.*, 1975, 53.
[146] D. F. Church and G. J. Gleicher, *J. Org. Chem.*, 1975, **40**, 536.
[147] A. Sugii, K. Harada, and K. Kitahara, *Chem. and Pharm. Bull. (Japan)*, 1975, **23**, 2415.

A detailed study of the free-radical addition of alkanethiols to (alkylthio)-vinylacetylenes [148] and their Se and Te analogues reveals interesting differences conferred by the presence of the different heteroatoms. While $MeSC\equiv CCH=CH_2$ gives $MeSC\equiv CCH_2CH_2SEt$, $MeSC\equiv CCHMeSEt$, and $MeSCH=C(SEt)-CH_2CH_2SEt$ with ethanethiol, the Se analogue gives $MeSeCH=C(SEt)CH_2-CH_2SEt$ (40%), $MeSeC(SEt)=CHCH=CH_2$ (*ca.* 20%), $MeSeC\equiv CCH_2CH_2SEt$ (25%), and $MeSeCH=C=CHCH_2SEt$ (11%); the Te analogue, in contrast, suffers oxidation by the t-butyl peroxide used to create the thiyl radicals, or C—Te bond cleavage in the absence of peroxide.[148]

The other studies of thiol addition to alkynes in the recent literature involve the ionic mechanism, the major interest being the stereoisomer ratios of the resulting vinyl sulphides.[149, 150] An interesting study [150] of the addition of 2 equivalents of $PhCH_2SH$ to acetylenedicarboxylic acid in alkaline solution includes an assessment of factors allowing control of the proportions of the resulting racemic and *meso*-1,2-bis(benzylthio)succinic acids.

Addition of C_6F_5SH to various unsaturated systems [151] and the kinetics of addition of propanethiol to benzylideneanilines in non-aqueous solvents [152] provide examples of related reactions.

Further Reactions of Thiols.—Treatment of $PhCH_2SLi$ with Bu^nLi at $-5\,^{\circ}C$ gives the species $(Ph\bar{C}H—S^-)2Li^+$, which is equally well named the thiobenzaldehyde dianion or the dianion of the thiol.[153] The carbanion centre is involved in the reaction of the dianion with electrophiles; *e.g.* benzaldehyde gives $PhCH(SH)CH(OH)Ph$.

1,2-Anionic rearrangement from sulphur to carbon under the influence of Bu^tLi is observed with benzylthiotrimethylsilane, $PhCH_2SSiMe_3 \rightarrow PhCH(SH)SiMe_3$.[154]

A form of anchimeric assistance is involved in the elimination of HCl from 3-chlorobutanethiols (10).[155] Two papers [156, 157] deal with differently based

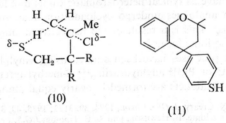

(10)   (11)

[148] S. I. Radchenko, I. G. Sulimov, and A. A. Petrov, *Zhur. org. Khim.*, 1974, **10**, 2456.
[149] M. T. Omar and M. N. Basyouni, *Bull. Chem. Soc. Japan*, 1974, **47**, 2325; L. V. Timokhina, A. S. Nakhmanovich, I. D. Kalikhman, and M. G. Voronkov, *Zhur. org. Khim.*, 1974, **10**, 2468; I. L. Mikhelashvili and E. N. Prilezhaeva, *ibid.*, p. 2524; I. L. Mikhelashvili-Fioliya, V. S. Bogdanov, N. D. Chuvylkin, and E. N. Prilezhaeva, *Izvest. Akad. Nauk S.S.S.R., Ser. khim.*, 1975, 890; H. A. Selling, *Tetrahedron*, 1975, **31**, 2387.
[150] E. Larsson, *Bull. Soc. chim. belges*, 1975, **84**, 697.
[151] T. S. Leong and M. E. Peach, *J. Fluorine Chem.*, 1975, **6**, 145.
[152] Y. Ogata and A. Kawasaki, *J.C.S. Perkin II*, 1975, 134.
[153] D. Seebach and K. H. Geiss, *Angew. Chem.*, 1974, **86**, 202.
[154] A. Wright and R. West, *J. Amer. Chem. Soc.*, 1974, **96**, 3222.
[155] C. A. Grob, B. Schmitz, A. Sutter, and A. H. Weber, *Tetrahedron Letters*, 1975, 3551.
[156] A. D. U. Hardy, D. D. MacNicol, J. J. McKendrick, and D. R. Wilson, *Tetrahedron Letters*, 1975, 4711.
[157] H. Maarse and M. C. Ten Noever De Brauw, *Chem. and Ind.*, 1974, 36.

physical properties of thiols; one describes CCl_4 clathrate formation by the sulphur analogue (11) of Dianin's compound.[156] Commercial t-dodecanethiol has a catty odour, which is detectable over several miles, and which has now been shown [157] to be due to the presence of an impurity, $Pr^iC(SH)Me_2$.

Protection of Thiols in Synthesis.—Commonly used protecting groups for cysteine (*S*-trityl, *S*-diarylmethyl, *S*-acetamidomethyl) are cleanly removed by 2-nitro-benzenesulphenyl chloride in AcOH.[158]

Generation of Thiyl Radicals and Thiolate-stabilized Radicals.—Oxidation of thiols by $Ti^{III}-H_2O_2$ or Ce^{IV} species gives products resulting from successively formed thiyl (RS·), sulphinyl (RSO·), and sulphonyl (RSO$_2$·) radicals.[159] A new outlook is provided by reassignment of e.s.r. spectra, previously interpreted in terms of thiyl radicals, to $RS\dot{-}SR_2$ σ*-radicals, structurally similar to disulphide anion radicals $RS\dot{-}\bar{S}R$.[160, 161] Low-temperature radiolysis of thiols, sulphides, and disulphides definitely involves the formation of thiyl radicals as the initial step;[161, 162] this was shown by spin-trapping with Bu^tNO.[161] Radiolysis of RSH at 77 K gives thiyl radicals,[162] and R_2S gives RS^- and R· through an electron-capture process during radiolysis.[162]

A long-lived alkylselenolate radical $R_nMSe\dot{C}Bu^t_2$ is formed [163] by the reaction of di-t-butyl selenoketone with organometalloid species R_nM· (M is an element of Group IV, V, or VI).

Thiols in Biochemistry.—As in previous volumes of these Reports, no attempt is made at comprehensive coverage in this section. However, there are relatively few topic areas linking to the general title of this section, and it is possible to indicate these with a few key references.

A review of cysteine proteinases [164] gives a perspective to one role of SH groups in proteins. The estimation of the reactivity of this functional group in proteins, as a function of the local charge distribution, is feasible using 5,5′-dithio-bis(2-nitro-*N*-trimethylbenzylammonium iodide) and the *N*-2′-hydroxy-ethylbenzamide as positively charged and neutral analogues, respectively, of Ellman's reagent.[165] Modified procedures for the determination of SH groups in proteins have been reported, employing 9-vinylacridine,[166] *N*-(7-dimethyl-amino-4-methyl-3-coumarinyl)maleimide and its 6-coumarinyl analogue,[167] and *p*-hydroxymercuribenzoate [168] as reagents. Estimation of disulphide groups in peptides and proteins employs a similar approach after first cleaving the disulphide.

[158] A. Fontana, *J.C.S. Chem. Comm.*, 1975, 976.
[159] B. C. Gilbert, H. A. H. Laue, R. O. C. Norman, and R. C. Sealy, *J.C.S. Perkin II*, 1975, 892.
[160] M. C. R. Symons, *J.C.S. Perkin II*, 1974, 1618.
[161] J. A. Wargon and F. Williams, *J.C.S. Chem. Comm.*, 1975, 947.
[162] D. Nelson and M. C. R. Symons, *Chem. Phys. Letters*, 1975, **36**, 340.
[163] J. C. Scaiano and K. U. Ingold, *J.C.S. Chem. Comm.*, 1976, 205.
[164] G. Lowe, *Tetrahedron*, 1976, **32**, 291.
[165] G. Legler, *Biochim. Biophys. Acta*, 1975, **405**, 136.
[166] Y. Nara and K. Tuzimura, *Agric. and Biol. Chem. (Japan)*, 1975, **39**, 7.
[167] M. Machida, N. Ushijima, M. I. Machida, and Y. Kanaoka, *Chem. and Pharm. Bull. (Japan)*, 1975, **23**, 1385.
[168] J. B. Carlsen, *Analyt. Biochem.*, 1975, **64**, 53.

Studies of addition of thiolates to flavans [169-171] showing that addition to a 4a- or 5-position of an isoalloxazine ring may be implicated in flavin catalysis of thiol–disulphide oxidation, are sufficiently broad to include dithiols [170] and arenesulphinic acids.[171]

Of a number of thiols tested,[172] mercaptoethanol was the most effective for the catalysis of the conversion of oleic acid → *trans*-9-octadecenoic acid. The compound (12; $n = 1$) was more effective than homologues ($n = 0, 2,$ or 3) in

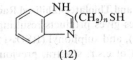

(12)

the catalysed hydrolysis of *p*-nitrophenyl acetate, indicating bifunctional catalysis.[173]

The importance of arene oxides in the metabolism of aromatic substrates is attested by weighty contributions to the literature; the mechanistic aspects of the thiolate–benzene oxide (and thiolate–Malachite Green) reactions have been studied.[174]

Cysteine and cytosine undergo acetone-sensitized photoreaction in water, to give 5,6-dihydrouracil and 5-*S*-cysteinyluracil.[175]

Thiol-acids and Thiol-esters.—Most of the references in the recent literature dealing with these classes of compound from the point of view of new chemistry concern thiol-esters, and apart from a study of the formation of diacyl sulphide using DCCI [176] and reaction with arene oxides [(13) → (14)],[177] the current literature on

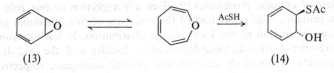

(13)   (14)

thiol-acids concerns mainly well-established reactions *e.g.*[178] homolytic addition of AcSH to methylenecyclohexanes.

Agathosma oils, previously shown to be rich in organosulphur compounds, contain *S*-prenyl thioisobutyrate, $Me_2C=CHCH_2SC(O)CHMe_2$.[179]

Novel synthetic routes to thiol-esters use carboxylic acids and thiols with diethyl phosphorocyanidate or diphenyl phosphorazidate, and Et_3N in DMF,[180]

[169] I. Yokoe and T. C. Bruice, *J. Amer. Chem. Soc.*, 1975, **97**, 450.
[170] E. L. Loechler and T. C. Hollocher, *J. Amer. Chem. Soc.*, 1975, **97**, 3235.
[171] B. R. Brown and M. R. Shaw, *J.C.S. Perkin I*, 1974, 2036.
[172] W. G. Niehaus, *Bio-org. Chem.*, 1974, **3**, 302.
[173] J. Schoenleber and P. Lochon, *Compt. rend.*, 1974, **278**, *C*, 1381.
[174] D. M. E. Reuben and T. C. Bruice, *J. Amer. Chem. Soc.*, 1976, **98**, 114.
[175] N. C. Yang, R. Okazaki, and F.-T. Lu, *J.C.S. Chem. Comm.*, 1974, 462.
[176] M. Mikolajczyk, P. Kielbasinski, and H. M. Schiebel, *J.C.S. Perkin I*, 1976, 564.
[177] A. M. Jeffrey, H. J. C. Yeh, D. M. Jerina, R. M. De Marinis, C. H. Foster, D. E. Piccolo, and G. A. Berchtold, *J. Amer. Chem. Soc.*, 1974, **96**, 6929.
[178] J. C. Richer and C. Lamarre, *Canad. J. Chem.*, 1975, **53**, 3005.
[179] D. E. A. Rivett, *Tetrahedron Letters*, 1974, 1253.
[180] S. Yamada, Y. Yokoyama, and T. Shioiri, *J. Org. Chem.*, 1974, **39**, 3302.

[RCO_2H + RSTl + $(EtO)_2P(O)Cl$];[181] acid chlorides ($RCOCl$ + Bu^tSTl);[182] aldehydes (R^1CHO + R^2SSR^3 → R^1COSR^2 by irradiation);[183] and ketones [R^1COR^2 converted into the tosylhydrazone, $TosNHN=CR^1R^2$ + HCN + EtSH + HCl → $TosNHNHCR^1R^2C(SEt)=NH_2^+$ Cl^- → $R^1R^2CHC(O)SEt$ on pyrolysis];[184] [R^1COR^2 + $R^3R^4CClCOCl$ + HCHO + Bu^n_2BSPh → R^1R^2C-$(OH)CR^3R^4C(O)SPh$].[185] The conventional route to thiol-esters from metal thio-carboxylates has been extended to t-amino-alkyl-Bunte salts as the source of the *S*-substituent (primary and secondary amino-alkyl-Bunte salts, however, yield acylamido-alkyl disulphides).[186]

Cleavage reactions of thiol-esters have been studied from both mechanistic and synthetic points of view. Methanolysis of thiol-esters is promoted by transition-metal ions, particularly by Cu^{2+},[187] while transesterification of *S*-t-butyl thiol-esters with Bu^tOH employs mercury(II) trifluoroacetate as catalyst[182] (thus pro-viding a novel route to t-butyl carboxylate esters). Photolysis of tolyl thio-benzoates involves S—acyl bond cleavage, leading to aldehydes, benzils, and di-tolyl disulphide,[188] while both S—acyl and S—alkyl bond cleavages occur on photolysis of 3-(trimethylsilyl)propyl thioacetate and 2-(trimethylsilyl)ethyl thioacetate.[189] S—Acyl cleavage is involved in the liberation of thiols from thiolacetates with cysteamine,[190] but S—alkyl cleavage is involved in the con-version of a thiazol-5-one–maleic anhydride adduct into an acyclic thiol-ester.[191] The reaction of organocopper(I) complexes R^1_2CuLi with thiol-esters $R^2C(O)SR^3$ to give ketones R^2COR^1 in high yields, with efficient use of the organometallic reagent (no tertiary alcohol is formed through competing carbonyl addition processes),[192] has been put to use in the synthesis of cockroach pheromone, $Me(CH_2)_{17}CO(CH_2)_6OCHMeOEt$.[193]

In rearrangements of glycidic thiol-esters (15) to formylthiolacetates (15a), the thiol-ester group shows a greater migratory aptitude than the carboxylate group in the analogous carboxylate esters.[194] Base-catalysed isomerization of

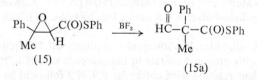

[181] S. Masamune, S. Kamata, J. Diakur, Y. Sugihara, and G. S. Bates, *Canad. J. Chem.*, 1975, **53**, 3693.
[182] S. Masamune, S. Kamata, and W. Schilling, *J. Amer. Chem. Soc.*, 1975, **97**, 3515.
[183] M. Takagi, S. Goto, and T. Matsuda, *J.C.S. Chem. Comm.*, 1976, 92.
[184] S. Cacchi, L. Caglioti, and G. Paolucci, *Synthesis*, 1975, 120.
[185] K. Inomata, T. Kawahara, and T. Mukaiyama, *Chem. Letters*, 1974, 245.
[186] D. L. Klayman and T. S. Woods, *Internat. J. Sulfur Chem.*, 1973, **8**, 5.
[187] C. L. Green, R. P. Houghton, and D. A. Phipps, *J.C.S. Perkin I*, 1974, 2623.
[188] J. Martens and K. Praefcke, *Chem. Ber.*, 1974, **107**, 2319.
[189] T. I. Ito and W. P. Weber, *J. Org. Chem.*, 1974, **39**, 1691.
[190] T. Endo, K. Oda, and T. Mukaiyama, *Chem. Letters*, 1974, 443.
[191] G. C. Barrett and R. Walker, *Tetrahedron*, 1976, **32**, 571.
[192] R. J. Anderson, C. A. Henrick, and L. D. Rosenblum, *J. Amer. Chem. Soc.*, 1974, **96**, 3654.
[193] L. D. Rosenblum, R. J. Anderson, and C. A. Henrick, *Tetrahedron Letters*, 1976, 419.
[194] D. J. Dagli, R. A. Gorski, and J. Wemple, *J. Org. Chem.*, 1975, **40**, 1741.

t-butyl thiolbut-3-enoate to t-butyl thiolcrotonate has been studied as a model for the transfer of an α-methyl proton from thiol-esters.[195]

Very little attention has been paid in the past to base-induced Claisen-type condensation reactions of thiol-esters, and related reactions. With LiPr$^i_2$N at −78 °C, ester enolates are formed from thiol-esters having an α-proton, and β-hydroxy-thiol-esters are readily formed when ketones are added.[196]

The synthesis and properties of thiol-ester *S*-oxides R^{1}C(O)S(O)R^2 have been investigated in a search for new syntheses of sulphenates by *O*-alkylation followed by nucleophilic cleavage [R^{1}C(O)$\overset{+}{S}$(OR^3)R^2 + Nu$^-$ → R^{3}OSR^2 + R^{1}C(O)Nu]. In the event, sulphenates were not formed from thiocarbonate *S*-oxides (R^1 = OEt) nor from N or S analogues,[197] or from derived oxo-sulphonium salts.

4 Sulphides

Preparation.—Methods employing simple sulphur reagents for the synthesis of sulphides continue in regular use. Primary and secondary dialkyl and aryl alkyl sulphides are conveniently prepared from alkyl or aryl halides with sodium sulphide in aqueous solution in the presence of a phase-transfer catalyst, *e.g.* hexadecyltributylphosphonium bromide.[198] Cyclohexene gives di-(2-chloro-cyclohexyl) sulphide with SCl$_2$,[199] and 1,3,5-trithiane reacts with dichlorodi-sulphane, S$_2$Cl$_2$, to give bis(mercaptomethyl) sulphide *via* (−SSCH$_2$SCH$_2$−)$_n$.[200] Highly specific formation of the unsymmetrical sulphide is observed when H$_2$S reacts with a mixture of Cl(CH$_2$)$_m$C(O)NHC(O)NHAr and Cl(CH$_2$)$_n$C(O)NHC-(O)NHAr.[201]

Classical synthetic routes employing metal thiolates and alkyl or aryl halides [198, 202] have been studied in conjunction with phase-transfer catalysts. Phosphorus reagents (RO)$_3$PO and Ph$_2$P(O)OCH$_2$Ph may be used in place of the alkyl halide in this route.[203] Analogous routes, starting from alcohols, proceed *via* ROPPh$_3$$^+$ NMePh$^-$,[204] or *via* (Me$_2$N)$_3$$\overset{+}{P}$(OR) X$^-$.[205] Conversion of MeS$^-$ into Me$_2$S by methylcobalamin is a model for the last step in the biosynthesis of methionine.[206]

Trimethylsilyl sulphides R^{1}SSiMe$_3$ give γ-alkoxy-allyl sulphides R^{1}SCHR^1-CH=CHOMe with propenyl acetals in the presence of AlCl$_3$,[207] and sulphides R^{1}SCH$_2$R^2 by their reaction with aldehydes R^{2}CHO followed by reduction with LiAlH$_4$–AlCl$_3$ of the resulting adduct.[208] Thiols can be converted into unsym-

[195] L. Fodor and P. H. Gray, *J. Amer. Chem. Soc.*, 1976, **98**, 783.

[196] J. Wemple, *Tetrahedron Letters*, 1975, 3255.

[197] D. H. R. Barton, D. P. Manly, and D. A. Widdowson, *J.C.S. Perkin I*, 1975, 1568.

[198] D. Landini and F. Rolla, *Synthesis*, 1974, 565.

[199] N. N. Novitskaya, R. G. Kantynkova, and G. A. Tolstikov, *Zhur. org. Khim.*, 1974, **44**, 2732.

[200] M. Schmidt and L. Endres, *Z. Naturforsch.*, 1975, **30b**, 632.

[201] T. Endo, S. Sato, and T. Mukaiyama, *Tetrahedron Letters*, 1974, 1195.

[202] A. W. Herriott and D. Picker, *Synthesis*, 1975, 447.

[203] P. Savignac and P. Coutret, *Synthesis*, 1974, 818.

[204] Y. Tanigawa, H. Kanamaru, and S. Murahashi, *Tetrahedron Letters*, 1975, 4655.

[205] I. M. Downie, H. Heaney, and G. Kemp, *Angew. Chem.*, 1975, **87**, 357.

[206] G. N. Schrauzer and E. A. Stadlbauer, *Bio-inorg. Chem.*, 1974, **3**, 353.

[207] T. Mukaiyama, T. Takeda, and K. Atsumi, *Chem. Letters*, 1974, 1013.

[208] R. S. Glass, *Synth. Comm.*, 1976, **6**, 47.

metrical sulphides by reaction with dialkyl carbonates in the presence of NaOEt.[209]

The route from thioketone to sulphide has become by now a standard preparative procedure. The addition of a Grignard reagent to thiobenzophenone leads to products of addition to S (*i.e.* Ph_2CHSR), and of double addition to C and S (*i.e.* Ph_2CRSR) as well as tetraphenylthiiran.[210] Analogous results are obtained using BuLi or PhLi, though products from double addition are not formed.[100] Di-t-butyl thioketone reacts anomalously in giving Bu^t_2CHSH with BuLi.[100] Methylation of the cyclobutene-1,2-dithione (16) gives the corresponding bis(methylthio)cyclobutene (17).[211]

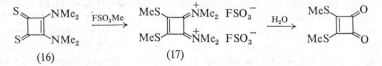

(16)   (17)

Disulphides and sulphenyl chlorides behave as sulphenylating agents, towards alkenes and carbanions in particular, leading to sulphides. A much broader range of substrates can be converted into sulphides by these reagents, as illustrated in recent papers: diaryl disulphides + SbF_5 + polyfluorobenzenes give aryl polyfluorophenyl sulphides;[212] $CF_2=CF_2$ + Me_2S_2 give $MeSCF=CF_2$,[213] ester enolates, $R^1R^2CCO_2R^3$,[214] and α-lithiated aliphatic nitriles [215] give α-sulphenylated products when a disulphide or a sulphenyl halide is used. Addition of 2,4-dinitrobenzenesulphenyl chloride to the allylic alcohol (18) gives the product arising from addition and a 1,2-aryl shift.[216] Benzvalene (19) adds PhSCl at −78 °C.[217]

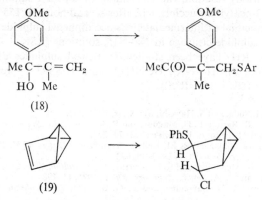

(18)

(19)

209 Y. Tamura, T. Saito, and H. Ishibashi, *Synthesis*, 1975, 641.
210 M. Dagonneau, *Compt. rend.*, 1974, **279**, C, 285; M. Dagonneau and J. Vialle, *Tetrahedron*, 1974, **30**, 3119.
211 G. Seitz, R. Schmiedel, and K. Mann, *Synthesis*, 1974, 578; G. Seitz, H. Morck, K. Mann, and R. Schmiedel, *Chem.-Ztg.*, 1974, **98**, 459.
212 G. G. Furin, L. N. Shchegoleva, and G. G. Yakobson, *Zhur. org. Khim.*, 1975, **11**, 1290.
213 R. N. Haszeldine, B. Hewitson, and A. E. Tipping, *J.C.S. Perkin I*, 1974, 1706.
214 T. J. Brocksom, N. Petragnani, and R. Rodrigues, *J. Org. Chem.*, 1974, **39**, 2114.
215 D. N. Brattesani and C. H. Heathcock, *Tetrahedron Letters*, 1974, 2279.
216 V. R. Kartashov, N. A. Kartashova, and E. V. Skorobogatova, *Zhur. org. Khim.*, 1974, **10**, 171.
217 T. J. Katz and K. C. Nicolaou, *J. Amer. Chem. Soc.*, 1974, **96**, 1948.

Sulphenylation with a sulphenamide occurs in the formation of sulphides from an *N*-alkylthio-phthalimide or -succinimide with a Grignard reagent or an alkyl-lithium compound, respectively,[218] and in the formation of α-phenylthio-ketones from enamines and $(PhS)_3N$.[219]

Unusual examples of sulphide formation have been reported. EtSCN gives $EtSCCl_3$ on reaction with $CHCl_3$ and aqueous NaOH in the presence of a phase-transfer catalyst,[220] and $Me_2CHSCMe_2OAc$ is obtained from tetramethyl-1,3,4-thiadiazoline on treatment with Ac_2O by way of a Pummerer rearrangement.[221]

Methods for the synthesis of selenides are often the analogues of those used in the sulphur series; examples are $PhCH_2Se(CH_2)_nCO_2H$ ($n = 1$—10), from the diselenide by reduction and alkylation;[222] $[MeSe(CH_2)_2SeCH_2]_2CH_2$ from chloro-alkyl intermediates;[223] $EtC{\equiv}C{-}C{\equiv}CSeMe$ from $EtC{\equiv}C{-}C{\equiv}C^-$ Na$^+$ and Se followed by methylation (the Te analogue may be prepared in the same way);[224] and reaction of Se with CF_3I or n-C_3F_7I at 330 °C to give 30% selenide and 70% diselenide.[225] A new selenide synthesis [226] uses SeO_2 suspended in THF and an alkylmagnesium halide [226a] or a trialkylborane.[226b] Photolysis (350 nm) of dibenzyl diselenide gives dibenzyl selenide.[227, 843] Unsymmetrical tellurides may be obtained from ArTeX and RMgX (X = halogen).[228]

Properties and Reactions of Saturated Sulphides.—Cleavage and displacement reactions of sulphides have applications in organic synthesis. Tungsten hexa- or penta-sulphide gives $WCl_6(SR)_2 + RCl$ *en route* to $(R_3S^+)_2[WCl_6]^{2-}$,[229] and glucopyranosyl-α-chlorides are formed from alkyl- or aryl-1-thio-β-D-gluco-pyranosides with $Cl_2CHOMe–ZnCl_2$.[230] A new method for the conversion of a sulphide into an ether ($R^1SR \to R^2OR$) uses an alcohol R^2OH with thallium(III) nitrate.[231] A curious reaction, the conversion of dipentyl sulphide and maleic anhydride into 2-pentylthiosuccinic acid after several hours at 155 °C, is believed to involve pentanethial as intermediate,[232] since diphenyl sulphide did not react. The kinetics of sulphide cleavage in Na–NH$_3$ solutions (RSR + 2e$^-$ + NH$_3$ $\to$ RS$^-$ + NH$_2^-$ + RH) have been studied.[233] Benzyl sulphides PhCH$_2$SR^1 give benzylideneamines PhCH=NR^2, (R^{2}NH)$_2$S, and R^{1}SSR^1 by reaction with a sulphur di-imide R^{2}N=S=NR^2.[234]

[218] M. Furukawa, I. Suda, and S. Hayashi, *Synthesis*, 1974, 282.
[219] J. Almog, D. H. R. Barton, P. D. Magnus, and R. K. Norris, *J.C.S. Perkin I*, 1974, 853.
[220] M. Makosza and M. Fedorynski, *Synthesis*, 1974, 274.
[221] J. Buter, P. W. Raynolds, and R. M. Kellogg, *Tetrahedron Letters*, 1974, 2901.
[222] A. Fredga, *Acta Chem. Scand.* (*B*), 1974, **28**, 692.
[223] W. Levason, C. A. McAuliffe, and S. G. Murray, *J.C.S. Dalton*, 1976, 269.
[224] S. I. Radchenko and A. A. Petrov, *Zhur. org. Khim.*, 1975, **11**, 1988.
[225] L. S. Koshcheeva, A. N. Lavrentev, and E. G. Sochilin, *Zhur. cobshhei Khim.*, 1974, **44**, 2103.
[226] A. Arase and Y. Masuda, *Chem. Letters*, 1975, (*a*) 1331; (*b*) 419.
[227] W. Stanley, M. R. Van De Mark, and P. L. Kumler, *J.C.S. Chem. Comm.*, 1974, 700.
[228] N. Petragnani, L. Torres, and K. J. Wynne, *J. Organometallic Chem.*, 1975, **92**, 185.
[229] P. M. Boorman, T. Chivers, and K. N. Mahadev, *J.C.S. Chem. Comm.*, 1974, 502; *Canad. J. Chem.*, 1975, **53**, 383.
[230] I. Farkas, R. Bognar, M. M. Menyhart, A. K. Tarnai, M. Bihari, and J. Tamas, *Acta Chim. Acad. Sci. Hung.*, 1975, **84**, 325.
[231] Y. Nagao, K. Kaneko, M. Ochiai, and E. Fujita, *J.C.S. Chem. Comm.*, 1976, 202.
[232] L. Field, *J. Org. Chem.*, 1974, **39**, 2110.
[233] R. L. Jones and R. R. Dewald, *J. Amer. Chem. Soc.*, 1974, **96**, 2315.
[234] G. Kresze and N. Schoenberger, *Annalen*, 1975, 1721.

Thiocarbenes formed from chloromethyl aryl sulphides yield cyclopropyl sulphides with alkenes in 50% aqueous NaOH–CH$_2$Cl$_2$–phase-transfer catalyst system,[235] and give amino-cyclopropyl sulphides with enamines.[236] 1-Arylthio-methylimidazoles are readily prepared from chloromethyl sulphides.[237] Heterolysis of the C—I bond in 2-iodoethyl *p*-tolyl sulphide at 100 °C accounts for the formation of I$_2$ + ArSCH$_2$CH$_2$SAr.[238] β-Phenylthio-radicals, formed by the reaction of tri-butyltin radicals on *erythro*- and *threo*-2-bromo-3-phenylthio-butanes, eliminate PhS· to give (*Z*)/(*E*)-mixtures of butenes, indicating that the PhS—CR$_2$—ĊR— radical is not stabilized by sulphur bridging.[239] Studies similarly directed towards elucidating the physical and electronic properties of divalent sulphur show participation by sulphur in the stabilization of cationic intermediates formed during addition reactions of *o*-allyl-thioanisole [240] and the influence of substituents on the stability of charge-transfer complexes formed between substituted diphenyl sulphides and I$_2$.[241] Revised data on the stabilities of carbanions XCH$_2^-$, taking account of internal return and ion-pairing, show that PhS is about as stabilizing as Ph$_2$P.[242] The interaction between the sulphur and cyano-groups in α-alkylthio-nitriles,[243] and the complexation propensities of MeS(CH$_2$)$_2$S(CH$_2$)$_3$S(CH$_2$)$_2$SMe [244] and of the 'octopus' molecules (20) [245] have also been studied.

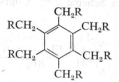

(20) R = MeS, PhS, Bu$-$(CH$_2$CH$_2$O$)-_3$S or Bu$-$(CH$_2$CH$_2$O$)-_2$S

The 1,2-shift in the dehydration of *trans*-1-hydroxy-2-ethylthio-1,2-dihydro-naphthalene, a simple model for the glutathione–naphthalene 1,2-oxide adduct (the so-called premercapturic acid which is receiving considerable attention because of the importance of arene oxides in metabolism), leads to ethyl 1-naphthyl sulphide *via* a cyclic sulphonium ion.[246]

Preparation of Vinyl Sulphides.—Improved procedures for certain standard methods have been described in the recent literature. Dehydration of hydroxy-sulphides without rearrangement can be brought about with SOCl$_2$ and pyridine at 0 °C.[247] Thiophilic addition of PhLi to thioketens R^{1}R^{2}C=C=S at −78 °C

[235] G. Boche and D. R. Schneider, *Tetrahedron Letters*, 1975, 4248.
[236] R. H. Rynbrandt and F. E. Dutton, *J. Org. Chem.*, 1975, **40**, 2282.
[237] K. H. Baggaley, S. G. Brooks, and R. M. Hindley, *J.C.S. Perkin I*, 1975, 1670.
[238] H. H. Radwan and L. R. Williams, *Austral. J. Chem.*, 1975, **28**, 1999.
[239] T. E. Boothe, J. L. Greene, and P. B. Shevlin, *J. Amer. Chem. Soc.*, 1976, **98**, 951.
[240] H. Kwart and D. Drayer, *J. Org. Chem.*, 1974, **39**, 2157.
[241] S. Santini, G. Reichenbach, and U. Mazzucato, *J.C.S. Perkin II*, 1974, 494.
[242] F. G. Bordwell, W. S. Matthews, and N. R. Vanier, *J. Amer. Chem. Soc.*, 1975, **97**, 442.
[243] M. T. Fabi, L. Marzorati, P. R. Olivato, R. Rittner, H. Viertler, and B. Wladislaw, *J.C.S. Perkin II*, 1976, 16.
[244] G. F. Smith and D. W. Margerum, *J.C.S. Chem. Comm.*, 1975, 807.
[245] F. Voegtle and E. Weber, *Angew. Chem.*, 1974, **86**, 896.
[246] A. M. Jeffery and D. M. Jerina, *J. Amer. Chem. Soc.*, 1975, **97**, 4427.
[247] B. M. Trost and D. E. Keeley, *J. Amer. Chem. Soc.*, 1976, **98**, 248.

followed by addition of MeOH gives $R^1R^2C=CHSPh$, while quenching with CO_2 gives the corresponding carboxylic acid $R^1R^2C=C(CO_2H)SPh$, and methylation gives $R^1R^2C=CMeSPh$ [248] (at room temperature the intermediate carbenoid gives the α-elimination product $R^1R^2C=C=C=CR^1R^2$).[248] Thiocamphor similarly gives alkyl vinyl sulphides.[470] Wittig synthesis using an alkylthio-substituted phosphorane ($R^1CHO + Ph_3\overset{+}{P}-\overset{-}{C}HSR^2 \rightarrow R^1CH=CHSR^2$) [249, cf. 467] and the condensation of thiols, or orthothio-esters, with ketones in the presence of $AlCl_3$ ($R^1R^2CHCOR^3 \rightarrow R^1R^2C=C(SR^4)R^3$) [250] also give vinyl sulphides. Addition of SCl_2 to allyl chloride [to give $S\{C(CH_2Cl)=CH_2\}_2$],[251] of thiols to alkynes (see also p. 12),[252] of a selenenyl halide to an alkene followed by dehydrohalogenation of the resulting 2-chloroalkyl selenide,[253] and of a sulphenyl halide to an alkyne [254] have been reported. Reduction of a 1-alkynyl sulphide with a copper(I) hydride complex gives the *cis*-1-alkenyl sulphide, and with $LiAlH_4$ gives the *trans*-isomer.[255]

Newer methods for the synthesis of vinyl sulphides, described in the recent literature, include procedures of potential general application. Successive alkylation of propenethiol dianion $(CH_2\!=\!\!=\!\!CH\!=\!\!=\!\!CH\!=\!\!=\!\!S)^{2-}$ $2Li^+$ with n-hexyl bromide and MeI gives $C_6H_{13}CH_2CH=CHSMe$ (78%) and $CH_2=CHCH-(SMe)C_6H_{13}$.[256] A method for the introduction of a thiovinyl group into a tertiary carbon atom is displayed in Scheme 2.[257] Copper(I)-promoted removal of a

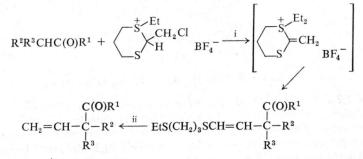

Reagents: i, $LiPr^i_2N$; ii, Raney Ni

Scheme 2

phenylthio-group from a diphenylthioacetal, followed by loss of a β-proton from the resulting carbonium ion, gives a phenyl vinyl sulphide.[258] Recently discovered alkyl bis(alkylthio)sulphonium salts [*e.g.* (21)] react with alkynes to give the long-lived thiirenium cations [*e.g.* (22)], which are susceptible to nucleophilic ring-opening to give β-substituted vinyl sulphides.[254, 612, 851]

[248] E. Schaumann and W. Walter, *Chem. Ber.*, 1974, **107**, 3562.
[249] H. J. Bestmann and J. Angerer, *Annalen*, 1974, 2085.
[250] F. Akiyama, *J.C.S. Chem. Comm.*, 1976, 208.
[251] E. L. Kalinina and P. S. Makovetskii, *Ukrain. khim. Zhur.*, 1975, **41**, 662.
[252] M. S. R. Naidu and S. G. Peeran, *Tetrahedron*, 1975, **31**, 465.
[253] E. G. Kataev, T. G. Mannafov, and Y. Y. Samitov, *Zhur. org. Khim.*, 1975, **11**, 2322.
[254] G. Capozzi, O. De Lucchi, V. Lucchini, and G. Modena, *J.C.S. Chem. Comm.*, 1975, 248.
[255] P. Vermeer, J. Meijer, C. Eylander, and L. Brandsma, *Rec. Trav. chim.*, 1976, **95**, 25.
[256] K. H. Geiss, B. Seuring, R. Pieter, and D. Seebach, *Angew. Chem.*, 1974, **86**, 484.
[257] T. Oishi, H. Takeuchi, and Y. Ban, *Tetrahedron Letters*, 1974, 3757.
[258] T. Cohen, G. Herman, J. R. Falck, and A. J. Mura, *J. Org. Chem.*, 1975, **40**, 812.

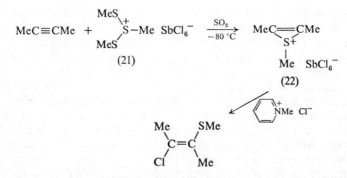

(21)

(22)

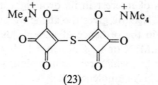

Photoaddition of bis(methylthio)acetylene to a thione gives an α-(thiomethoxy-thiocarbonyl) vinyl sulphide $R^1R^2C=C(CSSMe)SMe$.[259] Vinyl sulphides of particular interest, prepared by straightforward methods, include: optically active alkyl vinyl sulphides $EtCHMe(CH_2)_nSCH=CH_2$;[260] amides $R^1NHCMe=C(SR^2)CONHPh$;[261] and 2-pyridyl vinyl sulphides.[262] The novel sulphur-linked squaric acid analogue (23), prepared from the corresponding thiolate anion by

(23)

treatment with $CuCl_2$, has pK_1 probably less than zero (pK_1 of squaric acid = 0.5).[263] The vinyl sulphide rearrangement product from penicillin G (see Vol. 3, p. 32) has been synthesized, and thereby confirmed to be the (Z)-isomer.[264]

Properties of Vinyl and Allyl Sulphides.—3-Chloroalkenyl phenyl sulphides may be used in synthesis as acrolein synthons, reaction with an organometallic compound or an anion followed by hydrolysis ($TiCl_4$–PbO–H_2O) giving the homologated aldehyde.[265] The success or otherwise of the aqueous $HgCl_2$ reagent for converting vinyl sulphides into aldehydes is highly dependent on structure, and the $TiCl_4$ method is also somewhat erratic.[265] α-Alkylthio-aldehydes may be prepared from vinyl sulphides.[249] The kinetics of acid hydrolysis of $RSCH=CHC≡CH$, giving either $MeC(O)CH_2CHO$ or $RSCH=CHC(O)Me$, have been elucidated.[266]

[259] A. C. Brouwer and H. J. T. Bos, *Tetrahedron Letters*, 1976, 209.
[260] E. Chiellini, M. Marchetti, P. Salvadori, and L. Lardicci, *Internat. J. Sulfur Chem.*, 1973, **8**, 19.
[261] J. B. Pierce, R. J. Fanning, and R. A. Davis, *Canad. J. Chem.*, 1975, **53**, 1327.
[262] K. Undheim and L. A. Riege, *J.C.S. Perkin I*, 1975, 1493.
[263] D. Eggerding, J. L. Straub, and R. West, *Tetrahedron Letters*, 1975, 3589.
[264] C. J. Veal and D. W. Young, *J.C.S. Perkin I*, 1975, 2086.
[265] B. M. Trost, K. Hiroi, and S. Kurozumi, *J. Amer. Chem. Soc.*, 1975, **97**, 438; A. J. Mura, D. A. Bennett, and T. Cohen, *Tetrahedron Letters*, 1975, 4433; A. J. Mura, G. Majetich, P. A. Grieco, and T. Cohen, *Tetrahedron Letters*, 1975, 4437.
[266] A. N. Khudyatova, A. N. Volkov, B. A. Trofimov, L. V. Mametova, and R. N. Kudyakova, *Izvest. Akad. Nauk S.S.S.R., Ser. khim.*, 1975, 1975.

A condensation of vinyl sulphides of synthetic value involves the reaction of α-ethylthio-styrenes with an aldehyde in the presence of a Lewis acid catalyst to give 3-ethylthio-indenes.[267] Regiospecific and stereospecific pentannelation (*i.e.* fusion of a cyclopentane ring) of an alicyclic ketone involves arylthiomethylen-ation followed by base-catalysed addition to an αβ-unsaturated ester.[268] α-Phenyl-thiobutenolides are less reactive Michael acceptors than their sulphoxide counterparts,[269] and ββ-dichlorovinyl sulphides are less reactive in electrophilic reactions than their oxygen counterparts.[270] Dimerization of aryl α-cyanovinyl sulphides occurs more readily than that of alkyl analogues, since the intermediate 1,4-diradical ArSĊ(CN)CH$_2$CH$_2$Ċ(CN)SAr, which cyclizes to the cyclobutane, is more stable.[271] Products obtained by photolysis of methyl vinyl sulphide are accounted for by homolysis of both vinyl—S and methyl—S bonds.[272] The kinetics of catalysed *cis–trans* isomerization of aryl vinyl sulphides have been studied.[273]

γ-Addition is observed when allyl aryl sulphides react with aldehydes in the presence of base.[274] In the presence of AgSbF$_6$, *trans*-β-chlorovinyl selenides are in equilibrium with the corresponding selenenirenium cations.[610]

Acetylenic Sulphides.—Addition reactions of 1-alkynyl sulphides continue to provide chemistry that is of interest in its own right and of value in synthesis. Alkylation of MeSC≡CCH$_2$OMe gives the allene MeSCR=C=CHOMe, which on addition to a ketone in the presence of base yields a 3-furanone.[275] Stereo-specific addition of R$^2_2$CuMgX or R$^2_2$CuLi to a 1-alkynyl sulphide gives [R^{1}R^2-C=C(SMe)]$_2$CuMgX (or Li), which react anomalously on attempted alkylation to give exclusively the *cis*-cumulenes R^{1}R^{2}C=C=C=CR^{1}R^2, by carbene dimerization.[276] 3-Alkylthio-furans are accessible from the photo-adduct R^{1}SC(COR^2)=C(COR^2)R^3 of a 1,2-diketone to an alkyl alkynyl sulphide, after cyclization with SnCl$_2$.[277] HgO-Catalysed addition of a carboxylic acid to an alkyl alkynyl sulphide gives the corresponding *O*-acyl thiol-ester enolate R^{1}CH=C(SR^2)OCOR^3, which is readily converted into the thiol-ester.[278] Related reactions to which attention has been given are: hydration kinetics in aqueous HClO$_4$;[279] successive addition of a Grignard reagent and H$_2$O, CO$_2$, or I$_2$ to CH≡CSEt, giving RCH=CXSEt (X = H, CO$_2$H, or I) (the corresponding vinyl ether gives XCH=CROEt);[280] ionic addition of HCl in the presence of

[267] T. Mukaiyama, K. Kamio, and K. Narasaka, *Chem. Letters*, 1974, 565.
[268] J. P. Marino and W. B. Mesbergen, *J. Amer. Chem. Soc.*, 1974, **96**, 4050.
[269] K. Iwai, H. Kosugi, and H. Uda, *Chem. Letters*, 1974, 1237; 1975, 981; H. Hagiwara, K. Nakayama, and H. Uda, *Bull. Chem. Soc. Japan*, 1975, **48**, 3769.
[270] A. S. Atavin, A. N. Mirskova, E. F. Zorina, and Y. L. Frolov, *Zhur. org. Khim.*, 1974, **10**, 1157.
[271] K. D. Gundermann and E. Roehrl, *Annalen*, 1974, 1661.
[272] G. Leduc and Y. Rousseau, *Canad. J. Chem.*, 1975, **53**, 433.
[273] F. Turecek and M. Prochazka, *Coll. Czech. Chem. Comm.*, 1974, **39**, 2073.
[274] K. Kondo, K. Matsui, and A. Negishi, *Chem. Letters*, 1974, 1371.
[275] R. M. Carlson, R. W. Jones, and A. S. Hatcher, *Tetrahedron Letters*, 1975, 1741.
[276] H. Westmije, J. Meijer, and P. Vermeer, *Tetrahedron Letters*, 1975, 2923.
[277] A. Mosterd, H. J. Matser, and H. J. T. Bos, *Tetrahedron Letters*, 1974, 4179.
[278] M. L. Petrov, V. A. Bobylev, and A. A. Petrov, *Zhur. org. Khim.*, 1975, **11**, 267.
[279] W. F. Verhelst and W. Drenth, *J. Org. Chem.*, 1975, **40**, 130.
[280] J. F. Normant, A. Alexakis, A. Commercon, G. Cahiez, and J. Villieras, *Compt. rend.*, 1974, **279**, C, 763.

CuCl and hydroquinone to $MeSC\equiv C-C\equiv CPr$ to give *cis,cis-* and *cis,trans-*
$MeSCCl=CHCCl=CHPr$ in 1 : 2 ratio;[281] and formation of $EtC\equiv C-C\equiv CPr$
by the reaction of $MeSC\equiv C-C\equiv CEt$ with Pr^nLi.[281]

Phenylthio-ynamines, *e.g.* $PhSC\equiv CNEt_2$, are in a class apart, and the formation of the cyclobutadiene (24) by treatment with BF_3 is notable.[282]

$$PhSC\equiv CNEt_2 \longrightarrow \begin{array}{c} Et_2N SPh \\ \square \\ PhS NEt_2 \end{array}$$

(24)

Radchenko's group continues to report novel reactions of Se and Te analogues,
which provide further insight into the different behaviour of S, Se, and Te
functional groups. $MeSeC\equiv CCH=CH_2$ gives vinylacetylene by reaction with
BuONa in BuOH, whereas MeONa in MeOH in a sealed tube at 120 °C gives
$MeSeCH_2C(OMe)=CHMe$ together with $MeSeC(OMe)=CHCH_2CH_3$.[283] Addition of bromine at -60 °C to the same selenide gives $MeSeCBr=CBrCH=CH_2$,
whereas the tellurium analogue (EtTe in place of MeSe) gives $EtTeBr_2C\equiv$
$CCHBrCH_2Br$.[284] Acetic acid adds to $MeTeC\equiv CCH=CH_2$ at 50—60 °C to give
$MeTeC(OAc)=CHCH=CH_2$.[285]

Cleavage of Unsaturated Sulphides by Carbenes and Nitrenes; Photolysis.—
Continuing studies by Ando's group have clarified further the factors controlling
the course of the reaction between a carbene and an acetylenic sulphide ($RC\equiv$
$CSEt + N_2CHCO_2Me + CuSO_4 \rightarrow RC\equiv CSCH_2CO_2Me$),[286] and the insertion
reaction between the analogous nitrene and an allylic sulphide or ether.[287]
Photo-oxidation of vinyl sulphides in the presence of Rose Bengal might involve
1,2-dioxetan or ene-addition routes.[288] Support for the dioxetan route is provided by the rationalization of ratios of C—S and C—C cleavage products
resulting from photo-oxidation of 1-ethylthio-cycloalkenes.[289]

Rearrangements of Unsaturated Sulphides.—The 1,3-allylic shift exemplified in
the sequence $PhSCH_2C(O)R \rightarrow PhSCMe_2CH(OH)R \rightarrow CH_2=CMeCH(SPh)R \rightarrow$
(E/Z)-$PhSCH_2CMe=CHR$ may be brought about thermally or in daylight, and
the overall scheme allows the movement of the phenylthio-group one carbon
atom either way along a carbon chain.[290]

The Wittig rearrangement of allyl vinyl and benzyl vinyl ethers brought about
by strong bases is not shown by benzyl vinyl sulphides.[291] Base-catalysed
rearrangement of bis(propargyl) sulphides involves first the formation of the

[281] S. I. Radchenko, *Zhur. org. Khim.*, 1976, **12**, 229.
[282] R. Gompper, S. Mensch, and G. Seybold, *Angew. Chem.*, 1975, **87**, 711.
[283] S. I. Radchenko, *Zhur. org. Khim.*, 1975, **11**, 652.
[284] S. I. Radchenko and A. A. Petrov, *Zhur. org. Khim.*, 1974, **10**, 1986.
[285] S. I. Radchenko, I. G. Savich, and A. A. Petrov, *Zhur. org. Khim.*, 1974, **10**, 2269.
[286] W. Ando, H. Higuchi, and T. Migita, *J.C.S. Chem. Comm.*, 1974, 523.
[287] W. Ando, H. Fujii, I. Nakamura, N. Ogino, and T. Migita, *Internat. J. Sulfur Chem.*, 1973, **8**, 13.
[288] W. Ando, K. Watanabe, J. Suzuki, and T. Migita, *J. Amer. Chem. Soc.*, 1974, **96**, 6766.
[289] W. Ando, K. Watanabe, and T. Migita, *Tetrahedron Letters*, 1975, 4127; *J.C.S. Chem. Comm.*, 1975, 961.
[290] P. Brownbridge and S. Warren, *J.C.S. Chem. Comm.*, 1975, 820.
[291] V. Rautenstrauch, G. Buchi, and H. Wuest, *J. Amer. Chem. Soc.*, 1974, **96**, 2576.

isomeric bis(allenyl) sulphide and then its intramolecular rearrangement into a thienyl bis(methylene) diradical.[292] Base-catalysed [2,3]-sigmatropic rearrangements of allylic, allenic, and propargylic sulphides are described in this volume in the chapter dealing with ylides. Thermal [3,3]-sigmatropic rearrangement of allenyl vinyl sulphides $R^1CH=C=CR^2SCR^3=CHR^4$ in aqueous DMSO leads to the formation of $R^2C\equiv CCHR^1CHR^4C(O)R^3$.[293]

Examples of thio-Claisen rearrangements in the recent literature involve 3-substituted 4-allylthio-pyrimidin-2-ones,[294] 2-propargylthiobenzimidazoles,[295] and *S*-allylthiodimedones (25), which give (26) when Ac_2O is used as solvent, or

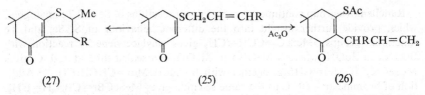

(27)   (25)   (26)

(27) when the rearrangement is conducted in a passive solvent.[296] Unlike the Claisen rearrangement itself, which is catalysed only by electrophilic species, the thio-Claisen rearrangement is catalysed only by nucleophilic species.[297]

Naturally occurring Sulphides, and Sulphides Related to Natural Products.— 5-Methylthiopentane-2,3-dione, the first α-diketone sulphide found in Nature, is produced in the anal scent gland of the striped hyena.[298] The structure, assigned through the interpretation of mass spectra, was readily confirmed by synthesis from biacetyl and chloromethyl methyl sulphide, or from biacetyl, formaldehyde, and MeSH. An unusual amino-acid derivative, 2,5-dicysteinyl-dopa, has been isolated from *Lepisosteus spatula*.[299]

A g.l.c.–atomic absorption combination has been developed for the detection and determination of dimethyl selenide and dimethyl diselenide in atmospheric samples.[300]

Acyclic sulphides handled during a continuing study of the reactions of penicillins and cephalosporins are the result of reaction of methyl fluorosulphonate,[301] methyl chloroformate,[264] and acetyl chloride with penicillin-derived thiazine *S*-oxides.[302] 4-Methylthioazetidinones formed by treatment of penicillin esters with MeI and strong base [303] suffer replacement of the MeS group by acetoxy on treatment with $Pb(OAc)_4$.[304] The MeS group turns up unexpectedly on the

292 P. J. Garratt and S. B. Neoh, *J. Amer. Chem. Soc.*, 1975, **97**, 3255.
293 L. Brandsma and H. D. Verkruijsse, *Rec. Trav. chim.*, 1974, **93**, 319.
294 J. L. Fourrey, E. Estrabaud, and P. Jouin, *J.C.S. Chem. Comm.*, 1975, 993.
295 K. K. Balasubramanian and B. Venugopalan, *Tetrahedron Letters*, 1974, 2643, 2645.
296 L. Dalgaard and S. O. Lawesson, *Acta Chem. Scand. (B)*, 1974, **28**, 1077.
297 H. Kwart and J. L. Schwartz, *J. Org. Chem.*, 1974, **39**, 1575.
298 J. W. Wheeler, D. W. von Endt, and C. Wemmer, *J. Amer. Chem. Soc.*, 1975, **97**, 441.
299 S. Ito and J. A. C. Nicol, *Tetrahedron Letters*, 1975, 3287.
300 Y. K. Chau, P. T. S. Wong, and P. D. Goulden, *Analyt. Chem.*, 1975, **47**, 2279.
301 D. K. Herron, *Tetrahedron Letters*, 1975, 2145.
302 R. J. Stoodley and R. B. Wilkins, *J.C.S. Perkin I*, 1975, 716.
303 E. G. Brain, I. McMillan, J. H. C. Nayler, R. Southgate, and P. Tolliday, *J.C.S. Perkin I*, 1975, 562.
304 E. G. Brain, A. J. Eglington, J. H. C. Nayler, M. J. Pearson, and R. Southgate, *J.C.S. Perkin I*, 1976, 447.

exocyclic amide nitrogen atom. Synthesis of 4-*p*-methoxybenzylthioazetidinones by addition of phthaloylglycyl chloride to the appropriate *S*-substituted thioimidate,[305] from the fused oxazoline-azetidinone obtainable from benzylpenicillin,[306] and through previously established methods,[307] has been described, and transformations of functional groups of the products have been studied.[305, 307]

Synthesis of monosaccharide analogues in which one or more oxygen atoms is replaced by an alkyl or arylthio-group is represented in the recent literature, as in that of several years past. 5-Thio-D-ribose and xylose derivatives have been converted into thiopyranoses, from which 4-thio-L-arabinofuranoses can be prepared.[308] The primary hydroxy-group of nucleosides and ribonucleosides is replaced by an alkylthio-group by treatment with the corresponding dialkyl disulphide and tri-n-butylphosphine.[309] Syntheses of 5-thio-D-fructofuranose [310] *via* acyclic sulphides and of maltose derivatives with sulphide groupings at C-1 and C-6 in the reducing residue [311] have been reported.

Thiocyclopropenium Salts.—Preparation of a tri(alkylthio)cyclopropenium perchlorate (28) from condensation of tetrachlorocyclopropene and a thiol in the

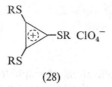

(28)

presence of $HClO_4$ (see Volume 3, p. 25), according to a prescribed procedure, led to a violent explosion.[312] A selenium analogue (29) has been prepared as shown in Scheme 3.[313]

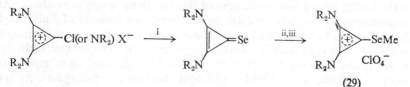

(29)

Reagents: i, NaSeH; ii, MeI in CH_2Cl_2; iii, $KClO_4$

Scheme 3

Heteroaryl Sulphides.—The earlier section dealing with 'Preparations of Sulphides' describes applications of general synthetic methods leading to heteroaryl sulphides, and this section deals with more specific syntheses and some properties of these compounds.

[305] M. D. Bachi and O. Goldberg, *J.C.S. Perkin I*, 1974, 1184.
[306] D. F. Corbett and R. J. Stoodley, *J.C.S. Chem. Comm.*, 1974, 438; *J.C.S. Perkin I*, 1975, 432.
[307] R. Lattrell and G. Lohaus, *Annalen*, 1974, 901; R. Lattrell, *ibid.*, p. 1361.
[308] N. A. Hughes and C. J. Wood, *J.C.S. Chem. Comm.*, 1975, 294; W. Clegg, N. A. Hughes, and C. J. Wood, *ibid.*, p. 300.
[309] I. Nakagawa and T. Hata, *Tetrahedron Letters*, 1975, 1409.
[310] M. Chmielewski and R. L. Whistler, *J. Org. Chem.*, 1975, **40**, 639.
[311] M. Mori, M. Haga, and S. Tejima, *Chem. and Pharm. Bull. (Japan)*, 1974, **22**, 1331.
[312] K. G. R. Sundelin, *Chem. Eng. News*, 1974, **52**, 3.
[313] Z. Yoshida, H. Konishi, and H. Ogoshi, *J.C.S. Chem. Comm.*, 1975, 359.

Pyridine 1-oxide reacts with adamantane-1-thiol in acetic anhydride to give 2- and 3-(1-adamantanethio)pyridines and their tetrahydropyridine analogues.[314] Corresponding butylthio-substituted pyridines are obtained using Bu^nSH or Bu^tSH in this reaction in the presence of a chloroformic acid derivative.[315] No complications arise in the use of 3-bromopyridine 1-oxide and KSH for the synthesis of 3,3'-dipyridyl sulphide.[316] 2-Methyl- or 2-phenyl-pyridine is converted into its 5-alkylthio-homologue by reaction of the pyridine–MeLi adduct with a dialkyl disulphide.[317]

2-Phenylindole reacts with dimethyl sulphoxide to give the 3-methylthio-homologue.[318] Bis(2- and 3-indolyl) sulphides are accessible from the indoles by reaction with SCl_2.[319]

A novel synthesis of *p*-tolylthio-imidazoles involves the addition of *p*-tolylthio-methyl isocyanide $TolSCH_2N{=}C$ to a nitrile.[320] Thermal rearrangement of 5-methylthio-imidazolines (30) to imidazoles (31) has been shown to be an intramolecular process.[321]

(30)   (31)

Sulphides in Synthesis.—Applications of sulphides in synthesis making use of the potential nucleophilic properties conferred by the sulphur atom in alkyl or aryl alkyl sulphides are reviewed in the following chapter.

The influence of a phenylthio-group in potentiating the oxidative ring-cleavage of a cycloalkanone,[322] illustrated in Scheme 4 [(32) → (33)], and in ring-expansion [(32) → (34)] adds a new application of this functional group in synthesis. An α-ethylthio-group usefully modifies the Michael donor tendency of β-nitroethyl benzenes.[323] α-Substitution of an *N*-nitrosopyrrolidine with an acetoxy-group is effected by first introducing an α-ethylthio-group and then treating the derivative with Cl_2 and NH_4OAc at $-78\ °C$.[324] The oxidative decyanation of aryl aceto nitriles, $ArCHRCN → ArC(O)R$, is brought about by introducing an α-phenyl-thio-group followed by treatment with *N*-bromosuccinimide.[325] Oxidative decarboxylation of carboxylic acids is similarly brought about *via* α-methylthiol-ation followed by formation of the labile *S*-chlorosulphonium salt.[326] This

314 B. A. Mikrut, K. K. Khullar, P. Y. P. Chan, J. M. Kokosa, L. Bauer, and R. S. Egan, *J. Heterocyclic Chem.*, 1974, **11**, 713.
315 L. Bauer, T. E. Dickerhofe, and K. Y. Tserng, *J. Heterocyclic Chem.*, 1975, **12**, 797.
316 T. Talik and Z. Talik, *Synthesis*, 1975, 499.
317 N. Finch and C. W. Gemenden, *J. Org. Chem.*, 1975, **40**, 569.
318 J. Hocker, K. Ley, and R. Merten, *Synthesis*, 1975, 334.
319 H. Piotrowska, B. Serafin, and K. Wejroch-Matacz, *Roczniki Chem.*, 1975, **49**, 635.
320 A. M. van Leusen and J. Schut, *Tetrahedron Letters*, 1976, 285.
321 J. Nyitrai, K. Lempert, and T. Cserfalvi, *Chem. Ber.*, 1974, **107**, 1645.
322 B. M. Trost and K. Hiroi, *J. Amer. Chem. Soc.*, 1975, **97**, 6911.
323 Y. Nagao, K. Kaneko, and E. Fujita, *Tetrahedron Letters*, 1976, 1215.
324 J. E. Baldwin, S. E. Branz, R. F. Gomez, P. L. Kraft, A. J. Sinskey, and S. R. Tannenbaum *Tetrahedron Letters*, 1976, 333.
325 S. J. Selikson and D. S. Watt, *Tetrahedron Letters*, 1974, 3029.
326 B. M. Trost and Y. Tamaru, *J. Amer. Chem. Soc.*, 1975, **97**, 3528, 3797.

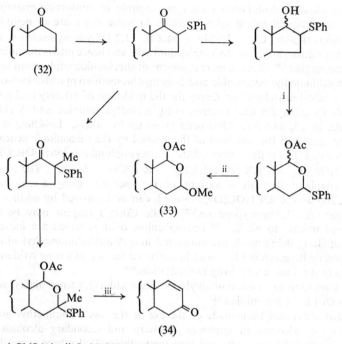

Reagents: i, Pb(OAc)₄; ii, I₂–MeOH; iii, KOH

Scheme 4

process can be applied to 2-(methylthio)acetic acid MeSCH₂CO₂H, from which an αα-disubstituted derivative MeSCR¹R²CO₂H may be formed which may itself be converted into a ketone R¹COR². A rival to diethyl malonate as a source of unsymmetrical ketones is therefore available. Further studies of the selective *ortho*-formylation procedure of aromatic and heteroaromatic amines (see Vol. 3, p. 26) involving *N*-chlorination, followed by reaction with a sulphide, rearrangement of the resulting aminosulphonium salt to the *o*-(alkyl- or aryl-thiomethyl)aniline, and conversion into the *o*-formylaniline (—CH₂SR → —CHClSR → —CHO), have been reported.[327, 328] When β-keto-sulphides are used, indoles are formed.[328, 329] Selective *ortho*-formylation of phenols can be performed similarly.[330] *N*-Chlorination of an amide, treatment with Me₂S, then with NEt₃ (Stevens rearrangement), gives the *O*-methylthiomethyl derivative, R¹NHC(O)R² → R¹N=CR²OCH₂SMe.[331] Conversely, *N*-methylthiomethylation of an amide requires chloromethyl methyl sulphide in TFA or in MeSO₃H, and conventional desulphurization with Raney nickel leads to the *N*-methyl-amide.[332]

[327] P. G. Gassman and H. R. Drewes, *J. Amer. Chem. Soc.*, 1974, **96**, 3002.
[328] P. G. Gassman, G. D. Gruetzmacher, and T. J. van Bergen, *J. Amer. Chem. Soc.*, 1974, **96**, 5512.
[329] P. G. Gassman and G. D. Gruetzmacher, *J. Amer. Chem. Soc.*, 1974, **96**, 5487.
[330] P. G. Gassman and D. R. Amick, *Tetrahedron Letters*, 1974, 3463.
[331] E. Vilsmaier and R. Bayer, *Synthesis*, 1976, 46.
[332] L. Bernardi, R. de Castiglione, and U. Scarponi, *J.C.S. Chem. Comm.*, 1975, 320.

Succinimido-dialkylsulphonium salts are capable of converting enamines into the β-dimethylsulphonium salts,[333] which in some cases are demethylated by distillation ($R\overset{+}{S}Me_2$ Cl^- → $RSMe$ + $MeCl$). 1,2-Dihydroxybenzenes may be converted into aryloxy dimethylsulphonium salts and hence into *ortho*-quinones by the same reagent.[334] Stevens rearrangement of succinimidosulphonium salts gives *N*-methylthiomethylsuccinimide and 2-methylthiomethoxypyrrol-2-in-5-one.[335]

The method introduced by Corey for the oxidation of primary and secondary alcohols to aldehydes and ketones using a dialkyl sulphide and *N*-chlorosuccinimide, or Cl_2 and NEt_3, has been taken up by others. Labelling studies [336] indicate uptake of the α-proton of the alcohol by the carbanionic centre of the ylide formed from the intermediate alkoxysulphonium ion; R^1R^2CHOH → $R^1R^2CHO\overset{+}{S}Me_2$ → $R^1R^2CHO\overset{+}{S}MeCH_2^-$ → R^1R^2CO + Me_2S. The moderate yields obtained through a variant of the method, using a chloroformate [R^1R^2CHOH → $R^1R^2CHOC(O)Cl$ → *etc.*], can be improved by adding an acid scavenger, *viz.* 1,2-epoxypropane.[337] While Corey's reagent may be used to dehydrate oximes to nitriles,[338] benzophenone oximes, which are incapable of being similarly dehydrated, are converted into *N*-methylthiomethyl nitrones.[339] Poly(*p*-methylthiostyrene) has been investigated for use in Corey oxidation, and appears to show some selectivity towards diols.[340]

The formation of t-butoxy-dialkylselenonium chlorides from dialkyl selenides and Bu^{t}OCl has been studied.[341]

Further study has been made of the use of the methylthiomethyl group for protection of alcohols in synthesis. Primary and secondary alcohols [342] and tertiary alcohols [343] are converted into methylthiomethyl ethers using dimethyl sulphoxide and Ac$_2$O, and de-protection is brought about under neutral conditions with a mercury(II) salt. Under these conditions, a 1,3-dithian moiety or a silyl ether grouping in the same molecule are unaffected. Protection of alcohols as 2-phenylselenoethyl ethers and their de-protection by oxidation has been proposed.[344]

Simpler sulphides also have uses in synthesis. Diphenyl sulphide is proposed as a reagent for the cleavage of 1,2-dioxetans with rearrangement,[345] and hydroboration by BH_3–Me_2S in Et_2O has been reported, hex-1-ene giving 94% hexan-1-ol and 6% hexan-2-ol after treatment of the intermediate trihexylboranes with alkaline H_2O_2.[346]

Sulphonium Salts.—Alkoxysulphonium salts $R\overset{+}{O}SR_2$ X^- and aminosulphonium salts are reactive intermediates, and were mentioned in this context in the

[333] E. Vilsmaier, W. Sprugel, and K. Gagel, *Tetrahedron Letters*, 1974, 2475.
[334] J. P. Marino and A. Schwartz, *J.C.S. Chem. Comm.*, 1974, 812.
[335] E. Vilsmaier, K. H. Dittrich, and W. Sprugel, *Tetrahedron Letters*, 1974, 3601.
[336] J. P. McCormick, *Tetrahedron Letters*, 1974, 1701.
[337] D. H. R. Barton and C. P. Forbes, *J.C.S. Perkin I*, 1975, 1614.
[338] T. L. Ho and C. M. Wong, *Synth. Comm.*, 1975, **5**, 423.
[339] W. M. Leyshon and D. A. Wilson, *J.C.S. Perkin I*, 1975, 1920.
[340] G. A. Crosby, N. M. Weinshenker, and H. S. Uh, *J. Amer. Chem. Soc.*, 1975, **97**, 2232.
[341] N. Y. Derkach and T. V. Lyapina, *Zhur. org. Khim.*, 1974, **10**, 1991.
[342] E. J. Corey and M. G. Bock, *Tetrahedron Letters*, 1975, 2643, 3269.
[343] K. Yamada, K. Kato, H. Nagase, and Y. Hirata, *Tetrahedron Letters*, 1976, 65.
[344] T. L. Ho and T. W. Hall, *Synth. Comm.*, 1975, **5**, 367.
[345] H. H. Wasserman and I. Saito, *J. Amer. Chem. Soc.*, 1975, **97**, 905.
[346] C. F. Lane, *J. Org. Chem.*, 1974, **39**, 1437.

preceding section. They receive further attention in later sections dealing with sulphoxides and their nitrogen analogues. Sulphonium ylides are covered in Chapter 2.

Synthesis of sulphonium salts from sulphides and alkyl halides is a time-honoured method requiring no citations from current literature, but interesting variations have been described involving: methyl fluorosulphonate for conversion of an *S*-methylcysteine peptide into the corresponding dimethyl-sulphonium salt;[347] alkylation of methionine with *N*-ethoxycarbonyl- or *N*-carboxamido-aziridines;[348] formation of $Me_2\overset{+}{S}Ph\ BF_4^-$, Ph_2S, and AcNHPh from $PhN_2^+\ BF_4^-$ and PhSMe in MeCN;[349] and formation of $Me_3\overset{+}{S}\ BF_4^-$ and PhSMe from $PhN_2^+\ BF_4^-$ and Me_2S.[349] The last two examples involve attack of Ph^+ at sulphur. Me_2S is very rapidly t-butylated by $Bu^t_2\overset{+}{S}Me$ to give $Bu^t\overset{+}{S}Me_2$, presumably *via* a quadricovalent sulphur intermediate, but the di-isopropyl analogue was an ineffective isopropylation reagent.[350] A kinetic study of the corresponding benzyl-transfer reaction of benzylsulphonium salts has been reported.[351] Alkylation of *cis*-1,2-bis(alkylthio)ethylene with trialkyloxonium tetrafluoroborates gives the mono-sulphonium salt *cis*-$R^1R^2\overset{+}{S}CH=CHSR$.[352] Studies[353] leading to the synthesis of the bis-sulphonium salt $R_2\overset{+}{S}CH_2C\equiv CCH_2\overset{+}{S}R_2\ 2X^-$ and its conversion into the cumulene $CH_2=C=C=CH\overset{+}{S}R_2\ X^-$ have been described.

The absolute configuration of a chiral trialkylsulphonium ion has been established 75 years after the first preparation of such species. (+)-Methylethyl-n-propylsulphonium 2,4,6-trinitrobenzenesulphonate has the (*S*)-configuration (35), which was established through correlation with lactic acid[354] and confirmed by *X*-ray analysis.

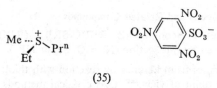

(35)

Mechanistic studies with sulphonium salts concern decomposition kinetics in hydroxylic solvents;[355] ^1H-^2H exchange of the α-protons of methyl and allyl-sulphonium salts, the relative rates of which are strongly dependent upon both solvent and micellar effects;[356] radical chain reactions of triarylsulphonium halides and sodium alkoxides, which lead to arenes, anisoles, diaryl sulphides, and aldehydes or ketones;[357] and the catalysis of hydrocarbon autoxidation by

[347] D. H. Rich and J. P. Tam, *Tetrahedron Letters*, 1975, 211.
[348] P. A. Capps and A. R. Jones, *J.C.S. Chem. Comm.*, 1974, 320.
[349] M. Kobayashi, H. Minato, J. Fukui, and N. Kamigata, *Bull. Chem. Soc. Japan*, 1975, **48**, 729.
[350] H. Minato, T. Miura, F. Takagi, and M. Kobayashi, *Chem. Letters*, 1975, 211.
[351] D. van Ooteghem, R. Deveux, and E. J. Goethals, *Internat. J. Sulfur Chem.*, 1973, **8**, 31.
[352] H. Braun and A. Amann, *Angew. Chem.*, 1975, **87**, 773, 775.
[353] H. Braun and G. Strobl, *Angew. Chem.*, 1974, **86**, 477; H. Braun, G. Strobl, and H. Gotzler, *ibid.*, 1974, **86**, 477.
[354] E. Kelstrup, A. Kjaer, S. Abrahamsson, and B. Dahlen, *J.C.S. Chem. Comm.*, 1975, 629.
[355] A. A. Sosunov, A. P. Kilimov, and V. V. Smirnov, *Zhur. obshchei Khim.*, 1975, **45**, 1533.
[356] T. Okonogi, T. Umezawa, and W. Tagaki, *J.C.S. Chem. Comm.*, 1974, 363.
[357] J. W. Knapczyk, C. C. Lai, W. E. McEwen, J. L. Calderon, and J. J. Lubinkowski, *J. Amer. Chem. Soc.*, 1975, **97**, 1188.

triphenylsulphonium salts.[358] In the last-mentioned study, the catalyst intervenes at the hydroperoxide homolysis step and not at the initial stage to activate the O_2 molecule, as claimed earlier.

Sulphuranes.—Continuing studies by Martin's group provide most of the material for this section. The diaryldialkoxysulphurane $Ph_2S[OC(CF_3)_2Ph]_2$ (see Vol. 3, p. 33) fails to give a sulphurane oxide on treatment with RuO_4, but nevertheless such species are implicated in the oxidation of sulphoxides to sulphones using Cl_2 and $PhC(CF_3)_2O^- K^+$.[359] Further uses for this diaryldialkoxysulphurane have been found. 1,2-Diols give oxirans in excellent yields by treatment with the reagent but higher homologues give only moderate yields of oxetans, tetra-hydrofurans, or tetrahydropyrans.[360] The sulphurane gives sulphimides with many amines, amides, and sulphonamides, while oxidation is observed in other cases ($PhCH_2NH_2 \rightarrow PhCN$; $RCH_2NHR \rightarrow RCH{=}NR$).[361, 567]

Alkoxysulphuranes are intermediates in certain preparative routes to sulphoxides (see p. 34) and cyclic tetra-alkoxysulphuranes result from the reaction of a 1,2-dioxetan (an energy-rich peroxide) with $(RO)_2S$.[362] Cyclic sulphuranes (36) are obtained through a new route, apparently of limited generality.[363]

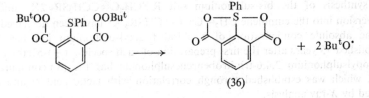

(36)

5 Thioacetals and Related Compounds

The general classes of compounds $R^1R^2C(SR^3)_2$, $(R^3X)CR^1R^2(SR^3)$, and $R^1C(SR^3)_3$ are covered by this section ($X = O$, Se, or NR^3).

Preparation.—Syntheses from ketones by reaction with thiols have been studied from a mechanistic point of view.[364] Less classical methods involve the use of methylthiotrimethylsilane, $MeSSiMe_3$, as a thioketalization reagent operating under neutral conditions;[365] preparation of O-trimethylsilyl hemithio-acetals and -ketals from an aldehyde or ketone, a thiol, pyridine, and Me_3SiCl;[366] and preparation of $\alpha\alpha$-bis(phenylthio)-γ-butyrolactone from the lactone and N-(phenylthio)phthalimide with Pr^i_2NLi.[367]

α-Bromo-sulphoxides undergo an unexpected cleavage reaction with $Mo(CO)_6$, R^1SOCHR^2Br giving $R^1SCHR^2SR^1$ and R^2CHBr_2.[368] Pummerer rearrangement

[358] W. J. M. van Tilborg, *Tetrahedron*, 1975, **31**, 2841.
[359] J. C. Martin and E. F. Perozzi, *J. Amer. Chem. Soc.*, 1974, **96**, 3155.
[360] J. C. Martin, J. A. Franz, and R. J. Arhart, *J. Amer. Chem. Soc.*, 1974, **96**, 4604.
[361] J. A. Franz and J. C. Martin, *J. Amer. Chem. Soc.*, 1975, **97**, 583.
[362] B. S. Campbell, D. B. Denney, D. Z. Denney, and L. S. Shih, *J. Amer. Chem. Soc.*, 1975, **79**, 3850.
[363] J. C. Martin and M. M. Chan, *J. Amer. Chem. Soc.*, 1974, **96**, 3319.
[364] L. Fournier, G. Lamaty, A. Natat, and J. P. Roque, *Tetrahedron*, 1975, **31**, 809.
[365] D. A. Evans, K. G. Grimm, and L. K. Truesdale, *J. Amer. Chem. Soc.*, 1975, **97**, 3229.
[366] T. H. Chan and B. S. Ong, *Tetrahedron Letters*, 1976, 319.
[367] M. Watanabe, K. Shirai, and T. Kumamoto, *Chem. Letters*, 1975, 855.
[368] H. Alper and G. Wall, *J.C.S. Chem. Comm.*, 1976, 263.

of alkylsulphinyl methylphosphonates $(EtO)_2P(O)CH_2S(O)R^1$ gives the *O,S*-thioacetal $(EtO)_2P(O)CH(OR^2)SR^1$ in the presence of an alcohol R^2OH.[369] Preparative procedures involving other simple sulphur functional groups include: carbene insertion into a limited range of disulphides $(RSSR \rightarrow RSCH_2SR)$;[370] alkylation of $LiCH(SMe)_2$ with an oxiran to give eventually a cyclopropanone thioacetal;[371] [2,3]-sigmatropic rearrangement of allylic thioacetals $[CH_2=CMeCH_2SCH_2SR + LiNPr^i_2 \rightarrow CH_2=CMeCH_2CH(SMe)SR$ (after methylation with MeI];[372] and ring-opening of cyclic *O,S*-thioacetals [2-acetylthio-tetra-hydropyran + RSH + $AlCl_3 \rightarrow HO(CH_2)_4CH(SR)SAc]$[373] and related keten thioacetals [(37) → (38)].[374]

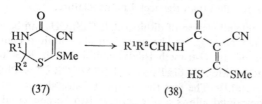

(37)   (38)

The formation of thioacetals from thiocarbonyl compounds is an area of study which has attracted considerably more interest recently. Thiophilic addition of a Grignard reagent to a dithioester has been established as a general route to dithio-acetals $[R^1C(S)SMe + R^2MgX \rightarrow R^1\bar{C}(SMe)SR^2 (MgX)^+]$.[375] Diels–Alder addition of a diene to methyl cyanodithioformate gives a cyclic dithioacetal from which novel routes to sulphur analogues of carbohydrates may be developed.[376] Alkyl dithiomalonates EtO_2CCH_2CSSR, and their thionester analogues, suffer attack at the thione grouping by acyl chlorides to give keten dithioacetals and *O,S*-keten acetals $EtO_2CCH=C(OR^2)SR^1$ respectively, and by alkyl dithiochloro-formates to give dithioacetal analogues.[377] Ethyl dithiophenylacetate behaves similarly[377] and alkyl chloroformates can be used as acylating agents. Alkyl chlorothionoformates react with malononitrile and NaH to give, after alkylation, keten *O,S*-acetals, and chlorodithioformates similarly give dithioacetals.[378] A novel example of a sulphide-contraction reaction $[RC(O)CH_2SC(S)OEt + NaH \rightarrow RC(O)CH_2C(S)OEt \rightarrow RC(O)CH=C(OEt)SMe$ (after methylation)] has been discovered.[379] A more conventional route to a keten dithioacetal starts with a ketone: $R^1COR^2 + LiC(SMe)_2SiMe_3 \rightarrow R^1R^2C=C(SMe)_2$.[380]

Selenium analogues of the various thioacetals discussed above have been prepared in analogous ways, though the various methods used turn out to be the

[369] M. Mikolajczyk, B. Costisella, S. Grzejszczak, and A. Zatorski, *Tetrahedron Letters*, 1976, 477.
[370] L. Field and C. H. Banks, *J. Org. Chem.*, 1975, **40**, 2774.
[371] M. Braun and D. Seebach, *Chem. Ber.*, 1976, **109**, 669.
[372] J. Sylvestre, V. Ratovelomanan, and C. Huynh, *Compt. rend.*, 1974, **278**, C, 371.
[373] M. Martin and L. Bassery, *Compt. rend.*, 1975, **281**, C, 571.
[374] M. Yokoyama, *J.C.S. Perkin I*, 1975, 1417.
[375] L. Leger and M. Saquet, *Compt. rend.*, 1974, **279**, C, 695; *Bull. Soc. chim. France*, 1975, 657; L. Leger, M. Saquet, A. Thuillier, and S. Julia, *J. Organometallic Chem.*, 1975, **96**, 313.
[376] D. M. Vyas and G. W. Hay, *J.C.S. Perkin I*, 1975, 180.
[377] K. Thimm and J. Voss, *Z. Naturforsch.*, 1975, **30b**, 932.
[378] N. H. Nilsson, *Tetrahedron*, 1974, **30**, 3181.
[379] A. J. Bridges and G. H. Whitham, *J.C.S. Perkin I*, 1975, 1603.
[380] D. Seebach and R. Buerstinghaus, *Synthesis*, 1975, 461.

first reported preparations of these compounds in some cases;[381—383] an example
is: $PhSeH + CH_2O + Et_2NH \rightarrow PhSeCH_2NEt_2$. Mixed *S,Se*-acetals, *e.g.*
$R^1SCR^2R^3SePh$, are now accessible through the reaction between $PhSe^- K^+$ and
an appropriate α-bromo-sulphide, or by addition of PhSeH to a vinyl sulphide.[381]
Diselenoacetals $R^1R^2C(SeR^3)_2$ are made conveniently from a selenol R^3SeH and
a ketone,[383] while a route to ditelluroacetals uses a 1,1-di-iodo-alkane [PhLi +
Te → $PhTeLi(+ CH_2I_2) \rightarrow PhTeCH_2TePh$].[138]

Incidentally, the nomenclature used by Krief[381—383] will doubtless be used
increasingly in this area in the future. Names such as α-seleno-sulphides, α-seleno-
ethers, α-seleno-silanes are preferable to α-selenonio-sulphides, *etc.*[381] and are
more readily appreciated than the acetal nomenclature.

Reactions.—Double alkylation of dithioacetals $(RS)_2CH_2$ can be effected with an
alkyl bromide and $NaNH_2$.[384] The use of an alkyl-lithium compound as base
causes cleavage of an *O*-trimethylsilyl hemithioacetal, $R^1SCR^2R^3OSiMe_3$ +
$R^4Li \rightarrow R^2R^3R^4COSiMe_3$,[366] a diselenoacetal, $(PhSe)_2CR^1R^2 \rightarrow PhSeCR^1R^2$,[381—383]
and a ditelluroacetal.[138] The fact that simple thioacetals are not cleaved by an
alkyl-lithium compound allows a diselenoacetal to be converted into an α-sul-
phenyl-carbanion $[R^1R^2C(SePh)_2 + BuLi \rightarrow R^1R^2CLi(SePh) \rightarrow PhSCR^1R^2SePh$
→ $Ph\overline{C}R^1R^2$].[385] Applications of bis(α-alkyl- or aryl-thio)-carbanions in
synthesis are reviewed in this volume in the following chapter.

Anodic oxidation of dithioacetals gives the disulphide, rather than the expected
sulphoxide, sulphone, or sulphonium salt (Scheme 5).[386] Further oxidation to the
thiolsulphonate can be brought about by prolonged electrolysis. Di-t-butyl
peroxide initiates a homolytic 1,2-shift so as to put another carbon atom between
the alkylthio-groups in a dithioacetal, resulting in a 1,3-bis(alkylthio)alkane

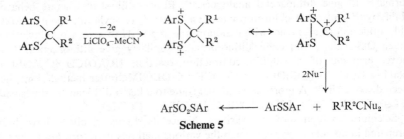

Scheme 5

$[(RCH_2)_2C(SBu)_2 \rightarrow RCH_2CH(SBu)CHRSBu]$; side-products include vinyl sulphides and butyl alkyl sulphides.[387]

[381] A. Anciaux, A. Eman, W. Dumont, D. van Ende, and A. Krief, *Tetrahedron Letters*, 1975, 1613.
[382] A. Anciaux, A. Eman, W. Dumont, and A. Krief, *Tetrahedron Letters*, 1975, 1617.
[383] J. N. Denis, W. Dumont, and A. Krief, *Tetrahedron Letters*, 1976, 453; D. van Ende and A. Krief, *ibid.*, p. 457.
[384] G. Schill and C. Merkel, *Synthesis*, 1975, 387.
[385] D. Seebach and A. K. Beck, *Angew. Chem.*, 1974, **86**, 859.
[386] J.-G. Gourcy, G. Jeminet, and J. Simonet, *J.C.S. Chem. Comm.*, 1974, 634.
[387] R. G. Petrova, T. D. Churkina, S. A. Karapetyan, and R. K. Freidlina, *Izvest. Akad. Nauk S.S.S.R., Ser. khim.*, 1974, 2517; R. G. Petrova and T. D. Churkina, *ibid.*, 1975, 716.

Kinetics of hydrolysis of thioacylals $RSCH_2OAc$ (R = Me or Ph) have been studied.[388]

α-Oxoketen dithioacetals [*e.g.* (39)] are useful three-carbon fragments for participation with cyanoacetamide in the synthesis of (2*H*)-pyridones (40).[389]

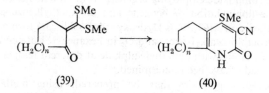

(39)   (40)

Thioacylketen dithioacetals participate with alkenes and sulphenes in 1,4-cyclo-addition reactions.[390] Thio-Claisen rearrangement of 1-allylthio-1-aminoalkenes to thioamides or isothiocyanates,[391] and of *S*-allyl-*S*-methyl-keten mercaptals to dithioesters,[392] has been described.

While orthoesters do not react with trifluoroacetic anhydride, aryltrithio-orthoacetates $(ArS)_3CMe$ give keten thioacetals $(ArS)_2C=CHC(O)CF_3$, by way of the parent keten dithioacetal $(ArS)_2C=CH_2$;[393] acetic anhydride does not react. Aryl trithio-orthoacetates readily exchange ^1H with $CF_3CO_2{}^2H$ as a consequence of the reversible loss of ArSH, and this can be exploited in the synthesis of deuteriated hydrocarbons, *e.g.* $(ArS)_2CArC^2H_3 \rightarrow ArCH_2C^2H_3$ (after treatment with Raney nickel).[394]

E.s.r. spectra of tris(organothio)methyl radicals, formed by heating tetrakis-(organothio)methanes and hexakis(alkylthio)ethanes, have been interpreted to show deviation from planarity of their tervalent carbon centre.[395]

6 Sulphoxides

Preparation.—The most straightforward route for the preparation of sulphoxides is the oxidation of the corresponding sulphide. A novel reagent for this purpose is SO_2Cl_2 together with wet silica gel;[396] if no water is present, an α-chloro-sulphide results. Oxidation of a sulphide with t-butyl hydroperoxide in the presence of bis(acetylacetonato)oxovanadium(IV) gives the sulphoxide at a rate that is convenient for kinetic study.[397] Dimethyl sulphoxide and hydrogen chloride oxidizes L-methionine to the sulphoxide,[398] and bis(alkylthio)alkanes $RS(CH_2)_nSR$ to the corresponding bis(alkylsulphinyl) analogues.[399] Both *cis*- and *trans*-1,2-bis(trifluoromethylthio)ethylenes give only 50% of the bis(trifluoro-methylthio)oxirans by oxidation with peroxytrifluoroacetic acid, the major

[388] R. A. McClelland, *Canad. J. Chem.*, 1975, **53**, 2772.
[389] R. R. Rastogi, H. Ila, and H. Junjappa, *J.C.S. Chem. Comm.*, 1975, 645.
[390] R. Okazaki, A. Kitamura, and N. Inamoto, *J.C.S. Chem. Comm.*, 1975, 257.
[391] F. C. V. Larsson and S.-O. Lawesson, *Tetrahedron*, 1974, **30**, 1283.
[392] L. Jensen, L. Dalgaard, and S.-O. Lawesson, *Tetrahedron*, 1974, **30**, 2413.
[393] M. Hojo, R. Masuda, and Y. Kamitori, *Tetrahedron Letters*, 1976, 1009.
[394] M. Hojo, R. Masuda, K. Yamane, and H. Takahashi, *Tetrahedron Letters*, 1975, 1899.
[395] H. B. Stegmann, K. Scheffler, and D. Seebach, *Chem. Ber.*, 1975, **108**, 64.
[396] M. Hojo and R. Masuda, *Tetrahedron Letters*, 1976, 613.
[397] R. Curci, F. Di Furia, R. Testi, and G. Modena, *J.C.S. Perkin II*, 1974, 752.
[398] S. H. Lipton and C. E. Bodwell, *J. Agric. Food Chem.*, 1976, **24**, 26.
[399] C. M. Hull and T. W. Bargar, *J. Org. Chem.*, 1975, **40**, 3152.

products being the bis(trifluoromethanesulphinyl) analogues.[400] Pulse radiolysis of aqueous solutions of dialkyl sulphides gives $R_2\dot{S}OH$, which reacts further to give the short-lived radical complex $(R_2S)_2OH\cdot$ from which $(R_2S)_2^+ OH^-$ and eventually the sulphoxide is formed.[401] A representative citation for the synthesis of an aromatic sulphoxide employing a general route for this class of compound describes arenesulphinylation of benzene and polymethylbenzenes with arene-sulphinyl chlorides and $AlCl_3$.[402] β-Hydroxy-sulphoxides are formed in irradiated mixtures of arenethiols, alk-1-enes, and O_2 in hexane.[403] Rate constants for the reaction of $O(^3P)$ atoms with dimethyl sulphide at different temperatures, to give dimethyl sulphoxide, have been determined.[404]

Optically active sulphoxides may be prepared, using methods involving asymmetric induction, starting with achiral sulphides and *N*-bromo-ε-capro-lactam together with an optically active alcohol [*e.g.* $(-)$-menthol], $R^1SR^2 \rightarrow R^1R^2\overset{+}{S}OR^3 \rightarrow R^1S(O)R^2$,[405] or employing the analogous combination of Bu^tOCl and $(-)$-menthol.[406] A strain of *Aspergillus niger* has been shown to be capable of oxidizing a sulphide to the corresponding optically active sulphoxide.[407] Optically active alkoxy diarylsulphonium salts react with halide ions to give sulphoxides with retention of configuration when Cl^-, Br^-, or I^- is used, but with inversion of configuration when F^- is used. Retention is considered to involve nucleophilic substitution at carbon, but inversion is tentatively suggested to involve nucleophilic substitution at sulphur, involving apical attack by F^- and equatorial departure of the leaving group [*e.g.* Scheme 6; (S)-(41) $\rightarrow$ (R)-(42)],

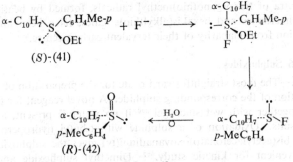

Scheme 6

or a difluorosulphurane intermediate, or Berry pseudorotation.[408] Menthyl sulphinates react with appropriately substituted propargyl Grignard reagents to give allenic sulphoxides which are asymmetric both at the allene moiety and at

[400] Y. V. Samusenko, A. M. Aleksandrov, and L. M. Yagupolskii, *Ukrain. khim. Zhur.*, 1975, **41**, 397.

[401] M. Bonifacic, H. Möckel, D. Bahnemann, and K.-D. Asmus, *J.C.S. Perkin II*, 1975, 675.

[402] G. A. Olah and J. Nishimura, *J. Org. Chem.*, 1974, **39**, 1203.

[403] S. Iriuchijima, K. Maniwa, T. Sakakibara, and G. Tsuchihashi, *J. Org. Chem.*, 1974, **39**, 1170.

[404] J. H. Lee, R. B. Timmons, and L. J. Stief, *J. Chem. Phys.*, 1976, **64**, 300.

[405] M. Kinoshita, Y. Sato, and N. Kunieda, *Chem. Letters*, 1974, 377.

[406] M. Moriyama, S. Oae, T. Numata, and N. Furukawa, *Chem. and Ind.*, 1976, 163.

[407] B. J. Auret, D. R. Boyd, H. B. Henbest, C. G. Watson, K. Balenovic, V. Polak, V. Johanides, and S. Divjak, *Phytochemistry*, 1974, **13**, 65.

[408] R. Annunziata, M. Cinquini, and S. Colonna, *J.C.S. Perkin I*, 1975, 404.

the sulphoxide sulphur centre.[409] The analogous preparation of (*R*)-PhCH$_2$S(O)-C^{2}H$_2$Ph has been described.[410] Arenesulphinate esters react with cyclopentanone and NaH, or with ArSO$_2$Me and base, to yield the arenesulphinyl ketone, or the sulphinylsulphone ArSO$_2$CH$_2$SOAr′, respectively.[411]

Properties and Reactions of Sulphoxides.—A kinetic study has been reported for the reduction of dialkyl sulphoxides with H$_2$S to give the sulphide, H$_2$O, and S$_8$.[412] Ionic intermediates are involved, and the catalytic role of sulphoxides in the reaction of H$_2$S with SO$_2$ is thus clarified. Laboratory methods for the reduction of sulphoxides to sulphides described in the recent literature include both new methods and developments of known methods. Mild conditions are used for deoxygenation with I$_2$–pyridine–SO$_2$ and quantitative conversion into the sulphide has been reported (Br$_2$ can be used in place of I$_2$).[413] Catalytic hydrogenation (H$_2$–Pd/C) is effective.[414] Hydride reagents, sodium bis-(2-methoxyethoxy)aluminium hydride,[415] and sodium cyanohydridoborate in conjunction with 18-crown-6-ether[416] have been used for the direct reduction of sulphoxides to sulphides and for the reduction of the methoxysulphonium salts formed by alkylation of the sulphoxide with methyl fluorosulphonate, respectively. 2-Phenoxy-1,3,2-benzodioxaphosphole deoxygenates sulphoxides under mild conditions,[417] and dibenzyl selenoxide is efficiently reduced with Ph$_3$P.[418]

Oxidation of sulphoxides to sulphones is readily accomplished using familiar oxidants; a novel observation, that *N*-oxides of tertiary and heterocyclic amines are reduced by sulphoxides (R$^1_3$N$-$Ō + R$^2_2$SO → R$^1_3$N$-$O$-$SR$^2_2-$Ō → R$^1_3$N + R$^2_2$SO$_2$),[419] may be as valuable in amine chemistry as in organosulphur chemistry. Its value is enhanced since the intermediate adduct may be captured by a nucleophile, thus giving a sulphinate as an oxidation product of the sulphoxide.[419]

An alkanesulphinyl substituent activates a benzene ring towards molecular halogenation,[420] and towards nitration with HNO$_3$–H$_2$SO$_4$.[421] The methane-sulphinyl group can be classified as a good leaving group in acid-catalysed reactions, illustrated by the formation of (43)[422] together with products from the decomposition of methanesulphenic acid.

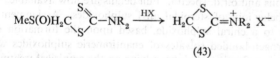

(43)

[409] M. Cinquini, S. Colonna, and C. J. M. Stirling, *J.C.S. Chem. Comm.*, 1975, 256.
[410] K. K. Andersen, M. Cinquini, S. Colonna, and F. C. Pilar, *J. Org. Chem.*, 1975, **40**, 3780.
[411] R. M. Coates and H. D. Pigott, *Synthesis*, 1975, 319.
[412] Y. Mehmet and J. B. Hyne, *Preprints, Div. Petrol. Chem., Amer. Chem. Soc.*, 1974, **19**, 187.
[413] M. Nojima, T. Nagata, and N. Tokura, *Bull. Chem. Soc. Japan*, 1975, **48**, 1343.
[414] K. Ogura, M. Yamashita, and G. Tsuchihashi, *Synthesis*, 1975, 385.
[415] T.-L. Ho and C. M. Wong, *Org. Prep. Proced. Internat.*, 1975, **7**, 163.
[416] H. D. Durst, J. W. Zubrick, and G. R. Kieczykowski, *Tetrahedron Letters*, 1974, 1777.
[417] M. Dreux, Y. Leroux, and P. Savignac, *Synthesis*, 1974, 506.
[418] S. Tamagaki, I. Hatanaka, and K. Tamura, *Chem. Letters*, 1976, 81.
[419] M. E. C. Biffin, G. Bocksteiner, J. Miller, and D. B. Paul, *Austral. J. Chem.*, 1974, **27**, 789.
[420] A. C. Boicelli, R. Danieli, A. Mangini, A. Ricci, and G. Pirazzini, *J.C.S. Perkin II*, 1974, 1343.
[421] N. C. Marziano, E. Maccarone, G. M. Cimino, and R. C. Passerini, *J. Org. Chem.*, 1974, **39**, 1098.
[422] Y. Ueno, Y. Masuyama, and M. Okawara, *Tetrahedron Letters*, 1974, 2577.

Reactions of sulphoxides leading to alkoxysulphonium salts have been augmented by a procedure using an alkyl chlorosulphinate and $SbCl_5$ [$MeOS(O)Cl + Me_2SO \rightarrow MeSCH_2\overset{+}{S}Me_2\ SbCl_6^-$, which with EtOH gives $Me_2\overset{+}{S}OEt\ SbCl_6^-$].[423]

Tetraco-ordinate sulphur(IV) species are increasingly coming to mind when hypothetical reaction mechanistic schemes are being devised. The formation of di(*p*-tolyl) sulphide, *p,p'*-bitolyl, and *m,p'*-bitolyl from di(*p*-tolyl) sulphoxide and *p*-tolyl-lithium suggests a 3-toluyne intermediate, formed from the sulphur(IV) precursor $Tol_3SO^-\ Li^+$.[424] The (*S*)-(−)-sulphoxide (44), on treatment with

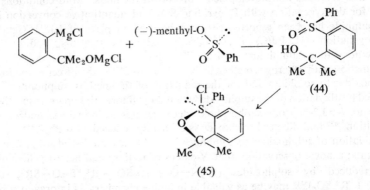

(44)

(45)

acetyl chloride, gives the first example of an optically active tetraco-ordinate sulphur(IV) species, *viz.* the chloro-sulphurane (45).[425] Of course, tellurium(IV) compounds form more readily [*e.g.* $Ar_2TeO + RC(O)OC(O)R \rightarrow Ar_2Te(OCOR)_2$].[426]

Reactions with some stereochemical significance have occupied a central position in sulphoxide chemistry in recent years. Straightforward methods are used for the resolution of carboxymethyl sulphoxides $RS(O)CH_2CO_2H$ (R = Ar or $ArCH_2$),[427] a point of interest in these compounds being their solvent-dependent optical rotations and o.r.d. spectra. Full details are now available[428] accounting for the c.d. behaviour of steroidal allyl sulphoxides. Assignment of absolute configuration to a chiral sulphoxide, based upon the formation of diastereoisomeric hydrogen-bonded solvates of enantiomeric sulphoxides with (*R*)-(−)-$CF_3CHPhOH$ is a simple technique, relying on the empirical treatment of n.m.r. data.[429] It is particularly useful for assessing the enantiomeric purity of a chiral sulphoxide. An alternative simple method for determining the absolute con-

[423] Y. Hara and M. Matsuda, *J.C.S. Chem. Comm.*, 1974, 919.
[424] B. K. Ackerman, K. K. Andersen, I. Karup-Nielsen, N. B. Peynircioglu, and S. A. Yeager, *J. Org. Chem.*, 1974, **39**, 964.
[425] T. M. Balthazor and J. C. Martin, *J. Amer. Chem. Soc.*, 1975, **97**, 5634.
[426] I. D. Sadekov, A. A. Maksimenko, A. I. Usachev, and V. I. Minkin, *Zhur. obshchei Khim.*, 1975, **45**, 2562.
[427] M. Janczewski and S. Dacka, *Roczniki Chem.*, 1974, **48**, 753; M. Janczewski and B. Dziurzynska, *ibid.*, p. 409.
[428] D. N. Jones, J. Blenkinsopp, A. C. F. Edmonds, E. Helmy, and R. J. K. Taylor, *J.C.S. Perkin I*, 1974, 937.
[429] W. H. Pirkle, S. D. Beare, and R. L. Muntz, *Tetrahedron Letters*, 1974, 2295.

figuration of enantiomers of a sulphoxide employs (S)-$(+)$-α-phenylbutyroyl chloride, which is allowed to undergo incomplete reaction with the racemic sulphoxide in the presence of a tertiary amine. An empirical relationship,[430] which states that the predominant enantiomer in the unreacted sulphoxide will be of absolute configuration (46), allows absolute configurational assignments

$$
\begin{array}{c}
O \\
\parallel \\
M \cdots S-: \\
L \nearrow
\end{array}
$$

(46)

to be made to the $(+)$- and $(-)$-enantiomers of the sulphoxide when the optical rotation of the recovered unreacted sulphoxide can be determined. It is necessary, of course, to separate the reaction products and reagent from the unreacted sulphoxide, and t.l.c. methods are suitable.[430]

β-Deuterium isotope effects on protonation equilibria and racemization kinetics of $(+)$-alkyl t-butyl sulphoxides in $HClO_4$ indicate the formation of the Bu^{t+}–$RSOH$ cation–molecule pair as the first step in cleavage of the protonated sulphoxide.[431] Electrophilic α-chlorination and α-bromination of benzyl methyl sulphoxide introduces the halogen atom through two pathways, one involving retention, the other inversion, at sulphur. Full details of earlier published work (see Vol. 3, p. 43) are now available.[432] (R)-$(+)$-$PhC^1H_2S(O)C^2H_2Ph$ gives one of the four possible diastereoisomeric α-halogeno-sulphoxides on treatment with $PhICl_2$ or Br_2–pyridine, and provides the first example of asymmetric induction promoted by an isotopically dissymmetric centre; ^1H is substituted in preference to ^2H.[433]

Allenmark's work on anchimerically assisted reactions of sulphoxides[434] continues with a study of the rates of reduction of the *syn*- and *anti*-forms of *endo-cis*-3-benzenesulphinylbicyclo[2,2,1]heptane-2-carboxylic acids. The *syn*-was more reactive than the *anti*-isomer by a factor of more than 3.2×10^3.[435] The relative ease of formation of the acyloxysulphonium ion resulting from nucleophilic attack by the carboxy-group on sulphur accounts for this result, and also for the relative racemization rates of β-carboxy-substituted sulphoxides.[436] Pummerer reaction of one enantiomer of a benzyl *o*-carboxyphenyl sulphoxide in the presence of DCCI leads to 2-phenyl-3,1-benzoxathian-4-one (47) in an optically active form, indicating the transfer of chirality from S to C.[437]

X-Ray analysis of 1-(*p*-bromophenyl)ethyl t-butyl sulphoxide confirms the *gauche* relationship between aryl and t-butyl groups implied by chemical studies.[438]

430   S. Juge and H. B. Kagan, *Tetrahedron Letters*, 1975, 2733.
431   P. Bonvicini, A. Levi, and G. Scorrano, *Gazzetta*, 1974, **104**, 1.
432   M. Cinquini, S. Colonna, and F. Montanari, *J.C.S. Perkin I*, 1974, 1719.
433   M. Cinquini and S. Colonna, *J.C.S. Chem. Comm.*, 1974, 769.
434   S. Allenmark, *Internat. J. Sulfur Chem.*, 1973, **8**, 127.
435   H. Johnsson and S. Allenmark, *Chem. Scripta*, 1975, **8**, 216, 223.
436   C. E. Hagberg and S. Allenmark, *Chem. Scripta*, 1974, **5**, 13.
437   B. Stridsberg and S. Allenmark, *Acta Chem. Scand. (B)*, 1974, **28**, 591; see also S. Oae and T. Numata, *Tetrahedron*, 1974, **30**, 2641.
438   Y. Iitaka. Y. Kodoma, K. Nishihata, and M. Nishio, *J.C.S. Chem. Comm.*, 1974, 389.

(47)

Pummerer rearrangement processes have been described in several papers.[437, 439-441] Typically,[439] $RS(O)(CH_2)_4SR \rightarrow RSCH=CH(CH_2)_2SR$ on treatment with Ac_2O, and the formation of the ring-expanded lactone *via* 2-hydroxy-2-methylsulphinylmethyltetrahydrofuran,[440] illustrate the possibilities of these processes in synthesis. A silicon-Pummerer rearrangement, $Me_3Si-CH_2S(O)Ph \rightarrow CH_2=S(OSiMe_3)Ph \rightarrow Me_3SiOCH_2SPh$, effected at 60 °C, has been studied further some seven years after the process was discovered;[441] substantial amounts of vinyl sulphides are also formed.

Alternative rearrangement processes open to sulphoxides have been illustrated recently. Thio-Claisen rearrangement of a simple allyl aryl sulphoxide (48) has

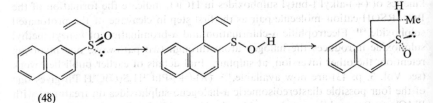

(48)

been reported for the first time.[442] Aryl prop-2-ynyl sulphoxides give condensed thiophens through consecutive [2,3]-sigmatropic and [3,3]-sigmatropic rearrangements[443] followed by ring-closure in acid media. The first of these rearrangement processes leads to an allenic sulphenate,[443] and the alternative version of this process, the equilibrium between an alk-2-ynyl sulphenate (49) and the isomeric allenic sulphoxide (50), is believed[409] to account for the mutarotation of

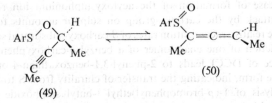

(49)   (50)

allenic sulphoxides in acetone at room temperature. This is considered a more likely explanation than pyramidal inversion.[409]

An unexpected thermal rearrangement of a 2-pyridylaldehyde dimethyl-dithioacetal *S*-oxide to a sulphinate has been encountered during the exploration

[439] G. M. Prokhorov, V. I. Dronov, R. V. Zainullina, E. E. Zaev, N. S. Lyubopytova, V. I. Khvostenko, and E. G. Galkin, *Zhur. org. Khim.*, 1974, **10**, 1852.
[440] B. M. Trost and C. H. Miller, *J. Amer. Chem. Soc.*, 1975, **97**, 7182.
[441] E. Vedejs and M. Mullins, *Tetrahedron Letters*, 1975, 2017.
[442] Y. Makisumi, S. Takada, and Y. Matsukura, *J.C.S. Chem. Comm.*, 1974, 850.
[443] Y. Makisumi and S. Takada, *J.C.S. Chem. Comm.*, 1974, 848.

of a route from 2-bromopyridines to 2-carbaldehydes making use of MeSCH$_2$-S(O)Me.[444]

The selenium equivalent of the known rearrangement of benzyl phenyl sulphoxides to benzaldehydes and diphenyl sulphide is brought about under milder conditions, 110—130 °C during 2—3 min., than those (210—280 °C) required for sulphoxides.[445]

The basicity–structure relationships applying to a series of diaryl sulphoxides are concluded to be dominated by the inductive effects of the aryl substituents.[446]

α-Functional Sulphoxides.—α-Halogeno-sulphoxides (see also preceding section) may be prepared using N-halogeno-succinimides,[447] N-chlorobenzotriazole,[448] SO$_2$Cl$_2$,[449] PhICl$_2$,[433] or Br$_2$–pyridine.[433] Haloform reactions with β-keto-sulphoxides have been used in the synthesis of a series of α-bromo- and α-iodo-sulphoxides.[450] Phenylsulphinyl chloride reacts with diazomethane to give a mixture of chloromethyl phenyl sulphoxide and diazomethyl phenyl sulphoxide.[451] The latter compound reacts as phenylsulphinylcarbene, and is the first example of this structural type. The high *anti*-stereospecificity obtained in this case is in contrast to the corresponding sulphenylcarbene, from which cyclopropyl sulphides are obtained practically non-stereoselectively.[451]

Sulphinyl-conjugated radicals R^{1}S(O)ĊR^{2}R^3, are formed from 1,3-bis(alkane-sulphinyl)alkanes and HO· radical, or from a sulphoxide with Ph· radical.[452]

Discussion of structures and synthetic application of α-sulphinyl carbanions is in Chapter 2 in this volume.

β-Keto-sulphoxides and Related Compounds.—Standard syntheses have been illustrated in the recent literature; methyl benzenesulphinate reacts with an active methylene compound in the presence of NaH to give a β-keto-sulphoxide,[453] and preparations from cyanohydrins have been described.[454]

pK_a Values of a series of substituted α-phenylsulphinylacetophenones are readily correlated with Hammett σ-values.[455]

New uses for β-keto-sulphoxides in synthesis are a one-step conversion into α-acetoxy-thiolesters with Ac$_2$O and NaOAc [PhC(O)CH$_2$S(O)Ar → PhCH(OAc)-C(O)SAr],[456] conversion of a 2-phenylsulphinylcyclohexanone into a 2-phenyl-thiocyclohex-2-en-1-one (Pummerer rearrangement),[457] and the acid-catalysed cyclization of β-keto-sulphoxides of general structure ArCHR^{1}CHR^{2}C(O)-

[444] G. R. Newkome, J. M. Robinson, and J. D. Sauer, *J.C.S. Chem. Comm.*, 1974, 410.
[445] I. D. Entwistle, R. A. W. Johnstone, and J. H. Varley, *J.C.S. Chem. Comm.*, 1976, 61.
[446] R. P. Fedoezzhina, E. P. Buchikhin, E. A. Kanevskii, and A. I. Zarubin, *Zhur. obshchei Khim.*, 1974, **44**, 877, 1351.
[447] F. Jung, K. C. Tin, and F. Durst, *Internat. J. Sulfur Chem.*, 1973, **8**, 1.
[448] D. Landini and A. Maia, *J.C.S. Perkin II*, 1975, 218.
[449] B. B. Jarvis and H. E. Fried, *J. Org. Chem.*, 1975, **40**, 1278.
[450] P. Del Buttero, S. Maiorana, and E. Marone, *Chimica e Industria*, 1974, **56**, 120.
[451] C. G. Venier, H. J. Barager, and M. A. Ward, *J. Amer. Chem. Soc.*, 1975, **97**, 3238.
[452] P. M. Carton, B. C. Gilbert, H. A. H. Lane, R. O. C. Norman, and R. C. Sealy, *J.C.S. Perkin II*, 1975, 1245.
[453] H. J. Monteiro and J. P. de Souza, *Tetrahedron Letters*, 1975, 921.
[454] S. Iriuchijima and G. Tsuchihashi, *Synthesis*, 1975, 401.
[455] N. Kunieda, Y. Fujiwara, J. Nokami, and M. Kinoshita, *Bull. Chem. Soc. Japan*, 1976, **49**, 575.
[456] S. Iriuchijima, K. Maniwa, and G. Tsuchihashi, *J. Amer. Chem. Soc.*, 1975, **97**, 596.
[457] H. J. Monteiro and A. L. Gemal, *Synthesis*, 1975, 437.

CHR³S(O)Me to give 2-methylthio-3-oxo-tetrahydronaphthalenes.[458] Synthesis of a heterocyclic analogue starting from the corresponding pyrrole has been reported.[459] Synthesis of a known intermediate for use in an oestrone synthesis starts from the homologue (51), which is subjected to thermal elimination in the presence of 2-methylcyclopentane-1,3-dione.[460]

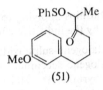

(51)

One equivalent of a Grignard reagent gives the unusual organomagnesium compound ArS(O)CH(MgX)C(O)Ar′, for which some uses in synthesis have been explored.[461, 462] Addition of a second equivalent of the Grignard reagent RMgX leads to the formation of the tertiary alcohol ArS(O)CH₂CR(OH)Ar′. When an optically active sulphoxide is used, *e.g.* (R)-(+)-TolS(O)CH₂C(O)Ph, the resulting diastereoisomer mixture with 2 equivalents of Grignard reagent contains 70% of the (RS)-isomer.[462] α-Phenylsulphonylacetates react similarly with RMgX.[461]

α-Sulphinyl esters may be used in the synthesis of α-keto-esters [R¹S(O)-CHR²CO₂Et + HCl in CHCl₃ → R¹SCR²ClCO₂Et → R²C(O)CO₂Et].[463] As expected from their propensity towards α-carbanion formation, α-sulphinyl esters can be α-alkylated, can be converted into αβ-unsaturated esters, and can participate in Michael addition reactions.[463] Treatment with NaH or LiPriₐN gives the ester enolate RS(O)CH=C(OEt)O⁻ Na⁺ (or Li⁺), which gives ethyl cinnamate with benzyl bromide.[464]

A crystalline, thermally stable, β-sultine is obtained by cyclization of the β-hydroxy-sulphoxide Ph₂C(OH)CMe₂S(O)Buᵗ with SO₂Cl₂, confirming and developing an earlier observation.[465]

Unsaturated Sulphoxides.—Variations of known methods for the synthesis of vinyl sulphoxides are described in reports of asymmetric selectivity in the elimination of HCl from β-halogenoethyl aryl sulphoxides using an optically active base, leading to partially resolved vinyl sulphoxides,[466] and in reports of the condensation of carbonyl compounds with (EtO)₂P(O)CH₂S(O)R in the presence of a phase-transfer catalyst.[467, 468] Vinyl sulphoxides are formed when sulphenic acids are trapped after generation in the presence of an acetylenic ester, and a full account is available of the trapping of the sulphenic acid of

⁴⁵⁸ Y. Oikawa and O. Yonemitsu, *Tetrahedron*, 1974, **30**, 2653; *J. Org. Chem.*, 1976, **41**, 1118.
⁴⁵⁹ Y. Oikawa, O. Setoyama, and O. Yonemitsu, *Heterocycles*, 1974, **2**, 21.
⁴⁶⁰ Y. Oikawa, T. Kurosawa, and O. Yonemitsu, *Chem. and Pharm. Bull.* (*Japan*), 1975, **23**, 2466
⁴⁶¹ N. Kunieda, J. Nokami, and M. Kinoshita, *Tetrahedron Letters*, 1974, 3997.
⁴⁶² N. Kunieda, J. Nokami, and M. Kinoshita, *Chem. Letters*, 1974, 369.
⁴⁶³ J. J. A. van Asten and R. Louw, *Tetrahedron Letters*, 1975, 671.
⁴⁶⁴ B. M. Trost, W. P. Conway, P. E. Strege, and T. J. Dietsche, *J. Amer. Chem. Soc.*, 1974, **96**, 7165.
⁴⁶⁵ T. Durst and B. P. Gimbarzersky, *J.C.S. Chem. Comm.*, 1975, 724.
⁴⁶⁶ G. Marchese, F. Naso, and L. Ronzini, *J.C.S. Chem. Comm.*, 1974, 830.
⁴⁶⁷ M. Mikolajczyk and S. Grzejszczak, and A. Zatorski, *J. Org. Chem.*, 1975, **40**, 1979.
⁴⁶⁸ M. Mikolajczyk, S. Grzejszczak, W. Midura, and A. Zatorski, *Synthesis*, 1975, 278.

greatest current interest, that derived from thermolysis of a penicillin sulphoxide.[469] A route of outstanding interest has been developed following the establishment of synthetic methods for the formation of sulphines. *S*-Alkylation of thiocamphor *S*-oxide (52) with an alkyl halide and TlOEt gives the vinyl sulphoxide (53).[470] Mono-alkyl copper(I) reagents R^3Cu undergo *cis*-addition to 1-alkynyl sulphoxides [$R^1C≡CS(O)R^2 → R^1R^3C=CHS(O)R^2$].[471]

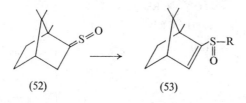

(52)   (53)

Apart from one citation,[472] all the papers collected here to illustrate the reactions of vinyl sulphoxides deal with addition reactions.[473-481] As would be expected from the nature of the sulphoxide functional group, there are several novel features in these addition reactions compared with those of other vinyl compounds. Pyrolysis of alkyl styryl sulphoxides gives benzothiophens. The proposed mechanism, which involves formation and homolysis of a sulphenic acid followed by cyclization, (54) → (55), is supported by the formation of benzothiophens from styryl sulphides and disulphides at 580 °C.[472] Addition of trialkylboranes to styryl sulphoxides gives *trans*-alkenes [$PhCH=CHS(O)Me + R_3B → PhCH=CHR$] *via* the vinyl radical,[473] while addition of PCl_5 to a styryl sulphoxide in CH_2Cl_2 gives the novel rearrangement products $PhC(O)CH_2SR$, $PhCH(SR)CHO$, and $PhCCl=CHSR$.[474] The fact that Michael addition to phenyl vinyl sulphoxide can be followed immediately by thermal *syn*-elimination provides a route for the introduction of a vinyl group into an activated carbon centre, $CH_2=CHS(O)Ph + R^1R^2R^3C^-(H^+) → R^1R^2R^3CCH_2CH_2S(O)Ph → R^1R^2R^3CCH=CH_2$.[475] Diels–Alder addition of cyclopentadiene to methyl 1-bromovinyl sulphoxide gives an adduct (56) which is susceptible to acid-catalysed rearrangement to the corresponding sulphenate and subsequent elimination of MeSBr, which then adds to the C=C grouping to give the bicyclo-[2,2,1]heptan-2-one (57).[476] In this overall route, the vinyl sulphoxide has acted as a keten synthon.

[469] D. H. R. Barton, I. H. Coates, P. G. Sammes, and C. M. Cooper, *J.C.S. Perkin I*, 1974, 1459.
[470] G. E. Veenstra and B. Zwanenburg, *Rec. Trav. chim.*, 1976, **95**, 37.
[471] W. E. Truce and M. J. Lusch, *J. Org. Chem.*, 1974, **39**, 3174; P. Vermeer, J. Meijer, and C. E. Eylander, *Rec. Trav. chim.*, 1974, **93**, 240.
[472] W. Ando, T. Oikawa, K. Kishi, T. Saiki, and T. Migita, *J.C.S. Chem. Comm.*, 1975, 704.
[473] N. Miyamoto, D. Fukuoka, K. Utimoto, and H. Nozaki, *Bull. Chem. Soc. Japan*, 1974, **47**, 503.
[474] N. Miyamoto, D. Fukuoka, K. Utimoto, and H. Nozaki, *Bull. Chem. Soc. Japan*, 1974, **47**, 1817.
[475] G. A. Koppel and M. D. Kinnick, *J.C.S. Chem. Comm.*, 1975, 473.
[476] J. C. Philips, M. Penzo, and G. T. S. Lee, *J.C.S. Chem. Comm.*, 1975, 107.
[477] D. J. Abbott, S. Colonna, and C. J. M. Stirling, *J.C.S. Perkin I*, 1976, 492.
[478] G. Tsuchihashi, S. Mitamura, and K. Ogura, *Tetrahedron Letters*, 1976, 855.
[479] J. J. Hansen and A. Kjaer, *Acta Chem. Scand. (B)*, 1974, **28**, 418.
[480] H. Kosugi, H. Uda, and S. Yamagiwa, *J.C.S. Chem. Comm.*, 1976, 71.
[481] H. Kosugi, H. Uda, and S. Yamagiwa, *J.C.S. Chem. Comm.*, 1975, 192.

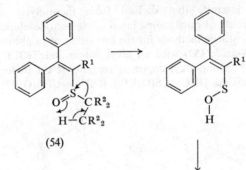

(54)

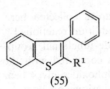

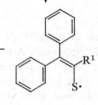

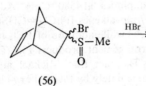

(55)

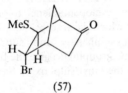

(56)

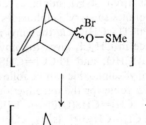

MeS

(57)

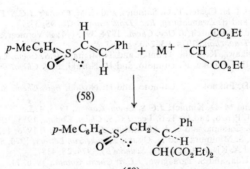

(58)

(59)

Asymmetric induction is a factor in nucleophilic and electrophilic additions to a chiral vinyl sulphoxide.[477–479] A full account is available [477] of asymmetric syntheses involving the addition of piperidine and of bromine to chiral propenyl and vinyl *p*-tolyl sulphoxides, respectively (see Vol. 2, p. 50). An observation enhancing the value of a vinyl sulphoxide enantiomer in organic synthesis concerns the control exerted by the metal counterion in Michael addition of the diethyl malonate carbanion to (*R*)-(+)-*trans*-β-styryl *p*-tolyl sulphoxide (58).[478] The K⁺ and Na⁺ salts in EtOH lead to a diastereoisomer mixture in which (59) predominates, while the Li⁺ salt in THF as the reaction medium leads to a preponderance of the epimeric adduct.[478]

The thiourea derived from the chiral vinyl sulphoxide sulphoraphene (60), a constituent of radish seeds, undergoes cyclization to give unequal amounts of diastereoisomer (61) and its epimer, a consequence of asymmetric induction in the intramolecular addition reaction.[479]

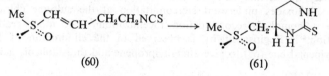

(60)   (61)

An additive Pummerer rearrangement is an apt description [480] of the course of the reaction between AcCl and the vinyl sulphoxide (62). Hot Ac₂O converts (62) into products (63) and (64) of a vinylogous Pummerer rearrangement.[481]

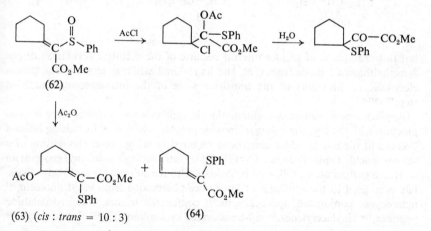

(62)

(63) (*cis* : *trans* = 10 : 3)   (64)

An outstanding new synthesis of amino-acids from nitriles and $MeSCH_2S(O)Me$ depends upon a Pummerer-type rearrangement of the intermediate $RC(NH_2)=C(SMe)S(O)Me$ with Ac_2O to give $MeSCR(NHAc)C(O)SMe$.[482]

Elimination Reactions of Aliphatic Sulphoxides.—The general process $R^1S(O)-CR^2R^3CR^4R^5H \rightarrow R^2R^3C=CR^4R^5 + R^1SOH$ represents the area in which current activity is mostly concentrated, though an interesting variation [483]

[482] K. Ogura and G. Tsuchihashi, *J. Amer. Chem. Soc.*, 1974, **96**, 1960.
[483] M. Naruse, K. Utimoto, and H. Nozaki, *Tetrahedron*, 1974, **30**, 2159.

permits the conversion of a terminal alkyne $HC\equiv CR^1$ into a homologue, $R^2C\equiv CR^1$, through treatment of the derived lithium trialkylalkynylborate with MeSOCl followed by *cis*-elimination of the methanesulphinyl and dialkylboron groups.

The regiospecificity of the sulphoxide elimination reaction leading to alkenes is subject to dipole–dipole interactions as well as to steric effects.[484] The predominant alkene formed from $R^1S(O)CMe(CHR^2R^3)CO_2Me$ is $CH_2=C(CHR^1R^2)CO_2Me$, indicating the Me group to be the preferred site for proton abstraction.[484] A preference for endocyclic $C=C$ bond formation in elimination from an α-methyl-α-phenylsulphinyl-γ-butyrolactone was dictated by the conformational rigidity of the particular substrate used.[484] The first example has been reported of regiospecific generation of isomeric alkenes from two sulphoxides that are diastereoisomeric at sulphur, *viz.* (*R*)- and (*S*)-3α-(1-adamantyl-sulphinyl)-5α-cholestanes, which give 5α-cholest-3-ene and 5α-cholest-2-ene, respectively, on thermolysis at 110 °C.[485] The corresponding steroidal diphenyl-methyl sulphoxides underwent stereomutation at the sulphur chiral centre, through a homolysis–recombination racemization path, more readily than elimination.[485] Regiospecificity is observed in the elimination of 5-t-butyl-sulphinylpent-1-ene (65) to give 2-methylpropene and the sulphenic acid (66),[486]

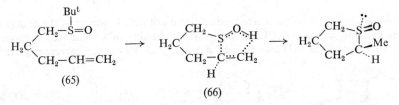

(65)   (66)

but this example is of greater interest because of the exclusive formation of *cis*-2-methylthiolan 1-oxide from (66), and its rationalization in terms of the stereo-electronic requirements of the transition state of the intramolecular addition step.[486, *cf.* 442]

Applications of sulphoxide elimination in synthesis are reviewed here in conjunction with the corresponding selenoxide process, which is of increasing interest because of the much milder conditions required to bring about elimination of a selenenic acid. Generation of a $C=C$ bond by introducing a sulphenyl group at an activated carbon atom, followed by oxidation to the sulphoxide and thermolysis, has been used in the synthesis of honey-bee pheromone from ethyl linoleate,[487] macrocyclic conjugated hydrocarbons,[488] conjugated trienes,[489] dehydroalanine peptides,[490] vinylacetylenes,[491] αβ-unsaturated cycloalkanones,[492] β-keto-esters,[493]

[484] B. M. Trost and K. K. Leung, *Tetrahedron Letters*, 1975, 4197.
[485] D. N. Jones, A. C. F. Edmonds, and S. D. Knox, *J.C.S. Perkin I*, 1976, 459.
[486] D. N. Jones and D. A. Lewton, *J.C.S. Chem. Comm.*, 1974, 457.
[487] B. M. Trost and T. N. Salzmann, *J. Org. Chem.*, 1975, **40**, 148.
[488] P. J. Jessup and J. A. Reiss, *Tetrahedron Letters*, 1975, 1453.
[489] A. S. Kende, D. Constantinides, S. J. Lee, and L. Liebeskind, *Tetrahedron Letters*, 1975, 405.
[490] D. H. Rich, J. Tam, P. Mathiaparanam, J. A. Grant, and C. Mabuni, *J.C.S. Chem. Comm.*, 1974, 897.
[491] D. K. Bates and B. S. Thyagarajan, *Internat. J. Sulfur Chem.*, 1973, **8**, 419.
[492] P. A. Grieco and C. S. Pogonowski, *J.C.S. Chem. Comm.*, 1975, 72.
[493] J. Nokami, N. Kunieda, and M. Kinoshita, *Tetrahedron Letters*, 1975, 2841.

$\alpha\beta$-unsaturated esters,[494] α-vinylketones,[475] and more general routes to alkenes.[323, 495] A detailed study of this method for the synthesis of α-methylene-lactones includes both sulphoxide [496, 497] and selenoxide [497] versions. Protection of an α-methylene-γ-lactone moiety by addition of PhS⁻ [498] or PhSe⁻ [499] and de-protection by oxidation followed by thermolysis has been proposed.

Synthesis of a butenolide from an α-alkyl-lactone by α-phenylselenylation, followed by oxidation and thermolysis, can be extended into a general furan synthesis.[500] A full account is given [501] of the use of the selenoxide route for conversion of ketones into $\alpha\beta$-unsaturated analogues, and also for the synthesis of $\alpha\beta$-unsaturated nitriles.[501] The fact that a vinyl sulphoxide is formed by thermolysis of an α-phenylsulphinyl α-phenylseleninyl ketone emphasizes the greater ease of selenoxide elimination compared with that of an analogous sulphoxide.[501] Notable syntheses of allylic alcohols by the selenoxide method [502, 503] have been described, and a synthesis of tetrasubstituted alkenes, which are difficult if not impossible to prepare by the Wittig synthesis, has been established.[504] In an account of the conversion of a CHCO$_2$Me group into a C=CH$_2$ grouping, *via* CH$_2$OH, CH$_2$OMs, and CH$_2$Se(o-NO$_2$-C$_6$H$_4$) stages, and selenoxide elimination, for a step in the total synthesis of moenocinol, details are relegated to the small print, since the method can now be regarded as having joined the stock of routine laboratory synthetic methods.[505]

Applications of Dimethyl Sulphoxide and Dimethyl Selenoxide as Reagents.— The use of these compounds in oxidative reactions is not new, but there are several novel applications in the recent literature. Oxidation of methyl 10,11-epoxyundecanoate with Me$_2$SO–BF$_3$ gives HOCH$_2$C(O)(CH$_2$)$_8$CO$_2$Me,[506] while an epoxide reacts with Me$_2$SO in acid solution to give a β-hydroxy-alkoxy-sulphonium salt.[507] Benzyl alcohols are oxidized to benzaldehydes by Me$_2$SO and O$_2$ if a strong acid is present. Once again an alkoxysulphonium salt is implicated.[508] Dehydration of tertiary alcohols in Me$_2$SO, and self-decomposition of Me$_2$SO to give MeSH + HCHO, also occurs in the presence of a strong acid.[508] I$_2$-Catalysed oxidation of a phosphine sulphide or selenide to the corresponding oxide R$_3$P=O, and of thioamides to amides, is readily effected in Me$_2$SO.[509] Me$_2$SeO converts a tertiary amine into its N-oxide.[510] The report that Me$_2$SO

[494] J. Nokami, N. Kunieda, and M. Kinoshita, *Tetrahedron Letters*, 1975, 2179.
[495] R. H. Mitchell, *J.C.S. Chem. Comm.*, 1974, 990.
[496] P. A. Grieco and J. J. Reap, *Tetrahedron Letters*, 1974, 1097.
[497] K. Yamakawa, K. Nishitani, and T. Tominaga, *Tetrahedron Letters*, 1975, 2829, 4137.
[498] P. A. Grieco and M. Miyashita, *J. Org. Chem.*, 1975, **40**, 1181.
[499] P. A. Grieco and M. Miyashita, *Tetrahedron Letters*, 1974, 1869.
[500] P. A. Grieco, C. S. Pogonowski, and S. Burke, *J. Org. Chem.*, 1975, **40**, 542.
[501] H. J. Reich, J. M. Renga, and I. L. Reich, *J. Org. Chem.*, 1974, **39**, 2133; *J. Amer. Chem. Soc.*, 1975, **97**, 5434.
[502] H. J. Reich and S. K. Shah, *J. Amer. Chem. Soc.*, 1975, **97**, 3250.
[503] W. Dumont, P. Bayet, and A. Krief, *Angew. Chem.*, 1974, **86**, 857.
[504] H. J. Reich and F. Chow, *J.C.S. Chem. Comm.*, 1975, 790.
[505] P. A. Grieco, Y. Masaki, and D. Boxler, *J. Amer. Chem. Soc.*, 1975, **97**, 1597.
[506] S. M. Osman and G. A. Qazi, *Fette, Seifen, Anstrichm.*, 1975, **77**, 106.
[507] T. M. Santosusso and D. Swern, *J. Org. Chem.*, 1975, **40**, 2764.
[508] T. M. Santosusso and D. Swern, *Tetrahedron Letters*, 1974, 4255.
[509] M. Mikolajczyk and J. Luczak, *Synthesis*, 1975, 114.
[510] M. Poje and K. Balenovic, *Bull. Sci., Cons. Acad. Sci. Arts R.S.F. Yougoslav., Sect. A*, 1975, **20**, 1.

oxidizes thiols (cysteine, penicillamine, and reduced glutathione) to disulphides has been corrected.[511] Routine use of Me₂SO–Ac₂O for oxidizing a secondary alcohol to a ketone is exemplified with methyl 4,6-*O*-benzylidene-α-D-gluco-pyranose.[512] The low yield (31%) of ketone and the relatively large proportions of side-products [the 3-methylthiomethyl ether (29%) and the 3-acetate (17%)] is not exceptional for this reagent.

More unusual properties of Me₂SO include its use in the conversion of di- and tri-nitrochlorobenzenes into the corresponding phenols together with methyl-thiomethyl- and formyl-substituted phenols and thioanisoles as side-products,[513] and the formation of ethylbenzene (25%), (*Z*)- and (*E*)-ethylidenecyclohex-2-ene (15%), and *o*-ethyltoluene (*ca.* 50%) from the cyclohexene–dichlorocarbene adduct and KOBu^t.[514] These last observations imply that Me₂SO can act as a methylating agent. The formation of 2-acetoxyalkyl methyl selenides from alkenes with Me₂SeO in AcOH–CHCl₃, possibly *via* MeSeOAc, has no precedent in the chemistry of dimethyl sulphoxide.[515]

Sulphoxonium salts Me₂$\overset{+}{\text{S}}$OSO₂Me MeSO₃⁻ [516] and Me₂$\overset{+}{\text{S}}$OSO₂CF₃ CF₃SO₃⁻ [517] have been tested as oxidizing agents towards primary and secondary alcohols, and, although effective, substantial amounts of methylthiomethyl ethers are formed also. Pummerer rearrangement of the corresponding salt Me₂$\overset{+}{\text{S}}$OC(O)CF₃-CF₃CO₂⁻ occurs when it is allowed to warm from −60 to 30 °C, giving MeSCH₂OC(O)CF₃. This salt gives sulphimides (see p. 51) with amines, amides, and sulphonamides and offers an improvement over known routes to these derivatives.[518, 571]

7 Sulphones

Preparation of Sulphones.—Routine oxidation methods give sulphones, starting from sulphides and sulphoxides, and, again, a warning has appeared about the use of H₂O₂ in acetone which led to a delayed explosion during an attempt to oxidize a sulphur heterocycle to the sulphone.[519]

The majority of papers concerned with preparations of sulphones in the recent literature deal with sulphinic acids as reagents. The reaction of a tetrabutyl-ammonium sulphinate with an alkyl halide has been developed further,[520] and the addition of a sulphinic acid to a Schiff base to give ArNHCH(Ar)SO₂Ar,[521] to a bromo-nitro-styrene (PhCH=CBrNO₂ + RSO₂H → RSO₂CHPhCHBrNO₂),[522] or to 3-phenylcyclobutene-1,2-dione (67),[523] has been studied.

[511] J. T. Snow, J. W. Finley, and M. Friedman, *Biochem. Biophys. Res. Comm.*, 1975, **64**, 441; *erratum, ibid.*, p. 1323.
[512] J. Defaye and A. Gadelle, *Carbohydrate Res.*, 1974, **35**, 264.
[513] M. E. C. Biffin and D. B. Paul, *Austral. J. Chem.*, 1974, **27**, 777.
[514] C. J. Ransom and C. B. Reese, *J.C.S. Chem. Comm.*, 1975, 970.
[515] N. Miyoshi, S. Furui, S. Murai, and N. Sonoda, *J.C.S. Chem. Comm.*, 1975, 293.
[516] J. D. Albright, *J. Org. Chem.*, 1974, **39**, 1977.
[517] J. B. Hendrickson and S. M. Schwartzman, *Tetrahedron Letters*, 1975, 273.
[518] A. K. Sharma and D. Swern, *Tetrahedron Letters*, 1974, 1503.
[519] A. D. Brewer, *Chem. in Britain*, 1975, **11**, 335.
[520] G. E. Vennstra and B. Zwanenburg, *Synthesis*, 1975, 519.
[521] P. Messinger, *Arch. Pharm.*, 1974, **307**, 348.
[522] D. I. Aleksiev, *Zhur. org. Khim.*, 1975, **11**, 211; D. I. Aleksiev and A. Zlatarov, *ibid.*, p. 908.
[523] W. Reid, A. H. Schmidt, and H. Knorr, *Chem. Ber.*, 1975, **108**, 533.

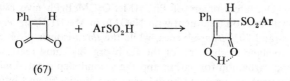

(67)

A chlorosulphine reacts with a sulphinate to give a sulphone after further reaction and hydrolysis of the initial substitution product ($RCCl{=}S{=}O +$ $ArSO_2^- \rightarrow ArSO_2CR{=}S{=}O \rightarrow ArSO_2CH_2R$).[524] Copper(I)-catalysed addition of a sulphonyl halide R^1SO_2X to a conjugated diene $R^2CH{=}CHCH{=}CH_2$ gives products of 1,2- and 1,4-addition,[525] while the corresponding reaction with $R^2C{\equiv}CCH{=}CH_2$ gives $R^2C{\equiv}CCHBrCH_2SO_2R^1$ and $R^2CBr{=}C{=}CHCH_2-$ SO_2R^1 when $R^2 = Me_3Si$, but $R^2C{\equiv}CCHBrCH_2SO_2R$ when $R^2 = Bu^t$.[526] Uncatalysed substitution of the acidic hydrogen atom in $PhC{\equiv}CH$ by $-SO_2CF_3$ is effected with trifluoromethanesulphonic anhydride.[527]

Allylic sulphones are available through the reaction of a t-butyl sulphoxide and an allyl alcohol with *N*-chlorosuccinimide,[528] and through the reaction of an allyl alcohol with di(*N*-phthalimido) sulphide or di(*N*-imidazolyl) sulphide to give a sulphoxylate, *e.g.* $CH_2{=}CHCH_2OSOCH_2CH{=}CH_2$, which rearranges to the bis(allyl) sulphone by way of the sulphinate.[529] Further study of the rearrangement of *N*-aryl arenesulphonamides to *N*-(*o*-aminoaryl) aryl sulphones has been described.[530]

General methods for the synthesis of α-arylsulphonyl- and α-alkanesulphonyl phenyldiazomethanes have been surveyed.[531]

α-Functional Sulphones.—A new synthesis of α-diazo-sulphones employs tosyl azide and $RSO_2CH_2C(O)CO_2Et$.[532]

α-Halogeno-sulphones are the α-functional sulphones of particular interest in general synthetic chemistry and in certain areas of mechanistic investigation. Chlorination of benzylthiomethyl methyl sulphones in aqueous AcOH gives $ClSO_2CCl_2SO_2Me$.[533] Bromomethyl sulphones $BrCH_2SO_2R^1$ give alkyl homologues $R^2CH_2SO_2R^1$ by their reaction with trialkylboranes and Bu^tOK.[534] Benzhydryl sulphones give 1,1-diarylalkenes rapidly and quantitatively, *via* the mono-chloro-derivatives, on treatment with CCl_4–$KOBu^t$–KOH.[535] More general studies of the Ramberg–Bäcklund reaction have been published with the objective of obtaining detailed information on mechanism rather than synthetic

[524] G. E. Vennstra and B. Zwanenburg, *Rec. Trav. chim.*, **95**, 28.
[525] M. M. Tanaskov and M. D. Stadnichuk, *Zhur. obshchei Khim.*, 1975, **45**, 843.
[526] M. D. Stadnichuk, T. B. Kryukova, and A. A. Petrov, *Zhur. obshchei Khim.*, 1975, **45**, 838.
[527] R. S. Glass and D. L. Smith, *J. Org. Chem.*, 1974, **39**, 3712.
[528] P. A. Grieco and D. Boxler, *Synth. Comm.*, 1975, **5**, 315.
[529] G. Buchi and R. M. Freidinger, *J. Amer. Chem. Soc.*, 1974, **96**, 3332.
[530] D. Hellwinkel and M. Supp, *Tetrahedron Letters*, 1975, 1499.
[531] A. M. van Leusen, B. A. Reith, and D. van Leusen, *Tetrahedron*, 1975, **31**, 597.
[532] A. L. Fridman, Yu. S. Andreichikov, V. L. Gein, and L. F. Gein, *Zhur. org. Khim.*, 1976, **12**, 463.
[533] W. R. Hardstaff, R. F. Langler, J. Leahy, and M. J. Newman, *Canad. J. Chem.*, 1975, **53**, 2664.
[534] W. E. Truce, L. A. Murall, P. J. Smith, and F. Young, *J. Org. Chem.*, 1974, **39**, 1449.
[535] C. Y. Meyers, W. S. Matthews, G. J. McCollum, and J. C. Bianca, *Tetrahedron Letters*, 1974, 1105.

applications. Diastereoisomers of PhCHMeSO$_2$CMeBrPh give *cis*- and *trans*-αα'-dimethylstilbenes on treatment with NaOMe in MeOH;[536] the *erythro*-isomer gives 93% *cis*- and 70% *trans*-alkenes, while the *threo*-isomer gives >95% *trans*-alkene. A two-stage mechanism for the Ramberg–Bäcklund rearrangement, the first step being carbanion formation and the second step a rate-limiting nucleophilic displacement, is suggested by ^{2}H-exchange studies [537] and by the relationship between solvent polarity and reaction rate.[538] Overall retention of configuration results from reduction of (±)-*erythro*-(α-bromo-α-methylbenzyl α'-methylbenzyl sulphone) with Na$_2$SO$_3$ in aqueous MeOH, or a phosphine, to give *meso*-bis(α-methylbenzyl) sulphone.[539] The possibility of an α-halogenosulphinate intermediate in the Ramberg–Bäcklund rearrangement is suggested by the ready elimination of SO$_2$ and halide ion from such compounds.[685] Dihalogeno-alkyl sulphones were prepared for these studies by halogenative decarboxylation of α-aryl- and α-alkyl-sulphonylalkanoic acids.[540] Steric and electronic factors control the rate of reduction of α-bromo- and α-iodo-sulphones by phosphines.[541] Ramberg–Bäcklund rearrangement of a vinylogous α-halogenosulphone, *viz.* *cis*- and *trans*-4-bromo-2-penten-2-yl benzyl sulphones, with Bu^{t}OK gives a mixture of equal amounts of the (*E*)- and (*Z*)-isomers of 1-phenyl-2-methyl-1,3-pentadiene.[542]

Radical intermediates are involved in reactions of α-nitro-sulphones with nucleophiles, leading to products from which the sulphone grouping has been displaced.[543] A useful method of synthesis of 1,2-dinitroalkanes emerges from this work [R^{1}R^{2}C(NO$_2$)SO$_2$R^3 + R^{4}R$^5\bar{C}$NO$_2$ → R^{1}R^{2}C(NO$_2$)CR^{4}R^{5}NO$_2$ + R^{3}SO$_2^-$]. Bis(trihalogenomethyl) sulphones add to alkenes *via* the radicals CX$_3$SO$_2$CX$_2$· to give mono- and symmetrical di-adducts.[544] Hydroxyl or phenyl radicals react with aliphatic sulphones to give stabilized radicals R^{1}R$^2\dot{C}$SO$_2$R^3.[452]

Condensation reactions resulting from the activation of the carbon atom adjacent to the sulphone grouping give vinyl sulphones from aldehydes and (PhSO$_2$)$_2$CH$_2$ [545] or RSO$_2$CH$_2$R (phase-transfer catalysis).[546] Analogues (MeSO$_2$)$_2$C=CHNR^{1}R^2 result from the reaction of 2,2-bis(methanesulphonyl)-vinyl ethyl ether with amines.[547]

Trifluoromethyl sulphones, neologized as 'triflones', are particularly reactive compared with other sulphones towards α-alkylation, and towards elimination reactions.[548]

[536] F. G. Bordwell and E. Doomes, *J. Org. Chem.*, 1974, **39**, 2526.
[537] F. G. Bordwell and J. B. O'Dwyer, *J. Org. Chem.*, 1974, **39**, 2519.
[538] F. G. Bordwell and M. D. Wolfinger, *J. Org. Chem.*, 1974, **39**, 2521.
[539] F. G. Bordwell and E. Doomes, *J. Org. Chem.*, 1974, **39**, 2298.
[540] F. G. Bordwell, M. D. Wolfinger, and J. B. O'Dwyer, *J. Org. Chem.*, 1974, **39**, 2516.
[541] B. B. Jarvis and B. A. Marien, *J. Org. Chem.*, 1975, **40**, 2587; B. B. Jarvis, R. L. Harper, and W. P. Tong, *ibid.*, p. 3778.
[542] R. B. Mitra, M. J. Natekar, and S. D. Virkar, *Indian J. Chem.*, 1975, **13**, 254.
[543] N. Kornblum, S. D. Boyd, and N. Duo, *J. Amer. Chem. Soc.*, 1974, **96**, 2580.
[544] C. J. Kelley and M. Carmack, *Tetrahedron Letters*, 1975, 3605.
[545] H. Stetter and K. Steinbeck, *Annalen*, 1974, 1315.
[546] J. J. Zeilstra and J. B. F. N. Engberts, *J. Org. Chem.*, 1974, **39**, 3215; G. Cardillo, D. Savoia, and A. Umani-Ronchi, *Synthesis*, 1975, 453; G. Ferdinand, K. Schank, and A. Weber, *Annalen*, 1975, 1484.
[547] A. R. Friedman and D. R. Graber, *J. Org. Chem.*, 1974, **39**, 1432.
[548] J. B. Hendrickson, A. Giga, and J. Wareing, *J. Amer. Chem. Soc.*, 1974, **96**, 2279.

Reactions of α-sulphonyl carbanions in synthesis are reviewed in the following chapter in this volume.

Properties of Sulphones.—The extreme difficulty experienced in achieving the reduction of open-chain and six-membered cyclic sulphones with sodium bis-(2-methoxyethoxy)aluminium hydride does not apply to five-membered sulphones.[549] Surprisingly, this reagent leads first to the α-monoanion or α,α-dianion, and LiAlH$_4$ also gives sulphides with some exchange of α-^{2}H; further details are to be elucidated.[549]

Highly selective β-chlorination results from the use of SO$_2$Cl$_2$ as chlorinating agent for aliphatic sulphones; this is a useful adjunct to well-known methods favouring α-chlorination.[550]

Formation of Meisenheimer complexes between 1,3,5-tris(trifluoromethane-sulphonyl)benzene and CN$^-$ or MeO$^-$ has been reported.[551] Photolysis of aromatic sulphones in C$_6$H$_6$ gives mixtures of biphenyls.[552]

Uses of Sulphones in Synthesis.—In acid-promoted ring-opening reactions of α-hydroxyalkylcyclopropanes (68), the concentrated bulk of the PhSO$_2$ group

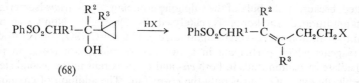

(68)

enhances the stereoselectivity of the reaction.[553] A similar principle is exploited in the stereocontrol of R^{1}SO$_2$CH$_2$CR^2(OH)CH=CH$_2$ + HX/ZnX$_2$ → R^{1}SO$_2$-CH$_2$CR^2=CHCH$_2$X in favour of the (E)-isomer.[554]

Mention has been made in a preceding section of the value of the Ramberg-Bäcklund reaction in synthesis. An example[528] is the conversion of allylic sulphones, with CCl$_4$ and KOH, into dienoic acids (CH$_2$=CHCR^{1}R^{2}SO$_2$CH$_2$-CO$_2$Me → R^{1}R^{2}C=CHCH=CHCO$_2$H), and this is similar to the conversion of a diallyl sulphone into a triene,[529] using CCl$_4$ and KOH, as illustrated by the conversion of vitamin A into β-retinyl sulphone and the contraction into β-carotene.[529]

Unsaturated Sulphones.—A sulphone group influences the properties of an adjacent multiple bond to a considerable extent, both by electronic and steric effects. This property can be potentiated further by an adjacent trifluoromethyl group, as illustrated in the exceptionally facile Diels–Alder reactions involving PhC≡CSO$_2$CF$_3$ and dienes.[527] Interesting stereochemical control of the addition of organocopper(I) reagents to 1-alkynyl sulphones R^{1}C≡CSO$_2$R^2 is found.[555] The proportions of the *cis-* and *trans*-adducts R^{1}R^{2}C=CHSO$_2$R^3 depend on the relative amounts of CuBr and RMgX used to prepare the reagent, excess CuBr

[549] W. P. Weber, P. Stromquist, and T. I. Ito, *Tetrahedron Letters*, 1974, 2595.

[550] I. Tabushi, Y. Tamaru, and Z. Yoshida, *Tetrahedron*, 1974, **30**, 1457.

[551] L. M. Yagupolskii, V. N. Boiko, G. M. Shchupak, N. V. Kondratenko, and V. P. Sambur *Tetrahedron Letters*, 1975, 4413.

[552] A. I. Khodair, T. Nakabayashi, and N. Kharasch, *Internat. J. Sulfur Chem.*, 1973, **8**, 37.

[553] M. Julia and J.-M. Paris, *Tetrahedron Letters*, 1974, 3445.

[554] M. Julia and D. Deprez, *Tetrahedron Letters*, 1976, 277.

[555] J. Meijer and P. Vermeer, *Rec. Trav. chim.*, 1975, **94**, 14.

leading to predominant *cis*-addition. Stereospecific coupling of organocopper(I) reagents to (*E*)-2-iodo-1-alkenyl sulphones is observed, the incoming group taking the position occupied by the halogen ($R^1SO_2CH=CR^2I + R^3Cu \rightarrow R^1SO_2CH=CR^2R^3$).[556] Substitution of the halogen atom of $PhCH=C(SO_2Ar)$-CH_2Br by reaction with a secondary amine leads to two products, one the direct substitution product and the other the rearranged substitution product $PhCH(NR_2)C(SO_2Ar)=CH_2$;[557] the arenesulphonyl group is believed to stabilize an S_N2'-type transition state in this reaction.

The non-stereospecific nature of addition of Br_2 to a vinyl sulphone followed by dehydrobromination with NEt_3 to give bromovinyl sulphones ($RCH=CBr$-SO_2Ph from $RCH=CHSO_2Ph$) is associated with failure of the $PhSO_2$ and the β-substituent to attain a *trans*-relationship before the elimination step.[558] A full account of the addition–elimination reactions of *cis*- and *trans*-phenoxyvinyl *p*-tolyl sulphones has been presented,[559] in continuation of extensive studies over the years by Stirling's group. Equilibration of $\beta\gamma$-unsaturated sulphones to $\alpha\beta$-unsaturated isomers leads to the *trans*-isomer predominantly. The equilibrium can, however, lie in favour of the $\beta\gamma$-isomer to a greater extent than might be expected, because of the bulk of the sulphone grouping.[560] *cis–trans* Isomerization of $\alpha\beta$-unsaturated sulphones is catalysed by Br_2.[558, 560] Non-stereospecific oxiran formation from *cis*-vinyl sulphones, *e.g.* $PhCH=CHSO_2Tol$, is associated with reagents which participate in a two-step reaction path (rotation in the intermediate carbanion leads to both *cis*- and *trans*-oxirans in the second step).[561] Hypochlorite ion, ClO^-, gives only the *cis*-oxiran. The addition of enamines to $\alpha\beta$-unsaturated sulphones is non-regiospecific.[562]

Michael addition reactions of allyl sulphones and $\alpha\beta$-unsaturated carbonyl compounds provide useful starting points in synthesis, illustrated in a synthesis of chrysanthemate ester [$Me_2C=CHCH_2SO_2Ph + Me_2C=CHCO_2R \rightarrow Me_2C=CHCH(SO_2Ph)CMe_2CH_2CO_2R \rightarrow$ *cis*- and *trans*-cyclopropane esters].[563]

8 Sulphimides and Sulphoximides

The introduction of a section on nitrogen analogues of sulphoxides and sulphones in Volume 3 of these Specialist Periodical Reports reflected the increasing activity in this area. The section is retained for this volume, and is supported by an even larger number of papers.

Sulphimides.—The reaction of a nitrene with a sulphide yields a sulphimide. This well-established route is illustrated in the recent literature with Chloramine-T ($R^1SR^2 + ArSO_2\overset{-}{N}Cl\ Na^+ \rightarrow R^1R^2S=NSO_2Ar$),[564, 565] and has been used[566] in

556 W. E. Truce, A. W. Borel, and P. J. Marek, *J. Org. Chem.*, 1976, **41**, 401.
557 E. Doomes, P. A. Thiel, and M. L. Nelson, *J. Org. Chem.*, 1976, **41**, 248.
558 J. C. Philips, M. Aregullin, M. Oku, and A. Sierra, *Tetrahedron Letters*, 1974, 4157.
559 M. J. van der Sluijs and C. J. M. Stirling, *J.C.S. Perkin II*, 1974, 1268.
560 I. Sataty and C. Y. Meyers, *Tetrahedron Letters*, 1974, 4161.
561 R. Curci and F. Di Furia, *Tetrahedron Letters*, 1974, 4085.
562 S. Fatutta and A. Risaliti, *J.C.S. Perkin I*, 1974, 2387.
563 R. V. M. Campbell, L. Crombie, D. A. R. Findley, R. W. King, G. Pattenden, and D. A. Whiting, *J.C.S. Perkin I*, 1975, 897.
564 F. Ruff and A. Kucsman, *J.C.S. Perkin II*, 1975, 509.
565 M. M. Campbell and G. Johnson, *J.C.S. Perkin I*, 1975, 1077.
566 S. Tamagaki, S. Oae, and K. Sakaki, *Tetrahedron Letters*, 1975, 649.

the first synthesis of a selenimide, $(PhCH_2)_2Se=NSO_2Tol$. A sulphinylnitrene undergoes 1,2-dipolar cycloaddition with a sulphoxide, to give a *N*-benzene-sulphonyl-sulphimide.[709] The sulphurane $Ph_2S[OC(CF_3)_2Ph]_2$ reacts with amides to give *N*-acylsulphimides.[567] The reaction of anilines with sulphides and *N*-chloro-succinimide, Bu^tOCl, or SO_2Cl_2 gives amino-sulphonium salts.[568, 327-330] Oxy-sulphonium salts derived from Me_2SO [518] and Lewis acids,[569-571] *e.g.* SO_3, P_4O_{10}, BF_3, H_2SO_4,[570] or $(CF_3CO)_2O$,[571] give sulphimides with amines and amides.

Sulphimides have uses in synthesis in addition to the selective *ortho*-formylation of amines and phenols discussed earlier in this chapter (p. 27).[327-330] Aryl and heteroaryl sulphimides $ArN=SMe_2$, have been used in reactions with cyanates leading to new annelation processes,[572, 573] and in reactions with diphenylcyclo-propenone in the synthesis of 4-pyrimidones and other ring-opened products.[573, 574] *N*-Phenybenzimidoylsulphimides give 2-substituted benzimidazoles under irradiation,[575] *via* cyclization of the derived nitrene.

The reactions of sulphimides studied recently concern the kinetics of the reaction between PhSH and *N*-toluene-*p*-sulphonyl-sulphimides [576, 577] to give PhSSPh,[577] or RSPh [576] (other differences exist between the results from these two studies); the formation of vinyl sulphides by treatment of *N*-toluene-*p*-sulphonyl aryl ethyl sulphimides with Bu^tOK [$TsN=S(Et)Ar \rightarrow CH_2=CHSAr$];[578] the halide-ion-induced Stevens-type rearrangement to the sulphen-amide [$TsN=SRAr \rightarrow RN(Ts)SAr$];[579] and the *S*-substitution or Pummerer rearrangement products from *N*-toluene-*p*-sulphonyl-sulphimides by cleavage with NaOH or NaOMe in MeOH.[580]

The recently opened route to the *N*-unsubstituted sulphimides $R^1S(=NH)R^2$ has permitted a study to be made of reactions at nitrogen. Acetic anhydride gives the *N*-acetyl-sulphimide,[578] while Michael addition to *trans*-dibenzoyl-ethylene gives R^1SR^2 and $PhC(O)CH=C(NH_2)C(O)Ph$, together with *trans*-1,2-dibenzoylaziridine.[581] Reaction with an alkene $RCH=CHR$ similarly gives the (E)-enamine and the aziridine.[582] Alkylation of diphenylsulphimide gives the *N*-alkyl analogue.[583] Rearrangement of phenyl propargyl sulphimide gives the sulphenamide $PhSN=CHCH=CH_2$,[584] and the same product, which could

[567] J. C. Martin and J. A. Franz, *J. Amer. Chem. Soc.*, 1975, **97**, 6137.
[568] P. K. Claus, W. Rieder, P. Hofbauer, and E. Vilsmeier, *Tetrahedron*, 1975, **31**, 505.
[569] G. F. Whitfield, H. S. Beilam, D. Saika, and D. Swern, *J. Org. Chem.*, 1974, **39**, 2148.
[570] T. E. Varkey, G. F. Whitfield, and D. Swern, *J. Org. Chem.*, 1974, **39**, 3365.
[571] A. K. Sharma, T. Ku, A. D. Dawson, and D. Swern, *J. Org. Chem.*, 1975, **40**, 2758.
[572] T. L. Gilchrist, C. J. Harris, and C. W. Rees, *J.C.S. Chem. Comm.*, 1974, 485.
[573] T. L. Gilchrist, C. J. Harris, C. J. Moody, and C. W. Rees, *J.C.S. Chem. Comm.*, 1974, 486; T. L. Gilchrist, C. J. Harris, and C. W. Rees, *ibid.*, p. 487.
[574] T. L. Gilchrist, C. J. Harris, C. J. Moody, and C. W. Rees, *J.C.S. Perkin I*, 1975, 1969.
[575] T. L. Gilchrist, C. J. Moody, and C. W. Rees, *J.C.S. Perkin I*, 1975, 1964.
[576] S. Oae, T. Aida, M. Nakajima, and N. Furukawa, *Tetrahedron*, 1974, **30**, 947.
[577] G. Guanti, G. Garbarino, C. Dell'Erba, and G. Leandri, *Gazzetta*, 1975, **105**, 849.
[578] N. Furukawa, S. Oae, and T. Masuda, *Chem. and Ind.*, 1975, 396.
[579] S. Oae, T. Aida, and N. Furukawa, *J.C.S. Perkin II*, 1974, 1231.
[580] N. Furukawa, T. Masuda, M. Yakushiji, and S. Oae, *Bull. Chem. Soc. Japan*, 1974, **49**, 2247.
[581] N. Furukawa, T. Yoshimura, T. Omata, and S. Oae, *Chem. and Ind.*, 1974, 702.
[582] N. Furukawa, S. Oae, and T. Yoshimura, *Synthesis*, 1976, 30.
[583] Y. Tamura, H. Matsushima, M. Ikeda, and K. Sumoto, *Tetrahedron*, 1976, **32**, 431.
[584] Y. Tamura, H. Matsushima, J. Minamikawa, M. Ikeda, and K. Sumoto, *Tetrahedron*, 1975, **31**, 3035.

otherwise be named an *S*-aryl thioaldoxime, is formed from *N*-phthalimido allyl aryl sulphimides.[585]

Alkyl phenyl sulphimides are found [586] to exchange an α-^{1}H for ^{2}H in alkaline ^{2}H$_2$O faster than the corresponding sulphoximide. This implies a higher degree of electron-withdrawal in the sulphimides.

N-Toluene-*p*-sulphonyl phenyl vinyl sulphimide gives Michael adducts RCH$_2$CH$_2$SPh=NTs, which give the vinyl compounds RCH=CH$_2$ on pyrolysis,[587] thus creating a useful vinylation method.

Stereochemical aspects of sulphimides covered in the recent literature concern the racemization mechanism [588] and the stereochemistry of substitution reactions involving the chiral sulphur centre [589] (see also p. 110). Racemization of an *N*-toluene-*p*-sulphonyl aryl methyl sulphimide involves pyramidal inversion rather than a cleavage–recombination mechanism.[588] The (+)-isomer of methyl *p*-tolyl sulphimide has the (*R*)-configuration, established by correlation with its (+)-*N*-toluene-*p*-sulphonyl derivative.[589] Resolution of a sulphimide R^{1}S-(=NH)R^2 can be achieved using (+)-α-bromo-π-camphorsulphonic acid.[589] Treatment of the (*R*)-sulphimide (69) with Chloramine-T gives the (−)-sulphurdi-imide, which on deamination gives the (−)-*N*-toluene-*p*-sulphonyl-sulphimide (*S*)-(70). The deamination step (caused by HNO$_2$) proceeds with retention of configuration, so that the stereochemical assignments shown in (69) and (70) are soundly based.[589]

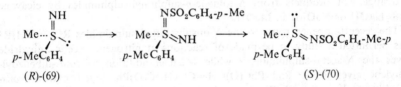

Sulphoximides.—Oxidation of *N*-arylsulphimides, or reaction of anilines with a dialkyl sulphoxide and Bu^{t}OCl or SO$_2$Cl$_2$, gives *N*-arylsulphoximines.[568] A route to the parent sulphoximides R^{1}S(O)(NH)R^2, starting from a sulphoxide, uses the hydroxylamine derivative ArSO$_2$ONH$_2$;[584, 590] a drawback to this method lies in the explosive nature of the reagent. Stereochemical aspects of preparations of sulphoximides by the standard nitrene + sulphoxide route have been studied. Retention of configuration accompanies the reaction of an (*R*)-sulphoxide with Chloramine-T or toluene-*p*-sulphonyl azide in the presence of a copper salt,[591] and of the chiral sulphoxide with *N*-amino-phthalimides.[592]

Alkaline hydrolysis of a sulphimide to the corresponding sulphoxide occurs with inversion of configuration, but treatment of a *N*-phthalimido-sulphoximide with NaOEt or NH$_2$NH$_2$ in EtOH leads to the sulphoxide, with retention of configuration.[592] Deamination of L-methionine (*R*)- or (*S*)-sulphoximine with

585 R. S. Atkinson and S. B. Awad, *J.C.S. Chem. Comm.*, 1975, 651.
586 M. Kobayashi, A. Mori, and H. Minato, *Bull. Chem. Soc. Japan*, 1974, **47**, 891.
587 T. Yamamoto and M. Okawara, *Chem. Letters*, 1975, 581.
588 D. Darwish and S. K. Datta, *Tetrahedron*, 1974, **30**, 1155.
589 B. W. Christensen and A. Kjaer, *J.C.S. Chem. Comm.*, 1975, 784.
590 C. R. Johnson, R. A. Kirchhoff, and H. G. Corkins, *J. Org. Chem.*, 1974, **39**, 2458.
591 M. Moriyama, T. Numata, and S. Oae, *Org. Prep. Proceed. Internat.*, 1974, **6**, 207.
592 S. Colonna and C. J. M. Stirling, *J.C.S. Perkin I*, 1974, 2120.

HNO_2 gives the corresponding L-methionine sulphoxide with retention of configuration (exclusive attack at the sulphoximide nitrogen is notable).[593]

The parent sulphoximides $R^1S(O)(NH)R^2$ react through nitrogen with S_2Cl_2, $SOCl_2$, or SO_2Cl_2 in the presence of Et_3N to give the sulphur di-imide derivatives $(R^1R^2S(O)N)_2X$ [X = SS, S(O), and SO_2, respectively].[594] Photolysis of *N*-aryl-sulphonyl-*SS*-dimethylsulphoximides in aromatic solvents gives biphenyls *via* aryl radicals, together with SO_2, Me_2SO_2, and the intact parent sulphoximine $Me_2S(O)(NH)$.[595] Sulphoximides derived from *N*-amino-oxazolidones (71) give

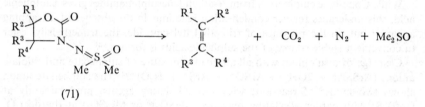

(71)

alkenes on thermolysis in DMSO at 110—130 °C.[596] Although this reaction is advocated as a new alkene synthesis, one route to the starting material involves an oxiran (see also Chapter 2, Part II, p. 105, for a more extensive review of sulphur-nitrogen chemistry).

9 Sulphenic Acids

Preparation.—It is possible to review syntheses of sulphenic acids and to refer now to stable products as well as fleeting species that need to be trapped in order to verify their existence. Further details of the stable penicillin-derived sulphenic acid (72) described in Volume 3 are available.[597, 598] The anion of (72) is, however,

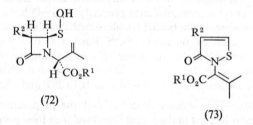

(72)

(73)

unstable if formed with 4-methylmorpholine and rearranges to the isothiazolone (73). Nevertheless it gives the methyl sulphenate when $LiPr^i_2N$ is used as base in THF at −126 °C and methyl fluorosulphonate is added (the exclusive *O*-alkylation in this reaction is notable).[598]

Elimination reactions leading to simple sulphenic acids have been described earlier in this chapter, in the review of recent literature covering properties of

[593] R. A. Stephani and A. Meister, *Tetrahedron Letters*, 1974, 2307.
[594] M. Okahara, E. Yoshikawa, I. Ikeda, and S. Komori, *Synthesis*, 1975, 521.
[595] R. A. Abramovitch and T. Takaya, *J.C.S. Perkin I*, 1975, 1806.
[596] J. D. White and M. Kim, *Tetrahedron Letters*, 1974, 3361.
[597] S. Kukolja, S. R. Lammert, M. R. Gleissner, and A. L. Ellis, *J. Amer. Chem. Soc.*, 1975, **97**, 3192; S. Kukolja, *Preprints, Div. Petrol. Chem., Amer. Chem. Soc.*, 1974, **19**, 300.
[598] G. A. Koppel and S. Kukolja, *J.C.S. Chem. Comm.*, 1975, 57.

sulphoxides (*e.g.* refs. 422, 486). In these cases the conditions applied for bringing about elimination invariably lead the sulphenic acid to familiar transformation products. Trapping techniques continue to be studied and arenesulphinic acids, which give arenethiolsulphonates, have proved to be effective [599] for trapping the penicillin-derived sulphenic acid. Intramolecular trapping of the sulphenic acid derived from 6-phenylthioacetamidopenicillin sulphoxide has been reported;[600] here the sulphenic acid acts as an electrophile and is attacked, as soon as it is formed, by the thiocarbonyl sulphur atom.

While Cope-type elimination from *N*-alkylidenesulphinamides gives a sulphenic acid, this undergoes further condensation reactions in the absence of a trapping agent,[704] but, in the presence of trimethylsilyl chloride, the trimethylsilyl ester (a convenient stable source of the sulphenic acid) is formed.[601]

Cleavage of disulphides with alkali gives a mixture of sulphenate and thiolate anions (ArSSAr + 2OH⁻ → ArSO⁻ + ArS⁻ + H₂O).[602] The sulphenate anion shows two modes of reaction; with soft alkylating agents, predominantly at (soft) S; with harder alkylating reagents (MeSO₃F or Me₂SO₄) at (harder) O, leading to sulphoxides and sulphenates, respectively, and revealing an ambident nucleophilic character to the sulphenate anion.[602] Thiolsulphinates are useful precursors of alkanesulphenic acids.[858]

Reactions of Sulphenic Acids.—In addition to the trapping reactions referred to in the preceding section, the transformation products of sulphenic acids are mentioned elsewhere in this chapter.

Sulphenates.—Thiocarbonate *S*-oxides fail to give sulphenates by pyrolysis or after conversion into oxosulphonium salts (see p. 16).[197]

Sulphenates react with phosphines *via* the oxythiophosphorane R¹OPR³₃SR², to give ethers and sulphides;[197] these results extend earlier studies.[603]

Solvolysis of benzyl arenesulphenates proceeds *via* S—O bond cleavage and is a second-order process, while benzyl trichloromethanesulphenates undergo first-order C—O bond cleavage on solvolysis.[604] Formation of a sulphoxide from a benzyl trichloromethanesulphenate is one of two reaction paths resulting from ionization in polar media, the alternative to this ionization–recombination process being further dissociation into dichloro-sulphine (Cl₂CSO) and a benzyl chloride.[605]

Sulphenyl Halides.—This section describes first the preparations of sulphenyl halides of particular interest in the recent literature, then their properties and their addition and substitution reactions. More space is allocated this year for results from studies of organo-sulphur, -selenium, and -tellurium halides involving higher valency states.

Chlorinolysis of disulphides is widely used as a direct synthesis of sulphenyl halides, and related cleavage by *N*-chloroformimidoyl chloride has been studied

[599] R. D. Allan, D. H. R. Barton, M. Girijavallabhan, and P. G. Sammes, *J.C.S. Perkin I*, 1974, 1456.
[600] H. Tanida, R. Muneyuki, and T. Tsushima, *Tetrahedron Letters*, 1975, 3063.
[601] F. A. Davis and A. J. Friedman, *J. Org. Chem.*, 1976, **41**, 897.
[602] D. R. Hogg and A. Robertson, *Tetrahedron Letters*, 1974, 3783.
[603] B. Krawiecka, J. Michalski, J. Mikolajczak, M. Mikolajczyk, J. Omelanczuk, and A. Skow-ranska, *J.C.S. Chem. Comm.*, 1974, 630.
[604] S. Braverman and D. Reisman, *Tetrahedron*, 1974, **30**, 3891.
[605] S. Braverman and B. Sredni, *Tetrahedron*, 1974, **30**, 2379.

$(R^1CCl=NCl + R^2SSR^2 \rightarrow R^1CCl=NSR^2 + R^2SCl$; also, $R^1CCl=NCl + S_8 \rightarrow R^1CCl=NSCl)$.[606] Cleavage of a disulphide with SO_2Cl_2-Ph_3P, urea, or DMF gives an arenesulphenyl chloride.[607] Syntheses and reactions of perfluoroalkane-sulphenyl halides have been reviewed.[608]

Sulphenyl halides give cationoid species RS^+Y^- ($Y = BF_4$ or SbF_6) on reaction with the corresponding Ag salt;[609] an areneselenenyl hexafluorophosphate and a corresponding hexafluoroantimonate have been described.[610] Further details of the equilibrium $2RSCl \rightleftharpoons R\overset{+}{S}(Cl)SR + Cl^-$, set up by an alkane-sulphenyl chloride in liquid SO_2 in the presence of a Lewis acid, or in H_2SO_4 or FSO_3H, are available.[611] Such cations have been suggested to be implicated in the reactions of disulphides and sulphenyl halides, and the reaction in these media gives the cation $R^1\overset{+}{S}(SR^2)SR^1$.[612, 254, 851]

Addition of a sulphenyl halide to an alkene gives a β-halogeno-sulphide or a thiiranium salt. More accurately, the latter, or a sulphurane equivalent, is established as an intermediate in analogous systems involving the addition to an alkene of a sulphenyl tetrafluoroborate or hexafluoroantimonate,[609] an areneselenenyl hexafluorophosphate,[610] or an areneselenenyl chloride.[613] The analogous selenenirenium cation results from the addition of benzeneselenenyl hexafluoroantimonate to but-2-yne.[610] However, the fact that the addition of PhSCl to a strained cyclobutene is non-stereospecific seems to discount the intermediacy of the thiiranium cation, and is more in keeping with a carbonium ion intermediate.[614] Furthermore, the cyclic selenuran structure assigned by Garratt and Schmid[613] to the adduct between toluene-p-selenenyl chloride and ethylene has been shown to be incorrect,[615] and the 2-chloroethyl dichloroselenuran structure $TolSeCl_2CH_2CH_2Cl$ has been established for the reaction product.[615] The reaction of 2-chloroethyl phenyl selenide with benzeneselenenyl chloride gives the dichloroselenenuran and diphenyl diselenide,[615] giving credence to the unexpected chemistry involved.

Routine studies of the addition of sulphenyl halides to alkenes have been reported, most studies dealing with the kinetics of the reaction as a function of structure and stereochemistry.[151, 217, 616–622] Among these reports is a notable

[606] J. Geevers and W. P. Trompen, *Tetrahedron Letters*, 1974, 1687, 1691.
[607] E. A. Parfenov and V. A. Fomin, *Zhur. obshchei Khim.*, 1975, **45**, 1129.
[608] G. Dahms, G. Diderrich, A. Haas, and M. Yazdanbakhsch, *Chem.-Ztg.*, 1974, **98**, 109.
[609] W. A. Smit, M. Z. Krimer, and E. A. Vorobeva, *Tetrahedron Letters*, 1975, 2451.
[610] G. H. Schmid and D. G. Garratt, *Tetrahedron Letters*, 1975, 3991.
[611] G. Capozzi, V. Lucchini, G. Modena, and F. Rivetti, *J.C.S. Perkin II*, 1975, 361.
[612] G. Capozzi, V. Lucchini, G. Modena, and F. Rivetti, *J.C.S. Perkin II*, 1975, 900.
[613] D. G. Garratt and G. H. Schmid, *Canad. J. Chem.*, 1974, **52**, 1027.
[614] G. Mehta and P. N. Pandey, *Tetrahedron Letters*, 1975, 3567.
[615] H. J. Reich and J. E. Trend, *Canad. J. Chem.*, 1975, **53**, 1922.
[616] K. Izawa, T. Okuyama, and T. Fueno, *Bull. Chem. Soc. Japan*, 1974, **47**, 1480.
[617] L. Rasteikiene, Z. Talaikyte, and V. Taliene, *Zhur. org. Khim.*, 1975, **11**, 920.
[618] W. M. Baluzow, G. Just, and W. Pritzkow, *J. prakt. Chem.*, 1974, **316**, 1051.
[619] G. H. Schmid and D. G. Garratt, *Canad. J. Chem.*, 1974, **52**, 1807.
[620] C. L. Dean, D. G. Garratt, T. T. Tidwell, and G. H. Schmid, *J. Amer. Chem. Soc.*, 1974, **96**, 4958.
[621] N. S. Zefirov, L. G. Gurvich, A. S. Shashkov, and V. A. Smit, *Zhur. org. Khim.*, 1974, **10**, 1786.
[622] V. R. Kartashov, I. V. Bodrikov, E. V. Skorobogatova, and N. S. Zefirov, *Zhur. org. Khim.*, 1976, **12**, 297; I. V. Bodrikov, T. S. Ganzhenko, N. S. Zefirov, and V. R. Kartashov, *Doklady Akad. Nauk S.S.S.R.*, 1976, **226**, 831.

56   *Organic Compounds of Sulphur, Selenium, and Tellurium*

result, the largest rate difference yet observed (a factor of 1.6×10^5) for an addition reaction to a pair of *cis*- and *trans*-isomers (4-chlorobenzenesulphenyl chloride to 1,2-di-t-butylethylene isomers).[620] The observation that the amount of the rearrangement product MeC(O)CMe(Ar)CH$_2$SPh formed together with direct adducts by the addition of PhSCl or ArCMe(OH)CMe=CH$_2$ is increased both by the presence of an electron-donating substituent in the aryl moiety and by the addition of LiClO$_4$ is an important clue to mechanism,[622] implying a balance between sulphurane and ion-pair intermediates. The addition of 2,4-dinitrobenzenesulphenyl chloride to an aliphatic allene RCH=C=CH$_2$ gives the adduct RCHClC(SAr)=CH$_2$, whereas phenylallene gives the oppositely oriented adduct PhCH=C(SAr)CH$_2$Cl.[623]

The intramolecular addition $(74) \rightarrow (75) + (76)$ gives mainly the anti-Markovnikoff adduct (76), although the two products are interconverted on heating [624] (see also Chapter 4, p. 193).

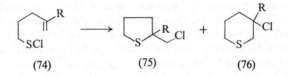

(74)   (75)   (76)

Substitution reactions of C$_6$F$_5$SCl [151] and p-MeC$_6$H$_4$SCl [813] with aromatic compounds [813] and active-methylene compounds [151] represent a separate area of sulphenyl halide chemistry. A related process, the ring-opening of the cyclopropane moiety of quadricyclanes by PhSCl, has been studied.[625]

Higher valency states are particularly easily reached in preparations of organoselenium and organotellurium halides, but special cases of sulphenyl perhalides are well known. Vinyl pentafluorosulphur, CH$_2$=CHSF$_5$, adds Cl$_2$ at the double bond,[626] gives BrCH(SF$_5$)CH$_2$SF$_5$ with the pseudohalogen BrSF$_5$,[627] and gives a 2 : 1 adduct (CF$_3$)$_2$NOCH$_2$CH(SF$_5$)ON(CF$_3$)$_2$ with di(trifluoromethyl)amine oxide.[626] Two new routes for the synthesis of diarylselenium dichlorides Ar$_2$SeCl$_2$ from SeCl$_4$,[628] and the broader use of phenyltin compounds for the synthesis of Ph$_2$SeCl$_2$, PhSeCl$_3$, and their tellurium analogues,[629] have been reported. TeCl$_4$ with more than one molar equivalent of AlCl$_3$ is an efficient reagent for the introduction of Te into aromatic compounds.[630, 631] Its use has been illustrated through the synthesis of PhTeCl$_3$, Ph$_2$TeCl$_2$, and Ph$_3$TeCl.[630] In addition to the arenetellurenyl halide ArTeX (X = Br or I), the tri-iodide ArTeI$_3$ and the diaryl ditellurium tetraiodide ArTeI$_2$TeI$_2$Ar are formed by treatment of a diaryl ditelluride with the halogens.[632] Mixtures of tellurated aromatic hydrocarbons

[623] T. L. Jacobs and R. C. Kammerer, *J. Amer. Chem. Soc.*, 1974, **96**, 6213.
[624] S. Ikegami, J. Ohishi, and Y. Shimizu, *Tetrahedron Letters*, 1975, 3923.
[625] T. C. Morrill, S. Malasanta, K. M. Warren, and B. E. Greenwald, *J. Org. Chem.*, 1975, **40**, 3032.
[626] M. D. Vorobev, A. S. Filatov, and M. A. Englin, *Zhur. org. Khim.*, 1974, **10**, 407.
[627] A. D. Berry and W. B. Fox, *J. Fluorine Chem.*, 1976, **7**, 449.
[628] E. R. Clark and M. A. Al-Turaihi, *J. Organometallic Chem.*, 1975, **96**, 251.
[629] R. C. Paul, K. K. Bhasin, and R. K. Chadha, *J. Inorg. Nuclear Chem.*, 1975, **37**, 2337.
[630] W. H. H. Guenther, J. Nepywoda, and J. Y. C. Chu, *J. Organometallic Chem.*, 1974, **74**, 79.
[631] M. Albeck and S. Shaik, *J.C.S. Perkin I*, 1975, 1223.
[632] P. Schulz and G. Klar, *Z. Naturforsch.*, 1975, **30b**, 40, 43.

$Ar'Te(Ar)_nCl_{3-n}$ and chlorinated aromatic hydrocarbons are formed by the reaction of $Ar'H$ with aryltellurium chlorides Ar_nTeCl_{4-n}.[631] Anisole undergoes *para*-substitution on reaction with $TeCl_4$ to give the organotellurium trichloride $4\text{-MeOC}_6H_4TeCl_3$.[633] The synthesis of the new anionic tellurium(IV) species $R^1{}^+ R^2TeX_4{}^-$ (X = Cl, Br, or I) from the corresponding trihalides has been reported.[634]

Addition and substitution reactions of the Se^{IV} and Te^{IV} halides are represented in the recent literature by a number of papers. *anti*-Addition has been established for the reaction of 2,4-dinitrophenylselenium trichloride with *cis*- and *trans*-1-phenylpropene, the initial adducts (*threo*- and *erythro*-$ArSeCl_2CHPhCHClMe$, respectively) giving the corresponding 1,2-dichloro-1-phenylpropanes and ArSeCl under mild conditions.[635] Diaryltellurium dichlorides give substitution products $Ar_2Te(O_2C)_2R$ with the sodium salt of a dicarboxylic acid $R(CO_2Na)_2$,[636] but the related selenium dihalides Ph_2SeX_2 (X = Cl or Br) give diphenyl selenide on treatment with NH_3, $MeNH_2$, or Me_2NH at $-70\,°C$, with no trace of Se^{IV} diamides.[637] The mono-amide $Ph_2SeClNMe_2$ is formed from diphenyl-selenium dichloride by reaction with Me_3SiNMe_2.[637]

Sulphenamides and Related Compounds.—Conventional synthetic routes to sulphenamides are represented with new systems in the recent literature. A sulphenyl chloride reacts with a hydrazone to give the sulphenamide $ArCR= NNHSAr$,[638] and with an acid hydrazide to give the analogous product.[639] Salts are formed between an arenesulphenyl chloride or a thiolsulphonate and a tertiary amine (*e.g.* $ArSCl + NEt_3 \rightarrow ArS\overset{+}{N}Et_3\ Cl^-$).[640]

Dialkylaminosulphenyl chlorides R_2NSCl give disulphides R_2NSSCF_3 with mercury(II) trifluoromethanethiolate.[641] Reduction of analogous disulphides gives the thermally labile thiohydroxylamines R_2NSH, which decompose above $-40\,°C$ into the secondary amine and sulphur.[642] These compounds have been postulated as reaction intermediates but not previously isolated. Thio-oximes $R_2C=NSH$, formed by reduction of an analogous disulphide, are unstable above $-70\,°C$.[643]

Base-catalysed elimination of thiolate anion from arenesulphenylhydrazones offers a novel route to aryl diazoalkanes.[638] Treatment of an arenesulphenyl-hydrazide $RC(O)NHNHSAr$ with base gives the corresponding aldehyde, providing overall a new, though long, route from an acid to the corresponding aldehyde.[639] A full account of new reactions of tribenzenesulphenamide $[(PhS)_3N]$ has been published.[219] Wittig-type reagents $R_3P=NSPh$ are formed

[633] F. J. Berry, E. H. Kustan, M. Roshani, and B. C. Smith, *J. Organometallic Chem.*, 1975, **99**, 115.
[634] N. Petragnani, J. V. Comasseto, and Y. Kawano, *J. Inorg. Nuclear Chem.*, 1976, **38**, 608.
[635] D. G. Garratt and G. H. Schmid, *Canad. J. Chem.*, 1974, **52**, 3599.
[636] N. Dance and W. R. McWhinnie, *J. Organometallic Chem.*, 1976, **104**, 317.
[637] V. Horn and R. Paetzold, *Z. anorg. Chem.*, 1974, **404**, 213.
[638] D. E. Dana and J.-P. Anselme, *Tetrahedron Letters*, 1975, 1565.
[639] S. Cacchi and G. Paolucci, *Gazzetta*, 1974, **104**, 221.
[640] J. H. Wevers and H. Kloosterziel, *J.C.S. Chem. Comm.*, 1975, 413.
[641] F. Bur-Bur, A. Haas, and W. Klug, *Chem. Ber.*, 1975, **108**, 1365.
[642] D. H. R. Barton, S. V. Ley, and P. D. Magnus, *J.C.S. Chem. Comm.*, 1975, 855.
[643] D. H. R. Barton, P. D. Magnus, and S. I. Pennanen, *J.C.S. Chem. Comm.*, 1974, 1007; C. Brown, B. T. Grayson, and R. F. Hudson, *J.C.S. Chem. Comm.*, 1974, 1007.

with phosphines, and on reaction with aromatic aldehydes give *S*-phenyl thio-
oximes. Tribenzenesulphenamide reacts with an enamine to give the α-sulphenylated ketone after acid hydrolysis of the intermediate, indicating that the tri-
sulphenamide undergoes ionic reactions as well as the radical reactions demonstrated earlier (see Vol. 3, p. 63).[219]

Cyclization of an *o*-methoxycarbonylbenzenesulphenamide to a benziso-
thiazolone is readily brought about by base.[644]

N,N-Dialkylbenzeneselenenamides are generally similar to the corresponding
sulphenamides in their selenenylation reactivity towards unsaturated systems.[645]

Dibenzenesulphenamide in benzene turns purple on oxidation with PbO_2, due
to the formation of a stable (7 days) N-centred radical.[646] An analogous radical
is formed by homolysis of the N—N bond in a substituted arenesulphenyl-
hydrazine.[647]

Barriers to rotation about the S—N bond in a sulphenamide derived from a
secondary amine are amenable to study by n.m.r., and Raban's pioneering
studies over recent years, published mainly in the form of preliminary accounts,
have now been fully described.[648] A similar study of an *N*-acyl-*N*-benzyl arene-
sulphenamide [649] leads to the conclusion that $p_\pi-d_\pi$ bonding is not a major factor
in determining the size of the rotation barrier. One possible alternative is a
gauche effect in which the polarity of the C—S bond of a sulphenamide is of
crucial importance. A similar conclusion that the electronic effects transmitted
to the bonds surrounding the sulphenamide grouping in *N*-(4,4′-dimethylbenzo-
phenylidene)arene-sulphenamides and -selenenamides are a major factor determining planar inversion barriers has been discussed.[650] A ¹H n.m.r. study of
analogous *S*-aryl thio-oximes and sulphenamides, demonstrating the transmission of electronic effects from the *S*-aryl substituent through the S—N bond,
implicates both *p*- and *d*-orbitals on sulphur in the process.[651]

10 Thiocyanates and Isothiocyanates

Preparation of Thiocyanates.—Variations of standard methods for the synthesis
of thiocyanates are illustrated in the addition of alkoxy- and thiocyanato-groups
to alkenes using KSCN, $CuCl_2$ (or other Cu^{II} salt), and an alcohol as solvent, to
give α-alkoxy-alkyl thiocyanates;[652] and in the addition of pseudohalogens
ClSCN or $(SCN)_2$ to chalcones.[653] A sulphonylthiocyanate RSO_2SCN, prepared
from $(SCN)_2$ and a sodium sulphinate, adds similarly to alkenes to give α-thio-
cyanato-alkyl sulphones.[654, 655] Aryl selenocyanates may be prepared from the

[644] J. C. Grivas, *J. Org. Chem.*, 1975, **40**, 2029.
[645] H. J. Reich and J. M. Renga, *J. Org. Chem.*, 1975, **40**, 3313.
[646] Y. Miura, N. Makita, and M. Kinoshita, *Tetrahedron Letters*, 1975, 127.
[647] R. S. Atkinson, S. B. Awad, E. A. Smith, and M. C. R. Symons, *J.C.S. Chem. Comm.*, 1976, 22.
[648] M. Raban, D. A. Noyd, and L. Bermann, *J. Org. Chem.*, 1975, **40**, 752; M. Raban, S. K. Lauderback, and D. Kost, *J. Amer. Chem. Soc.*, 1975, **97**, 5178.
[649] D. Kost and A. Zeichner, *Tetrahedron Letters*, 1975, 3239.
[650] F. A. Davis and E. W. Kluger, *J. Amer. Chem. Soc.*, 1976, **98**, 302.
[651] F. A. Davis, J. M. Kaminski, E. W. Kluger, and H. S. Freilich, *J. Amer. Chem. Soc.*, 1975, **97**, 7085.
[652] A. Onoe, S. Uemura, and M. Okano, *Bull. Chem. Soc. Japan*, 1974, **47**, 2818.
[653] F. G. Weber, A. Holzenger, G. Westphal, and U. Pusch, *Pharmazie*, 1975, **30**, 800.
[654] G. C. Wolf, *J. Org. Chem.*, 1974, **39**, 3454.
[655] S. C. Olsen and C. Christophersen, *Acta Chem. Scand.* (*B*), 1975, **29**, 717.

corresponding arylthallium derivative and KSeCN,[656] and aliphatic examples from an alkene, Tl(OAc)$_3$, and KSeCN.[656] The synthesis of a thiocyanate from a thiol using a 1-cyano-4-dimethylaminopyridinium salt is rapid in neutral or acidic aqueous solutions, offering a valuable method for modifying thiol groups in proteins.[657] Conversely, the synthesis of thiocyanates from primary alkoxy-carbonyl thiocyanates or isothiocyanates by heating at 240—375 °C at *ca.* 5 Torr provides a method for the synthesis of thiols from primary alcohols [ROH → ROC(O)SCN → RSCN → RSH].[658] Routes to aroyl thiocyanates have been established.[659]

Reactions of Thiocyanates.—[35]S-Labelled 1-methyl-4-thiocyanatouridine reacts with SH$^-$ to give the corresponding thiol, which, from the distribution of the label, is formed mainly through R—S bond fission, but the alternative S—CN fission occurs to a significant extent (30%).[660] Elimination of HSCN is readily achieved from the styrene–sulphonylthiocyanate adduct ArSO$_2$CH$_2$CHPhSCN, using a tertiary amine.[655] 2,4-Dinitro-1-thiocyanatobenzene undergoes true S_NAr exchange with N$_3$$^-$ or $^-$SCN, and in this respect is similar to the iodo-analogue.[661] A β-keto-thiocyanate ArC(O)CH$_2$SCN reacts as an active methylene compound, undergoing the Japp–Klingemann reaction with aryldiazonium cations to give thiadiazoles (77).[662]

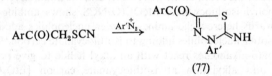

(77)

Oxidation of a thiocyanate with *m*-chloroperoxybenzoic acid gives the sulphinyl cyanate RS(O)CN. Although the product could not be isolated, it could be converted back into the thiocyanate using Ph$_3$P.[663] Cleavage by cyanide has been assessed.[835]

Preparation of Isothiocyanates.—Syntheses from amines *via* dithiocarbamates [664–666] include conventional [R^{1}NHC(S)SR^2 → R^{1}NCS at 145—185 °C] [664] and more interesting procedures [RNH$_2$ + CS$_2$ + EtMgBr → RN(MgBr)C(S)SMgBr → RNCS,[666] or the equivalent procedure using Bu^{n}Li [665]]. Pyrolysis of secondary or tertiary alkoxycarbonyl-thiocyanates or -isothiocyanates at 240—375 °C at *ca.* 5 Torr gives the corresponding isothiocyanates.[658] This

[656] S. Uemura, A. Toshimitsu, M. Okano, and K. Ichikawa, *Bull. Chem. Soc. Japan*, 1975, **48**, 1925.
[657] M. Wakselman, E. Guibé-Jampel, A. Raoult, and W. D. Busse, *J.C.S. Chem. Comm.*, 1976, 21.
[658] D. Liotta and R. Engel, *Canad. J. Chem.*, 1975, **53**, 907.
[659] C. Christophersen and P. Carlsen, *Tetrahedron*, 1976, **32**, 745.
[660] B. C. Pal and D. G. Schmidt, *J. Amer. Chem. Soc.*, 1974, **96**, 5943.
[661] J. Miller and F. H. Kendall, *J.C.S. Perkin II*, 1974, 1645.
[662] A. S. Shawali and A. O. Abdelhamid, *Tetrahedron Letters*, 1975, 163.
[663] A. Boerma-Markerink, J. C. Tagt, H. Meyer, J. Wildeman, and A. M. van Leusen, *Synth. Comm.*, 1975, **5**, 147.
[664] Y. E. Moharir, *J. Indian Chem. Soc.*, 1975, **52**, 148.
[665] S. Sakai, T. Aizawa, and T. Fujinami, *J. Org. Chem.*, 1974, **39**, 1970.
[666] S. Sakai, T. Fujinami, and T. Aizawa, *Bull. Chem. Soc. Japan*, 1975, **48**, 2981.

provides a method for the conversion of a primary or a secondary alcohol into the corresponding amine *via* the isothiocyanate.

Cleavage of pyridine with thiophosgene and $BaCO_3$ gives a mixture of *trans,cis*- and *trans,trans*-5-isothiocyanatopenta-2,4-dienals SCNCH=CHCH=CHCHO.[667] The corresponding reaction with 4,7-dichloroquinoline gives the expected *ortho*-substituted phenyl isothiocyanate.[668] *N*-Trimethylsilyl-lactams give ω-isothiocyanatoalkanoyl chlorides $SCN(CH_2)_nC(O)Cl$, *via* *N*-chlorothiocarbonyl derivatives, by treatment with thiophosgene.[669]

Although a good deal of variety in structure is conceivable with bifunctional isothiocyanates, preparative methods are generally routine. An example is the conversion of Cl_2FCSCl through several steps into $Cl_2FCSNCS$.[670]

Reactions of Isothiocyanates.—The volume of work reported on the addition reactions of isothiocyanates is, as usual, large but based on well-established chemistry. Among the many papers covering the synthesis of thioureas by addition of an amine to an isothiocyanate, one extraordinary account is to be found describing the conversion of an isothiocyanate into a thiourea by reaction with 3 equivalents of Me_2SO.[671] The reaction presumably involves conversion of isothiocyanate into amine, since the corresponding reaction with 1 equivalent each of isothiocyanate and benzaldehyde, together with 3 equivalents of Me_2SO, gives a mixture of the Schiff base PhCH=NR and the thiourea RNHC(S)NHR.[671] Methoxycarbonyl isothiocyanate MeOC(O)NCS shows ambident nucleophilic reactivity towards heterocyclic amines, harder nucleophiles adding to carbonyl carbon, and softer nucleophiles adding to isothiocyanate carbon.[672] α-Metallated α-isothiocyanato-alkanoates react with an alkyl halide to give products of both α-alkylation and alkylation at isothiocyanate carbon [$EtO_2CCHR^1NCS \rightarrow$ $(EtO_2CCR^1NCS)^- \rightarrow EtO_2CCR^1R^2NCS + EtO_2CCR^1=NC(S)R^2$].[673]

Chlorination of a vinyl isothiocyanate gives the corresponding imine (Me_2C=CPhNCS $\rightarrow Me_2CClCPhClN$=$CCl_2$).[674] The formation of the thiocarbamoyl chloride (78) from phenyl isothiocyanate and $SbCl_5$ offers a mechanistic puzzle for further study.[675]

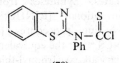

(78)

11 Sulphinic Acids

Preparation.—Steady work over recent years has led to the accumulation of a substantial body of information on the factors determining C—S *versus* C—O

[667] F. T. Boyle and R. Hull, *J.C.S. Perkin I*, 1974, 1541.
[668] R. Hull, P. J. van den Broek, and M. L. Swain, *J.C.S. Perkin I*, 1975, 922.
[669] H. R. Kricheldorf, *Angew. Chem.*, 1975, **87**, 517.
[670] E. Kuehle, H. Hagemann, and L. Oehlmann, *Angew. Chem.*, 1975, **87**, 707.
[671] J. B. Chattopadhyaya and A. V. R. Rao, *Synthesis*, 1974, 289.
[672] T. Matsui and M. Nagano, *Chem. and Pharm. Bull. (Japan)*, 1974, **22**, 2123.
[673] I. Hoppe, D. Hoppe, and U. Schollkopf, *Tetrahedron Letters*, 1976, 609.
[674] V. I. Gorbatenko and L. I. Samarai, *Zhur. org. Khim.*, 1974, **10**, 1785.
[675] D. Herrmann and D. Keune, *Annalen*, 1975, 1025.

bond formation during the insertion of SO_2 into metal–alkyl complexes. Reactions of this type leading to sulphinates have been reviewed.[676]

Conversions of sulphonyl chlorides into sulphinic acids have been illustrated in the aromatic series,[677] and an interesting aliphatic case (Cl_2CHSO_2Cl + Et_3N or pyridine $\rightarrow$ Cl_3CSO_2H, *via* $Cl_2C{=}SO_2$) has been reported.[678]

The isolation of 3-sulphinopropionic acid from *Pseudomonas fluorescens* has been reported [679] and the structure confirmed by synthesis involving cleavage of bis(2-carboxyethyl)sulphone. Cleavage of thiiran 1,1-dioxides with metal halides [680] or metal thiolates [681] gives β-halogeno- and β-alkylthio-alkane-sulphinates, respectively.

Properties and Reactions of Sulphinic Acids.—Addition reactions of sulphinic acids to alkenes lead to sulphones; studies of the addition of $PhSO_2H$ to maleimides [682] and flavan-4-ols [171] have been reported. Arenesulphinic acids catalyse the *cis–trans* equilibration of disubstituted alkenes, leading to a large number of sulphur-containing side-products, none of which were identified, although sulphones must be included amongst these compounds.[683] Methoxymethylation of sodium 4-chlorobenzenesulphinate with $MeOCH_2Br$ or $MeOCH_2SO_2Me$ gives either the methoxymethyl sulphone or the unstable methoxymethanesulphinate, depending on the reaction conditions.[684] β-Halogenoalkanesulphinates fragment readily, giving SO_2, an alkene, and halide ion.[680, 685]

There are two reaction pathways to choose between for the classical oxidation with SeO_2 of aliphatic ketones or aldehydes to α-diketones or glyoxals, respectively, and a β-keto-seleninic acid intermediate $R^1C(O)CHR^2SeOOH$ is preferred [686] to the enol ester intermediate $[R^1C(O)CH_2R^2 + H_2SeO_3 \rightarrow R^1C(OSeOOH){=}CHR^2]$.

Diphenylseleninic anhydride, $PhSeOOSeOPh$, has been found to be a useful oxidizing agent, giving *ortho*- and *para*-hydroxylated products with phenols but only *ortho*-hydroxylation with phenolate anions.[687] Benzylamines are oxidized by the reagent to benzaldehydes,[688] and primary amines of general structure R_2CHNH_2 to ketones R_2CO.[688]

Sulphinyl Halides.—The isolation of an α-cyanoalkanesulphinyl chloride from a reaction mixture containing thionyl chloride and a secondary alkyl cyanide has been reported, although the yield was low.[689] Primary alkyl cyanides RCH_2CN give α-chloro-α-cyanoalkanesulphenyl chlorides with $SOCl_2$.[689] Cleavage of

[676] A. Wojcicki, *Adv. Organometallic Chem.*, 1974, **12**, 31.
[677] E. C. Dart and G. Holt, *J.C.S. Perkin I*, 1974, 1403.
[678] T. Kempe and T. Norin, *Acta Chem. Scand.* (*B*), 1974, **28**, 609.
[679] B. Jolles-Bergeret, *European J. Biochem.*, 1974, **42**, 349.
[680] E. Vilsmaier, R. Tropitsch, and O. Vostrowsky, *Tetrahedron Letters*, 1974, 3275.
[681] E. Vilsmaier and G. Becker, *Synthesis*, 1975, 55.
[682] I. Matsuda, K. Akiyama, T. Toyoshima, S. Kato, and M. Mizuta, *Bull. Chem. Soc. Japan*, 1975, **48**, 3675.
[683] T. W. Gibson and P. Strassburger, *J. Org. Chem.*, 1976, **41**, 791.
[684] K. Schank and H. G. Schmitt, *Chem. Ber.*, 1974, **107**, 3026.
[685] T. Kempe and T. Norin, *Acta Chem. Scand.* (*B*), 1974, **28**, 613.
[686] K. B. Sharpless and K. M. Gordon, *J. Amer. Chem. Soc.*, 1976, **98**, 300.
[687] D. H. R. Barton, P. D. Magnus, and M. N. Rosenfeld, *J.C.S. Chem. Comm.*, 1975, 301.
[688] M. R. Czarny, *J.C.S. Chem. Comm.*, 1976, 81.
[689] M. Ohoka, T. Kojitani, S. Yanagida, M. Okahara, and S. Komori, *J. Org. Chem.*, 1975, **40**, 3540.

aralkyl methanesulphonylmethyl sulphides with Cl_2 in aqueous AcOH leads eventually to Cl_2CHSO_2Me *via* the sulphinyl chloride intermediates $ClS(O)$-$CHClSO_2Me$ and $ClS(O)CCl_2SO_2Me$.[533]

Sulphinate Esters.—A wide variety of synthetic routes to sulphinate esters is represented in the recent literature. A one-step synthesis [690] of alkyl t-alkanesulphinates from a Grignard reagent and a sulphite has been explored. Among the reports of sulphinate syntheses from sulphinyl halides are three accounts of studies with unusual interest, *viz.* $CF_3SOF \rightarrow CF_3S(O)OR$,[691] *N*-t-butylsulphinyloxycarbamates from Bu^tSOCl and $HONR^1C(O)OR^2$,[692] and an asymmetric synthesis involving a sulphinyl chloride, an achiral alcohol, and an optically active tertiary amine.[693] In the best experiment an optical purity of 45% was achieved in the asymmetric synthesis.[693]

The reaction of an amine oxide with a sulphoxide gives the betaine $R^1_3\overset{+}{N}OSR^2_2O^-$, from which sulphinates may be obtained by nucleophilic attack.[419] Another novel synthesis, of methyl prop-2-enesulphinates from fluoromethane, SbF_5, and a vinyl chloride $RCH_2CCl=CH_2$, in liquid SO_2, is formally an ene reaction involving the species $MeO\overset{+}{S}=O$ as enophile.[694] Photolysis of toluene-*p*-sulphinamides in alcohol solvents gives the corresponding sulphinate esters.[695]

Oxidation of trichloromethanesulphenates with *m*-chloroperoxybenzoic acid gives sulphinates, but arenesulphenates give sulphonates under the same conditions.[696]

Cleavage of sulphinate esters in aqueous acidic [697] or alkaline [697, 698] media follows second-order kinetics, hydrolysis in alkali involving a trigonal-bipyramidal intermediate.[698] Exclusive C—O bond fission is involved in ethanolysis of trichloromethanesulphinates ($ArCH_2OSOCCl_3 + EtOH \rightarrow ArCH_2OEt + Cl_3CSO_2H$) [696] and of furfuryl arenesulphinates,[699] while benzyl arenesulphinates undergo exclusive S—O bond fission in EtOH at much lower rates.[696] Apparently,[696] trichloromethanesulphinates behave as if their sulphur atom possesses lower nucleophilicity than the corresponding sulphur atom in aryl analogues. Fission of furfuryl arenesulphinates in hydroxylic solvents is accompanied by rearrangement to the corresponding sulphone, while in non-hydroxylic solvents the rearrangement pathway is the sole reaction.[699] [2,3]-Sigmatropic rearrangement of propargyl benzenesulphinates to allenyl phenyl sulphones [700] and the rearrangement of sulphinyl carbamates $Bu^tS(O)$-$ONR^1CO_2R^2$ to *N*-alkoxycarbonylsulphonamides [692] have been thoroughly studied.

Reactions of sulphinates and sulphoxylates with FSO_3Me and CF_3SO_3Me proceed *via* dialkoxysulphonium ions produced, in the former case, by methylation

690 M. Mikolajczyk and J. Drabowicz, *Synthesis*, 1974, 124.
691 A. Majid and J. M. Shreeve, *Inorg. Chem.*, 1974, **13**, 2710.
692 W. J. Bouma and J. B. F. N. Engberts, *J. Org. Chem.*, 1976, **41**, 143.
693 M. Mikolajczyk and J. Drabowicz, *J.C.S. Chem. Comm.*, 1974, 547.
694 P. E. Peterson, R. Brockington, and M. Dunham, *J. Amer. Chem. Soc.*, 1975, **97**, 3517.
695 H. Tsuda, H. Minato, and M. Kobayashi, *Chem. Letters*, 1976, 149.
696 S. Braverman and Y. Duar, *Tetrahedron Letters*, 1975, 343.
697 M. Kobayashi, R. Nishi, and H. Minato, *Bull. Chem. Soc. Japan*, 1974, **47**, 888.
698 A. A. Najam and J. G. Tillett, *J.C.S. Perkin II*, 1975, 858.
699 S. Braverman and T. G. Globerman, *Tetrahedron*, 1974, **30**, 3873.
700 S. Braverman and H. Mechoulam, *Tetrahedron*, 1974, **30**, 3883.

of sulphinyl oxygen.[701] Reversal of the ene reaction by which methyl prop-2-enesulphinates MeOS(O)CH$_2$CCl=CHR are formed from the vinyl halide is achieved by base hydrolysis;[694] an ingenious suggestion follows from this observation, that sulphinate ester groups may be useful in activating a vinyl carbon atom towards substitution reactions.[694]

Inversion of configuration accompanies nucleophilic substitution of (−)-(*S*)-(−)-menthyl toluene-*p*-sulphinate by the magnesium enolate of t-butyl acetate.[702]

Sulphinamides and Related Compounds.—A standard route to sulphinamides is illustrated by the reaction of toluene-*p*-sulphinyl chloride with (−)-menthyl-carbamate;[703] separation of the resulting diastereoisomers could not be achieved. The reaction of the sulphinyl chloride with (−)-menthyl *N*-(*p*-nitrobenzene-sulphonyloxy)carbamate, however, gave a chiral arenesulphonimidoyl chloride TolSO(Cl)NHCO$_2$R [R = (−)-menthyl],[703] from which both enantiomers of a chiral sulphinamide could be obtained. Conversion of the sulphinamide enantiomers into the known (−)-(*S*)- and (+)-(*R*)-methyl tolyl sulphoxides by reaction with MeMgBr, which is known to proceed with inversion of configuration, was used to establish absolute configurations. Several new relatives of chiral sulphinamides have been prepared through a study of the nucleophilic substitution reactions of chiral arylsulphonimidoyl chlorides (79) displayed in Scheme 7.[703] An efficient synthesis of alkanesulphinylanilides involves the addition of RMgX to an *N*-sulphinylaniline.[704]

Oxidation of a sulphenamide to a sulphinamide has been described in two recent papers;[705] ArSN=CR$_2$ gives ArS(O)N=CR$_2$ with *m*-chloroperoxy-benzoic acid,[705a] and C$_6$F$_5$SNH$_2$ gives C$_6$F$_5$S(O)NH$_2$ with active MnO$_2$ at room temperature. In the latter reaction, oxidation to the sulphonamide takes place at 70 °C.[705b]

As with the analogous sulphenamides,[638] base-induced α-elimination from methanesulphinylhydrazones gives aryldiazoalkanes.[706] Cope-type elimination from *N*-alkylidene-arenesulphinamides gives the corresponding sulphenic acids,[704] which undergo further condensation reactions of a familiar type. Sulphenate esters ArS(O)SiMe$_3$ are formed when the elimination reaction is conducted in the presence of Me$_3$SiCl.[601] Photolysis of arenesulphinamides in alcohol solvents gives corresponding sulphinate esters.[695] Stannyl sulphinamides R^{1}S(O)NR^2-SnR$^3_3$ are formed by transamination of a sulphinamide with R$^3_3$SnNMe$_2$.[707]

Powerful methylating agents, *e.g.* CF$_3$SO$_3$Me, convert a sulphinamide into the corresponding methoxyaminosulphonium salt;[708] the reaction may be reversed with Me$_2$S.[708]

Benzenesulphinyl azides do not undergo Curtius rearrangement, in contrast to their sulphonyl analogues.[709] They react with sulphoxides through a 1,2-dipolar

[701] H. Minato, K. Yamaguchi, and M. Kobayashi, *Chem. Letters*, 1975, 307.
[702] C. Mioskowski and G. Solladie, *Tetrahedron Letters*, 1975, 3341.
[703] M. R. Jones and D. J. Cram, *J. Amer. Chem. Soc.*, 1974, 96, 2183.
[704] S. B. Bowlus and J. A. Katzenellenbogen, *Synth. Comm.*, 1974, 4, 137.
[705] (a) F. A. Davis, A. J. Friedman, and E. W. Kluger, *J. Amer. Chem. Soc.*, 1974, 96, 5000; (b) I. Glander and A. Golloch, *J. Fluorine Chem.*, 1975, 5, 83.
[706] J. G. Shelnut, S. Mataka, and J.-P. Anselme, *J.C.S. Chem. Comm.*, 1975, 114.
[707] E. Wenschuh, W. D. Riedmann, L. Korecz, and K. Burger, *Z. anorg. Chem.*, 1975, 413, 143.
[708] H. Minato, K. Yamaguchi, and M. Kobayashi, *Chem. Letters*, 1975, 991.
[709] (a) T. J. Maricich and V. L. Hoffman, *J. Amer. Chem. Soc.*, 1974, 96, 7770; (b) G. de Luca, G. Renzi, V. Bartocci, and C. Panattoni, *Chem. and Ind.*, 1975, 1054.

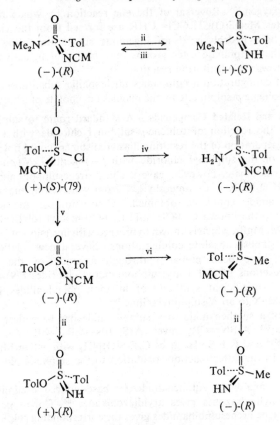

NCM = NCO₂[(−)-menthyl]

$$\text{NCM} = \text{NCO}_2[(-)\text{-menthyl}]$$

Reagents: i, Me₂NH; ii, H₂SO₄; iii, ClCM; iv, NaNH₂–NH₃; v, KOTol; vi, MeMgBr

Scheme 7

cycloaddition process involving the sulphinylnitrene, and give *N*-benzene-sulphonylsulphimides rather than *N*-sulphinylsulphoximides.[709]

Diazotization of primary sulphinamides RS(O)NH₂ with isopentyl nitrite gives mainly the sulphinic acids, together with isopentyl sulphones and sulphoxides.[709a]

12 Sulphonic Acids

Preparation.—An increase in the number of papers dealing with the preparation of aliphatic sulphonic acids is noticeable in the period under review. The first synthesis of cyclopropanesulphonic acid has been achieved through a novel reaction sequence [RSiMe₃ + ClSO₃SiMe₃ → RSO₃SiMe₃ → RSO₃H,H₂O (R = cyclopropyl)].[710] Addition of NaHSO₃ to an alkene under radical-forming

710 M. Grignon-Dubois, J. Dunogues, and R. Calas, *Tetrahedron Letters*, 1976, 1197.

conditions [711, 712] involves H-abstraction from the bisulphite radical anion by the intermediate β-sulphonylalkyl radical.[711] Addition of SO_3 to but-1-ene and homologues gives 1,3-propanesultones or but-2-ene-1-sulphonic acids, depending on the degree of substitution of the double bond.[712] Addition of Me_2S-SO_3 to alk-1-enes gives betaines $R^1R^2C(\overset{+}{S}Me_2)CH_2SO_3^-$, from which $\alpha\beta$-unsaturated sulphonic acids $R^1R^2C{=}CHSO_3H$ are obtained by treatment with alkali.[713] The product of reaction of *trans,trans*-dibenzylideneacetone with H_2SO_4 and Ac_2O is the cyclopentanone-2-sulphonic acid (80), correcting an earlier structure assignment.[714]

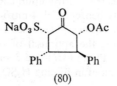

(80)

Aminolysis of 1,3-propanesultones gives 3-aminoalkylalkanesulphonic acids.[715]

Useful procedures have been worked out for the oxidation of alkanethiols [716] and aromatic disulphides [717] to sulphonic acids. Optimum conditions for oxidation of alkanethiols by HNO_3 involves cooling to $1-2\,°C$ above the freezing point of the thiol in a N_2 atmosphere.[716]

Substantial contributions continue to be made concerning the addition of bisulphite to uracil and its 5-halogeno-derivatives, and the dehalogenation of 5-halogenouracil-6-sulphonates.[718-720] A two-step mechanism operates in the elimination of bisulphite ion from uracil-6-sulphonates,[719a] and attack by bisulphite ion on a 5-halogeno-5,6-dihydro-6-methoxy-uracil is the rate-determining step [719b] in the dehalogenation of these compounds, involving either S_N2-displacement of the halogen or an $E1cb$ route.[719b, c; cf. 720a] Related studies, but with synthetic objectives, have provided a route to purine-6-sulphonic acids by the addition of $KHSO_3$ or H_2SO_3 to 6-substituted purines.[721]

Sulphonation of [(cyclobutadiene)Fe(CO)₃] [722] and of [(cyclohexa-1,3-diene)-Fe(CO)₃] [723] has been described, leading to the Fe(CO)₃ complex of the 5-sulphonic acid in the latter case.[723]

[711] T. Miyata, A. Sakumoto, M. Washino, and T. Abe, *Chem. Letters*, 1975, 367; A. Sakumoto, T. Miyata, and M. Washino, *ibid.*, p. 563.
[712] M. D. Robbins and C. D. Broaddus, *J. Org. Chem.*, 1974, **39**, 2459.
[713] M. Nagayama, O. Okumura, K. Yaguchi, and A. Mori, *Bull. Chem. Soc. Japan*, 1974, **47**, 2473.
[714] C. W. Shoppee and B. J. A. Cooke, *J.C.S. Perkin I*, 1974, 189.
[715] I. Zeid and I. Ismail, *Annalen*, 1974, 667.
[716] W. G. Filby, *Lab. Practice*, 1974, **23**, 355.
[717] W. H. Dennis, D. H. Rosenblatt, and J. R. Simcox, *Synthesis*, 1974, 295.
[718] H. Hayatsu, T. Chikuma, and K. Negishi, *J. Org. Chem.*, 1975, **40**, 3862.
[719] G. S. Rork and I. H. Pitman, *J. Amer. Chem. Soc.*, (*a*) 1974, **96**, 4654; (*b*) 1975, **97**, 5566; (*c*) 1975, **97**, 5559; (*d*) *J. Pharm. Sci.*, 1975, **64**, 216.
[720] (*a*) F. A. Sedor, D. G. Jacobson, and E. G. Sander, *Bio-org. Chem.*, 1974, **3**, 221; (*b*) F. A. Sedor, D. G. Jacobson, and E. G. Sander, *J. Amer. Chem. Soc.*, 1975, **97**, 5572; (*c*) D. G. Jacobson, F. A. Sedor, and E. G. Sander, *Bio-org. Chem.*, 1975, **4**, 72; (*d*) F. A. Sedor and E. G. Sander, *Arch. Biochim. Biophys.*, 1974, **161**, 632.
[721] W. Prendergast, *J.C.S. Perkin I*, 1975, 2240.
[722] P. Marcincal and E. Cuinguet, *Tetrahedron Letters*, 1975, 1223.
[723] A. J. Birch, I. D. Jenkins, and A. J. Liepa, *Tetrahedron Letters*, 1975, 1723.

Aromatic sulphonation studies are again well represented in the recent literature, and novel observations in areas of methodology and mechanism have been made. Photo-induced sulphonation of aromatic halides or sulphones [724] and of 1-aminoanthraquinones to give the 2-sulphonates [725] has been described. Brief treatment of 1,4-di(phenylsulphonyl)-2,3,5,6-tetrachlorobenzene with sodium sulphite in aqueous dioxan gives the disodium 1,4-disulphonate, while longer reaction (6 h) in the presence of a copper(II) salt leads to an 80% yield of hexasodium benzenehexasulphonate.[726] Tetrachloropyridines give pentasodium pyridinepentasulphonate in the same way.[727] Sulphonation of anthracenes,[728] *meso*-methylated anthracenes,[728, 729] acenaphthenes,[730] 1,2-dihydrobenzocyclobutene and di- and tri-t-butylbenzenes,[731] 4-nitro-, 4-amino-, and 4-halogenotoluenes,[732] and simple alkylbenzenes [733] represents a familiar area of study, and some of these projects include the appraisal of ring-sulphonation *versus* side-chain-substitution processes. Pyran-(4*H*)-one 3-sulphonic acid is formed from the parent heterocyclic compound in fuming H_2SO_4.[734]

Properties of Sulphonic Acids.—In the course of a study of the kinetics of nitration of arenesulphonic acids in H_2SO_4 [735] it was shown that the sulphonate anion is a deactivating substituent even though it carries a negative charge.

Some ^{1}H and [13]C n.m.r. studies of alkane- and arene-sulphonic acids in H_2SO_4 yielded pK_{BH} values for representative compounds.[736] Electric dipole-moment data for potassium toluene-*p*-sulphonate, in the presence of LiCl and dibenzo-18-crown-6 ether, in octanoic acid indicate a symmetrical structure for the ion-pair, with K^+ adjacent to the tetrahedral sulphur atom.[737]

The discovery that an arenesulphonic acid is in equilibrium with its fluoride in FSO_3H solution ($ArSO_3H + FSO_3H \rightleftharpoons ArSO_2F + H_2SO_4$) [738] may open up a new field of study because of its implications in mechanistic and synthetic aspects.

Hydrogenation of arenesulphonic acids to *cis*-cyclohexanesulphonic acids is feasible using $Rh-Al_2O_3$ or $Ru-Al_2O_3$ catalysts.[739]

[724] A. N. Frolov, E. V. Smirnov, and A. V. Eltsov, *Zhur. org. Khim.*, 1974, **10**, 1686.
[725] J. O. Morley, *J.C.S. Chem. Comm.*, 1976, 88.
[726] N. S. Dokunikhin and G. A. Mezentseva, *Zhur. Vsesoyuz Khim. obshch. im. D.I. Mendeleeva.*, 1974, **19**, 356.
[727] S. D. Moshchitskii, G. A. Zalesskii, V. P. Kukhar, and L. M. Yagupolskii, *Zhur. org. Khim.*, 1975, **11**, 1134.
[728] H. Cerfontain, A. Koeberg-Telder, C. Ris, and C. Schenk, *J.C.S. Perkin II*, 1975, 966.
[729] A. Koeberg-Telder and H. Cerfontain, *Tetrahedron Letters*, 1974, 3535.
[730] H. Cerfontain and Z. R. H. Schassberg-Nienhuis, *J.C.S. Perkin II*, 1974, 989.
[731] A. Koeberg-Telder and H. Cerfontain, *J.C.S. Perkin II*, 1974, 1206; C. Ris and H. Cerfontain, *ibid.*, 1975, 1438.
[732] Y. Muramoto and H. Asakura, *Nippon Kagaku Kaishi*, 1975, 1070 (*Chem. Abs.*, 1975, **83**, 113 296).
[733] B.-G. Gnedin and N. I. Rudakova, *Izvest. V. U. Z., Khim. i khim. Tekhnol.*, 1974, **17**, 1820; V. A. Kozlov, A. A. Spryskov, and E. N. Krylov, *ibid.*, p. 710; B. G. Gnedin, G. P. Chudinova, T. A. Baranova, and V. V. Bykova, *ibid.*, p. 769; B. G. Gnedin, A. A. Spryskov, and G. P. Chudinova, *ibid.*, p. 388; B. G. Gnedin, T. A. Baronova, and G. P. Chudinova, *ibid.*, p. 552; V. P. Leschev, V. V. Kharitonov, A. A. Spryskov, and V. S. Stepanova, *ibid.*, p. 621.
[734] C. C. McCarney, R. S. Ward, and D. W. Roberts, *J.C.S. Perkin I*, 1974, 1381.
[735] R. B. Moodie, K. Schofield, and T. Yoshida, *J.C.S. Perkin II*, 1975, 788.
[736] H. Cerfontain, A. Koeberg-Telder, and C. Kruk, *Tetrahedron Letters*, 1975, 3639; A. Koeberg-Telder and H. Cerfontain, *J.C.S. Perkin II*, 1975, 226.
[737] Ting-Po I and E. Grunwald, *J. Amer. Chem. Soc.*, 1974, **96**, 2879.
[738] B. G. Gnedin, S. N. Ivanov, N. I. Rudakova, and A. A. Spryskov, *Izvest. V. U. Z., Khim. khim. Tekhnol.*, 1975, **18**, 1039.
[739] R. Egli and C. H. Eugster, *Helv. Chim. Acta*, 1975, **58**, 2321.

Sulphonic Anhydrides.—Unstable mixed sulphonic anhydrides $ArSO_2OSO_2CF_3$, prepared from the arenesulphonyl bromide and silver trifluoromethane-sulphonate,[740] are useful reagents for Friedel–Crafts sulphonoxylation.[740] Aromatic amines are some ten times more reactive than MeOH towards benzene-sulphonic anhydride in the presence of pyridine.[741]

Sulphonyl Peroxides.—Continuing interest in this class of compound is a consequence of their value in aromatic sulphonoxylation [742, 764] and in their addition to alkenes to give sulphonate esters. A useful electrochemical preparation of bis(methanesulphonyl) peroxide [743] from sodium methanesulphonate makes available a reagent for the synthesis of methanesulphonates of less reactive aromatic hydrocarbons.

p-Nitrobenzenesulphonyl peroxide labelled with ^{18}O at the sulphonyl oxygen atoms gives sulphonates with *p*-xylene and with benzene, the oxygen atom that links the sulphonyl group to the aromatic moiety being derived exclusively from the peroxide oxygen atoms by clean first-order reactions.[742] However, under different conditions, up to 41% of the product contains ^{18}O in this position.

Stereoselective addition of nitrobenzene-*p*-sulphonyl peroxide to *cis*- and *trans*-stilbenes has been studied. Random ratios of *meso*- and *dl*-adducts were obtained, and it was concluded that the carbonium ion intermediate in the electrophilic addition process is sufficiently long-lived in this case to undergo rotation before capture.[744]

Sulphonyl Halides.—Novel preparative procedures have been described; a sulphonyl chloride can be converted into a sulphonyl fluoride with XeF_2,[745] and chlorinolysis of perfluoroalkanesulphinylmagnesium bromides, from the Grignard reagent and SO_2, yields the corresponding sulphonyl chlorides.[746] Other methods employing organometallic or organometalloid reagents [RLi + SO_2Cl_2 at -20 to $-65\,°C$ gives RSO_2Cl,[747] and RSO_3SiMe_3 + $Cl_2CHOMe \rightarrow Me_3SiCl$ + RSO_2Cl [748]] have been reported. Azulenes react easily, and surprisingly, with thionyl chloride to give a mixture of azulenesulphonyl chlorides and bis(azulenyl) sulphides.[749]

Migration of Cl from S to C is observed in the radical addition of thiols to vinylsulphonyl chloride (CH_2=$CHSO_2Cl$ + RSH $\rightarrow$ $ClCH_2CH_2SR$ as a minor product, together with SO_2 and R_2S).[750] Other cleavage reactions of sulphonyl halides, deferring discussion of nucleophilic substitution of halogen to the next paragraph, lead to sulphinic acids [677, 678, 751, 752] through elimination [677] and

[740] F. Effenberger and K. Huthmacher, *Angew. Chem.*, 1974, **86**, 409; K. Huthmacher, G. Koenig, and F. Effenberger, *Chem. Ber.*, 1975, **108**, 2947.
[741] N. T. Maleeva, V. A. Savelova, L. M. Litvinenko, and L. G. Kuryakova, *Zhur. org. Khim.*, 1975, **11**, 1015.
[742] R. L. Dannley, R. V. Hoffman, P. K. Tornstrom, R. L. Waller, and R. B. Srivastava, *J. Org. Chem.*, 1974, **39**, 2543.
[743] C. J. Myall and D. Pletcher, *J.C.S. Perkin I*, 1975, 953.
[744] R. V. Hoffman and R. D. Bishop, *Tetrahedron Letters*, 1976, 33.
[745] S. A. Volkova, Z. M. Sinyntina, and L. N. Nikolenko, *Zhur. obshchei Khim.*, 1974, **44**, 2592.
[746] P. Moreu, G. Dalverny, and A. Commeyras, *J. Fluorine Chem.*, 1975, **5**, 265.
[747] H. Quast and F. Kees, *Synthesis*, 1974, 489.
[748] P. Bourgeois, *Compt. rend.*, 1974, **279**, C, 585.
[749] J. A. Billow and T. J. Speaker, *J. Pharm. Sci.*, 1975, **64**, 862.
[750] I. I. Kandror and R. K. Friedling, *Izvest. Akad. Nauk S.S.S.R.*, *Ser. khim.*, 1974, 1890.
[751] A. Horowitz and L. A. Rajbenbach, *J. Amer. Chem. Soc.*, 1975, **97**, 10.
[752] I. I. Kandror, R. G. Gasanov, and R. K. Freidlina, *Tetrahedron Letters*, 1976, 1075.

displacement [678] routes, or radical pathways.[751, 752] While photolysis of PhSO$_2$Cl gives products resulting from both C—S and S—Cl bond homolysis, PhSO$_2$F suffers only C—S cleavage.[752] Radical chain processes are involved in thermal and copper(II)-catalysed addition of RSO$_2$Br to phenylacetylene. Exclusive *trans*-addition occurs in the uncatalysed route, and this is accounted for by equilibration of the intermediate styrylsulphonyl radical.[753]

Nucleophilic substitution reactions of sulphonyl halides continue to receive detailed mechanistic study.[754—759, 775] Points of interest from hydrolysis studies [754—758] include the smaller sensitivity to substituent effects of hydrolysis rates of ArSO$_2$Cl compared with corresponding ArCH$_2$SO$_2$Cl,[754] catalysis by AcO$^-$ and NEt$_3$ of the hydrolysis of arenesulphonyl fluorides,[755] and a large ρ-value for the alkaline hydrolysis of substituted benzenesulphonyl fluorides,[758] which gives some support to the postulate of an addition–elimination mechanism for nucleophilic substitution at sulphur. Since *ortho*-substituents accelerate the rate of Cl exchange between ArSO$_2$Cl and Et$_4$N$^+$ Cl$^-$,[760] further support for an initial addition step (ArSO$_2$Cl + Cl$^-$ $\rightleftharpoons$ ArSO$_2$Cl$_2^-$) is provided since the trigonal-bipyramidal structure of the adduct involves smaller steric interactions, particularly those between sulphonyl oxygen atoms and the *ortho*-substituents. Several journals carry the results of a study of the aminolysis of 2- and 3-thiophensulphonyl halides by anilines.[759] Deuteration studies reveal the operation of the sulphene mechanism (PhCH$_2$SO$_2$Cl → PhCH=SO$_2$ → PhCH$_2$SO$_2$OR or PhCH$_2$SO$_2$NHAr), rather than direct nucleophilic substitution of halide, in the tertiary-amine-catalysed reaction with Pr^{i}OH and *p*-toluidine mixtures [761] (see Chapter 3).

Full details of the use of mesitylenesulphonyl chloride as a selective sulphonylation reagent for carbohydrates have been published.[762]

Chlorinolysis of pyrazole-4-sulphonyl chlorides with Cl$_2$ in aqueous AcOH causes displacement of the chlorosulphonyl group and cleavage of the ring.[763]

Sulphonates.—Synthesis of aryl arenesulphonates through arenesulphonoxylation of unactivated aromatic hydrocarbons can be accomplished using arenesulphonyl peroxides.[743, 764] Although other serviceable routes to sulphonate esters do not need to be documented here, since they are so well known, a procedure for the

[753] Y. Amiel, *J. Org. Chem.*, 1974, **39**, 3867.
[754] R. V. Visgert, I. M. Tuchapski, and Y. G. Skrypnik, *Reakts. spos. org. Soedinenii*, 1974, **11**, 21.
[755] J. L. Kice and E. A. Lunney, *J. Org. Chem.*, 1975, **40**, 2125.
[756] A. R. Haughton, R. M. Laird, and M. J. Spence, *J.C.S. Perkin II*, 1975, 637.
[757] L. Senatore, L. Sagramora, and E. Ciuffarin, *J.C.S. Perkin II*, 1974, 722.
[758] E. Ciuffarin and L. Senatore, *Tetrahedron Letters*, 1974, 1635.
[759] E. Maccarone, G. Musumarra, and G. A. Tomaselli, *Ann. Chim. (Italy)*, 1973, **63**, 861; A. Arcoria, E. Maccarone, G. A. Tomaselli, R. Cali, and S. Gurrieri, *J. Heterocyclic Chem.*, 1975, **12**, 333; A. Arcoria, E. Maccarone, G. Musumarra, and G. A. Tomaselli, *J. Org. Chem.*, 1974, **39**, 1689; E. Maccarone, G. Musumarra, and G. A. Tomaselli, *ibid.*, p. 3286; A. Arcoria, E. Maccarone, G. Musumarra, and G. A. Tomaselli, *ibid.*, p. 3595.
[760] M. Mikolajczyk, M. Gajl, and W. Reinischüssel, *Tetrahedron Letters*, 1975, 1325.
[761] J. F. King and Y. I. Kang, *J.C.S. Chem. Comm.*, 1975, 52; see also J. F. King, *Accounts Chem. Res.*, 1975, **8**, 10.
[762] S. E. Creasey and R. D. Guthrie, *J.C.S. Perkin I*, 1974, 1373.
[763] R. J. Alabaster and W. J. Barry, *J.C.S. Perkin I*, 1976, 428.
[764] R. L. Dannley and P. K. Tornstrom, *J. Org. Chem.*, 1975, **40**, 2278; R. L. Dannley and R. V. Hoffman, *ibid.*, p. 2426.

synthesis of 3-methylbut-2-en-2-yl trifluoromethanesulphonate should be mentioned.[765]

Elimination reactions of aryl arylmethanesulphonates leading to phenyl-sulphenes have been shown to involve $E1cb$ characteristics [766-768] where the leaving group is a phenol with pK_a less than 6,[768] otherwise a concerted $E2$ mechanism applies.

A broadening variety of uses is being found for vinyl trifluoromethane-sulphonates ('vinyl triflates'). Benzyl triflate, though unstable at temperatures above -60 °C, is a powerful benzylating agent.[769] Primary vinyl triflates $R^1R^2C=CHOSO_2CF_3$ are excellent precursors of vinylidenecarbenes $R^1R^2C=C$: and give alkylidenecyclopropanes with alkenes,[770] and alkynes when $R^1 = H$.[771] Allenes $R^1R^2C=C=CR^3R^4$ are formed from trisubstituted vinyl triflates $R^1R^2C=C(CHR^3R^4)OSO_2CF_3$ on treatment with quinoline at 100 °C.[772] The reaction of alkyl triflates with nucleophiles involves C—O bond cleavage but the $1H,1H$-perfluoroalkyl analogues predominantly undergo S—O bond fission in their reactions with alkoxides.[773]

Hammett and Taft substituent constants have been determined for mesylate, tosylate, and triflate groups.[774]

Solvolysis studies reported during the period under review [775-780] are all of a type familair to readers of earlier volumes of these Reports. Substrates used in these studies are allyl α-toluenesulphonates [775] and benzenesulphonates,[776] 2,4,6-trimethoxy- and pentamethyl-benzenesulphonates,[777] and 2-substituted ethyl [778, 779] and vinyl [780] arenesulphonates. Points of interest from these studies are the low rates of hydrolysis of the *ortho*-disubstituted benzenesulphonates [777] and 1,2-migration accompanying solvolysis of 2-chloroethyl arenesulphonates.[778] The (Z)- and (E)-isomers of oxime sulphonates PhC($=$NO Tos)CMeRSMe behave very differently under Beckmann rearrangement conditions, the (Z)-isomer undergoing rearrangement but the (E)-isomer undergoing fragmentation, indicating a powerful anchimeric assistance by the divalent sulphur atom.[781]

Sulphonamides.—Preparations of sulphonamides from sulphonyl halides are illustrated with examples of particular interest from the recent literature, *viz.* α-bromoalkanesulphonamides RCHBrSO$_2$NHBu^t [782] and *N*-alkyl-*NN*-

[765] P. J. Stang and T. E. Deuber, *Org. Synth.*, 1974, **54**, 79.
[766] A. Williams, K. T. Douglas, and J. S. Loran, *J.C.S. Chem. Comm.*, 1974, 689.
[767] J. F. King and R. P. Beatson, *Tetrahedron Letters*, 1975, 973.
[768] K. T. Douglas, A. Steltner, and A. Williams, *J.C.S. Chem. Comm.*, 1974, 353.
[769] R. U. Lemieux and T. Kondo, *Carbohydrate Res.*, 1974, **35**, C4.
[770] P. J. Stang, M. G. Mangum, D. P. Fox, and P. Haak, *J. Amer. Chem. Soc.*, 1974, **96**, 4562.
[771] P. J. Stang, J. Davis, and D. P. Fox, *J.C.S. Chem. Comm.*, 1975, 17.
[772] P. J. Stang and J. R. Hargrove, *J. Org. Chem.*, 1975, **40**, 657.
[773] P. Johncock, *J. Fluorine Chem.*, 1974, **4**, 25.
[774] P. J. Stang and A. G. Anderson, *J. Org. Chem.*, 1976, **41**, 781.
[775] R. V. Vizgert and I. M. Tuchapskii, *Zhur. org. Khim.*, 1975, **11**, 1886, 1890; R. V. Vizgert, E. P. Panov, Y. G. Skrypnik, and M. P. Starodubtseva, *ibid.*, p. 1894.
[776] N. G. Gorbatenko and R. V. Sendega, *Reakts. spos. org. Soedinenii*, 1973, **10**, 673, 691, 707.
[777] C. M. Paleos, F. S. Varreri, and G. A. Gregoriou, *J. Org. Chem.*, 1974, **39**, 3594.
[778] T. A. Smolina, E. D. Gopius, and O. A. Reutov, *Zhur. org. Khim.*, 1974, **10**, 1121.
[779] T. Ando, Y. Saito, J. Yamawaki, H. Morisaki, M. Sawada, and Y. Yukawa, *J. Org. Chem.*, 1974, **39**, 2645.
[780] P. Bassi and U. Tonellato, *J.C.S. Perkin II*, 1974, 1283.
[781] C. A. Grob and J. Ide, *Helv. Chim. Acta*, 1974, **57**, 2571.
[782] J. C. Sheehan, U. Zoller, and D. Ben-Ishai, *J. Org. Chem.*, 1974, **39**, 1817.

disulphonimides $(R^1SO_2)_2NR^2$.[783] Addition of an arenesulphinate salt to an azo-dicarbonyl compound gives $RC(O)N(SO_2Ar)NHC(O)R$.[784]

N-Sulphinylsulphonamides can be used for the preparation of N-aryl-sulphonamides, through attack on an unactivated benzene ring.[785] Alternative procedures for N-substitution of a sulphonamide involve N-sodio-derivatives $[BrCH_2SO_2NHBu^t \rightarrow BrCH_2SO_2N(CH_2SO_2NHBu^t)Bu^t]^{782}$ and a stable thallium(I) derivative $(ArSO_2NTlR \rightarrow ArSO_2NRSO_2CH_2CF_3)$.[786] An N-aryl sulphonamide can be converted into an NN-diarylsulphonamide through copper(II)-catalysed reaction with an aryl bromide.[787] Further condensation products than those already known have been obtained from the reaction of a primary sulphonamide with formaldehyde.[788] A simpler reaction course is followed in the reaction between Chloramine-T and an aromatic aldehyde which gives the N-aroyl-sulphonamide.[789] N-(2-Hydroxyalkyl)toluene-p-sulphonamides are formed through the reaction of Chloramine-T with an alkene in the presence of OsO_4,[790] and the 2-bromoalkyl analogues are obtainable from an NN-dibromosulphon-amide and an alkene.[791] Full details have now been published of the N-alkylation of N-substituted trifluoromethanesulphonamides ('triflamides') and the parallel N- and C-alkylation of phenacyl analogues.[792] The products can be cleaved to give secondary amines, and a viable modification of the Gabriel synthesis has been established as a result of this work. Forced methylation of NN-disubstituted methanesulphonamides gives salts, *e.g.* $MeSO_2\overset{+}{N}Et_2Me\ FSO_3^-$, which are effective methanesulphonylating agents towards amines and alcohols under mild conditions (this earns the salt the name 'Easy Mesyl').[793]

Nucleophilic attack by OH^-, SH^-, and CN^- on N-alkyl-NN-disulphonamides gives N—S bond fission, but with other nucleophiles these compounds behave like simple alkyl sulphonates; *i.e.* the grouping $\overline{N}(SO_2R)_2$ is a good leaving group in certain nucleophilic substitution reactions.[794] Substitution of the sulphinate anion by reaction with isopropenyl-lithium represents an unusual nucleophilic aromatic substitution, discovered during studies of sulphon-hydrazones (81).[795] Photorearrangement of N-vinylsulphonamides gives β-sulphonylvinylamines.[796] Continuing studies of the desulphonylative double Smiles rearrangement of N-(2-hydroxyalkyl)sulphonamides $(ArSO_2NR^1CR^2R^3\text{-}CHR^4OH \rightarrow ArNR^1CR^2R^3CHR^4OH)$ have been described.[797]

783 P. J. De Christopher, J. P. Adamek, G. D. Lyon, S. A. Klein, and R. J. Baumgarten, *J. Org. Chem.*, 1974, **39**, 3525.
784 J. E. Herweh and R. M. Fantazier, *J. Org. Chem.*, 1976, **41**, 116.
785 T. Minami, Y. Tsumori, K. Yoshida, and T. Agawa, *J. Org. Chem.*, 1974, **39**, 3412.
786 H. L. Pan and T. L. Fletcher, *Synthesis*, 1975, 39.
787 I. G. C. Coutts and M. Hamblin, *J.C.S. Perkin I*, 1975, 2445.
788 O. O. Orazi and R. A. Corral, *J.C.S. Perkin I*, 1975, 772.
789 V. G. Dolyuk, M. M. Kremlev, and M. S. Rovinskii, *Voprosy Khim. i khim. Tekhnol.*, 1975, **38**, 22.
790 K. B. Sharpless, A. O. Chong, and K. Oshima, *J. Org. Chem.*, 1976, **41**, 177.
791 H. Terauchi and S. Takemura, *Chem. and Pharm. Bull. (Japan)*, 1975, **23**, 2410.
792 J. B. Hendrickson, R. Bergeron, and D. D. Sternbach, *Tetrahedron*, 1975, **31**, 2517.
793 J. F. King and J. R. du Manoir, *J. Amer. Chem. Soc.*, 1975, **97**, 2566.
794 V. A. Curtis, A. Raheja, J. E. Rejowski, R. W. Majewski, and R. J. Baumgarten, *Tetrahedron Letters*, 1975, 3107.
795 K. B. Tomer and A. Weiss, *Tetrahedron Letters*, 1976, 231.
796 W. R. Hertler, *J. Org. Chem.*, 1974, **39**, 3219.
797 A. C. Knipe, *Tetrahedron Letters*, 1975, 3563.

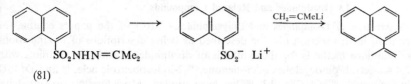

(81)

A detailed study of the intramolecular catalysis of the normally very slow acid hydrolysis of a sulphonamide has been described.[798] Where a carboxy-group is suitably placed, *e.g.* in *cis*-$HO_2CCH=CHSO_2NMePh$, substantial rate enhancement is observed.

Conjugation between the nitrogen atom and an aromatic system in toluene-*p*-sulphonamides appears to affect the co-ordination sphere of the sulphur atom, as judged by bond lengths and bond angles established through *X*-ray crystal analysis.[799] Further work is promised [799] to define the geometrical and structural factors determining d_π-p_π interaction in sulphonamides.

Sulphonyl Azides.—The generation of a sulphonylnitrene from the corresponding sulphonyl azide in solution, followed by attack on the solvent, accounts for the various mixtures of reaction products obtained. The simplest sulphonyl azides continue to be used, though more complex examples, *e.g.* *o*-dialkylaminobenzene-sulphonyl azide [800] and *o*- and *p*-benzenedisulphonyl azides,[801] have been studied.

Thermolysis of benzenesulphonyl azide gives azobenzene, diphenylamine, and biphenyl, but no benzenesulphonanilide.[802] Related products are obtained by similar treatment of *o*- and *p*-benzenesulphonyl azides in hydrocarbon solvents [801] and of arenesulphonyl azides in substituted benzenes, particularly those with electron-withdrawing substituents.[803]

Addition of an aromatic sulphonyl azide to an unstrained alkene gives the corresponding *N*-arenesulphonylimine, which is tautomeric with the corresponding enamine, and which can be hydrolysed easily to a mixture of a ketone and a sulphonamide.[804] Arenesulphonylimino-derivatives are formed, amongst other products, when an arenesulphonyl azide reacts with 1,4-dihydroquinolines and isoquinolines, indoles, and carbazoles.[805]

Complex formation between a metal carbonyl and methane-, benzene-, or toluene-*p*-sulphonyl azide leads to CO replacement; $Fe(CO)_5$, for example, gives $[Fe(RSO_2N)(CO)_2(H_2O)]_n$.[806]

[798] A. Wagenaar, A. J. Kirby, and J. B. F. N. Engberts, *Tetrahedron Letters*, 1974, 3735; 1976, 489.

[799] T. S. Cameron, K. Prout, B. Denton, R. Spagna, and E. White, *J.C.S. Perkin II*, 1975, 176.

[800] J. Martin, O. Meth-Cohn, and H. Suschitzky, *J.C.S. Perkin I*, 1974, 2451.

[801] R. A. Abramovitch and G. N. Knaus, *J. Org. Chem.*, 1975, **40**, 883.

[802] W. B. Renfrow and M. Devadoss, *J. Org. Chem.*, 1975, **40**, 1525.

[803] R. A. Abramovitch, G. N. Knaus, and V. Uma, *J. Org. Chem.*, 1974, **39**, 1101.

[804] R. A. Abramovitch, G. N. Knaus, M. Pavlin, and W. D. Holcomb, *J.C.S. Perkin I*, 1974, 2169.

[805] A. S. Bailey, P. A. Hill, and J. F. Seager, *J.C.S. Perkin I*, 1974, 967; A. S. Bailey, C. J. Barnes, and P. A. Wilkinson, *ibid.*, p. 1321; A. S. Bailey, J. F. Seager, and Z. Rashid, *ibid.*, p. 2384; A. S. Bailey, T. Morris, and Z. Rashid, *ibid.*, 1975, 420.

[806] R. A. Abramovitch, G. N. Knaus, and R. W. Stowe, *J. Org. Chem.*, 1974, **39**, 2513.

13 Disulphides and Related Compounds

Preparation of Disulphides and Polysulphides.—Most of the papers eligible for citation in this section may be described as being descriptions of modifications of standard methods for the synthesis of disulphides. Oxidation of thiols with 2,4,4,6-tetrabromocyclohexa-2,5-dienone,[807] 2-iodosobenzoic acid,[808] and DMSO in the presence of HCl and I_2 [809] has been studied. More familiar reagents have been employed for the oxidation of thiolate anions resulting from ring-opening of benzothiazoles with methoxide ion,[810a] and of 2-nitrothiophen with secondary amines [810b] (see Vol. 3, p. 79), and also for the oxidation of PhC(O)CPh=CHSeH to the diselenide.[811] A carefully elaborated route to unsymmetrical disulphides [812] involves successive treatment of a silver thiolate with 2,4-dinitrobenzene-sulphenyl chloride, then with a thiol together with AgOAc. Examples of disulphide formation with simple sulphur reagents include the use of S_2Cl_2 combined with silica gel for the conversion of aromatic hydrocarbons into diaryl disulphides,[813] and S_2Cl_2 alone for the synthesis of bis(2- and 3-indolyl) disulphides.[319]

Synthesis of disulphides from thiocarbonyl compounds can be achieved through photoreduction of a thione by a thiol,[814] or through $AlCl_3$-catalysed addition of a thiol to thiophosgene ($Bu^tSH + CSCl_2 \rightarrow Bu^tSSCHCl_2$).[815] 1-(Methylthio)-propyl propenyl disulphide and s-butyl 3-(methylthio)allyl disulphide, which have been isolated from asafoetida, have been synthesized from methyl dithio-propionate and from 1-methylthio-3-chloropropene, respectively.[816] Oxidation of thiohydrazides $R^1C(S)NHNHR^2$ with O_2, or with I_2,[817, 818] or their treatment with $MeSO_2Cl$ and pyridine,[819] gives the corresponding hydrazonyl disulphides. The more obvious route from a thiocarbonyl compound, reduction to the thiol, followed by oxidative coupling is illustrated for the selenium analogue $[Bu^t_2C=Se \rightarrow (Bu^t_2CH)_2Se_2]$.[820]

An interesting crop of papers dealing with polysulphides contains a variety of preparative procedures. Treatment of C^2H_3SH with NN'-thiobisphthalimide gives the trisulphide $C^2H_3S_3C^2H_3$,[821] and a selenium analogue $(HO_2CCMe_2S)_2Se$ has been prepared [822] from the thiol and selenous or selenic acid. Cysteine trisulphide is a minor product of the radiolysis of cysteine solutions at pH 5.3.[823] The use

[807] T. L. Ho, T. W. Hall, and C. W. Wong, *Synthesis*, 1974, 872.
[808] F. Garcia, J. Galvez, J. Vera, and A. Serna, *Anales de Quim.*, 1975, **71**, 277.
[809] O. G. Lowe, *J. Org. Chem.*, 1975, **40**, 2096.
[810] (a) G. Bartoli, M. Fiorentino, F. Ciminale, and P. E. Todesco, *J.C.S. Chem. Comm.*, 1974 732; (b) G. Guanti, C. Dell'Erba, G. Leandri, and S. Thea, *J.C.S. Perkin I*, 1974, 2357.
[811] G. Wilke and E. Uhlemann, *Z. Chem.*, 1975, **15**, 453.
[812] T. Endo, H. Tasai, and T. Ishigami, *Chem. Letters*, 1975, 813.
[813] M. Hojo and R. Masuda, *Synth. Comm.*, 1975, **5**, 173.
[814] J. R. Bolton, K. S. Chen, A. H. Lawrence, and P. de Mayo, *J. Amer. Chem. Soc.*, 1975, **97** 1832.
[815] N. H. Nilsson, *J.C.S. Perkin I*, 1974, 1308.
[816] J. Meijer and P. Vermeer, *Rec. Trav. chim.*, 1974, **93**, 242.
[817] D. H. R. Barton, J. W. Ducker, W. A. Lord, and P. D. Magnus, *J.C.S. Perkin I*, 1976, 38.
[818] P. Wolkoff, S. Hammerum, P. D. Callaghan, and M. S. Gibson, *Canad. J. Chem.*, 1974, **52** 879.
[819] P. Wolkoff and M. S. Gibson, *J.C.S. Perkin I*, 1974, 1173.
[820] T. G. Back, D. H. R. Barton, M. R. Britten-Kelly, and F. S. Guziec, *J.C.S. Chem. Comm.* 1975, 539.
[821] D. N. Harpp and T. G. Back, *J. Labelled Compounds*, 1975, **11**, 95.
[822] E. R. Clark and A. J. Collett, *J. Inorg. Nuclear Chem.*, 1974, **36**, 3860.
[823] G. C. Goyal and D. A. Armstrong, *Canad. J. Chem.*, 1975, **53**, 1475.

of S_2Cl_2 and SCl_2 in the synthesis of dipentafluorophenyl tri- and tetra-sulphides has been reported.[824] Moderate yields of trisulphides ($RNHCOCH_2CH_2)_2S_3$ are obtained by treatment of the corresponding disulphides with amines, or from the corresponding thiolsulphonate salts by reaction with an amine and sulphur.[825] The reaction of CF_3SCl with H_2S gives trifluoromethyldisulphane CF_3SSH from which symmetrical tri- and tetra-sulphides have been obtained.[826] Correspondingly, CF_3SSCl gives the trisulphane with H_2S,[827] and trisulphides with thiols.[828] Selenaditellurides have been obtained [228, 829] from an arenetellurenyl halide and RSeMgBr.

Properties of Disulphides and Polysulphides.—Cleavage processes of the simplest type, *i.e.* reduction to thiols, can be brought about with NaSeH,[829] chromium(II) salts,[830] or formamidinesulphinic acid.[831] Cleavage by tetramethyldistibine [832] or trialkyltin hydrides [833] has also been reported. More familiar processes, disproportionation of unsymmetrical disulphides (a chain-type mechanism involving ionic intermediates),[834] exchange with thiolate anion [835-837] [these papers range from quantitative assessment of exchange rate of bis(2-nitro-4-trifluoromethylphenyl) disulphide using ^{19}F n.m.r.[835] to the chemistry of analytical procedures for the determination of disulphide groups in peptides [836, 837]], and alkali cleavage [602, 838] have been further studied.

Several papers have appeared describing the cleavage of disulphides and diselenides with phosphines. A two-step mechanism applies to the reduction of a disulphide with Ph_3P in aqueous media, and the intermediate $Ph_3\overset{+}{P}SAr\ ArS^-$ hydrolyses to $ArSH + Ph_3PO$.[839] More complex disulphides containing acetamido-alkyl and αα-di(ethoxycarbonyl)alkyl substituents give oxazole and trisulphide side-products during reduction with Ph_3P.[840] Conversion of dimethyl diselenide with $(R_2N)_3P$ into dimethyl selenide,[841] and photochemical deselenation of diethyl diselenide to diethyl selenide with Ph_2PMe, which involves an EtSe· radical chain mechanism,[842] have been described.

Photolysis of disulphides in aldehyde solvents gives thiol-esters.[183] Irradiation of dibenzyl diselenides at wavelengths longer than *ca.* 300 nm gives selenides [227, 843]

[824] M. E. Peach, *Internat. J. Sulfur Chem.*, 1973, **8**, 27.
[825] M. Furukawa, S. Kato, and S. Hayashi, *Chem. and Pharm. Bull. (Japan)*, 1974, **22**, 2987.
[826] W. Gombler and F. Seel, *Z. Naturforsch.*, 1975, **30b**, 169.
[827] C. A. Burton and J. M. Shreeve, *Inorg. Nuclear Chem. Letters*, 1976, **12**, 373.
[828] N. R. Zack and J. M. Shreeve, *J.C.S. Perkin I*, 1975, 614.
[829] T. S. Woods and D. L. Klayman, *J. Org. Chem.*, 1974, **39**, 3716.
[830] L. E. Asher and E. Deutsch, *Inorg. Chem.*, 1975, **14**, 2799.
[831] G. Borgogno, S. Colonna, and R. Fornasier, *Synthesis*, 1975, 529.
[832] C. L. Baimbridge, C. D. Mickey, and R. A. Zingaro, *J.C.S. Perkin I*, 1975, 1395.
[833] W. P. Neumann and J. Schwindt, *Chem. Ber.*, 1975, **108**, 1339, 1346.
[834] J. L. Kice and G. E. Ekman, *J. Org. Chem.*, 1975, **40**, 711.
[835] D. R. Hogg and J. Stewart, *J.C.S. Perkin II*, 1974, 1040.
[836] J. Carlsson, M. Kierstan, P. J. Marek, and K. Brocklehurst, *Biochem. J.*, 1974, **139**, 221.
[837] P. D. J. Weitzman, *Biochem. J.*, 1975, **149**, 281.
[838] W. L. Anderson and D. B. Wetlaufer, *Analyt. Biochem.*, 1975, **67**, 493.
[839] L. E. Overman, D. Matzinger, E. M. O'Connor, and J. D. Overman, *J. Amer. Chem. Soc.*, 1974, **96**, 6081; L. E. Overman and S. T Petty, *J. Org. Chem.*, 1975, **40**, 2779; L. E. Overman and E. M. O'Connor, *J. Amer. Chem. Soc.*, 1976, **98**, 771.
[840] T. Sato and T. Hino, *Tetrahedron*, 1976, **32**, 507.
[841] I. A. Nuretdinov, E. V. Bayandina, and G. M. Vinokurova, *Zhur. obshchei Khim.*, 1974, **44**, 2588.
[842] R. J. Cross and D. Millington, *J.C.S. Chem. Comm.*, 1975, 455.
[843] J. Y. C. Chu, D. G. Marsh, and W. H. H. Gunther, *J. Amer. Chem. Soc.*, 1975, **97**, 4905.

4

if O_2 is excluded, otherwise an aldehyde and elementary selenium are formed.[227] Storage of organoselenium and organotellurium compounds often results in deposition of Se or Te. Other processes leading to radical formation are thermal cleavage,[844] radiolysis (R_2S + HO· → RSSR[+]·),[845] and flash photolysis[846] leading to a transient disulphide radical anion. Dipropyl disulphide is a scavenger for electrons, positive ions, and free radicals in γ-irradiated cyclohexane.[847] Facile photorearrangement of bis(o-acetylaminophenyl) disulphides arylthiobenzothiazoles proceeds through recombination of arylthiyl radicals; previous ambiguous results in this area are clarified by this study.[848]

Alkylated disulphide salts $R_2\overset{+}{S}SR$ X^- are often encountered as intermediates in cleavage reactions of disulphides by electrophiles. The study of the simplest system (Me_2S + $Me_2\overset{+}{S}SMe$) had earlier shown the existence of reversible exchange with a sulphide and now the existence of an irreversible process on heating or prolonged standing (RSMe + $Me_2\overset{+}{S}SMe$ → $R\overset{+}{S}MeSMe$ → $R\overset{+}{S}Me_2$ + Me_2S_2), involving a 1,2-shift in the intermediate ion, has been established through ^{2}H-labelling.[849] A synthesis of aryl analogues $ArS\overset{+}{S}Me_2$ ClO_4^- has been established, employing an arenesulphenyl chloride, $AgClO_4$, and Me_2S in MeCN.[850] Preliminary studies[849-852] indicate that nucleophilic attack on these cations occurs at the sulphenyl centre; consequently they are useful reagents for the synthesis of unsymmetrical disulphides.

Methyl bis(methylthio)sulphonium hexachloroantimonate, $Me\overset{+}{S}(SMe)_2$ $SbCl_6^-$,[254, 612] prepared by methylthiolation of the disulphide, shows useful methylthiolating properties; *e.g.* alkenes give *S*-methylthiiranium salts.[851] Methylation of a trisulphide with Me_3O^+ BF_4^- gives the analogous salt $Me\overset{+}{S}(SR)_2$ BF_4^-.[852]

Reference has been made in a preceding paragraph to disulphides of biological interest. Additional papers in this area describe the metabolic formation of dimethyl disulphide in the ant *Paltothyreus tarsatus*,[853] discuss the apparent anomaly that cystine is the only natural amino-acid which is susceptible to racemization in 6N-HCl (this is ascribed to stabilization by the β-heteroatom of the enol tautomer of the carboxy-group),[854] and display a growing realization that polysulphides have important functions *in vivo*. Thus cystine trisulphide, glutathione trisulphide, and the mixed cystine–glutathione trisulphide are activators of 5-aminolaevulinate synthase.[855] The trisulphide $[NaO_2S(CH_2)_4S]_2S$ and a naphthalene analogue have been proposed for use as anti-radiation agents.[856]

[844] T. Miyashita, M. Matsuda, and M. Iino, *Bull. Chem. Soc. Japan*, 1975, **48**, 3230; R. D. Costa, J. Tanaka, and D. E. Wood, *J. Phys. Chem.*, 1976, **80**, 213.
[845] M. Bonifacic, K. Schaefer, H. Moeckel, and K. D. Asmus, *J. Phys. Chem.*, 1975, **79**, 1496.
[846] T. L. Tung and J. A. Stone, *Canad. J. Chem.*, 1975, **53**, 3153.
[847] J. A. Stone and J. Esser, *Canad. J. Chem.*, 1974, **52**, 1253.
[848] Y. Maki and M. Sako, *Tetrahedron Letters*, 1976, 851.
[849] J. K. Kim and M. C. Caserio, *J. Amer. Chem. Soc.*, 1974, **96**, 1930.
[850] H. Minato, T. Miura, and M. Kobayashi, *Chem. Letters*, 1975, 1055.
[851] G. Capozzi, O. de Lucchi, V. Lucchini, and G. Modena, *Tetrahedron Letters*, 1975, 2603.
[852] P. Dubs and R. Stüssi, Paper read at VIIth International Symposium on Organosulphur Chemistry, Hamburg, July 12—16th, 1976.
[853] R. M. Crewe and F. P. Ross, *Nature*, 1975, **254**, 443.
[854] S. J. Jacobson, C. G. Willson, and H. Rapoport, *J. Org. Chem.*, 1974, **39**, 1074.
[855] J. D. Sandy, R. C. Davies, and A. Neuberger, *Biochem. J.*, 1975, **150**, 245.
[856] P. K. Srivastava, L. Field, and M. M. Grenan, *J. Medicin. Chem.*, 1975, **18**, 798.

Thiolsulphinates.—The literature of the period under review provides an excellent survey of the chemistry of the thiolsulphinate functional group, and is a reminder that most of this knowledge has been gathered only recently.

A standard synthesis [$R^1S(O)Cl + R^2SH \rightarrow R^1S(O)SR^2$] requires the presence of a base, and when this is a chiral tertiary amine the route provides a small preponderance of one enantiomer.[857] The greatest optical purity (*ca.* 10%) was obtained for t-butyl toluene-*p*-thiolsulphinate ($R^1 = p$-tolyl, $R^2 = Bu^t$), supporting the notion that the low optical stability of these compounds is due to an equilibrium involving a sulphenic acid[858] (see Vol. 3, p. 83), and that thiolsulphinates with a bulky sulphenyl substituent (hindering attack at this sulphur atom) would be expected to show greater optical stability. Steroidal thiolsulphinates ($R^1 = R^2 = 2$- or 3-cholestanyl moiety) have been prepared through the reaction of the corresponding episulphoxides in EtOH containing a little H_2SO_4, and the resulting diastereoisomer mixture has been separated.[859] The latter route to a thiolsulphinate has also been studied with the simplest system, ethylene episulphoxide in MeOH at 90 °C giving $MeOCH_2CH_2S(O)SCH_2$-CH_2OMe.[860]

Asymmetric induction is involved in the synthesis of a novel thiolsulphinate derivative, *viz.* an amidothiosulphite $Me_2NS(O)SBu^t$, through the reaction between a chiral isothiocyanate (+)-PhCHMeNCS, Bu^tSH, and thionyl dimethylamide $SO(NMe_2)_2$.[861] The reaction of the amidothiosulphite with $HgCl_2$ to give $ClS(O)NMe_2$ proceeds with retention of configuration, probably through a concerted four-centre transition state.[861] This is only the third reported example of retention of configuration at sulphinyl sulphur during nucleophilic substitution. This configurational assignment is confirmed by the reaction of the chlorosulphinyl derivative with Bu^tSH to give the enantiomer of the starting amidothiosulphite.[861]

Some representative reactions have been studied using partially resolved (−)-(*S*)-t-butyl toluene-*p*-thiolsulphinate;[857] MeMgI gives (+)-(*R*)-methyl *p*-tolyl sulphoxide, Et_2NMgBr gives the (+)-(*S*)-sulphinamide, and MeOH with NBS gives (−)-(*R*)-methyl toluene-*p*-sulphinate, with the expected inversion of configuration at each reaction implied in the absolute configurations stated. Block and O'Connor[858] have provided fascinating and very readable accounts of their thiolsulphinate studies, mostly reported previously in preliminary form. The SO—S bond is some 29 kcal mol⁻¹ weaker than a disulphide bond; oxidation of disulphides to thiolsulphinates is not regiospecific; and $MeS(O)SMe$ is a methanesulphinylating agent and gives methyl vinyl sulphoxides with alkynes as a result of the liberation of methanesulphenic acid.[858] The t-butyl analogue $Bu^tS(O)SBu^t$, in contrast, gives the thiolsulphinate, $C_5H_{11}C(=CH_2)S(O)SBu^t$ with $C_5H_{11}C\equiv CH$.[858] Pummerer-type rearrangements leading to α-methoxy-,[860] α-alkanesulphinyl-,[858] and α-alkanesulphonyl-alkyl disulphides[858] have been observed with sulphenyl oxysulphonium salts derived from thiolsulphinates.

[857] M. Mikolajczyk and J. Drabowicz, *J.C.S. Chem. Comm.*, 1976, 220.
[858] E. Block and J. O'Connor, *J. Amer. Chem. Soc.*, 1974, **96**, 3921, 3929.
[859] M. Kishi, S. Ishihara, and T. Komeno, *Tetrahedron*, 1974, **30**, 2135.
[860] K. Kondo and A. Negishi, *Chem. Letters*, 1974, 1525.
[861] M. Mikolajczyk and J. Drabowicz, *J.C.S. Chem. Comm.*, 1974, 775.

Kinetics of the alkaline hydrolysis of phenyl benzenethiolsulphinate and the corresponding thiolsulphonate support a mechanism proposed earlier (Savige, 1964) and contradict an alternative suggestion by Oae (see Vol. 1, p. 106). Hydrolysis of a thiolsulphonate proceeds through the thiolsulphinate, and since the latter is hydrolysed more slowly, its concentration, as an intermediate, builds up during the hydrolysis.[862] Nucleophilic attack at both sulphenyl and sulphinyl centres is observed in the reaction of OH⁻ with a thiolsulphinate.[862] Thiolsulphinates and thiolsulphonates react more rapidly with RS⁻ than with RSH, to an unexpectedly high degree, but, as with hydrolysis, the thiolsulphonate reacts faster than a corresponding thiolsulphinate.[863] Formation of thiocyanates by treatment of thiolsulphinates with CN⁻ has been studied.[864]

Thiolsulphonates.—Reference is also made in the preceding and following sections to these, the most easily prepared and commonly encountered disulphide oxides.

The reaction of a Grignard reagent at $-70\,^{\circ}C$ with phenyl chlorothiolsulphonate, $PhSSO_2Cl$, gives the corresponding thiolsulphonate together with several side-products, *e.g.* SO_2, disulphide, and hydrocarbons.[865] The aryl arenethiolsulphonate is formed as a result of N—O bond homolysis in the reaction of an *N*-aroyl-*N*-arylhydroxylamine with *o*-nitrobenzenesulphenyl chloride.[866] Trapping a sulphenic acid with an arenesulphinic acid gives the arenethiolsulphonate.[599]

Thiolsulphonates are well known as thioalkylating or thioarylating agents, as illustrated by the reaction with ethyl acetoacetate [$MeC(O)CH_2CO_2Et \rightarrow MeC(O)CH(SR)CO_2Et$].[867] An equilibrium between a mixture of aryl arenethiolsulphonate with a tertiary amine and the corresponding arylthiotrialkylammonium arenesulphinate salt $Ar^1SNR_3^+\ Ar^2SO_2^-$ explains[640] the beneficial effect of NEt_3 on the formation of sulphenate esters from thiolsulphonates and alcohols.[868] Conversion of a thiolsulphonate into a disulphide with NaN_3 in aqueous dioxan involves a sulphenyl azide as an intermediate.[599] Acetylacetone gives 2-acetyl-1,3-dithian by reaction with bis(toluene-*p*-sulphonylthio)-1,3-propane.[869] This cleavage of 1,3-diketones under mild conditions suggests a novel synthetic use and a further bonus arises from the fact that the dithian is a synthon for the $MeC(O)CO$-group.[869]

A detailed study of the reactions of the novel grouping $RS(O)(OH)(OMe)SR$, the mono-hemiketal of a thiolsulphonate, derived from the disulphide by reaction with Bu^tOCl and MeOH at $-80\,^{\circ}C$,[870] has been reported. Acid hydrolysis gives the sulphonic acid, and treatment with anhydrous toluene-*p*-sulphonic acid gives the parent thiolsulphonate.[870]

α-Disulphones.—[(Cyclohexa-1,3-diene)Fe(CO)₃] gives the α-disulphone RSO_2-SO_2R {$R = $[(cyclohexa-1,3-dien-5-yl)Fe(CO)₃]} by reaction with sodium dithionite.[723]

[862] J. L. Kice and T. E. Rogers, *J. Amer. Chem. Soc.*, 1974, **96**, 8009.
[863] J. L. Kice and T. E. Rogers, *J. Amer. Chem. Soc.*, 1974, **96**, 8015.
[864] T. Anstad, *Acta Chem. Scand. (A)*, 1975, **29**, 241.
[865] J. Lazar and E. Vinkler, *Acta Chim. Acad. Sci. Hung.*, 1974, **82**, 87.
[866] S. Oae and T. Sakurai, *Bull. Chem. Soc. Japan*, 1975, **48**, 3759.
[867] B. G. Boldyrev and L. N. Aristarkhova, *Zhur. org. Khim.*, 1975, **11**, 454, 455.
[868] B. G. Boldyrev, L. C. Vid, and S. A. Kolesnikova, *Zhur. org. Khim.*, 1974, **10**, 405.
[869] R. J. Bryant and E. McDonald, *Tetrahedron Letters*, 1975, 3841.
[870] A. I. Scott and E. M. Gordon, *Tetrahedron Letters*, 1974, 3507.

α-Disulphones are useful substrates for comparisons of nucleophilicities of common species, whose order of reactivity towards sulphonyl sulphur has recently been shown (see Vol. 3, p. 84) to be the same as that towards carbonyl carbon.[871] Corresponding data on phenyl benzenethiolsulphonate have been collected, and a generally similar electrophilic reactivity is shown by sulphonyl sulphur in the two series.[871] Earlier results suggesting that Et₃N catalyses the hydrazinolysis of diphenyl α-disulphone have been reviewed;[872] the amine exerts no effect.

[1] J. L. Kice, T. E. Rogers, and A. C. Warheit, *J. Amer. Chem. Soc.*, 1974, **96**, 8020.
[2] J. L. Kice and E. A. Lunney, *J. Org. Chem.*, 1975, **40**, 2128.

2
Ylides of Sulphur, Selenium, and Tellurium, and Related Structures

BY E. BLOCK AND M. HAAK

PART I: Ylides and Carbanionic Compounds of Sulphur, Selenium, and Tellurium
by *E. Block*

1 Introduction

This Report covers the literature of the topic which has appeared since Volume
of this series up until March 1976. Readers will note a number of changes in the
organization and nomenclature employed in this chapter compared to previous
volumes. Johnson's [1] definition of ylides as 'substances in which a carbanion is
attached directly to a heteroatom carrying a high degree of positive charge . .
[including] those molecular systems whose heteroatoms carry less than a full
formal positive charge' is not entirely satisfactory since it groups together
zwitterions, organometallic substances [*e.g.* RS(O)CH₂Li], and compounds such
as sulphine (1) in which ylidic resonance hybrids are now thought to make little
contribution.[2] We prefer Trost and Melvin's [3] simpler and more restrictive
definition of sulphur ylides as 'zwitterions in which a carbanion achieves
stabilization by interaction with an adjacent sulphonium center'. Developments
in the chemistry of sulphur-containing organometallics of the type RS(O)ₙCR
metal will then be reviewed under the headings sulphenyl carbanions (n = 0
sulphinyl carbanions (n = 1), and sulphonyl carbanions (n = 2). We shall also

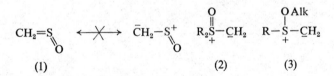

$$CH_2=\underset{O}{S} \quad \longleftrightarrow\!\!\!\times\!\!\!\longrightarrow \quad \bar{C}H_2-\underset{O}{\overset{+}{S}} \qquad R_2\overset{+}{\underset{}{S}}-\underset{O}{\overset{\parallel}{\underset{}{}}}\bar{C}H_2 \qquad R-\overset{O\,Alk}{\underset{+}{S}}-\bar{C}H_2$$

(1)   (2)   (3)

follow the nomenclature used by Trost and Melvin [3] in distinguishing between
oxosulphonium and alkoxysulphonium ylides [*e.g.* (2) and (3), respectively
Ylides containing nitrogen directly bonded to sulphur will all be treated in the
subsection on azasulphonium ylides.

A number of reviews dealing with sulphur ylides and carbanions have appeared
within the past two years. These include a major book by Trost and Melvin
on sulphur ylides, covering the literature through 1974,[3] and reviews of

[1] A. Wm. Johnson, 'Ylid Chemistry', Academic Press, New York and London, 1966.
[2] E. Block, H. Bock, S. Mohmand, P. Rosmus, and B. Solouki, *Angew. Chem. Internat. Edn*, 1976, **15**, 383.
[3] B. M. Trost and L. S. Melvin, jun., 'Sulfur Ylides', Academic Press, New York, 1975.

azasulphonium ylides,[4a] thiabenzenes (cyclic ylides),[4b] sulphinyl carbanions,[4c] sulphonyl carbanions,[4d] thienyl-lithium derivatives,[4e] various carbanionic and ylidic species,[4f, 4g] and carbanions of allylic sulphides and sulphoxides.[5] Several general reviews deal completely or in part with ylides and carbanions.[6-10] An article on 'Umpolung (dipole inversion) of Carbonyl Reactivity'[11] contains much sulphur chemistry.

The period covered by this Report has seen some ingenious applications of ylidic and carbanionic species in organic synthesis. There has also been an extensive search for new reagents incorporating sulphur-stabilized carbanions adjacent to a wide variety of other functional groups. Considerable attention has rightfully been directed toward the metallic counterion in carbanions with regard to chelation, solvation, and ion-pairing effects, as the reactivity of carbanions and ylides can be significantly modified by the presence of a transition-metal ion such as copper. Such effects may well be important in biochemical processes involving sulphur ylides. Striking effects can even be obtained by adding an ionic salt such as LiBr to a solution of a lithio-sulphur compound.

There has been considerable activity during the period of this Report in the area of theoretical calculations on sulphur carbanions and ylides. Interest has concentrated on the carbon–sulphur bond and the question of whether vacant 3d-orbitals on sulphur stabilize the adjacent anionic centre by d–$p\pi$-bonding. The bias of these studies is indicated by the title of one[12] of the papers, 'The Irrelevance of d-Orbital Conjugation . . .'. Non-empirical (*ab initio*) computational methods have been employed in two[12, 13] independent comparisons of the model thiocarbanion $^-CH_2SH$ with the first-row counterparts $^-CH_2OH$[12] and $^-CH_2CH_3$.[13] Both groups conclude that the gas-phase acidity of the C—H bond adjacent to sulphur is significantly greater than that of a C—H bond adjacent to oxygen or carbon *whether or not* d-*type functions are employed.* Further arguing against d–$p\pi$ conjugation in the thiocarbanion are the findings that the optimized C—S bond length in $^-CH_2SH$ is *longer* than in CH_3SH[12] and the barrier to rotation about the C—X bond is lower in $^-CH_2SH$ than in $^-CH_2OH$. In the case of $^-CH_2SH$ and CH_3SH, introduction of 3d functions is said to lower the overall energy by increasing the flexibility of the basis set but *to the same degree* for both the carbon acid and its conjugate anion.[13] Both groups offer the suggestion that it is the greater polarizability of sulphur and the longer C—S

[4] 'Organic Sulphur Chemistry', ed. C. J. M. Stirling, Butterworths, London, 1975 (proceedings of the VIth International Conference on Organic Sulphur Chemistry, Bangor, Wales, 1974): (a) C. R. Johnson, R. A. Kirchhoff, E. V. Jonsson, and J. C. Saukaitis, pp. 95–113; (b) G. H. Senkler, jun., B. E. Maryanoff, J. Stackhouse, J. D. Andose, and K. Mislow, pp. 157–179; (c) F. Montanari, pp. 181–202; (d) M. J. Janssen, pp. 19–42; (e) S. Gronowitz, pp. 203–228; (f) B. M. Trost, pp. 237–263; (g) T. Mukaiyama, pp. 265–284.

[5] D. A. Evans and G. C. Andrews, *Accounts Chem. Res.*, 1974, **7**, 147.

[6] K. Kondo, *Yuki Gosei Kagaku Kyokai Shi*, 1973, **31**, 1011 (*Chem. Abs.*, 1974, **81**, 135 210).

[7] P. Neumann and F. Vögtle, *Chem. Ztg.*, 1974, **98**, 138; *ibid.*, 1975, **99**, 308.

[8] R. K. Olsen and J. O. Currie, jun., in 'The Chemistry of the Thiol Group', Vol. II, ed. S. Patai, John Wiley and Sons, London, 1974.

[9] I. Fleming, *Chem. and Ind.*, 1975, 449.

[10] A. H. Davidson, P. K. G. Hodgson, D. Howells, and S. Warren, *Chem. and Ind.*, 1975, 455.

[11] D. Seebach and M. Kolb, *Chem. and Ind.*, 1974, 687.

[12] F. Bernardi, I. G. Csizmadia, A. Mangini, H. B. Schlegel, M.-H. Whangbo, and S. Wolfe, *J. Amer. Chem. Soc.*, 1975, **97**, 2209.

[13] A. Streitwieser, jun. and J. E. Williams, jun., *J. Amer. Chem. Soc.*, 1975, **97**, 191.

bond length that is responsible for the superior carbanion-stabilizing ability of sulphur compared to oxygen or carbon. Additional calculations comparing $^-CH_2SH$ and $^-CH_2OH$ have been published.[14]

The question of d–$p\pi$ conjugation in ylides has also been explored.[15, 16] The calculated [16] (*ab initio*) C—X bond lengths of the anions, cations, and radicals CH_2XH and CH_3X (X = S or O) do not correlate with the overlap population of the bond but correlate very well with the *ionic* bond orders of the C—X bond, defined by the coulombic term $-q_Aq_B/r_{AB}$.[16] This result implies that the observed shortening of the C–heteroatom distance on deprotonation of onium salts to give ylides need not be ascribed to an ylene formulation involving d–$p\pi$ conjugation but rather can be attributed to coulombic effects.[16]

Further quantitative data on carbanion (and ylide) stability have been published.[17-21] Bordwell [17] is developing an accurate scale of equilibrium acidities determined in DMSO. The pK's have been found to be correlated with heats of deprotonation in DMSO by dimsylpotassium, and pK measurements in DMSO were shown to be free from ion-association effects.[17] Equilibrium pK data are suggested [18] to be a much more reliable indicator of carbanion stability than are kinetic acidity data. Representative pK_a values have been given for a variety of organosulphur (and other) compounds and the relative stability of the corresponding carbanions has been discussed (pK values in parentheses):[17, 18] DMSO (35.1), Ph_3CH_2 (~32.3), Me_2SO_2 (31.1), $(PhS)_2CH_2$ (30.8), $(Ph_2P)_2CH_2$ (29.9), $PhSO_2Me$ (29.0), $PhCH_2SOMe$ (29.0), $PhCH_2SO_2Ph$ (23.4), $PhSCH_2SO_2Ph$ (20.3), Ph_2PCH_2-SO_2Ph (20.2). It has been concluded that the order of acidifying effect is $Ph_2P > PhS > Ph$. The pK_a values of a series of trifluoromethyl sulphones and phenyl ketones have similarly been determined: $MeSO_2CF_3$ (18.76), $Pr^iSO_2CF_3$ (21.80), cyclo-$C_3H_5SO_2CF_3$ (26.60), $MeCOPh$ (24.70), Pr^iCOPh (26.26), and cyclo-C_3H_5COPh (28.18).[19] Bordwell concludes that the CF_3SO_2 group is exerting a large conjugative effect rather than just a polar effect since the acidification is much larger than that seen with the Me_3N^+ group.[19] It is also inferred that the CF_3SO_2 group imposes *the same kind* of electronic demand on the cyclopropane ring as do the PhCO or NO_2 groups, *e.g.* with CF_3SO_2 there is conjugative d–$p\pi$ overlap.[19] In contrast, a study of substituent effects on the equilibrium acidity of contact ion-pairs of dithians in cyclohexylamine was rationalized on the basis of localized pyramidal carbanions, and no significant role was assigned to delocalized (d–$p\pi$) structures.[20]

The dialogue between the proponents and critics of sulphur d–$p\pi$ conjugation in carbanions and ylides will no doubt continue. It may be anticipated that the properties of organoselenium and organotellurium carbanions and

[14] I. G. Csizmadia in 'The Chemistry of the Thiol Group', Vol. I, ed. S. Patai, John Wiley and Sons, London, 1974.

[15] J. I. Musher, *Tetrahedron*, 1974, **30**, 1747.

[16] M.-H. Whangbo, S. Wolfe, and F. Bernardi, *Canad. J. Chem.*, 1975, **53**, 3040.

[17] W. S. Matthews, J. E. Bares, J. E. Bartmess, F. G. Bordwell, F. J. Cornforth, G. E. Drucker Z. Margolin, R. J. McCallum, G. J. McCollum, and N. R. Vanier, *J. Amer. Chem. Soc.*, 1975 **97**, 7006.

[18] F. G. Bordwell, W. S. Matthews, and N. R. Vanier, *J. Amer. Chem. Soc.*, 1975, **97**, 442.

[19] F. G. Bordwell, N. R. Vanier, W. S. Matthews, J. B. Hendrickson, and P. L. Skipper, *J. Amer Chem. Soc.*, 1975, **97**, 7160.

[20] A. Streitwieser, jun. and S. P. Ewing, *J. Amer. Chem. Soc.*, 1975, **97**, 190.

[21] N. Kunieda, Y. Fujiwara, J. Nokami, and M. Kinoshita, *Bull. Chem. Soc. Japan*, 1976, **49**, 575

ylides will be relevant to this discussion since polarizability and bond-length effects should be enhanced in systems containing these heavy elements.

2 Sulphonium Ylides

Synthesis and Properties.—The two most general methods for preparing sulphonium ylides continue to be α-deprotonation of a sulphonium salt [21] and the reaction of a sulphide (or disulphide [22]) with a carbene. A new development involves [23] the reaction of the thianthrene or phenoxathiin cation radical with a dicarbonyl compound, *e.g.* ethyl benzoylacetate, to give (4). An infrequently used but useful route to sulphonium ylides involves reaction of sulphides (*e.g.* dithia[3,3]cyclophanes [24]) with benzyne. A detailed description of a preparation of the sulphonium salt precursor to Trost's diphenylsulphonium cyclopropylide has appeared.[25] The selectivity of ylide formation in the reaction of cyclic and acyclic sulphides with carbenes has been examined and compared [26] with the much

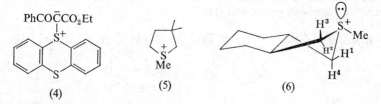

(4)   (5)   (6)

lower selectivity seen in nitrene addition. A full report [27] of a study of ylide formation by intramolecular trapping of carbenes by arylthio- or allylthio-groups on the β-carbon has appeared.

Fava and co-workers [28] have examined H–D exchange in (5) and (6). Exchange occurs in a highly stereospecific manner, *e.g.* in (6) the relative rates in D_2O are H-1 $\leqslant$ 1; H-2 = 200; H-3 = 3; H-4 = 3. The value of k_{H-2}/k_{H-1} being >200 (estimated to be as large as 1300), this exchange is suggested to be one of the most stereospecific non-enzymatic reactions yet reported. The exchange data are inconsistent with any theory (*e.g.* the *gauche*-effect theory) based on the idea that the all-important factor in determining anion stability is the C_α—S dihedral angle.

Thiabenzenes are best represented as cyclic ylides rather than species with aromatic ring currents involving through-sulphur delocalization.[4b, 29, 30] Thia-

[22] L. Field and C. H. Banks, *J. Org. Chem.*, 1975, **40**, 2774.
[23] (*a*) K. Kim and H. J. Shine, *Tetrahedron Letters*, 1974, 4413; (*b*) K. Kim, S. R. Mani, and H. J. Shine, *J. Org. Chem.*, 1975, **40**, 3857.
[24] T. Otsubo and V. Boekelheide, *Tetrahedron Letters*, 1975, 3881.
[25] M. J. Bogdanowicz and B. M. Trost, *Org. Synth.*, 1974, **54**, 27.
[26] D. C. Appleton, D. C. Bull, J. McKenna, J. M. McKenna, and A. R. Walley, *J.C.S. Chem. Comm.*, 1974, 140.
[27] K. Kondo and I. Ojima, *Bull. Chem. Soc. Japan*, 1975, **48**, 1490.
[28] G. Barbarella, A. Garbesi, and A. Fava, *J. Amer. Chem. Soc.*, 1975, **97**, 5883; U. Folli, D. Iarossi, I. Moretti, F. Taddei, and G. Torre, *J.C.S. Perkin II*, 1974, 1655; U. Folli, D. Iarossi, and F. Taddei, *ibid.*, p. 1658.
[29] A. G. Hortmann, R. L. Harris, and J. A. Miles, *J. Amer. Chem. Soc.*, 1974, **96**, 6119.
[30] (*a*) J. Stackhouse, B. E. Maryanoff, G. H. Senkler, jun., and K. Mislow, *J. Amer. Chem. Soc.*, 1974, **96**, 5650; (*b*) B. E. Maryanoff, G. H. Senkler, jun., J. Stackhouse, and K. Mislow, *ibid.*, p. 5651; (*c*) B. E. Maryanoff, J. Stackhouse, G. H. Senkler, jun., and K. Mislow, *ibid.*, 1975, **97**, 2718.

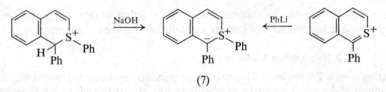

(7)

naphthalenes such as (7) may be generated either by deprotonation or by addition of an organolithium to thiochromenium salts.[29, 30] Indicative of the ylidic character of the thiabenzenes and thianaphthalenes is the occurrence of Stevens rearrangement, the ease of protonation, substituent and solvent effects, the [13]C n.m.r. spectra, and the pyramidal geometry at sulphur.[29, 30] Hori[31] has synthesized the two thiabenzenes 1-cyano- and 1-benzoyl-2-methyl-2-thianaphthalene, and has reported certain of their reactions.

Among the novel sulphonium ylides synthesized in this period are (8),[32] (9)[33] (which is unusually basic), (10),[34] $Me_2\overset{+}{S}-\overline{C}HPO(OMe)_2$,[35] $EtMe\overset{+}{S}-\overline{C}=CHS-MeEt$[36] (a cumulated ylide), and (11).[37]

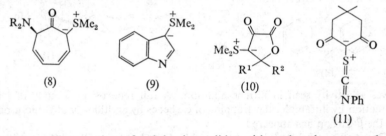

(8)   (9)   (10)

(11)

Reactions.—The reaction of sulphonium ylides with carbonyl compounds to form oxirans is now a standard reaction. Among the more unusual examples reported are an intramolecular version directed towards the preparation of arene epoxide (12)[38] (not isolable), the synthesis of the epoxide of 8-phenylthio-methylenecycloheptane,[39] and the direct conversion of a diketone by dimethyl-sulphonium methylide into a doubly homologated hydroxymethyl aldehyde (13) by the rearrangement shown.[40]

A second common reaction of sulphonium ylides is Michael addition to $\alpha\beta$-unsaturated carbonyl compounds, giving cyclopropanes. This reaction forms the basis of a cyclopentene synthesis starting from sulphonium allylide (14).[41] Other recent examples include the preparation of spiro[cyclopropane-1,4-$\Delta^{2'}$-pyrazolin]-5'-one derivatives (15)[42] and cyclopropyl-ulose (16).[43] Un-

[31] M. Hori, T. Kataoka, H. Shimizu, K. Narita, S. Ohno, and H. Aoki, *Chem. Letters*, 1974, 1101.
[32] M. Cavazza, C. A. Veracini, and F. Pietra, *Tetrahedron Letters*, 1975, 2085.
[33] G. D. Daves, jun., W. R. Anderson, jun., and M. V. Pickering, *J.C.S. Chem. Comm.*, 1974, 301.
[34] K. Hagio, N. Yoneda, and H. Takei, *Bull. Chem. Soc. Japan*, 1974, 47, 909.
[35] K. Kondo, Y. Liu, and D. Tunemoto, *J.C.S. Perkin I*, 1974, 1279.
[36] H. Braun and A. Amann, *Angew. Chem. Internat. Edn.*, 1975, 14, 756.
[37] G. F. Koser and S.-M. Yu, *J. Org. Chem.*, 1976, 41, 125.
[38] M. S. Newman and L.-F. Lee, *J. Org. Chem.*, 1974, 39, 1446; *ibid.*, 1975, 40, 2650.
[39] T. Cohen, D. Kuhn, and J. R. Falck, *J. Amer. Chem. Soc.*, 1975, 97, 4749.
[40] R. H. Mitchell, I. Calder, H. Huisman, and V. Boekelheide, *Tetrahedron*, 1975, 31, 1109.
[41] J. R. Neff, R. Gruetzmacher, and E. J. Nordlander, *J. Org. Chem.*, 1974, 39, 3814.
[42] P. D. Croce and D. Pocar, *J.C.S. Perkin I*, 1976, 620.
[43] J. M. J. Tronchet and H. Eder, *Helv. Chim. Acta*, 1975, 58, 1799.

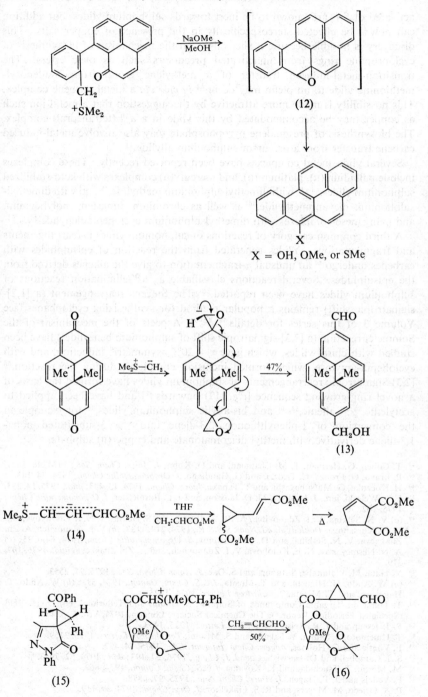

activated olefins are known to be inert towards sulphonium ylides, but addition can now [44] be effected stereospecifically in the presence of copper salts. This discovery is notable because it sheds light on the mechanism of biosynthesis of cyclopropane rings from unactivated precursors such as oleic esters. The transition-metal-induced transfer of a methylene group from *S*-adenosylmethionine ylide to an olefin may occur [44] *in vivo via* a metal–carbene complex. This possibility is made more attractive by the suggestion that a metal ion such as copper may be accommodated by this ylide in a d^{10} (tetrahedral) complex. The biosynthesis of presqualene pyrophosphate may also involve metal-induced carbene transfer from a precursor sulphonium allylide.[44]

Several ylide–metal complexes have been reported recently. These complexes include palladium(II), platinum(II), and mercury(II) complexes with keto-stabilized sulphonium ylides,[45–47] with dimethylsulphonium methylide,[45] and with dimethylsulphonium dicyanomethylide,[46] as well as chromium, tungsten, molybdenum, and manganese complexes with dimethylsulphonium cyclopentadienylide.[48]

A third common category of reactions of sulphonium ylides is rearrangements and fragmentations. Ylides generated from the reaction of episulphides with carbenes undergo [49] an unusual α-fragmentation to give the alkenes derived from the episulphides. Several reactions classifiable as $\alpha\beta'$-elimination reactions of sulphonium ylides have been reported.[50] The Stevens rearrangement (a [1,2]-sigmatropic shift) remains a popular method for synthesizing cyclophanes (see Volume 3 of this series for details [51]).[52–54] Aspects of the mechanism of the Sommelet reaction (a [2,3]-sigmatropic shift of sulphonium benzylides) have been studied with chiral ylides, which gave [55] ~20% asymmetric induction, and with cyclohepta-amylose, which formed inclusion complexes during the reaction.[56] [2,3]-Sigmatropic rearrangements of sulphonium ylides have formed the basis of a novel ring-growing sequence [*e.g.* (17) onwards [57]] and have been applied to acetylenic,[58, 59] allenic,[58, 59] and bis-allylic sulphonium ylides,[60] for example in the conversion of 1-phenylthionona-2,3-diene into [58] a 3-substituted nona-1,3-diene derivative with methyl diazomalonate and copper(II) sulphate.

[44] T. Cohen, G. Herman, T. M. Chapman, and D. Kuhn, *J. Amer. Chem. Soc.*, 1974, **96**, 5627.
[45] P. Bravo, G. Fronza, C. Ticozzi, and G. Gaudiano, *J. Organometallic Chem.*, 1974, **74**, 143.
[46] H. Koezuka, G. Matsubayashi, and T. Tanaka, *Inorg. Chem.*, 1974, **13**, 443; *ibid.*, 1975, **14**, 253.
[47] E. T. Weleski, jun., J. L. Silver, M. D. Jansson, and J. L. Burmeister, *J. Organometallic Chem.*, 1975, **102**, 365.
[48] (*a*) V. N. Setkina, V. I. Zdanovitch, A. Zh. Zhakaeva, Yu. S. Nekrasov, N. I. Vasyukova, and D. N. Kursanova, *Doklady Akad. Nauk S.S.S.R.*, 1974, **219**, 1137; (*b*) V. I. Zdanovitch, A. Zh. Zhakaeva, V. N. Setkina, and D. N. Kursanov, *J. Organometallic Chem.*, 1974, **64**, C25; (*c*) A. N. Nesmeyanov, N. E. Kolobova, V. I. Zdanovitch, and A. Zh. Zhakaeva, *ibid.*, 1976, **107**, 319.
[49] Y. Hata, M. Watanabe, S. Inoue, and S. Oae, *J. Amer. Chem. Soc.*, 1975, **97**, 2553.
[50] (*a*) W. Ando, H. Higuchi, and T. Migata, *J.C.S. Chem. Comm.*, 1974, 523; (*b*) W. Ando, T. Hagiwara, and T. Migata, *Tetrahedron Letters*, 1974, 1425.
[51] T. Durst, in 'Organic Compounds of Sulphur, Selenium, and Tellurium', ed. D. H. Reid (Specialist Periodical Reports), The Chemical Society, London, 1975, Vol. 3, p. 178.
[52] P. J. Jessup and J. R. Reiss, *Tetrahedron Letters*, 1975, 1453.
[53] T. Umemoto, S. Satani, Y. Sakata, and S. Misumi, *Tetrahedron Letters*, 1975, 3159.
[54] F. Vögtle and G. Hohner, *Angew. Chem. Internat. Edn.*, 1975, **14**, 497.
[55] S. J. Campbell and D. Darwish, *Canad. J. Chem.*, 1976, **54**, 193; *ibid.*, 1974, **52**, 2953.
[56] M. Mitani, T. Tsuchida, and K. Koyama, *J.C.S. Chem. Comm.*, 1974, 869.
[57] E. Vedejs and J. P. Hagen, *J. Amer. Chem. Soc.*, 1975, **97**, 6878.
[58] P. A. Grieco, M. Meyers, and R. S. Finkelhor, *J. Org. Chem.*, 1974, **39**, 119.

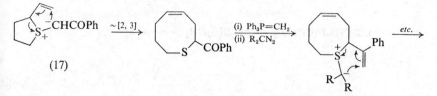

(17)

Notable in the category of miscellaneous reactions of sulphonium ylides is the use of an ylide as a *blocking group*, as shown for the stable ylide (18).[61] The hydrogenation [62] and alkaline methanolysis [63] of sulphonium ylides have been

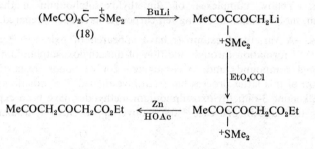

reported, as has the use of stabilized sulphonium ylides to synthesize tri-azolines,[64] isoxazoles,[65] and furans.[66] The reaction of a sulphonium ylide with trimethylsilylketen has also been described.[67]

3 Oxosulphonium Ylides

Synthesis and Properties.—A number of heavily functionalized oxosulphonium ylides have been prepared, *e.g.* (19),[68] which is converted by base into a thia-

$$Me_2\overset{+}{S}(O)\overset{-}{C}HCOPh + EtOCH=C(COMe)_2 \longrightarrow Me_2\overset{+}{S}(O)\overset{-}{C}CH=C(COMe)_2$$
$$|$$
$$COPh$$
$$(19)$$

$$Me_2\overset{+}{S}(O)\overset{-}{C}H_2 + (EtO)_2P(O)H \xrightarrow[-CHCl_3]{CCl_4} (EtO)_2P(O)\overset{-}{C}H\overset{+}{S}(O)Me_2$$
$$(20)$$

[59] G. Pourcelot, L. Veniard, and P. Cadiot, *Bull. Soc. chim. France*, 1975, 1275, 1281.
[60] A. J. H. Labuschagne, C. J. Meyer, H. S. C. Spies, and D. F. Schneider, *J.C.S. Perkin I*, 1975, 2129.
[61] M. Yamamoto, *J.C.S. Chem. Comm.*, 1975, 289.
[62] H. Wittmann and F. A. Petio, *Z. Naturforsch.*, 1975, **30b**, 763.
[63] N. Furukawa, T. Masuda, M. Yakushiji, and S. Oae, *Bull. Chem. Soc. Japan*, 1974, **49**, 2247.
[64] G. L'Abbe, G. Mathys, and S. Toppet, *Chem. and Ind.*, 1975, 278.
[65] P. Bravo and C. Ticozzi, *Gazzetta*, 1975, **105**, 91.
[66] A. R. Katritzky, S. Q. A. Rizvi, and J. W. Suwinski, *Heterocycles*, 1975, **3**, 379; *J.C.S. Perkin I*, 1975, 2489.
[67] R. A. Ruden, *J. Org. Chem.*, 1974, **39**, 3607.
[68] M. Watanabe, T. Kinoshita, and S. Furukawa, *Chem. and Pharm. Bull. (Japan)*, 1975, **23**, 82, 258.

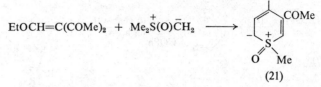

(21)

benzene 1-oxide, and (20),[69] which can be further functionalized. The synthesis and properties of cyclic oxosulphonium ylides, *e.g.* (21),[70] have been described. This thiabenzene 1-oxide undergoes H–D exchange and electrophilic substitution (bromination, nitration) but the ylide character is clearly indicated by its n.m.r. spectrum. Yellow complexes of dimethyloxosulphonium methylide with chromium, molybdenum, and tungsten carbonyls have been prepared.[71]

Reactions.—A variety of examples have appeared of oxiran [72, 73] and cyclopropane [74–77] formation through addition of dimethyloxosulphonium methylide to carbonyl compounds and to conjugated double bonds, respectively. The mechanism of this latter process has been investigated.[78] Synthetic applications of cycloalkenone 3-dimethyloxosulphonium methylides have been explored,[79, 80] providing the basis in one case [79] for a novel approach to hydroazulenes (22).

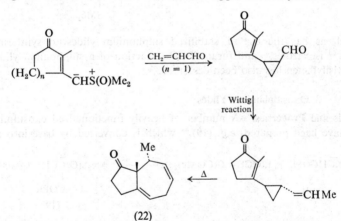

(22)

[69] (a) V. P. Lysenko, I. E. Boldeskul, R. A. Loktionova, and Yu. G. Gololobov, *Zhur. obshchei Khim.*, 1975, **45**, 2341; (b) *Zhur. org. Khim.*, 1975, **11**, 2440.
[70] Y. Tamura, H. Taniguchi, T. Miyamoto, M. Tsunekawa, and M. Ikeda, *J. Org. Chem.*, 1974, **39**, 3519.
[71] L. Weber, *J. Organometallic Chem.*, 1976, **105**, C9.
[72] G. I. Dmitrienko, A. Szakolcai, and S. McLean, *Tetrahedron Letters*, 1974, 2599.
[73] J. A. Donnelly, S. O'Brien, and J. O'Grady, *J.C.S. Perkin I*, 1974, 1674.
[74] R. M. Pagni and C. R. Watson, jun., *J.C.S. Chem. Comm.*, 1974, 224.
[75] J. A. Marshall and R. H. Ellison, *J. Org. Chem.*, 1975, **40**, 2070.
[76] R. E. Ireland, M. I. Dawson, C. J. Kowalski, C. A. Lipinski, D. R. Marshall, J. W. Tilley, J. Bordner, and B. L. Trus, *J. Org. Chem.*, 1975, **40**, 973.
[77] G. D. Anderson, T. J. Powers, C. Djerassi, J. Fayos, and J. Clardy, *J. Amer. Chem. Soc.*, 1975, **97**, 388.
[78] F. Rocquet and A. Sevin, *Bull. Soc. chim. France*, 1974, 881, 888.
[79] J. P. Marino and T. Kaneko, *J. Org. Chem.*, 1974, **39**, 3175.
[80] (a) Y. Tamura, T. Miyamoto, H. Kiyokawa, and Y. Kita, *J.C.S. Perkin I*, 1974, 1125; (b) Y. Tamura, T. Miyamoto, and Y. Kita, *J.C.S. Chem. Comm.*, 1974, 531.

Other substrates which have been allowed to react with oxosulphonium ylides are diphenylcyclopropenone [81] and azomethine ylides.[82]

4 Alkoxy- and Aza-sulphonium and Related Ylides

Alkoxysulphonium Ylides.—Alkoxysulphonium ylides are thought to be involved in the oxidation of alcohols with DMSO-related reagents, as demonstrated [83] for the deuteriated form of Corey's *N*-chlorosuccinimide–dimethyl sulphide reagent (23). Chlorosulphonium or chloro-oxosulphonium reagents $R_2\overset{+}{S}(O)_n$-

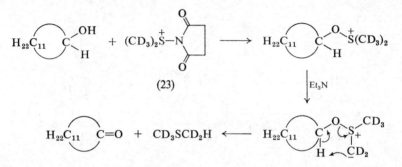

(23)

Cl Cl$^-$ ($n = 0$ or 1) have been used [84] to convert the tricyclic *vic*-diol (24) into an α-hydroxy-ketone. Most inorganic oxidants would have caused C—C bond cleavage, giving dicarbonyl compounds, in this situation. The polymer version (25) of Corey's reagent has been recommended [85] for the high-yield oxidation of

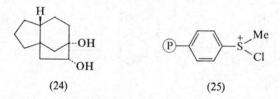

(24)   (25)

sensitive alcohols to labile aldehydes. Barton [86] reports a modification of his chloroformate–DMSO oxidation procedure in which the yield of secondary alcohols is improved through the use of 1,2-epoxypropane as an acid scavenger. Applications of Barton's original procedure in a prostaglandin synthesis have appeared.[87] The use of DMSO with trifluoromethanesulphonic anhydride as an oxidant has been described.[88] The synthesis of nitriles from oximes is suggested to involve an oximinosulphonium ylide intermediate $RCH{=}NO\overset{+}{S}(Me)\overset{-}{C}H_2$.[89]

[81] L. Salisbury, *J. Org. Chem.*, 1975, **40**, 1340.
[82] M. Vaultier, R. Danion-Bougot, D. Danion, J. Hamelin, and R. Carrie, *J. Org. Chem.*, 1975, **40**, 2990.
[83] J. P. McCormick, *Tetrahedron Letters*, 1974, 1701.
[84] E. J. Corey and C. U. Kim, *Tetrahedron Letters*, 1974, 287.
[85] G. A. Crosby, N. M. Weinshenker, and H.-S. Uh, *J. Amer. Chem. Soc.*, 1975, **97**, 2232.
[86] D. H. R. Barton and C. P. Forbes, *J.C.S. Perkin I*, 1975, 1614.
[87] N. Finch, J. J. Fitt, and I. H. S. Hsu, *J. Org. Chem.*, 1975, **40**, 206.
[88] J. B. Hendrickson and S. M. Schwartzman, *Tetrahedron Letters*, 1975, 273.
[89] T.-L. Ho and C. M. Wong, *Synthetic Comm.*, 1975, **5**, 423.

Aryloxysulphonium Ylides.—Aryloxysulphonium ylides are known to undergo [2,3]-sigmatropic (Sommelet) rearrangements.[90] Gassman and Amick [91] have exploited this reaction for the *ortho*-functionalization of phenols, as illustrated by their synthesis of (26).

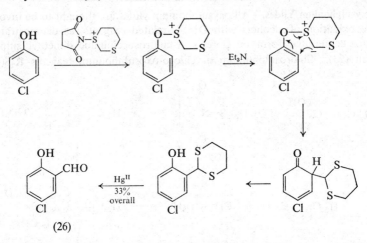

(26)

Azasulphonium Ylides.—A valuable series of papers has appeared [92] on the azasulphonium route to *ortho*-alkylated aromatic amines, indoles, and oxindoles. This procedure involves (i) generation of an azasulphonium salt such as (27),

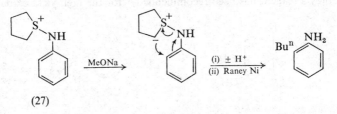

(27)

either through mono-*N*-chlorination of the aniline followed by reaction with a dialkyl sulphide, or by the reaction of a chlorosulphonium salt with the aniline; (ii) treatment with base, giving an azasulphonium ylide; (iii) Sommelet-type rearrangement followed by hydrogen transfer and re-aromatization, giving *o*-alkyl-α-thioalkoxy-substituted anilines; and (iv) reduction with Raney nickel. The yields are generally quite good. The use of β-keto-sulphides or β-formyl-sulphides, instead of the dialkyl sulphides, allows the preparation of indoles, while the use of α-alkoxycarbonyl sulphides gives oxindoles. Extensive examples have been given. A method for the selective *ortho*-formylation of aromatic

[90] A. W. Johnson in 'Organic Compounds of Sulphur, Selenium, and Tellurium', ed. D. H. Reid (Specialist Periodical Reports), The Chemical Society, London, 1970, Vol. 1, p. 276.

[91] P. G. Gassman and D. R. Amick, *Tetrahedron Letters*, 1974, 889, 3463.

[92] (a) P. G. Gassman and G. D. Gruetzmacher, *J. Amer. Chem. Soc.*, 1974, 96, 5487; (b) P. G. Gassman, T. J. van Bergen, D. P. Gilbert, and B. W. Cue, jun., *ibid.*, p. 5495; (c) P. G. Gassman and T. J. van Bergen, *ibid.*, p. 5508; (d) P. G. Gassman, G. Gruetzmacher, and T. J. van Bergen, *ibid.*, p. 5512.

amines, which mechanistically is very similar to the *ortho*-formylation of phenols [*cf.* (26)], has also been published.[93]

Vilsmaier[94] has reported that succinimido-sulphonium salts may undergo [1,2]- and [2,3]-sigmatropic rearrangement, giving *N*-methylthiomethyl succinimide and the appropriate methylthiomethyl ether, respectively.

5 Thiocarbonyl Ylides

Thiocarbonyl ylides (29) are unstable ring-opened valence isomers of thiirans (30), often generated by pyrolysis of Δ^3-1,3,4-thiadiazolines (28). When this pyrolysis is conducted in the presence of triaryl- or trialkyl-phosphines, the

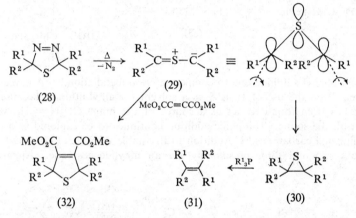

intermediate thiirans are desulphurized, affording olefins (31).[95, 96] This reaction is of considerable synthetic utility since it is stereospecific, ylide (29) cyclizing in a conrotatory manner [*cf.* (29) → (30)] in accord with theoretical predictions.[97] The sequence (28) → (30) has been used[98] to make unusual thiirans such as *cis*- and *trans*-2,3-di-t-butylthiiran, and the overall route (28) → (31), which has been called[96] a 'two-fold extrusion process', has been used to prepare some novel[98] and some very hindered olefins containing a *gem*-t-butyl group.[96] The ylide (29) can be diverted by addition of an acid AH, to give $R^1R^2CHSCAR^1R^2$ (A = ArS, ArO, or AcO),[99] and by addition of dimethyl acetylenedicarboxylate to give the 2,5-dialkyl 2,5-dihydrothiophens (32).[100] Other reactions in which it has been suggested that thiocarbonyl ylides are intermediates are a photo-arylation process involving a substituted 2-arylthiocyclohex-2-enone,[101] the deprotonation of thiuronium salts,[102] and the addition of carbenes to tetra-methylthiourea.[102] Finally, it should be noted that thiocarbonyl ylide structures

[93] P. G. Gassman and H. R. Drewes, *J. Amer. Chem. Soc.*, 1974, **96**, 3002.
[94] E. Vilsmaier, K. H. Dittrich, and W. Sprügel, *Tetrahedron Letters*, 1974, 3601.
[95] R. M. Kellogg, M. Noteboom, and J. M. Kaiser, *J. Org. Chem.*, 1975, **39**, 2573.
[96] D. H. R. Barton, F. S. Guziec, jun., and I. Shahak, *J.C.S. Perkin I*, 1974, 1794.
[97] J. P. Snyder, *Angew. Chem. Internat. Edn.*, 1974, **13**, 351.
[98] P. Raynolds, S. Zonnebelt, S. Bakker, and R. M. Kellogg, *J. Amer. Chem. Soc.*, 1974, **96**, 3147.
[99] J. Buter, P. W. Raynolds, and R. M. Kellogg, *Tetrahedron Letters*, 1974, 2901.
[100] R. M. Kellogg and W. L. Prins, *J. Org. Chem.*, 1974, **39**, 2366.
[101] A. G. Schultz, *J. Org. Chem.*, 1974, **39**, 3185.
[102] S. Mitamura, M. Takaku, and H. Nozaki, *Bull. Chem. Soc. Japan*, 1974, **47**, 3152.

may be important resonance forms for such non-classical heterocycles as thieno-[3,4-*c*]thiophens.[103]

6 Sulphenyl Carbanions

Synthesis and Properties.—Seebach [104] and Krief [105] and their co-workers have published details of a useful new approach to reactive α-sulphenyl carbanions involving the RSe–Li exchange reaction, as exemplified by the preparation of (34).[105] This sequence is valuable because (33) cannot be prepared directly from

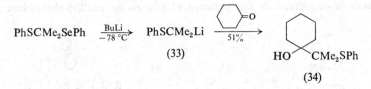

$$PhSCMe_2SePh \xrightarrow[-78\,°C]{BuLi} PhSCMe_2Li \xrightarrow{51\%}$$

(33)

(34)

isopropyl phenyl sulphide due to competing *ortho*-metallation.[104] A caveat has appeared [106] on the use of HMPA as a solvent for alkyl-lithiums because this solvent has been found to act as a source of the imine $CH_3N=CH_2$, which reacts with the solute. Thiophilic addition continues to be explored as a route to α-sulphenyl carbanions.[107] Addition to the stable thioketen (35) gives [108] the bicyclic compound (36), presumably *via* an alkylidenecarbene. Apparently

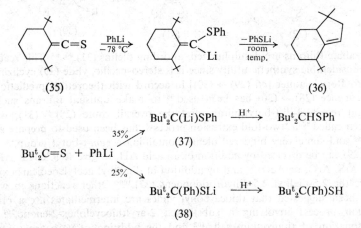

$$\xrightarrow[-78\,°C]{PhLi}$$

$$\xrightarrow[\text{room}\atop\text{temp.}]{-PhSLi}$$

(35)   (36)

$$Bu^t_2C(Li)SPh \xrightarrow{H^+} Bu^t_2CHSPh$$

(37)

$$Bu^t_2C=S + PhLi$$

$$Bu^t_2C(Ph)SLi \xrightarrow{H^+} Bu^t_2C(Ph)SH$$

(38)

the instability of α-sulphenyl carbanion (37) is only slightly less than the instability due to steric factors of the lithium thiolate (38), since the extent of thiophilic addition in this system is much less than is usually observed.[109]

[103] M. P. Cava and M. V. Lakshmikantham, *Accounts Chem. Res.*, 1975, 139.
[104] D. Seebach and A. K. Beck, *Angew. Chem. Internat. Edn.*, 1974, **13**, 806.
[105] A. Anciaux, A. Eman, W. Dumont, and A. Krief, *Tetrahedron Letters*, 1975, 1617.
[106] A. G. Abatjoglou and E. L. Eliel, *J. Org. Chem.*, 1974, **39**, 3043.
[107] P. Beak, J. Yamamoto, and C. J. Upton, *J. Org. Chem.*, 1975, **40**, 3052.
[108] E. Schaumann and W. Walter, *Chem. Ber.*, 1974, **107**, 3562.
[109] A. Ohno, K. Nakamura, M. Uohama, and S. Oka, *Chem. Letters*, 1975, 983.

There is a striking contrast in behaviour of the 10π-electron 1,4-dithiepinide ion (39) and the 1,3-dithiepinide ion (42). Heterocycle (39) is apparently *not* formed on reaction of 5*H*-1,4-dithiepin (40) with base, since alkylation occurs at the 2- and not at the 5-position.[110] In contrast, 1,3-dithiacyclohepta-4,6-diene

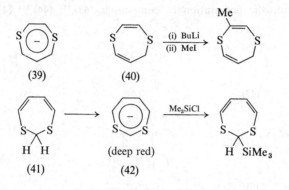

(39) (40) (41) (42) (deep red)

(41) is significantly more acidic than its open-chain analogue, and a moderately stable deep-red aromatic 1,3-dithiepinide anion (42) can be prepared.[111] α-Disulphide carbanions have proven to be very elusive, undoubtedly due to the sensitivity of the S—S linkage to nucleophilic attack and the tendency of activated disulphides to undergo α-elimination or rearrangement. Base-catalysed incorporation of deuterium in t-butylsulphonylmethyl methyl disulphide, $Bu^tSO_2CH_2SSMe$, is presented[112] as the first unequivocal evidence for the intermediacy of a discrete α-disulphide carbanion.

Reactions.—1,3-*Dithianyl Anions.* The Corey–Seebach 1,3-dithianyl anion rates as one of the most popular sulphur anions and nucleophilic acylating agents, as demonstrated by its recent use in the syntheses of potential precursors of the macrocyclic antitumour agent maytansine,[113, 114] of cyclopentenones[115, 116] and other unsaturated ketones,[117] of octoses,[118] of 11α-hydroxyprogesterone,[119] of deuteriated aldehydes,[120, 121] of alnusone,[122] of the sex-attractant of the Douglas fir tussock moth,[123] of 1,4-diketones,[124] of 3-s-butylglutaraldehyde,[125] of linaloyl

110 I. Murata, K. Nakasuji, and Y. Nakajima, *Tetrahedron Letters*, 1975, 1895.
111 C. L. Semmelhack, I.-C. Chiu, and K. G. Grohmann, *J. Amer. Chem. Soc.*, 1976, **98**, 2005.
112 E. Block and J. O'Connor, *J. Amer. Chem. Soc.*, 1974, **96**, 3929; E. Block, *Preprints, Div. Petrol. Chem., Amer. Chem. Soc.*, 1974, **19**, 205.
113 E. J. Corey and M. G. Bock, *Tetrahedron Letters*, 1975, 2643.
114 A. I. Meyers and R. S. Brinkmeyer, *Tetrahedron Letters*, 1975, 1749.
115 I. Kawamoto, S. Muramatsu, and Y. Yura, *Tetrahedron Letters*, 1974, 4223.
116 R. A. Ellison, E. R. Lukenbach, and C. Chiu, *Tetrahedron Letters*, 1975, 499.
117 R. H. Fischer, M. Baumann, and G. Köbrich, *Tetrahedron Letters*, 1974, 1207.
118 A. Gateau-Olesker, S. D. Gero, C. Pascard-Billy, C. Riche, A.-M. Sepulchre, G. Vass, and N. A. Hughes, *J.C.S. Chem. Comm.*, 1974, 811.
119 W. S. Johnson, S. Escher, and B. W. Metcalf, *J. Amer. Chem. Soc.*, 1976, **98**, 1039.
120 W. T. Wipke and G. L. Goeke, *J. Org. Chem.*, 1975, **40**, 3242.
121 F. Mutterer and J. P. Fluery, *J. Org. Chem.*, 1974, **39**, 640.
122 M. F. Semmelhack and L. S. Ryono, *J. Amer. Chem. Soc.*, 1975, **97**, 3873.
123 R. G. Smith, G. D. Daves, jun., and G. E. Daterman, *J. Org. Chem.*, 1975, **40**, 1593.
124 W. B. Sudweeks and H. S. Broadbent, *J. Org. Chem.*, 1975, **40**, 1131.
125 D. Tatore, T. C. Dich, R. Nacco, and C. Botteghi, *J. Org. Chem.*, 1975, **40**, 2987.

oxide,[126] and of certain substituted hydronaphthalenes [127] and hydroanthra-
cenes.[128] Corey and Seebach [129] have published further details on lithiodithian,
Eliel [130] has reviewed his studies on the stereochemistry of 1,3-dithianyl carbanions,
and reports have appeared on the addition of lithiodithian to nitro-olefins [131]
and iminium salts.[132, 133] Some of the more unusual applications of the dithianyl
anions include the preparation of compounds (43),[134] (44),[134] (45),[135] (46),[136]

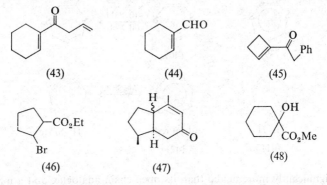

(43)   (44)   (45)

(46)   (47)   (48)

(47),[137] and (48).[138] It should be noted that one of the reagents recommended for
the hydrolysis of 1,3-dithians [135, 139] has proven to be explosive.[140]

Thioallyl and Related Anions. Alkylation of thioallyl anions, often termed
Biellmann alkylation, represents a useful method for forming new C—C bonds,
as illustrated by the synthesis of dendrolasin (49),[141] which is a furanoterpene

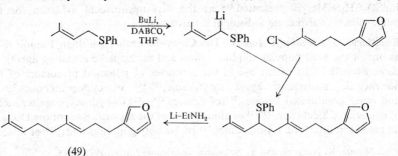

(49)

126 S. Torii, K. Uneyama, and M. Isihara, *J. Org. Chem.*, 1974, **39**, 3645.
127 R. Sarges, *J. Org. Chem.*, 1975, **40**, 1216.
128 W. Amrein and K. Schaffner, *Helv. Chim. Acta*, 1975, **58**, 380.
129 D. Seebach and E. J. Corey, *J. Org. Chem.*, 1975, **40**, 231.
130 E. L. Eliel, *Tetrahedron*, 1974, **30**, 1503.
131 D. Seebach, H. F. Leitz, and V. Ehrig, *Chem. Ber.*, 1975, **108**, 1924.
132 L. Duhamel, P. Duhamel, and N. Mancelle, *Bull. Soc. chim. France*, 1974, 331.
133 D. Seebach, V. Ehrig, H. F. Leitz, and R. Henning, *Chem. Ber.*, 1975, **108**, 1946.
134 E. J. Corey and A. P. Kozikowski, *Tetrahedron Letters*, 1975, 925.
135 D. Seebach, M. Kolb, and B.-T. Gröbel, *Tetrahedron Letters*, 1974, 3171.
136 B.-T. Gröbel, R. Bürstinghaus, and D. Seebach, *Synthesis*, 1976, 121.
137 N. H. Andersen, Y. Yamamota, and A. D. Denniston, *Tetrahedron Letters*, 1975, 4547.
138 W. D. Woessner, *Chem. Letters*, 1976, 43.
139 Y. Tamura, K. Sumoto, S. Fujii, H. Satoh, and M. Ikeda, *Synthesis*, 1973, 312.
140 R. Y. Ning, *Chem. Eng. News* (Letter to Editor), Dec. 17, 1973.
141 K. Kondo and M. Matsumoto, *Tetrahedron Letters*, 1976, 391.

found in ants, and related syntheses of geranylgeraniol.[142] An intramolecular Biellmann alkylation of an epoxy-function forms the basis for a clever synthesis of cembrene-A,[143] a trail-making pheromone of termites.

The regioselectivity of alkylation and carbonyl addition of various thioallyl anions has been examined, particularly with regard to chelation and ion-pairing effects.[5] Thus, Biellmann and co-workers[144] have found that thioallyl-lithium (50), in the presence of [2,2,2]cryptate, adds to acetone exclusively from the

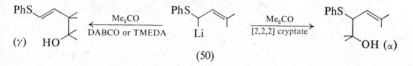

(50)

α-position of (50), while in the presence of DABCO or TMEDA the addition occurs exclusively from the γ-position. It is argued that, by complexing with lithium, cryptate favours the formation of a free anion, which adds at the α-position due to a combination of steric and charge-concentration factors. Where steric factors are less important, as with methyl iodide, a 60/40 mixture of α-/γ-alkylated products is formed in the presence of the cryptate.[144] Under conditions favouring intimate or solvated ion pairs (DABCO, TMEDA, or no added reagent), complex formation involving acetone and lithium is thought to lead to bonding at the available γ-position.

The regioselectivity of alkylation has also been studied with cyclic thioallyl anions,[145] with thioallyl–copper systems[146] which involve S_N2' reactions, and with the dianion of prop-2-enethiol (51).[147, 148] The products of addition of aldehyde at the α-position of (50) have been converted into cyclopropane

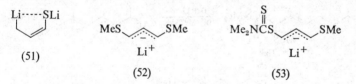

(51)   (52)   (53)

carboxyaldehydes.[149] Interest continues in Corey's[150] reagent 1,3-bis(methylthio)-allyl-lithium (52), which functions as a 1-oxopropene synthon ($^{-}COCH{=}CH_2$), and which is now described[151] in 'Organic Synthesis'. Applications in connection with a new synthesis of macrocycles,[152] and in the synthesis of yomogi alcohol, using the copper complex[146] of (52), have been reported. Anion (53) is recommended[153]

[142] L. J. Altman, L. Ash, and S. Marson, *Synthesis*, 1974, 129.
[143] M. Kodama, Y. Matsuki, and S. Ito, *Tetrahedron Letters*, 1975, 3065.
[144] P. M. Atlanti, J. F. Biellmann, S. Dube, and J. J. Vicens, *Tetrahedron Letters*, 1974, 2665.
[145] S. Torii, H. Tanaka, and Y. Tomotaki, *Chem. Letters*, 1974, 1541.
[146] K. Oshima, H. Yamamoto, and H. Nozaki, *Bull. Chem. Soc. Japan*, 1975, **48**, 1567.
[147] K. Geiss, B. Seuring, R. Pieter, and D. Seebach, *Angew. Chem. Internat. Edn.*, 1974, **13**, 479.
[148] J. Hartmann, R. Muthukrishnan, and M. Schlosser, *Helv. Chim. Acta*, 1974, **57**, 2261.
[149] K. Kondo, K. Matsui, and A. Negishi, *Chem. Letters*, 1974, 1371.
[150] E. J. Corey, B. W. Erickson, and R. Noyori, *J. Amer. Chem. Soc.*, 1971, **93**, 1724.
[151] B. W. Erickson, *Org. Synth.*, 1974, **54**, 19.
[152] R. G. Carlson and W. S. Mardis, *J. Org. Chem.*, 1975, **40**, 817.
[153] T. Nakai, H. Shiono, and M. Okawara, *Tetrahedron Letters*, 1974, 3625.

as being superior to (52) because of its more facile hydrolysis, and synthetic applications involving the related system $Et_2NC(S)SCH\colon CHCHRSC(S)NEt_2$ have been described.[154]

Thioallyl anions such as (55) may participate [155] in $\pi 2s + \pi 4s$ cycloadditions as the 4π component. The [2,3]-sigmatropic rearrangements of anions of

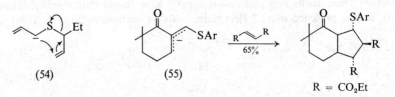

(54)   (55)

$R = CO_2Et$

bis(allyl) sulphides [*e.g.* (54)] and propargylallyl sulphides have been used as the key step in an olefin synthesis.[156] The anions of *S*-allyldithiocarbamates, *e.g.* (56), are of synthetic interest because of the possibility of [3,3]-sigmatropic rearrangement following alkylation, as utilized in the synthesis of nuciferal (57),[157a] among other examples.[153, 154, 157] A group of compounds related to the

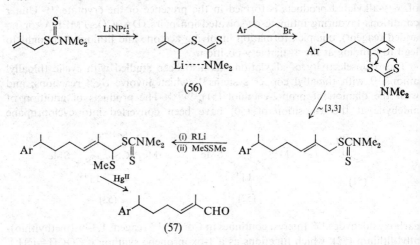

(56)

~ [3,3]

(57)

thioallyl anions, whose chemistry has been of interest,[158] are the ene-sulphide anions, such as the chelated 1-ethoxy-2-lithio-2-phenylthioethene $EtOCH= CLiSPh$, used in the synthesis of 9-deoxo-9-thiaprostaglandins.[158a]

Acyclic α-Thiocarbanions, Thioacetal Anions, and Trithioformate Anions. A new episulphide synthesis has been reported independently by Meyers [159] and

[154] I. Hori, T. Hayashi, and H. Midorikawa, *Synthesis*, 1975, 727.
[155] J. P. Marino and W. B. Mesbergen, *J. Amer. Chem. Soc.*, 1974, **96**, 4052.
[156] W. Kreiser and H. Wurziger, *Tetrahedron Letters*, 1975, 1669.
[157] (*a*) T. Nakai, H. Shiono, and M. Okawara, *Chem. Letters*, 1975, 249; (*b*) *Tetrahedron Letters*, 1975, 4027; T. Hayashi, *ibid.*, 1974, 339; T. Hayashi and H. Midorikawa, *Synthesis*, 1975, 100; I. Vlattas, L. Della Vecchia, and A. O. Lee, *J. Amer. Chem. Soc.*, 1976, **98**, 2008.
[158] (*a*) R. Muthukrishnam and M. Schlosser, *Helv. Chim. Acta*, 1976, **59**, 13; (*b*) H. Westmijze, J. Meijer, and P. Vermeer, *Tetrahedron Letters*, 1975, 2923.
[159] A. I. Meyers and M. E. Ford, *Tetrahedron Letters*, 1975, 2861.

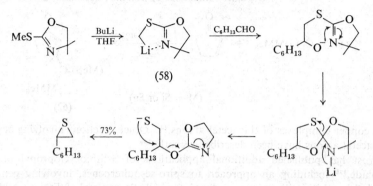

(58)

Johnson,[160] utilizing the lithio-derivative (58) and 2-(thiomethyl)-Δ^2-oxazoline or -thiazoline. A full report on the Wittig rearrangement ([1,2]-anionic rearrangement) of alkylthiotrialkyl-silanes and -germanes [*e.g.* (59) → (60)] has been

$$CH_3SSiMe_2Bu^t \xrightarrow[THF]{Bu^tLi} LiCH_2SSiMe_2Bu^t \xrightarrow[35\%]{[1,2]} Bu^tMe_2SiCH_2SLi$$

(59)   (60)

published,[161] together with similar reactions of the anions of thiacyclophanes[162] and dithiocarbamates.[163] Thiophenylmethyl-lithium has been used in a synthesis of (±)-laurene.[164]

A new procedure for the alkylation of α-thiocarbanions and thioacetal anions involves treatment of the lithio salt with a trialkylborane,[165] as illustrated with the reaction of (61).[165a]

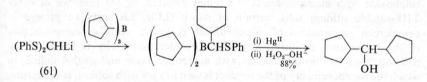

(61)

With αβ-unsaturated carbonyl compounds, 1,2- rather than 1,4-addition is generally the rule with thioacetal anions. Seebach[166] now finds that lithiated bis(methylthio)(silyl)- and bis(methylthio)(stannyl)-methanes undergo 1,4-addition in good yield, affording the Michael adduct, *e.g.* (62). Other anionic sulphur systems reported to undergo 1,4-addition are trithioformate anions[167]

[160] C. R. Johnson, A. Nakanishi, N. Nakanishi, and K. Tanaka, *Tetrahedron Letters*, 1975, 2865.
[161] A. Wright and R. West, *J. Amer. Chem. Soc.*, 1974, **96**, 3222.
[162] (a) R. H. Mitchell, T. Otsubo, and V. Boekelheide, *Tetrahedron Letters*, 1975, 219; (b) R. H. Mitchell and R. J. Carruthers, *ibid.*, 1975, 4331.
[163] T. Hayashi and H. Baba, *J. Amer. Chem. Soc.*, 1975, **97**, 1608.
[164] J. E. McMurray and L. A. von Beroldingen, *Tetrahedron*, 1974, **30**, 2027.
[165] (a) R. J. Hughes, A. Pelter, K. Smith, E. Negishi, and T. Yoshida, *Tetrahedron Letters*, 1976, 87; R. J. Hughes, A. Pelter, and K. Smith, *J.C.S. Chem. Comm.*, 1974, 863; (b) E. Negishi, T. Yoshida, A. Silveira, jun., and B. L. Chiou, *J. Org. Chem.*, 1975, **40**, 814; E. J. Corey, R. L. Danheiser, and S. Chandrasekaran, *ibid.*, 1976, **41**, 260.
[166] D. Seebach and R. Bürstinghaus, *Angew. Chem. Internat. Edn.*, 1975, **14**, 57.
[167] A.-R. B. Manas and R. A. J. Smith, *J.C.S. Chem. Comm.*, 1975, 216.

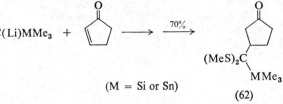

$$(\text{MeS})_2\text{C(Li)MMe}_3$$

(M = Si or Sn)

(62)

and copper complexes of thioacetal anions.[168] Other reactions involving acyclic thioacetal anions have been described recently.[169]

Trost has published additional applications of α-lithiocyclopropyl phenyl sulphide,[170] including an approach to spiro-sesquiterpenes, involving geminal alkylation, which utilizes a process that is formally the reversal of dithian-carbonyl addition [170a] (also see related studies by Marshall [171]). Synthetic applications of the highly nucleophilic methoxy-phenylthio-methyl-lithium [172] PhSCH(Li)OMe and of the acyl anion equivalent methylthioacetic acid dianion RSCH(Li)CO₂Li (R = Me [173] or Ph [174]) have also been described, as have synthetic procedures based on anions of α-alkylthio- or α-phenylthio-ketones [175] and methylthiomethyl dithiocarbamates.[176]

7 Sulphinyl Carbanions

Synthesis and Properties.—Potassium hydride is recommended [177] as a base for the generation of dimsylpotassium. The sensitivity of the preferred conformation of α-sulphinyl carbanions to ion-pairing effects has been neatly exploited in an asymmetric synthesis based on addition of carbanions to vinyl sulphoxides.[178] The ratio of diastereomers produced by quenching α-lithiobenzyl methyl sulphoxide with electrophiles is [179] strongly affected by the presence of other THF-soluble lithium salts, certain of which (LiBr, LiI) may be present in commercial preparations of organolithium reagents. Another demonstration of the importance of ion-pairing effects on stereoselectivity is found in the reaction [180] of α-lithio-sulphoxides with deuterium oxide and methyl iodide, in which the stereochemistry of the product is seen to vary with solvent, temperature, and with the presence or absence of agents that chelate lithium, such as cryptates. It is argued that cation-oxygen chelation (63) may control the conformation.

168   T. Cohen, G. Herman, J. R. Falck, and A. J. Mura, jun., *J. Org. Chem.*, 1975, **40**, 812.
169   L. Leger, M. Saquet, A. Thuillier, and S. Julia, *J. Organometallic Chem.*, 1975, **96**, 313; G. Schill and C. Merkel, *Synthesis*, 1975, 387; T. Cohen, D. Kuhn, and J. R. Falck, *J. Amer. Chem. Soc.*, 1975, **97**, 4749.
170   (a) B. M. Trost, M. Preckel, and L. M. Leichter, *J. Amer. Chem. Soc.*, 1975, **97**, 2224; (b) B. M. Trost and D. E. Keeley, *ibid.*, 1976, **98**, 248.
171   J. A. Marshall and D. E. Seitz, *J. Org. Chem.*, 1974, **39**, 1814.
172   B. M. Trost and C. H. Miller, *J. Amer. Chem. Soc.*, 1975, **97**, 7182.
173   B. M. Trost and Y. Tamaru, *Tetrahedron Letters*, 1975, 3797.
174   K. Iwai, M. Kawai, H. Kosugi, and H. Uda, *Chem. Letters*, 1974, 385.
175   R. F. Romanet and R. H. Schlessinger, *J. Amer. Chem. Soc.*, 1974, **96**, 3701; R. M. Coates, H. D. Pigott, and J. Ollinger, *Tetrahedron Letters*, 1974, 3955.
176   I. Hori, T. Hayashi, and H. Midorikawa, *Synthesis*, 1974, 705.
177   C. A. Brown, *J. Org. Chem.*, 1974, **39**, 3913.
178   G. Tsuchihashi, S. Mitamura, and K. Ogura, *Tetrahedron Letters*, 1976, 855.
179   T. Durst and M. Molin, *Tetrahedron Letters*, 1975, 63.
180   J. F. Biellmann and J. J. Vicens, *Tetrahedron Letters*, 1974, 2915.

(63)

A ¹H n.m.r. study [181] of α-lithio-sulphoxides leads to the conclusion that the possibility of a planar metallated α-carbon should be strongly considered. Fava and co-workers [28] have demonstrated the essential non-stereospecificity of exchange in D_2O of some thiolan 1-oxide systems, and they concluded that, contrary to the *gauche*-effect theory, there seems to be no obvious logical connection between kinetic acidity and geometrical factors.

Reactions.—A variety of interesting and useful syntheses have been published involving the reaction of dimsyl anion $[MeS(O)CH_2^-]$ with esters and lactones,[172, 182] with disulphides,[183] with chlorosilanes,[184] with sulphinate esters,[185] with organoboranes,[186] and with stilbenes.[187] Simple and functionalized α-sulphinyl carbanions can be condensed with carbonyl compounds or alkylated, often in a stereocontrolled manner, as in a nicely conceived synthesis of biotin.[188] Considerable attention has been given to methods for the removal of the sulphoxide function following carbon–carbon bond formation. Among the methods used are reduction by aluminium amalgam (with β-keto-sulphoxides),[189] reduction with Raney nickel,[190] pyrolytic elimination of sulphenic acid,[191] elimination of sulphur dioxide from sultines, *e.g.* (64),[192] and sulphoxide–

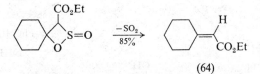

(64)

sulphenate interconversion (a [2,3]-sigmatropic process) with reduction of sulphenate.[193] Elimination of sulphenic acid, following carbonyl addition by α-sulphinyl ester anions, provides a synthesis of β-keto-esters such

[181] R. Lett and A. Marquet, *Tetrahedron Letters*, 1975, 1579.
[182] S. Iriuchijima, K. Maniwa, and G. Tsuchihashi, *J. Amer. Chem. Soc.*, 1975, **97**, 596; Y. Oikawa and O. Yonemitsu, *J. Org. Chem.*, 1976, **41**, 1118; S. F. Dyke and E. P. Tiley, *Tetrahedron*, 1975, **31**, 561.
[183] B. M. Trost and J. L. Stanton, *J. Amer. Chem. Soc.*, 1975, **97**, 4018.
[184] E. Vedejs and M. Mullins, *Tetrahedron Letters*, 1975, 2017.
[185] N. Kunieda, J. Nokami, and M. Kinoshita, *Bull. Chem. Soc. Japan*, 1976, **49**, 256
[186] E. Negishi, K.-W. Chiu, and T. Yosida, *J. Org. Chem.*, 1975, **40**, 1677.
[187] B. G. James and G. Pattenden, *J.C.S. Perkin I*, 1974, 1195, 1204.
[188] S. Bory, M. J. Luche, B. Moreau, S. Lavielle, and A. Marquet, *Tetrahedron Letters*, 1975, 827.
[189] P. A. Grieco and C. S. Pogonowski, *J. Org. Chem.*, 1974, **39**, 732.
[190] N. Kunieda, J. Nokami, and M. Kinoshita, *Chem. Letters*, 1974, 369; *Tetrahedron Letters*, 1974, 3997.
[191] B. M. Trost and A. J. Bridges, *J. Org. Chem.*, 1975, **40**, 2014; B. M. Trost, W. P. Conway, P. E. Strege, and T. J. Dietsche, *J. Amer. Chem. Soc.*, 1974, **96**, 7165; B. M. Trost and K. K. Leung, *Tetrahedron Letters*, 1975, 4197; P. A. Grieco, D. Boxler, and C. S. Pogonowski, *J.C.S. Chem. Comm.*, 1974, 497; P. A. Grieco and C. S. Pogonowski, *ibid.*, 1975, 72.
[192] J. Nokami, N. Kunieda, and M. Kinoshita, *Tetrahedron Letters*, 1975, 2179.
[193] D. A. Evans, T. C. Crawford, T. T. Fujimoto, and R. C. Thomas, *J. Org. Chem.*, 1974, **39**, 3176.

98   *Organic Compounds of Sulphur, Selenium, and Tellurium*

as (65).[194] Other reactions involving α-sulphinyl ester anions have been reported,[190, 191b, c, 192, 195] as have reactions of α-phosphoryl α-sulphinyl carbanions.[196]

The methyl methylthiomethyl sulphoxide anion and its derivatives continue to attract attention as useful synthetic intermediates. Examples include the syntheses

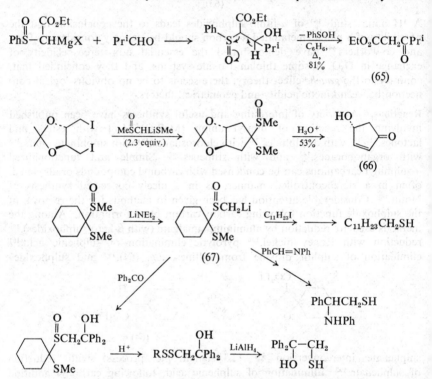

of the chiral prostaglandin precursor (*R*)-4-hydroxycyclopent-2-enone (66),[197] of α-amino-acids,[198] of cyclobutanones[199] and other cycloalkanones,[200] of pyridine carboxyaldehydes,[201] of vinyl sulphides,[202] of ketones,[203] of homologated esters,[204] and, by a clever sequence involving (67), of monothio-glycols, α-amino-thiols, and homologated thiols.[205]

[194] J. Nokami, N. Kunieda, and M. Kinoshita, *Tetrahedron Letters*, 1975, 2841.
[195] J. J. A. von Asten and R. Louw, *Tetrahedron Letters*, 1975, 671.
[196] M. Mikolajczyk, S. Grzejszczak, and A. Zatorski, *J. Org. Chem.*, 1975, **40**, 1979.
[197] K. Ogura, M. Yamashita, and G. Tsuchihashi, *Tetrahedron Letters*, 1976, 759.
[198] K. Ogura and G. Tsuchihashi, *J. Amer. Chem. Soc.*, 1974, **96**, 1960.
[199] K. Ogura, M. Yamashita, M. Suzuki, and G. Tsuchihashi, *Tetrahedron Letters*, 1974, 3653.
[200] K. Ogura, M. Yamashita, S. Furukawa, M. Suzuki, and G. Tsuchihashi, *Tetrahedron Letters*, 1975, 2767.
[201] G. R. Newkome, J. M. Robinson, and J. D. Sauer, *J.C.S. Chem. Comm.*, 1974, 410.
[202] H. Tsukasa and S. Saito, *Yukagaku*, 1975, **24**, 659; C. Huynh, V. Ratovelomanana, and J. Sylvestre, *Compt. rend.*, 1975, **280**, C, 1231.
[203] G. Schill and P. R. Jones, *Synthesis*, 1974, 117.
[204] K. Ogura, S. Furukawa, and G. Tsuchihashi, *Chem. Letters*, 1974, 659.
[205] K. Ogura, S. Furukawa, and G. Tsuchihashi, *Synthesis*, 1976, 202.

8 Sulphonyl Carbanions

Synthesis and Properties.—The preparation of sulphonyl-stabilized alkylidene-phosphoranes (68) has been reported.[206] Further investigations have been made [207] into the generation of α-sulphonyl carbanions through the reaction of α-halogeno-sulphones with phosphines and other reducing agents. The generation of lithio-compounds (69) and (70) has been reported.[208] Acidity measurements on dinitro-

$$R^1SO_2F + R^2\bar{C}H\overset{+}{P}Ph_3 \longrightarrow R^1SO_2\bar{C}R^2\overset{+}{P}Ph_3$$

$$(68)$$

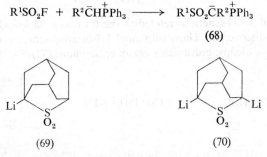

(69)   (70)

methyl sulphones and sulphides have been compared.[209] Two separate examples of differential kinetic acidity of geminal hydrogens α to a sulphonyl group have been reported.[28, 210] Chiral n.m.r. solvents have been utilized in a study of the stereochemistry of α-sulphonyl carbanions.[211]

Reactions.—The mechanism and synthetic applications of the Ramberg–Bäcklund reaction have been areas of active interest. Bordwell [212] has published a series of papers dealing with a detailed mechanistic study of the Ramberg–Bäcklund reaction, including a study of the stereochemistry in both acyclic and cyclic systems, the kinetics, deuterium exchange, salt and solvent effects, and a search for a concerted mechanism which was not found even in favourable cases. Among the olefins prepared by Ramberg–Bäcklund reaction of α-chlorosulphones are pro-pellanes [213] [*e.g.* (71) [213b]] and various bicyclic structures.[214] The one-step Meyers modification of the Ramberg–Bäcklund reaction (sulphone, potassium hydroxide, carbon tetrachloride) has been used to prepare 1,1-diarylalkenes,[215] cyclophanes,[216] silacycloheptenes,[217] dienoic and trienoic acids,[218] and polyolefins.[219]

[206] B. A. Reith, J. Strating, and A. M. van Leusen, *J. Org. Chem.*, 1974, **39**, 2728.
[207] F. G. Bordwell and E. Doomes, *J. Org. Chem.*, 1974, **39**, 2298; B. B. Jarvis, R. L. Harper, jun., and W. P. Tong, *ibid.*, 1975, **40**, 3778; B. B. Jarvis and B. A. Marien, *ibid.*, p. 2587.
[208] J. Janku, J. Burkhard, and L. Vodicka, *Z. Chem.*, 1975, **15**, 397.
[209] V. I. Erashko, A. V. Sultanov, S. A. Shevelev, T. I. Rozhkova, and A. A. Fainsilberg, *Izvest. Akad. Nauk S.S.S.R., Ser. khim.*, 1974, 1289.
[210] J. Kattenberg, E. R. de Waard, and H. O. Huisman, *Rec. Trav. chim.*, 1975, **94**, 89.
[211] T. A. Whitney and W. H. Pirkle, *Tetrahedron Letters*, 1974, 2299.
[212] F. G. Bordwell and M. D. Wolfinger, *J. Org. Chem.*, 1974, **39**, 2521; F. G. Bordwell and J. B. O'Dwyer, *ibid.*, p. 2519; F. G. Bordwell and E. Doomes, *ibid.*, pp. 2526, 2531.
[213] (*a*) K. Weinges and K. Klessing, *Chem. Ber.*, 1974, **107**, 1915, 1925; 1976, **109**, 793; (*b*) L. A. Paquette, R. E. Wingard, jun., and J. M. Photis, *J. Amer. Chem. Soc.*, 1974, **96**, 5801.
[214] R. G. Carlson and K. D. May, *Tetrahedron Letters*, 1975, 947; J. Kattenberg, E. R. de Waard, and H. O. Huisman, *Tetrahedron*, 1974, **30**, 3177.
[215] C. Y. Meyers, W. S. Matthews, G. J. McCollum, and J. C. Branca, *Tetrahedron Letters*, 1974, 1105.
[216] H. Bestmann and W. Schaper, *Tetrahedron Letters*, 1975, 3511.
[217] K. E. Koenig, R. A. Felix, and W. P. Weber, *J. Org. Chem.*, 1974, **39**, 1539.
[218] P. Grieco and D. Boxler, *Synthetic Comm.*, 1975, **5**, 315.
[219] G. Büchi and R. M. Freidinger, *J. Amer. Chem. Soc.*, 1974, **96**, 3332.

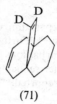

(71)

A number of novel reactions related to the Ramberg–Bäcklund reaction have recently been discovered. Thus, sulphonyl 1,3-dicarbanions, on oxidation with $CuCl_2$ or I_2, give olefins, presumably *via* an episulphone (72).[219, 220] This reaction

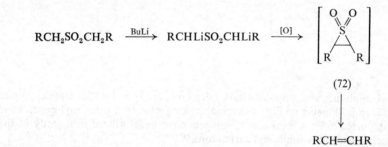

(72)

$$RCH=CHR$$

has been used [219] in an elegant synthesis of all-*trans*-β-carotene. A similar transformation, particularly useful in the synthesis of cyclobutenes, can be effected through the action of lithium aluminium hydride on α-sulphonyl carbanions.[221] In some instances this reagent alone can effect desulphonylation.[221] Other studies suggest that it can convert sulphones into α-monoanions and αα'-dianions.[222] Alkylation of sulphones followed by base-catalysed elimination of sulphinate anion, or replacement of the sulphonyl group with hydrogen using lithium in ethylamine, represents a very useful approach to carbon–carbon bond formation. Recent examples include the synthesis of the sesquiterpene deoxytrisporone (73) [223] and related syntheses of 2,2'-dinor-carotenoids,[224] vitamin A,[225] apo-carotenoids,[226] β-carotene,[227] sesquifenchene,[228] diumycinol,[229] squalene,[230] and geranylgeraniol.[231] An intramolecular alkylation of a sulphonyl anion has been used to make a 1-phenylsulphonylbicyclo[1,1,0]butane.[232] Dialkylation of an αα'-sulphonyl dianion, derived from a 1,4-sulphone-bridged cyclo-octatriene,

[220] J. S. Crossert, J. Buter, E. W. H. Asveld, and R. M. Kellogg, *Tetrahedron Letters*, 1974, 2805.
[221] J. M. Photis and L. A. Paquette, *J. Amer. Chem. Soc.*, 1974, **96**, 4715.
[222] W. P. Weber, P. Stronquist, and T. I. Ito, *Tetrahedron Letters*, 1974, 2595.
[223] K. Uneyama and S. Torii, *Tetrahedron Letters*, 1976, 443.
[224] F. Kienzle and R. E. Minder, *Helv. Chim. Acta*, 1976, **59**, 439.
[225] P. S. Manchand, M. Rosenberger, G. Saucy, P. A. Wehrli, H. Wong, L. Chambers, M. P. Ferro, and W. Jackson, *Helv. Chim. Acta*, 1976, **59**, 387; A. Fischli, H. Mayer, W. Simon, and H.-J. Stoller, *ibid.*, p. 397.
[226] A. Fischli and H. Mayer, *Helv. Chim. Acta*, 1975, **58**, 1492.
[227] A. Fischli and H. Mayer, *Helv. Chim. Acta*, 1975, **58**, 1586.
[228] P. A. Grieco and Y. Masaki, *J. Org. Chem.*, 1975, **40**, 150.
[229] P. A. Grieco, Y. Masaki, and D. Boxler, *J. Org. Chem.*, 1975, **40**, 2261.
[230] P. A. Grieco and Y. Masaki, *J. Org. Chem.*, 1974, **39**, 2135.
[231] B. M. Trost and L. Weber, *J. Org. Chem.*, 1975, **40**, 3617.
[232] Y. Gaoni, *Tetrahedron Letters*, 1976, 503.

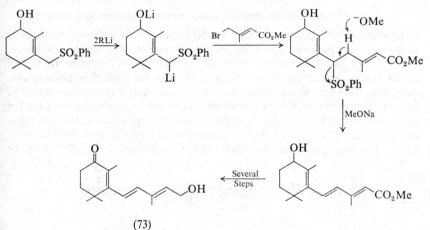

(73)

followed by photochemical extrusion of SO_2 affords 1,4-disubstituted cyclo-octatetraenes.[233]

Michael addition of sulphonyl carbanions followed by 1,3-elimination of sulphinate anion provides a route to cyclopropanes utilized in the total synthesis of presqualene alcohol,[234] prephytoene alcohol,[234] and the preparation of the simple ketone (74).[235] The palladium complex of an α-sulphonyl carbanion has been reported [236] to add to isolated double bonds.

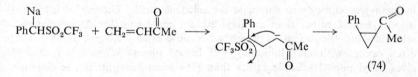

(74)

A variety of α-functionalized α-sulphonyl carbanions have been studied. Phenyl(phenylsulphonyldichloromethyl)mercury, $PhHgCCl_2SO_2Ph$, on prolonged heating at 140 °C (for 8 days), acts as an agent that transfers the $PhSO_2CCl$ group and which undergoes homolytic decomposition.[237] Several reactions of α-halogeno-α-sulphonyl carbanions have been reported,[238] as have syntheses based on β-keto-sulphone anions,[239] α-ethoxy-β-keto-sulphone anions,[240] and anions of α-keto-sulphone acetals.[241]

[233] L. A. Paquette, S. V. Ley, R. H. Meisinger, R. K. Russell, and M. Oku, *J. Amer. Chem. Soc.*, 1974, **96**, 5806.
[234] R. V. M. Campbell, L. Crombia, D. A. R. Findley, R. W. King, G. Pattenden, and D. A. Whiting, *J.C.S. Perkin I*, 1975, 897.
[235] J. B. Hendrickson, A. Giga, and J. Wareing, *J. Amer. Chem. Soc.*, 1974, **96**, 2275.
[236] M. Julia and L. Saussine, *Tetrahedron Letters*, 1974, 3443.
[237] D. Seyferth and R. A. Woodruff, *J. Organometallic Chem.*, 1974, **71**, 335.
[238] W. E. Truce, L. A. Mura, P. J. Smith, and F. Young, *J. Org. Chem.*, 1974, **39**, 1449; A. Jonczyk, K. Banko, and M. Makosza, *ibid.*, 1975, **40**, 266.
[239] B. Koutek, L. Pavlickova, and M. Soucek, *Coll. Czech. Chem. Comm.*, 1974, **39**, 192.
[240] G. Ferdinand, K. Schank, and A. Weber, *Annalen*, 1975, 1484.
[241] K. Kondo and D. Tunemoto, *Tetrahedron Letters*, 1975, 1007, 1397; K. Kondo, E. Saito, and D. Tunemoto, *ibid.*, p. 2275; M. Julia and B. Badet, *Bull. Soc. chim. France*, 1975, 1363.

9 Organoselenium and Organotellurium Carbanions and Ylides

Organoselenium chemistry has recently experienced a healthy growth, particularly with regard to applications of organoselenium reagents in synthesis. Following the discovery that RSe groups stabilize adjacent carbanion centres nearly as well as RS groups, researchers have sought to develop the area of organoselenium carbanion chemistry along similar lines to organosulphur carbanion chemistry. Part of the motivation behind this effort was to use to synthetic advantage the subtle differences in reactivity of organoselenium compounds at the carbanion and at the carbanion adduct stage, compared with their organosulphur counterparts. For example,[242] α-selenoalkyl-lithium compounds, *e.g.* (75), may be

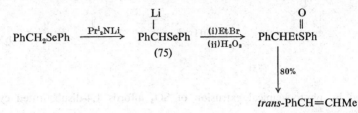

alkylated and the resulting selenide treated with hydrogen peroxide to give a selenoxide which spontaneously eliminates PhSeOH, giving an olefin. This sequence capitalizes on the well-known [243] ability of selenoxides to undergo cyclo-elimination to olefins under far milder conditions than sulphoxides. A more direct reaction [244] utilizes α-selenoxyalkyl-lithium reagents, *e.g.* PhSe(O)CH(Li)Ph, which are alkylated, or added to carbonyl groups at -78 °C, and then refluxed in methylene chloride to eliminate the selenenic acid. The use of lithium di-isopropylamide rather than an alkyl-lithium compound for deprotonation of selenides (and selenoxides) is prescribed because extensive Se—C bond cleavage often occurs with alkyl-lithiums;[242, 244] benzyl phenyl selenide, for example, gives butyl phenyl selenide rather than (75) when butyl-lithium is employed. The RSe–Li exchange reaction [104, 105] is comparable to Br–Li exchange [104] as a means of generating α-selenoalkyl-lithiums. When applied to seleno- or mixed sulpho-seleno-acetals it offers a kinetically controlled method for the preparation of secondary and tertiary phenylseleno- (or alkylseleno-), and phenylthioalkyl-lithium compounds (76) and (77), respectively. When applied to methyl (α-silyl-

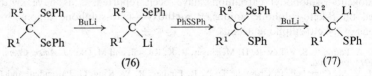

methyl) selenides, formed by the reaction of α-selenoalkyl-lithium compounds with silyl chlorides, α-silylcarbanions can be generated.[245] Two sequences

[242] R. H. Mitchell, *J.C.S. Chem. Comm.*, 1974, 990.
[243] D. N. Jones, D. Mundy, and R. D. Whitehouse, *Chem. Comm.*, 1970, 86; H. J. Reich, I. L. Reich, and J. M. Renga, *J. Amer. Chem. Soc.*, 1973, **95**, 5813; K. B. Sharpless, R. F. Lauer, and A. Y. Teranishi, *J. Amer. Chem. Soc.*, 1973, **95**, 6137.
[244] H. J. Reich and S. K. Shah, *J. Amer. Chem. Soc.*, 1975, **97**, 3250; H. Gilman and R. L. Bebb, *ibid.*, 1939, **61**, 1110; H. Gilman and F. J. Webb, *ibid.*, 1949, **71**, 4064.
[245] W. Dumont and A. Krief, *Angew. Chem.*, 1976, **88**, 184.

utilizing the RSe–Li exchange reaction are the synthesis of allyl alcohols [104, 246] by subsequent reaction of α-selenoalkyl-lithium with an aldehyde and the synthesis of an ester by subsequent reaction with a chloroformate.[246] In these reactions the α-phenylselenocyclohexyl-lithium can be regarded as a masked vinyl-lithium compound. α-Selenoalkyl-lithium compounds have also been employed [247] in the synthesis of epoxides from carbonyl compounds by a sequence involving conversion of the intermediate β-hydroxy methyl (or phenyl) selenides into β-hydroxyalkylselenonium salts, and hence, *via* the betaine (79), into the epoxide. This same betaine can also be formed directly [248] from the α-selenonio-alkylide

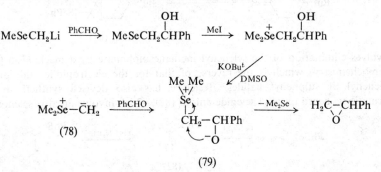

(78). The advantage of the α-selenoalkyl-lithium compounds is their very high nucleophilicity, which, for example, permits reaction with enolizable or hindered carbonyl compounds, such as deoxybenzoin, or 2,2,6,6-tetramethylcyclohexanone, respectively,[247] and which favours 1,2- rather than 1,4-addition, allowing the preparation of (80) from cyclohexenone.[249] A novel olefin synthesis has been

$$(MeSe)_2CMe_2 \xrightarrow{BuLi} MeSeCMe_2Li \longrightarrow$$

(80)

developed from β-hydroxy-selenides, *e.g.* (81), which can be prepared from α-selenoalkyl-lithium compounds with carbonyl compounds, or from α-selenoxy-alkyl-lithium compounds, as shown, *via* reduction.[250] The reaction probably

[246] W. Dumont, P. Bayet, and A. Krief, *Angew. Chem. Internat. Edn.*, 1974, **13**, 804; J. N. Denis, W. Dumont, and A. Krief, *Tetrahedron Letters*, 1976, 453.

[247] W. Dumont and A. Krief, *Angew. Chem. Internat. Edn.*, 1975, **14**, 350; D. Van Ende, W. Dumont, and A. Krief, *ibid.*, p. 700.

[248] W. Dumont, P. Bayet, and A. Krief, *Angew. Chem. Internat. Edn.*, 1974, **13**, 274.

[249] D. Van Ende and A. Krief, *Tetrahedron Letters*, 1976, 457.

[250] H. J. Reich and F. Chow, *J.C.S. Chem. Comm.*, 1975, 790.

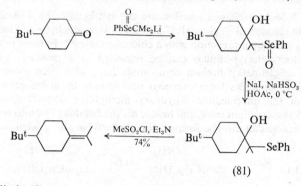

(81)

involves elimination of the selenenyl methanesulphonate by a mechanism (*anti* stereochemistry) which is the reverse of that for the electrophilic addition of selenenyl or sulphenyl halides. Reich [251] has also devised synthetic transformations, based on allyl selenide anions (82), which involve an allyl selenoxide

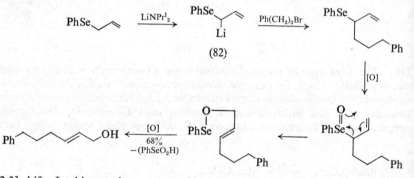

(82)

[2,3] shift. In this reaction excess oxidant is used, so that no volatile selenium-containing compounds remain after oxidation, and no trapping agent is needed to cleave the allyl selenate. With (82) and its homologues, problems of γ- *vs.* α-alkylation are encountered which are similar to those found for related sulphur systems. [251]

A limited number of studies of selenium ylides have appeared during the period covered by this Report. The preparation of unstabilized α-selenonio-alkylides, *e.g.* (78), by deprotonation of the corresponding selenonium salts has been reported, [248] and the formation of selenonium ylides through the reaction of selenoxides or selenonium imides, *e.g.* (83), with active-methylene compounds has been described. [252] A report [253] of the preparation and *X*-ray structural analysis of selenium ylide (84) argues, on the basis of the long Se—C bond [1.906(8) Å], that there is a large contribution from the ylide (dipolar) resonance hybrid. Dibenzylselenonium ylides (85) are reported [254] to undergo deselenization

[251] H. J. Reich, *J. Org. Chem.*, 1975, **40**, 2570.
[252] S. Tamagaki and K. Sakaki, *Chem. Letters*, 1975, 503.
[253] K.-T. H. Wei, I. C. Paul, M.-M. Y. Chang, and J. I. Musher, *J. Amer. Chem. Soc.*, 1974, **96**, 4099.
[254] S. Tamagaki, I. Hatanaka, and K. Tamura, *Chem. Letters*, 1976, 81.

$$\text{(ring)}Se{=}X + CH_2(CN)_2 \xrightarrow{82-84\%} \text{(ring)}\overset{+}{Se}{-}\bar{C}(CN)_2$$

(83) X = O or NTs

$$Ph_2SeCl_2 + 2NaCH(CMe)_2 \parallel O \longrightarrow Ph_2\overset{+}{Se}\bar{C}(CMe)_2 \parallel O$$

(84)

$$(PhCH_2)_2\overset{+}{Se}\bar{C}R_2 \xrightarrow[25\ °C]{Ph_3P} (PhCH_2)_2CR_2 + Ph_3PSe + (PhCH_2Se{-})_2$$

(85)

with triphenylphosphine. Syntheses and reactions of stable selenonium and telluronium ylides of type (86) have been published.[255]

(86) X = Se or Te

PART II: Compounds with S=N Functional Groups *by M. Haake*

1 Di-co-ordinate Sulphur

Sulphinylamines and Sulphinylamides.—The $-N{=}S{=}O$ functional group can be generated from silylated amines and amides by reaction with thionyl chloride, as in the preparation of (1),[1] (3),[2] (5),[3] or (6).[4] The potassium salt of tricyano-

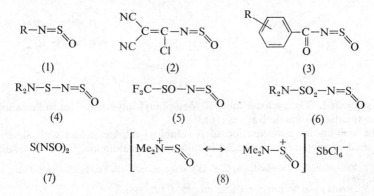

(1)   (2)   (3)

(4)   (5)   (6)

(7)   (8)

[255] N. N. Magdesiera, R. A. Kyandzhetsian, and O. A. Rakitin, *Zhur. org. Khim.*, 1975, **11**, 2562; 1976, **12**, 36; I. D. Sadekov, A. I. Usachev, A. A. Maksimenko, and V. I. Minkin, *Zhur. obshchei Khim.*, 1975, **45**, 2563.
[1] E. M. Dorokhova, E. S. Levchenko, and N. P. Pel'kis, *Zhur. org. Khim.*, 1975, **11**, 762.
[2] H. W. Roesky and G. Holtschneider, *J. Fluorine Chem.*, 1976, **7**, 77.
[3] R. Appel and M. Montenarh, *Chem. Ber.*, 1975, **108**, 2340.
[4] D. A. Armitage and A. W. Sinden, *J. Inorg. Nuclear Chem.*, 1974, **36**, 993.

methane was similarly converted into (2).[5] Thionyl chloride reacts with *NN'*-tetra(silyl)-sulphamide to give the reduced bifunctional derivative (7),[4] which is converted by silyl-dialkylamines into *N*-sulphinyl compounds of type (4).[6] These were also accessible from reactions of aminosulphinyl chlorides with silyl-sulphinylamine.[1] Considerable double-bond character was suggested for the sulphur–nitrogen bond of salt (8), which was obtained in high yields from reactions of $SbCl_5$ with dimethylaminosulphinyl chloride or tetramethyl-sulphurous diamide.[7]

The first examples of *N*-thiosulphinylanilines (12) were synthesized from (9) and from (10).[8, 9] They undergo cycloaddition reactions, for example with

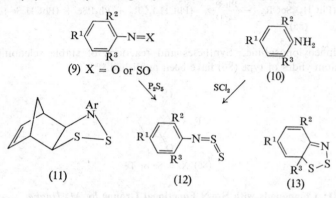

(9) X = O or SO

(10)

(11)

(12)

(13)

norbornadiene to give (11).[8] Compound (12; R^1, R^2, R^3 = Bu^t) was shown to be in equilibrium with (13).[9]

Diels–Alder reactions to give thiazine oxides (14) and (15) were reported for *N*-sulphinyl-benzamides and *N*-sulphinyl-sulphonamides.[10, 11] As the four diastereomers of (15) were formed, a non-concerted (two-step) dipolar mechanism

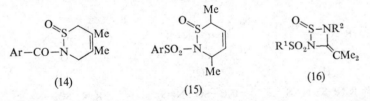

(14)

(15)

(16)

was suggested. The cycloaddition of *N*-sulphinyl-sulphonamides to ketenimines led to thiadiazetidine derivatives (16).[12]

The well-known condensation of *N*-sulphinyl-sulphonamides with aldehydes was used to prepare *N*-tosyl-2,3,3-trichloropropionaldimine,[13] and the addition

[5] N. S. Zefirov, N. K. Chapovskaya, L. Y. D'yachkova, and S. S. Trach, *Zhur. org. Khim.*, 1975, **11**, 1981.
[6] H. W. Roesky and W. Schaper, *Chem. Ber.*, 1974, **107**, 3451.
[7] W. Warthmann and A. Schmidt, *Z. anorg. Chem.*, 1975, **418**, 61.
[8] D. H. R. Barton and M. J. Robson, *J.C.S. Perkin I*, 1974, 1245.
[9] Y. Inagaki, R. Okazaki, and N. Inamoto, *Tetrahedron Letters*, 1975, 4575.
[10] E. M. Dorokhova, E. S. Levchenko, and T. Y. Lavrenyuk, *Zhur. org. Khim.*, 1974, **10**, 1865.
[11] W. L. Mock and R. M. Nugent, *J. Amer. Chem. Soc.*, 1975, **97**, 6521.
[12] T. Minami, F. Takimoto, and T. Agawa, *Bull. Chem. Soc. Japan*, 1975, **48**, 3259.
[13] H. Zinner, W. E. Siems, and G. Erfurt, *J. prakt. Chem.*, 1974, **316**, 698.

of n-butyl-lithium to *N*-sulphinyl-hydrazones produced diazoalkanes.[14] Alcoholysis studies of the *N*-sulphinyl-*p*-nitro- and *N*-sulphinyl-*p*-chloro-aniline –copper(II) chloride–ethanol system implied that the complex RNSO:CuCl$_2$ is not involved in the reaction mechanism.[15] In an analogous manner to the Strecker degradation, α-amino-acids were converted into carbonyl derivatives by treatment with *N*-sulphinylaniline.[16] In contrast to earlier theoretical calculations, it was concluded, on the basis of electric dipole moments and the i.r. and n.m.r. spectra of a series of substituted *N*-sulphinyl-amines (1; R = alkyl or aryl), that these compounds adopt preferentially the *cis*-structure (*Z*-configuration).[17] Low-energy photolysis of *cis*-thionylimide (1; R = H) led to formation of *trans*-thionylimide, both of which were obtained in an argon matrix and characterized by their i.r. spectra.[18]

Sulphurdi-imides.—Perhaps the most interesting findings in this area were concerned with the synthesis, structure, and properties of the polymeric metal (SN)$_x$ (17).[19, 20] This material, although prepared as early as 1910, has now been obtained analytically pure by slowly growing crystals of S$_2$N$_2$ from the vapour of pyrolysed S$_4$N$_4$, followed by solid-state polymerization. The single-crystal *X*-ray study revealed that (SN)$_x$ consists of an almost planar chain of alternating sulphur and nitrogen atoms. The crystals showed conductivities as high as 2.5 × 10^3 (Ω cm)$^{-1}$ at room temperature, and superconductivity at 0.26 K.

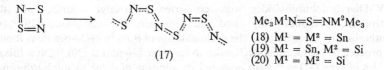

$$Me_3M^1N=S=NM^2Me_3$$

(18) M^1 = M^2 = Sn
(19) M^1 = Sn, M^2 = Si
(20) M^1 = M^2 = Si

S$_4$N$_4$ was further shown to be a versatile starting material, *e.g.* for the preparation of the sulphurdi-imides (18) and (19).[21] Reactions of *N*-silyl-imines with S$_4$N$_4$ [22] or S$_3$N$_3$Cl$_3$ [23] led to the known di-imide (20). The *N*-functional di-imides (18) and (20) were used for the synthesis of numerous novel heterocycles containing the —N=S=N— functional group, *e.g.* (21),[24, 25] (22),[26] (23),[27] (24),[28] (25),[29–31]

[14] J. G. Shelnut, S. Mataka, and J. P. Anselme, *J.C.S. Chem. Comm.*, 1975, 114.
[15] N. C. Collins and W. K. Glass, *J.C.S. Perkin II*, 1974, 713.
[16] T. Taguchi, S. Morita, and Y. Kawazoe, *Chem. and Pharm. Bull.* (*Japan*), 1975, **23**, 2654.
[17] H. F. van Woerden and S. H. Bijl-Vlieger, *Rec. Trav. chim.*, 1974, **93**, 85.
[18] P. O. Tchir and R. D. Spratley, *Canad. J. Chem.*, 1975, **53**, 2311, 2331.
[19] A. G. MacDiarmid, C. M. Mikulski, P. J. Russo, M. S. Saran, A. F. Garito, and A. J. Heeger, *J.C.S. Chem. Comm.*, 1975, 476.
[20] C. M. Mikulski, P. J. Russo, M. S. Saran, A. G. MacDiarmid, A. F. Garito, and A. J. Heeger, *J. Amer. Chem. Soc.*, 1975, **97**, 6358.
[21] H. W. Roesky and H. Wiezer, *Chem. Ber.*, 1974, **107**, 3186.
[22] I. Ruppert, V. Bastian, and R. Appel, *Chem. Ber.*, 1974, **107**, 3426.
[23] A. Golloch and M. Kuss, *Z. Naturforsch.*, 1974, **29b**, 320.
[24] H. W. Roesky and B. Kuhtz, *Chem. Ber.*, 1974, **107**, 1.
[25] R. Appel, H. Uhlenhaut, and M. Montenarh, *Z. Naturforsch.*, 1974, **29b**, 799.
[26] R. Appel, I. Ruppert, R. Milker, and V. Bastian, *Chem. Ber.*, 1974, **107**, 380.
[27] H. W. Roesky and B. Kuhtz, *Chem. Ber.*, 1975, **108**, 2536.
[28] H. W. Roesky and H. Wiezer, *Angew. Chem.*, 1974, **86**, 130.
[29] R. Neidlein and P. Leinberger, *Chem. Ztg.*, 1975, **99**, 433.
[30] H. W. Roesky and H. Wiezer, *Angew. Chem.*, 1975, **87**, 254.
[31] H. W. Roesky and E. Wehner, *Angew. Chem.*, 1975, **87**, 521.

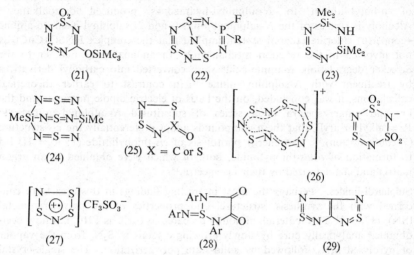

(21)   (22)   (23)

(24)   (25) X = C or S   (26)

(27)   (28)   (29)

or (26).[32] The radical cation (27),[33] as well as novel sulphonylisocyanate cyclo-adducts,[34] were obtained directly from S_4N_4. The reaction of *N*-trimethylsilyl-*N'*-t-butylsulphurdi-imide [35, 36] in MeOH led to the bicyclic anion $S_4N_5^-$.[36] *NN'*-Diarylsulphurdi-imides were converted by oxalyl chloride into thia-diazolidine-3,4-diones (28),[37] and the heteroaromatic compound (29) [38] was synthesized by three routes. The cycloaddition of S_4N_4 to cyclo-octene is thought to proceed by a concerted mechanism to form the bis-adduct (30).[39] New thiazine imides of type (31) [11, 40] were prepared by addition of diene to sulphurdi-imides, and the [2 + 2]-cycloaddition with metathiophosphoramides afforded the heterocycles (32).[41]

NN'-Bis(tosyl)sulphurdi-imide (33) was shown to be a useful reagent for the dehydrogenation of hydrazo- to azo-compounds [42] as well as for the allylic

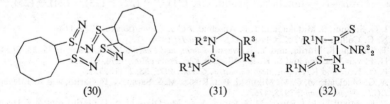

(30)   (31)   (32)

[32] H. W. Roesky, W. G. Bowing, I. Rayment, and H. M. M. Shearer, *J.C.S. Chem. Comm.*, 1975, 735.
[33] H. W. Roesky and A. Hazza, *Angew. Chem.*, 1976, **88**, 226.
[34] R. Appel, M. Montenarh, and I. Ruppert, *Chem. Ber.*, 1975, **108**, 582.
[35] R. Appel and M. Montenarh, *Z. Naturforsch.*, 1975, **30b**, 847.
[36] O. J. Scherer and G. Wolmershäuser, *Angew. Chem.*, 1975, **87**, 485.
[37] R. Neidlein and P. Leinberger, *Angew. Chem.*, 1975, **87**, 811.
[38] A. P. Komin, R. W. Street, and M. Carmack, *J. Org. Chem.*, 1975, **40**, 2749.
[39] W. L. Mock and I. Mehrotra, *J.C.S. Chem. Comm.*, 1976, 123.
[40] E. S. Levchenko and E. I. Slyusarenko, *Zhur. org. Khim.*, 1975, **11**, 871.
[41] N. T. Kulbach and O. J. Scherer, *Tetrahedron Letters*, 1975, 2297.
[42] G. Kresze and N. Schönberger, *Annalen*, 1974, 847.

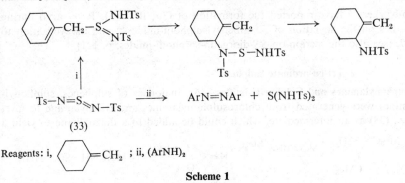

(33)

Reagents: i, 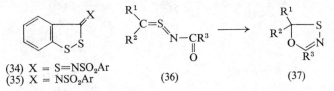 ; ii, (ArNH)$_2$

Scheme 1

amination of alkenes or alkynes [43] (Scheme 1). Cleavage reactions of (20) with chlorine and HCl afforded the highly explosive $HN=S=NCl$.[44] NN'-Dialkyl-sulphurdi-imides were prepared from SF_4 with primary amines,[45] and their metal complexing ability with a series of transition metals [45-49] as well as with organo-metallic reagents [50] was demonstrated.

The molecular structure of NN'-dimethylsulphurdi-imide was studied in the gas phase.[51] The preparation, properties, and structures of a selection of di-, tri-, and tetra-co-ordinate cyclic sulphur–nitrogen compounds have been discussed.[52]

Thione-S-imides.—This type of heterocumulene is characterized by the $>C=S=N-$ functional group, and has been discovered only very recently. The desulphurization of 1,2-dithiole-3-thione-S-imides (34) to (35) not only

(34) X = S=NSO$_2$Ar
(35) X = NSO$_2$Ar

(36)

(37)

occurs thermally,[53a] but also with phosphines,[53b] amines,[53c] acyl halides,[53d] thiol-catalysts,[53e] or halides.[53f] The generation and cycloaddition of thione-S-imides of type (36; R^1, R^2 = aryl) have been described,[54] and Haake and

[43] K. B. Sharpless and T. Hori, *J. Org. Chem.*, 1976, **41**, 176.
[44] W. Lidy, W. Sundermeyer, and W. Verbeck, *Z. anorg. Chem.*, 1974, **406**, 228.
[45] J. Kuyper and K. Vrieze, *J. Organometallic Chem.*, 1974, **74**, 289.
[46] J. D. Wilkins, *J. Inorg. Nuclear. Chem.*, 1976, **38**, 673.
[47] E. W. Lindsell and G. R. Faulds, *J.C.S. Dalton*, 1975, 40.
[48] R. Meij, J. Kuyper, D. J. Stufkens, and K. Vrieze, *J. Organometallic Chem.*, 1976, **110**, 219.
[49] U. Wannagat and M. Schlingmann, *Z. anorg. Chem.*, 1974, **406**, 312.
[50] J. Kuyper and K. Vrieze, *J.C.S. Chem. Comm.*, 1976, 64.
[51] J. Kuyper, P. H. Isselmann, F. C. Mijlhoff, A. Spelbos, and G. Renes, *J. Mol. Structure*, 1975, **29**, 247.
[52] H. W. Roesky, *Chem.-Ztg.*, 1974, **98**, 121.
[53] S. Tamagaki, K. Sakaki, and S. Oae, (a) *Heterocycles*, 1974, **2**, 39; (b) *ibid.*, p. 631; (c) *Bull. Chem. Soc. Japan*, 1974, **47** 3084; (d) *ibid.*, **48**, 1975, 2983; (e) *ibid.* p. 2985; (f) *ibid.*, p. 2987.
[54] E. M. Burgess and H. R. Penton, jun., *J. Amer. Chem. Soc.*, 1973, **95**, 279; *J. Org. Chem.*, 1974, **39**, 2885.

co-workers[55] have reported the formation of (36; R^1 = H, R^2 = aryl) during studies on the acylation of *SS*-dibenzylsulphodi-imides[55] and its cyclization to (37) (see also the section on *SS*-diorgano-sulphodi-imides, p. 121).

2 Tri-co-ordinate Sulphur

Sulphonylamines and Sulphurtri-imides.—As analogues of sulphenes, sulphonyl-amines were generated from chlorosulphonylamines and triethylamine to give, *e.g.*, (38) as an intermediate, which could be added to a diazoalkane to yield a

$$Me_3CN\overset{SO_2}{\underset{CMe_3}{\diagdown}} \xleftarrow{Me_3CCHN_2} \overset{Me_3C}{\diagdown}N=S\overset{O}{\underset{O}{\diagup}} \xrightarrow{RNH_2} Me_3CNH-SO_2-NHR$$

$$(39) \qquad\qquad\qquad (38) \qquad\qquad\qquad (40)$$

stable thiaziridine dioxide (39),[56] or to a primary amine to give the sulphamide (40).[57] As nitrogen analogues of SO_3, stable sulphurtri-imides were synthesized from the sulphurdi-imide (20) with trifluoroacetic anhydride.[58]

***SS*-Diorgano-sulphimides and Azasulphonium Salts.**—(See also Chapter 1, p. 50.) For efficient *S*-amination of sulphides under very mild conditions *O*-mesitylene-sulphonylhydroxylamine (MSH) appears to be the reagent of choice. Thus various types of sulphides were converted into a series of novel *S*-amino-sulphonium mesitylenesulphonates (41) (Scheme 2), the deprotonation of which afforded the more or less stable 'free' sulphimides (44).[59, 60] Reactions of allyl

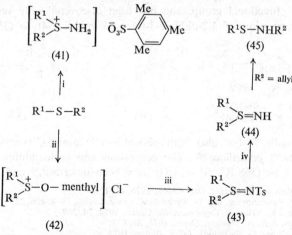

Reagents: i, MSH; ii, Me₃COCl–py–menthol; iii, NaNHTs; iv, H₂SO₄

Scheme 2

[55] M. Haake, B. Eichenauer, and K. H. Ahrens, *Z. Naturforsch.*, 1974, **29b**, 284.
[56] H. Quast and F. Kees, *Angew. Chem.*, 1974, **86**, 816.
[57] J. D. Catt and W. L. Matier, *J. Org. Chem.*, 1974, **39**, 566.
[58] R. Höfer and O. Glemser, *Z. Naturforsch.*, 1975, **30b**, 460.
[59] Y. Tamura, H. Matsushima, M. Ikeda, and K. Sumoto, *Synthesis*, 1974, 277.
[60] Y. Tamura, H. Matsushima, J. Minamikawa, M. Ikeda, and K. Sumoto, *Tetrahedron*, 1975, **31**, 3035.

sulphides with MSH led directly, in good yields, to salts of allylamines, presumably *via* a [2,3]-sigmatropic rearrangement of the un-isolable allyl sulphimides (44) to (45) followed by S—N bond cleavage.[60] In the case of 2-(dimethylamino)-ethyl phenyl sulphide the N-amino-ammonium salt was formed in 72% yield, indicating the limited scope for S-amination if other groups are present which are susceptible to attack by MSH. The addition of MSH to dibenzyl sulphide only afforded the thiolsulphonate ester $PhCH_2SO_2SCH_2Ph$, presumably *via* hydrolysis of the S-aminosulphonium salt to the S-oxide followed by disproportionation.

Resolution of the racemic mixture of S-methyl-S-p-tolyl-sulphimides produced by reaction of the sulphide with MSH gave a chiral 'free' sulphimide (120), *i.e.* (+)-(R)-(44; R^1 = Me, R^2 = p-tolyl), the configuration of which was assigned by conversion into the previously known N-tosylsulphimide enantiomer.[61] Another synthetic route for the preparation of optically active *ortho*-substituted diarylsulphimides directly from the sulphides involved the formation of the (−)-menthoxysulphonium salts (42) in solution and their conversion into the optically active (43), which on hydrolysis afforded the chiral 'free' sulphimides (44).[62]

In contrast to numerous novel reports on N-substituted sulphimides, few papers have appeared on reactions of 'free' sulphimides. These were concerned, for example, with the Michael-type addition of diphenylsulphimide $Ph_2S=NH$ to acrylonitrile and phenyl vinyl sulphone to give the N-(β-substituted)alkyl-sulphimides (46) and (47), respectively.[63] With carbonyl-activated alkenes,

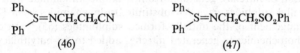

(46)   (47)

e.g. PhCOCH=CHCOPh, diphenylsulphimide was shown to undergo ylide-type reactions to give aziridines (49; R = H) together with amino-ethenes (50; R = H).[63] Similar reactions were achieved with novel N-alkylated diphenyl-sulphimides (48), which were synthesized from diphenylsulphimide with various alkyl bromides.[64]

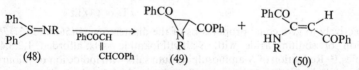

(48)   (49)   (50)

A variety of known and new methods was employed for the preparation of N-substituted sulphimides, using sulphides or sulphoxides as starting materials. From reactions involving activated heterocumulenes [65–69] (Scheme 3), interesting

[61] B. W. Christensen and A. Kjaer, *J.C.S. Chem. Comm.*, 1975, 784.
[62] S. Oae, T. Numata, and N. Furukawa, *Chem. and Ind.*, 1976, 163.
[63] N. Furukawa, S. Oae, and T. Yoshimura, *Synthesis*, 1976, 30.
[64] Y. Tamura, H. Matsushima, M. Ikeda, and K. Sumoto, *Tetrahedron*, 1976, 32, 431.
[65] B. A. Arbusov, N. N. Zobova, and O. V. Sofranova, *Izvest. Akad. Nauk S.S.S.R., Ser. khim.*, 1975, 1206.
[66] E. Behrend and A. Haas, *J. Fluorine Chem.*, 1974, 4, 83.
[67] K. Wagner, K. Ley, and L. Oehlmann, *Chem. Ber.*, 1974, 107, 414.
[68] T. Minami, Y. Tsumori, K. Yoshida, and T. Agawa, *J. Org. Chem.*, 1974, 39, 3412.
[69] G. Kresze and N. Schönberger, *Annalen*, 1975, 1721.

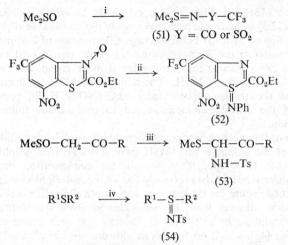

(51) Y = CO or SO$_2$

(52)

MeSO—CH$_2$—CO—R $\xrightarrow{iii}$ MeS—CH—CO—R
                                                  |
                                               NH—Ts

(53)

R^{1}SR^2 $\xrightarrow{iv}$ R^1—S—R^2
                                  ‖
                                 NTs

(54)

Reagents: i, CF$_3$YNCO; ii, PhNCO–PhNSO; iii, TsNSO; iv, TsN=S=NTs

Scheme 3

findings were the conversion of sulphides into (52)[62] and (54),[69] and the instability of S-(β-carbonyl)alkyl-S-methylsulphimides, which rapidly rearranged to (53).[68]

The addition of nitrenes, generated from N-aminophthalimide with lead tetra-acetate, to allyl sulphides gave N-(substituted)sulphenamides by a [2,3]-sigma-tropic rearrangement of the initially formed sulphimides (55).[70] In contrast to carbenes, photochemically generated nitrenes added to 4-t-butylthiacyclohexane

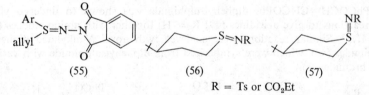

(55)   (56)   (57)

R = Ts or CO$_2$Et

unselectively to give equal proportions of the diastereomers (56) and (57).[71] The reaction of sodium azide with S-alkylthioamide salts afforded 1,2,3,4-thia-triazoles.[72] Reduction of S-aminosulphonium salts with iodide ion or thiourea led to the sulphides.[73] The Chloramine-T method, which has been shown to involve ion-pair intermediates,[74] was used for the conversion of various sulphides into S-functionalized sulphimides such as (58),[75] (59),[76] (60),[77] (61),[78] (62),[78] or (63).[79]

[70] R. S. Atkinson and S. B. Awad, *J.C.S. Chem. Comm.*, 1975, 651.
[71] D. C. Appleton, D. C. Bull, J. McKenna, and A. R. Walley, *J.C.S. Chem. Comm.*, 1974, 140.
[72] S. I. Mathew and F. Stansfield, *J.C.S. Perkin I*, 1974, 540.
[73] J. H. Krueger and A. J. Kiyokane, *Inorg. Chem.*, 1974, **13**, 2522.
[74] F. Ruff and A. Kucsman, *J.C.S. Perkin II*, 1975, 509.
[75] L. N. Markowskii, V. V. Vasilev, A. D. Sinitsa, and Y. A. Nuzhina, *Zhur. org. Khim.*, 1975, **11**, 2567.
[76] S. Colonna and C. J. M. Stirling, *J.C.S. Perkin I*, 1974, 2120.
[77] T. Yamamoto and M. Okawara, *Chem. Letters*, 1975, 581.
[78] R. B. Greenwald, D. H. Evans, and J. R. DeMember, *Tetrahedron Letters*, 1975, 3885.
[79] Y. Ueno and M. Okawara, *Bull. Chem. Soc. Japan*, 1974, **47**, 1033.

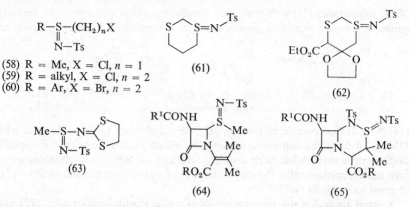

(58) R = Me, X = Cl, n = 1
(59) R = alkyl, X = Cl, n = 2
(60) R = Ar, X = Br, n = 2

(61)

(62)

(63)

(64)

(65)

As indicated by structures (64)[80] and (65),[81] the method was also employed for the *S*-imidation of β-lactam sulphides, including transformations of penicillins, which reacted with two mole equivalents of Chloramine-T.

N-Functionalized sulphimides and azasulphonium salts, *e.g.* (66),[82] (67),[83] (68),[84] and (69),[85] were obtained from sulphides with *N*-chloro-carbamates,

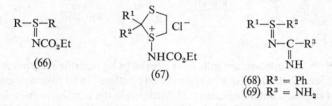

R—S—R
‖
NCO$_2$Et

(66)

(67)

R^1—S—R^2
‖
N—C—R^3
‖
NH

(68) R^3 = Ph
(69) R^3 = NH$_2$

N-chloro-amidines, or *N*-chloro-guanidines, and shown to be useful intermediates for further syntheses. The reaction sequence (70) → (71) → (72) was employed for the synthesis of novel cyclic sulphimides, which could be oxidized further with permanganate to give sulphoximides.[86]

N-Arylsulphimides were accessible from sulphides and anilines, using t-butyl hypochlorite or sulphuryl chloride as oxidants,[87] but more simply by treatment

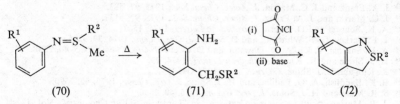

(70)

(71)

(72)

[80] M. M. Campbell and G. Johnson, *J.C.S. Perkin I*, 1975, 1077.
[81] M. M. Campbell, G. Johnson, A. F. Cameron, and I. R. Cameron, *J.C.S. Chem. Comm.*, 1974, 868; *J.C.S. Perkin I*, 1975, 1208.
[82] G. F. Whitfield, H. S. Beilan, D. Saika, and D. Swern, *J. Org. Chem.*, 1974, **39**, 2148.
[83] H. Yoshino, Y. Kawazoe, and T. Taguchi, *Synthesis*, 1974, 713.
[84] T. Fuchigami and K. Odo, *Chem. Letters*, 1974, 247.
[85] A. Heesing and G. Imsieke, *Chem. Ber.*, 1974, **107**, 1536.
[86] P. K. Claus, P. Hofbauer, and W. Rieder, *Tetrahedron Letters*, 1974, 3319.
[87] P. K. Claus, W. Rieder, P. Hofbauer, and E. Vilsmaier, *Tetrahedron*, 1975, **31**, 505.

of the sulphuranes (73) with anilines.[88] Similar reactions with ammonia, primary amines, or amides provide a general route to *N*-substituted *SS*-diarylsulphimides

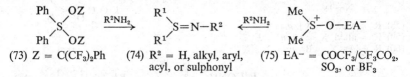

(73) Z = C(CF$_3$)$_2$Ph   (74) R^2 = H, alkyl, aryl,   (75) EA$^-$ = COCF$_3$/CF$_3$CO$_2$,
                                  acyl, or sulphonyl   SO$_3$, or BF$_3$

(74).[88, 89] Trifluoroacetic anhydride, SO$_3$, BF$_3$, and other Lewis acids react with DMSO to form the sulphonium ions (75), which readily undergo nucleophilic displacement on sulphur with amides, aromatic amines, or sulphonamides to give azasulphonium salts, the deprotonation of which gives the sulphimides (74) in good to excellent yields.[90-92]

A novel route for the preparation of tricyclic azasulphonium salts (77) and sulphimides (78) was found by trapping the sulphide radical cations (76) with

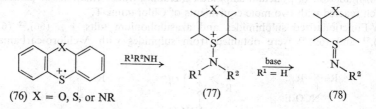

(76) X = O, S, or NR   (77)   (78)

amines.[93-96] Azasulphonium salts have proved to be versatile reagents in organic synthesis, and their chemistry has been briefly outlined in an overview on various classes of sulphur-containing cations.[97] Their use in the selective *ortho*-substitution of amines and phenols[98, 101] and in the preparation of indoles[99, 101] and oxindoles[100, 101] has been discussed elsewhere in this volume (Chapter 1, p. 27) and earlier volumes (see Vol. 3, p. 26). A series of papers has appeared on the synthetic uses of azasulphonium salts of type (79) formed as intermediates from sulphides and *N*-chlorosuccinimide. The reactions studied include Stevens-type rearrangements,[102] nucleophilic displacement on sulphur by enamines,[103]

[88] J. A. Franz and J. C. Martin, *J. Amer. Chem. Soc.*, 1975, **97**, 583.
[89] J. C. Martin and J. A. Franz, *J. Amer. Chem. Soc.*, 1975, **97**, 6137.
[90] A. K. Sharma and D. Swern, *Tetrahedron Letters*, 1974, 1503.
[91] T. E. Varkey, G. F. Whitfield, and D. Swern, *J. Org. Chem.*, 1974, **39**, 3365.
[92] A. K. Sharma, T. Ku, A. D. Dawson, and D. Swern, *J. Org. Chem.*, 1975, **40**, 2758.
[93] H. J. Shine and K. Kim, *Tetrahedron Letters*, 1974, 99.
[94] K. Kim and H. J. Shine, *J. Org. Chem.*, 1974, **39**, 2537.
[95] B. K. Bandlish, A. G. Padilla, and H. J. Shine, *J. Org. Chem.*, 1975, **40**, 2590.
[96] S. R. Mani and H. J. Shine, *J. Org. Chem.*, 1975, **40**, 2756.
[97] J. P. Marino, 'Sulfur-containing cations', in 'Topics in Sulfur Chemistry', Vol. I, ed. A. Senning, Thieme, Stuttgart, 1976.
[98] P. G. Gassmann and G. D. Gruetzmacher, *J. Amer. Chem. Soc.*, 1974, **96**, 5487.
[99] P. G. Gassmann, T. J. van Bergen, D. P. Gilbert, and B. W. Cue, jun., *J. Amer. Chem. Soc.*, 1974, **96**, 5495.
[100] P. G. Gassmann and T. J. van Bergen, *J. Amer. Chem. Soc.*, 1974, **96**, 5508.
[101] P. G. Gassmann, G. D. Gruetzmacher, and T. J. van Bergen, *J. Amer. Chem. Soc.*, 1974, **96**, 5512.
[102] E. Vilsmaier, K. H. Dittrich, and W. Sprügel, *Tetrahedron Letters*, 1974, 3601.
[103] E. Vilsmaier, W. Sprügel, and K. Gagel, *Tetrahedron Letters*, 1974, 2475.

amines,[87] indoles,[104] catechols,[105] phenols,[106] or aldehyde-alcohols.[107] Also, novel mono-, di-, and tri-azasulphonium chlorides such as (80),[108] (81),[109] and (82)[110] were prepared, and shown to be useful synthetic intermediates.

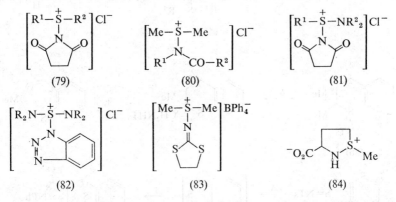

(79)   (80)   (81)

(82)   (83)   (84)

Methylation of the thio-oxime gave the salt (83),[79] and the structure of dehydro-methionine was confirmed by X-ray analysis to be azasulphonium betaine (84).[111] The synthetic utility of N-substituted sulphimides has been the subject of increasing attention. Photolysis of (85)[112] and thermolysis of (85),[112] (86),[113] and (87)[59] afforded various types of heterocycles, as indicated in Scheme 4. Reactions of sulphimides (88)[114, 115] and (89)[116, 117] with 1,3-dipoles or activated multiple bonds provided novel routes to the indicated heterocycles.

N-Tosylsulphimides (90) are reduced to sulphides (92) in DMF in high yields by the action of phosphines or cyanide ion[118] (Scheme 5). This reaction proceeds by initial attack of the nucleophile on tervalent sulphur (S_N2-S mechanism). The intermediate 1,3-dipoles (91) were shown to be powerful dehydrating reagents which converted carbonic acids into anhydrides, or, in the presence of alcohols and amines, into esters and amides, respectively.[119] The use of alcohols alone led to alkyl-exchanged sulphides (93). Thiophenolate ion and (90) gave quantitatively the sulphide arising by a S_N2-type reaction on the carbon atom

[104] K. Tomita and A. Terada, *Heterocycles*, 1975, **3**, 1110.
[105] J. P. Marino and A. Schwartz, *J.C.S. Chem. Comm.*, 1974, 812.
[106] P. G. Gassmann and D. R. Amick, *Tetrahedron Letters*, 1974, 889; *Synth. Comm.*, 1975, **5**, 325.
[107] J. C. Depezay and Y. Le Merrer, *Tetrahedron Letters*, 1975, 2755.
[108] E. Vilsmaier and R. Bayer, *Synthesis*, 1976, 46.
[109] M. Haake and H. Benack, *Synthesis*, 1976, 306, 310.
[110] H. Minato, K. Okuma, and M. Kobayashi, *J.C.S. Chem. Comm.*, 1975, 868.
[111] R. S. Glass and J. R. Duchek, *J. Amer. Chem. Soc.*, 1976, **98**, 965.
[112] T. L. Gilchrist, C. J. Moody, and C. W. Rees, *J.C.S. Perkin I*, 1975, 1964.
[113] M. Kise, M. Murase, T. Tomita, M. Kitano, and H. Murai, *Heterocycles*, 1975, **3**, 1120; *Tetrahedron Letters*, 1976, 691.
[114] T. L. Gilchrist, C. J. Harris, and C. W. Rees, *J.C.S. Chem. Comm.*, 1974, 485, 486, 487.
[115] T. L. Gilchrist, C. J. Harris, C. J. Moody, and C. W. Rees, *J.C.S. Perkin I*, 1975, 1969.
[116] Y. Hayashi, Y. Iwagami, A. Kadoi, T. Shono, and D. Swern, *Tetrahedron Letters*, 1974, 1071.
[117] P. Barraclough, E. M. Michael, T. L. Gilchrist, and C. J. Harris, *J.C.S. Perkin I*, 1976, 716.
[118] T. Aida, M. Nakajima, T. Inoue, N. Furukawa, and S. Oae, *Bull. Chem. Soc. Japan*, 1975, **48**, 723.
[119] T. Aida, N. Furukawa, and S. Oae, *Chem. Letters*, 1974, 121; 1975, 29.

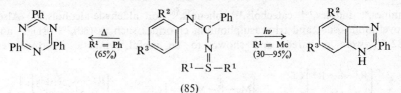

(85)

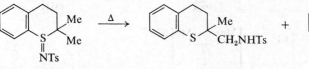

(86)

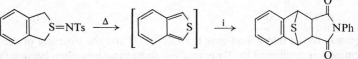

(87)

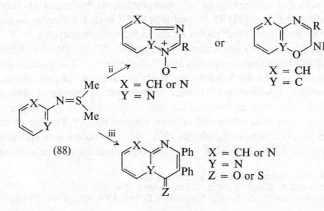

(88)

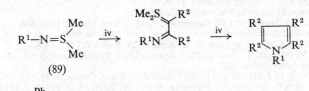

(89)

Reagents: i, 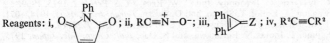 ; ii, RC≡N⁺–O⁻; iii, Ph/Ph C=Z ; iv, R²C≡CR²

Scheme 4

adjacent to the tervalent sulphur (S_N2-C mechanism).[120] Similarly, the halide-ion-catalysed rearrangement of (90) to the sulphenamide (94) was stated to proceed through initial formation of an alkyl halide followed by alkylation of the sulphenamide anion (Scheme 5).[121] The reactions of dialkyl- and alkyl-aryl-

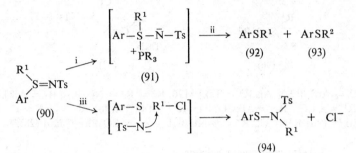

Reagents: i, PR$_3$; ii, R^{2}OH; iii, Cl$^-$

Scheme 5

N-tosylsulphimides with hydroxide or methoxide in methanol afforded *S*-substitution and Pummerer-type products.[122]

Free-radical cleavage of the S=N linkage of (90) with tributyltin hydrides,[123] and the reduction with formamidinesulphinic acid under phase-transfer conditions,[124] led to the sulphides. The reaction products of *SS*-diaryl-*N*-tosyl-sulphimides with aryl-lithium reagents were proposed to arise from the initially formed sulphurane.[125] With sodium hydride, *N*-tosylsulphimides (95) were converted into carbanions (Scheme 6), which were shown to be alkylidene-transfer reagents comparable to those generated from the corresponding sulphoximides.[126] The complexing ability of *N*-acylsulphimides was demonstrated

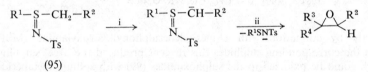

Reagents: i, NaH; ii, R^{3}R^{4}CO

Scheme 6

with Pt^{II} and Pd^{II} salts.[127, 128] Novel i.r.- and u.v.-spectroscopic studies on *N*-tosylsulphimides have been reported.[129]

[120] S. Oae, T. Aida, M. Nakajima, and N. Furukawa, *Tetrahedron*, 1974, **30**, 947.
[121] S. Oae, T. Aida, and N. Furukawa, *J.C.S. Perkin II*, 1974, 1231.
[122] N. Furukawa, T. Matsuda, M. Yakushiji, and S. Oae, *Bull. Chem. Soc. Japan*, 1974, **47**, 2247.
[123] S. Kozuka, S. Furumai, and T. Akasaka, *Chem. and Ind.*, 1974, 496.
[124] G. Borgogno, S. Colonna, and R. Fornasier, *Synthesis*, 1975, 529.
[125] B. K. Ackerman, K. K. Andersen, I. K. Nielsen, N. B. Peynircioglu, and S. A. Yeager, *J. Org. Chem.*, 1974, **39**, 964.
[126] Y. Tamura, H. Matsushima, M. Ikeda, and K. Sumoto, *Synthesis*, 1976, 35.
[127] G. Matsubayashi, M. Toriuchi, and T. Tanaka, *Bull. Chem. Soc. Japan*, 1974, **47**, 765.
[128] M. Toriuchi, G. Matsubayashi, E. Koezuka, and T. Tanaka, *Inorg. Chim. Acta*, 1976, **17**, 253.
[129] J. J. Shah, *Canad. J. Chem.*, 1957, **53**, 2381.

Out of numerous papers which appeared on the preparation and reactions of various *S*-heterosubstituted sulphimides, some interesting examples are (96; $R^1 = F$, $R^2 = Cl$, $R^3 = COF$),[130] (97),[131] (96; $R^1 = R^2 = Cl$, $R^3 = COAlk$),[132]

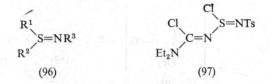

(96)   (97)

(96; $R^1 = ArS$, $R^2 = Ar$, $R^3 = Ts$),[133] [96; $R^1 = R^2 = N(Alk)_2$, $R^3 = Alk$],[134, 135] (96; $R^1 = Cl_3C$, $R^2 = AlkCONH$, $R^3 = Ts$),[136] (96; $R^1 = R^2 = F$, $R^3 = CH_2Ph$),[137] (96; $R^1 = Cl$, $R^2 = p\text{-}NO_2C_6H_4$, $R^3 = SO_2Ph$),[138] and (28).[37]

3 Tetra-co-ordinate Sulphur

SS-Diorgano-sulphoximides.—(See also Chapter 1, p. 50.) This type of compound has recently been reviewed.[139] Interest in their pharmacological activity has led to the preparation of some novel cyclic sulphoximides. Thus 2-(alkyl-sulphinyl)benzamides were cyclized with hydrazoic acid to the benzisothiazole oxides (98).[140] Tricyclic sulphoxides were converted into the free sulphoximides (99) either by the tosyl azide–copper method with subsequent hydrolysis, or

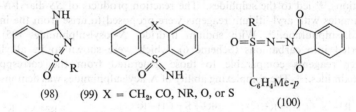

(98)   (99) X = CH₂, CO, NR, O, or S
                                              (100)

directly by amination with *O*-mesitylenesulphonylhydroxylamine (MSH).[141] With the corresponding sulphides this reagent produced the 'free' sulphimides, which could be oxidized to the sulphoximides (99) with sodium metaperiodate. This route had not been applied previously for the preparation of 'free' sulphoximides. Optically active sulphoxides were converted into optically

130 R. Mews, *J. Fluorine Chem.*, 1974, **4**, 445.
131 L. N. Markovskii, Y. G. Shermolovich, Y. A. Nuzhdina, and V. I. Shevchenko, *Zhur. org. Khim.*, 1974, **10**, 1000.
132 L. N. Markovskii, G. S. Fedyuk, and Y. A. Nuzhdina, *Zhur. org. Khim.*, 1974, **10**, 992.
133 E. S. Levchenko and L. V. Budnik, *Zhur. org. Khim.*, 1975, **11**, 2293.
134 L. N. Markovskii, G. S. Fedyuk, and Y. G. Balon, *Zhur. org. Khim.*, 1974, **10**, 1434.
135 L. N. Markovskii, V. E. Pashinnik, and N. A. Kirsanova, *Zhur. org. Khim.*, 1975, **11**, 74.
136 M. M. Kremlev, A. I. Tarasenko, I. V. Koval, and V. V. Ryabenko, *Zhur. org. Khim.*, 1974, **10**, 2320.
137 J. R. Grunwell and S. L. Dye, *Tetrahedron Letters*, 1975, 1739.
138 S. S. Trach and N. S. Zefirov, *Zhur. org. Khim.*, 1974, **10**, 219.
139 P. D. Kennewell and J. B. Taylor, *Chem. Soc. Rev.*, 1975, **4**, 189.
140 P. Stoss and G. Satzinger, *Chem. Ber.*, 1975, **108**, 3855.
141 P. Stoss and G. Satzinger, *Tetrahedron Letters*, 1974, 1973.

active 'free' sulphoximides *via* the tosyl azide or Chloramine-T method in the presence of copper catalyst, followed by acid hydrolysis of the resulting N-tosyl-sulphoximides.[142] The amination of optically active sulphoxides by means of the MSH reagent was recommended for the direct synthesis of 'free' sulphoximides of high optical purity. It occurs with retention of configuration at the sulphur atom.[143] This has also been observed for the conversion of optically active sulphoxides into N-phthalimidosulphoximides (100) by treatment with N-aminophthalimide and lead tetra-acetate.[76] The sulphoximides (101) were prepared by this method from 3-amino-2-oxazolidones, and underwent efficient and stereospecific thermal fragmentation (Scheme 7) *via* the diazenes to the corresponding alkenes.[144, 145]

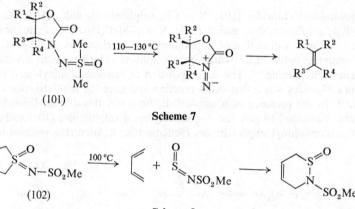

Scheme 7

Scheme 8

The N-mesylsulphoximide (102) rearranged on heating by a dissociation–recombination mechanism (Scheme 8) to a thiazine oxide.[146] Photolysis of N-arylsulphonyl sulphoximides gave only biphenyls in high yields; the formation of sulphonylnitrenes was not observed.[147] Optically pure 'free' sulphoximides, including those of amino-acids such as of L-methionine or S-methyl-L-cysteine, were stereospecifically converted by nitrous acid into the sulphoxides with complete retention of configuration at the sulphur atom.[142, 148] Some stereochemical aspects of tri- and tetra-co-ordinate sulphur compounds of natural derivation were discussed[149] with reference to the biologically active diastereoisomer of L-methionine sulphoximide having the (S)-configuration at sulphur.

The stereochemical course of substitution at sulphur in the 'sulphon' oxidation state, attached to four different ligands, was examined by Cram and Jones.[150]

142 M. Moriyama, T. Numata, and S. Oae, *Org. Prep. Proceed. Internat.*, 1974, **6**, 207.
143 C. R. Johnson, R. A. Kirchhoff, and H. G. Corkins, *J. Org. Chem.*, 1974, **39**, 2458.
144 J. D. White and M. Kim, *Tetrahedron Letters*, 1974, 3361.
145 M. Kim and J. D. White, *J. Amer. Chem. Soc.*, 1975, **97**, 451.
146 W. L. Mock and R. M. Nugent, *J. Amer. Chem. Soc.*, 1975, **97**, 6526.
147 R. A. Abramovitch and T. Takaya, *J.C.S. Perkin I*, 1975, 1806.
148 R. A. Stephani and A. Meister, *Tetrahedron Letters*, 1974, 2307.
149 A. Kjaer, *Tetrahedron*, 1974, **30**, 1551.
150 M. R. Jones and D. J. Cram, *J. Amer. Chem. Soc.*, 1974, **96**, 2183.

They described a number of new reactions, and classes of compound never prepared before in an optically active state (Scheme 9), including diastereomeric

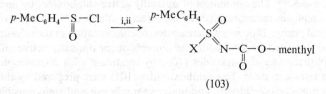

(103)

Reagents: i, p-NO$_2$C$_6$H$_4$SO$_2$ONHCO$_2$ menthyl; ii, nucleophile, X$^-$

Scheme 9

sulphonimidoyl chlorides (103; X = Cl), sulphonimidamides (104; X = NH$_2$ or NR$_2$), sulphonimidate esters (103; X = p-MeC$_6$H$_4$O), and sulphoximides (103; X = Me). Formation of sulphonimidoyl chlorides was also reported from sulphonamides with PCl$_5$ [151] and from sulphinyl chlorides with NN-dichloro-benzenesulphonamide.[152] The decomposition of benzenesulphinyl- and toluene-p-sulphinyl-azides was a first-order reaction and gave the trithiatriazine trioxide (105).[153] In the presence of a sulphoxide, however, the sulphinylnitrene inter-mediates combined to give the N-arenesulphonyl sulphimides (104) rather than the N-arenesulphinyl sulphoximides (Scheme 10). Related ring systems to (105)

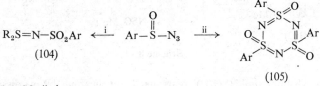

(104)   (105)

Reagents: i, R$_2$SO; ii, Δ

Scheme 10

with various kinds of substituents on sulphur were described by Glemser's group,[154–156] who also reported the formation of (106) from the reaction of sulphonyl isocyanate with bis(trifluoromethyl)diazomethane.[157] From per-fluorinated sulphimides, novel N-trifluoromethyl-S-fluoro-sulphoximides were prepared.[168, 169]

N-Aryl sulphoximides (107) could be obtained either by oxidation of N-aryl sulphimides with potassium permanganate, or directly from reactions of anilines with Me$_3$COCl–DMSO or SO$_2$Cl$_2$–DMSO.[87] The formation of N-perfluoroaryl sulphoximides by thermal decomposition of azides in DMSO [158] and the synthesis

[151] E. S. Levchenko and L. V. Budnik, *Zhur. org. Khim.*, 1975, **11**, 2044.
[152] L. N. Markovskii, Y. G. Shermolovich, V. I. Gorbatenko, and V. I. Shevchenko, *Zhur. org. Khim.*, 1975, **11**, 751.
[153] T. J. Maricich and V. L. Hoffmann, *J. Amer. Chem. Soc.*, 1974, **96**, 7770.
[154] H. Wagner, R. Mews, T. P. Lin, and O. Glemser, *Chem. Ber.*, 1974, **107**, 584.
[155] W. Heider, U. Klingebiel, T. P. Lin, and O. Glemser, *Chem. Ber.*, 1974, **107**, 592.
[156] D. C. Wagner, H. Wagner, and O. Glemser, *Chem. Ber.*, 1976, **109**, 1224.
[157] H. Steinbeisser, R. Mews, and O. Glemser, *Z. anorg. Chem.*, 1974, **406**, 299.
[158] R. E. Banks and A. Prakash, *J.C.S. Perkin I*, 1974, 1365.

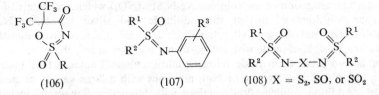

(106)   (107)   (108) X = S_2, SO, or SO_2

of novel bis(sulphoximido)-compounds (108) from condensation reactions of the 'free' sulphoximides, $R^1R^2S(:O)NH$, with sulphur chlorides and oxychlorides have been described.[159] The pyridine-catalysed α-hydrogen–deuterium exchange for *S*-alkyl-*S*-phenyl-*N*-tosylsulphoximides is much slower than that for the corresponding sulphimides, although the sulphoximide functional group is expected to be more acidifying. The findings were explained by greater resonance stabilization of the carbanion by a sulphimide group.[160]

SS-Diorgano-sulphodi-imides.—The first review of this relatively new class of compound has appeared.[161] It covers the literature since the detection of (109) in 1964 and many recent results obtained in the author's laboratory. The review

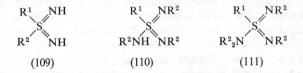

(109)   (110)   (111)

also briefly includes the chemistry of bis(alkylimide)sulphonamides (110) and (111). Experimental details are given for the preparation of (109) by the oxidation of sulphides with t-butyl hypochlorite in the presence of ammonia, which is more convenient than the previously used gaseous mixtures of chloramine–ammonia generated in special equipment. With Chloramine-T in liquid ammonia the 'free' (*R*)-sulphimide enantiomer (112) could be converted into an optically active sulphodi-imide (113), the (*S*)-configuration of which was established on

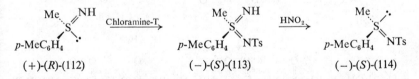

(+)-(*R*)-(112)   (−)-(*S*)-(113)   (−)-(*S*)-(114)

the basis of retention, upon conversion with nitrous acid into the sulphimide (114), of the defined chirality.[61] The reactions of sulphodi-imides (109) so far described have mainly been concerned with the substitution of the imide-hydrogens by metals, carbon, silicon, phosphorus, sulphur, and halogens. The metal-complexing ability of the sulphodi-imide functional group previously shown for the silylated derivatives [162] was further demonstrated by a crystal-

[159] M. Okahara, E. Yoshikawa, I. Ikeda, and S. Komori, *Synthesis*, 1975, 521.
[160] M. Kobayashi, A. Mori, and H. Minato, *Bull. Chem. Soc. Japan*, 1974, 47, 891.
[161] M. Haake, 'The chemistry of S,S-diorgano-sulfodi-imides', in 'Topics in Sulfur Chemistry', Vol. 1, ed. A. Senning, Thieme, Stuttgart, 1976.
[162] W. Wolfsberger and H. Försterling, *J. Organometallic Chem.*, 1973, 56, C17.

structure investigation of the complex Ag₃N₂SEt₂(NO₃) which precipitated from aqueous solutions of diethyl sulphodi-imide and silver nitrate.[163] Novel *NN'*-bis(silyl) and -bis(germyl) sulphodi-imides were prepared from dimethyl- and diethyl-sulphodi-imide with chloro-silanes and -germanes.[164]

In accord with their basic bifunctionality, sulphodi-imides (109) undergo substitution reactions on one or both nitrogens with a large variety of electrophiles. Addition–condensation reactions with bifunctional reagents, including reactive dichlorides, isocyanates, and multiple-bond systems, have afforded a series of novel heterocycles (115)—(123).[161] Although most of the reported

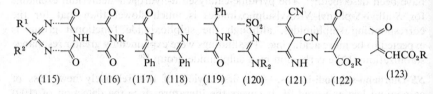

(115)   (116)   (117)   (118)   (119)   (120)   (121)   (122)   (123)

N-sulphonyl-, *N*-acyl-, and *NN'*-bis(acyl)-sulphodi-imides were found to be fairly stable compounds, a benzyl group attached to the sulphur atom was easily removed in nucleophilic protic solvents or in the presence of other nucleophiles to give sulphinamidines in high yields.[161] Similarly, sulphinamidines were readily formed from *SS*-dibenzyl sulphodi-imides with benzoyl chlorides in the presence of triethylamine. Rather surprisingly, formation of 1,3,4-oxathiazoles (37) was also observed, indicating the generation of unstable thione-*S*-imides (34).[55] Thermally induced cleavage of a benzyl group, and 1,2- and 1,4-shifts from sulphodi-imide sulphur to nitrogen and oxygen, respectively, were established by isolation of products (124)—(128) after boiling (125) in water or toluene.[161]

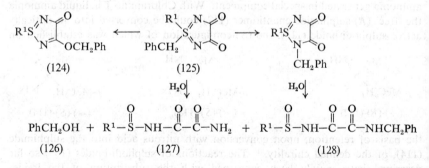

(124)   (125)

(126)   (127)   (128)

N-Alkylation of sulphodi-imides can be effected by trialkyloxonium salts, or by conversion into *N*-alkali-metal salts and reaction with alkyl halides.[161] Reactions of sulphur difluoride di-imides[165] afforded nitrogen analogues of sulphuric acid, *e.g.* (129).[166] Photoelectron spectra and molecular properties of

[163] D. Hass and G. Bergerhoff, *Acta Cryst.*, 1974, **B30**, 1361.
[164] W. Wolfsberger and H. Försterling, *Chem.-Ztg.*, 1976, **100**, 35.
[165] O. Glemser and R. Höfer, *Z. Naturforsch.*, 1974, **29b**, 121.
[166] R. Höfer and O. Glemser, *Z. Naturforsch.*, 1975, **30b**, 460.

sulphuric acid derivatives $X_2S(=Y)_2$, including sulphodi-imides (X = alkyl, Y = NH), were discussed on the basis of M.O. models and CNDO calculations.[167]

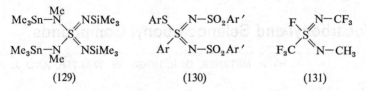

(129)   (130)   (131)

Further examples of *S*-heterosubstituted sulphodi-imides (130) and (131) were obtained by disproportionation of iminothiolsulphinates [168] and from reactions of *S*-perfluoro-sulphimides.[169]

4 Selenium Analogues

Selenium di-imides (132) were formed in methylene chloride solution from selenium tetrachloride with t-butylamine or sulphonamides and also from selenium metal with anhydrous Chloramine-T.[170] They were found to be efficient reagents for the allylic amination of alkenes and alkynes. The first examples of selenium imides (133) were synthesized from diaryl selenides and Chloramine-T,[171] from

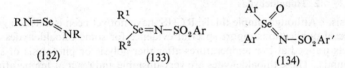

(132)   (133)   (134)

various selenides with Bu^tOCl–$NaHNSO_2C_6H_4Me$-*p*,[172] or by addition of iminoseleninyl chlorides (133; R^1 = aryl, R^2 = Cl) to styrene in the presence of Cu^ICl.[173] Condensation reactions of *N*-sulphinylarenesulphonamides with diaryl selenones afforded iminoselenones of type (134) in excellent yields,[171] the structures of which were confirmed by i.r. and electron spectroscopic studies.[174]

[167] B. Salouki, H. Bock, and R. Appel, *Chem. Ber.*, 1975, **108**, 897.
[168] S. L. Yu and J. M. Shreeve, *J. Fluorine Chem.*, 1976, **7**, 85.
[169] I. Stahl, R. Mews, and O. Glemser, *J. Fluorine Chem.*, 1976, **7**, 55.
[170] K. B. Sharpless, T. Hori, L. K. Truesdale, and C. O. Dietrich, *J. Amer. Chem. Soc.*, 1976, **98**, 269.
[171] N. Y. Derkach, T. V. Lyapina, and N. A. Pasmurtseva, *Zhur. org. Khim.*, 1974, **10**, 807.
[172] S. Tamagaki, S. Oae, and K. Sakaki, *Tetrahedron Letters*, 1975, 649.
[173] N. Y. Derkach, N. A. Pasmurtseva, T. V. Lyapina, and E. S. Levchenko, *Zhur. org. Khim.*, 1974, **10**, 1873.
[174] Y. A. Nuzhdina, Y. P. Egorov, L. V. Mironyuk, and N. Y. Derkach, *Teor. i eksp. Khim.*, 1975, **11**, 605.

3
Thiocarbonyl and Selenocarbonyl Compounds

BY P. METZNER, D. R. HOGG, W. WALTER, AND J. VOSS

PART I: **Thioaldehydes, Thioketones, Thioketens, and their Selenium Analogues**
by P. Metzner

1 Introduction

Reviews.—The chemistry of fluorinated thiocarbonyl compounds [1] and enamino-thioketones [2] has been reviewed. An important paper deals with the reaction of organomagnesium reagents with thiocarbonyl compounds.[3] Thione photochemistry has been covered in other reviews,[4, 5] and the synthesis of thioketens and oligomers has been surveyed.[6]

2 Thioaldehydes

Synthesis.—Although simple thials, RCHS, have not yet been isolated, a growing number of publications report spectral evidence for some thioaldehydes among products trapped at low temperatures after thermolysis or photolysis of sulphur compounds. These thioaldehydes are very unstable and react at temperatures far below ambient. Thus thioformaldehyde has been detected during the pyrolysis of thietan [7] and of trithian,[8] and during the photolysis of methanethiol [8] and of thietan.[9] Spectral evidence for the presence of thioformaldehyde in interstellar space has also been reported.[10] Thioacetaldehyde has been detected after thermolysis of trimethyltrithian.[11] de Mayo and his co-workers [12] successfully used flash thermolysis of allyl alkyl sulphides to obtain propene and two previously uncharacterized conjugated molecules, thioacrolein and thiobenzaldehyde.

The thioaldehyde (1), associated with penicillin synthesis, has been trapped chemically, but not isolated.[13] It undergoes polymerization or thioenolization,

[1] R. E. Banks, in 'Fluorocarbon and Related Chemistry', ed. R. E. Banks and M. G. Barlow (Specialist Periodical Reports), The Chemical Society, London, 1974, Vol. 2, p. 124.
[2] Y. F. Freimanis, 'Khimiya Enaminoketonov, Enaminoiminov i Enaminotionov', Zinatne, Riga,U.S.S.R., 1974.
[3] D. Paquer, *Bull. Soc. chim. France*, 1975, 1439.
[4] P. de Mayo, *Accounts Chem. Res.*, 1976, 52.
[5] J. D. Coyle, *Chem. Soc. Rev.*, 1975, 4, 523.
[6] R. Mayer and H. Kröber, *Z. Chem.*, 1975, 15, 91.
[7] R. H. Judge and G. W. King, *Canad. J. Phys.*, 1975, 53, 1927.
[8] M. E. Jacox and D. E. Milligan, *J. Mol. Spectroscopy*, 1975, 58, 142.
[9] D. R. Dice and R. P. Steer, *Canad. J. Chem.*, 1974, 52, 3518.
[10] L. H. Doherty, J. M. MacLeod, and T. Oka, *Astrophys. J.*, 1974, 192, L, 157.
[11] H. W. Kroto, B. M. Landsberg, R. J. Suffolk, and A. Vodden, *Chem. Phys. Letters*, 1974, 29, 265.
[12] H. G. Giles, R. A. Marty, and P. de Mayo, *J.C.S. Chem. Comm.*, 1974, 409.
[13] J. Cheney, C. J. Moores, J. A. Raleigh, A. I. Scott, and D. W. Young, *J.C.S. Chem. Comm.*, 1974, 47; *J.C.S. Perkin I*, 1974, 986.

favoured by the presence of a β-imino-group. Cycloaddition of 1,2-dithiole-3-thiones, not substituted in the 5-position, to acetylenic compounds [14] gave stable conjugated thioaldehydes (2). Preparation of a thioaldehyde (3; R^1 = H,

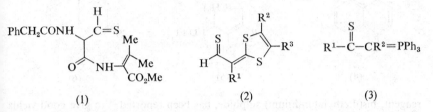

(1)   (2)   (3)

R^2 = alkyl) substituted with a phosphonium ylide has been reported.[15] Syntheses of thioaldehyde dianions [16, 17] and of iron complexes of thioacrolein and 2-ethyl-crotonthioaldehyde [18] have appeared.

Reactions.—Owing to the low thermal stability of thioaldehydes obtained by flash thermolysis, their reactivity has not yet been investigated. However, one paper reports that thioformaldehyde can be trapped with cyclopentadiene to give Diels–Alder cycloadducts.[9]

3 Thioketones and Selenoketones

Synthesis of Thioketones.—The classical synthesis of thioketones involving sulphurization of corresponding ketones is still being used. Treatment with hydrogen sulphide and hydrogen chloride yields alkylcycloalkyl thioketones,[19] bicyclic thioketones,[19] alkyl aryl thioketones,[20] polycyclic aromatic thioketones, *e.g.* (4), in very good yields,[21] fluorinated thio-β-diketones,[22] and 2-arylmono-thioindane-1,3-diones.[23] Applied to cyclic ketones bearing a β-ester group this method gives, according to Duus,[24] mixtures of thioketone and *cis*- and *trans*-conjugated enethiols and, according to Paquer,[25] mixtures of thioketone, conjugated enethiol, and non-conjugated enethiol (5).

Treatment of steroid dienones with tetraphosphorus decasulphide gave the previously unreported purple-blue dienethiones (6).[26] An analogous reaction is claimed to yield 2-arylindane-1,3-dithione.[23] Sulphurization with boron sulphide or silicon sulphide was applied to the synthesis of pyran-4-thione rings.[27] A new

[14] H. Davy and J. Vialle, *Bull. Soc. chim. France*, 1975, 1435.
[15] H. Yoshida, H. Matsuura, T. Ogata, and S. Inokawa, *Bull. Chem. Soc. Japan*, 1975, **48**, 2907.
[16] D. Seebach and K. H. Geiss, *Angew. Chem. Internat. Edn.*, 1974, **13**, 202.
[17] K. Geiss, B. Seuring, R. Pieter, and D. Seebach, *Angew. Chem. Internat. Edn.*, 1974, **13**, 479.
[18] A. I. M. Tsai, *Diss. Abs. (C)*, 1974, **35**, 743.
[19] C. Fournier, D. Paquer, and M. Vazeux, *Bull. Soc. chim. France*, 1975, 2753.
[20] P. de Mayo and R. Suau, *J.C.S. Perkin I*, 1974, 2559.
[21] A. Cox, D. R. Kemp, R. Lapouyade, P. de Mayo, J. Joussot-Dubien, and R. Bonneau, *Canad. J. Chem.*, 1975, **53**, 2386.
[22] C. S. Saba and T. R. Sweet, *Analyt. Chim. Acta*, 1974, **69**, 478.
[23] V. A. Usov, N. A. Korchevin, Ya. S. Tsetlin, and M. G. Voronkov, *Zhur. org. Khim.*, 1975, **11**, 410.
[24] F. Duus, *Tetrahedron*, 1974, **30**, 3753.
[25] D. Paquer, S. Smadja, and J. Vialle, *Compt. rend.*, 1974, **279**, C, 529.
[26] D. H. R. Barton, L. S. L. Choi, R. H. Hesse, M. M. Pechet, and C. Wilshire, *J.C.S. Chem. Comm.*, 1975, 557.
[27] F. M. Dean, J. Goodchild, A. W. Hill, S. Murray, and A. Zahman, *J.C.S. Perkin I*, 1975, 1335.

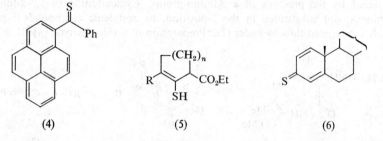

(4)   (5)   (6)

reagent, bis(diethylaluminium) sulphide, has been reported [28] to give good yields of various thioketones.

Barton and his group [29] have prepared the interesting new di-t-butyl thioketone (7) in high yield by reaction of carbon disulphide with the lithium salt of the corresponding ketone imine. Independently, Ohno and his co-workers [30] have published an analogous synthesis of this non-enethiolizable thioketone (7). Aliphatic thiones and, possibly, enethiols have been obtained by reaction of arylimines with benzoic anhydride and a stream of hydrogen sulphide.[31] Treatment of a vinyl chloride with sodium monosulphide led to the isolation of 'thiodimedone' (8; R = H), which has an enethiol ketone structure.[32] Basic hydrolysis

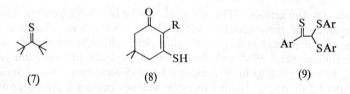

(7)   (8)   (9)

of 2-aryl-3-alkylaminoindene-1-thiones gave keto-enethiols.[23] Reaction of aromatic Grignard reagents with chlorodithioformates gave, among other products, thioketones (9).[33]

Thermolysis techniques have been investigated in the field of thioketone synthesis with moderate success. Flash thermolysis of allyl bornyl sulphide gave thiocamphor.[12] Pyrolysis of 1-methylallyl sulphide gave, after collection at −196 °C, a deep-blue liquid, presumably methyl vinyl thioketone, which could not be characterized or isolated because it underwent polymerization.[34] Preparation of cyclic thioketones by heating spirotrithians, under reduced pressure, is applicable to the relatively stable norbornanethione (10) but did not give reasonable yields of cycloalkanethiones, owing partially to their instability.[35] Thioacetone has been detected during the pyrolysis of hexamethyltrithian.[11]

[28] Y. Ishii, T. Hirabayashi, H. Imaeda, and K. Ito, Jap. P. 40 441/1974, App. 79 983/1970.
[29] D. H. R. Barton, F. S. Guziec, jun., and I. Shahak, *J.C.S. Perkin I*, 1974, 1794.
[30] A. Ohno, K. Nakamura, Y. Nakazima, and S. Oka, *Bull. Chem. Soc. Japan*, 1975, **48**, 2403.
[31] E. Ziegler, C. Mayer, and J. G. Zwainz, *Z. Naturforsch.*, 1975, **30b**, 760.
[32] L. Dalgaard and S. O. Lawesson, *Acta Chem. Scand. (B)*, 1974, **28**, 1077.
[33] F. C. V. Larsson and S. O. Lawesson, *Rec. Trav. chim.*, 1975, **94**, 1.
[34] W. J. Bailey and M. Isogawa, *Polymer Preprints, Amer. Chem. Soc., Div. Polymer Chem.*, 1973, **14**, 300.
[35] P. S. Fraser, L. V. Robbins, and W. S. Chilton, *J. Org. Chem.*, 1974, **39**, 2509.

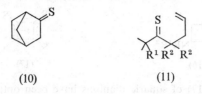

(10)   (11)

Retro-Diels–Alder reactions yielding thioketones did not occur, contrary to the previous related results.[36]

[3,3]-Sigmatropic rearrangement (thio-Claisen) of allyl enesulphides, obtained after basic alkylation of thioketones or enethiols, has led to γ-unsaturated thioketones (11),[37] to an enethiol (8; R = allyl),[32] and to an enethiol in the indole series.[38]

A new type of thioketone (12), conjugated with an α-ester group, has been obtained[39] by basic cleavage of a thiosulphate (Bunte salt) obtained from the α-chloro-ester. This unstable deep-blue thioketone dimerizes easily to a dithietan. Preparation of the α-dithione (13; R = p-Me₂NC₆H₄) has been reported[40] in detail. In the solid state it exists in the dithione form and in solution in

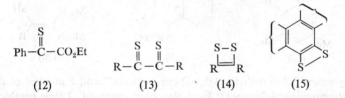

(12)   (13)   (14)   (15)

equilibrium with the dithiet form (14). Photolytic elimination of ethylene from a dihydrobenzodithiin allowed[41] the isolation of a stable dithiet (15), which is a tautomer of a dithio-o-quinone. The previously described structure of the dithiet (14; R = CF₃) has been confirmed and the dithione form (13; R = CF₃) disproved.[42]

A number of mostly novel thio-derivatives of squaric acid have been synthesized,[43-48] mainly by Seitz *et al.* The monothione (16; X = O)[45] and the dithione (16; X = S)[44, 45] were prepared by treatment of the corresponding carbonyl compounds with tetraphosphorus decasulphide. Monothio-, dithio-, and

[36] B. König, J. Martens, K. Praefcke, A. Schönberg, H. Schwartz, and R. Zeisberg, *Chem. Ber.*, 1974, **107**, 2931.
[37] L. Morin and D. Paquer, *Compt. rend.*, 1976, **282**, C, 353.
[38] H. Plieninger, H. P. Kraemer, and H. Sirowej, *Chem. Ber.*, 1974, **107**, 3915.
[39] K. Thimm and J. Voss, *Tetrahedron Letters*, 1975, 537.
[40] W. Küsters and P. de Mayo, *J. Amer. Chem. Soc.*, 1974, **96**, 3502.
[41] R. B. Boar, D. W. Hawkins, J. F. McGhie, S. C. Misra, D. H. R. Barton, M. F. C. Ladd, and D. C. Povey, *J.C.S. Chem. Comm.*, 1975, 756.
[42] J. L. Hencher, Q. Shen, and D. G. Tuck, *J. Amer. Chem. Soc.*, 1976, **98**, 899.
[43] G. Seitz, R. Schmiedel, and K. Mann, *Synthesis*, 1974, 578.
[44] G. Seitz, K. Mann, and R. Schmiedel, *Chem.-Ztg.*, 1975, **99**, 332.
[45] G. Seitz, H. Morck, K. Mann, and R. Schmiedel, *Chem.-Ztg.*, 1974, **98**, 459.
[46] G. Seitz, K. Mann, R. Schmiedel, and R. Matusch, *Chem.-Ztg.*, 1975, **99**, 90.
[47] D. Coucouvanis, F. J. Hollander, R. West, and D. Eggerding, *J. Amer. Chem. Soc.*, 1974, **96**, 3006.
[48] D. Coucouvanis, D. G. Holah, and F. J. Hollander, *Inorg. Chem.*, 1975, **14**, 2657.

(16)   (17)

tetrathio-analogues (17) of squaric dianions have been obtained by reaction of
hydrosulphide anion with squaric acid derivatives.[46-48]

The synthesis of enamino-thioketones has been pursued mainly by the groups
of Quiniou [49-52] and Usov.[53-55] Traditional methods employed include reaction
of P_4S_{10} with enamino-ketones,[49] reaction of amines with 1,2-dithiolylium salts,[56]
and transamination of enamino-thioketones,[49, 50] which was used [49] to prepare
the previously unknown simple compound (18). Access to thioamide vinylogues
has also been achieved through the reaction of a monothio-β-diketone with
ammonia,[52] and, for 3-amino-2-arylindene-1-thiones (19), by reaction of
enamino-imines with P_4S_{10} or sodium hydrosulphide [53] or by reaction of 1-imino-
2-aryl-2-alkyloxyindene tetrafluoroborates with sodium hydrosulphide.[54]

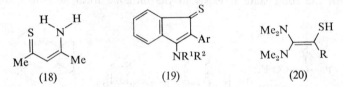

(18)   (19)   (20)

3-Phenylamino-5,5-dimethylcyclohex-2-ene-1-thione [54] and a number of deuteriated bis(enamino)thioketones [51] have also been prepared. Labile enethiols (20)
bearing two β-amino-groups have been reported.[57] Electrolytic reduction of
1,2-dithiolylium salts produced a propane dithionate.[58]

Several thioketones bearing heterocyclic groups have been synthesized, including a number of thienyl aryl thioketones [59] (21), and a thioketone and some
enethiols in the dihydrofuran [60] and dihydrothiophen [61] series. Basic cleavage of
6a-thiathiophthens affords thioketones (22) or the corresponding enethiols,
depending on the substituents.[62, 63] 1,2-Dithiole-3-thiones and their derivatives
are useful starting materials for the preparation of thiones (23) by cycloaddition

[49] G. Duguay, C. Metayer, H. Quiniou, and J. Bourrigaud, *Bull. Soc. chim. France*, 1974, 2507.
[50] A. Reliquet and F. Reliquet-Clesse, *Compt. rend.*, 1975, **280**, C, 145.
[51] G. Duguay, C. Metayer, and H. Quiniou, *Bull. Soc. chim. France*, 1974, 2853.
[52] G. Duguay, *Compt. rend.*, 1975, **281**, C, 1077.
[53] Ya. S. Tsetlin, V. A. Usov, and M. G. Voronkov, *Zhur. org. Khim.*, 1975, **11**, 1945.
[54] V. A. Usov, Ya. S. Tsetlin, and M. G. Voronkov, *Izvest. sibirsk. Otdel. Akad. Nauk, Ser. khim.
     Nauk*, 1975, 159.
[55] V. A. Usov, Ya. S. Tsetlin, N. A. Korchevin, and M. G. Voronkov, *Izvest. sibirsk. Otdel. Akad.
     Nauk, Ser. khim. Nauk*, 1974, 99.
[56] E. Uhlemann and B. Zöllner, *Z. Chem.*, 1974, **14**, 245.
[57] S. Mitamura, M. Takaku, and H. Nozaki, *Bull. Chem. Soc. Japan*, 1974, **47**, 3152.
[58] C. Th. Pedersen, *Angew. Chem. Internat. Edn.*, 1974, **13**, 349.
[59] K. S. N'Guyen, R. Pinel, and Y. Mollier, *Bull. Soc. chim. France*, 1974, 471, 1359.
[60] A. N. Volkov, Y. M. Skvortsov, Y. V. Kind, and M. G. Voronkov, *Zhur. org. Khim.*, 1974, **10**,
     174.
[61] G. A. Van den Ouweland and H. G. Peer, *J. Agric. Food Chem.*, 1975, **23**, 501.
[62] A. Josse, M. Stavaux, and N. Lozac'h, *Bull. Soc. chim. France*, 1974, 1723.
[63] A. Josse and M. Stavaux, *Bull. Soc. chim. France*, 1974, 1727.

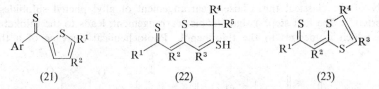

(21)   (22)   (23)

with acetylenic esters [64] or with benzyne.[65, 66] Alternatively, dithiolylium ions may be treated [67] successively with amines and hydrogen sulphide. Analogous products are obtained by irradiation with olefins.[68] Pedersen has established that the reaction product of thioacetic acid with phenylacetylene has the structure (23) and has also reported a propane dithionate.[69] Other miscellaneous heterocyclic thioketones have been reported.[27, 70—73] A thioketone (3; R^1 = Ph, R^2 = H) in which the thione group is conjugated with a phosphonium ylide has been prepared.[15]

Synthesis of Selenoketones.—Two new simple monomeric selenoketones, selenofenchone (24) and the blue di-t-butyl selenoketone, were prepared [74] by heating the ketone phosphoranylidene hydrazones with selenium. Treatment of a

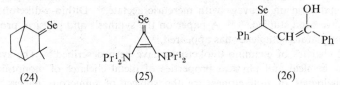

(24)   (25)   (26)

cyclopropenylium salt with hydrogen selenide gave [75] an excellent yield of the cyclopropene selenoketone (25). Monoselenodibenzoylmethane (26) was obtained by reaction of benzoylphenylacetylene with selenourea.[76] Its structure is best represented as an enol-selenoketone. These selenoketones are stable in the absence of air.

Transient Species.—Treatment of carbonyl compounds with hydrogen sulphide under pressure in the presence of molybdenum sulphide leads to reduction, and involves the formation of thioketonic intermediates.[77] Aliphatic thioketones [78] and a β-unsaturated thioketone (27) [79] are proposed as intermediates in the reaction of Grignard reagents with various dithioesters. A mechanistic

[64] H. Davy and J. M. Decrouen, *Bull. Soc. chim. France*, 1976, 115.
[65] D. Paquer and R. Pou, *Bull. Soc. chim. France*, 1976, 120.
[66] J. M. Decrouen, D. Paquer, and R. Pou, *Compt. rend.*, 1974, **279**, C, 259.
[67] E. Fanghänel, *J. prakt. Chem.*, 1975, **317**, 137.
[68] P. de Mayo and H. Y. Ng, *J.C.S. Chem. Comm.*, 1974, 877.
[69] C. Th. Pedersen, *Acta Chem. Scand.* (B), 1974, **28**, 367.
[70] C. J. Grol, *Tetrahedron*, 1974, **30**, 3621.
[71] K. S. N'Guyen, R. Pinel, and Y. Mollier, *Bull. Soc. chim. France*, 1974, 1356.
[72] Th. Eicher and V. Schäfer, *Tetrahedron*, 1974, **30**, 4025.
[73] K. T. Potts, J. Baum, and E. Houghton, *J. Org. Chem.*, 1976, **41**, 818.
[74] T. G. Back, D. H. R. Barton, M. R. Britten-Kelly, and F. S. Guziec, jun., *J.C.S. Chem. Comm.*, 1975, 539.
[75] Z. I. Yoshida, H. Konishi, and H. Ogoshi, *J.C.S. Chem. Comm.*, 1975, 359.
[76] G. Wilke and E. Uhlemann, *Z. Chem.*, 1975, **15**, 66.
[77] T. Takido, Y. Yamane, and K. Itabashi, *J. Synthetic Org. Chem., Japan*, 1975, **33**, 694.
[78] L. Leger and M. Saquet, *Bull. Soc. chim. France*, 1975, 657.
[79] L. Leger, M. Saquet, A. Thuillier, and S. Julia, *J. Organometallic Chem.*, 1975, **96**, 313.

study of the classical thio-Claisen rearrangement of allyl phenyl sulphides is reported.[80] The first step, a sigmatropic rearrangement, leads to the thioketone (28), which isomerizes to the thiophenol. Photochemical cleavage of a thiol

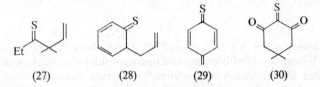

(27)   (28)   (29)   (30)

ester is reported to proceed *via* a conjugated thioketone (29).[81] A [1,3] sigma-tropic reaction, possibly involving a transient thioketone, is observed [32] with a bis(enesulphide). Evidence for the formation of monothiotrione (30) from the reaction of phenyldimedonyl iodone with phenyl isothiocyanate is presented.[82] Dithioketonate anion species have been suggested as possible intermediates in the photolysis of 1,2-dithiolylium salts.[83]

Metal Complexes.—Thiobenzophenones give complexes with metals [84] and *ortho*-metallated complexes with metal carbonyls.[85, 86] The latter complexes give *ortho*-metallation on cleavage with mercuric acetate.[85] Dithio-α-diketone com-plexes have been studied.[87, 88] A paper on the synthesis and physical properties of dithiosquarate complexes has appeared.[48]

Metal chelates of enamino-thioketones have been described.[55] The synthesis and the chemical and physical properties of metal chelates of monothio- and dithio-analogues of β-diketones are the subject of numerous papers.[89] The

[80] H. Kwart and J. L. Schwartz, *J. Org. Chem.*, 1974, **39**, 1575.
[81] J. Martens and K. Praefcke, *Chem. Ber.*, 1974, **107**, 2319.
[82] G. F. Koser and S. M. Yu, *J. Org. Chem.*, 1976, **41**, 125.
[83] C. T. Pedersen and C. Lohse, *Acta Chem. Scand. (B)*, 1975, **29**, 831.
[84] A. T. Pilipenko, O. P. Ryabushko, and G. S. Matsibura, *Ukrain. khim. Zhur.*, 1975, **41**, 664.
[85] H. Alper and W. G. Root, *Tetrahedron Letters*, 1974, 1611.
[86] H. Alper, *J. Organometallic Chem.*, 1974, **73**, 359.
[87] A. Z. Ryzhmanova, A. D. Troitskaya, Y. V. Yablokov, and B. V. Kudryavtsev, *Zhur. neorg. Khim.*, 1975, **20**, 165.
[88] A. V. Ryzhmanova, Y. V. Yablokov, and A. D. Troitskaya, *Zhur. neorg. Khim.*, 1975, **20**, 1911.
[89] M. Das and S. E. Livingstone, *Austral. J. Chem.*, 1974, **27**, 53, 749; 1177; 2109, 2115; 1975, **28**, 513; S. E. Livingstone and N. Saha, *ibid.*, 1975, **28**, 1249; S. E. Livingstone, J. H. Mayfield, and D. S. Moore, *ibid.*, 1975, **28**, 2531; S. E. Livingstone and D. S. Moore, *ibid.*, 1976, **29**, 283; E. Uhlemann and U. Eckelmann, *Z. Chem.*, 1974, **14**, 66; G. Engelhardt, B. Schuknecht, and E. Uhlemann, *ibid.*, 1975, **15**, 367; E. Uhlemann and B. Schuknecht, *Analyt. Chim. Acta*, 1974, **69**, 79; B. Schuknecht, G. Robisch, and E. Uhlemann, *ibid.*, p. 329; C. G. MacDonald, R. L. Martin, and A. F. Masters, *Austral. J. Chem.*, 1976, **29**, 257; C. Cauletti and C. Furlani, *J. Electron Spectroscopy Related Phenomena*, 1975, **6**, 465; M. H. Dhingra, B. Maiti, and R. M. Mathe, *Indian J. Chem.*, 1975, **13**, 359; O. Siiman, D. D. Titus, C. D. Cowman, J. Fresco, and H. B. Gray, *J. Amer. Chem. Soc.*, 1974, **96**, 2353; R. C. Burton, *Diss. Abs. (B)*, 1974, **34**, 5356; M. McPartlin, G. B. Robertson, G. H. Barnett, and M. K. Cooper, *J.C.S. Chem. Comm.*, 1974, 305; R. Beckett and B. F. Hoskins, *J.C.S. Dalton*, 1974, 622; B. E. Reichert and B. O. West, *J. Organometallic Chem.*, 1974, **71**, 291; D. R. Dakternieks and D. P. Graddon, *Austral. J. Chem.*, 1974, **27**, 1351; K. Nag and M. Chaudhury, *J. Inorg. Nuclear Chem.*, 1976, **38**, 309; K. Nag and M. Chaudhury, *Inorg. Nuclear Chem. Letters*, 1976, **12**, 307; M. Chikuma, A. Yokoyama, Y. Ooi, and H. Tanaka, *Chem. and Pharm. Bull. (Japan)*, 1975, **23**, 507; M. Chikuma, A. Yokoyama, and H. Tanaka, *ibid.*, 1974, **22**, 1378; N. Nakanishi, A. Yokoyama, and H. Tanaka, *ibid.*, 1975, **23**, 1677, 1684; K. R. Solanke and S. M. Khopkar, *Talanta*, 1974, **21**, 245; S. M. Khopkar, *Z. analyt. Chem.*, 1974, **272**, 283; R. R. Mulye and S. M. Khopkar, *Analyt. Chim. Acta*, 1975, **76**, 204; R. Hendrickson, R. K. Y. Ho, and R. L. Martin, *Inorg. Chem.*, 1974, **13**, 1279.

formation of chelates of thio-β-dicarbonyl compounds with various metals, as previously reported, is the method of choice for the extraction and precise spectrophotometric determination of these metals, even in very low concentration. Novel metal chelates of seleno-β-diketones are prepared by treatment of acetylenic ketones with selenourea to give salts, which are then treated with the appropriate metal salts.[90]

Reactions.—The chemical behaviour of thioketones has mainly been investigated in four directions; reaction with organometallic compounds, reaction with nucleophiles, thermal cycloadditions, and photochemical reactions.

The thiophilic addition of organomagnesium and organolithium derivatives is still the subject of interesting research, particularly from a mechanistic viewpoint. Dagonneau and Vialle[91] report further data on the Grignard reaction with thiobenzophenone. Thiophilic addition of the organic group is general, but carbophilic addition, formation of thiirans, and double addition are also observed. The mechanism is suggested to involve homolytic cleavage of the organomagnesium intermediate to give a reactive radical, which can be monitored by e.s.r. Genesis of double addition products is discussed.[92] Reaction of various thioketones with vinylic organomagnesium compounds leads to kinetically controlled *C*-addition products (31) or thermodynamically controlled *S*-addition products (32) (Scheme 1).[93] A radical rearrangement of the corresponding

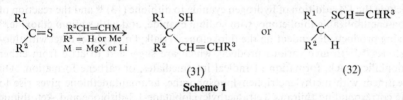

(31)   (32)

Scheme 1

intermediates is considered. An important and detailed study[94] has shown that isomerically pure 1-propenyl-lithium or 1-propenylmagnesium bromide leads to a thiophilic addition product (32) with retention of configuration, which leads the authors to rule out a free-radical path, in disagreement with the mechanism suggested by Dagonneau.[40] A series of organometallic compounds bearing phenyl groups give *C*- and *S*-additions and double addition. The yield of *S*-addition depends on the organic portion of the organometallic compound and appears to be inversely proportional to its ability to transfer an electron to the thioketone as in path *b* (Scheme 2). Beak *et al.*[94] prefer a mechanism which involves the transfer of two electrons in the formation of the carbon–sulphur bond and thus gives an anion directly (path *a*). The stability of the anion does not seem to be a dominant factor in determining the reaction course. The reaction of thiobenzophenone and of di-t-butyl thioketone (7) with butyl-lithium or phenyl-lithium has been shown[95] to involve thiophilic and carbophilic

[90] G. Wilke and E. Uhlemann, *Z. Chem.*, 1974, **14**, 288; 1974, **15**, 453.
[91] M. Dagonneau and J. Vialle, *Tetrahedron*, 1974, **30**, 3119.
[92] M. Dagonneau, *Compt. rend.*, 1974, **279**, C, 285.
[93] M. Dagonneau, *J. Organometallic Chem.*, 1974, **80**, 1.
[94] P. Beak, J. Yamamoto, and C. J. Upton, *J. Org. Chem.*, 1975, **40**, 3052.
[95] A. Ohno, K. Nakamura, M. Uohama, and S. Oka, *Chem. Letters*, 1975, 983; A. Ohno, K. Nakamura, M. Uohama, S. Oka, T. Yamabe, and S. Nagata, *Bull. Chem. Soc. Japan*, 1975, **48**, 3718.

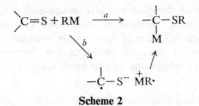

Scheme 2

additions and reduction. The reaction course is considered to proceed through a charge-transfer mechanism (path *b*), in contrast with the preceding arguments. As a result of theoretical calculations, two different intermediates (33) and (34), depending on the thioketone, were proposed.

$$\overset{+}{M}\ \overset{\cdot}{R}\qquad\qquad \overset{\cdot}{R}\ \overset{+}{M}$$

$$Ph_2\bar{C}-S\cdot\qquad\qquad Bu^t_2\overset{\cdot}{C}-\bar{S}\qquad\qquad Bu^t_2CH-Se-Se-CHBu^t_2$$

(33)   (34)   (35)

Reduction of di-t-butyl selenoketone with sodium borohydride gives the diselenide (35).[74] Reactions of thioketones with nucleophiles reported recently include the 1,4-addition of hydrogen cyanide to dithione (13)[40] and the reaction of non-enethiolizable thioketones with sodium cyanide, and with sodium ethoxide,[95] giving products dependent on the thioketone. Trialkyl phosphites react with thio-ketones[96, 97] to give various products which are suggested to arise from either thiophilic attack, formation of radical intermediates, or carbene formation. On treatment with methylenetriphenylphosphorane, adamantanethione gives rise to the corresponding thiiran. Tetramethylcyclobutane-1,3-dithione and -ketothione give novel thiocarbonyl-stabilized ylides.[98] Alkylation of aliphatic thio-ketones, in a basic medium, normally yields enesulphides derived from *S*-alkylation,[32, 37, 99] but in one case *C*-alkylation is observed.[100] Amines or ammonia attack enamino-thioketones[50, 52] and a monothio-β-diketone[52] to yield different enamino-thioketones.

Squaric dithiones (16; X = S) react with an amine to give transamination with intramolecular migration of a sulphur atom.[44] *S*-Methylated cations are obtained with methyl fluorosulphonate,[43] and on hydrolysis lose a dimethylamino-group to give a ketone. Reactions of a cyclobutenedithione with a thiocyanate[101] and reaction of a monothiosquarate dianion[102] have been reported.

A number of thermal cycloadditions have been reported. The thiocarbonyl group may behave as a dienophile with butadiene[36] or as a 1,3-dipolarophile with benzonitrile oxide[103] and, especially, with diazoalkanes. Reaction of

[96] Z. I. Yoshida, T. Kawase, and S. Yoneda, *Tetrahedron Letters*, 1975, 235.
[97] Y. Ogata, M. Yamashita, and M. Mizutani, *Tetrahedron*, 1974, **30**, 3709.
[98] A. P. Krapcho, M. P. Silvon, and S. D. Flanders, *Tetrahedron Letters*, 1974, 3817.
[99] G. E. Veenstra and B. Zwanenburg, *Rec. Trav. chim.*, 1976, **95**, 37.
[100] R. Couturier, D. Paquer, and A. Vibet, *Bull. Soc. chim. France*, 1975, 1670.
[101] G. Seitz, R. Schmiedel, and K. Mann, *Chem.-Ztg.*, 1975, **99**, 463.
[102] D. Eggerding, J. L. Straub, and R. West, *Tetrahedron Letters*, 1975, 3589.
[103] N. A. Korchevin, V. A. Usov, A. Tokareva, and M. G. Voronkov, *Khim. geterotsikl. Soedinenii*, 1975, 278.

(36)   (37)

thioketones with diazoalkanes afforded normal products, *i.e.* thiadiazolines (36) and (37),[29, 104] thiirans,[29, 105] alkenes,[26, 29, 105] and dithiolans.[105] Reaction of di-t-butyl thioketone (7) with diphenyldiazomethane[29] yielded the same thiadiazoline as that obtained by the reaction of thiobenzophenone with di-t-butyldiazomethane, thus proving elegantly and unequivocally that the product had the Δ^3-1,3,4-thiadiazoline structure (36; $R^1 = R^2 = Bu^t$, $R^3 = Ph$). On the other hand, addition of diazomethane to adamantanethione led to a mixture of the corresponding Δ^2-1,2,3-thiadiazoline (36) and Δ^3-1,3,4-thiadiazoline (37). The product ratio is highly dependent on the nature of the reaction solvent.[104] Such a dependence is not observed, however, with some of the other thioketones. Loss of nitrogen from these heterocycles and desulphurization of thiirans leads to a successful synthesis of new highly hindered alkenes, starting from hindered thioketones and hindered diazoalkanes.[29] Reaction of di-t-butyl selenoketone with diphenyldiazomethane gave a Δ^3-1,3,4-selenodiazoline. A competitive reaction with the corresponding thioketone (7) showed the greater reactivity of the selenoketone.[74] Conjugated thioketones, *e.g.* 1,4-dithione (13)[40] and enamino-thioketones,[106, 107] may also react as 1,4-heterodienes with dienophiles to give heterocycles. The enethiol form of an enamino-thioketone is reported to add to acetylenic ketones.[108] Diphenylcyclopropenethione has found applications as a versatile intermediate in organic synthesis. Reactions with enamines,[109] imines,[110] keten acetals,[111, 112] azomethine ylides,[72] and meso-ionic systems[73] have been studied. In general, the first step in the reaction is a cycloaddition.

(38)

Oxidation of thioketones with peroxy-acids gives sulphines, *e.g.* (38);[26, 95, 99] in one example, the *syn-* and *anti-*isomer could be separated.[26] In contrast, di-t-butyl selenoketone yields di-t-butyl ketone and selenium. The hydrolysis of mono-thioacetylacetone and monothiobenzoylacetone in acidic solution leads to the

[104] A. P. Krapcho, M. P. Silvon, I. Goldberg, and E. G. E. Jahngen, *J. Org. Chem.*, 1974, **39**, 860.
[105] N. A. Korchevin, V. A. Usov, and M. G. Voronkov, *Khim. geterotsikl. Soedinenii*, 1974, 713, 714.
[106] J. C. Meslin, Y. T. N'Guessan, H. Quiniou, and F. Tonnard, *Tetrahedron*, 1975, **31**, 2679.
[107] J. P. Pradere, Y. T. N'Guessan, and H. Quiniou, *Tetrahedron*, 1975, **31**, 3059.
[108] N. A. Korchevin, V. N. Elokhina, V. A. Usov, and M. G. Voronkov, *Zhur. org. Khim.*, 1975, **11**, 1532.
[109] T. Eicher and S. Boehm, *Chem. Ber.*, 1974, **107**, 2238.
[110] T. Eicher and J. L. Weber, *Tetrahedron Letters*, 1974, 3409.
[111] M. Hirth, H. Krapf, R. Riedl, J. Sauer, and E. Oeser, *Chem. Ber.*, 1976, **109**, 562.
[112] H. Krapf, P. Riedl, and J. Sauer, *Chem. Ber.*, 1976, **109**, 576.

corresponding ketones.[113] The hydrolysis of aryl-trifluoromethyl-monothio-
β-diketones results in cleavage to hydrogen sulphide, probably trifluoroacetic
acid, and a methyl ketone. The kinetics of this reaction have been studied.[114]
3-Aminoindene-1-thiones are hydrolysed in a basic medium to give 3-mercapto-
inden-1-ones.[23, 54] Treatment of thioketones with hydridotetracarbonylferrate
anion, $[HFe(CO_4)]^-$, gives the desulphurized hydrocarbons.[115, 116]

The photochemical reactions of thioketones have previously been the subject
of a large number of papers, due mostly to Ohno, who pioneered this field, and
to de Mayo. Current research involves more complete investigations of earlier
work as well as the study of novel reactions. It has been confirmed that, on
photochemical excitation, diaryl thioketones react in the (n,π^*) state with
electron-rich olefins to give thietans or 1,4-dithians, and in the (π,π^*) state with
electron-deficient olefins to give thietans.[116] Irradiation of aromatic thioketones
at $\lambda > 540$ nm with bis(methylthio)ethyne gives the previously unknown
αβ-unsaturated dithio-esters (39), *via* the decomposition of an unstable thiet.[117]

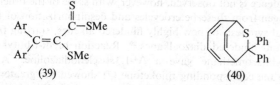

(39)   (40)

Irradiation of thiobenzophenone with cyclo-octatetraene at $\lambda > 340$ nm gives
a 1,4-adduct (40), but, in contrast, 6,6-diphenylfulvene yields[118] a 1,1-adduct.
The photochemical reaction of *o*-benzylthiobenzophenone has been examined
because of its relevance to the reduction of thiobenzophenone.[119] An intramolecular transfer of hydrogen was postulated to give an enethiol, which could
be trapped. It is suggested that the photoreduction of thiobenzophenone involves
the initial formation of a radical, $Ph_2\dot{C}SH$. The photochemical behaviour of
adamantanethione (41) has also been studied. In the (n,π^*) state it gives the

$$R^1R^2C{=}S \xrightarrow{\ h\nu\ } R^1R^2CH{-}S{-}S{-}CHR^1R^2$$

(41)   (42)

disulphide (42).[120] During irradiation in the presence of adamantanethiol, an
e.s.r. signal was observed. Spin-trapping of the radical showed that its structure
was consistent with the formation of $R^1R^2\dot{C}SSCDR^1R^2$, when using the
deuteriated thiol R^1R^2CDSH. Excitation of various thioketones has been shown
to give cyclization. Alkyl aryl thioketones (43), having an activated β-position,
ring-close to give substituted cyclopropanes (44),[121] polycyclic aromatic thio-

[113] M. Leban, J. Fresco, and S. E. Livingstone, *Austral. J. Chem.*, 1974, **27**, 2353.
[114] M. Leban, J. Fresco, M. Das, and S. E. Livingstone, *Austral. J. Chem.*, 1974, **27**, 2357.
[115] H. Alper, *J. Org. Chem.*, 1975, **40**, 2694.
[116] H. Gotthardt, *Chem. Ber.*, 1974, **107**, 1856.
[117] A. C. Brouwer and H. J. T. Bos, *Tetrahedron Letters*, 1976, 209.
[118] T. S. Cantrell, *J. Org. Chem.*, 1974, **39**, 853.
[119] N. Kito and A. Ohno, *Internat. J. Sulfur Chem.*, 1973, **8**, 427.
[120] J. R. Bolton, K. S. Chen, A. H. Lawrence, and P. de Mayo, *J. Amer. Chem. Soc.*, 1975, **97**, 1832.
[121] A. Couture, M. Hoshino, and P. de Mayo, *J.C.S. Chem. Comm.*, 1976, 131.

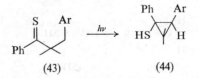

(43)　　　　　　　(44)

ketones having a free *peri*-position nearby give a thiophane ring,[21] and alkyl aryl thioketones (45) give cyclopentanethiols (46).[122] The synthetic interest of this reaction has been demonstrated by its application as a step in the preparation of

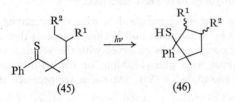

(45)　　　　　　　(46)

the sesquiterpenoid (±)-cuparene.[20] In an important paper, de Mayo has examined and rationalized all of the results of thione photochemistry, which appears to have distinctive characteristics of its own.[4] The formation of several products is interpreted as involving a higher state of the excited species. A study [32] of the irradiation of enethiols (8), yielding disulphides and cyclic products, has been reported.

4 Thioketens and Selenoketens

Synthesis.—Owing to their reactivity and instability, few thioketens were isolated until very recently.[6] A remarkable method [123] for the synthesis of thioketens (48) in good yields by the flash thermolysis of 1,2,3-thiadiazoles (47) has now

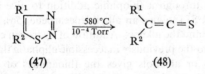

(47)　　　　　　　(48)

been reported. The stabilities of the products, which were trapped at −196 °C, have been discussed. Diarylthioketens dimerize to give the bis(methylene)-dithietans. The solid thioketen (49) is made relatively stable by neighbouring bulky substituents, and its properties have been studied.[124] I.r. spectroscopy provides evidence that the unsubstituted thioketen (50) is produced by the decomposition of 1,2,3-thiadiazole,[125] and that the isomeric ethynyl thiol (51) is also formed. Microwave techniques allowed the detection of thioketen (50) during the pyrolysis of hexamethyltrithian.[126] Addition of thiophosgene to

[122] P. de Mayo and R. Suau, *J. Amer. Chem. Soc.*, 1974, **96**, 6807.
[123] G. Seybold and C. Heibl, *Angew. Chem. Internat. Edn.*, 1975, **14**, 248.
[124] E. Schaumann, S. Harto, and G. Adiwidjaja, *Angew. Chem. Internat. Edn.*, 1976, **15**, 40.
[125] A. Krantz and J. Laureni, *J. Amer. Chem. Soc.*, 1974, **96**, 6768.
[126] K. Georgiou, H. W. Kroto, and B. M. Landsberg, *J.C.S. Chem. Comm.*, 1974, 739.

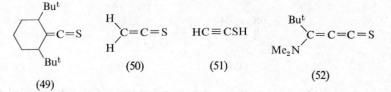

(49)   (50)   (51)   (52)

a butene, *e.g.* 2-dimethylamino-3,3-dimethylbut-1-ene, in ether containing triethylamine gave [127] the alkylidenethioketen (52). Two thioketens conjugated with phosphonium ylides have been described.[128]

Reactions.—Three interesting papers deal with the reactivity of thioketens. A noteworthy paper by Schaumann and Walter [129] reports that thiophilic addition of organometallic compounds is observed with thioketens. At $-78\,°C$, dialkyl-thioketens (53) react with phenyl-lithium or methyl-lithium by 1,2-thiophilic addition to give enesulphides (55). At room temperature the main products

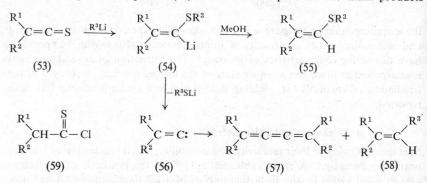

(53)   (54)   (55)

(59)   (56)   (57)   (58)

are the butatrienes (57) and alkenes (58). The formation of these products is considered to occur through a thiophilic addition to give an intermediate (54) and elimination of R^3SLi to give an alkylidenecarbene (56), which reacts further to give the cumulene and the alkene. Addition of hydrogen chloride to thioketens at $-80\,°C$ leads [130] to the previously inaccessible aliphatic thioacyl chlorides (59). Addition of amines or alcohols gives the thioamides or thiono-esters, either reaction being suitable for the quantitative determination of thioketens.[123]

PART II: Sulphines and Sulphenes *by D. R. Hogg*

1 Sulphines

Although many stable sulphines are known, the parent compound, CH_2SO, has not been isolated. It is generated [1] cleanly by flash vacuum pyrolysis of 1,3-di-thietan 1-oxide above $300\,°C$, and of thietan *S*-oxide or methanesulphinyl

[127] M. Parmantier, J. Galloy, M. Van Meersche, and H. G. Viehe, *Angew. Chem. Internat. Edn.*, 1975, **14**, 53.

[128] H. T. Bestmann and G. Schmid, *Angew. Chem. Internat. Edn.*, 1974, **13**, 273; H. J. Bestmann, G. Schmid, and D. Sandmeier, *ibid.*, 1975, **14**, 53.

[129] E. Schaumann and W. Walter, *Chem. Ber.*, 1974, **107**, 3562.

[130] G. Seybold, *Angew. Chem. Internat. Edn.*, 1975, **14**, 703.

[1] E. Block, R. E. Penn, R. J. Olsen, and P. F. Sherwin, *J. Amer. Chem. Soc.*, 1976, **98**, 1264.

chloride above 600 °C in the gas phase. Less efficient sulphine precursors include 1,3,5-trithian 1-oxide and DMSO. A microwave study [1, 2] of the pyrolysis products enabled bond lengths and angles to be calculated, and indicated that sulphine is planar, with a dipole moment of 2.994 D, oriented at an angle of 25.50° to the S—O bond. The C=S bond is physically similar to that in thio-formaldehyde. This is in broad agreement with *ab initio* LCAO–SCF–MO calculations [3] which suggest that the most stable conformation is planar, with a bent C=S=O group, a full C=S double bond, and a partial S—O double bond. Similar results on the C=S=O portion of the molecules were obtained [4] from *X*-ray diffraction studies on (*E*)- and (*Z*)-mesityl(phenylsulphinyl)sulphine and (*E*)- and (*Z*)-mesityl(phenylsulphonyl)sulphine. Ethane- and propane-2-sulphinyl chlorides did not give higher sulphines on flash vacuum pyrolysis, but dimethyl-sulphine was obtained from 2,2,4,4-tetramethyl-1,3-dithietan 1-oxide.

Dichlorosulphine is formed [5] in good yield by the thermal decomposition of *p*-anisyl trichloromethanesulphenate in acetonitrile or chloroform. The reaction is very sensitive to solvent polarity, and, in non-polar solvents, rearrangement of the sulphenate to the sulphoxide occurs by a polar mechanism. Since the tri-chloromethanesulphenate anion is known to give dichlorosulphine, an inter-mediate ion-pair is postulated, *i.e.*

$$ArCH_2OSCCl_3 \xrightarrow[\Delta]{MeCN} ArCH_2^+ \ \bar{O}SCCl_3 \longrightarrow ArCH_2Cl + Cl_2C{=}S{=}O$$

A new type of sulphine, phenylcyanosulphine, $PhC(CN){=}S{=}O$, has been reported [6] to be produced in low yield from the reaction of benzonitrile with thionyl chloride in the presence of hydrogen chloride. It gradually decomposes on standing to give *trans*-αβ-dicyanostilbene and sulphur dioxide.

Studies of dipole moments [7] of diarylsulphines, fused aromatic sulphines, adamantanethione *S*-oxide, and dichlorosulphine give a value of 3.84 D for the group dipole moment of the CSO system. It is directed towards the oxygen atom at an angle of 20° with the C—S bond. The CSO group releases electron density to aromatic systems, showing the importance of resonance hybrid (1), in which the carbon atom has a negative charge. Other studies and calculations [3] are in agreement with there being a slightly negative carbon atom.

Rotational isomerism in 3,3′-disubstituted diphenylsulphines has been studied [8] by n.m.r. The deshielding of the *ortho*-protons *syn* to the CSO system depends upon the rotational equilibrium and hence on the solvent polarity. In apolar solvents the rotamer populations were approximately equal, or the rotamer with the lower dipole moment was slightly favoured. The more polar rotamer [*e.g.* (2)] was favoured in polar solvents. Rotational barriers about the aryl–sulphine bond have been determined,[9] using a coalescence method, for

[2] R. E. Penn and R. J. Olsen, *J. Mol. Spectroscopy*, 1976, **61**, 21.

[3] F. Bernardi, G. Maccagnani, and A. Mangini, *Anales de Quím.*, 1974, **70**, 1199.

[4] Th. W. Hummelink, *J. Cryst. Mol. Structure*, 1974, **4**, 87, 373: *Cryst. Struct. Comm.*, 1975, **4**, 441; 1976, **5**, 169.

[5] S. Braverman and B. Sredni, *Tetrahedron*, 1974, **30**, 2379.

[6] M. Ohoka, T. Kojitani, S. Yanagida, M. Okahara, and S. Komori, *J. Org. Chem.*, 1975, **40**, 3540.

[7] A. Tangerman and B. Zwanenburg, *J.C.S. Perkin II*, 1974, 1413.

[8] A. Tangerman and B. Zwanenburg, *J.C.S. Perkin II*, 1974, 1141.

[9] A. Tangerman and B. Zwanenburg, *J.C.S. Perkin II*, 1975, 916.

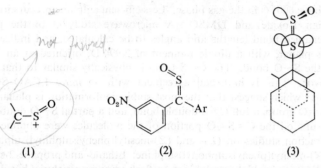

(1)   (2)   (3)

substituted 2,6-dimethylphenylsulphines having arylthio-, arylsulphinyl-, or arylsulphonyl-groups as the other component, *e.g.* ArC(=SO)SO$_n$Ar (n = 0—2). Barriers for the (E)- and the (Z)-isomers are greater than those for the corresponding thiocarboxylates and dithiocarboxylates. The preferred conformations of a similar series of mesityl aryl- or alkyl-thio-, -sulphinyl-, or -sulphonyl-sulphines 2,4,6-Me$_3$C$_6$H$_2$C(=SO)SO$_n$R (n = 0—2; R = alkyl or aryl) have been established [10] by n.m.r. and dipole-moment studies. Unlike carboxylates, thiocarboxylates, and dithiocarboxylates, none of the sulphines possesses a linear *s-trans*-conformation. Rather rigid folded and *gauche* conformations predominate. For example, the mesityl-arylthio-(Z)-sulphines have a preferred *s-cis*-conformation (3) in which the aromatic rings lie face to face. The (E)-isomers predominantly exist as the *gauche* conformations, in which the arylthio-group has rotated through ~90° about the C—S bond. *X*-Ray studies on (E)-mesityl(phenylsulphonyl)sulphine [4] indicate a torsion angle of 82°.

Sulphines are formed by the oxidation of thiones with peroxy-acids, but they then react further, albeit much more slowly, to give the ketone, sulphur, and sulphur dioxide. This latter reaction [11] is kinetically similar to many other oxidations with peroxy-acids, and the results are consistent with electrophilic attack of the peroxy-acid on the C=S double bond to give a cyclic sulphinate ester (4), which rapidly decomposes to give the ketone and sulphur monoxide. In contrast, oxidation of thiobenzanilide *S*-oxide with *m*-chloroperoxybenzoic acid gives [12] products derived from the sulphene PhC(NHPh)=SO$_2$. Further examples of the reaction between diarylsulphines and aryldiazomethanes to give episulphoxides have been reported.[13]

Toluene-*p*-sulphinate anion reacts with chloro-*p*-tolylsulphine to give [14] *p*-methylbenzyl *p*-tolyl sulphone instead of the expected substitution product, *p*-tolyl(toluene-*p*-sulphonyl)sulphine:

$$p\text{-MeC}_6\text{H}_4\text{SO}_2^- + p\text{-MeC}_6\text{H}_4\text{C(Cl)=S=O} \xrightarrow{\text{MeCN}} p\text{-MeC}_6\text{H}_4\text{SO}_2\text{CH}_2\text{C}_6\text{H}_4\text{Me-}p$$

The expected substitution product is considered to react further with the nucleophile at the sulphine sulphur atom to give a sulphinyl-sulphone, which then

[10] A. Tangerman and B. Zwanenburg, *J.C.S. Perkin II*, 1975, 352.
[11] A. Battaglia, A. Dondoni, G. Maccagnani, and G. Mazzanti, *J.C.S. Perkin II*, 1974, 609.
[12] W. Walter and O. H. Bauer, *Annalen*, 1975, 305.
[13] B. F. Bonini, A. Capelli, G. Maccagnani, and G. Mazzanti, *Gazzetta*, 1975, **105**, 827.
[14] G. E. Veenstra and B. Zwanenburg, *Rec. Trav. chim.*, 1976, **95**, 28.

hydrolyses and loses SO_2 to give the product. Other chloro-sulphines react similarly. Methylation of sulphines with methyl-lithium gives the methyl sulphoxide. The reaction of the salt formed from thiocamphor *S*-oxide and thallium(I) ethoxide with electrophilic alkylating agents gives [15] the unsaturated sulphoxide (5) derived from the vinylsulphenate. Diarylsulphines are

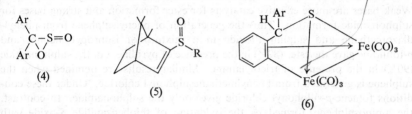

(4)

(5)

(6)

deoxygenated [16] by dimanganese decacarbonyl and with di-iron enneacarbonyl. In the latter case, *ortho*-metallated complexes (6) derived from the thioketone were also obtained.

2 Sulphenes

The chemistry of sulphenes and the evidence for their existence as short-lived intermediates has been reviewed.[17] The generation of sulphenes from alkane-sulphonyl chlorides and tertiary amines is generally considered [17] to be an E2 process with E1cb character. Sulphonyl derivatives having poorer leaving groups have now been shown by kinetic [18, 19] and hydrogen–deuterium exchange studies [20] to follow either a reversible or an irreversible E1cb mechanism, depending on the leaving group:

$$RCH_2SO_2X + Et_3N \; \rightleftharpoons \; R\bar{C}HSO_2X + Et_3\overset{+}{N}H$$
(7)

$$R\bar{C}HSO_2X \longrightarrow RCH{=}SO_2 + X^-$$

$$RCH{=}SO_2 + D_2O + Et_3N \longrightarrow RCHDSO_3^- + Et_3\overset{+}{N}D$$

2,4-Dinitrophenyl phenylmethanesulphonate followed the irreversible mechanism; it gave a monodeuteriated product, and the unreacted ester was not exchanged. The 4-nitrophenyl ester followed the reversible mechanism, and both the product and the recovered ester were dideuteriated. With triethylamine in DME–D_2O, the phenyl ester was not hydrolysed but the recovered ester was deuteriated. The methylsulphonylammonium ion (7; R = H, X = $\overset{+}{N}RR'Me$), prepared from the *NN*-disubstituted methanesulphonamide and methyl fluorosulphonate, also reacted [20] by the reversible mechanism. These quaternary ammonium salts mesylate amines and alcohols in high yield under very mild conditions, and have

[15] G. E. Veenstra and B. Zwanenburg, *Rec. Trav. chim.*, 1976, **95**, 37.
[16] H. Alper, *J. Organometallic Chem.*, 1975, **84**, 347.
[17] J. F. King, *Accounts Chem. Res.*, 1975, **8**, 10.
[18] A. Williams, K. T. Douglas, and J. S. Loran, *J.C.S. Chem. Comm.*, 1974, 689.
[19] J. F. King and R. P. Beatson, *Tetrahedron Letters*, 1975, 973.
[20] J. F. King and J. R. du Manoir, *J. Amer. Chem. Soc.*, 1975, **97**, 2566.

considerable synthetic potential. In the absence of other nucleophiles, sulphene reacts[21] with the triethylamine complex $^-CH_2SO_2\overset{+}{N}Et_3$ to form the dimeric zwitterion $^-CH_2SO_2CH_2SO_2\overset{+}{N}Et_3$, which can tautomerize and hence form the dimeric sulphene $MeSO_2CH{=}SO_2$. The tetrameric zwitterion is formed similarly.

The reactions of sulphenes with alcohols and amines are catalysed[22] by bases. Weak bases are more effective catalysts for ester formation and strong bases for sulphonamide formation. Thus the generation of phenylsulphene from phenyl-diazomethane and sulphur dioxide in a mixture of isopropyl alcohol and *p*-toluidine gave the ester (82%) in the presence of pyridine, but the sulphonamide (90%) in the presence of triethylamine. Similar results are obtained when the sulphene is generated from phenylmethanesulphonyl chloride. Under these conditions toluene-*p*-sulphonyl chloride gives only the sulphonamide. In contrast, the aminosulphene, formed by the oxidation of thiobenzanilide *S*-oxide with peroxy-acid, is considered[12] to react with alcohols at the sulphene carbon atom to give imidates $PhC(OR){=}NPh$. Photolysis of 3-phenyl-2*H*-thiopyran 1,1-dioxide in methanol gives[23] the sulphene, which reacts with the solvent to give methyl 4-phenylpenta-2,4-diene-1-sulphonate. The sulphene obtained[24] from the photolysis of 2*H*-1-benzothiopyran 1,1-dioxide is not trapped by methanol, but gives a mixture of cyclic sulphinate esters. These observations support the generality of the cycloreversion (8) to (9), and show that the presence of an atom in the ring with an unshared electron pair is not a necessary condition for reaction.

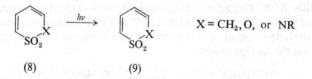

$$X = CH_2, O, \text{ or } NR$$

(8)   (9)

The extrusion of sulphur dioxide from a sulphene to give a carbene has been postulated previously, but never authenticated. Flash pyrolysis of α-(toluene-*p*-sulphonyl)phenyldiazomethane gave[25] *p*-tolyl phenyl ketone (15%) and 2-methylfluorene (5%). The former could arise from phenyl-*p*-tolylsulphene by loss of sulphur monoxide and the latter by loss of sulphur dioxide to give the carbene, which rearranges to the fluorene. Sulphene is postulated[26] as an intermediate in the formation of carbonyl sulphide by the solid-state photolysis of dimethyl sulphone and by various other high-energy processes.

Reactions of sulphenes with *NN*-disubstituted 2-methylprop-1-enylamines[27] and 1-(*N*-morpholino)propenes[28] give largely the thermodynamically less stable *cis*-adducts (10). The *trans*-adduct is the major product with alkylsulphenes, where the atom β to the sulphonyl group cannot develop appreciable negative

21   J. S. Grossert and M. M. Bharadwaj, *J.C.S. Chem. Comm.*, 1974, 144.
22   J. F. King and Y. I. Kang, *J.C.S. Chem. Comm.*, 1975, 52.
23   J. F. King, E. G. Lewars, D. R. K. Harding, and R. M. Enanoza, *Canad. J. Chem.*, 1975, **53**, 3656.
24   C. R. Hall and D. J. H. Smith, *Tetrahedron Letters*, 1974, 3633.
25   B. E. Sarver, M. Jones, jun., and A. M. van Leusen, *J. Amer. Chem. Soc.*, 1975, **97**, 4771.
26   H. Hiraoka, *J.C.S. Chem. Comm.*, 1974, 1014.
27   W. E. Truce and J. F. Rach, *J. Org. Chem.*, 1974, **39**, 1109.
28   V. N. Drozd, V. V. Sergeichuk, and V. A. Moskalenko, *Zhur. org. Khim.*, 1975, **11**, 135.

character, or when the products readily isomerize. These results can be explained [27] on the basis of a concerted $[_\pi 2s + _\pi 2s]$ process or by an attraction of opposite charges in a 1,4-dipolar intermediate. The latter explanation is supported [29] by an increasing tendency to form the *trans*-adduct in more polar solvents. Acyclic [29] and di-adducts [30] [*e.g.* (11)] have been reported.

CNDO/2 calculations suggest [31] that sulphenes should not undergo concerted additions to alkenes, but that more nucleophilic alkenes may add stepwise.

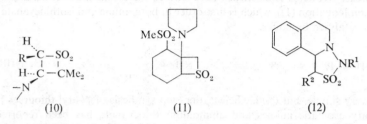

(10)   (11)   (12)

Alkyl or cycloalkyl vinyl sulphides, which are substantially less nucleophilic than enamines, react [32] with methylsulphonylsulphene, but not with sulphene or phenylsulphene. Additions to Schiff bases to give 1,2-thiazetidine 1,1-dioxides,[33] to *NN*-dialkyl-2-aminomethylene-1-tetralones and their heterocyclic analogues to give [34] oxathiin 2,2-dioxides, and to azomethine imines by a concerted $[_\pi 2s + _\pi 4s]$ process to give [35] 1,2,3-thiadiazolidine derivatives (12) have been reported. 1,3-Cycloadditions of sulphenes, as in the last reaction, are uncommon. The reaction of phenyl benzoylmethanesulphonate with substituted salicylaldehydes in basic media to give sultones is proposed [36] to involve transesterification, *via* a sulphene intermediate, followed by a Knoevenagel condensation.

PART III: Thioureas, Thiosemicarbazides, Thioamides, Thiono- and Dithio-carboxylic Acids, their Derivatives, and their Selenium Analogues
by W. Walter and J. Voss

1 Reviews

A review on *N*-functional derivatives of thiocarboxylic acids, covering the synthesis of thioamides, thioureas, and thiocarbamates, has appeared in *Methodicum Chimicum*.[1a]

[29] T. Tanabe, T. Shingaki, and T. Nagai, *Chem. Letters*, 1975, 679.
[30] B. Lamm, *Acta Chem. Scand. (B)*, 1975, **29**, 332.
[31] K. N. Houk, R. W. Strozier, and J. A. Hall, *Tetrahedron Letters*, 1974, 897.
[32] E. N. Prilezhaeva, N. P. Petukhova, V. I. Kurilkin, A. U. Stepanyants, and V. P. Lezina, *Izvest. Akad. Nauk S.S.S.R., Ser. khim.*, 1974, 1827.
[33] T. Hiraoka and T. Kobayashi, *Bull. Chem. Soc. Japan*, 1975, **48**, 480.
[34] P. Schenone, L. Mosti, G. Bignardi, and A. Tasca, *Ann. Chim. (Italy)*, 1974, **64**, 603: P. Schenone, L. Mosti, and G. Bignardi, *J. Heterocyclic Chem.*, 1976, **13**, 225.
[35] W. E. Truce and J. R. Allison, *J. Org. Chem.*, 1975, **40**, 2260.
[36] B. E. Hoogenboom, M. S. El-Faghi, S. C. Fink, P. J. Ihrig, A. N. Langsjoen, C. J. Linn, and K. L. Maehling, *J. Org. Chem.*, 1975, **40**, 880.
[1] (a) W. Walter, in 'Methodicum Chimicum,' Vol. 6, ed. F. Zymalkowski, Academic Press, New York, and Georg-Thieme-Verlag, Stuttgart, 1975, p. 752; (b) G. Zolyomi, *J. Labelled Compounds*, 1974, **10**, 361.

2 Thioureas and Selenoureas

Synthesis.—Unsubstituted ^{35}S-labelled thiourea has been synthesized in 88% yield by treating K[35]SCN (obtained from [35]S_8 and KCN) with benzoyl chloride, aminolysis of the resulting benzoyl isothiocyanate, and subsequent hydrolysis of the benzoyl[^{35}S]thiourea.[1b] The addition of ammonia or amines to isothiocyanates has continued to be the main method for the preparation of substituted thioureas. Williams and Jencks[2] have studied the mechanism of this important reaction [equation (1)], which is first-order in both amine and isothiocyanate and is not catalysed by buffer.

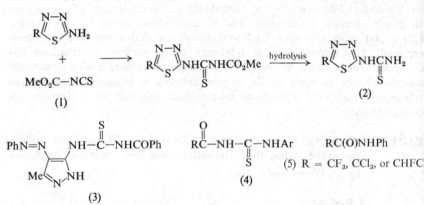

$$\text{\textbackslash{}N-H} + R-N=C=S \rightleftharpoons \text{\textbackslash{}NH-C} \longrightarrow \text{\textbackslash{}N-C-NHR} \quad (1)$$

A very simple, but quite efficient, one-step synthesis of *N*-aryl-thioureas from primary aromatic amines and ammonium thiocyanate has been reported.[3, 4] Activated isothiocyanates, such as methoxycarbonyl isothiocyanate (1) and acyl or aroyl isothiocyanates, have been applied in the preparation of the *N*-heteroarylthioureas (2)[5] and (3),[6] or acyl-thioureas (4).[7–9] Chloroacetyl isothiocyanate

yields thiourea (4; R = CH_2Cl, Ar = Ph) on reaction with aniline, whereas the acyl group is split off and anilides (5) are formed from the di- and tri-substituted acetyl isothiocyanates.[10]

[2] A. Williams and W. P. Jencks, *J.C.S. Perkin II*, 1974, 1753.
[3] C. P. Joshua and K. N. Rajasekharan, *Chem. and Ind.*, 1974, 750.
[4] T. E. Achary, C. S. Panda, and A. Nayak, *J. Indian Chem. Soc.*, 1975, **52**, 1065.
[5] F. Russo, M. Santagati, and M. Alberghina, *Farmaco, Ed. Sci.*, 1975, **30**, 1031 (*Chem. Abs.*, 1976, **84**, 74 187).
[6] M. H. Elnagdi, M. M. M. Sallam, H. M. Fahmy, S. A.-M. Ibrahim, and M. A. M. Elias, *Helv. Chim. Acta*, 1976, **59**, 551.
[7] C.-C. Yu and R. J. Kuhr, *J. Agric. Food Chem.*, 1976, **24**, 134.
[8] K. Oyamada, J. Tobitsuka, M. Saito, and M. Nagano, *Sankyo Kenkyusho Nempo*, 1975, **27**, 85 (*Chem. Abs.*, 1976, **84**, 150 274).
[9] M. O. Lazinskii, A. F. Shivanyuk, V. N. Bodnar, and P. S. Pel'kis, *Zhur. org. Khim.*, 1975, **11**, 1983.
[10] A. V. Fokin, A. F. Kolomiets, Yu. N. Studnev, A. I. Rapkin, and V. I. Yakutin, *Izvest. Akad. Nauk S.S.S.R., Ser. khim.*, 1974, 2349.

Numerous alkyl[11-21] as well as aryl and heteroaryl[12, 16, 18, 19, 22-31] isothio-cyanates have been used for the preparation of thioureas. Phenylazoalkylthiourea (6) is obtained from the corresponding isothiocyanate, which can be prepared by oxidation of 2-phenyl-5,5-dimethyl-1,2,4-triazolidine-3-thione.[15] The unsaturated isothiocyanate $MeSOCH=CH(CH_2)_2NCS$, a constituent of radish seeds, has

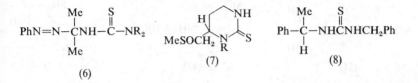

(6)   (7)   (8)

been converted, on aminolysis, into a mixture of two epimeric hexahydro-1,3-diazine-2-thiones (7; R = H, Me, or Ph).[13] The optically active thiourea (8) has been synthesized from (R)-α-phenylethyl bromide and $Hg(SCN)_2$ followed by aminolysis of the resulting isothiocyanate. This method is especially convenient for the preparation of s- and t-alkyl isothiocyanates.[17]

1,2-Dichloroethyl isothiocyanate reacts with thioureas $R^1NHC(S)NR^2_2$ at 100 °C to yield thiazolyl-2-thioureas (9), or at 20 °C to give the thiadiazines (10), which may be rearranged, on heating or with base, to the thiazoles (9).[20]

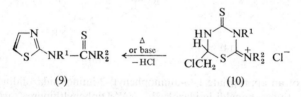

(9)   (10)

[11] H. Böhme and F. Ziegler, *Annalen*, 1974, 1474.
[12] A. Kreutzberger and H. H. Schröders, *Arch. Pharm.*, 1975, **308**, 748.
[13] J. J. Hansen and A. Kjaer, *Acta Chem. Scand.* (*B*), 1974, **28**, 418.
[14] W. Walter and C. Rohloff, *Annalen*, 1975, 295.
[15] J. Schantl, *Monatsh.*, 1974, **105**, 427.
[16] G. Crank, *Tetrahedron Letters*, 1974, 4537.
[17] N. Watanabe, M. Okano, and S. Uemura, *Bull. Chem. Soc. Japan*, 1974, **47**, 2745.
[18] F. T. Boyle and R. Hull, *J.C.S. Perkin I*, 1974, 1541.
[19] V. J. Ram, *J. Indian Chem. Soc.*, 1975, **52**, 240.
[20] R. Lantzsch and D. Arlt, *Synthesis*, 1975, 675.
[21] J. E. Oliver and J. L. Flippen, *J. Org. Chem.*, 1974, **39**, 2233.
[22] M. Furukawa, M. Goto, and S. Hayashi, *Bull. Chem. Soc. Japan*, 1974, **47**, 1977.
[23] R. G. Levit, M. M. Donskaya, and B. V. Unkovskii, *Trudy Mosk. Inst. Tonkoi Khim. Tekhnol.*, 1974, **4**, 50 (*Chem. Abs.*, 1976, **84**, 16 913).
[24] N. B. Galstukhova, Z. A. Pankina, I. M. Berzina, and M. N. Shchukina, *Zhur. org. Khim.*, 1974, **10**, 804.
[25] S. P. Singh, A. Chaudhari, S. S. Parmar, and W. E. Cornatzer, *Canad. J. Pharm. Sci.*, 1974, **9**, 110.
[26] A. M. M. E. Omar, M. S. Ragab, and A. A. B. Hazzaa, *Pharmazie*, 1974, **29**, 445.
[27] D. N. Dhar and R. C. Munjal, *Z. Naturforsch.*, 1974, **29b**, 408.
[28] T. Jen, H. van Hoeven, W. Groves, R. A. McLean, and B. Loev, *J. Medicin. Chem.*, 1975, **18**, 90.
[29] A. Nabeya, T. Shigemoto, and Y. Iwakura, *J. Org. Chem.*, 1975, **40**, 3536.
[30] N. A. Barba and Ya. E. Gutsu, *Zhur. Vsesoyuz. Khim. obshch. im. D. I. Mendeleeva*, 1974, **19**, 109 (*Chem. Abs.*, 1974, **80**, 108 146).
[31] R. E. Manning and F. M. Schaefer, *Tetrahedron Letters*, 1974, 3343.

Adamantylthioureas, which exhibit marked antiviral activity, are obtained from 1-aminoadamantane[12] and isothiocyanates. Compound (11) is similarly obtained from 4-aminophenyl phosphate.[24] Oxazolyl-2-thioureas (12) are only obtained as minor products from 2-amino-oxazoles, the main components of the reaction mixture being the thioamides (13; R = Me, Et, allyl, or Ph).[16]

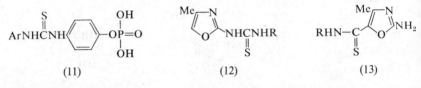

(11)   (12)   (13)

A somewhat surprising reaction takes place if isothiocyanates are treated with DMSO: symmetrically substituted thioureas RNHC(S)NHR (R = Me or Ar) are formed in high yield.[32]

Thioureas have also been synthesized by aminolysis of other thiocarbonic acid derivatives. Carbon disulphide reacted with guanidines to yield $RNHC(S)NH_2$ along with 1,3,5-thiadiazines (14).[33] The tricyclic compound (15) was formed by

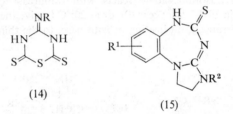

(14)

(15)

cyclization of an appropriate 1-(*o*-aminophenyl)-2-iminoimidazolidine with CS_2, $CSCl_2$, or thiocarbonyldi-imidazole.[34] *NN'*-Diphenylthiourea was formed directly from CS_2 and $PhNLi_2$,[35] and trisubstituted thioureas can be obtained [36] according to equation (2) (R = Ph, $PhCH_2$, or cyclohexyl). The reaction of

$$P(NMe_2)_3 + CS_2 + RNH_2 \longrightarrow Me_2NC(S)NHR \qquad (2)$$

iminotriphenylphosphorane with dimethylthiocarbamyl chloride gives the phosphoranylidenethiourea $Me_2NC(S)N=PPh_3$.[37] Dithiocarbamates have been used by several authors,[38-40] and Doyle and Kurzer[41] have published a valuable review of the synthesis of activated thiocarbonic acid derivatives $R_2NC(S)SCH_2$-

[32] J. B. Chattopadhyaya and A. V. R. Rao, *Synthesis*, 1974, 289.
[33] Y. E. Moharir, *Indian J. Chem.*, 1974, **12**, 490.
[34] G. Doleschall, G. Hornyák, B. Agái, G. Simig, J. Fetter, and K. Lempert, *Tetrahedron*, 1976, **32**, 57.
[35] S. Sakai, T. Aizawa, and T. Fujinami, *J. Org. Chem.*, 1974, **39**, 1970.
[36] N. Yamazaki, T. Tomioka, and F. Higashi, *Synthesis*, 1975, 384.
[37] A. S. Shtepanek, V. V. Doroshenko, V. A. Zasorina, and L. M. Tochilkina, *Zhur. obshche. Khim.*, 1974, **44**, 2130.
[38] K. T. Potts and J. Kane, *J. Org. Chem.*, 1974, **39**, 3783.
[39] S. Palazzo, L. I. Giannola, S. Caronna, and M. Neri, *Atti Accad. Sci., Lett. Arti Palermo Parte* 1, 1973, **33**, 411 (*Chem. Abs.*, 1975, **83**, 114 317).
[40] T. D. Duffy and D. G. Wibberley, *J.C.S. Perkin I*, 1974, 1921.
[41] K. M. Doyle and F. Kurzer, *Chem. and Ind.*, 1974, 803.

CO_2H and $S:C(SCH_2CO_2H)_2$ and their application to the preparation of thioureas. Benzothiazolyl-2-thioureas (16) are obtained by the reaction of 2-aminobenzothiazoles with benzoylthiourea and subsequent hydrolysis.[42] Trimethylthiourea is formed by the action of elemental sulphur on DMF in the

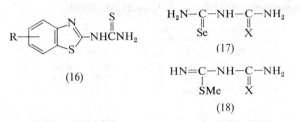

(16)

(17)

(18)

presence of an alkene (*e.g.* isobutene) in a somewhat curious reaction.[43] Seleno-biurets (17; X = O, S, or Se) have been synthesized by selenolysis of the corresponding *S*-methylisothiobiurets, *e.g.* (18).[44]

Physical Properties.—The molecular and electronic structures of thioureas have been studied by various methods.

X-Ray diffraction of (19) has shown that the four sulphur atoms lie on a straight line, but that the S—S- and C=S-distances are not equal, indicating the

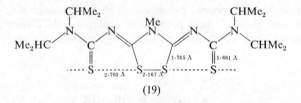

(19)

independent existence of thiourea groups and a 1,2,4-dithiazolidine system in the molecule.[45] Photoelectron spectra and CNDO calculations [46-48] show that the non-bonded *n*-orbital, mostly localized on the sulphur atom of thioureas, has the lowest ionization potential, but is nearly degenerate with the first π-orbital, and the bonding MO's are markedly dependent on the substituents on nitrogen. The *X*-ray photoelectron spectrum of selenourea has been compared with that of thiourea.[49] 3*d*-Orbital interaction in the MOs of several thioureas has been ruled out by measurements of *X*-ray-fluorescence and -absorption spectra.[50]

Thorough computer simulations and CNDO calculations of the u.v. spectra have revealed that, unlike thiophosgene and thioacetamide, thiourea shows no

[42] P. N. Dhal, T. E. Achary, and A. Nayak, *J. Indian Chem. Soc.*, 1974, **51**, 931.
[43] J. P. Brown and M. Thompson, *J.C.S. Perkin I*, 1974, 863.
[44] T. S. Griffin and D. L. Klayman, *J. Org. Chem.*, 1974, **39**, 3161.
[45] J. Sletten, *Acta Chem. Scand.* (*A*), 1974, **28**, 989.
[46] T. P. Debies and J. W. Rabelais, *J. Electron Spectroscopy Related Phenomena*, 1974, **3**, 315 (*Chem. Abs.*, 1974, **80**, 132 346).
[47] C. Guimon, D. Gonbeau, G. Pfister-Guillouzo, L. Åsbrink, and J. Sandström, *J. Electron Spectroscopy Related Phenomena*, 1974, **4**, 49 (*Chem. Abs.*, 1974, **81**, 113 377).
[48] G. W. Mines and H. W. Thompson, *Spectrochim. Acta*, 1975, **31A**, 137.
[49] H. Rupp and U. Weser, *Bioinorg. Chem.*, 1975, **5**, 21 (*Chem. Abs.*, 1975, **83**, 205 426).
[50] A. I. Shpon'ko, L. N. Mazalov, A. P. Sadovskii, S. V. Larionov, and G. K. Parygina, *Izvest. sibirsk. Otdel. Akad. Nauk, Ser. khim. Nauk*, 1975, 138 (*Chem. Abs.*, 1975, **83**, 210 875).

$n \to \pi^*$ excitation.[51] Normal co-ordinate analysis of the i.r. spectra of thiourea,[52] selenourea,[52] and thiobiuret [53] has been carried out. Table 1 shows the decrease of k_{CX} in the order X = O, S, Se, while k_{CN} remains unchanged.

Table 1 *Force constants, k/mdyn Å^{-1}, for $H_2NC(X)NH_2$*

X	k_{CX}		k_{CN}	
	solid	soln.	solid	soln.
O	6.62	8.25	5.22	4.83
S	2.90	3.20	5.23	4.84
Se	2.44		5.23	

Restricted rotation about the CN bond of alkylthioureas ($\Delta G^{\ddagger} = 53.3 \pm 1$ kJ mol^{-1} in NN'-di-t-butylthiourea [54]),[54, 55] and diselenobiuret (17; X = Se [44]), and *syn–anti*-isomerization of *S*-methylisothioureas by an inversion mechanism [56, 57] has been studied by ¹H n.m.r. spectroscopy. The extent and site of protonation (namely at the S atom) of thioureas have been studied in HSO$_3$Cl,[58] and in aqueous HCl,[59] by conductivity, n.m.r., and u.v. spectroscopic methods. Aryl derivatives are weaker bases than alkyl-thioureas.[59]

Analytical Properties and Applications.—Thioureas may be determined quantitatively by titration with phenyl iodosoacetate.[60] Chromatography on SiO$_2$ is suitable for analytical [61, 62] and preparative separation of thioureas, *e.g.* the antiviral agent (20) from its isomer (21).[63] Thioureas have found widespread

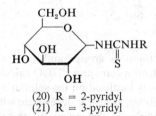

(20) R = 2-pyridyl
(21) R = 3-pyridyl

applications, but their detailed discussion is not within the scope of this Report. Some aspects are, however, mentioned in the text, and may be extended by

[51] J. Barrett and F. S. Deghaidy, *Spectrochim. Acta*, 1975, **31A**, 707.
[52] K. Geetharani and D. N. Sathyanarayana, *Spectrochim. Acta*, 1976, **32A**, 227.
[53] D. Hadži, J. Kidrič, Ž. U. Knezevic, and B. Barlič, *Spectrochim. Acta*, 1976, **32A**, 693.
[54] C. D. Freeman and D. L. Hooper, *J. Phys. Chem.*, 1974, **78**, 961.
[55] R. H. Sullivan and E. Price, *Org. Magn. Resonance*, 1975, **7**, 143.
[56] A. P. Engoyan, T. F. Vlasova, Yu. N. Sheinker, and L. A. Ignatova, *Doklady Akad. Nauk S.S.S.R.*, 1975, **225**, 1329.
[57] G. Tóth, L. Toldy, I. Tóth, and B. Rezessy, *Tetrahedron*, 1974, **30**, 1219.
[58] S. A. A. Zaidi and Z. A. Siddiqi, *J. Inorg. Nuclear Chem.*, 1975, **37**, 1806.
[59] J. Maslowska and R. Soloniewicz, *Roczniki Chem.*, 1974, **48**, 1391.
[60] K. K. Verma, *Z. analyt. Chem.*, 1975, **275**, 287.
[61] H. Nakamura and Z. Tamura, *J. Chromatog.*, 1974, **96**, 195.
[62] M. B. Devani, C. J. Shishoo, and B. K. Dadia, *J. Chromatog.*, 1975, **105**, 186.
[63] M. Szczesniak and W. Wieniawski, *Acta Polon. Pharm.*, 1974, **31**, 27 (*Chem. Abs.*, 1975, **82**, 106 087).

references concerning radio-protective properties,[64, 65] stimulation of plant growth,[66-68] or use as solvent for the selective extraction of metal ions.[69]

Chemical Reactions.—Two typical features of thioureas are the reason for most of their reactions: the nucleophilicity (mainly of the sulphur atom), which is responsible for alkylation, oxidation, and related reactions; and the bi- or tri-functionality of thiourea molecules, which leads to the formation of a vast number of heterocycles. These two reaction types are not clearly distinguishable, but may be used as a method of classification.

Alkylation. Theoretical treatment by the CNDO method, and the analysis of orbital interactions, show a frontier orbital control for the *S*-methylation of MeNHC(S)NHMe.[70] Kinetic measurements of the alkylation of thioureas with alkyl halides[71] or tosylates[72] have shown that this reaction is first-order in thiourea and alkylating agent, implying an S_N2 mechanism, with direct attack by the sulphur atom.

S-Alkylisothiouronium salts of various types,[73-80] *e.g.* the ^{14}C-labelled (22),[73] (23), which exhibits surface activity,[74] the glucosamine (24),[75] and the seleno-sugar (25),[76] have been prepared from the corresponding alkyl halides, thiirans,[81] or 4-hydroxyphosphonolactone ('phostone'),[82] whereas the *N*-alkylated thio-urea $H_2NC(S)NH(CH_2)_4SO_3H$ is formed from butanesultone.[83] 2-Halogeno-pyridines,[84, 85] and -indoles[86, 87] yield the corresponding *S*-heteroaryl-isothioureas and isoselenuronium salts. Compound (26), which exhibits antitumour activity against leukaemia cells,[88] is derived from the corresponding halogeno-purine.[88, 89]

[64] R. Badiello, O. Sapora, G. Simone, and M. Tamba, *Ann. Ist. Super Sanità*, 1974, **10**, 147 (*Chem. Abs.*, 1976, **84**, 101 606).

[65] J. Barnes, G. Esslemont, and P. Holt, *Makromol. Chem.*, 1975, **176**, 275.

[66] N. Mashev and G. Vasilev, *Fiziol. Rast.*, 1974, **1**, 19 (*Chem. Abs.*, 1975, **83**, 92 190).

[67] R. Radnev, N. Mashev, and G. Vasilev, *Rastenievud. Nauki*, 1975, **12**, (8), 21 (*Chem. Abs.*, 1976, **84**, 100 722).

[68] N. Shimizu, H. Takahashi, and K. Tajima, *Nippon Sochi Gakkai-Shi*, 1974, **20**, 173.

[69] Yu. A. Zolotov, I. V. Seryakova, and G. A. Vorob'eva, *Proc. Internat. Solvent Extract. Conf.*, 1974, **2**, 1923 (*Chem. Abs.*, 1975, **83**, 121 576).

[70] D. Gonbeau and G. Pfister-Guillouzo, *Canad. J. Chem.*, 1976, **54**, 118.

[71] A. M. Bhatti and N. K. Ralhan, *Indian J. Chem.*, 1974, **12**, 969.

[72] D. J. McLennan, *Tetrahedron Letters*, 1975, 4689.

[73] L. Kronrad and I. Kozak, *Radioisotopy*, 1974, **15**, 135.

[74] I. P. Komkov and I. M. Kuleshova, *Izvest. V.U.Z., Khim. i khim. Tekhnol.*, 1974, **17**, 820 (*Chem. Abs.*, 1975, **82**, 32 763).

[75] M. E. Rafestin, A. Obrenovitch, A. Oblin, and M. Monsigny, *F.E.B.S. Letters*, 1974, **40**, 62.

[76] G. C. Chen, R. A. Zingaro, and C. R. Thompson, *Carbohydrate Res.*, 1975, **39**, 61.

[77] W. Heffe, R. W. Balsiger, and K. Thoma, *Helv. Chim. Acta*, 1974, **57**, 1242.

[78] Y. Ueno, Y. Masuyama, and M. Okawara, *Tetrahedron Letters*, 1974, 2577.

[79] R. E. Ardrey and L. A. Cort, *J.C.S. Perkin I*, 1976, 121.

[80] A. V. Fokin, A. F. Kolomiets, and T. I. Fedyushina, *Doklady Akad. Nauk S.S.S.R.*, 1976, **227**, 104.

[81] E. Vilsmaier and G. Becker, *Synthesis*, 1975, 55.

[82] H. Stutz and H.-G. Henning, *Z. Chem.*, 1975, **15**, 52.

[83] I. Zeid, H. Moussa, and I. Ismail, *Annalen*, 1974, 1816.

[84] T. Zawisza and S. Respond, *Roczniki Chem.*, 1975, **49**, 743.

[85] B. Iddon, H. Suschitzky, A. W. Thompson, and E. Ager, *J.C.S. Perkin I*, 1974, 2300.

[86] T. Hino, M. Endo, and M. Nakagawa, *Chem. and Pharm. Bull. (Japan)*, 1974, **22**, 2728.

[87] T. Hino, M. Tonozuka, and M. Nakagawa, *Tetrahedron*, 1974, **30**, 2123.

[88] S.-H. Chu, C.-Y. Shiue, and M. Y. Chu, *J. Medicin. Chem.*, 1975, **18**, 559.

[89] S.-H. Chu, C.-Y. Shiue, and M. Y. Chu, *J. Pharm. Sci.*, 1975, **64**, 1343.

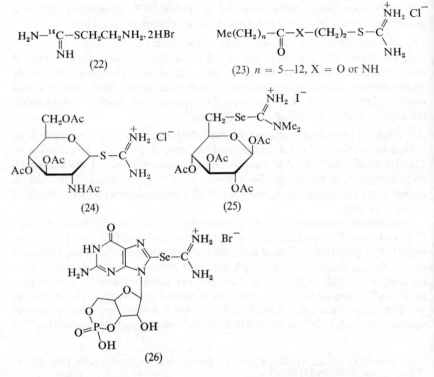

(22)

(23) $n = 5—12$, X = O or NH

(24)

(25)

(26)

The reaction of halides and subsequent alkaline hydrolysis without isolation of the intermediate isothiouronium salt has been used for the synthesis of thiols.[90—94] $\alpha\alpha'$-Dimercapto-*o*-xylene has been obtained from the corresponding bromide in 97% yield.[90]

Addition to Multiple Bonds. Kinetic measurements of the addition of acrylic acid and its methyl ester have shown this reaction to be first-order in both thiourea and alkenes.[95] Divinyl sulphides (27)[96] and (28)[97] are formed from the corresponding alkynes and thioureas, but rearrangement is observed[98] in the reaction with diphenylacetylene to give $(Ph_2C=CH)_2S$. *S*-Vinyl-isoselenourea (29) is obtained from the addition of selenourea to 1,3-diphenylprop-2-yn-1-one,

[90] J. J. Mayerle, S. E. Denmark, B. V. DePamphilis, J. A. Ibers, and R. H. Holm, *J. Amer. Chem. Soc.*, 1975, **97**, 1032.
[91] P. Cagniant and G. Kirsch, *Compt. rend.*, 1974, **279**, C, 829.
[92] M. Cirule and Yu. A. Bankovskii, *Latv. P.S.R., Zinat. Akad. Vestis, Kim. Ser.*, 1974, 496 (*Chem. Abs.*, 1974, **81**, 169 417).
[93] R. M. Titkova, A. S. Elina, E. N. Padeiskaya, and L. M. Polukhina, *Khim. Farm. Zhur.*, 1975, **9**, 10 (*Chem. Abs.*, 1975, **82**, 139 983).
[94] K. Schulze, E.-M. Dietrich, and M. Mühlstädt, *Z. Chem.*, 1975, **15**, 302.
[95] A. I. Konovalov, L. K. Konovalova, and E. G. Kataev, *Zhur. org. Khim.*, 1974, **10**, 1580.
[96] S. V. Amosova, O. A. Tarasova, B. A. Trofimov, G. A. Kalabin, V. M. Bzhezovskii, and V. B. Modonov, *Zhur. org. Khim.*, 1975, **11**, 2026.
[97] M. N. Basyouni and M. T. Omar, *Austral. J. Chem.*, 1974, **27**, 1585.
[98] B. A. Trofimov, S. V. Amosova, O. A. Tarasova, V. M. Bzhezovskii, V. B. Modonov, and G. A. Kalabin, *Zhur. org. Khim.*, 1975, **11**, 657.

$(RCH{=}CH)_2S$ $(ArCOCH{=}CPh)_2S$ $PhCOCH{=}C-Se-C-NH_2$

$\qquad\qquad\qquad\qquad\qquad\qquad\qquad\qquad\qquad\qquad\quad$ $\underset{Ph}{|}\qquad\underset{NH}{\|}$

$\qquad$ (27) $\qquad\qquad\qquad$ (28) $\qquad\qquad\qquad\qquad$ (29)

$\underset{NAr}{\overset{Ar}{ArNHC-N-CNHAr,HCl}}$ $\underset{NR\quad NH}{R_2NC-S-CNH_2,HCl}$ $\underset{NR\quad CO_2CH_2Ph}{RNHC-S-N-NHCO_2CH_2Ph}$

$\qquad$ (30) $\qquad\qquad\qquad\qquad$ (31) $\qquad\qquad\qquad\qquad$ (32)

$PhCOC{\equiv}CPh$.[99] $C{=}N-$, $C{\equiv}N-$, and $-N{=}N-$ bonds also undergo addition reactions with thioureas, yielding formamidino-thioureas (30),[100] thio-bis-(formamidines) (31),[101] and the sulphenylhydrazine (32),[102] from $ArN{=}C{=}NAr$, $H_2NC{\equiv}N$, and $PhCH_2O_2CN{=}NCO_2CH_2Ph$, respectively.

Oxidation. Formamidino-disulphides (33) are formed by oxidation of thioureas with 2 equivalents of Ce^{IV} [103] or $PhSMe\overset{+}{N}H_2\ X^-$,[104] ureas being the products with an excess of Ce^{IV}.[103] Hydrogen peroxide yields thiourea *SS*-dioxides.[105] Tri- and

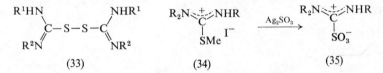

$\qquad$ (33) $\qquad\qquad\qquad\qquad$ (34) $\qquad\qquad\qquad\qquad$ (35)

tetra-substituted thiourea *SSS*-trioxides (35) are not obtainable by oxidation, but can be prepared from isothiouronium (34) or chloro-formamidinium salts.[14]

The decomposition of thioureas by nitrous acid has been studied kinetically, and S-to-N migration of the NO group in the intermediate [equation (3)], which

$$(H_2N)_2C{=}S + HO{-}NO \underset{}{\overset{fast}{\rightleftharpoons}} [(H_2N)_2C{=}\overset{+}{S}{-}NO]$$

$$[(H_2N)_2C{=}\overset{+}{S}{-}NO] \xrightarrow{slow} HSCN + N_2 + H_3O^+ \qquad\qquad (3)$$

may be detected by u.v. spectroscopy, has been found to be the rate-determining step.[106, 107]

Hydrolysis and Related Reactions. These reactions involve nucleophilic substitution and not electrophilic attack on sulphur. The alkaline decomposition of thiourea and *N*-ethylthiourea, which is first order in thiourea and NaOH, has

[99] G. Wilke and E. Uhlemann, *Z. Chem.*, 1974, **14**, 288.
[100] R. Evers and E. Fischer, *Z. Chem.*, 1976, **16**, 15.
[101] R. Evers, G. Rembarz, and E. Fischer, *Z. Chem.*, 1976, **16**, 101.
[102] K. K. De, G. T. Shiau, and R. E. Harmon, *J. Carbohydrates, Nucleosides, Nucleotides*, 1975, **2**, 259.
[103] N. M. Turkevich and R. T. Dmitrishin, *Visn. L'viv. Politekh. Inst.*, 1974, **82**, 38 (*Chem. Abs.*, 1975, **83**, 52 982).
[104] H. J. Krueger and R. J. Kiyokane, *Inorg. Chem.*, 1974, **13**, 2522.
[105] J. J. Havel and R. Q. Kluttz, *Synthetic Comm.*, 1974, **4**, 389.
[106] K. Al-Mallah, P. Collings, and G. Stedman, *J.C.S. Dalton*, 1974, 2469.
[107] P. Collings, K. Al-Mallah, and G. Stedman, *J.C.S. Perkin II*, 1975, 1734.

been studied kinetically, and the activation energies have been measured.[108-110] Exchange reactions with $Na^{35}SH$ and $^{35}S_8$ have been used for the preparation of ^{35}S-labelled thioureas.[111] The elimination of the elements of hydrogen sulphide from several thioureas, to give carbodi-imides $RN=C=NR$, has been achieved by *N*-phenylbenzimidoyl chloride,[112] $PhN=C(Cl)Ph$, or chloro-hetero-aromatics.[113]

The addition of $SO_3^{\overline{\cdot}}$ to form the persistent radical anions $R_2NC(NR_2)$-$(SO_3^-)S\cdot$ ($g = 2.0051$ for $R = H$) has been studied by e.s.r. spectroscopy.[114]

Cyclization. Formation of heterocycles is the most frequently used reaction of thioureas, and the literature until 1973 has been reviewed.[115] This section deals with the different types of heterocycles derived from thioureas.

The thiiran groups of thia-tris-*σ*-homobenzenes (37) and (38) are formed from the tris(oxiran) (36) *via* isothiouronium salts.[116-118] Interconversion of epoxides into episulphides has also been reported in the carbohydrate series.[119]

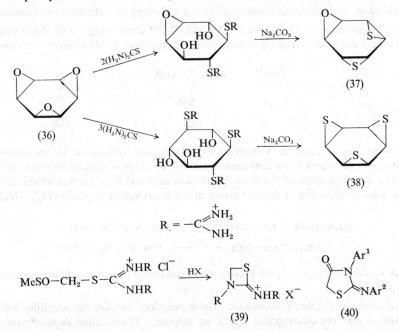

108 G. A. Kitaev, and I. T. Romanov, *Izvest. V.U.Z., Khim. i khim. Tekhnol.*, 1974, **17**, 1427 (*Chem. Abs.*, 1975, **82**, 30 741).
109 G. Marcotrigiano, R. Battistuzzi, and G. Peyronel, *Gazzetta*, 1974, **104**, 781.
110 R. Battistuzzi, G. Marcotrigiano, and G. Peyronel, *J.C.S. Perkin II*, 1975, 169.
111 T. Sato, *Radioisotopes*, 1974, **23**, 145.
112 S. Furumoto, *Yuki Gosei Kagaju Kyokai Shi*, 1975, **33**, 748 (*Chem. Abs.*, 1976, **84**, 59 367).
113 S. Furumoto, *Yuki Gosei Kagaku Kyokai Shi*, 1974, **32**, 727 (*Chem. Abs.*, 1975, **82**, 125 361).
114 T. Ozawa, M. Setaka, H. Yamamoto, and T. Kwan, *Chem. and Pharm. Bull.* (*Japan*), 1974, **22**, 962.
115 T. S. Griffin, T. S. Woods, and D. L. Klayman, *Adv. Heterocyclic Chem.*, 1975, **18**, 99.
116 H. Prinzbach, C. Kaiser, and H. Fritz, *Angew. Chem.*, 1975, **87**, 249.
117 S. Kagabu and H. Prinzbach, *Angew. Chem.*, 1975, **87**, 248.
118 S. Kagabu and H. Prinzbach, *Tetrahedron Letters*, 1975, 29.
119 M. V. Jesudason and L. N. Owen, *J.C.S. Perkin I*, 1974, 2019.

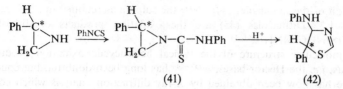

(41)   (42)

Cyclization of S-methylsulphinylmethyl-isothiouronium salts yields azathiets (39).[78]

Thioureas react with α-halogenocarboxylic acids and their derivatives to give 2-aminothiazolin-4-ones,[42, 120-124] *e.g.* (40), which exhibits fungicidal activity.[120] The reaction of α-bromo-ketones[125-135] and analogous compounds[26, 136, 137] with thioureas gives 2-amino-thiazoles by the Hantzsch synthesis. Ring-expansion of N-thiocarbamoylaziridine (41) yields the aminothiazoline (42), with 60% retention of configuration,[29] while oxidation of arylthioureas gives 2-amino-benzothiazoles by the Hugershoff reaction.[138] 2-Aminoselenazole has been obtained from selenourea.[139] (See also Chapter 12, p. 384, and Chapter 13, p. 387.)

Besides thiazoles, five-membered heterocycles of different types are obtainable from thioureas. Total loss of sulphur takes place in the condensation with arylacetonitriles, 2-amino-imidazoles (43) being formed,[140] whereas in the

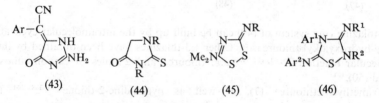

(43)   (44)   (45)   (46)

[120] S. R. Singh, *J. Indian Chem. Soc.*, 1975, **52**, 734.
[121] A. A. Shaikh, *J. Inst. Chem. Calcutta*, 1974, **46**, 135 (*Chem. Abs.*, 1975, **82**, 125 312).
[122] A. Singh and A. S. Uppal, *Austral. J. Chem.*, 1975, **28**, 1049.
[123] M. Augustin, W.-D. Rudorf, and R. Pasche, *Z. Chem.*, 1974, **14**, 434.
[124] P. N. Dhal, T. E. Achary, and A. Nayak, *Indian J. Chem.*, 1975, **13**, 753.
[125] J. Bödeker, H. Pries, D. Rösch, and G. Malewski, *J. prakt. Chem.*, 1975, **317**, 953.
[126] B. V. Passet, G. N. Kul'bitskii, V. Ya. Samarenko, and L. I. Vekshina, *Khim. Farm. Zhur.*, 1974, **8** (11), 48 (*Chem. Abs.*, 1975, **82**, 57 596).
[127] S. N. Sawhney, J. Singh, and O. P. Bansal, *J. Indian Chem. Soc.*, 1974, **51**, 566.
[128] S. N. Sawhney, J. Singh, and O. P. Bansal, *J. Indian Chem. Soc.*, 1975, **52**, 561.
[129] J. Mohan and H. K. Pujari, *Indian J. Chem.*, 1975, **13**, 871.
[130] W. A. Remers and G. S. Jones, jun., *J. Heterocyclic Chem.*, 1975, **12**, 421.
[131] H. K. Gakhar, V. Parkash, and K. Bhushan, *J. Indian Chem. Soc.*, 1974, **51**, 941.
[132] J. Ashby and D. Griffiths, *J.C.S. Perkin I*, 1975, 657.
[133] J.-F. Robert and J. J. Panouse, *Compt. rend.*, 1974, **278**, C, 1289.
[134] B. Arena, R. Gulbe, and A. Arens, *Latv. P.S.R. Zinat. Akad. Vestis, Kim. Ser.*, 1975, 600 (*Chem. Abs.*, 1976, **85**, 21 184).
[135] W. Ried and L. Kaiser, *Annalen*, 1976, 395.
[136] V. Barkane, E. Gudriniece, and D. Skerite, *Latv. P.S.R. Zinat. Akad. Vestis., Kim. Ser.*, 1975, 345 (*Chem. Abs.*, 1975, **83**, 178 899).
[137] K. S. Dhaka, J. Mohan, V. K. Chadha, and H. K. Pujari, *Indian J. Chem.*, 1974, **12**, 966.
[138] M. S. A. El-Meligy and S. A. Mohamed, *J. prakt. Chem.*, 1974, **316**, 154.
[139] A. Shafiee and I. Lalezari, *J. Heterocyclic Chem.*, 1975, **12**, 675.
[140] L. A. Grigorian, R. G. Mirzoyan, M. A. Kaldrikyan, and A. A. Aroyan, *Armyan. khim. Zhur.*, 1975, **28**, 564 (*Chem. Abs.*, 1976, **84**, 4890).

152   *Organic Compounds of Sulphur, Selenium, and Tellurium*

formation of the imidazoles (44) [141, 142] the sulphur is retained in a thiocarbonyl group. 1,2,4-Dithiazoles (45) are the oxidation products of dithiobiurets $Me_2NC(S)NHC(S)NHR$.[21, 143, 144]

The molecular structure of the typical heterocyclic oxidation products of thioureas, *i.e.* the Hector bases,[38, 145-148] has long been doubtful, but conclusive evidence has now been obtained by *X*-ray diffraction analysis which confirms that they are 3,5-di-imino-1,2,4-thiadiazoles (46).[145] Simultaneous oxidation of different thioureas has been studied, and uniform thiadiazoles (46), not mixtures, have been isolated.[146-148] The isomeric 1,3,4-thiadiazoles may be obtained from hydrazonyl chlorides $ArNHN=C(Cl)Ph$[149] (see also Chapter 14, p. 423).

Various types of six-membered rings are formed from thioureas. Elimination of isothiocyanate from the butadienylthiourea $ArNHC(S)NH(CH=CH)_2CHO$ yields pyridine,[18] whereas the oxazine (48) arises from ketone (47).[150] The

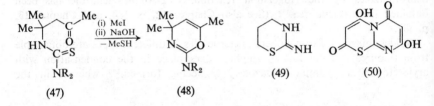

(47)   (48)   (49)   (50)

1,3-thiazine ring system of (49) can be built up by the intramolecular cyclization of γ-hydroxyalkyl-thioureas.[28] Other 1,3-thiazines have been obtained by intermolecular reactions,[39, 84, 151-155] *e.g.* thiourea and 2 moles of carbon suboxide yield (50).[151]

Trimethylenethiourea (7), as well as pyrimidine-2-thiones [40, 156-161] and

141 B. Behura, G. Panda, A. C. Rath, and G. N. Mahapatra, *Indian J. Chem.*, 1974, **12**, 781.
142 M. Behera, P. N. Dhal, and A. Nayak, *J. Indian Chem. Soc.*, 1975, **52**, 1067.
143 J. E. Oliver and A. B. DeMilo, *J. Org. Chem.*, 1974, **39**, 2225.
144 J. E. Oliver and R. T. Brown, *J. Org. Chem.*, 1974, **39**, 2228.
145 C. Christophersen, T. Øttersen, K. Seff, and S. Treppendahl, *J. Amer. Chem. Soc.*, 1975, **97**, 5237.
146 C. P. Joshua and P. N. K. Nambisan, *Indian J. Chem.*, 1974, **12**, 962.
147 P. N. K. Nambisan, *Tetrahedron Letters*, 1974, 2907.
148 C. P. Joshua and P. N. K. Nambisan, *Indian J. Chem.*, 1975, **13**, 241.
149 P. Wolkoff, S. T. Nemeth, and M. S. Gibson, *Canad. J. Chem.*, 1975, **53**, 3211.
150 L. A. Ignatova, A. E. Gekhman, M. A. Spektor, P. L. Ovechkin, and B. V. Unkovskii, *Khim. geterotsikl. Soedinenii*, 1974, 764.
151 T. Kappe, G. Lang, and E. Ziegler, *Z. Naturforsch.*, 1974, **29b**, 258.
152 N. Cohen, B. L. Banner, J. F. Blount, G. Weber, M. Tsai, and G. Saucy, *J. Org. Chem.*, 1974, **39**, 1824.
153 E. Akerblom, *Chem. Scripta*, 1974, **6**, 35 (*Chem. Abs.*, 1975, **82**, 43 301).
154 M. Fuertes, M. T. García-López, G. García-Muñoz, and R. Madroñero, *J. Carbohydrates, Nucleosides, Nucleotides*, 1975, **2**, 277.
155 A. N. Mirskowa, G. G. Levkovskaya, and A. S. Atavin, *Zhur. org. Khim.*, 1976, **12**, 904.
156 L. Pritasil and J. Filip, *Radioisotopy*, 1975, **16**, 297.
157 V. Krchnak and Z. Arnold, *Coll. Czech. Chem. Comm.*, 1974, **39**, 3327.
158 H. Bredereck, G. Simchen, and W. Griebenow, *Chem. Ber.*, 1974, **107**, 1545.
159 R. Nutiŭ, I. Sebe, and M. Nutiŭ, *Rev. Roumaine Chim.*, 1974, **19**, 679.
160 S. M. S. Chauhan and H. Junjappa, *Synthesis*, 1974, 880.
161 S. M. S. Chauhan and H. Junjappa, *Tetrahedron*, 1976, **32**, 1779.

dihydropyrimidine-2-thiones,[162-167] is prepared from appropriate bifunctional reagents. For instance, ^{14}C-labelled thiourea and ethyl acetoacetate have been used for the synthesis of totally ^{14}C-labelled 6-methyluracil (52) *via* (51),[156] and cyclization of RNHC(S)NR(CH$_2$)$_2$CN provides a route to dihydro-2-thiouracils.[162]

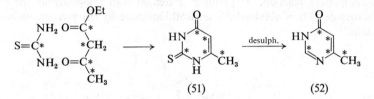

(51)   desulph.   (52)

The 1,2,4-selenadiazine (53) is formed from allylthiourea and Se$_2$Cl$_2$.[168]
Symmetrical 4-imino-1,3,5-thiadiazines (54) [169] or (55) [170] result from thioureas and sodium dicyanoamide, and 1,3,5-thiadiazinium salts (58) are formed from

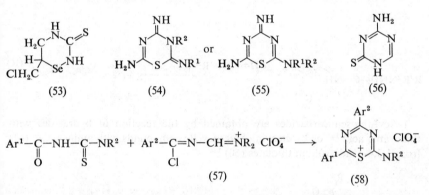

(53)   (54)   or   (55)   (56)

$$Ar^1-C(=O)-NH-C(=S)-NR^2 + Ar^2-C(=N-CH=\overset{+}{N}R_2)Cl \; ClO_4^- \longrightarrow$$

(57)   (58)

the imidoyl chlorides (57).[171] 1,3,5-Triazine-2-thiones,[22, 172, 173] *e.g.* 2-thioxo-5-azacytosine (56), from which the nucleotide analogue 5-azacytidine can be prepared,[172] are obtained from the reaction of ethyl orthoformate with, for example, thiocarbamoylguanidine, NH$_2$C(S)N=C(NH$_2$)$_2$, in DMF at 150 °C.

3 Thiosemicarbazides and Selenosemicarbazides

Synthesis.—The principal synthetic method in the field of thiosemicarbazide chemistry is the reaction of isothiocyanates with hydrazines, and new thiosemi-

[162] M. Dembecki and T. Pyl, *Z. Chem.*, 1976, **16**, 148.
[163] Y. Yuki and K. Inoue, *Nippon Kagaku Kaishi*, 1974, 2140 (*Chem. Abs.*, 1975, **82**, 140 061).
[164] M. I. Ali, A.-E. M. Abd-Elfattah, and H. A. Hammouda, *Z. Naturforsch.*, 1976, **31b**, 254.
[165] M. I. Ali, M. A. F. Elkaschef, and A.-E. G. Hammam, *J. Chem. and Eng. Data*, 1975, **20**, 128.
[166] R. Neidlein and H.-G. Hege, *Chem.-Ztg.*, 1974, **98**, 513.
[167] F. Hofmann, D. Heydenhauss, G. Jaenecke, L. Meister, and H. Voig, *Z. Chem.*, 1975, **15**, 441.
[168] M. Apostolescu, *Bull. Inst. Politeh. Iasi*, 1974, **20**, 9 (*Chem. Abs.*, 1974, **82**, 112 049)
[169] E. Fischer, R. Evers, and G. Rembarz, *Chimia (Switz.)*, 1974, **28**, 388.
[170] E. Fischer, R. Evers, and G. Rembarz, *Chimia (Switz.)*, 1974, **28**, 390.
[171] J. Liebscher and H. Hartmann, *Z. Chem.*, 1975, **15**, 438.
[172] U. Niedballa and H. Vorbrüggen, *J. Org. Chem.*, 1974, **39**, 3672.
[173] C. P. Joshua and V. P. Rajan, *Austral. J. Chem.*, 1975, **28**, 427.

carbazides have been prepared by this method.[174-180] An interesting modification by Neidlein and Hege [181] involves the use of trimethylsilyl isothiocyanate. The thiosemicarbazide was prepared *in situ* from Me$_3$SiNCS and phenylhydrazine, and immediately combined with a ketone to form the thiosemicarbazones in a three-component reaction. If pyridine is treated with thiophosgene the thio-semicarbazide (59) is obtained *via* an isothiocyanate by the given sequence of reactions.[182]

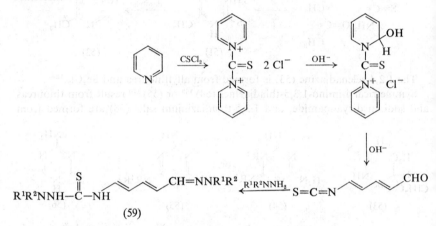

(59)

1-Acyl-thiosemicarbazides are obtained by the reaction of hydrazides with thiocyanic acid [183] or by cleaving heterocyclic compounds with amines, *e.g.* (61) from 1,3,4-oxadiazolium thiolates (60).[184]

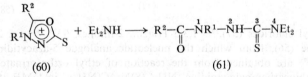

(60)   (61)

1-Oxamoyl-4-phenylthiosemicarbazide (62) is obtained either from oxamoyl-hydrazide and phenyl isothiocyanate or from 3-thioxohexahydro-1,2,4-triazine-5,6-dione (63) with aniline.[185] 2-Amino-4-phenyl-1,3,4-thiadiazepin-5-one, on acid hydrolysis, gives 1-phenylthiosemicarbazide and its *S*-carboxyethyl-

174 H. Böhme and F. Ziegler, *Annalen*, 1974, 1474.
175 D. Twomey, *Proc. Roy. Soc. Irish Acad., Sect. B*, 1974, **74**, 37 (*Chem. Abs.*, 1974, **81**, 3864).
176 J. H. Wikel and C. J. Paget, *J. Org. Chem.*, 1974, **39**, 3506.
177 G. J. Ikeda, C. B. Estep, L. H. Wiemeler, and A. Alter, *J. Medicin. Chem.*, 1974, **17**, 1079.
178 A. K. Sen Gupta, K. Avasthi, and P. K. Seth, *J. Indian Chem. Soc.*, 1975, **52**, 1084.
179 S. P. Singh, R. S. Misra, S. S. Parmar, and S. J. Brumleve, *J. Pharm. Sci.*, 1975, **64**, 1245.
180 G. N. Vasilev and N. P. Maschev, *Biochem. Physiol. Pflanz.*, 1974, **165**, 49 (*Chem. Abs.*, 1974, **80**, 104 762).
181 R. Neidlein and H. G. Hege, *Chem.-Ztg.*, 1974, **98**, 512.
182 F. T. Boyle and R. Hull, *J.C.S. Perkin I*, 1974, 1541.
183 J.-P. Hénichart, B. Lablanche, and R. Houssin, *Compt. rend.*, 1976, **282**, C, 857.
184 A. R. McCarthy, W. D. Ollis, and C. A. Ramsden, *J.C.S. Perkin I*, 1974, 627.
185 U. Anthoni, B. Mynster Dahl, H. Eggert, C. Larsen, and P. H. Nielsen, *Acta Chem. Scand.* (*B*), 1976, **30**, 71.

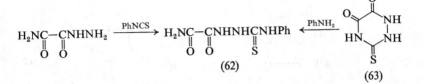

$$\text{H}_2\text{NC}-\text{CNHNH}_2 \xrightarrow{\text{PhNCS}} \text{H}_2\text{NC}-\text{CNHNHCNHPh} \xleftarrow{\text{PhNH}_2}$$

(62)

(63)

derivative.[186] Some bis(thiocarbamoyl)hydrazines have been described[187] and the synthesis of 4-hydroxy-thiosemicarbazides has been reported.[188, 189] [*Cf.* (61) for the numbering.]

Physical Properties.—Unsubstituted isatin-2-thiosemicarbazones exist primarily as iminohydrazine tautomers (64), although with increasing substitution in the side-chain the tautomer (65) is favoured.[190] 3-Thiosemicarbazones and their

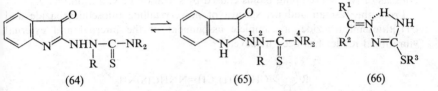

(64)

(65)

(66)

4,4-dialkyl-derivatives predominantly exist as (*Z*)-isomers,[191] but isothiosemi-carbazones with an unsubstituted 2-position exist exclusively in the amino-form (66), with the (*E*)-configuration. Stabilization by an intramolecular hydrogen-bond, as shown in (66), occurs when N-4 is only bonded to hydrogen, but if N-4 is monosubstituted, an equilibrium mixture of (*E*)- and (*Z*)-configurations is observed, the position of equilibrium depending on the solvent used.[192] *X*-Ray diffraction analysis has shown that 5-hydroxy-2-formylpyridine thiosemicarbazone sesquihydrate and acetone thiosemicarbazone exist in the (*E*)-configuration, in the solid state. The N—N bond in the pyridine derivative is significantly shorter (1.379 Å) than that in the acetone thiosemicarbazone (1.398 Å).[193] In the *NN'*-bis(thiocarbamoyl)hydrazine [H$_2$NC(S)NH]$_2$, each of the central N atoms has 3 intramolecular hydrogen-bonds (2.0—2.5 Å), which may be responsible for the twisting of the molecule; the bonds of each C atom have almost regular sp^2 symmetry, the C—S (1.691—1.701 Å) and C—N distances (1.31—1.34 Å)

[186] V. I. Pleshnev, *Farm. Zhur.* (*Kiev*), 1974, **29**, 88 (*Chem. Abs.*, 1975, **82**, 112 053).
[187] Z. Budesinsky, F. Roubinek, J. Kral, D. Nemcova, and V. Janata, *Cesk. Farm.*, 1974, **23**, 200 (*Chem. Abs.*, 1975, **82**, 42 936); R. G. Dubenko and E. F. Gorbenko, *Khim. geterotsikl. Soedinenii*, 1975, 346 (*Chem. Abs.*, 1975, **83**, 28 144).
[188] P. Gröbner and E. Müller, *Monatsh.*, 1974, **105**, 969.
[189] E. Müller and P. Gröbner, *Monatsh.*, 1975, **106**, 27.
[190] A. B. Tomchin and G. A. Shirokii, *Zhur. org. Khim.*, 1974, **10**, 2465; A. B. Tomchin, G. A. Shirokii, and V. S. Dmitrukha, *Khim. geterotsikl. Soedinenii*, 1976, 83 (*Chem. Abs.*, 1976, **84**, 179 383).
[191] A. B. Tomchin, I. S. Ioffe, A. J. Kol'tsov, and Yu. V. Lepp, *Khim. geterotsikl. Soedinenii*, 1974, 503 (*Chem. Abs.*, 1974, **81**, 37 145).
[192] C. Yamazaki, *Canad. J. Chem.*, 1975, **53**, 610.
[193] G. J. Palenik, D. F. Rendle, and W. S. Carter, *Acta Cryst.*, 1974, **30B**, 2390.

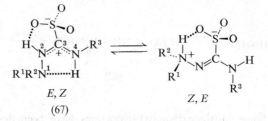

E,Z   Z, E

(67)

indicating a high bond order.[194] In the thiosemicarbazide S-trioxides an equilibrium between a 2,4-betaine with (E,Z)-configuration (67) and a 1-betaine with (Z,E)-configuration can be observed.[195] These studies of tautomerism and configuration are based mainly on n.m.r. and i.r. spectra.

For analytical reasons, the u.v. spectrum of (68) has been investigated in several solvents and in water at different pH values. The long-wavelength band consists of two overlapping bands caused by solvation or vibration.[196]

From the Raman and i.r. spectra of polycrystalline thiosemicarbazide and [^{2}H$_5$]thiosemicarbazide, a complete assignment of the internal and external vibrational modes has been achieved.[197]

$$R\text{—}\underset{O}{\overset{}{\boxed{}}}\text{—}(CH=CH)_nCH=NNHC(S)NH_2$$

(68)

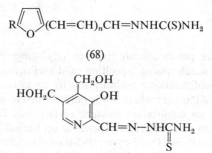

(69)

The pK_a values of β-resorcylidene thiosemicarbazone [198] and of picoline-2-aldehyde thiosemicarbazone [199] have been determined, and the reactions of the latter with various metal ions studied. Phenanthraquinone monothiosemicarbazone can be used as indicator for chelatometric determinations of copper, zinc, cadmium, mercury, and nickel ions.[200] Compound (69) was prepared in an attempt to utilize its chelating potential in studies connected with the zinc-requiring enzyme pyridoxalphosphokinase.[201] Their chelating properties are one of the reasons why thiosemicarbazones are of considerable pharmacological

[194] A. Pignedoli, G. Peyronel, and L. Anatolini, *Acta Cryst.*, 1975, **B31**, 1903.
[195] W. Walter and C. Rohloff, *Annalen*, 1975, 1563.
[196] L. I. Mas'ko, V. P. Kerentseva, and M. D. Lipanova, *Zhur. analit. Khim.*, 1974, **29**, 1490 (*Chem. Abs.*, 1974, **81**, 151 158).
[197] G. Keresztury and M. P. Marzocchi, *Spectrochim. Acta*, 1975, **31A**, 275.
[198] S. Stankoviansky, A. Beno, and J. Carsky, *Chem. Zvesti*, 1974, **28**, 614 (*Chem. Abs.*, 1975, **82**, 77 700).
[199] I. J. Leggett and W. A. E. McBryde, *Talanta*, 1974, **21**, 1005.
[200] A. K. Singh, K. C. Trikha, R. P. Singh, and M. Katyal, *Talanta*, 1975, **22**, 551.
[201] K. C. Agrawal, S. Clayman, and A. C. Sartorelli, *J. Pharm. Sci.*, 1976, **65**, 297.

interest. Much work in this field has involved the thiosemicarbazones of isatins [*cf.* (64)] and related compounds.[190, 191, 202]

Reactions.—For the above reasons, thiosemicarbazones have been prepared from apionylacetaldehydes,[203] flavonoids,[204] α-methylchalcone,[205] *m*-aminophenyl-substituted 2-formylpyridines,[206] substituted 1-formylquinolines,[207] D-*threo*-pentulose and other keto-sugars,[208] substituted butyrophenones,[209] substituted acetophenones,[210] furfural derivatives,[211] 2-formyl-3-methylthiophen,[212] 3-formylrifamicin SV,[213] the aglycone of leucomycin A₃,[213] and barbital.[214] In the last case the corresponding selenosemicarbazone was obtained by heating barbital with acetone selenosemicarbazone in butanol containing acetic acid.[214] Thio- and seleno-semicarbazones of 5-hydroxy-2-formylpyridine have been described,[215] in addition to the corresponding 4-hydroxy-thiosemicarbazones.[188, 189] Sayer and co-workers have continued their mechanistic studies on this important reaction.[216]

A number of mono-, di-, and tri-substituted thiosemicarbazide *S*-trioxides (α-hydrazino-α-iminomethanesulphonic acid betaines) have been prepared by oxidation with peroxyacetic acid at low temperature [*cf.* (67)].[195]

[202] J. Borysiewicz, Z. Potec, B. Luka-Sobstel, and A. Zejc, *Acta Pharm. Jugoslavia*, 1975, **25**, 165 (*Chem. Abs.*, 1976, **84**, 12 965); V. Cavrini, G. Giovanninetti, A. Chiarini, P. A. Mannini, and M. Borgatti, *Farmaco, Ed. Sci. Tossicol.*, 1975, **30**, 974 (*Chem. Abs.*, 1976, **84**, 121 703); G. Giovanninetti, V. Cavrini, A. Chiarini, L. Garuti, and A. Mannini-Palenzona, *ibid.*, 1974, **29**, 375 (*Chem. Abs.*, 1974, **81**, 25 480); V. S. Misra, R. S. Varma, and S. Agarwal, *J. Indian Chem. Soc.*, 1975, **52**, 981; M. N. Preobazhenskaya, I. V. Yartseva, and L. V. Ektova, *Doklady Akad. Nauk S.S.S.R.*, 1974, **215**, 873; G. Saint-Ruf and J. C. Bourgeade, *Chim. Ther.*, 1973, **8**, 447 (*Chem. Abs.*, 1974, **81**, 25 487); A. B. Tomchin and E. A. Rusakov, *Khim. Farm. Zhur.*, 1974, **8**, 23 (*Chem. Abs.*, 1975, **82**, 111 889); M. Tonew, E. Tonew, and L. Heinisch, *Acta Virol.*, 1974, **18**, 17 (*Chem. Abs.*, 1974, **80**, 104 425); I. V. Yartseva, L. V. Ektova, M. N. Preobrazhenskaya, N. A. Lesnaya, N. P. Yavorskaya, G. N. Platonova, and Z. P. Sof'ina, *Bioorg. Khim.*, 1975, **1**, 1589 (*Chem. Abs.*, 1976, **84**, 105 995); I. Zawadowska, *Acta polon. Pharm.*, 1975, **32**, 33 (*Chem. Abs.*, 1975, **83**, 131 399).
[203] F. Dallacker and H. van Wersch, *Chem. Ber.*, 1975, **108**, 561.
[204] F. Kallay and G. Janzso, *Kem. Kozlem*, 1974, **42**, 213 (*Chem. Abs.*, 1975, **83**, 9701).
[205] B. Prescott, *Internat. J. Clin. Pharmacol. Biopharm.*, 1975, **11**, 332 (*Chem. Abs.*, 1975, **83**, 126 292).
[206] K. C. Agrawal, A.-J. Lin, B. A. Booth, J. R. Wheaton, and A. C. Sartorelli, *J. Medicin. Chem.*, 1974, **17**, 631.
[207] K. C. Agrawal, P. D. Mooney, and A. C. Sartorelli, *J. Medicin. Chem.*, 1976, **19**, 970; P. D. Mooney, B. A. Booth, E. C. Moore, K. C. Agrawal, and A. C. Sartorelli, *J. Medicin. Chem.*, 1974, **17**, 1145; A. S. Tsiftsoglou, K. M. Hwang, K. C. Agrawal, and A. C. Sartorelli, *Biochem. Pharmacol.*, 1975, **24**, 1631.
[208] F. H. H. Carlsson, A. J. Charlson, and E. C. Watton, *Carbohydrate Res.*, 1974, **36**, 359.
[209] V. S. Misra and S. Prakash, *J. Indian Chem. Soc.*, 1974, **51**, 715.
[210] S. S. Tiwari and A. Sharma, *J. Indian Chem. Soc.*, 1975, **52**, 153.
[211] N. Saldabols, A. Cimanis, S. Hillers, J. Popelis, L. N. Alekseeva, A. Zile, and A. K. Yalynskaya, *Khim. Farm. Zhur.*, 1974, **8**, 12; Sh. Yoshina, A. Tanaka, Chiing-Hsing Wu, and Hsim-Saw Kuo, *Yakugaku Zasshi*, 1975, **95**, 883 (*Chem. Abs.*, 1975, **83**, 206 077).
[212] H. R. Wilson, G. R. Revankar, and R. L. Tolman, *J. Medicin. Chem.*, 1974, **17**, 760.
[213] R. Cricchio, G. Cietto, E. Rossi, and V. Arioli, *Farmaco, Ed. Sci.*, 1975, **30**, 695 (*Chem. Abs.*, 1975, **83**, 192 973); S. Omura, A. Nakagawa, K. Suzuki, and T. Hata, *J. Antibiotics*, 1974, **27**, 370 (*Chem. Abs.*, 1974, **81**, 37 770).
[214] A. A. Tsurkau and V. S. Bazalitskaya, *Farm. Zhur.* (*Kiev*), 1975, **30**, 45 (*Chem. Abs.*, 1976, **84**, 164 719).
[215] K. C. Agrawal, B. A. Booth, R. L. Michand, E. C. Moore, and A. C. Sartorelli, *Biochem. Pharmacol.*, 1976, **23**, 2421.
[216] J. M. Sayer, B. Pinsky, A. Schonbrunn, and W. Washtien, *J. Amer. Chem. Soc.*, 1974, **96**, 7998.

Acylation of thiosemicarbazides is reported to occur at the terminal nitrogen,[217] whereas alkylation normally occurs at the sulphur. With triphenylphosphine the alkylation product of 2,2-diethyl-1,1-dimethylthiosemicarbazide (70) undergoes a 'sulphide contraction' to give the corresponding acylmethylamidrazone.[218] Thiosemicarbazide reacts with ketones and cyanide ion under acid conditions to give the cyanothiosemicarbazones, which are converted into the dihydrotriazine-3-thiones (71) by heating with concentrated hydrochloric acid.[219] Benzil monothiosemicarbazone exists from the moment of its formation as the dihydrotriazine-3-thione (72), which on heating or treatment with acid

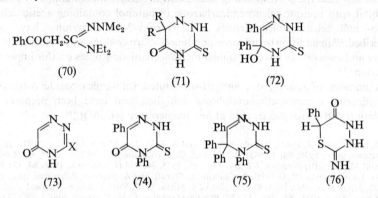

(70)   (71)   (72)

(73)   (74)   (75)   (76)

loses water to form the corresponding triazine-3-thione.[220] An analogous reaction has been reported for arylglyoxal [221] and oxaloacetic acid esters,[222] and has been applied to the synthesis of 5-cyclopropyl-6-azauracil.[223] The energies of activation for closing the analogous ring in the β-thiosemicarbazones of isatin have been found [224] to be 10.5—30.7 kcal mol^{-1}. On reaction of glyoxylic acid monohydrate with S-alkylthiosemicarbazide, the triazine derivative (73; X = SR) is obtained, in which the alkylthio-group can be exchanged by a substituted amino-group (73; X = NRAr).[225] A thiosemicarbazone is an intermediate when the triazine-3-thione (74) is converted into (75) by reaction with phenylmagnesium bromide.[226] 1,3,4-Thiadiazin-5-ones (76) are formed from PhCH(OH)CCl$_3$ and thiosemicarbazide under basic conditions.[227] The first step in the reaction is the formation of an oxiran, which undergoes ring-opening with the thiosemicarbazide, acting as a S-nucleophile, giving an acid chloride, which cyclizes.

[217] T. Yabuuchi, M. Hisaki, and R. Kimura, *Chem. and Pharm. Bull. (Japan)*, 1975, **23**, 668 (*Chem. Abs.*, 1975, **83**, 58 385).
[218] W. Heffe, R. W. Balsinger, and K. Thoma, *Helv. Chim. Acta*, 1974, **57**, 1242.
[219] L. C. March, K. Wasti, and M. M. Joullié, *J.C.S. Perkin I*, 1976, 83.
[220] A. B. Tomchin, Yu. V. Lepp, and T. N. Timofeeva, *Zhur. org. Khim.*, 1974, **10**, 2002.
[221] S. S. Smagin, V. E. Bogachev, V. V. Yakubovskii, S. E. Metkalova, T. P. Privol'neva, and E. F. Lavretskaya, *Khim. Farm. Zhur.*, 1975, **9**, 11 (*Chem. Abs.*, 1975, **83**, 108 311).
[222] D. Maudet, R. Granet, and S. Piekarski, *Bull. Soc. chim. France*, 1975, 2696.
[223] I. Basnak and J. Farkas, *Coll. Czech. Chem. Comm.*, 1975, **40**, 1038.
[224] A. B. Tomchin, I. S. Ioffe, and G. A. Shirokii, *Zhur. org. Khim.*, 1974, **10**, 1962; A. B. Tomchin and Yu. V. Lepp, *ibid.*, p. 1972.
[225] L. Heinisch, *J. prakt. Chem.*, 1974, **316**, 667.
[226] A. K. Mansour, Y. A. Ibrahim, and M. M. Eid, *Indian J. Chem.*, 1974, **12**, 301.
[227] C. N. O'Callaghan, *Proc. Roy. Irish Acad., Sect. B*, 1976, **76**, 37 (*Chem. Abs.*, 1976, **85**, 21 243); W. Reeve and E. R. Barron, *J. Org. Chem.*, 1975, **40**, 1917.

4-Alkyl-1-cyanoacetyl-thiosemicarbazides were cyclized by hydrazine hydrate to give the triazoline-3-thiones (77).[227] Analogous products were obtained with other 1-acyl-thiosemicarbazides.[228] Cyclization of 1-thioacylthiocarbonohydrazides in alkaline media gave the corresponding 3-mercapto-1,2,4-triazole (78) together with other products.[229] In contrast, thiobenzoylation of S-methyl

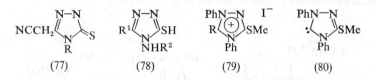

(77) (78) (79) (80)

isothiocarbonohydrazide gave 2,3-dihydro-1-methylthio-4-phenyl-2,3,5,6-tetrazine.[229] If S-methyl-1,4-diphenylthiosemicarbazide is treated with ethyl orthoformate, the triazolium ion (79; R = H) is formed, which is converted by sodium hydride into the nucleophilic carbene (80) (or its dimer, respectively). The carbene can be alkylated with alkyl halides to give the alkylated triazolium ion (79; R = alkyl), which is converted into the aldehyde RCHO by reduction with sodium borohydride and hydrolysis. In total, this is a further example of an acylation with reversed polarity.[230, 231]

4-(Alkyl)thiosemicarbazide, $R^1NHC(S)NHNH_2$, reacts with isocyanate and methyl iodide to give the triazolinones (81); on the other hand, the diaminotriazoles (82) are obtained on reaction with a carbodi-imide $R^2N:C:NR^2$.[232] Oxadiazoles, *e.g.* (83), are formed from 1-acylthiosemicarbazides on reaction with

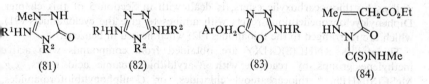

(81) (82) (83) (84)

iodine and sodium hydroxide in ethanol.[178] The 4-methylthiosemicarbazone of diethyl acetylsuccinate, on reaction with ammonia, gives the pyrazolone (84).[233]

Limitations of space preclude the treatment of heterocyclic systems, which are covered in other chapters of this book.

4 Thioamides and Selenoamides

Synthesis.—The preparation of thioamides and selenoamides is dominated by standard methods of long standing. Many amides of various types, open-chain

[228] C. N. O'Callaghan, *Proc. Roy. Irish Acad., Sect. B*, 1974, **74**, 455 (*Chem. Abs.*, 1975, **82**, 140 031).
[229] R. Esmail and F. Kurzer, *J.C.S. Perkin I*, 1975, 1787.
[230] G. Doleschall, *Tetrahedron Letters*, 1975, 1889.
[231] D. Seebach and D. Enders, *Angew. Chem.*, 1975, **87**, 1.
[232] R. Sunderdiek and G. Zinner, *Arch. Pharm.*, 1974, **307**, 504.
[233] A. A. Santilli, B. R. Hofmann, and D. H. Kim, *J. Heterocyclic Chem.*, 1974, **11**, 879.

160   *Organic Compounds of Sulphur, Selenium, and Tellurium*

aliphatic,[234-236] aromatic,[237-243] heteroaromatic,[244] and cyclic amides,[245] have been converted into the corresponding thioamides by treatment with phosphorus pentasulphide in solvents such as benzene, toluene, dichloromethane, dioxan, or pyridine. In the cephalosporin (85), the amide group in the side-chain was exclusively attacked by phosphorus pentasulphide to give (86), whereas boron sulphide in chloroform gave a mixture of (87) and (88).[246]

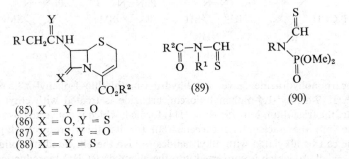

(85) X = Y = O
(86) X = O, Y = S
(87) X = S, Y = O
(88) X = Y = S

A new thionation reagent, $(EtAlS)_n$, is the first product of the reaction of triethylaluminium with hydrogen sulphide; it reacts with various dimethylamides to form the corresponding thioamides.[247]

Thiocarboxylic acids react with isocyanides to give *N*-thioformyl-*N*-acylamides (89);[248] with the esters of phosphorus dithio-acids, the novel *N*-thioformyl-*N*-phosphoramides (90) are similarly formed.[248]

The thioacylation of NH groups by dithiocarboxylic esters and their salts, as well as by thionocarboxylic esters, is dealt with in Section 5 of this chapter. Diphenylcyclopropenethione reacts with amines to give the cycloadducts (91), which are rearranged to the thioamides (92) when heated in benzene.[249]

Thioamides RNHC(S)CHXY are obtained from compounds with active methylene groups by reaction with *N*-aryldithiocarbamic acid esters, e.g. MeSC(S)NHPh,[250] thiocarbamoyl chlorides, or *C*-sulphonylthioformamides,

234 I. I. Ershova, V. I. Staninets, and T. A. Degurko, *Dopovidi Akad. Nauk Ukrain. R.S.R.,* Ser. B, 1975, 1097 (*Chem. Abs.*, 1976, **84**, 58 199).
235 B. Zeeh and H. Kiefer, *Annalen*, 1975, 1984.
236 G. Ewin and J. O. Hill, *Austral. J. Chem.*, 1975, **28**, 909.
237 M. Jancevska, B. Prisaganec, and M. Lazarevic, *God. Zb. Prir.-Mat. Fak. Univ. Skopje, Mat., Fiz. Hem.*, 1974, **24**, 65 (*Chem. Abs.*, 1975, **82**, 125 022).
238 B. Prisaganec and M. Lazarevic, *God. Zb. Prir.-Mat. Fak. Univ. Skopje, Mat., Fiz. Hem.,* 1974, **24**, 65 (*Chem. Abs.*, 1975, **82**, 125 022).
239 R. Grashey, E. Jänchen, and J. Litzke, *Chem.-Ztg.*, 1973, **97**, 657.
240 L. Heinisch, *J. prakt. Chem.*, 1975, **317**, 435.
241 J. Liebscher and H. Hartmann, *Z. Chem.*, 1975, **15**, 438.
242 W. Walter, Th. Fleck, J. Voss, and M. Gerwin, *Annalen*, 1975, 275.
243 D. Petrova and K. Jakopcic, *Croat. Chem. Acta*, 1976, **48**, 49 (*Chem. Abs.*, 1976, **85**, 5334).
244 W. Christ, D. Rakow, and S. Strauss, *J. Heterocyclic Chem.*, 1974, **11**, 397.
245 P. Nuhn, H.-J. Nitzsche, and G. Wagner, *Pharmazie*, 1974, **29**, 267 (*Chem. Abs.*, 1974, **81**, 3734).
246 P. W. Wojtkowski, J. E. Dolfini, O. Kocy, and C. M. Cimarusti, *J. Amer. Chem. Soc.*, 1975, **97**, 5628.
247 T. Hirabayashi, K. Inoue, K. Yokota, and Y. Ishii, *J. Organometallic Chem.*, 1975, **92**, 139.
248 J. P. Chupp and K. L. Leschinsky, *J. Org. Chem.*, 1975, **40**, 66.
249 Th. Eicher and S. Böhm, *Chem. Ber.*, 1974, **107**, 2238.
250 I. M. Bazavova, R. G. Dubenko, and P. S. Pel'kis, *Zhur. org. Khim.*, 1974, **10**, 1992.

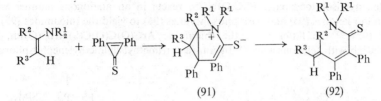

(91)   (92)

ArSO$_2$C(S)NR$_2$.[251] This is a direct introduction of the thioamido-group, which is normally achieved by means of isothiocyanates.[252-257] Ethoxycarbonyl isothiocyanate SCNCO$_2$Et reacts with aromatic compounds and Friedel–Crafts catalysts to give the ethoxycarbonylthioamides ArC(S)NHCO$_2$Et, which are easily hydrolysed and decarboxylated to the corresponding thioamides.[258] *para*-Substituted thioamides are obtained when toluene or anisole reacts with potassium thiocyanate in hydrogen fluoride.[259] The preference for reaction at carbon in competition with reaction at nitrogen is also demonstrated by a reaction already mentioned, which gives the thioamides (13) as the major and the thioureas (12) as the minor products.[16]

Thioacylimines (94) are obtained from α-metallated α-isothiocyanatoalkanoic acid esters (93; R = H or Me) with alkyl halides. They rearrange on heating

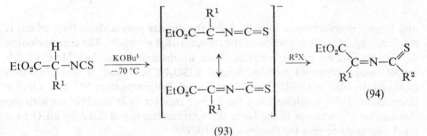

(93)   (94)

to give dialkyl isothiocyanates EtO$_2$CCR^{1}R^{2}NCS, presumably by a sigmatropic [2,3]shift.[260] Diethyl cyanomethylphosphonate, (EtO)$_2$P(O)CH$_2$CN, and similar compounds react with aryl isothiocyanates to give the thioamides (EtO)$_2$P(O)-CH(CN)C(S)NHAr in a normal reaction.[261] The parent diester (EtO)$_2$P(O)H forms a thioamide with vicinal dialkylphosphonate and thiocarbamoyl groups, (EtO)$_2$P(O)C(S)NHAr;[262] a similar type of compound is obtained on reaction of trimethylsilylphosphonamide, Me$_3$SiP(O)(NR$_2$)$_2$, with phenyl isothiocyanate.[263]

[251] N. H. Nilsson and J. Sandström, *Synthesis*, 1974, 433.
[252] F. Darré, A. M. Lamazouère, and J. Sotiropoulos, *Bull. Soc. chim. France*, 1975, 829.
[253] A. Étienne, G. Longchambon, and P. Giraudeau, *Compt. rend.*, 1974, **279**, C, 659.
[254] J. Goerdeler, A. Laqua, and C. Lindner, *Chem. Ber.*, 1974, **107**, 3518.
[255] W. Walter and A. Röhr, *Annalen*, 1975, 41.
[256] K. Gewald and M. Hentschel, *J. prakt. Chem.*, 1976, **318**, 343.
[257] W. Ried and L. Kaiser, *Annalen*, 1976, 395.
[258] E. P. Papadopoulos, *J. Org. Chem.*, 1974, **39**, 2540; 1976, **41**, 962.
[259] A. E. Feiring, *J. Org. Chem.*, 1976, **41**, 148.
[260] I. Hoppe, D. Hoppe, and U. Schöllkopf, *Tetrahedron Letters*, 1976, 609.
[261] G. Barnikow and G. Saeling, *J. prakt. Chem.*, 1974, **316**, 534.
[262] V. V. Alekseev and M. S. Malinovskii, *Zhur. obshchei Khim.*, 1975, **45**, 1484.
[263] E. S. Batyeva, Yu. N. Girfanova, G. U. Zamaletdinova, and A. N. Pudovik, *Izvest. Akad. Nauk S.S.S.R., Ser. khim.*, 1976, 455.

Benzoyl isothiocyanate, PhC(O)NCS, reacts in an analogous manner with enamino-ketones (95) and with phosphoranes (96) to yield the thioamides (97) [264] and (98); [265] the latter gives the thioamide ArC(O)C(O)C(S)NHC(O)Ph, by oxidation with selenous acid. [265] In contrast, trimethylsilylmethylenephosphorane

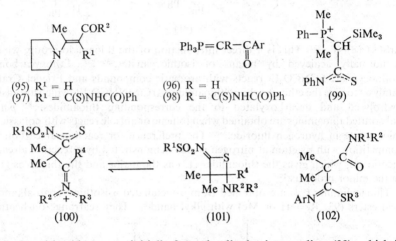

(95) R¹ = H
(97) R¹ = C(S)NHC(O)Ph

(96) R = H
(98) R = C(S)NHC(O)Ph

(99)

(100)   (101)   (102)

and phenyl isothiocyanate initially form the dipolar intermediate (99), which is stabilized by a C to S migration of the trimethylsilyl group. [266] The corresponding dipoles are formed by the reaction of the disubstituted enamines $Me_2C=CR^4-NR^2R^3$ with sulphonyl isothiocyanates R^1SO_2NCS, [267] and of dimethylketen O,N-acetals $Me_2C=C(OR^3)NR^1R^2$ with aryl isothiocyanates. [268] The dipolar intermediate (100) resulting from the former reaction is in equilibrium with the thietan (101), [267] whereas in the latter the intermediate is stabilized by an O to S alkyl migration to give the thiolimidate (102). [268]

Thioamides have been obtained by allowing aromatic hydrocarbons to react with thiocarbamoyl chloride in carbon disulphide in the presence of aluminium trichloride. [269] When 1-alkynyl propargyl sulphides are dissolved in a solution of a dialkylamine in DMSO or methanol, the thioamides of penta-3,4- and penta-2,4-dienoic acids are formed, as shown in equation (4). [270]

$R^1C\equiv CSCH_2C\equiv CR^2$

$\downarrow R^3_2NH$

(4)

$CH_2=C=CR^2CHR^1C(S)NR^3_2 + CH_2=CHCR^2=CR^1C(S)NR^3_2$

[264] O. Tsuge and A. Inaba, *Heterocycles*, 1975, **3**, 1081 (*Chem. Abs.*, 1976, **84**, 164 570).
[265] A. F. Tolochko, I. V. Megera, L. V. Zykova, and M. I. Shevchuk, *Zhur. obshchei Khim.*, 1975, **45**, 2150.
[266] K. Itoh, H. Hayashi, M. Fukui, and Y. Ishii, *J. Organometallic Chem.*, 1974, **78**, 339.
[267] E. Schaumann, S. Sieveking, and W. Walter, *Tetrahedron*, 1974, **30**, 4147.
[268] E. Schaumann, S. Sieveking, and W. Walter, *Chem. Ber.*, 1974, **107**, 3589.
[269] J. Goerdeler and K. Nandi, *Chem. Ber.*, 1975, **108**, 3066.
[270] J. Meijer, P. Vermeer, H. J. T. Bos, and L. Brandsma, *Rec. Trav. chim.*, 1974, **93**, 26.

The addition of hydrogen sulphide or hydrogen selenide to the nitrile group has been used to prepare aromatic thioamides,[242, 271, 272] heteroaromatic thioamides,[244] and selenoamides,[244, 273] as well as aliphatic thioamides.[274-277] The nitrile groups in polyacrylonitrile were found to be more reactive than in acetonitrile, 1,3-dicyanopropane, and 1,3,5-tricyanopentane;[278] even more reactive were copolymers with acrylonitrile.[279] Dicyandiamides RNHC(NH)NHCN and thiobenzamide or thioacetamide in the presence of hydrogen chloride give *C*-amidinothioformamides, amongst other products;[280] the thioamide acts as a sulphur-transfer reagent. A review of some thioimides which are useful in this respect has appeared.[281] *NN*-Dimethylthioformamide has been used to prepare *N*-(*trans-p*-menthan-*cis*-3-yl)thioacetamide (103) from the corresponding nitrilium salt

$$(103) \quad R = -NHCMe$$
$$\qquad\qquad\quad \overset{\|}{S}$$
$$(104) \quad R = -\overset{+}{N}{\equiv}CMe \quad SbCl_6{}^-$$

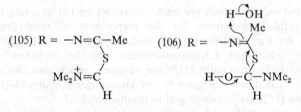

$$(105) \quad R = -N{=}C{-}Me \qquad (106) \quad R = -N{=}$$

(104).[282] A mechanism involving the intermediate structures (105) and (106) has been proposed.[282]

Imidoyl chlorides are useful starting materials for the preparation of selenoamides[283] and for labile thioamides, *e.g.* (107), prepared from the amide (108),

[271] V. I. Cohen, N. Rist, and S. Clavel, *European J. Med. Chem.-Chim. Ther.*, 1975, **10**, 134, 140 (*Chem. Abs.*, 1975, **83**, 96 669, 113 884).

[272] J. S. Walia, S. N. Bannore, A. S. Walia, and L. Guillot, *Chem. Letters*, 1974, 1005.

[273] W. Christ, D. Rakow, and S. Strauss, *J. Heterocyclic Chem.*, 1974, **11**, 397.

[274] K. Gewald and U. Hain, *J. prakt. Chem.*, 1975, **317**, 329.

[275] V. M. Neplyuev, M. G. Lekar, R. G. Dubenko, and P. S. Pel'kis, *Zhur. org. Khim.*, 1974, **10**, 2172.

[276] J. Poupaert and A. Bruylants, *Bull. Soc. chim. belges*, 1975, **84**, 61.

[277] C. Ressler and S. N. Banerjee, *J. Org. Chem.*, 1976, **41**, 1336.

[278] L. M. Levites, G. A. Gabrielyan, M. V. Shablygin, G. I. Kudryavtsev, and Z. A. Rogovin, *Vysokomol. Soedineniya, Ser. B*, 1974, **16**, 268 (*Chem. Abs.*, 1974, **81**, 121 334).

[279] L. M. Levites, G. A. Gabrielyan, G. I. Kudryavtsev, and Z. A. Rogovin, *Faserforsch. Textiltech.*, 1974, **25**, 153 (*Chem. Abs.*, 1974, **81**, 171 017).

[280] C. P. Joshua and V. P. Rajan, *Austral. J. Chem.*, 1974, **27**, 2627.

[281] P. T. S. Lau, *Eastman Org. Chem. Bull.*, 1975, **46**, 1 (*Chem. Abs.*, 1976, **84**, 150 417).

[282] D. H. R. Barton, P. D. Magnus, J. A. Garbarino, and R. N. Young, *J.C.S. Perkin I*, 1974, 2101.

[283] M. P. Cava and L. E. Saris, *J.C.S. Chem. Comm.*, 1975, 617.

which was converted (with phosphorus pentachloride) into the imidoyl chloride (109), and hence into (107) on treatment with hydrogen sulphide.[284] When perfluoro-*N*-isopropylacetimidoyl chloride is treated with hydrogen sulphide, the thioamide $CF_3C(S)NHCH(CF_3)_2$ and the Δ^3-1,2,4-dithiazoline (110) are obtained.[285] The reaction of *N*-phenylbenzimidoyl chloride with thiourea gave the corresponding carbodi-imide and thiobenzanilide.[286]

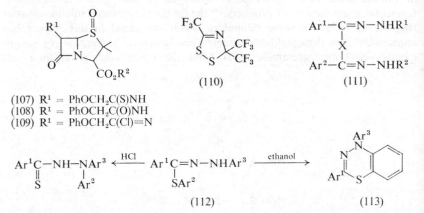

(107) R^1 = $PhOCH_2C(S)NH$
(108) R^1 = $PhOCH_2C(O)NH$
(109) R^1 = $PhOCH_2C(Cl)=N$

Hydrazonyl halides, which are readily available by halogenating hydrazones, may be useful intermediates for the preparation of thiohydrazides.[287, 288] Depending on the conditions, *sym*- and *unsym*-hydrazonyl sulphides (111; X = S), or substituted 1,3,4-thiadiazolines, may be prepared.[287, 288] With thiophenols, thiohydrazonates (112) are obtained, which give the thiadiazines (113) when heated with ethanol.[289] In concentrated hydrochloric acid the hydrazonates (112) are [289] rearranged to thiohydrazides.

A review of the Willgerodt–Kindler reaction covering the literature through 1972 is available.[290] Using DMF as solvent, the reaction has been applied to benzylamines and benzyl chlorides, giving *NN*-dimethylthiobenzamides;[43] the yields have been found to be different in the absence and presence of DMF.[291] In the absence of DMF, the reaction shown in equation (5) was observed.[291]

$$ArC(O)Me + PhCH_2NH_2 \xrightarrow{S_8} ArCH_2C(S)NHPh + PhC(S)NHCH_2Ph \quad (5)$$

[284] R. G. Micetich, C. G. Chin, and R. B. Morin, *Tetrahedron Letters*, 1976, 967, 975.
[285] K. E. Peterman and J. M. Shreeve, *J. Fluorine Chem.*, 1975, **6**, 83 (*Chem. Abs.*, 1975, **83**, 193 180).
[286] S. Furumoto, *Yuki Gosei Kagaku Kyokai Shi.*, 1975, **33**, 748 (*Chem. Abs.*, 1976, **84**, 59 376).
[287] P. Wolkoff, S. T. Nemeth, and M. S. Gibson, *Canad. J. Chem.*, 1975, **53**, 3211.
[288] P. Wolkoff, S. Hammerum, P. D. Callaghan, and M. S. Gibson, *Canad. J. Chem.*, 1974, **52**, 879.
[289] A. J. Elliott, P. D. Callaghan, M. S. Gibson, and S. T. Nemeth, *Canad. J. Chem.*, 1975, **53**, 1484.
[290] E. V. Brown, *Synthesis*, 1975, 358.
[291] T. Hisano and M. Ichikawa, *Chem. and Pharm. Bull.* (*Japan*), 1974, **22**, 2051 (*Chem. Abs.*, 1975, **82**, 57 598); E. P. Nakova, O. N. Tolchakev, and R. P. Evstigneeva, *Zhur. org. Khim.*, 1975, **11**, 2585.

Diphenylguanidine was successfully used as starting material for the Willgerodt–Kindler reaction, giving polythioamides.[292] Carbon disulphide reacted with triethylamine and DMF to give *NN*-diethyl- and *NN*-dimethyl-thioformamide. Under the same conditions, dithiocarbamates decompose to give *NN*-dialkyl-thioformamide and sulphur.[293] HMPT and sulphur have been shown to be a useful system for the oxidation of a variety of aromatic and heteroaromatic compounds to the corresponding *N*-methyl- and *NN*-dimethyl-thioamides;[294] in the case of phenyl benzyl ketone, *trans*-stilbene and the heterocyclic compounds (114) and (115) have been obtained as by-products.[295] Whereas the benzylic

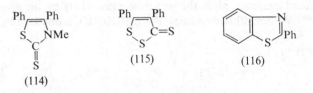

group in aliphatic benzyl esters is oxidized similarly to *NN*-dimethylthiobenz-amide, 2-phenylbenzothiazole (116) is obtained with benzyl benzoate. The same compound is formed when thiobenzanilide is heated with the same reagents in the presence of aniline.[296] Thioamides are modified in the carbon chain by the reaction sequence of Scheme 1, containing a hetero-Cope rearrangement as the last

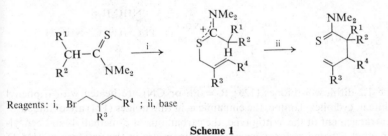

Reagents: i, Br$\diagup$$\diagdown$R^4 ; ii, base
R^3

Scheme 1

step.[297] The cycloaddition product (117) of thiobenzoyl isocyanate to 2,3-di-phenylaziridine gave the thioamide (118) on acid hydrolysis.[298]

Cleavage by potassium cyanide of bis(hydrazonyl) disulphides (111; X = S$_2$) gives the corresponding thiohydrazide and the 1,3,4-thiadiazoline (119).[287] Selenoformamides HC(Se)NR^{1}R^2 are obtained by reduction of the triseleno-carbonate (PhCH$_2$Se)$_2$C=Se with a selenol PhCH$_2$SeH and then allowing reaction of the resulting product with s-amines to take place.[299]

[292] Yu. N. Zafranskii, A. N. Semenova, K. E. Zhukova, and Z. M. Ivanova, *Zhur. Vsesoyuz. Khim. obshch. im. D. I. Mendeleeva*, 1974, **19**, 233 (*Chem. Abs.*, 1974, **80**, 145 695).
[293] J. Petermann and H. Plieninger, *Tetrahedron*, 1975, **31**, 1209.
[294] J. Perregaard, I. Thomsen, and S.-O. Lawesson, *Acta Chem. Scand.* (*B*), 1975, **29**, 538.
[295] J. Perregaard, I. Thomsen, and S.-O. Lawesson, *Acta Chem. Scand.* (*B*), 1975, **29**, 599.
[296] J. Perregaard and S.-O. Lawesson, *Acta Chem. Scand.* (*B*), 1975, **29**, 604.
[297] R. Gompper and W.-R. Ulrich, *Angew. Chem.*, 1976, **88**, 300.
[298] V. Nair and K. H. Kim, *Tetrahedron Letters*, 1974, 1487.
[299] L. Henriksen, *Synthesis*, 1974, 501.

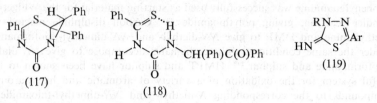

(117)   (118)   (119)

Several papers describing the preparation of thioamides by opening heterocyclic rings have appeared.[242, 300-306] The thiazole derivative (120), on reaction
with Grignard reagents, gives the oxo-thioamides (121) by the given reaction
sequence.[306] *N*-Formylthiobenzamide, PhC(S)NHCHO, is obtained by hydrolysis

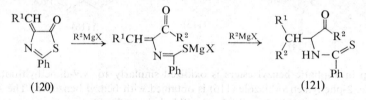

(120)   (121)

of 5-phenyl-1,2,4-dithiazolium tetrafluoroborate (122), which is available from
hydroxymethylthiobenzamide, PhC(S)NHCH$_2$OH, by successive cyclization
with sulphuryl chloride and treatment with triphenylmethyl fluoroborate.[300]

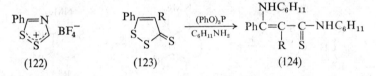

(122)   (123)   (124)

If the 1,2-dithiole-3-thiones (123; R = Ph or CN) are heated with triphenyl
phosphite in cyclohexylamine, the cinnamic acid thioamides (124) are obtained.[305]
In a rearrangement of the Wittig type, the carbanions of β-methallyl- and benzyl-
NN-dimethylcarbamates form α-mercapto-thioamides in the presence of HMPT
(Scheme 2).[307]

$$R\bar{C}HSC(S)NMe_2 \xrightarrow{\quad i \quad} \begin{array}{c} R-CHS^- \\ | \\ C(S)NMe_2 \end{array} \xrightarrow{\quad ii \quad} \begin{array}{c} R-CHSH \\ | \\ C(S)NMe_2 \end{array}$$

R = CH:CMe$_2$ or Ph

Reagents: i, HMPT–THF; ii, AcOH

Scheme 2

[300] H. Böhme and K. H. Ahrens, *Arch. Pharm.*, 1974, **307**, 828.
[301] A. Holm, N. H. Toubro, and N. Harrit, *Tetrahedron Letters*, 1976, 1909.
[302] L. Legrand and N. Lozac'h, *Bull. Soc. chim. France*, 1974, 1194.
[303] T. Matsuura and Y. Ito, *Tetrahedron*, 1975, **31**, 1245.
[304] G. Wagner, S. Leistner, and K. Winkler, *Pharmazie*, 1974, **29**, 681 (*Chem. Abs.*, 1975, **82**, 43 295).
[305] J. Goerdeler, J. Haag, C. Lindner, and R. Losch, *Chem. Ber.*, 1974, **107**, 502.
[306] M. A. F. Elkaschef, M. E. Abdel-Megeid, and S. M. A. Yassin, *Acta Chim. Acad. Sci. Hung.*, 1974, **80**, 119 (*Chem. Abs.*, 1974, **80**, 95 819).
[307] T. Hayashi and H. Baba, *J. Amer. Chem. Soc.*, 1975, **97**, 1608.

Physical Properties.—A tautomeric equilibrium between the iminomethane disulphides (125) and the aminovinyl disulphides (126) has been demonstrated.[308] The position of the equilibrium depends on the substituents; (125) is stabilized when R^3 is electron-withdrawing, whereas (126) is stabilized when R^1 and R^2

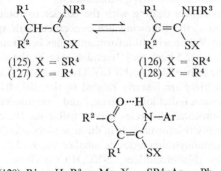

(125) X = SR⁴   (126) X = SR⁴
(127) X = R⁴   (128) X = R⁴

(129) R^1 = H, R^2 = Me, X = SR⁴, Ar = Ph
(130) R^1 = CO₂Et, R^2 = OEt, X = Me, Ar = *p*-BrC₆H₄

have this property; hydrogen-bonds may exert additional stabilization, as in (129). A Hammett correlation ($\rho = 0.278$) was achieved for a group of compounds (127; $R^1 = R^2 = CO_2Et$, $R^3 = 4\text{-}XC_6H_4$).[308] An analogous equilibrium has been observed between the thioimidates (127) and the keten *S,N*-acetals (128) ($\rho = -0.61$); in this case, the influence of R^4 on the equilibrium could be investigated, and it was found to be smaller than that of $R^1 \rightarrow R^3$; $\rho = 0.24$ for (127; $R^1 = R^2 = CO_2Et$, $R^3 = Ph$, $R^4 = 4\text{-}XC_6H_4$).[309] The quantity of the chelate keten *S,N*-acetal (130) decreases with increasing temperature, with increasing solvent polarity, and with the ability of the solvent to form intermolecular hydrogen-bonds.[310] In the *N*-(trimethylsilyl)thioamides (131) and (132), tautomerism resulting from a [1,3-N,S] migration of the trimethylsilyl group has been observed; the $\Delta G^{\ddagger}$ values being higher for this shift [(131), 22.6; (132), 21.6 kcal

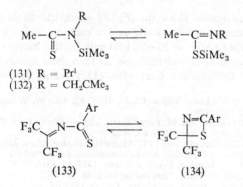

(131) R = Pr^i
(132) R = CH₂CMe₃

(133)   (134)

[308] W. Walter and H.-W. Meyer, *Annalen*, 1974, 776.
[309] W. Walter and H.-W. Meyer, *Annalen*, 1975, 19.
[310] W. Walter and H.-W. Meyer, *Annalen*, 1975, 36.

mol⁻¹] than the $\Delta G^{\ddagger}$ values of the hindered rotation about the C—N bond [(131), 11.4; (132), 13.9 kcal mol⁻¹].[311] Valence tautomerism between the *N*-alkylidenethioamide (133) and 2*H*-1,3-thiazete (134) has been demonstrated at elevated temperatures; (133) underwent capture reactions, *e.g.* with norbornene, to give the appropriate 4,5-dihydro-6*H*-1,3-thiazines.[312]

Among the numerous papers dealing with the barrier to rotation about the C—N bond in thioamides, the following are concerned with special aspects. The barrier to rotation in *NN*-dimethylthioformamide has been found to be lower in the gas phase ($\Delta G^{\ddagger}$ = 22.5 kcal mol⁻¹) and higher in the liquid ($\Delta G^{\ddagger}$ = 25.5 kcal mol⁻¹).[313] In the thioamides Me₂NC(S)CO₂H and Me₂NC(S)CN, derived from oxalic acid, the barriers are linearly related to the dielectric constants of the solvents nitrobenzene, *o*-dichlorobenzene, and naphthalene.[314] Kinetic studies show that an *N*-protonated species is responsible for the catalysis of the isomerization of *N*-neopentylthioformamide in dilute acid solution.[315] The barrier to rotation in the *N*-sulphonylthioamides is smaller than 12 kcal mol⁻¹;[255] in *ortho*-substituted tertiary thiobenzamides, 2-XC₆H₄C(S)NR¹R², not only the C—N barrier rises, but also the barrier to rotation about the C_{aryl}—C bond.[242] The same type of barrier was observed in *para*-substituted thioamides at low temperature.[316] The influence of substituents on the potential function in thio-amides such as (135) has been investigated.[317] There is a controversy between Fulea and Krueger[318] on one side and Jennings and Tolley[319, 320] on the other about the cause of the non-equivalence of the geminal protons in the methylene groups bonded to nitrogen. Molecular mechanics calculations of factors influencing the barriers to internal rotation in thioamides show that the torsional energy provides the principal contribution to the total energy.[321]

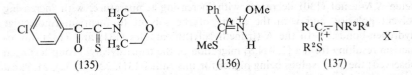

(135)   (136)   (137)

Thiobenzohydroximates show no (*E*)/(*Z*) isomerization in neutral solvents, but in the presence of trifluoroacetic acid a rotation occurs in the protonated molecule, *e.g.* (136) ($\Delta G^{\ddagger}$ = 23—25 kcal mol⁻¹); the barriers to inversion in these molecules must be considerably higher than these figures.[322] Likewise, the isomerization of thioimidium ester salts (137) occurs by rotation ($\Delta G^{\ddagger}$ =

311 W. Walter and H.-W. Lüke, *Angew. Chem.*, 1975, **87**, 420; W. Walter, H.-W. Lüke, and J. Voss, *Annalen*, 1975, 1808.
312 K. Burger, J. Albanbauer, and M. Eggersdorfer, *Angew. Chem.*, 1975, **87**, 816.
313 T. Drakenberg, *J. Phys. Chem.*, 1976, **80**, 1023.
314 J. A. Lepoivre, H. O. Desseyn, and F. C. Alderweireldt, *Org. Magn. Resonance*, 1974, **6**, 284.
315 W. Walter, M. Franzen-Sieveking, and E. Schaumann, *J.C.S. Perkin II*, 1975, 528.
316 A. O. Fulea and P. J. Krueger, *Canad. J. Chem.*, 1975, **53**, 3315.
317 P. J. Krueger and A. O. Fulea, *Tetrahedron*, 1975, **31**, 1813.
318 A. O. Fulea and P. J. Krueger, *Tetrahedron Letters*, 1975, 3135.
319 W. B. Jennings and M. S. Tolley, *Tetrahedron Letters*, 1976, 695.
320 W. B. Jennings, *Chem. Rev.*, 1975, **75**, 1307.
321 W. Walter and J. P. Imbert, *J. Mol. Structure*, 1975, **29**, 253.
322 W. Walter, C. O. Meese, and B. Schröder, *Annalen*, 1975, 1455.

89.9—117.6 kJ mol⁻¹).[323] For the secondary salts (137; R^3 = H) the rotation mechanism is only operative in strongly acidic solvents ($pK_a \leqslant 1.4$), otherwise the isomerization proceeds by planar inversion *via* a protonation–deprotonation mechanism.[323] Inversion is the normal way in which stereomutation occurs in the thioimidates RSC(R)=NR themselves, but there are cases in which an imine–enamine tautomerism, with fast rotation about the single C—N bond in the enamine, is operative, *e.g.* (127) ⇌ (128) ($R^1 = R^2 = H$, $R^3 = Me$, $R^4 = 2,6\text{-Pr}^i_2C_6H_3$).[324] This has to be taken into account when considering the $\Delta G^{\neq}$ values of 51.5—92.9 kJ mol⁻¹. The barriers to inversion in some selenoimidates have been found to be a little lower than those of the corresponding thio-imidates.[325] Hydrogen-bonds have been found to raise [324] or to lower [242] the barriers to isomerization in thioamides. In the thioamides (138) the NHN hydrogen-bond has been found to be stronger if R is phenyl or *o*-tolyl than if it is benzyl or α-methylbenzyl.[326] Several examples of the influence of hydrogen-bonds on the configuration of thioamides have been given.[254, 257, 327] I.r. data have been given for hydrogen-bonded complexes of thioacetamide and thio-benzamide with acetonitrile, THF–dimethylacetamide, triethylamine, and pyridine.[328]

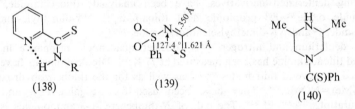

(138)   (139)   (140)

Ratios of (*E*)-/(*Z*)-isomers have been reported for thioamides,[329] thioimidates,[324] their salts,[323] and selenoimidates.[325] *N*-(Phenylsulphonyl)thiopropionic acid amide exists in the (*Z*)-configuration (139) in the crystal, as shown by *X*-ray diffraction.[330] This compound provides the first observed example of a deviation from the thioamide plane, which corresponds with the extremely low barrier of rotation about the C—N bond characteristic for this class of thioamides.[255] The bond length of C=S and C—N varies from 1.626 to 1.672 Å and from 1.30 to 1.358 Å, the NCS bond angles from 122.8° to 126.9°;[331] for the dipolar compound (100; R^1 = Ph, $R^2 = R^3$ = Me, R^4 = H) the figures are C=S 1.659 Å, C—N 1.318 Å.[331]

[323] W. Walter and C. O. Meese, *Chem. Ber.*, 1976, **109**, 947.
[324] W. Walter and C. O. Meese, *Chem. Ber.*, 1976, **109**, 922.
[325] C. O. Meese, W. Walter, H. Mrotzek, and H. Mirzai, *Chem. Ber.*, 1976, **109**, 956.
[326] S. D. Nasirdinov, N. A. Parpiev, Z. M. Musaev, and Yu. E. Il'ichev, *Uzbek. khim. Zhur.*, 1975, **19** (5), 11 (*Chem. Abs.*, 1976, **84**, 73 226).
[327] I. M. Ginzburg, L. B. Dashkevich, P. V. Kuznetsov, and P. B. Tarasov, *Zhur. obshchei Khim.*, 1975, **45**, 2665.
[328] I. M. Ginzburg and N. N. Bessonova, *Zhur. obshchei Khim.*, 1975, **45**, 622.
[329] D. L. Bate and I. D. Rae, *Austral. J. Chem.*, 1974, **27**, 2611.
[330] W. Walter, J. Holst, and A. Röhr, *Annalen*, 1975, 54.
[331] E. Schaumann, A. Röhr, S. Sieveking, and W. Walter, *Angew. Chem.*, 1975, **87**, 486; M. Cannas, G. Carta, A. Cristini, and G. Marongiu, *Acta Cryst.*, 1975, **B31**, 2909; A. Christensen, H. J. Geise, and B. J. Van der Veken, *Bull. Soc. chim. belges*, 1975, **84**, 1173; R. Sugisaki, T. Tanaka, and E. Hirota, *J. Mol. Spectroscopy*, 1974, **49**, 241.

The bulk of the evidence for the above physical properties is derived from
n.m.r., i.r., and Raman spectral studies. The following additional results are
also based on these methods. The Hammett-type dependence of the chemical
shifts of the thioamide group has been utilized in the assignment of the (E)-/(Z)-
isomers.[332] The influence of protonation of the thioamide group on the aniso-
tropic shielding [333] and on the distortion of the piperazine ring in the [²H₆]acetone
complex of (140) [334] has been investigated. The n.m.r. spectra of the two isomers
of NN-di-isopropylthioacetamide have been assigned on the assumption that
there is a plane of symmetry through the heavy atoms.[335] In substituted phos-
phinoylthioformamides R₂P(X)C(S)NHR, the ³J(PCNH) coupling constants can
be used to distinguish between the (E)- and (Z)-configurations; the configuration
has been shown to be (Z) in every example tested.[336] A reliable assignment of
signals has been performed for numerous thioamides derived from oxalic acid in
various solvents, using the ASIS effect.[337] The thiocarbonyl group of thioamides
has been found to be less complexing with lanthanide shift reagents than the
amide carbonyl group, but more effective than the sulphur of a thioether. Small
but useful shifts have been observed for selenoamides.[338] On the basis of i.r. and
Raman spectra, the fundamental vibrations of the following compounds, mostly
including deuteriated derivatives, have been analysed; thioacetamide, thioacetamide oxide,[339, 340] propionic acid thioamide,[341] NN-dimethylcyanothio-
formamide,[342] and NN-dimethylselenoformamide.[343]

The deuterium and nitrogen (¹⁴N) nuclear quadrupole resonance in deuteriated thioacetamide has been measured at 77 K.[344] Dipole moments have been
reported for several thioamides [345, 346] as well as for the thiobenzohydroximate
PhC(SMe)=NOMe.[322] They have been used for assigning (E)- and (Z)-
configurations.[255, 322, 324, 346] The c.d. of N-thiobenzoyl-L-α-amino-acids is very
useful for indicating the degree of resolution on separation of the enantiomers of
DL-amino-acids.[347] The mass spectra of N-acyl-thioamides have been investigated
systematically,[348] the e.s.r. spectra of the radical anions of mono- and di-thio-
oxamide have been recorded, and conclusions about the spin-density distribution
and the conformation of the N-alkyl group have been drawn from the g values
and from the hyperfine structure coupling.[349]

[332] M. Nicolaisen, *J. Mol. Structure*, 1975, **29**, 379.
[333] M. Yamazaki and J. Niwa, *Nippon Kagaku Kaishi*, 1974, 1501 (*Chem. Abs.*, 1974, **81**, 151 343).
[334] I. D. Rae, *Austral. J. Chem.*, 1974, **27**, 2621.
[335] A. O. Fulea and P. J. Krueger, *Canad. J. Chem.*, 1976, **54**, 566.
[336] O. Dahl and S. A. Laursen, *Org. Magn. Resonance*, 1976, **8**, 1.
[337] J. A. Lepoivre, H. O. Desseyn, and F. C. Alderweireldt, *Org. Magn. Resonance*, 1974, **6**, 279.
[338] I. D. Rae, *Austral. J. Chem.*, 1975, **28**, 2537.
[339] A. Ray and D. N. Sathyanarayana, *Bull. Chem. Soc. Japan*, 1974, **47**, 729.
[340] W. Walter and P. Stäglich, *Spectrochim. Acta*, 1974, **30A**, 1739.
[341] P. U. Bai and K. V. Ramiah, *Indian J. Pure Appl. Phys.*, 1976, **11**, 769.
[342] H. O. Desseyn and J. A. Le Poivre, *Spectrochim. Acta*, 1975, **31A**, 635, 647.
[343] U. Anthoni, L. Henriksen, P. H. Nielsen, G. Borch, and P. Klaboe, *Spectrochim. Acta*, 1974, **30A**, 1351.
[344] M. J. Hunt and A. L. Mackay, *J. Magn. Resonance*, 1974, **15**, 402.
[345] J. W. Diggle and D. Bogsanyi, *J. Phys. Chem.*, 1974, **78**, 1018.
[346] G. C. Pappalardo and S. Gruttadauria, *J.C.S. Perkin II*, 1974, 1441.
[347] G. C. Barrett and P. R. Cousins, *J.C.S. Perkin I*, 1975, 2313.
[348] T. N. Sumarokova, A. E. Lyuts, R. A. Slavinskaya, and V. A. Solomin, *Zhur. fiz. Khim.*, 1975, **49**, 2824 (*Chem. Abs.*, 1976, **84**, 37 798).
[349] J. Voss, *Annalen*, 1974, 1220, 1231.

Photoelectron spectra of thioamides have been recorded and interpreted; the non-bonded orbital mostly localized on the S atom was found to have the lowest ionization potential.[47] In the charge-transfer complexes between thio-amides and iodine the thioamides act as electron donors, thiobenzoyl piperidide, $PhC(S)NC_5H_{10}$, being a stronger donor than the corresponding derivative of phenylacetic acid.[350] Waisser and co-workers have found that the tuberculostatic activity of thiobenzamides and thionaphthamides increases with increasing electron density and superdelocalizability for electrophilic substitution on the sulphur atom, and with decreasing bond-order of the C—S bond.[351]

The thermodynamic acidities have been measured for N-sulphonyl-thioamides ($pK_a = 3.31—4.99$)[255] and for thioamides with electron-withdrawing groups substituted at the α-carbon atom [$XYCHC(S)NR_2$; X,Y = $ArSO_2$, H_2NCO, PhCO, etc.]; in these cases one has to discriminate among CH-, NH-, OH-, and SH-acidity.[352]

Reactions.—Many examples of the normal S-alkylation of thioamides have been reported [261, 277, 297, 309, 322—325, 353, 354] as well as the Se-alkylation of selenobenz-anilide with α-bromophenylacetic acid to yield (141), which cyclizes with acetic anhydride and base to give a meso-ionic selenium heterocycle.[283] The thio-acetimidates $R^2SC(Me)=NR^1$, obtained by alkylation of thioacetamides by successive reactions with phosgene and bases, give the N-(chlorocarbonyl)keten

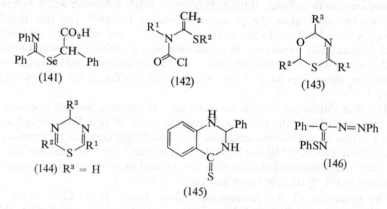

S,N-acetals (142), which are very useful starting materials for synthetic work.[235] The N-hydroxyalkylation and amino-alkylation with aldehydes and amines has found further application.[355, 356] The resulting N-hydroxyalkylthioamides $R^1C(S)NHCH(OH)R^3$ may be treated with another aldehyde to give the 6H-2,3,5-

[350] F. Cornea, C. Fulea, and S. Moldoveanu, *Rev. Roumaine Chim.*, 1974, **19**, 429 (*Chem. Abs*, 1974, **81**, 168 959).

[351] K. Waisser, M. Celadnik, J. Sova, and K. Palat, *Cesk. Farm.*, 1975, **24**, 423 (*Chem. Abs.*, 1976, **85**, 752).

[352] W. Walter, H.-W. Meyer, and A. Lehmann, *Annalen*, 1974, 765; V. M. Neplyuev, V. P. Kukhar, and R. G. Dubenko, *Zhur. org. Khim.*, 1974, **10**, 765.

[353] A. M. Lamazouère, F. Darré, and J. Sotiropoulos, *Bull. Soc. chim. France*, 1975, 2269.

[354] N. H. Leon, *J. Pharm. Sci.*, 1976, **65**, 146 (*Chem. Abs.*, 1976, **84**, 895 629).

[355] H. Böhme, K. H. Ahrens, and H. H. Hotzel, *Arch. Pharm.*, 1974, **307**, 748.

[356] J. Curtze and K. Thomas, *Annalen*, 1975, 2318.

oxathiazines (143).[357] Alternatively, on reaction with nitriles in the presence of boron trifluoride, the 4*H*-1,3,5-thiadiazines (144) are obtained.[358] One of the products obtained by allowing anthranilic acid thioamide to react with potassium cyanide and benzaldehyde in acidic solution is (145), which means that the reaction occurred at the nitrogen atom of the thioamide group.[359] An exchange between thiocarbonyl and disulphide structures involving thioamides has been observed (Scheme 3).[357] This reaction may be classified as a sulphenylation of the thioamide

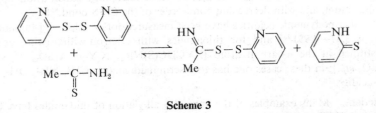

Scheme 3

group, which may likewise be performed by reaction with sulphenyl chlorides to yield, *e.g.*, (125),[308] or with tribenzenesulphenamide, N(SPh)₃, a new reagent introduced by Barton and co-workers, which gives the expected imino-disulphide [*cf.* (125)] with thiobenzanilide. This reagent gives some novel reactions. With thiobenzamide the sulphur di-imide PhSN=S=NSPh is formed, and N^2-phenyl-thiobenzhydrazide gives the phenylazothio-oxime (146).[360] The trimethylsilyl group is introduced on to the nitrogen of primary and secondary thioamides with hexamethyldisilazane or *via* metallation with potassium or n-butyl-lithium and reaction with chlorotrimethylsilane.[311] Triphenylstannyl iodide reacts with secondary thioamides to give the *S*-triphenylstannylthioamide Ph₃SnSC(R)=NR.[361]

The *N*-acylation of thioamides with aryl isocyanates has been shown to proceed smoothly in the presence of cuprous oxide to give *N*-thioacylureas.[362] The reaction of α-cyanothioacetamide with acetyl chloride gives the product of *C*-acylation MeC(O)CH(CN)C(S)NH₂.[363] The same type of reaction occurs between α-arylsulphonylthioacetamide and an aryl isocyanate in the presence of base to yield Ar¹NHC(O)CH(Ar²SO₂)C(S)NH₂.[364]

The formation of the resonance-stabilized anions R₂N—C̄=S (147), by metallation of *NN*-dialkylthioformamides, and their use as efficient thio-carbamoylation reagents, has been further elucidated.[365-367] Their reaction with

[357] J. Carlsson, M. P. J. Kierstan, and K. Brocklehurst, *Biochem. J.*, 1974, **139**, 221; C. Giordano and A. Belli, *Synthesis*, 1975, 789.
[358] C. Giordano, A. Belli, and V. Bellotti, *Synthesis*, 1975, 266.
[359] J. S. Walia, S. N. Bannore, A. S. Walia, and L. Guillot, *Chem. Letters*, 1974, 1005.
[360] J. Almog, D. H. R. Barton, P. D. Magnus, and R. K. Norris, *J.C.S. Perkin I*, 1974, 853.
[361] E. J. Kupchik and H. E. Hanke, *J. Organometallic Chem.*, 1975, **97**, 39.
[362] V. I. Cohen, *J. Org. Chem.*, 1974, **39**, 3043; *Helv. Chim. Acta*, 1976, **59**, 350.
[363] H. Schäfer and K. Gewald, *J. prakt. Chem.*, 1975, **317**, 771.
[364] V. M. Neplyuev, M. G. Lekar, R. G. Dubenko, and P. S. Pel'kis, *Zhur. org. Khim.*, 1974, **10**, 2172 (*Chem. Abs.*, 1975, **82**, 16 504).
[365] E. Campaigne and J. Beckman, 167th National Meeting of the American Chemical Society, Los Angeles, 1974 (Abstr. Bioorg. Chem. No. 90).
[366] D. Seebach, W. Lubosch, and D. Enders, *Chem. Ber.*, 1976, **109**, 1309.
[367] D. Seebach and W. Lubosch, *Angew. Chem.*, 1976, **88**, 339.

chlorotrimethylsilane gave $Me_2NC(S)SiMe_3$, while the anion of NN-diphenylthio-formamide dimerized to give tetraphenyldithio-oxamide, $Ph_2NC(S)C(S)NPh_2$. Because the sulphur of the nucleophilic thioacylation products may be easily removed by lithium aluminium hydride, the reaction sequence of Scheme 4

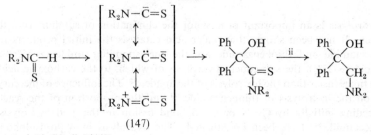

(147)

Reagents: i, Ph_2CO; ii, $LiAlH_4$

Scheme 4

represents a Mannich reaction with reversed polarity. This depends on the transformation of a secondary amine into a nucleophilic reagent for amino-methylation, by thioformylation and metallation.[366]

A reduction of the thiocarbonyl group to a methylene group can also be achieved by the hydridotetracarbonylferrate anion, $[HFe(CO)_4]^-$.[368] Thio-benzamide was found to be useful as a source of sulphur in converting 1-aroyl-2-phenylethyne into bis(2-aroyl-1-phenylethenyl) sulphide,[369] and as an alternative to thioacetamide in preparing the 6a-thiathiophthens (149) from the dithiolylidene-thiones (148).[370] In losing sulphur, primary thioamides are frequently trans-formed into nitriles;[357, 361] thioformanilides give aromatic nitriles in good yields

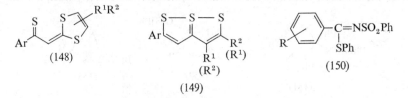

by flash thermolysis at 700 °C.[371] Bis(triphenylstannyl)carbodi-imide gives the cyano-amidine $RNHC(R)=NCN$ on reaction with secondary thioamides.[361] Amidrazones are formed from thioamides, or the thioimidates of primary amines, on reaction with hydrazines or substituted hydrazines.[372] The whole thioamide group is removed under the conditions of the Japp–Klingemann reaction (Scheme 5).[275, 364]

[368] H. Alper, *J. Org. Chem.*, 1975, **40**, 2694.
[369] M. N. Basyouni and M. T. Omar, *Austral. J. Chem.*, 1974, **27**, 1585.
[370] H. Davy and J. M. Decronen, *Bull. Soc. chim. France*, 1976, 115.
[371] R. F. C. Brown, I. M. Coddington, I. D. Rae, and G. J. Wright, *Austral. J. Chem.*, 1976, **29**, 931.
[372] F. H. Case, A. A. Schilt, and T. A. Fang, *J. Heterocyclic Chem.*, 1974, **11**, 463; K. M. Doyle and F. Kurzer, *Synthesis*, 1974, 583.

$$p\text{-}R^1C_6H_4SO_2-\underset{\underset{C(S)NH_2}{|}}{C}=N-NHPh \xrightarrow{\quad i \quad} p\text{-}R^1C_6H_4SO_2-\underset{\underset{N=NC_6H_4R^2\text{-}p}{|}}{C}=N-NHPh$$

Reagents: i, $p\text{-}R^2C_6H_4\overset{+}{N}_2$

Scheme 5

Hydrolysis is an important process for the elimination of sulphur from thio-amides. It has been shown that under acid catalysis the initial products from thioacetamide and thiobenzamide are the corresponding amide ($\sim 75\%$) and thioacid ($\sim 25\%$), the subsequent hydrolysis of which, to the carboxylic acid, is three times faster than the hydrolysis of the amide.[373] The influence of mercury(II) ions on the hydrolysis of thiobenzamide[374] and the proportion of the reaction proceeding initially by C—N or C—S bond fission in the acid hydrolysis of thioacetamide[375] have been investigated. The hydrolysis of N-(arylsulphonyl)-thiobenzimidates (150) at pH > 11 was initiated by attack of hydroxide ions, but at pH < 9 hydrolysis began by addition of water to the C=N group.[376] During the hydrolysis of N-acyl-thiobenzamides at pH 5.1 it was shown that the influence of a substituent was larger in the thioacyl group than in the acyl group.[377]

Evidence for the intermediate formation of an aminosulphene during the oxidative elimination of sulphur from thioamides is provided by the production of imidates on oxidation of thioamide S-oxides in the presence of alcohols.[378] (See Section 2, p. 138.) Oxidation of thiohydrazides gave the disulphides (111; X = S₂, Ar¹ = Ar² = R¹ = R² = Ph), which on further oxidation with potassium ferricyanide or dichlorodicyanobenzoquinone produced the dithia-diazine (151), which may lose sulphur to give the corresponding thiadiazole. Treatment of (111; X = S₂, Ar¹ = Ar² = R¹ = R² = Ph) with acetic anhydride gives the thioacylpyrazolone (152), which is also formed directly from the thio-

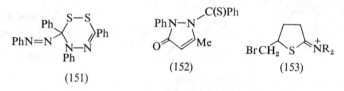

(151)   (152)   (153)

hydrazide and acetic anhydride.[379] The thioamides of pent-4-enoic acid, on oxidation with bromine or iodine, give thiolans, *e.g.* (153),[234] and the dithio-anilides of malonic acid give the 1,2-dithiole derivatives (155).[380] The meso-ionic compound (154) has been prepared by the given reaction sequence.[381] By reaction

[373] A. J. Hall and D. P. N. Satchell, *J.C.S. Perkin II*, 1974, 1077.
[374] A. J. Hall and D. P. N. Satchell, *J.C.S. Perkin II*, 1975, 778.
[375] O. M. Peeters and C. J. de Ranter, *J.C.S. Perkin II*, 1974, 1832.
[376] T.-S. Huh and T.-R. Kim, *Taehan Hwahak Hoechi*, 1976, **20**, 73 (*Chem. Abs.*, 1976, **85**, 5037).
[377] J. Mirek and B. Nawalek, *Roczniki Chem.*, 1974, **48**, 243 (*Chem. Abs.*, 1974, **81**, 24 638).
[378] W. Walter and O. H. Bauer, *Annalen*, 1975, 305.
[379] D. H. R. Barton, J. W. Ducker, W. A. Lord, and P. D. Magnus, *J.C.S. Perkin I*, 1976, 38.
[380] A. D. Grabenko, L. N. Kulaeva, and P. S. Pel'kis, *Khim. geterotsikl. Soedinenii*, 1974, 924 (*Chem. Abs.*, 1974, **81**, 120 520); L. Menabue and G. C. Pellacani, *J.C.S. Dalton*, 1976, 455.
[381] H. Gotthardt, M. C. Weisshuhn, and B. Christl, *Chem. Ber.*, 1976, **109**, 740.

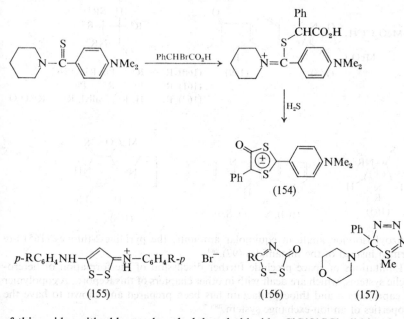

(154)

(155)   (156)   (157)

of thioamides with chlorocarbonylsulphenyl chloride, ClC(O)SCl, dithiazolones (156) are obtained (in ~60% yield), which give thioacyl isocyanates RC(S)NCO on treatment with triphenylphosphine.[269] From the quaternary salts of *NN*-disubstituted thioimidates and sodium azide, the 1,5-dihydro-1,2,3,4-thia(S^{IV})-triazoles, *e.g.* (157), are obtained.[382] Arylthiohydroxamic acids react with dimethyl acetylenedicarboxylate to give 1,2,4-oxathiazoles (158).[383]

The following heterocyclic compounds, which do not contain sulphur in the ring but contain exocyclic sulphur derived from a thioamide structure, are worthy of mention. Heating the phenylnitrones[240] of glyoxylthioanilides, $Ar^1NHC(S)CH=N(O)Ar^2$, in concentrated HCl gives the thioisatins (159). β-Thiolactams are formed in [2 + 2] cycloadditions of diarylthioketens with Schiff bases.[384] The isocyanoketen C=N—CH=C=O, formed from isocyano-acetyl chloride and triethylamine, reacts with alkyl thioimidates to give the isocyano-substituted β-lactams (160);[385] (161) and (162) are formed from the appropriate acid chlorides.[386, 387] Meso-ionic imidazole-4-thiones (163) are obtained by allowing substituted aminoacetic acid thioamides to react with ethyl orthoformate.[239] Triazoles (164), with various combinations of the furan, thiophen, and selenophen rings, are obtained by cyclization of the appropriate thioamides and hydrazides.[388] In the reaction of the enamino-ketones (95) with

[382] S. I. Mathew and F. Stansfield, *J.C.S. Perkin I*, 1974, 540.
[383] P. Rajagopalan and C. N. Talaty, *Heterocycles*, 1975, **3**, 563 (*Chem. Abs.*, 1975, **83**, 147 432).
[384] E. Schaumann, *Chem. Ber.*, 1976, **109**, 906.
[385] I. Hoppe and U. Schöllkopf, *Chem. Ber.*, 1976, **109**, 482.
[386] A. K. Bose, M. S. Manhas, J. S. Chib, H. P. S. Chwala, and B. Dayal, *J. Org. Chem.*, 1974, **39**, 2877.
[387] R. Lattrell and G. Lohaus, *Annalen*, 1974, 870.
[388] P. Dubus, B. Decroix, J. Morel, and C. Paulmier, *Compt. rend.*, 1974, **278**, *C*, 61.

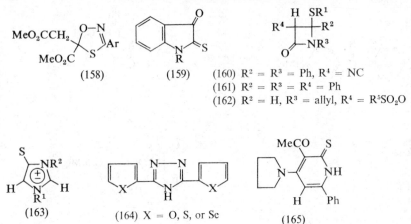

(158)   (159)   (160) $R^2 = R^3 = Ph$, $R^4 = NC$
                               (161) $R^2 = R^3 = R^4 = Ph$
                               (162) $R^2 = H$, $R^3 = $ allyl, $R^4 = R^5SO_2O$

(163)   (164) X = O, S, or Se   (165)

benzoyl isothiocyanate in equimolar amounts, the pyridine-2-thiones (165) are formed instead of the thioamides (97).[264]

Limitations of space preclude further discussion of the formation of heterocyclic systems, which are dealt with in other chapters of this Report. A copolymer of caprolactam and thiocaprolactam has been prepared and shown to have the properties of an ion-exchange system.[389]

5 Thiono- and Dithio-carboxylic Acids, their Derivatives, and their Selenium Analogues

Synthesis of Dithiocarboxylic Acids.—Several methods of long standing have been applied recently, although a procedure which is appropriate in all, or even most, cases is still lacking.

Thiolysis of phenyl trichloroacetate, Cl_3CO_2Ph, yields potassium trithiooxalate, $K_2[S_2CCOS]$,[390] and the reaction of benzyl chlorides or bromides with elemental sulphur and methoxide ion has proved to be a very convenient method for the preparation of pure dithiocarboxylic acids.[381]

The most important synthetic route now, as before, is the addition of Grignard reagents to carbon disulphide.[381, 391—396]

Activated methylene,[397] or methyl groups, *e.g.* in *N*-methylpicolinium iodide and related compounds,[398] are readily attacked by CS_2. The resulting dithiocarboxylate ion is methylated by excess of *N*-methylpicolinium iodide to yield (166).[398]

[389] L. A. Anan'eva, G. A. Gabrielyan, T. V. Druzhinina, and Z. A. Rogovin, *Khim. Volokna*, 1975, 22 (*Chem. Abs.*, 1976, **84**, 106 099).
[390] W. Stork and R. Mattes, *Angew. Chem.*, 1975, **87**, 452.
[391] M. Gisin and J. Wirz, *Helv. Chim. Acta*, 1975, **58**, 1768.
[392] D. H. R. Barton and S. Prabhakar, *J.C.S. Perkin I*, 1974, 781.
[393] R. Haraoubia, J. C. Gressier, and G. Levesque, *Makromol. Chem.*, 1975, **176**, 2143.
[394] J. Meijer, P. Vermeer, and L. Brandsma, *Rec. Trav. chim.*, 1975, **94**, 83.
[395] J. Voss and K. Schlapkohl, *Tetrahedron*, 1975, **31**, 2982.
[396] D. Paquer, *Bull. Soc. chim. France*, 1975, 1439.
[397] T. Takeshima, N. Fukada, E. Okabe, F. Mineshima, and M. Muraoka, *J.C.S. Perkin I*, 1975, 1277.
[398] K. Mizuyama, Y. Tominaga, Y. Matsuda, and G. Kobayashi, *Yakugaku Zasshi*, 1974, **94**, 702 (*Chem. Abs.*, 1974, **81**, 135 915).

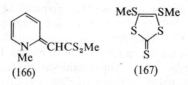

(166)   (167)

Various attempts to achieve reductive dimerization of CS_2 in order to obtain tetrathio-oxalates have been unsuccessful, the product isolated after methylation always being the 1,3-dithiole-2-thione (167).[399-403] Carbon monosulphide, 'CS', is involved in the formation of (167), and has been detected as the stable complex $Rh(CS)(PPh_3)_2$.[402]

It has been emphasized by Kato and his co-workers[404] that di- and tri-alkyl-ammonium dithiocarboxylates are readily obtained as stable crystals, which are more useful than metal salts in the purification and preparation of derivatives.

Synthesis of Thioacyl Halides.—Aromatic thioacyl chlorides, but not the unknown thioacetyl chloride, have been prepared in a convenient way, using phosgene as the chlorinating agent.[405]

Aliphatic thioacyl chlorides, *e.g.* $Bu^tCH_2C(S)Cl$ and $Bu^tCHClC(S)Cl$, can be obtained by addition of hydrogen chloride or chlorine, respectively, to the thioketen $Bu^tCH=C=S$, generated by flash pyrolysis of 4-t-butyl-1,2,3-thiadi-azoles or by direct reaction with the heated thiadiazole. The addition of hydrogen chloride to the alkynethiolate ion $Bu^tC\equiv CS^-$ also gives the thioacyl chloride $Bu^tCH_2C(S)Cl$.[406] Addition of thiophosgene to the enamine $Me_2NC(Bu^t)=CH_2$ gives the corresponding thioacyl chloride $Me_2\overset{+}{N}=C(Bu^t)CH_2C(S)Cl$.[407]

When trifluoroiodoethylene was bubbled through boiling sulphur, tetra-fluorodithiosuccinyl difluoride, $FC(S)(CF_2)_2C(S)F$, was formed.[408]

Synthesis of Thiono- and Dithio-esters.—*Alkylation of Thio- and Dithio-carb-oxylates.* Alkylation of thiocarboxylates takes place at the sulphur atom, and thiolesters are the reaction products in almost all cases. The reaction of thio-benzoic acid with diazomethane yields, however, some methyl thionobenzoate, along with a ten-fold amount of the thiolobenzoate, from which it may be cleanly separated by chromatography.[409] On the other hand, silylation of monothio-[410] or monoseleno-carboxylates[411] affords the *O*-silyl esters, in contrast to germylation or stannylation, which occur at the sulphur atom.

[399] K. Hartke and H. Hoppe, *Chem. Ber.*, 1974, **107**, 3121.
[400] G. Kiel, U. Reuter, and G. Gattow, *Chem. Ber.*, 1974, **107**, 2569.
[401] G. Steimecke, R. Kirmse, and E. Hoyer, *Z. Chem.*, 1975, **15**, 28.
[402] G. Bontempelli, F. Magno, G.-A. Mazzocchin, and R. Seeber, *J. Electroanalyt. Chem. Inter-facial Electrochem.*, 1975, **63**, 231.
[403] U. Reuter and G. Gattow, *Z. anorg. Chem.*, 1976, **421**, 143.
[404] S. Kato, T. Mitani, and M. Mizuta, *Internat. J. Sulfur Chem.*, 1973, **8**, 359.
[405] H. Viola and R. Mayer, *Z. Chem.*, 1975, **15**, 348.
[406] G. Seybold, *Angew. Chem.*, 1975, **87**, 710.
[407] M. Parmantier, J. Galloy, M. Van Meerssche, and H. G. Viehe, *Angew. Chem.*, 1975, **87**, 33.
[408] W. J. Middleton, *J. Org. Chem.*, 1975, **40**, 129.
[409] J. Rullkötter and H. Budzikiewicz, *Org. Mass Spectrometry*, 1976, **11**, 44.
[410] M. G. Voronkov, R. G. Mirskov, O. S. Ishchenko, and S. P. Sitnikowa, *Zhur. obshchei Khim.*, 1974, **44**, 2462.
[411] H. Ishihara and Y. Hirabayashi, *Chem. Letters*, 1976, 203.

A great number of dithioesters have been prepared by alkylation of dithio-carboxylates.[41, 381, 391-396, 398, 404, 409, 412] The carboxymethyl dithiocarboxylates $RC(S)SCH_2CO_2H$,[41, 381, 391] including the pentadeuteriophenyl derivative,[392] which are important thioacylation agents, can be obtained in this way. The use of ω-chloroalkyl bromides yields ω-chlorodithiocarboxylates $RC(S)S(CH_2)_nCl$,[394] and electrochemical reduction of benzo-1,2-dithiole-3-thione in the presence of CH_3I produces methyl 2-(methylthio)dithiobenzoate (168).[412] Triphenylplumbyl

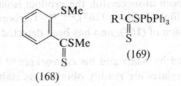

(168)

(169)

dithiocarboxylates (169), which are also thioacylating agents, have been prepared from the dialkylammonium salts of the dithiocarboxylates and triphenylplumbyl chloride, Ph_3PbCl.[413]

Thiolysis and Thionation of Carboxylic Acid Derivatives. Thiolysis of imidates $R^1C(OR^2)=NR^3$, or their salts $R^1C(OR^2)=\overset{+}{N}R^3_2 X^-$, respectively, has remained the method of choice for the synthesis of thionocarboxylates.[271, 399, 409, 414-421] The starting materials can be obtained by addition of alcohols to nitriles, by alcoholysis of imido-chlorides, or by *O*-alkylation of amides. Dialkyl mono-(170)[414] and bis-thiono-oxalates (171)[399] are prepared in this way, in which case it is especially advantageous to use the imidates as free bases instead of their

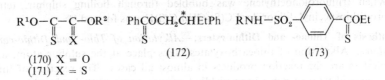

(170) X = O
(171) X = S

(172)

(173)

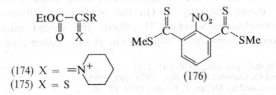

(174) X = $=N^+$
(175) X = S

(176)

[412] P. E. Iversen and H. Lund, *Acta Chem. Scand.* (*B*), 1974, **28**, 827.
[413] T. Katada, S. Kato, and M. Mizuta, *Chem. Letters*, 1975, 1037.
[414] P. Stäglich, K. Thimm, and J. Voss, *Annalen*, 1974, 671.
[415] Y. Ogata, K. Takagi, and S. Ihda, *J.C.S. Perkin I*, 1975, 1725.
[416] V. I. Cohen, N. Rist, and S. Clavel, *European J. Medicin. Chem.-Chim. Ther.*, 1975, **10**, 137 (*Chem. Abs.*, 1975, **83**, 126 074).
[417] D. H. R. Barton and S. W. McCombie, *J.C.S. Perkin I*, 1975, 1574.
[418] K. Hartke and H. Hoppe, *Chem.-Ztg.*, 1974, **98**, 618.
[419] J. Ellis and R. A. Schibeci, *Austral. J. Chem.*, 1974, **27**, 429.
[420] K. Thimm and J. Voss, *Z. Naturforsch.*, 1974, **29b**, 419.
[421] H. Alper and C. K. Foo, *Inorg. Chem.*, 1975, **14**, 2928.

hydrochlorides.[399] This method is quite appropriate for the synthesis of chiral (172),[415] or physiologically active (tuberculostatic) (173) [271, 416] thione esters, and has also been applied to the preparation of selone esters $R^1C(Se)OR^2$, but attempts to obtain tellurone esters have not been successful.[417]

Thiolysis of imidothiolates $R^1C(SR^2)=NR^3$ or the salts $R^1C(SR^2)=\overset{+}{N}R^3_2$ X^- affords dithiocarboxylates.[41, 276, 354, 381, 395, 414, 420, 422, 423] This method has been reviewed by Doyle and Kurzer,[41] and by Leon,[354] and it is suitable for the preparation of dithio-oxalates (175) from (174) [414] and α-acylamino-dithiocarboxylates $R^1C(O)NHCHR^2C(S)SR^3$.[276]

Thionation of esters [421, 424] and thiolesters [424] with phosphorus pentasulphide yields thiono- or dithio-esters, *e.g.* (176).[424]

Thioacylation of Alcohols and Thiols. Thiobenzoyl chloride [392] and Bu^tCH_2-$C(S)Cl$ [406] have been used to prepare various thiono- and dithio-esters, *e.g.* the chiral (177).[392] Dialkyl tetrafluorobis(thiono)succinates $[CF_2C(S)OR]_2$ result from alcoholysis of $\text{--}[CF_2C(S)F]_2$,[408] whereas the ortho-thioester (178) is obtained from $\text{--}[CF_2C(S)F]_2$ and methanethiol.[408]

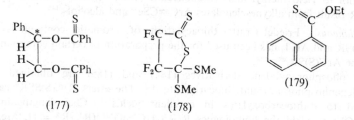

(177)   (178)   (179)

Carboxymethyl dithiocarboxylates are well-known thioacylating reagents, and have been the starting materials for the preparation of thionocarboxylates of different types,[391, 392, 395, 421, 425] *e.g.* (179), which has proved to be an efficient sensitizer for photochemical reactions in the visible region ($\lambda < 500$ nm),[391] and thioacylcholines $ArC(S)O(CH_2)_2\overset{+}{N}Me_3$ I^-.[425]

The thioacylating properties of (169) have already been mentioned, and thiono-esters are formed from it by reaction with alkoxides.[413] Thioacylation of methoxide ion can also be achieved by $R^1C(S)SSR^2$, which is obtained from $R^1CS_2^-$ $R_2NH_2^+$ and R^2SCl.[426] Dithiocarboxylates may be prepared in an analogous manner from mercaptides.[392, 395]

Thioketens, which are now more readily available [406, 427-429] than some years ago,[430] are valuable reagents for the thioacylation of alcohols and thiols at room temperature. The thioketen (183), which is formed from the thioaldehyde (182)

[422] A. Previero, A. Gourdol, J. Derancourt, and M.-A. Coletti-Previero, *F.E.B.S. Letters*, 1975, **51**, 68.
[423] K. Thimm and J. Voss, *Z. Naturforsch.*, 1975, **30b**, 932.
[424] M. S. Chauhan and D. M. McKinnon, *Canad. J. Chem.*, 1975, **53**, 1336.
[425] G. C. Barrett and P. H. Leigh, *F.E.B.S. Letters*, 1975, **57**, 19.
[426] T. Katada, S. Tsuji, T. Sugiyama, S. Kato, and M. Mizuta, *Chem. Letters*, 1976, 441.
[427] E. Schaumann and W. Walter, *Chem. Ber.*, 1974, **107**, 3562.
[428] G. Seybold and C. Heibl, *Angew. Chem.*, 1975, **87**, 171.
[429] H. Meier and H. Bühl, *J. Heterocyclic Chem.*, 1975, **12**, 605.
[430] R. Mayer and H. Kröber, *Z. Chem.*, 1975, **15**, 91.

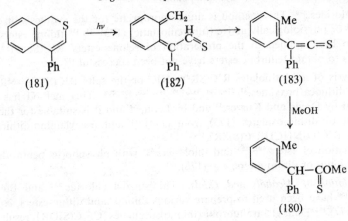

by a 1,5-hydrogen shift, is an intermediate in the photolysis of (181), and it adds methanol to yield methyl diarylthionoacetate (180).[431] Selone esters ArCH$_2$C-(Se)OR result from alkyneselenolates ArC$\equiv$CSe$^-$ and alcohols.[432]

Miscellaneous. Friedel–Crafts thioacylation of aromatic compounds with ClC(S)SR and AlCl$_3$ has been used for the preparation of certain compounds of the type ArC(S)SR.[395, 433]

The phosphoranylidene derivatives (184) and (185) are obtained from methylenephosphoranes and thiocarbonates.[434] The esters R^{1}C(S)SSR^2 can be reduced to dithiocarboxylates in excellent yields.[42] Chain-lengthening of HC(S)OEt to yield the homologues R^{1}R^{2}CHC(S)OEt (R^1, R^2 = H, Me, Ph, PhCH$_2$, or CO$_2$Et) can be performed by means of diazo-compounds R^{1}R^{2}CN$_2$.[435]

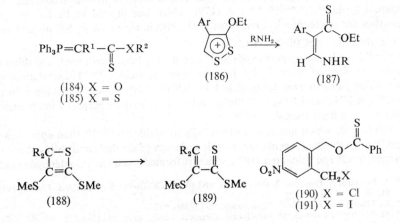

[431] A. Padwa, A. Au, G. A. Lee, and W. Owens, *J. Org. Chem.*, 1975, **40**, 1142.
[432] F. Malek-Yazdi, and M. Yalpani, *J. Org. Chem.*, 1976, **41**, 729.
[433] A. Tangerman and B. Zwanenburg, *J.C.S. Perkin II*, 1975, 916.
[434] H. Yoshida, H. Matsuura, T. Ogata, and S. Inokawa, *Bull. Chem. Soc. Japan*, 1975, **48**, 2907.
[435] H. Kröber and R. Mayer, *Internat. J. Sulfur Chem.*, 1976, **8**, 611.

Ring-opening of the dithiolium cation (186) by amines leads to ethyl β-amino-thionoacrylates (187).[436] The thiet (188), which is formed by photochemical addition of thioketones to bis(methylthio)acetylene, easily undergoes ring-opening to (189).[437] The thionobenzoate (190) can be interconverted into the iodo-derivative (191) by NaI.[392]

Physical Properties.—*X*-Ray diffraction studies have been performed on potassium[438] and rubidium[438, 439] dithioacetate, and have shown that both C—S distances are equal (d = 1.671 Å in MeCS$_2$K). Methyl thionoformate has been studied in the gaseous phase by electron diffraction. The molecule exists in the planar (*Z*)-configuration ($d_{C=S}$ = 1.612 Å, d_{C-O} = 1.369 Å).[440]

The u.v.—visible spectrum of the persistent aliphatic thioacyl chloride Bu^{t}CHClC(S)Cl [λ_{max}/nm, (log ε): 480 (1.08); 304 (2.24); 257 (3.74); 207 (4.10)],[406] agrees well with that predicted for MeC(S)Cl.[441] Methyl benzoylthionoacetate (192) exhibits an $n \rightarrow \pi^*$ absorption band at λ_{max} = 415 nm, which has been quoted as a proof for its molecular structure.[420] The characteristic absorptions

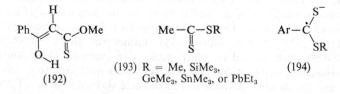

(192)

(193) R = Me, SiMe$_3$,
GeMe$_3$, SnMe$_3$, or PbEt$_3$

(194)

of RC(S)SSR are found at λ_{max} (log ε) = 534 nm (1.99); 310 nm (4.14).[426] Vibrational spectra of dithiocarboxylic acids have been measured by various groups,[390, 442–444] and a strong absorption at 1215/1230 cm^{-1} has been assigned to the C=S stretching vibration of Bu^{t}CHClC(S)Cl.[406] The i.r. spectra of carboxymethyl dithiocarboxylates and diselenocarboxylates are cited in the review by Doyle and Kurzer.[41] Thiono- and dithio-oxalates have been studied thoroughly.[399, 414] It has been shown that the latter exist as a mixture of (*E*)- and (*Z*)-isomers, in agreement with dipole-moment studies.[414] The C=S stretching frequency of organometallic dithioacetates (193) has been located at 1150—1200 cm^{-1} on the basis of i.r. and Raman spectroscopic evidence;[445] the corresponding value for RC(S)SSR is 1245 cm^{-1}.[426]

In order to elucidate the molecular structure, ^{1}H n.m.r. spectra of thio-esters have been measured in numerous cases, which are not cited here. ^{13}C n.m.r. spectra of thioacyl chlorides, *e.g.* Bu^{t}CHClC(S)Cl,[406] and thiono-esters[429] have also been reported. The barriers to internal rotation about the C—S bond are

[436] J. Faust, *Z. Chem.*, 1975, **15**, 478.
[437] A. C. Brouwer and H. J. T. Bos, *Tetrahedron Letters*, 1976, 209.
[438] M. M. Borel and M. Ledesert, *Z. anorg. Chem.*, 1975, **415**, 285.
[439] M. A. Bernard, M. M. Borel, and G. Dupriez, *Rev. Chim. minérale*, 1975, **12**, 181.
[440] J. DeRooij, F. C. Mijlhoff, and G. Renes, *J. Mol. Structure*, 1975, **25**, 169.
[441] J. Fabian, H. Viola, and R. Mayer, *Tetrahedron*, 1967, **23**, 4323.
[442] R. Mattes and W. Stork, *Spectrochim. Acta*, 1974, **30A**, 1385.
[443] N. V. Mel'nikova and A. T. Pilipenko, *Ukrain. khim. Zhur.*, 1974, **40**, 269 (*Chem. Abs.*, 1974, **80**, 132 286).
[444] H. S. Randhawa and C. N. R. Rao, *J. Mol. Structure*, 1974, **21**, 123.
[445] S. Kato, A. Hori, H. Shiotani, M. Mizuta, N. Hayashi, and T. Takakuwa, *J. Organometallic Chem.*, 1974, **82**, 223.

18—20 kcal mol^{-1} in sterically hindered dithiobenzoates, as determined by n.m.r. spectroscopy.[433]

The electronic structure of alkylthiono- and dithio-acetates has been compared with that of the corresponding thioloacetates and acetates, using photoelectron spectroscopy and CNDO/2 calculations.[446] Electron capture yields radical anions (194), which were studied by mass [409] and e.s.r. spectroscopy.[395] Decomposition of (194) takes place with rearrangement of the intermediate dithiocarboxylate anion, and loss of CS to form ArS$^-$.[409] The spin-density distribution of (194) has been determined from the h.f.s. coupling constants and g values.[395]

The pK_a values of dithio-acids have been measured spectrophotometrically.[439, 447] CH_3CS_2H is a stronger acid (pK_a = 2.57) than CH_3COSH (pK_a = 3.35).[439]

Reactions.—The kinetics of the reaction of PhC(S)OEt with Hg^{2+} to give PhCO$_2$Et have been studied. Loss of thioester is first-order, and Hg^{2+} is found to be much more efficient than other cations, e.g. Ag$^+$ or H$^+$.[448]

Interconversion of thiono- and dithio-esters has been considered earlier (p. 179). The reaction of these compounds with amines has remained one of the most important and widely applied methods for the preparation of thioamides and related substances. This is especially true for carboxymethyl dithiobenzoates,[41, 228, 449, 450] but other alkyl dithiocarboxylates [276, 422, 451–453] or RC(S)SSR [426] work as well in many cases. For example, MeCS$_2$Me reacted with amino-acids, free peptides, and peptides which were fixed to a polystyrene matrix, and the terminal N-thioacetyl amino-acid was split off by CF$_3$CO$_2$H to give 2-substituted thiazol-5-ones, which served to identify the N-terminal amino-acid.[422, 453] 3-Nitrothiobenzoylcholine, 3-NO$_2$C$_6$H$_4$C(S)O(CH$_2$)$_2\overset{+}{N}$Me$_3$ I$^-$, can also be applied advantageously in peptide sequence analysis, because it reacts faster with amino-groups than carboxymethyl dithiobenzoates, and excess reagent and choline can be removed from the reaction mixture by ion exchange.[425]

Sodium,[454] potassium,[397] dialkylammonium,[404] and triphenylstannyl dithiocarboxylates [455] yield thioamides on aminolysis. Thionocarboxylates have been used to prepare (195),[456] (196),[456] (197),[399] (198),[408] and (199).[457]

Paquer [396] has published a review on the Grignard reaction of dithiocarboxylates (and other thiocarbonyl compounds) (see Scheme 6). Dithioacetals (201), which result from thiophilic addition of the reagent, are the main products, and dithioketals (202) can be isolated after methylation of the intermediate (200),

[446] A. Flamini, E. Semprini, and G. Condorelli, *Chem. Phys. Letters*, 1975, **32**, 365.
[447] G. Mezaraups, L. D. Kulikova, E. Jansons, and T. Gutenko, *Latv. P.S.R. Zinat. Akad. Vestis, Kim. Ser.*, 1976, 196 (*Chem. Abs.*, 1976, **85**, 32 175).
[448] D. P. N. Satchell, M. N. White, and T. J. Weil, *Chem. and Ind.*, 1975, 791.
[449] R. Esmail and F. Kurzer, *J.C.S. Perkin I*, 1975, 1781.
[450] P. D. Callaghan, M. S. Gibson, and A. J. Elliott, *J.C.S. Perkin I*, 1975, 1386.
[451] V. A. Pechenyuk, L. B. Dashkevich, and P. V. Kuznetsov, *Zhur. org. Khim.*, 1975, **11**, 1201.
[452] V. A. Pechenyuk, P. V. Kuznetsov, and L. B. Dashkevich, *Zhur. org. Khim.*, 1975, **11**, 1345.
[453] A. Previero and J. C. Cavadore, *Phase Methods Protein Sequence Anal. Proc. Internat. Conf. 1st*, 1975, 63 (*Chem. Abs.*, 1976, **84**, 147 268).
[454] R. Zielke and H. Mägerlein, *Synthesis*, 1975, 47.
[455] S. Kato, T. Kato, T. Yamauchi, Y. Shibahashi, E. Kakuda, M. Mizuta, and Y. Ishii, *J. Organometallic Chem.*, 1974, **76**, 215.
[456] H. Hoppe and K. Hartke, *Arch. Pharm.*, 1975, **308**, 526.
[457] P. Jakobsen, *Acta Chem. Scand.* (*B*), 1975, **29**, 281.

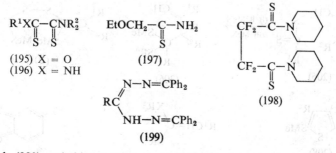

(195) X = O
(196) X = NH

(197)

(198)

(199)

but thiols (203) and thiones (204) are also obtained in several cases [458, 459] (see Section 1, p. 131).

S-Allyl dithiocarboxylates (205) form $\beta\gamma$-unsaturated dithioketals (206) after addition of R^5MgX and subsequent methylation; the reaction sequence pre-

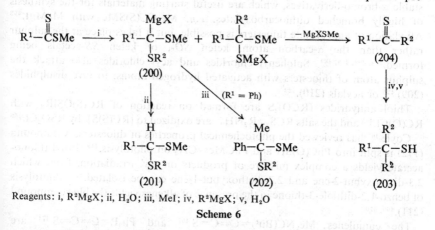

Reagents: i, R^2MgX; ii, H_2O; iii, MeI; iv, R^2MgX; v, H_2O

Scheme 6

sumably involves a [2,3]sigmatropic shift.[460] The reduction of methyl dithio-propionate by $NaBH_4$ yields the dithiosemiacetal $EtCH(SMe)SH$, from which $EtCH(SMe)SSCH=CHMe$, a constituent of asafoetida resin, can be prepared.[461] Thionobenzoates PhC(S)OR are reduced to the hydrocarbons RH by Bu^n_3SnH, which provides a method for the conversion of alcohols into hydrocarbons.[417] Alkyl thionocarboxylates $PhCH_2C(S)OR$ (R = Bu^n, cyclohexyl, or cholestanyl), on the other hand, are desulphurized by means of Raney nickel to form the ethers $PhCH_2CH_2OR$.[419] One of the C=S groups of (171; R = Et) is reduced by KSH to yield $EtOC(S)CH_2OEt$.[399]

Sterically hindered *S*-oxides of dithiocarboxylates, *viz.* sulphines, have been prepared by oxidation with metachloroperoxybenzoic acid [433] (see Section 2). Oxidation of $R^1R^2CHC(S)SMe$ with iodine results in the formation of disulphides (207), which, if R^2 = H, undergo cyclization to thiophens (208) on heating or

[458] L. Léger and M. Saquet, *Compt. rend.*, 1974, **279**, C, 695.
[459] L. Léger and M. Saquet, *Bull. Soc. chim. France*, 1975, 657.
[460] L. Léger, M. Saquet, A. Thuillier, and S. Julia, *J. Organometallic Chem.*, 1975, **96**, 313.
[461] J. Meijer and P. Vermeer, *Rec. Trav. chim.*, 1974, **93**, 242.

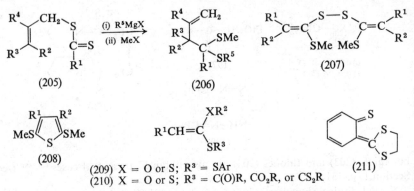

(205)   (206)   (207)

(208)

(209) X = O or S; R³ = SAr   (211)
(210) X = O or S; R³ = C(O)R, CO₂R, or CS₂R

in the presence of base.[462] The reaction of Me₂CHC(S)SMe with bromine yields stable α-bromo-derivatives, which are useful starting materials for the synthesis of highly branched dithiocarboxylates, *e.g.* Me₃CC(S)SMe with MeMgI.[463] Alkylation of appropriate thioesters is possible, but it takes place at the sulphur rather than the α-carbon atom, keten *SO*-, or keten *SS*-acetals being formed.[394, 398, 434, 464] Sulphenyl chlorides and acyl chlorides also attack the sulphur atom of thioesters with activated hydrogen atoms, to give disulphides (209)[420] or acylals (210).[423]

Thiol anhydrides (RCO)₂S are formed on reaction of RC(S)OSiR₃ with RC(O)Cl,[465] and the salts RCS₂⁻ R₂NH₂⁺ are oxidized to [RC(S)S]₂ by RSO₂Cl.[466]

Coyle[467] has reviewed the photochemical properties of thioesters. Compound (172) is split into PhC(O)SH and PhCMe=CH₂ on photolysis.[415] Ethyl thiono-acetate yields a complex mixture of products on u.v. irradiation, from which 2,3-diethoxybut-2-ene and 2,3-diethoxybut-1-ene can be isolated.[468] Photolysis of benzo-1,2-dithiole-3-thione in the presence of alkenes yields the blue compound (211).[469, 470]

The cumulenes Me₂NC(Buᵗ)=C=C=S[407] and Ph₃P=C=C=S[471] are formed by elimination of HCl from the thioacyl chloride or MeXH from Ph₃P=CHC(S)XMe (X = O or S), respectively.

Rearrangement reactions of alkyl thionocarboxylates[392, 399] and the polymerization of methyl 4-vinyldithiobenzoate have been studied.[393]

Various heterocycles can be obtained from thio- and dithio-carboxylic acids and their esters. *o*-Phenylenediamine reacts with (170) and (171) to give the quinoxalines (212)[472] and (213),[456] whereas substituted benzothiazoles[456, 472] are

[462] F. C. V. Larsson, L. Brandsma, and S.-O. Lawesson, *Rec. Trav. chim.*, 1974, **93**, 258.
[463] J. C. Wesdorp, J. Meijer, P. Vermeer, H. J. T. Bos, L. Brandsma, and J. F. Arens, *Rec. Trav. chim.*, 1974, **93**, 184.
[464] H. Yoshida, H. Matsuura, T. Ogata, and S. Inokawa, *Chem. Letters*, 1974, 1065.
[465] M. Mikołajczyk, P. Kiełbasiński, and H. M. Schiebel, *J.C.S. Perkin I*, 1976, 564.
[466] S. Kato, T. Kato, T. Kataoka, and M. Mizuta, *Internat. J. Sulfur Chem.*, 1973, **8**, 437
[467] J. D. Coyle, *Chem. Soc. Rev.*, 1975, **4**, 523.
[468] R. Jahn and U. Schmidt, *Chem. Ber.*, 1975, **108**, 630.
[469] P. DeMayo and H. Y. Ng, *J.C.S. Chem. Comm.*, 1974, 877.
[470] R. Okazaki and N. Inamoto, *Chem. Letters*, 1974, 1439.
[471] H. J. Bestmann and D. Sandmeier, *Angew. Chem.*, 1975, **87**, 630.
[472] K. Thimm and J. Voss, *Z. Naturforsch.*, 1975, **30b**, 292.

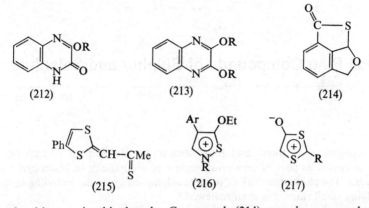

(212)   (213)   (214)

(215)   (216)   (217)

formed with *o*-aminothiophenol. Compound (214) can be prepared from thionophthalide by reaction with $Fe_2(CO)_9$ and oxidation of the intermediate iron complex with Ce^{IV}.[421] Two molecules of $MeCS_2H$ add to $PhC{\equiv}CH$, yielding (215).[473] 3-Amino-1,2,4-thiadiazoles are formed from thionocarboxylates and guanidines, and subsequent oxidation with bromine,[474] and (187) is oxidized to (216).[436] Meso-ionic dithiolium-4-olates (217) result from $RC(S)SCH_2CO_2H$ and $Ac_2O{-}NR_3$.[381] Diels–Alder addition of dimethyl acetylenedicarboxylate to (166) occurs with ring-opening, to yield thiopyran (218),[475] whereas ring-contraction takes place during the addition of (219), spiro-cyclopentadiene (220) being the product.[476]

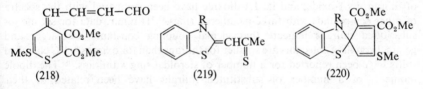

(218)   (219)   (220)

[473] C. T. Pedersen, *Acta Chem. Scand. (B)*, 1974, **28**, 367.
[474] B. Junge, *Annalen*, 1975, 1961.
[475] Y. Tominaga, K. Mizuyama, and G. Kobayashi, *Chem. and Pharm. Bull. (Japan)*, 1974, **22**, 1670.
[476] G. Kobayashi, Y. Matsuda, Y. Tominaga, and K. Mizuyama, *Heterocycles*, 1974, **2**, 309.

4

Small Ring Compounds of Sulphur and Selenium

BY D. C. DITTMER

The aromaticity of three- and four-membered cyclic sulphur compounds has been discussed as part of an overall review of aromaticity in heterocyclic compounds.[1] The photochemistry of organic sulphur compounds, including reactions involving small rings, has been reviewed.[2]

1 Thiirans (Episulphides)

Physical Properties and Theoretical Treatments.—The photoelectron spectra of cis- and trans-2,3-diethynylthiiran have been analysed and a comparison has been made with that of thiiran itself.[3] The photochemistry of thiirans has been reviewed,[4] and the optically active electronic transition at 260 nm of chiral thiirans has been discussed.[5] The microwave spectrum of 6-thiabicyclo[3,1,0]-hexane has supported the proposal that there is a single, stable, boat conformation for the molecule.[6] An X-ray analysis of $2\alpha,3\alpha$-epithio-5α-androst-6-en-17β-yl p-bromobenzoate has been accomplished.[7] The vibrational spectra of thiiran, its 1-oxide, and its 1,1-dioxide have been compared with the spectra of other compounds with three-membered rings.[8] ^{1}H n.m.r. data for six monosubstituted thiirans indicate that vicinal coupling constants are larger and geminal coupling constants are smaller for thiirans than for oxirans.[9] ^{13}C chemical shifts have been reported for a number of steroidal ring A thiirans.[10] The dipole moments of a number of substituted thiirans have been related to their conformations.[11]

Ab initio MO–SCF calculations have been done for thiiran, thiiran 1-oxide, and thiiran 1,1-dioxide.[12a] Variations in carbon–carbon bond distances are

[1] M. J. Cook, A. R. Katritzky, and P. Linda, in 'Advances in Heterocyclic Chemistry', ed. A. R. Katritzky and A. J. Boulton, Academic Press, New York and London, Vol. 17, 1975, p. 289.

[2] J. D. Coyle, *Chem. Soc. Rev.*, 1975, **4**, 323.

[3] F. Brogli, E. Heilbronner, J. Wirz, E. Kloster-Jensen, R. G. Bergman, K. P. C. Vollhardt, and A. J. Ashe, tert., *Helv. Chim. Acta*, 1975, **58**, 2620.

[4] A. Padwa, *Internat. J. Sulfur Chem.* (B), 1972, **7**, 331.

[5] G. Gottarelli, B. Samori, and G. Torre, *J.C.S. Chem. Comm.*, 1975, 398.

[6] D. J. Mjöberg, W. M. Ralowski, S. O. Ljunggren, and J. E. Backvall, *J. Mol. Spectroscopy*, 1976, **60**, 179.

[7] K. Utsumi-Oda and H. Koyama, *J.C.S. Perkin II*, 1975, 993.

[8] V. T. Aleksanyan and E. R. Razumova, *J. Struct. Chem.*, 1974, **15**, 955.

[9] K. J. Ivin, E. D. Lillie, and I. H. Petersen, *Internat. J. Sulfur Chem.*, 1973, **8**, 411.

[10] K. Tori and T. Komeno, *Tetrahedron Letters*, 1975, 135.

[11] E. N. Guryanova, A. M. Kuliev, K. Byashimov, and F. N. Mamedov, *J. Gen. Chem.* (*U.S.S.R.*), 1974, **44**, 1350.

[12] (a) M. Rohmer and B. Roos, *J. Amer. Chem. Soc.*, 1975, **97**, 2025; (b) R. Hoffmann, H. Fujimoto, J. R. Swenson, and C. Wan, *ibid.*, 1973, **95**, 7644.

explained by variation of the electron donor–acceptor strength of the sulphide, sulphoxide, and sulphone functions as they interact with the C_2H_4 moiety. These results were compared with those obtained previously by extended Hückel calculations,[12b] and they appear to explain more satisfactorily the variations in carbon–carbon bond lengths. Singlet transition energies and oscillator strengths have been calculated for thiiran.[13] A theoretical study of 2-phenylthiiran indicates that the planes of the phenyl ring and the thiiran ring should be orthogonal, and conjugation of the two ring systems was considered.[14] The barriers to rotation of methyl groups attached to thiirans and other three-membered heterocyclic rings have been discussed.[15]

Formation.—The methods of synthesis of thiirans have been reviewed.[16] A brief discussion of the preparation of thiirans is part of a review on the reactions of alkali-metal sulphides with organic halides and toluene-*p*-sulphonates.[17]

Oxirans are converted into thiirans stereospecifically and in high yield by treatment with 3-methylbenzothiazole-2-thione (Scheme 1).[18] Quantitative

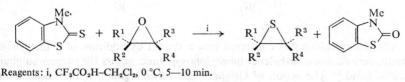

Reagents: i, CF_3CO_2H–CH_2Cl_2, 0 °C, 5—10 min.

Scheme 1

yields were reported for reactions with cyclohexene oxide, styrene oxide, epi-chlorohydrin, and *cis*-stilbene oxide. The reagent is said to be superior to thiourea (see Chapter 13). Phosphole and phospholen sulphides, *e.g.* 1-phenyl-3,4-dimethyl-Δ^3-phospholen sulphide, are claimed to be superior [19] to phosphine, sulphides [20] for the conversion of oxirans into thiirans. The reaction was exemplified with the epoxides of styrene, cyclopentene, cyclohexene, and cyclo-octene. A number of thiirans, among which are the homobenzene systems (1), (2), and (3), have been prepared from oxirans by treatment with thiocyanate ion or thiourea.[21] The episulphide of 3-carene was obtained from the adduct of the epoxide with diethyldithiophosphoric acid on addition of ethanolic potassium hydroxide.[22]

[13] H. Sakai, T. Yamabe, H. Kato, S. Nagata, and K. Fukui, *Bull. Chem. Soc. Japan*, 1975, **48**, 33.

[14] S. Sorriso, F. Stefani, E. Semprini, and A. Flamini, *J.C.S. Perkin II*, 1976, 374.

[15] W. J. Hehre, J. A. Pople, and A. J. P. Devaquet, *J. Amer. Chem. Soc.*, 1976, **98**, 664.

[16] A. V. Fokin and A. F. Kolomiets, *Russ. Chem. Rev.*, 1975, **44**, 138.

[17] D. Martinetz, *Z. Chem.*, 1976, **16**, 1.

[18] V. Calo, L. Lopez, L. Marchese, and G. Pesce, *J.C.S. Chem. Comm.*, 1975, 621.

[19] F. Mathey and G. Muller, *Compt. rend.*, 1975, **281**, C, 881.

[20] T. H. Chan and J. R. Finkenbine, *J. Amer. Chem. Soc.*, 1972, **94**, 2880; J. R. Finkenbine, *Diss. Abs.*, 1975, **35B**, 4833.

[21] (a) H. Prinzbach, C. Kaiser, and H. Fritz, *Angew. Chem.*, 1975, **87**, 249; (b) S. Kagabu and H. Prinzbach, *ibid.*, p. 248; *Tetrahedron Letters*, 1975, 29; (c) Z. Z. Zhumabaev, A. D. Aliev, and B. A. Krentsel, *Chem. Heterocyclic Compounds*, 1975, **10**, 743; (d) M. V. Jesudason and L. N. Owen, *J.C.S. Perkin I*, 1974, 2019; (e) A. M. Jeffrey, H. J. Yeh, and D. M. Jerina, *J. Org. Chem.*, 1974, **39**, 1405; (f) L. P. Vakhrushev, E. F. Filippov, N. F. Chernov, and V. P. Ageev, *J. Gen. Chem. (U.S.S.R.)*, 1975, **45**, 1878.

[22] O. N. Nuretdinova, G. A. Bakaleinik, and B. A. Arbuzov, *Bull. Acad. Sci., U.S.S.R.*, 1975, **24**, 878.

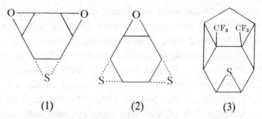

(1)   (2)   (3)

Thiirans are obtained in good yields (61—78%) by treatment of aldehydes and ketones with the lithium salt of 4,4-dimethyl-2-(methylthio)-Δ^2-oxazoline (4).[23] Other sulphur-stabilized carbanions may similarly be used, *e.g.* the lithium salts of 2-(alkylthio)thiazolidines, 2-(benzylthio)pyridine, and benzyl diethyl-dithiocarbamate.[24]

$$\text{LiCH}_2\text{S} - \underset{\underset{\text{Me}_2}{|}}{\overset{\text{O}\overline{}}{\diagdown}} \text{N} \quad + \quad \overset{\text{O}}{\overset{\|}{\text{R}^1\text{CR}^2}} \quad \xrightarrow{-78\ ^\circ\text{C}} \quad \overset{\text{R}^1}{\underset{\text{R}^2}{\diagup}}\overset{\text{S}}{\text{C}} - \text{CH}_2$$

(4)

Adamantanethione was converted into a thiiran by addition of a methylene group, derived from methylenetriphenylphosphorane, across the carbon–sulphur double bond.[25] The action of Grignard and other organometallic reagents on thiones gives thiirans together with other products,[26] and thiirans also are obtained by treatment of 3-piperidino-2-phenylindene-1-thione with several diazoalkanes.[27]

Intermediates in Reactions.—The desulphurization of thiepins is believed to proceed by ring contraction to a thiiran.[28] Treatment of benzene oxide–oxepin with thiocyanate ion did not yield benzene sulphide–thiepin, but rather gave benzene, possibly by desulphurization of the unstable benzene sulphide.[29] The synthesis of 1,6:8,13-bismethano[14]annulene involves extrusion of a sulphur atom *via* an intermediate thiiran,[30] and the photolysis of 4-phenylisothio-chromene to 3-phenylindene likewise involves a thiiran (Scheme 2).[31] The

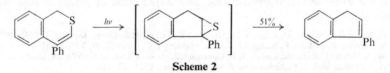

Scheme 2

[23] A. I. Meyers and M. E. Ford, *Tetrahedron Letters*, 1975, 2861.
[24] C. R. Johnson, A. Nakanishi, N. Nakanishi, and K. Tanaka, *Tetrahedron Letters*, 1975, 2865.
[25] A. P. Krapcho, M. P. Silvon, and S. D. Flanders, *Tetrahedron Letters*, 1974, 3817.
[26] M. Dayonneau and J. Vialle, *Tetrahedron*, 1974, **30**, 3119; this reaction has been reviewed by D. Paquer, *Bull. Soc. chim. France*, 1975, 1439.
[27] N. A. Korchevin, V. A. Usov, and M. G. Voronkov, *Chem. Heterocyclic Compounds*, 1975, **10**, 623.
[28] D. N. Reinhoudt and D. C. Kouwenhoven, *Tetrahedron*, 1974, **30**, 2093, 2431.
[29] A. M. Jeffrey, H. J. C. Yeh, D. M. Jerina, R. M. De Marinis, C. H. Foster, D. E. Piccolo, and G. A. Berchtold, *J. Amer. Chem. Soc.*, 1974, **96**, 6929.
[30] E. Vogel, J. Sombroek, and W. Wagemann, *Angew. Chem.*, 1975, **87**, 591.
[31] A. Padwa, A. Au, G. A. Lee, and W. Owens, *J. Org. Chem.*, 1975, **40**, 1142.

photochemical desulphurization of a substituted isothiazolone may possibly involve a thiiran.[32] The easy loss of sulphur from anions of certain 1,3,4-thia-diazines and 1,3-thiazines is suggested to proceed by a thiiran intermediate,[33] *e.g.* as shown in Scheme 3. Treatment of 2-nitrophenyl 2-pyridyl sulphide with

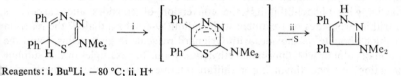

Reagents: i, Bu^{n}Li, -80 °C; ii, H$^+$

Scheme 3

triethyl phosphite resulted in a 2% yield of desulphurized product, believed to be formed *via* a thiiran produced when an intermediate nitrene attacked the nitrogen atom of the pyridine ring.[34]

The stereoselective isomerization of disubstituted olefins *via* thiirans has been reported.[35] *O*-Alkyl *S*-β-keto-alkyl dithiocarbonates (5) eliminate sulphur on

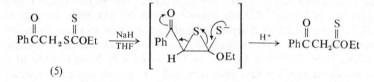

treatment with sodium hydride; a thiiran was suggested as an intermediate.[36] Treatment of 3*H*-4,5-benzo-1,2-dithiole-3-thione with diphenyldiazomethane proceeds *via* an intermediate thiiran to give 4,5-benzo-3-diphenylmethylene-1,2-dithiole.[37] Proton abstraction from thiuronium salts (6) yields enethiols (7) and enediamines (8), presumably by way of thiirans.[38] An enediamine is obtained by a similar reaction path when *NNN'N'*-tetramethylthiourea reacts with diethyl-azomalonate.[38]

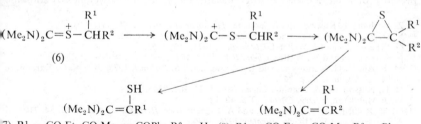

(7) R^1 = CO$_2$Et, CO$_2$Me, or COPh; R^2 = H   (8) R^1 = CO$_2$Et or CO$_2$Me; R^2 = Ph

[32] M. Maki and M. Sako, *Tetrahedron Letters*, 1976, 375.
[33] R. R. Schmidt and H. Huth, *Tetrahedron Letters*, 1975, 33; R. R. Schmidt and M. Dimmler, *Chem. Ber.*, 1975, **108**, 6.
[34] J. I. G. Cadogan and B. S. Tait, *J.C.S. Perkin I*, 1975, 2396.
[35] D. Van Ende and A. Krief, *Tetrahedron Letters*, 1975, 2709.
[36] A. J. Bridges and G. H. Whitham, *J.C.S. Perkin I*, 1975, 1603.
[37] S. Tamagaki, R. Ichihara, and S. Oae, *Bull. Chem. Soc. Japan*, 1975, **48**, 355.
[38] S. Mitamura, M. Takaku, and H. Nozaki, *Bull. Chem. Soc. Japan*, 1974, **47**, 3152.

Thiirans are believed to be intermediates in the reaction of 4,5-dicyano-1,3-dithiole-2-thione with trimethyl or triphenyl phosphite to give tetracyano-tetrathiafulvalene[39] and in the reactions of thiobenzophenone[40] or benzo-1,3-dithiole-2-thione[41] with phosphites to yield olefinic derivatives. Other reactions in which thiiran intermediates were suggested are the reaction of a thioketen with phenyl-lithium,[42] the conversion of epoxides and carbonates into trithiocarbonates by treatment with sodium *O*-ethyl xanthate,[43] the reaction of trithiocarbonate *SS*-dioxides with Grignard reagents to give olefins,[44] and the reaction of sulphur with fluorinated ethylenes.[45] Further examples of the 'sulphur contraction' method (involving a thiiran intermediate) for the synthesis of a carbon–carbon bond, discovered by Eschenmoser and co-workers,[46a] have been reported in studies on the synthesis of bile pigments.[46b]

Ions having a thiiran structure have been reported to occur in the mass spectra of various sulphur-containing compounds.[47]

Reactions.—Stereospecific desulphurization of several thiirans occurs in high yield on treatment with 3-methylbenzothiazole-2-selone,[48] which acts as a selenium-transfer reagent (*cf.* Scheme 1), giving the unstable seleniran, which readily undergoes deselenization. The desulphurization of thiirans to olefins by carbenes, derived from aliphatic diazo-compounds, was also reported to be essentially stereospecific.[49] Further examples of the formation of olefins by the extrusion of sulphur from thiirans by treatment with tervalent phosphorus derivatives,[21a, 21b, 21e, 22, 23] or by heating,[21b, 27] or even by being allowed to stand at room temperature,[22, 24] have been reported. Hydrogen atoms remove sulphur from thiiran to give ethylene.[50]

The isomerization of a 'Dewar thiophen' (9) to the thiophen structure has a half-life of 5.1 h at 160 °C. At room temperature the isomerization occurs in the presence of triphenylphosphine or phenylphosphine chlorides.[51] Thermolysis of 1,2-epithiocyclohexane at 210 °C yields a variety of products, bis(cyclohexyl) sulphide and disulphide, cyclohexyl phenyl sulphide, mercaptocyclohexane, cyclohexyl cyclohex-2-enyl sulphide, dodecahydrothianthrene, and H_2S. In the presence of a large excess of cyclohexane the yield of bis(cyclohexyl) sulphide

[39] Z. Yoshida, T. Kawase, and S. Yoneda, *Tetrahedron Letters*, 1975, 331.

[40] Y. Ogata, M. Yamashita, and M. Mizutani, *Tetrahedron*, 1974, **30**, 3709.

[41] G. Scherowsky and J. Weiland, *Chem. Ber.*, 1974, **107**, 3155.

[42] E. Schaumann and W. Walter, *Chem. Ber.*, 1974, **107**, 3562.

[43] M. E. Ali, N. G. Kardouche, and L. N. Owen, *J.C.S. Perkin I*, 1975, 748; N. G. Kardouche and L. N. Owen, *ibid.*, p. 754.

[44] N. H. Nilsson and A. Senning, *Chem. Ber.*, 1974, **107**, 2345.

[45] W. J. Middleton, *J. Org. Chem.*, 1975, **40**, 129.

[46] (*a*) M. Roth, P. Dubs, E. Götschi, and A. Eschenmoser, *Helv. Chim. Acta*, 1971, **54**, 710; (*b*) A. Gossauer and W. Hirsch, *Annalen*, 1974, 1496.

[47] (*a*) K. Hartke and H. Hoppe, *Chem. Ber.*, 1974, **107**, 3121; (*b*) J. B. Chattopadhyaya and A. V. R. Rao, *Org. Mass Spectrometry*, 1974, **9**, 649; (*c*) G. Cauquis, B. Divisia, and J. Ulrich, *ibid.*, 1975, **10**, 1021; (*d*) M. M. Vestling and R. L. Ogren, *J. Heterocyclic Chem.*, 1975, **12**, 243.

[48] V. Calo, L. Lopez, A. Mincuzzi, and G. Pesce, *Synthesis*, 1976, 200.

[49] Y. Hata, M. Watanabe, S. Inoue, and S. Oae, *J. Amer. Chem. Soc.*, 1975, **97**, 2553.

[50] T. Yokota, M. G. Ahmed, I. Safarik, O. P. Strausz, and H. E. Gunning, *J. Phys. Chem.*, 1975, **79**, 1758.

[51] Y. Kobayashi, I. Kumadaki, A. Ohsawa, and Y. Sekine, *Tetrahedron Letters*, 1975, 1639.

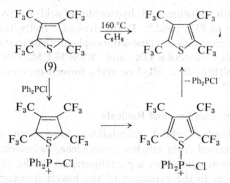

(9)

increased from 22 to 63%. The same products were also obtained by thermolysis of other thiirans in the presence of excess cyclohexane.[52]

The chemical reactions of *cis*- and *trans*-2,3-di-t-butyl-thiirans have been investigated.[53] The *trans*-isomer is more inert to ring-opening because of the equal shielding of both sides of the ring by the t-butyl groups. The *S*-oxides could be prepared by oxidation with *m*-chloroperoxybenzoic acid, but the sulphones could not be obtained. Treatment of the *cis*-isomer with methyl fluorosulphonate gave the *S*-methylthiiranium salt, which underwent ring opening in the presence of various nucleophiles (*trans* attack). [1]H n.m.r. and [13]C n.m.r. spectroscopy were used to study the protonation of the thiirans by fluorosulphonic acid. Treatment of the *cis*-isomer with chlorine or t-butyl hypochlorite gives different ring-opened products.

The reaction of thiiran with aryldiazonium chlorides in the presence of copper chloride yields β-chloroethyl aryl sulphides and diaryl disulphides *via* radical intermediates (Scheme 4).[54] Treatment of thiirans with sodium methyl xanthate

$$ ArN_2^+ \ Cl^- \xrightarrow[-N_2]{i,} [Ar^\cdot] \xrightarrow{ii} ArSCH_2CH_2^\cdot $$

$$ \downarrow \text{-(CH}_2\text{=CH}_2\text{)} \qquad \downarrow iii $$

$$ ArSSAr \longleftarrow ArS^\cdot \qquad ArSCH_2CH_2Cl $$

Reagents: i, CuCl₂; ii, $\triangle^S$; iii, CuCl₃⁻

Scheme 4

gives trithiocarbonates.[55] With several thiirans, prolonged treatment with potassium thiocyanate resulted in the partial stereomutation of the thiiran ring.[21d] Optically active polymers of substituted thiirans have been prepared.[56]

[52] S. Inoue and S. Oae, *Bull. Chem. Soc. Japan*, 1975, **48**, 1665.

[53] P. Raynolds, S. Zonnebelt, S. Bakker, and R. M. Kellogg, *J. Amer. Chem. Soc.*, 1974, **96**, 3146.

[54] B. V. Kopylova, L. V. Yashkina, S. A. Karapet'yan, and R. K. Freidlina, *Bull. Acad. Sci., U.S.S.R.*, 1975, **24**, 1090; B. V. Kopylova, S. A. Karapet'yan, and R. K. Freidlina, *ibid.*, 1974, **23**, 1839.

[55] M. V. Jesudason and L. N. Owen, *J.C.S. Perkin I*, 1974, 2024.

[56] P. Dumas, N. Spassky, and P. Sigwalt, *J. Polymer Sci., Part A*-1, *Polymer Chem.*, 1974, **12**, 1001; M. Sepulchre, N. Spassky, D. von Ooteghem, and E. J. Goethals, *ibid.*, p. 1683; N. Spassky, P. Dumas, M. Moreau, and J. P. Vacron, *Macromolecules*, 1975, **8**, 956.

Thiiran reacts with chlorine and hydrochloric acid to give 2-chloroethane-sulphonyl chloride in 85% yield.[57] Derivatives of α-methylthioglycidic acid are cleaved readily by methane- or arene-sulphenyl chlorides to mixtures of isomeric disulphides, $ClCH_2CMe(SSR)COX$ and $RSSCH_2CMe(Cl)COX$ (X = OMe, NH_2, NMe_2, or NHPh; R = alkyl or aryl), formed presumably *via* thiiranium ion intermediates.[58]

2 Thiiranium Ions and Radicals

Theoretical.—Molecular orbital calculations have been reported for the thiiranium ions derived from ethylene, propylene, isobutene, and butadiene.[59] The most stable system involves a *p*-configuration of the valence electrons of sulphur. Differences in the extension of the lowest unoccupied MO's on the ethylene carbon atoms were said to be important in determining the regiospecificity of ring-opening addition reactions with chloride ion. Theoretical calculations on $C_2H_4SH^+$ indicate that a bridged thiiranium structure would be more stable as an intermediate than an open structure in electrophilic additions of RSX to olefins.[60] The barrier to pyramidal inversion at the sulphur atom was calculated to be 78.11 kcal mol^{-1}.

Formation and Properties.—A number of relatively stable (at −30 °C) thiiranium ions have been reported, *e.g.* (10).[61a] (See also Chapter 1, p. 55.) Nucleophiles attack these ions in a *trans* fashion, and both Markovnikov and

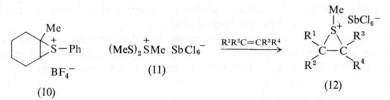

(10)   (11)   (12)

anti-Markovnikov additions occur.[61a, b] Thiiranium ions (12), stable for weeks at −10 °C, have also been prepared by treatment of alkenes with methyl bis-(methylthio)sulphonium hexachloroantimonate (11).[62] Only one isomer is obtained from *cis*-alkenes, *i.e.* the one with the *S*-methyl group *trans* to the alkyl groups of the alkene. The direction of ring-opening of 2,2-dimethyl-1-phenyl-thiiranium hexachloroantimonate differed according to reaction conditions (kinetic *vs.* thermodynamic control).[63] The synthesis and properties of 1-methyl *cis*-2,3-di-t-butylthiiranium fluorosulphonate have been described.[53]

[57] A. V. Fokin, A. F. Kolomiets, N. K. Bliznyuk, and R. N. Golubeva, *Bull. Acad. Sci., U.S.S.R.*, 1975, **24**, 2014.

[58] N. M. Karimova, M. G. Lin'kova, O. V. Kil'disheva, and I. L. Knunyants, *Chem. Heterocyclic Compounds*, 1974, **9**, 6.

[59] Y. Kikuzono, T. Yamabe, S. Nagata, H. Kato, and K. Fukui, *Tetrahedron*, 1974, **30**, 2197.

[60] J. W. Gordon, G. H. Schmid, and I. G. Csizmadia, *J.C.S. Perkin II*, 1975, 1722.

[61] (a) W. A. Smit, M. Z. Krimer, and E. A. Vorob'eva, *Tetrahedron Letters*, 1975, 2451; (b) E. A. Vorob'eva, L. G. Gurvich, N. S. Zefirov, M. Z. Kriner, and V. A. Smit, *J. Org. Chem. (U.S.S.R.)*, 1974, **10**, 891.

[62] G. Capozzi, O. De Lucchi, V. Lucchini, and G. Modena, *Tetrahedron Letters*, 1975, 2603.

[63] M. Oki, W. Nakanishi, M. Fukunaga, G. D. Smith, W. L. Duax, and Y. Osawa, *Chem. Letters*, 1975, 1277.

Intermediates in Reactions.—The role of thiiranium ions as intermediates in the addition of sulphenyl and selenenyl derivatives to alkenes [64-69] is discussed in Chapter 1 (p. 55). Thermolysis of the *N*-tosyl-sulphimide (13) is suggested [70] to involve formation of the sulphenamide (14), which gives (15) *via* a thiiranium ion intermediate.

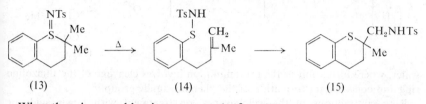

(13)   (14)   (15)

When there is a good leaving group at the β-position to a sulphur atom or the possibility of generation of a carbonium ion at that position, the formation of thiiranium ions is frequently proposed to account for the course of a reaction. The retention observed in the reduction of the halide *endo,endo*-2,6-dichloro-9-thiabicyclo[3,3,1]nonane with LiAlD₄ is nicely explained on the basis of an intermediate thiiranium cation. [71] The lack of retardation of the solvolysis of 2-*endo*-chloro-7-thianorbornane (16) by chloride ions, however, suggested that the chloride ion might be tightly bound, as in the sulphurane (17). [72]

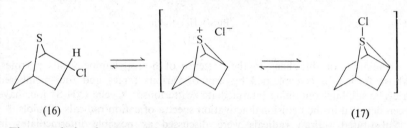

(16)   (17)

The acetolysis of chloromethylthiiran or 2-chloroethyl methyl sulphide does show a common-ion rate depression, and hence thiiranium ions were proposed as intermediates. [73] Thiiranium ions may be intermediates in the annelation of the steroid precursor (18), [74] and in the reaction of azetidinone (19) with iodine and

[64] N. S. Zefirov, N. K. Sadovaja, A. M. Maggerramov, I. V. Bodrikov, and V. R. Kartashov, *Tetrahedron*, 1975, **31**, 2948; I. V. Bodrikov, L. G. Gurvich, N. S. Zefirov, V. R. Kartashov, and A. L. Kurts, *J. Org. Chem. (U.S.S.R.)*, 1974, **10**, 1553.

[65] G. H. Schmid and V. J. Nowlan, *Canad. J. Chem.*, 1976, **54**, 695; G. Schmid and D. G. Garratt, *ibid.*, 1974, **52**, 1807; C. L. Dean, D. G. Garratt, T. T. Tidwell, and G. H. Schmid, *J. Amer. Chem. Soc.*, 1974, **96**, 4958; H. Kwart and D. Drayer, *J. Org. Chem.*, 1974, **39**, 2157.

[66] B. Giese, *Chem. Ber.*, 1975, **108**, 2978, 2998; *Tetrahedron Letters*, 1974, 3579.

[67] (a) D. J. Pasto, R. L. Smoroda, B. L. Turini, and D. J. Wampfler, *J. Org. Chem.*, 1976, **41**, 432; (b) S. Ikegami, J. Ohishi, and Y. Shimizu, *Tetrahedron Letters*, 1975, 3923; (c) E. A. Vorob'eva, M. Z. Krimer, and V. A. Smit, *Bull. Acad. Sci., U.S.S.R.*, 1974, **23**, 2564; (d) V. R. Kartashov, N. A. Kartashova, and E. V. Skorobogatova, *J. Org. Chem. (U.S.S.R.)*, 1974, **10**, 173.

[68] K. T. Burgoine, S. G. Davies, M. J. Peagram, and G. H. Whitham, *J.C.S. Perkin I*, 1974, 2629.

[69] Y. Makisumi, T. Takada, and Y. Matsukura, *J.C.S. Chem. Comm.*, 1974, 850.

[70] M. Kise, M. Murase, M. Kitano, T. Tomita, and H. Murai, *Tetrahedron Letters*, 1976, 691.

[71] J. N. Labows, jun., and N. Landmesser, *J. Org. Chem.*, 1975, **40**, 3798.

[72] I. Tabushi, Y. Tamaru, Z. Yoshida, and T. Sugimoto, *J. Amer. Chem. Soc.*, 1975, **97**, 2886.

[73] I. Tabushi, Y. Tamaru, and Z. Yoshida, *Bull. Chem. Soc. Japan*, 1974, **47**, 1455.

[74] P. T. Lansbury, T. R. Demmin, G. E. DuBois, and V. R. Haddon, *J. Amer. Chem. Soc.*, 1975, **97**, 394.

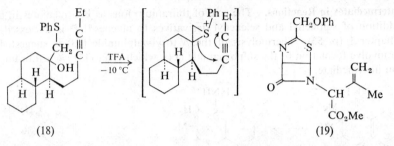

(18)   (19)

water, where formation of the thiiranium ion involves cleavage of the thiazoline ring and concomitant formation of the phenoxy-amide group.[75]

Ring contractions of thiacyclohexanes, on treatment with Lewis acids or hydride-ion-acceptors, are suggested to occur *via* thiiranium ions.[76] The thiiranium ion (20) is invoked to explain the formation of disulphide (21) when 5,6-dihydro-thiazolo[2,3-*b*]thiazolium salts are treated with sodium thiophenoxide.[77]

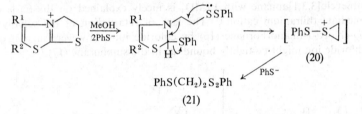

(20)

PhS(CH$_2$)$_2$S$_2$Ph

(21)

Irradiation of thiophens in the presence of n-propylamine yields pyrroles, whose formation is explained by way of various three- and four-membered tricyclic sulphur-containing intermediate zwitterions.[78] Cyclic $C_2H_5S^+$ ions have been detected in the collisional activation spectra of cation radicals of thiols.[79]

Substituted thiiran radicals were discussed as possible intermediates in rearrangements of β-halogeno-sulphides on treatment with reducing metals such as zinc.[80] The loss of stereochemistry in the formation of olefins by treatment of β-phenylthioethyl bromides with tributyltin radicals does not favour the intermediacy of an *S*-phenylthiiran radical.[81]

3 Thiiran 1-Oxides (Episulphoxides)

A stepwise process is favoured for the addition of aryldiazomethanes to diaryl-sulphines to give thiiran 1-oxides (21—68% yield), which readily decompose to olefins.[82] The reaction of dihalogeno-carbene precursors with diarylsulphines

[75] R. G. Micetich and R. B. Morin, *Tetrahedron Letters*, 1976, 979.
[76] A. K. Yus'kovich, T. A. Danilova, and L. M. Petrova, *Chem. Heterocyclic Compounds*, 1974, 9, 657; E. A. Viktorova, A. A. Freger, A. V. Egorov, and L. M. Petrova, *ibid.*, p. 132.
[77] H. Ohtsuka, T. Miyasaka, and K. Arakawa, *Chem. and Pharm. Bull.* (*Japan*), 1975, 23, 3243.
[78] A. Couture, A. Delevallee, A. Lablache-Combier, and C. Parkanyi, *Tetrahedron*, 1975, 31, 785.
[79] B. van de Graaf, P. P. Dymerski, and F. W. McLafferty, *J.C.S. Chem. Comm.*, 1975, 978.
[80] V. M. Fedoseev, V. S. Churilin, V. I. Mal'ko, M. N. Semenko, Y. I. Lys, and V. A. Nikanorov, *Proc. Acad. Sci. U.S.S.R.*, *Chem. Sect.*, 1974, 218, 751.
[81] T. E. Boothe, J. L. Greene, jun., and P. B. Shevlin, *J. Amer. Chem. Soc.*, 1976, 98, 951.
[82] B. F. Bonini, A. Cappelli, G. Maccagnani, and G. Mazzanti, *Gazzetta*, 1975, 105, 827.

yields *gem*-dihalogeno-alkenes and 2-halogeno-benzothiophen 1-oxides, which are also obtained by oxidation of the thiiran.

In alcoholic solvents thiiran 1-oxides yield thiolsulphinates, which may undergo further reaction with solvent to give α-alkoxy-disulphides,[83] as shown in Scheme 5.[83a] Oxidation of 2-methylthiiran 1-oxide by titanium(III) ion and hydrogen

Reagents: i, MeOH, 90 °C

Scheme 5

peroxide gave propene and the sulphur dioxide radical anion, which was detected by e.s.r. spectroscopy.[84] The 'ene' reactions of substituted thiiran 1-oxides have been discussed briefly in the context of a theoretical consideration of these reactions and [2 + 2] cycloadditions of heteronuclear systems.[85]

The photoelectron spectrum of thiiran 1-oxide has been reported, and orbital-correlation diagrams have been constructed.[86] The c.d. spectra of some chiral thiiran 1-oxides have been determined.[87]

4 Thiiran 1,1-Dioxides (Episulphones)

Quantitative M.O. models were used in a discussion of the photoelectron spectrum of thiiran 1,1-dioxide.[88] As compared with dimethyl sulphone, all antibonding S—C σ-orbitals increase in energy and bonding S—C σ-orbitals decrease in energy because of the relative compression of the S—C bonds in the cyclic sulphone. The chelotropic reactions of thiiran 1,1-dioxide have been compared with the formation of alkene–oxygen π-complexes.[89] Activation parameters for the extrusion of sulphur dioxide from thiiran 1,1-dioxide have been obtained.[90]

The chemistry of sulphenes, including their role in the formation of episulphones,[91] and reactions of diazo-compounds with sulphur dioxide to yield thiiran 1,1-dioxides [92a] have been reviewed. The reaction of sulphur dioxide with a number of substituted diazoalkanes was investigated.[92b] In most cases, no evidence for the formation of an episulphone was observed. 1-(4-Methoxyphenyl)diazoethane gave 19% of a mixture of isomeric 4,4'-dimethoxy-α,α'-

[83] (a) K. Kondo and A. Negishi, *Chem. Letters*, 1974, 1525; (b) M. Kishi, S. Ishihara, and T. Komeno, *Tetrahedron*, 1974, **30**, 2135.
[84] B. C. Gilbert, R. O. C. Norman, and R. C. Sealy, *J.C.S. Perkin II*, 1975, 308.
[85] S. Inagaki, T. Minato, S. Yamabe, H. Fujimoto, and K. Fukui, *Tetrahedron*, 1974, **30**, 2165.
[86] H. Bock and B. Solouki, *Chem. Ber.*, 1974, **107**, 2299.
[87] I. Moretti, G. Torre, and G. Gottarelli, *Tetrahedron Letters*, 1976, 711.
[88] B. Solouki, H. Bock, and R. Appel, *Chem. Ber.*, 1975, **108**, 897.
[89] M. J. S. Dewar and W. Thiel, *J. Amer. Chem. Soc.*, 1975, **97**, 3978.
[90] W. L. Mock, *J. Amer. Chem. Soc.*, 1975, **97**, 3673.
[91] J. F. King, *Accounts Chem. Res.*, 1975, **8**, 10.
[92] (a) T. Nagai and N. Tokura, *Internat. J. Sulfur Chem. (B)*, 1972, **7**, 207; (b) G. C. Brophy, D. J. Collins, J. J. Hobbs, and S. Sternhell, *Austral. J. Chem.*, 1975, **28**, 151.

dimethylstilbenes, probably formed *via* a thiiran dioxide. The reaction of bis-(trifluoromethyl)keten with bis(trifluoromethyl)methionic anhydride proceeds to give a number of products depending on conditions. A 15% yield of 2,2,3,3-tetra(trifluoromethyl)thiiran 1,1-dioxide was obtained.[93]

Thiiran 1,1-dioxides give good yields of ring-opened sulphinates, *e.g.* (22), when treated with mercaptide ions, sodium sulphide, or thiourea,[94] or with metal

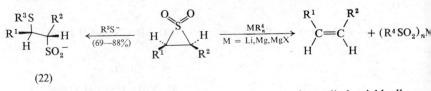

(22)

halides.[95a] Grignard reagents and lithium or magnesium alkyls yield alkenes and metal sulphinates.[95b]

Data have been obtained to support a two-stage mechanism for the Ramberg–Bäcklund reaction, which involves reversible carbanion formation followed by a rate-determining intramolecular nucleophilic displacement reaction to give a thiiran 1,1-dioxide, which rapidly extrudes sulphur dioxide.[96] This stereospecific loss of sulphur dioxide is said also to be possible *via* an intermediate alkane-sulphinate derived from attack of hydroxide ion on the episulphone.[97] The oxidation of sulphonyl 1,3-dicarbanions, $R^1\bar{C}HSO_2\bar{C}HR^2$, with copper(II) chloride yields alkenes, possibly *via* an intermediate diradical $R^1\dot{C}HSO_2\dot{C}HR^2$, which forms a thiiran 1,1-dioxide and extrudes sulphur dioxide.[98] Electrochemical reduction in DMF of 2,2,4,4-tetrabromo- or -tetrachloro-1,3-dithietan 1,1,3,3-tetroxide on a polished platinum cathode gives relatively stable radical anions, which lose halide ions stepwise to yield a postulated bicyclic disulphone (23), which was not isolated.[99]

(23)

5 Thiiren Derivatives

The thiiren 1,1-dioxide system has been investigated theoretically and by u.v. photoelectron spectroscopy. Charge transfer from the carbon–carbon double bond to the SO_2 unit is about one-third that predicted for the cyclopropenone system. 'Aromaticity' decreases in the order cyclopropenone > tropone >

[93] G. A. Sokol'skii, V. M. Pavlov, V. M. Golovkin, V. F. Gorelov, and I. L. Knunyants, *Chem. Heterocyclic Compounds*, 1975, **10**, 34.

[94] E. Vilsmaier and G. Becker, *Synthesis*, 1975, 55.

[95] (*a*) E. Vilsmaier, R. Tropitzsch, and O. Vostrowsky, *Tetrahedron Letters*, 1974, 3275; (*b*) *ibid.*, p. 3987.

[96] F. G. Bordwell and J. B. O'Dwyer, *J. Org. Chem.*, 1974, **39**, 2519; F. G. Bordwell and M. D. Wolfinger, *ibid.*, p. 2521; F. G. Bordwell and E. Doomes, *ibid.*, pp. 2526, 2531.

[97] T. Kempe and T. Norin, *Acta Chem. Scand.* (*B*), 1974, **28**, 613.

[98] J. S. Grossert, J. Buter, E. W. H. Asveld, and R. M. Kellogg, *Tetrahedron Letters*, 1974, 2805.

[99] J. G. Goury, G. Jeminet, and J. Simonet, *Bull. Soc. chim. France*, 1975, 1713.

thiiren dioxide.[100] Chemical ionization mass spectrometry has been applied to substituted thiiren 1-oxides and 1,1-dioxides, and molecular ions were observed, which was not possible with conventional electron-impact mass spectrometry.[101] A quantum chemical study of $C_2H_3S^+$ indicated that the thiirenium ion structure was more stable (1—14 kcal mol⁻¹) than the β-thiovinyl cation.[102] The barrier to interconversion of these valence tautomers was calculated as 12.8 kcal mol⁻¹. The barrier to pyramidal inversion at sulphur in the thiirenium ion was calculated as 72.9 kcal mol⁻¹.

Thiirenium ions (24), stable at low temperatures, have been obtained by addition of methyl bis(methylthio)sulphonium hexafluoroantimonate (11) to excess but-2-yne or hex-3-yne at −80 °C.[103] The ¹H n.m.r. spectrum is unchanged after several hours at −50 to −70 °C. Addition of *N*-methylpyridinium

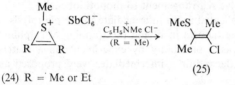

(24) R = Me or Et

(25)

chloride to the trimethyl cation gives 60—80% of (*E*)-1,2-dimethyl-2-(methylthio)-vinyl chloride (25). The trimethyl cation was also observed by n.m.r. spectroscopy as a transient species in the reaction of methanesulphenyl chloride with but-2-yne in liquid SO_2. Dialkylthiiren 1,1-dioxides have been obtained by treatment of bis(α,α-dibromoalkyl) sulphones with phosphines.[104] Although some evidence was obtained for the formation of a thiiren 1-oxide in an analogous way, the sulphoxide could not be isolated.[104b] 2,3-Diarylthiiran 1,1-dioxides are obtained in high yield by treatment of α,α-dichlorobenzyl sulphones with 1,4-diazabicyclo-octane in DMSO.[105] The thermal extrusion of sulphur dioxide from these episulphones is believed to involve a slow homolytic cleavage of one carbon–sulphur bond.

The reaction of 2,3-diphenyl- or 2,3-bis-(4-chlorophenyl)-thiiren 1,1-dioxide with enamines $R^1R^2C{=}CR^3NR^4_2$ yields a variety of heterocyclic compounds, including medium- and large-ring sulphur heterocycles, *via* decomposition of a bicyclic thiiran dioxide intermediate (26).[106] Treatment of these thiiran dioxides with 2-pyrrolidinyl- or 2-piperidinyl-2-norbornene gave adducts which possessed antifertility activity.[107] The reaction of 2,3-diphenylthiiren 1,1-dioxide with several meso-ionic compounds, *e.g.* (27), gave 2,3,5,6-tetraphenyl-4*H*-1,4-thiazine 1,1-dioxides by loss of carbon dioxide from the intermediate adduct (28).[108]

[100] C. Müller, A. Schweig, and H. Vermeer, *J. Amer. Chem. Soc.*, 1975, **97**, 982.
[101] P. Vouros and L. A. Carpino, *J. Org. Chem.*, 1974, **39**, 3777.
[102] I. G. Csizmadia, A. J. Duke, V. Lucchini, and G. Modena, *J.C.S. Perkin II*, 1974, 1808.
[103] G. Capozzi, O. De Lucchi, V. Lucchini, and G. Modena, *J.C.S. Chem. Comm.*, 1975, 248.
[104] (*a*) L. A. Carpino and J. R. Williams, *J. Org. Chem.*, 1974, **39**, 2320; J. R. Williams, *Diss. Abs.*, 1974, **34B**, 4890.
[105] O. Morales-Garzia, *Diss. Abs.*, 1975, **36B**, 1255.
[106] M. H. Rosen and G. Bonet, *J. Org. Chem.*, 1974, **39**, 3805.
[107] M. H. Rosen, I. Fengler, G. Bonet, T. Giannina, M. C. Butler, F. R. Popick, and B. G. Steinetz, *J. Medicin. Chem.*, 1976, **19**, 414.
[108] H. Matsukubo, M. Kojima, and H. Kato, *Chem. Letters*, 1975, 1153.

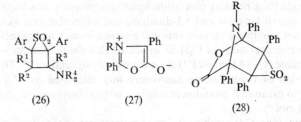

(26)   (27)   (28)

A variety of nucleophiles add to the double bond of 2,3-diphenylthiiren
1,1-dioxide to give mainly ring-opened products.[109] The principal mode of
fragmentation of thiiren oxides and dioxides in mass spectrometry involves
elimination of SO or SO_2. Fragmentation of 2,3-diphenylthiiren 1-oxide is also
believed to involve rearrangement to monothiobenzil.[110]

Thiiren is implicated as an intermediate in the photolysis of 1,2,3-thiadiazole;
an alternative mechanism involving the trapping of sulphur atoms by acetylene
was eliminated as a possibility by experiments in a matrix of [$^2H_2$]acetylene
(Scheme 6).[111] Benzothiiren intermediates were proposed as being involved in

$$R^1 = H, R^2 = D; R^1 = D, R^2 = H$$

Scheme 6

the thermolysis of sodium *o*-bromobenzenethiolate, which gives thianthrenes,[112a]
and in the decomposition of 1,2,3-benzothiadiazole 1,1-dioxide.[112b] The reactions
of sulphones with carbon tetrachloride in t-butyl alcohol containing powdered
potassium hydroxide were suggested to involve a thiiren.[113] A thiirenium ion
was proposed as an intermediate in the photochemical rearrangement of 3-methyl-
2-methylthio-5-phenyl-(1,3-thiazol-4-ylio)oxide.[114] Thiazole and isothiazole complexes of the Group VI transition metals undergo fragmentation in the mass
spectrometer to give ions which may be complexes of thiirens.[115]

6 Three-membered Rings containing Sulphur and One or Two Other Heteroatoms

A theoretical study of the oxathiiran ring system has indicated that the introduction of heteroatoms into a three-membered cyclic carbon framework can

[109] B. B. Jarvis, W. P. Tong, and H. L. Ammon, *J. Org. Chem.*, 1975, **40**, 3189; B. B. Jarvis and W. P. Tong, *Synthesis*, 1975, 102.
[110] P. Vouros, *J. Heterocyclic Chem.*, 1975, **12**, 21.
[111] A. Krantz and J. Laureni, *J. Amer. Chem. Soc.*, 1974, **96**, 6768.
[112] (*a*) J. I. G. Cadogan, J. T. Sharp, and M. J. Trattles, *J.C.S. Chem. Comm.*, 1974, 900; (*b*) K. Ramanathan, *Diss. Abs.*, 1975, **36B**, 243.
[113] L. L. Ho, *Diss. Abs.*, 1975, **35B**, 5817.
[114] O. Buchardt, J. Domanus, N. Harrit, A. Holm, G. Isaksson, and J. Sandström, *J.C.S. Chem. Comm.*, 1974, 376.
[115] K. H. Pannell, C. C. Lee, C. Parkanyi, and R. Redfearn, *Inorg. Chim. Acta*, 1975, **12**, 127.

lead to a severe re-ordering of energy levels.[116] The potential surface for CS_3, 2,3-dithiacyclopropanethione, has three minima, for the three possible valence tautomers, separated by barriers of 1.4 eV.[117]

The first example of a thiaziridine 1,1-dioxide (30) was obtained by addition of a diazo-compound to an *N*-sulphonyl-amine (29).[118] More details of the

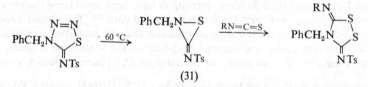

synthesis and properties of 2,3-di-t-butylthiadiaziridine 1,1-dioxide have been reported.[119]

α-Halogeno-benzylic sulphonamides, when treated with bases, gave products suggesting the intermediacy of a thiaziridine 1,1-dioxide (Scheme 7).[120]

Reagents: i, Et₃N or KOBuᵗ

Scheme 7

N-Tosyliminothiaziridines, *e.g.* (31), are possible intermediates in the thermolysis of 4-alkyl-5-tosylimino-1,2,3,4-thiatriazolines.[121] These intermediates may be trapped by electron-rich olefins, ketens, isocyanates, carbodi-imides, and iso-thiocyanates.

Thiaziridinimines may be intermediates in the reaction of alkyl azides with aryl isothiocyanates.[122] Thiaziridine intermediates have been suggested in the reaction of *N*-tosyl-sulphilimines with cyanide ion,[123a] in the decomposition of

[116] J. P. Snyder, *J. Amer. Chem. Soc.*, 1974, **96**, 5005.
[117] G. Calzaferri and R. Gleiter, *J.C.S. Perkin II*, 1975, 559.
[118] H. Quast and F. Kees, *Angew. Chem.*, 1974, **86**, 816.
[119] H. Chang, *Diss. Abs.*, 1975, **35B**, 3240.
[120] J. C. Sheehan, U. Zoller, and D. Ben-Ishai, *J. Org. Chem.*, 1974, **39**, 1817.
[121] G. L'abbé, E. Van Loock, R. Albert, S. Toppet, G. Verhelst, and G. Smets, *J. Amer. Chem. Soc.*, 1974, **96**, 3973; G. L'abbé, G. Verhelst, C. Yu, and S. Toppet, *J. Org. Chem.*, 1975, **40**, 1728; G. L'abbé, *Bull. Soc. chim. France*, 1975, 1127; R. Neidlein and K. Salzmann, *Synthesis*, 1975, 52.
[122] G. L'abbé, E. Van Loock, G. Verhelst, and S. Toppet, *J. Heterocyclic Chem.*, 1975, **12**, 607.
[123] (*a*) S. Oae, T. Aida, and N. Furukawa, *Internat. J. Sulfur Chem.*, 1975, **8**, 410; (*b*) S. Tamagaki, K. Sakaki, and S. Oae, *Heterocycles*, 1974, **2**, 39; (*c*) H. P. Braun, K. P. Zeller, and H. Meier, *Annalen*, 1975, 1257; (*d*) C. Riou, J. C. Poite, G. Vernin, and J. Metzger, *Tetrahedron*, 1974, **30**, 879; M. Maeda, A. Kawahara, M. Kai, and M. Kojima, *Heterocycles*, 1975, **3**, 389.

4,5-benzo-1,2-dithiole-3-thione imide,[123b] and in the scrambling of ring atoms in 1,2,3-thiadiazole 2-oxides[123c] and in thiazoles.[123d] A thiaziridinone was suggested as an intermediate in the reaction of a thiadiazolopyrimidinone with ethanolic hydrogen chloride,[124a] and a thiaziridinethione was proposed as an intermediate in the reaction of oxaziridines with carbon disulphide or phenyl isothiocyanate.[124b] Photolysis of the meso-ionic 4-phenyl-1,3,2-oxathiazolylio-5-oxide at 10 K in a nitrogen matrix may yield an azathiiren intermediate.[125] cis-Azoisopropane, $Pr^iN{=}NPr^i$, is obtained in high purity on treatment of di-isopropyl sulphamide with sodium hypochlorite and potassium carbonate; the reaction is believed to involve a thiadiaziridine 1,1-dioxide intermediate.[126] The oxidation of sulphines (thiocarbonyl S-oxide derivatives) by peroxy-acids may involve intermediate oxathiiran 1-oxides,[127] and such an intermediate may be formed in the reaction of methylene with sulphur dioxide, which gives carbonyl sulphide.[128] Photolysis of 8-methylthiochroman-4-one 1-oxide yields a disulphide, possibly via an oxathiiran.[129]

Charged dithiiran intermediates have been suggested to be involved in the rearrangement of 6-phenylthioacetamidopenicillin sulphoxide p-nitrobenzyl ester,[130a] the anodic oxidation of gem-disulphides (thioketals),[130b] and in the decomposition of a sulphonium salt containing a (phenylthio)methyl substituent.[130c]

7 Thietans

Physical Properties.—The potential function for the ring-puckering vibration of thietan in its lowest excited singlet state has been obtained.[131] The structure and ring-puckering of thietan have been investigated by far-i.r. spectroscopy,[132a] by 220 MHz n.m.r. spectroscopy on a nematic liquid crystal,[132b] and by electron diffraction and microwave spectroscopy.[132c] The conformations of thietan, thietan 1-oxide, and their 3-chloro-derivatives have been considered theoretically, the results agreeing well with known experimental data.[133] The crystal structure of cis-2,2-diphenyl-3,4-dichlorothietan indicates that the preferred conformation of the 3-chlorine atom is equatorial and that of the 4-chlorine is axial, in a puckered ring.[134] The structure of 5-thiabicyclo[2,1,1]hexane, which contain

124 (a) D. Baldwin and P. Van den Broek, J.C.S. Perkin I, 1975, 375; (b) M. Komatsu, Y. Ohshiro, K. Yasuda, S. Ichijima, and T. Agawa, J. Org. Chem., 1974, 39, 957.
125 I. R. Dunkin, M. Poliakoff, J. J. Turner, N. Harrit, and A. Holm, Tetrahedron Letters, 1976, 873.
126 P. S. Engel, Tetrahedron Letters, 1974, 2301.
127 W. Walter and O. H. Bauer, Annalen, 1975, 305; A. Battaglia, A. Dondoni, G. Maccagnani and G. Mazzanti, J.C.S. Perkin II, 1974, 609.
128 H. Hiraoka, J.C.S. Chem. Comm., 1974, 1014.
129 I. W. J. Still, P. C. Arora, M. S. Chauhan, M. H. Kwan, and M. T. Thomas, Canad. J. Chem. 1976, 54, 455.
130 (a) H. Tanida, R. Muneyuki, and T. Tsushima, Tetrahedron Letters, 1975, 3063; (b) J. G Gourcy, G. Jeminet, and J. Simonet, J.C.S. Chem. Comm., 1974, 634; (c) H. Matsuyama H. Minato, and M. Kobayashi, Bull. Chem. Soc. Japan, 1975, 48, 3287.
131 F. Momicchioli, G. Di Lonardo, and G. Galloni, J. Mol. Spectroscopy, 1974, 51, 273.
132 (a) H. Wieser, J. A. Duckett, and R. A. Kydd, J. Mol. Spectroscopy, 1974, 51, 115; C. Yamada T. Shigemune, and E. Hirota, ibid., 1975, 54, 261; (b) K. C. Cole and D. F. R. Gilson, Mol Phys., 1975, 29, 1749; (c) K. Karakida and K. Kuchitsu, Bull. Chem. Soc. Japan, 1975, 48, 1691.
133 C. Guimon, D. Liotard, and G. Pfister-Guillouzo, Canad. J. Chem., 1975, 53, 1224.
134 S. Kumakura and T. Kodama, Bull. Chem. Soc. Japan, 1975, 48, 2339.

a thietan ring, has been determined [135a] and its heat of formation calculated.[135b] The ring puckering of 3-thietanone has been investigated by i.r.[136a] and micro-wave spectroscopy.[136b] The ring of 2-thietanone (β-propiothiolactone) was concluded to be planar from investigations of its vibrational spectrum.[137]

Formation.—The acetylenic thietan derivative (32) and related compounds occur naturally. The structures have been established by independent synthesis.[138]

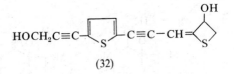

(32)

The use of thioureas in the preparation of thietans has been reviewed.[139] Thietans have been prepared by treatment of 1,3-dihalides with thiourea and base,[140a] or with hydrogen sulphide–aluminium chloride;[140b] by treatment of disulphonate esters of 1,3-glycols with sodium sulphide;[140c] and by treatment of 2-chloromethyl-3-phenyloxiran with sodium hydrosulphide.[140d] αα-Diphenyl-β-propiothiolactone was obtained by cyclization of the bromomethyl ester of diphenylthiolacetic acid.[141]

The photochemistry of thiones, including the formation of thietans and dithietans, has been reviewed.[142a] A number of investigations of the formation of thietans by the photochemical cyclization of thiones and alkenes have been reported.[142] Variations in the amount of thietan depend on whether or not the $n \rightarrow \pi^*$ or $\pi \rightarrow \pi^*$ excited state of the thioketone is involved.[142b]

Photolysis of 3,3,6,6-tetramethyl-5-acetoxy-1-thiacycloheptan-4-one (33) yields a small amount of 3,3-dimethyl-2-acetoxythietan.[143] This thietan could be obtained in 30% yield by treatment of 3,3-dimethylthietan with lead tetra-acetate in pyridine. The lack of reaction of the 5-hydroxy-derivative of (33) is explained by an electron-transfer quenching process involving a thietanium ion intermediate (34).

The mechanism of thietan-3-one formation by oxidation of 4-arylbutan-2-ones with thionyl chloride involves oxidation exclusively at the α-methylene

135 (a) T. Fukuyama, K. Oyanagi, and K. Kuchitsu, *Bull. Chem. Soc. Japan*, 1976, **49**, 638; (b) N. L. Allinger and M J. Hickey, *J. Amer. Chem. Soc.*, 1975, **97**, 5167.
136 (a) C. S. Blackwell and R. C. Lord, *J. Mol. Spectroscopy*, 1975, **55**, 460; (b) T. K. Avirah, R. L. Cook, and T. B. Malloy, jun., *ibid.*, p. 464.
137 G. M. Kuz'yants, *J. Struct. Chem.*, 1974, **15**, 696.
138 (a) F. Bohlmann and A. Suwita, *Chem. Ber.*, 1975, **108**, 515; (b) F. Bohlmann and J. Kocur, *ibid.*, 1974, **107**, 2115.
139 T. S. Griffin, T. S. Woods, and D. L. Klayman, in 'Advances in Heterocyclic Chemistry', ed. A. R. Katritzky and A. J. Boulton, Vol. 18, Academic Press, New York, 1975, p. 99.
140 (a) N. A. Nesmeyanov, V. A. Kalyavin, and O. A. Reutov, *Proc. Acad. Sci.* (*U.S.S.R.*), 1975, **223**, 510; (b) C. Mayer, *Helv. Chim. Acta*, 1974, **57**, 2514; (c) G. Seitz and W.-D. Mikulla, *Annalen*, 1974, 1328; M. Buza, *Diss. Abs.*, 1975, **36B**, 1704; (d) K. Haya, *ibid.*, 1974, **34B**, 5922.
141 L. I. Gapanovich, M. G. Lin'kova, O. V. Kil'disheva, and I. L. Knunyants, *Bull. Acad. Sci.*, *U.S.S.R.*, 1974, **23**, 142.
142 (a) P. de Mayo, *Accounts Chem. Res.*, 1976, **9**, 52; (b) H. Gotthardt, *Chem. Ber.*, 1974, **107**, 1856; (c) *ibid.*, p. 2544; (d) H. Gotthardt and M. Listl, *ibid.*, p. 2552; (e) J. L. Fourrey, P. Jouin, and J. Moron, *Tetrahedron Letters*, 1974, 3005; (f) H. Gotthardt and S. Nieberl, *ibid.*, p. 3397.
143 P. Y. Johnson and M. Berman, *J. Org. Chem.*, 1975, **40**, 3046.

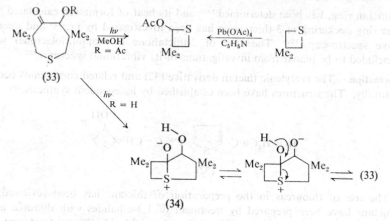

position.[144] Irradiation of (35) gave the bicyclic 3-thietanone (36).[145] α-Thietanones [β-propiothiolactones] (37) are obtained by a sulphur-transfer reaction involving 1,2-dithiolan-3-ones and triphenylphosphine.[146]

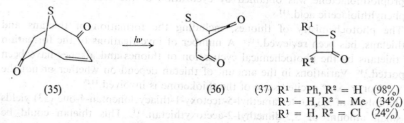

(35)   (36)   (37) R¹ = Ph, R² = H (98%)
                                                  R¹ = H, R² = Me (34%)
                                                  R¹ = H, R² = Cl (24%)

The reaction of bisthiomalonic acids $R^1R^2C(COSH)_2$ with isopropenyl acetate, acetone, and other carbonyl compounds in the presence of boron trifluoride etherate yields thietan-2,4-diones.[147] 2,2,4,4-Tetramethylcyclobutane-1,3-dithione reacts with methylenetriphenylphosphorane to yield a 2-thioxo-thietan (38; X = S).[25] The monothiodiketone yields a 2-keto-thietan (38; X = O).

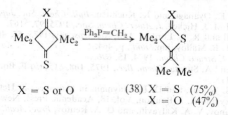

X = S or O   (38) X = S (75%)
                       X = O (47%)

A 2-thioxo-thietan, the dimer of dimethylthioketen, is obtained by treatment of isobutyric acid with hydrogen sulphide in the presence of titanium tetra-

[144] A. J. Krubsack, R. Sehgal, W. Loong, and W. E. Slack, *J. Org. Chem.*, 1975, **40**, 3179.
[145] H. Tsuruta, M. Ogasawara, and T. Mukai, *Chem. Letters*, 1974, 887.
[146] I. L. Knunyants, O. V. Kil'disheva, M. G. Lin'kova, and T. P. Vasil'eva, *Proc. Acad. Sci.* (*U.S.S.R.*), 1975, **224**, 521.
[147] J. H. Schauble, W. A. Van Saun, jun., and J. D. Williams, *J. Org. Chem.*, 1974, **39**, 2946.

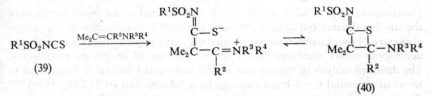

chloride.[148] Sulphonylisothiocyanates (39) and $\beta\beta$-dimethyl-enamines react to yield dipolar intermediates in equilibrium with thietan structures (40).[149]

Thietanonium ylides (42) are obtained from β-arylthio- and β-allylthio-alkylcarbenes (41).[150] These ylides undergo rearrangements to various structures (43)—(45), depending on the nature of the substituent on sulphur.

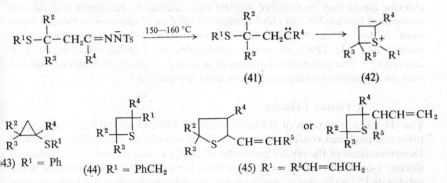

Cyclodehydration of isoquinoline derivative (46) may proceed *via* a thietanium ion,[151] and thermolysis of thiabicyclo-octene (47) may involve a 2-thietanone intermediate.[152]

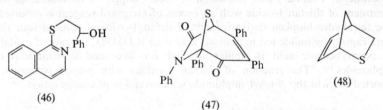

(46)

(47)

(48)

Chemical Properties.—Thietan is a useful starting material in the synthesis of macrocyclic compounds; the first step in the sequence is the reaction of thietan with allyl bromide to give an 85% yield of allyl 3-bromopropyl sulphide.[153] When *cis*- and *trans*-3-ethyl-2-propylthietan were irradiated in the presence of mercury vapour they gave but-1-ene and *cis*- and *trans*-hept-3-ene.[154a] At low

[148] F. Sannicolo, *Ann. Chim. (Italy)*, 1973, **63**, 825.
[149] E. Schaumann, S. Sieveking, and W. Walter, *Tetrahedron*, 1974, **30**, 4147.
[150] K. Kondo and I. Ojima, *Bull. Chem. Soc. Japan*, 1975, **48**, 1490.
[151] H. Singh, V. K. Vij, and K. Lal, *Indian J. Chem.*, 1974, **12**, 1242.
[152] K. T. Potts, J. Baum, and E. Houghton, *J. Org. Chem.*, 1976, **41**, 818.
[153] E. Vedejs and J. P. Hagen, *J. Amer. Chem. Soc.*, 1975, **97**, 6878.
[154] (*a*) D. R. Dice and R. P. Steer, *J. Amer. Chem. Soc.*, 1974, **96**, 7361; (*b*) *Canad. J. Chem.*, 1975, **53**, 1744; (*c*) *ibid.*, 1974, **52**, 3518.

pressures the ratio of *trans*- to *cis*-hept-3-ene is related to the stereochemistry of the starting thietan, but at high pressures the ratio is independent of the stereochemistry of the starting material, suggesting that energetic triplet biradicals can decompose rather stereospecifically in the absence of deactivating collisions. The direct photolysis of thietan and several substituted thietans is interpreted in terms of an initial C—S bond cleavage to a 1,4-biradical (·CH$_2$CH$_2$CH$_2$S·).[154b] Thioformaldehyde, the decomposition product of the diradical from thietan, can be trapped by cyclopentadiene to give the bicyclic adduct (48).[154c] A spirocyclic thietan also is reported to undergo photofragmentation to thioformaldehyde and an olefin.[142f] The interconversion of 2-methylthietan and tetrahydrothiophen *via* a carbonium ion is effected by triphenylmethyl fluoroborate or chloranil.[155] 2,2-Dichlorothietans may be hydrolysed on silica gel to β-thiolactones or the chlorine atoms may be removed without ring opening by reduction with lithium aluminium hydride.[142c] An aldol condensation of an aldehyde with thietan-2-one went in low yield.[138b] β-Propiothiolactones (thietan-2-ones) are cleaved by MeSCl and MeCOSCl to mixed disulphides of β-mercaptocarboxylic acid chlorides.[156] The polymerization rates of several β-propiothiolactones substituted with an arylsulphonamido-group have been determined.[157]

8 Thietan 1-Oxides

The [1]H n.m.r. spectrum of thietan 1-oxide at 300 and at 100 MHz indicates a preferred puckered conformation, with the oxygen in an equatorial position.[158] Determinations of the crystal structures of both *cis*- and *trans*-3-*p*-bromophenylthietan 1-oxide show that the oxygen atom is equatorial in both isomers in the solid state,[159] and further evidence for a puckered conformation has been obtained from microwave spectroscopy.[160]

Thermolysis of thietan 1-oxide, 1,3-dithietan 1-oxide, or 2,2,4,4-tetramethyl-1,3-dithietan 1-oxide yields sulphines.[161] (See Chapter 3, Section 2, p. 136.) Treatment of thietan 1-oxide with an excess of Grignard reagent is reported to give tetrahydrothiophen derivatives.[140a] Thietan 1-oxide is reduced four times more rapidly by iodide ion in aqueous acid than is DMSO; the rate of oxidation by peroxybenzoic acid is similar to that for five- and six-membered-ring sulphoxides.[162] The reaction of thietan 1-oxides with *p*-tosyl isocyanate is reported to yield the *N*-tosyl sulphimide with inversion of configuration.[163]

9 Thietan 1,1-Dioxides

Physical Properties.—*X*-Ray analysis of *cis*-2,2-diphenyl-3,4-dichlorothietan 1,1-dioxide indicates that the thietan ring is puckered, and that the 3-chlorine

[155] L. M. Petrova, A. A. Freger, and E. A. Viktorova, *Moscow University Chem. Bull.*, 1974, **29**, No. 4, p. 89.
[156] N. M. Karimova, M. G. Lin'kova, O. V. Kil'disheva, and I. L. Knunyants, *Chem. Heterocyclic Compounds*, 1974, **8**, 432.
[157] D. Krilov, Z. Veksli, and D. Fles, *J. Polymer Sci., Part A-1, Polymer Chem.*, 1976, **14**, 777.
[158] C. Cistaro, G. Fronza, R. Mondelli, S. Bradamante, and G. A. Pagani, *J. Magn. Resonance*, 1974, **15**, 367.
[159] J. H. Barlow, C. R. Hall, D. R. Russell, and D. J. H. Smith, *J.C.S. Chem. Comm.*, 1975, 133.
[160] J. W. Bevan, A. C. Legon, and D. J. Millen, *J.C.S. Chem. Comm.*, 1974, 659.
[161] E. Block, R. E. Penn, R. J. Olsen, and P. F. Sherwin, *J. Amer. Chem. Soc.*, 1976, **98**, 1264.
[162] R. Curci, F. DiFuria, A. Levi, and G. Scorrano, *J.C.S. Perkin II*, 1975, 408.
[163] C. R. Hall and D. J. H. Smith, *Tetrahedron Letters*, 1974, 1693.

atom is axial and the 4-chlorine equatorial, a conformation opposite to that in the unoxidized thietan.[164] The favoured axial conformation for the 3-chlorine atom may indicate a $3d$-$3p$ interaction with the sulphur atom; this suggestion was supported by MO calculations. [1]H n.m.r. measurements and X-ray analysis of 3-substituted thietan 1,1-dioxides also indicate a preference for the 3-substituent (OH, Cl, OAc) to be axial. Increasing the temperature to 150 °C increases the proportion of the less stable 3-equatorial conformation.[165] Thietan 1,1-dioxide itself is either planar or a mixture of rapidly interconverting equivalent conformers.[158] An X-ray analysis of *cis*-2-chloro-3-morpholino-4,4-dimethylthietan 1,1-dioxide shows a puckered ring, with the 2-chloro-substituent causing a rotation of the O—S—O group with respect to the C—S—C plane.[166]

Formation.—The use of sulphenes in the synthesis of thietan 1,1-dioxides has been reviewed.[91, 92a] The formation of thietan 1,1-dioxides from enamines and sulphenes and the stereoselectivity of the reaction is discussed in Chapter 3, Section 2 (p. 140).[167-170] When a mixture of *cis*- and *trans*-2-chloro-2,4,4-trimethyl-3-dialkylaminothietan 1,1-dioxides is treated with aqueous alkali, the *cis*-isomer is converted into the thietan-3-one, leaving the pure *trans*-isomer untouched.[167] Cyclic α-amino-ketoximes can be converted in one step into 2-(ω-cyanoalkyl)-2-dialkylamino-thietan 1,1-dioxides by treatment with methanesulphonyl chloride and pyridine.[171] Thietan dioxides may be obtained by addition of sulphenes to αβ-unsaturated sulphides.[172] A number of thietan dioxides were prepared as possible inhibitors of mono-amine oxidase or as analgesics.[140d]

Reactions.—Thermolysis of thietan 1,1-dioxide yields mainly cyclopropane and sulphur dioxide, with some propene.[173] Flash vacuum thermolysis of (49) gives 3-isopropyl-2,4-dimethylpenta-1,3-diene (50),[174] while irradiation of thietan-3-one 1,1-dioxide (51) yields keten and sulphene.[175]

Treatment of 2-halogeno-3-morpholinothietan derivatives (52) with base yields a variety of products. With triethylamine in aqueous dioxan (52; R^1 = Ph, R^2 = H) gives the oxathiole dioxide (53) and products derived from the thiiran dioxide (54) after extrusion of SO_2. The reaction is suggested[176] to involve ring-opening to the enamine, which hydrolyses to give PhCOC̄MeSO₂CH₂Cl and

[164] S. Kumakura, *Bull. Chem. Soc. Japan*, 1975, **48**, 2164.
[165] C. Cistaro, G. Fronza, R. Mondelli, S. Bradamante, and G. A. Pagani, *J. Magn. Resonance*, 1975, **17**, 219.
[166] G. D. Andreetti, G. Bocelli, and P. Sgarabotto, *Gazzetta*, 1974, **104**, 1207.
[167] W. E. Truce and J. F. Rach, *J. Org. Chem.*, 1974, **39**, 1109.
[168] V. N. Drozd, V. V. Sergeichuk, and N. D. Antonova, *J. Org. Chem.* (*U.S.S.R.*), 1974, **10**, 1508; V. N. Drozd, V. V. Sergeichuk, and V. A. Moskalenko, *ibid.*, 1975, **11**, 132.
[169] T. Tanabe, T. Shingaki, and T. Nagai, *Chem. Letters*, 1975, 679.
[170] M. Riviere, N. Paillous, and A. Lattes, *Bull. Soc. chim. France*, 1974, 1911.
[171] S. Chen and Y. L. Chow, *Canad. J. Chem.*, 1974, **52**, 2283.
[172] E. N. Prilezhaeva, N. P. Petukhova, V. I. Kurilkin, A. U. Stepanyants, and V. P. Lezina, *Bull. Acad. Sci. U.S.S.R.*, 1974, **23**, 1745.
[173] D. Cornell and W. Tsang, *Internat. J. Chem. Kinetics*, 1975, **7**, 799.
[174] R. J. Bushby, *J.C.S. Perkin I*, 1975, 2513.
[175] R. Langendries, F. C. De Schryver, P. de Mayo, R. A. Marty, and J. Schutyser, *J. Amer. Chem. Soc.*, 1974, **96**, 2964.
[176] P. Del Buttero, S. Maiorana, and M. Trautluft, *J.C.S. Perkin I*, 1974, 1411.

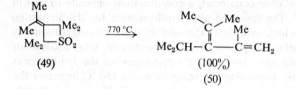

(49)   (50)

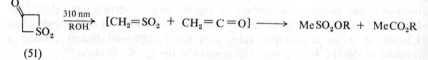

(51)

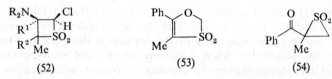

(52)   (53)   (54)

hence the cyclic compounds. With sodium hydroxide (52; $R^1 = H$, $R^2 = Me$)
eliminated HCl to give the enamine, and hence the ketone.[176]

Refluxing an ethanolic solution of 2-chloro-4-methyl-3-morpholino-3-phenyl-
thietan 1,1-dioxide (55) yields the thiazoline dioxide (56).[177]

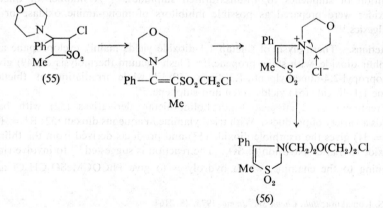

(55)

(56)

The kinetics of the base-catalysed H–D exchange and of *cis–trans* isomerization
of a series of 2-halogeno-3-morpholino-4,4-dimethylthietan 1,1-dioxides indicate
that for a given halogen the rate of exchange and the rate of isomerization are
essentially identical, demonstrating the high configurational instability of
strained α-sulphonyl-carbanion intermediates.[178] Acylation of the anion of
2,2-dimethylthietan 1,1-dioxide with ethyl acetate followed by a haloform
cleavage of the resulting methyl ketone is a convenient way of introducing an
α-halogen atom.[179] Stereospecific metallation by n-butyllithium of 3-dialkylamino-

[177] P. Del Buttero, S. Maiorana, D. Pocar, G. D. Andreetti, G. Bocelli, and P. Sgarabotto, *J.C.S.
       Perkin II*, 1974, 1483.
[178] S. Bradamante, P. Del Buttero, D. Landini, and S. Maiorana, *J.C.S. Perkin II*, 1974, 1676.
[179] P. Del Buttero and S. Maiorana, *Synthesis*, 1975, 333.

thietan dioxides, followed by addition of electrophilic reagents, gives *cis*-2-substituted 3-dialkylamino-thietan 1,1-dioxides.[180] The amine oxide elimination on a mixture of *cis*- and *trans*-2,2-dimethyl-3-morpholino-4-phenylthietan 1,1-dioxides left the *cis*-isomer unchanged.[180]

10 Thiet Derivatives

The crystal structure of *trans*-2,5-dibromo-7-thiabicyclo[4,2,0]oct-1(6)-ene 7,7-dioxide indicates that the cyclohexene ring is in a half-chair conformation.[181]

Methyl 2*H*-benzothiet-2-carboxylate (57) has been prepared by photolysis of 3-diazo-2-oxo-2,3-dihydro-benzo[*b*]thiophen; it is thermally stable and does not undergo valence tautomerism.[182a]

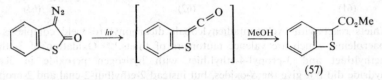

(57)

No evidence could be obtained for the formation of 2*H*-benzo[*b*]thiet or its valence tautomer by irradiation of 3*H*-1,2-benzodithiole 1,1-dioxide.[182b] Naphtho-[1,8-*bc*]thiet (58) has been prepared and its *S*-oxide, and *SS*-dioxide, have been obtained by oxidation with *m*-chloroperoxybenzoic acid. All three compounds react with lithium aluminium hydride to give ring-opened products as the result of hydride attack on sulphur.[183] After reduction and methylation, (58) gave the methyl sulphide (59), and the *S*-oxide gave bis(α-naphthyl) disulphide (75%) together with (58) (6%) and the corresponding sulphone (4%), which was the major product (85%) from the *SS*-dioxide (Scheme 8). Thiets are believed to be

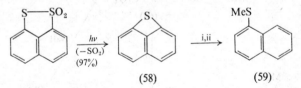

Reagents: i, LiAlH₄; ii, MeI–OH⁻

Scheme 8

intermediates in the addition of thiones to heteroatom-substituted acetylenes MeSC≡CSMe or MeC≡CNEt₂, to give αβ-unsaturated thioesters (60) and thioamides (R¹R²C=CMeCSNEt₂), respectively.[184] Treatment of the tetra-sulphide (61) with triethyl phosphite yields an orange-red material, suggested to have a bis(thiet) structure (62), which decomposes to the thiophen derivative (63).[185]

180 V. N. Drozd and V. V. Sergeichuk, *J. Org. Chem. (U.S.S.R.)*, 1975, **11**, 1301.
181 E. Ljungström, *Acta Chem. Scand. (B)*, 1975, **29**, 1071.
182 (*a*) E. Voight and H. Meier, *Angew. Chem.*, 1976, **88**, 94; (*b*) A. K. Bhattacharya, *Diss. Abs.*, 1976, **36B**, 4477.
183 J. Meinwald and S. Knapp, *J. Amer. Chem. Soc.*, 1974, **96**, 6532.
184 A. C. Brouwer and H. J. T. Bos, *Tetrahedron Letters*, 1976, 209.
185 H. Behringer and E. Meinetsberger, *Tetrahedron Letters*, 1975, 3473.

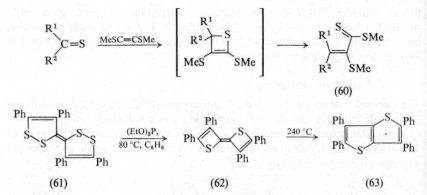

(60)

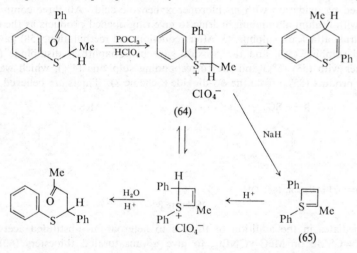

(61)   (62)   (63)

Thiets react with cyclopentadienylcobalt dicarbonyl to yield complexes of thioacroleins, which are valence tautomers of thiets.[186] Oxidation of 3-ethyl-4-methylthiet and 3-propyl-4-ethylthiet with hydrogen peroxide in acetic anhydride did not give the *S*-oxides, but instead 2-ethylbut-2-enal and 2-propyl-pent-2-enal, respectively.[187]

Stable thiet cations (64) reportedly were isolated by treatment of certain β-keto-sulphides with phosphorus oxychloride.[188] Addition of sodium hydride

(64)

(65)

to these cations gave green materials for which thiacyclobutadiene structures (65) were suggested.

Some improvements in the synthesis of thiet 1,1-dioxide have been reported,[189] and 2-phenylthiet 1,1-dioxide, 2-benzylthiet 1,1-dioxide, and 4-methyl-2-phenyl-thiet 1,1-dioxide have been prepared.[140d] The Michael addition of HCN to

[186] D. C. Dittmer, K. Takahashi, M. Iwanami, A. I. Tsai, P. L. Chang, B. B. Blidner, and I. K. Stamos, *J. Amer. Chem. Soc.*, 1976, **98**, 2795.

[187] A. I. Tsai, *Diss. Abs.*, 1974, **35B**, 743.

[188] R. S. Devdhar, V. N. Gogte, and B. D. Tilak, *Tetrahedron Letters*, 1974, 3911.

[189] B. Lamm and K. Gustafsson, *Acta Chem. Scand. (B)*, 1974, **28**, 701.

several thiet sulphones followed by reduction and methylation gave 3-dimethyl-aminothietan derivatives.[140d] Two improved syntheses, one new, of 2H-benzo-[b]thiet 1,1-dioxide have been reported.[190a, b] The compound yields o-thiocresol on reduction with lithium aluminium hydride and gives a sultine, 3H-2,1-benzoxa-thiole 1-oxide, on thermolysis.[190b] Flash thermolysis of sulphone (66) gave the benzothiet 1,1-dioxide (67).[191]

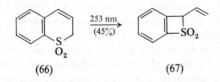

(66)   (67)

11 Four-membered Rings containing One Sulphur Atom and One or More Heteroatoms

Thiazetidines and Thiazetes.—Kinetic data for the cycloadditions of dicyclo-hexylcarbodi-imide with *para*-substituted phenyl isothiocyanates to give 1,3-thiazetidine derivatives (68) (yields of 31—92%) have been obtained for both

(68)

the forward and reverse reactions.[192a] The exocyclic double bonds have the (Z, E) configuration, and the stereochemistry of the cyclization is controlled by steric factors.[192b] Addition of phenylsulphene to Schiff bases of aromatic aldehydes and methylamine (Scheme 9) yields *cis*- and *trans*-2-methyl-4-phenyl-

$$PhCH{=}SO_2 \ + \ ArCH{=}NMe \xrightarrow{(52{-}88\%)} \underset{Ar}{\overset{Ph}{\diagdown}}\begin{matrix}{-}SO_2 \\ {-}NMe\end{matrix} \xrightarrow{i,ii} \underset{(Ar\,=\,Ph)}{PhCH} \underset{\ }{\overset{SO_3^-}{\underset{|}{-}}} \underset{Ph}{\overset{+}{\underset{|}{CHNH_2Me}}}$$

(69)

Reagents: i, NaOMe; ii, HCl

Scheme 9

3-aryl-1,2-thiazetidine 1,1-dioxide (69).[193] Thermolysis of the 3,4-diphenyl derivative gave *trans*-stilbene, and ring-opening occurred on treatment with sodium methoxide.

190 (a) B. Lamm, *Acta Chem. Scand.* (B), 1975, **29**, 332; (b) T. R. Nelsen and D. C. Dittmer, Abstracts of Papers, 168th National Meeting, Amer. Chem. Soc., Atlantic City, N.J., Sept. 1974, ORGN 74.
191 C. R. Hull and D. J. H. Smith, 'Organic Sulphur Chemistry', ed. C. J. M. Stirling, Butter-worths, London, 1975, p. 364.
192 (a) A. Dondoni and A. Battaglia, *J.C.S. Perkin II*, 1975, 1475; (b) O. Exner, V. Jehlicka, and A. Dondoni, *Coll. Czech. Chem. Comm.*, 1976, **41**, 562.
193 T. Hiraoka and T. Kobayashi, *Bull. Chem. Soc. Japan*, 1975, **48**, 480.

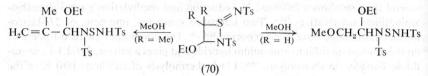

(70)

Cycloaddition of *NN'*-ditosyl sulphur di-imide, TsN=S=NTs, with vinyl ethers R_2C=CHOEt gives 1,2-thiazetidine derivatives (70), which decompose readily.[194] Cycloaddition of dialkylthioketens to azomethines may yield 1,3-thiazetidine intermediates, which decompose to thiocarbonyl compounds,[195] and the reaction of *SS*-diphenylmethylsulphilimine with phenyl isocyanate may be rationalized as occurring *via* a four-membered cyclic intermediate containing one sulphur atom and two nitrogen atoms.[196] 1,2-Thiazetidin-3-one 1,1-dioxides are obtained by treatment of α-chlorosulphonyl acid chlorides, $ClSO_2CR_2COCl$, with ammonia or primary amines, by treatment of α-chlorosulphonyl carboxylic acid esters with ammonia, or by treatment of α-chlorosulphonyl carboxylic acid amides with triethylamine.[197] Elimination of methanesulphenic acid from methylsulphinylmethyl isothiuronium salts (71) yields 1,3-thiazetidine derivatives.[198]

$$\underset{\text{(71)}}{\overset{\overset{\text{O}}{\|}\quad\overset{\overset{+}{N}HR}{\|}}{MeSCH_2SC-NHR}} \quad\xrightarrow[\substack{-\,MeSOH\\R=H,\,84\%\\R=Me,\,91\%}]{H^+}\quad RN{\Big<}{\overset{\displaystyle S}{\underset{+}{}}}{\overset{}{NHR}}$$

The action of hydrogen peroxide–acetic acid on 2-imino-1,3-thiazetidines gives ring-expanded products, *viz.* 3-oxo-1,2,4-thiadiazolidine 1-oxides, presumably *via* the thiazetidine 1-oxide.[199] The polymerization of 1,2-thiazetidine 1,1-dioxide has been studied.[200]

2*H*-1,3-Thiazetes (73) have been prepared by thermolysis of oxathiazines (72) obtained from hexafluoroacetone and thioamides of aromatic carboxylic acids (Scheme 10).[201a] These thiazetes undergo thermal ring opening to isopropylidene-thioamides, which can be trapped by various reagents, *e.g.* RNC, P_2S_5, and norbornene.[201b] Heating to 140 °C causes a change in colour to yellow-brown; on cooling, the colour reverts to yellow. This behaviour is attributed to an equilibrium between the thiazete and its ring-opened tautomer.

Fragmentation of α-alkylthio-ketoximes may proceed *via* 1,2-thiazetin-1-ium ions,[202] as shown in Scheme 11.

[194] N. Schönberger and G. Kresze, *Annalen*, 1975, 1725.
[195] E. Schaumann, *Chem. Ber.*, 1976, **109**, 906.
[196] J. A. Franz and J. C. Martin, *J. Amer. Chem. Soc.*, 1975, **97**, 583.
[197] A. Le Berre, A. Étienne, and B. Desmazières, *Bull. Soc. chim. France*, 1976, 277; *ibid.*, 1975, 807.
[198] Y. Ueno, Y. Masuyama, and M. Okawara, *Tetrahedron Letters*, 1974, 2577.
[199] W. Ried, O. Mösinger, and W. Schuckmann, *Angew. Chem.*, 1976, **88**, 120.
[200] Y. Imai, H. Hirukawa, and M. Ueda, *Kobunshi Ronbunshu*, 1974, **3**, 2044.
[201] (a) K. Burger, J. Albanbauer, and M. Eggersdorfer, *Angew. Chem.*, 1975, **87**, 816; (b) K. Burger, J. Albanbauer, and W. Foog, *ibid.*, p. 816.
[202] C. A. Grob and J. Ide, *Helv. Chim. Acta*, 1974, **57**, 2562, 2571.

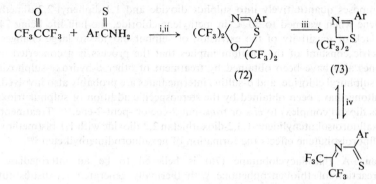

Reagents: i, −20 °C; ii, POCl₃–C₅H₅N, −30 °C (52—58 %); iii, 120—140 °C (65—78 %); iv, heat

Scheme 10

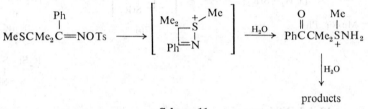

products

Scheme 11

Thiazete ring systems have also been proposed as being intermediates in the reaction of guanidines with carbon disulphide [203] and in the reaction of carboxylic acid amides with thionyl chloride and pyridine.[204]

In addition to the above examples involving nitrogen (in Group V of the Periodic Table), a phosphorane analogue of a 1,3-thiazetidine 1,1-dioxide has been suggested as an intermediate in the reaction of alkylidene- and arylidene-triphenylphosphoranes with sulphenes derived from alkanesulphonyl fluorides.[205]

Sultines, Sultones, and Cyclic Sulphates (1,2-Oxathietan 2-Oxides, 1,2-Oxathietan 2,2-Dioxides, and 1,3,2-Dioxathietan 2,2-Dioxides).—3,3-Dimethyl-4,4-diphenyl-1,2-oxathietan 2-oxide (75), in crystalline form, has been obtained by treatment of the β-hydroxy-sulphoxide (74) with sulphuryl chloride or *N*-chlorosuccinimide.[206] It is stable at room temperature for several days but

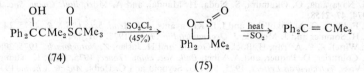

[203] Y. E. Moharir, *Indian J. Chem.*, 1974, **12**, 490.
[204] E. M. Dorokhova, E. S. Levchenko, and N. P. Pel'kis, *J. Org. Chem.* (*U.S.S.R.*), 1975, **11**, 755.
[205] B. A. Reith, J. Strating, and A. M. van Leusen, *J. Org. Chem.*, 1974, **39**, 2728.
[206] T. Durst and B. P. Gimbarzevsky, *J.C.S. Chem. Comm.*, 1975, 724.

decomposes quantitatively into sulphur dioxide and 1,1-diphenyl-2,2-dimethyl-ethylene when warmed to 30 °C in methylene chloride, the half-life being 24 h. The lack of sensitivity of the extrusion of sulphur dioxide to large changes in dielectric constant of the medium implies that the process is a concerted one. Alkenes also have been obtained by treatment of other β-hydroxy-sulphoxides with sulphuryl chloride, and β-sultine intermediates are probably also involved.[207]

Sultones have been obtained by the stereospecific addition of sulphur trioxide (as its dioxan complex) to *cis*- or *trans*-but-2-ene or -pent-2-ene.[208] Treatment of 4-hexafluoroisobutenylidene-1,3,2-dioxathietan 2,2-dioxide with 1,1-bis(methoxy)-hexafluoroisobutene effects the formation of hexafluorodimethylketen.[209]

Silicon.—A thiasilacyclobutane (76) is believed to be an intermediate in the reaction of thiobenzophenone with thermally generated 1,1-disubstituted sila-ethenes.[210]

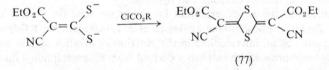

$$R = Me \text{ or } Ph \qquad\qquad (76) \qquad\qquad (R = Me)$$

$$\begin{array}{c} R_2Si-S \\ |\quad\ | \\ S-SiR_2 \end{array}$$

12 1,3-Dithietans

The photoelectron spectrum of 2-thioxo-1,3-dithietan has been determined.[211]

The dimerization of thiocarbonyl compounds yields 1,3-dithietans, and some recent examples have been reported.[195, 212] 1,3-Dithietans (77) also are obtained

$$\underset{NC}{\overset{EtO_2C}{>}}C=C\underset{S^-}{\overset{S^-}{<}} \quad\xrightarrow{ClCO_2R}\quad \underset{NC}{\overset{EtO_2C}{>}}C\underset{S}{\overset{S}{<}}C\underset{CN}{\overset{CO_2Et}{<}}$$

$$(77)$$

from 1,1-dimercapto-alkenes by treatment with methyl or ethyl chloro-carbonate,[213] by heating,[214] or by treatment with dihalogenomethanes.[215]

Refluxing *N*-alkyl-2-cyano-3-mercapto-3-(methylthio)acrylamides with acetic anhydride yields desaurins (78).[216]

[207] J. Nokami, N. Kunieda, and M. Kinoshita, *Tetrahedron Letters*, 1975, 2179.
[208] M. Nagayama, O. Okumura, S. Noda, H. Mandai, and A. Mori, *Bull. Chem. Soc. Japan*, 1974, **47**, 2158.
[209] A. F. Eleev, G. A. Sokol'skii, and I. L. Knunyants, *Bull. Acad. Sci., U.S.S.R.*, 1975, **24**, 1277.
[210] L. H. Sommer and J. McLick, *J. Organometallic Chem.*, 1975, **101**, 171.
[211] K. Wittel, E. E. Astrup, H. Bock, G. Graeffe, and H. Juslen, *Z. Naturforsch.*, 1975, **30b**, 862.
[212] R. Couturier, D. Paquer, and A. Vibet, *Bull. Soc. chim. France*, 1975, 1670; K. Thimm and J. Voss, *Tetrahedron Letters*, 1975, 537; G. Seybold and C. Heibl, *Angew. Chem.*, 1975, **87**, 171.
[213] K. Peseke, *Z. Chem.*, 1975, **15**, 19.
[214] A. Tajana, D. Nardi, and R. Cappelletti, *Ann. Chim. (Italy)*, 1974, **64**, 305.
[215] L. Henriksen, 'Organic Sulphur Chemistry', ed. C. J. M. Stirling, Butterworths, London, 1975, p. 337.
[216] M. Yokoyama, *J.C.S. Perkin I*, 1975, 1417.

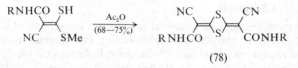

(78)

2-Isopropylidene-1,3-dithiacyclobutane was obtained by the thermal elimination of hydrogen chloride from chloromethyl 2-methyldithiopropanoate.[217] Grignard reagents convert 5-chloro-4-phenyl-3*H*-1,2-dithiol-3-one into desaurin (79).[218]

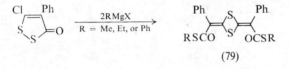

(79)

1,3-Dithietan-2-ylidenepyrazolines are prepared in 27—85% yield by treatment of a pyrazolinedithiocarboxylic acid with aromatic aldehydes in the presence of ethanolic sulphuric acid.[219] Sulphur dichloride converts β-keto-esters or β-diketones into 1,3-dithietans.[220] 2,4-Dibenzylidene-1,3-dithietan is obtained by treatment of benzylidenetriphenylphosphorane with carbon disulphide,[221] and difluorenylidene-1,3-dithietan is obtained by photolysis of 3,6-difluorenylidene-1,2,4,5-tetrathiacyclohexane.[222]

Ring-opening of 2,2,4,4-tetrakis(trifluoromethyl)-1,3-dithietan occurs on treatment with potassium fluoride, hexafluoropropene, and cuprous bromide, or with potassium fluoride, alcohols, and copper(I) or silver salts.[223] Treatment of the desaurin (80) with *o*-amino-phenols gives oxazepine derivatives,[224a] and with *o*-phenylenediamines, diazepines.[224b]

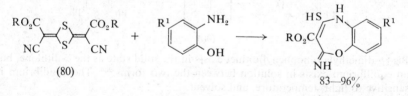

(80)

83—96%

Various ring-opened products are obtained from desaurins and nucleophiles (amines, mercaptides, hydrazine).[225] Several thioketens are formed in the loss of carbonyl sulphide from 4-dialkylidene-1,3-dithietan-2-ones.[226]

[217] J. Meijer, P. Vermeer, and L. Brandsma, *Rec. Trav. chim.*, 1975, **94**, 83.
[218] F. Boberg, M. Ghoudikian, and M. H. Khorgami, *Annalen*, 1974, 1261.
[219] T. Takeshima, N. Fukada, E. Okabe, M. Mineshina, and M. Muraoka, *J.C.S. Perkin I*, 1975, 1277.
[220] S. K. Gupta, *J. Org. Chem.*, 1974, **39**, 1944.
[221] G. Purrello and P. Fiandaca, *J.C.S. Perkin I*, 1976, 692.
[222] H. J. Kyi and K. Praefcke, *Tetrahedron Letters*, 1975, 555.
[223] T. Kitazume and N. Ishikawa, *Bull. Chem. Soc. Japan*, 1975, **48**, 361.
[224] (*a*) K. Peseke, *J. prakt. Chem.*, 1975, **317**, 648; (*b*) *Tetrahedron*, 1976, **32**, 483.
[225] K. Peseke, *Z. Chem.*, 1976, **16**, 16; W. Schroth, H. F. Uhlig, H. Bahn, A. Hildebrandt, and D. Schmiedl, *ibid.*, 1974, **14**, 186.
[226] E. Schaumann, E. Kausch, and W. Walter, *Chem. Ber.*, 1974, **107**, 3574.

13 1,2-Dithiets

Theoretical considerations indicate that the thermal conversion of an α-dithione into a 1,2-dithiet is allowed photochemically; the thermal processes, both forward and reverse, are forbidden.[227] Model calculations on the dithioglyoxal–dithiet equilibrium show that the Hückel method predicts the dithioglyoxal to be more stable than the dithiet, whereas the CNDO/2 method predicts the opposite.[228] The potential surface for C_2S_4 was calculated, and energy minima were predicted for (81), (82), (83), and (84).[228]

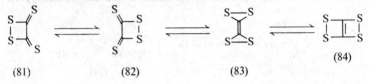

(81)   (82)   (83)   (84)

The structure of 3,4-bis(trifluoromethyl)-1,2-dithiet has been determined by electron diffraction.[229] Structures (85), (86), and (87) were said to contribute to the hybrid, but (88) does not.

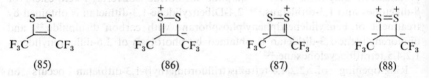

(85)   (86)   (87)   (88)

Photolysis of diphenylvinylene dithiocarbonate (89) yields diphenyldithiet (90), which is too reactive to be isolated, although it can be trapped as a molybdenum or a nickel complex in the presence of the metal carbonyl.[230]

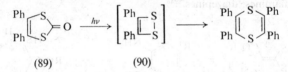

(89)   (90)

Bis-(*p*-dimethylaminophenyl)dithiet exists in the solid state as the α-dithione, but an equilibrium exists in solution between the two forms.[230] The equilibrium is sensitive to light, temperature, and solvent.

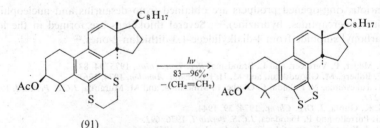

(91)

[227] J. P. Snyder, 'Organic Sulphur Chemistry', ed. C. J. M. Stirling, Butterworths, London, 1975, p. 307.
[228] G. Calzaferri and R. Gleiter, *J.C.S. Perkin II*, 1975, 559.
[229] J. L. Hencher, Q. Shen, and D. G. Tuck, *J. Amer. Chem. Soc.*, 1976, **98**, 899.
[230] W. Kusters and P. de Mayo, *J. Amer. Chem. Soc.*, 1974, **96**, 3502.

appropriate enamines. It was suggested that compounds (9) were formed *via* a [2 + 2] cycloaddition between an enamine and the thiiren 1,1-dioxide, to give (11), followed by opening of the cyclobutane ring, possibly with participation of nitrogen. Other modes of decomposition of (11) involving base-catalysed elimination of the amine function followed by opening of the three-membered ring and disproportionation were used to explain the formation of 4,5-dihydro-thiophen 1,1-dioxide derivatives (12), which were also obtained in many instances. When the compounds (9) were heated in the presence of proton-donating solvents, transannular cyclizations to (13) occurred. This cyclization can be explained by assuming an initial migration of the enamine double bond to the $\beta\gamma$-position relative to the SO_2 group, thus rendering it nucleophilic and able to attack the remaining electrophilic $\alpha\beta$-unsaturated sulphone group on the other side of the ring.

With the exception of the above routes, other reported preparations of these heterocycles followed the more conventional routes, such as intramolecular displacement by a thiolate ion or other nucleophilic sulphur function, intramolecular addition of a thiol radical to a carbon–carbon double bond, and transannular addition of SCl_2 to dienes.

Treatment of the *cis*-dibromide (14) with Na_2S, Na_2Se, or Na_2Te gave the 9-thia-, 9-selena-, or 9-tellura-[3,3,1]nona-2,6-dienes (15), in 22, 13, and 18% yields, respectively.[6]

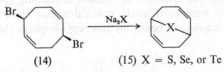

(14)   (15) X = S, Se, or Te

5-Bromomethylthiolan-2-one was obtained in about 75% overall yield by the reaction of the thioamide of pent-4-enoic acid with bromine, followed by hydrolysis of the intermediate immonium salt. The 5-iodo-analogue was also prepared.[7]

Radical-induced cyclizations have been used to prepare thiolans of the type (16),[8] the unsaturated thiol precursors being prepared by the reaction of allyl-

(16) $R^1 = R^2 = Ph$, $R^3 = Me$
$R^1 = Bu^t$, $R^2 = R^3 = Me$

magnesium bromide with thioketones. Photolytic cyclization of 1-mercapto-2-phenylpent-4-ene gave a mixture containing 56% of 3-phenylthian and 44% of 2-methyl-4-phenylthiolan.[9]

[6] (a) E. Cuthbertson and D D. MacNicol, *J.C.S. Perkin I*, 1974, 1893; (b) *J.C.S. Chem. Comm.*, 1974, 498.
[7] I. I. Ershova, V. I. Staninets, and T. A. Degurko, *Dopovidi Akad. Nauk Ukrain. R.S.R.*, *Ser. B*, 1975, 1097 (*Chem. Abs.*, 1976, **84**, 58 199).
[8] M. Dagonneau and J. Vialle, *Tetrahedron*, 1974, **30**, 415.
[9] V. P. Krivonogov, V. I. Dronov, and N. K. Pokoneshchiva, *Khim. geterotsikl. Soedinenii*, 1975, 1204 (*Chem. Abs.*, 1976, **84**, 59 113).

The 8-thiabicyclo[3,2,1]octanes (17) and (18), which are potential precursors to thia-analogues of tropane alkaloids, were obtained in near quantitative yield by addition of SCl_2 to cyclohepta-1,3-dien-6-yl benzoate and cyclohepta-1,4-dien-3-one, respectively.[10] In contrast, cyclohepta-3,5-dienone gave the unsaturated derivative (19) and HCl. Compound (19) readily lost a second mole of

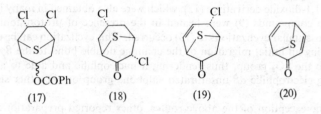

(17)   (18)   (19)   (20)

HCl to give the novel dienone (20), presumably *via* a thiiranium ion intermediate. The addition of SCl_2 to cyclododeca-1,5,9-triene has been reported,[11a] and reactions of the bicyclic system thus obtained, including solvolytic rearrangements and cleavage of the remaining double bond, have been described.[11b] Ionic hydrogenation of 3-phenylthiophen with trimethylsilane in trifluoroacetic acid furnished 3-phenylthiolan in 80% yield.[12]

Reports describing the synthesis of sulphur-containing compounds of pharmacological interest were particularly noticeable over the past two years (see Chapter 16 for an account of penicillin and cephalosporin chemistry). Included

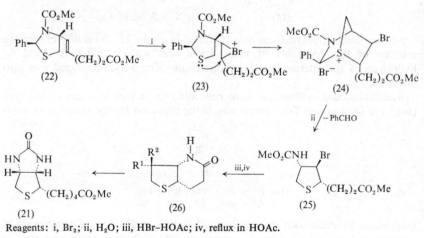

Reagents: i, Br_2; ii, H_2O; iii, HBr–HOAc; iv, reflux in HOAc.

Scheme 1

[10] P. H. McCabe and W. H. Routledge, *Tetrahedron Letters*, 1976, 85.
[11] (a) N. N. Novitskaya, G. K. Samirkanova, N. N. Pervushina, R. V. Kunakova, A. M. Shakirova, and G. A. Tolstikov, *Khim. geterotsikl. Soedinenii*, 1974, 360 (*Chem. Abs.*, 1974, **81**, 24 481); (b) N. N. Novitskaya, R. V. Kunakova, E. E. Zaev, G. A. Tolstikov, and L. V. Spirkhin, *Zhur. org. Khim.*, 1975, **11**, 1434.
[12] Z. N. Parnes, G. I. Bolestova, S. P. Dolgova, V. Udre, M. G. Voronkov, and D. N. Kursanov, *Izvest. Akad. Nauk S.S.S.R., Ser. khim.*, 1974, 1834 (*Chem. Abs.*, 1975, **81**, 16 635).

are three new routes to biotin; the preparation of the biotin ring-system containing selenium in place of sulphur, three syntheses of prostaglandins in which
a sulphide or sulphone grouping has replaced a ring carbon, and five steroidal
systems containing sulphur at various positions.

An elegant total synthesis of '*d*'-biotin (21), starting with L-(+)-cysteine, is
due to a group at Hoffmann–LaRoche (Nutley).[13] The thiolan ring-system,
appropriately substituted for further elaboration, was obtained (Scheme 1) in one
step upon bromination of (22). The reaction is presumed to involve opening of
the bromonium ion (23) by sulphur to generate the bicyclic sulphonium ion (24),
which was hydrolysed during work-up to give (25) and benzaldehyde. Hydrolysis
of the urethane function in (25), followed by refluxing in acetic acid, produced
via an aziridine intermediate, the *trans*-bromolactam (26; R^1 = H, R^2 = Br),
which was converted into the *cis*-azido-lactam (26; R^1 = N_3, R^2 = H) upon
heating with NaN_3 in DMF at 140 °C. The latter compound was transformed,
via d-bisnorbiotin, into d-biotin, using conventional reactions.

A second stereospecific route to *d*-biotin commenced with 2,3,5,6-di-*O*-
isopropylidene-α-mannofuranose.[14] This substance was converted, *via* a series of
steps, into the dimesylate (27) and then cyclized with Na_2S to give the thiolan (28).

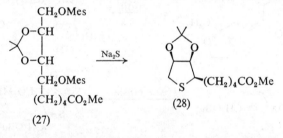

(27)   (28)

The third synthesis [15] differs from all earlier ones in that the side-chain is introduced
after the biotin ring-system has been constructed. The reaction of the sulphoxide
(29) with n-butyl-lithium followed by $I(CH_2)_4CO_2Bu^t$ gave *d,l*-biotin in 30%
yield (Scheme 2). The stereospecific introduction of the alkyl group *trans* to the

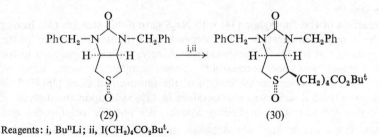

(29)   (30)

Reagents: i, Bu^nLi; ii, $I(CH_2)_4CO_2Bu^t$.

Scheme 2

[13] P. N. Confalone, G. Pizzolato, E. G. Baggiolini, D. Lollar, and M. R. Uskoković, *J. Amer. Chem. Soc.*, 1975, **97**, 5936.
[14] H. Ohru and S. Emoto, *Tetrahedron Letters*, 1975, 2765.
[15] S. Bory, M. J. Luche, B. Moreau, S. Lavielle, and A. Marquet, *Tetrahedron Letters*, 1975, 827.

S=O bond had been expected on the basis of previous work.[16] Fortunately, (29), having the desirable S=O stereochemistry, as shown, was the major isomer obtained upon oxidation of the corresponding sulphide with NaIO$_4$. The *trans*-fused biotin ring-system has been synthesized, 2,5-dihydrothiophen 1,1-dioxide and *N,N*-dichlorourethane being utilized as starting materials.[16] Treatment of *meso*-2,3-diazido-1,4-dimesyloxybutane with Na$_2$Se furnished *cis*-3,4-diamino-selenolan, which was converted into the selenium analogue of the biotin ring-system upon reaction with phosgene.[17]

The ring-systems required for the preparation of 9-[18, 19] and 11-thiaprosta-glandin[20] derivatives were constructed from their precursors as shown. The subsequent conversion of compounds (31)—(33) into prostaglandins followed

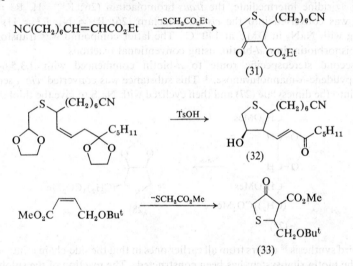

well-worked procedures. In a second route to the 9-thiaprostaglandin system, the condensation of (EtO)$_2$CHCH$_2$SC(Li)=CHOEt with NC(CH$_2$)$_6$CHO con-stituted the key step in the assembling of the precursor required for the ring-forming step.[19b]

The reaction of the dimesylate (34) with Na$_2$S gave 6-thia-steroids (35) having either the α- or β-configuration at C-5, depending on the stereochemistry of the mesyl group at C-5 in (34).[21] Somewhat surprisingly, a similar approach to the 4-thiacholestanes was not successful.[22] These compounds were, however, obtained by a photochemical cyclization of the unsaturated thiol (36).[22] This ring system in the 5-β series was also obtained in 48% yield upon thermolysis of the t-butyl sulphoxide (37) in boiling xylene. All possible sulphoxides and

[16] K. Ohba, T. Kitahara, K. Mori, and M. Matsui, *Agric. and Biol. Chem. (Japan)*, 1974, **38**, 1679 (*Chem. Abs.*, 1975, **82**, 140 011).
[17] R. L. Martin and B. E. Norcross, *J. Org. Chem.*, 1975, **40**, 523.
[18] I. Vlattas and L. DellaVecchia, *Tetrahedron Letters*, 1974, 4267.
[19] (a) I. Vlattas and L. DellaVecchia, *Tetrahedron Letters*, 1974, 4459; (b) I. Vlattas, L. Della Vecchia, and A. O. Lee, *J. Amer. Chem. Soc.*, 1976, **98**, 2008.
[20] I. T. Morrison, K. J. R. Taylor, and J. H. Fried, *Tetrahedron Letters*, 1975, 1165.
[21] W. N. Speckamp and H. Kesselaar, *Tetrahedron Letters*, 1974, 3405.
[22] D. N. Jones, D. A. Lewton, J. D. Msonthi, and R. J. K. Taylor, *J.C.S. Perkin I*, 1974, 2637.

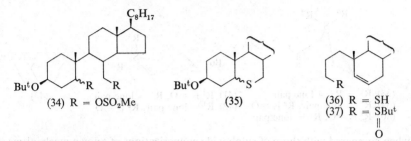

(34) R = OSO$_2$Me   (35)   (36) R = SH
(37) R = SBu^t
‖
O

sulphones in the 5-α and -β series were prepared. Structure assignments were made on the basis of n.m.r. data, base-catalysed interconversions, modes of synthesis, and chromatographic mobility. The sulphoxides were found to be thermally stable, but could be epimerized photochemically at sulphur without isomerization at C-5. This result argues against C—S bond cleavage to a diradical intermediate in the epimerization step, and suggests a simple photochemical inversion at sulphur. The preparation of A-nor-3-thia-5β-pregnane [23] has been described.

Properties and Reactions.—The conformational properties of heterocyclohexanes, including thians, have been reviewed.[24] I.r. spectra have been obtained for thian,[25a] selenan,[25a] and sulpholen [25b] in the vapour, liquid, and solid phases (−170 °C). Peak assignments were made with the help of studies of polarization in Raman spectra. The photoelectron spectra of dihydrofuran and its sulphur, selenium, and tellurium analogues were interpreted on the basis of local C_{2v} symmetry of the CH$_2$XCH$_2$ fragment, including mixing of the non-bonding electrons of the heteroatom system;[26] the mixing being greatest for the oxygen and least for the tellurium heterocycle. Photoelectron spectroscopy has been used to probe for interactions between the sulphur non-bonding electrons and the two π-electron systems of (38).[27] Interactions were observed between sulphur and the ethylene, but not the butadiene, bridge. These results differ from the corresponding nitrogen analogue, in which interaction among all π systems (bicycloconjugation) exists. The bicyclic sulphide (38) was prepared by reduction with LiAlH$_4$ of the sulphoxide (39) obtained by the interaction of SO with cyclo-octatetraene.[28] Oxidation of (38) with *m*-chloroperoxybenzoic acid (MCPBA) gave the isomeric sulphoxide (40), having the S=O bond *syn* to the C$_2$ bridge; similar stereochemical results were obtained upon oxidation of di-, tetra-, and hexa-hydro-(38), regardless of the degree of unsaturation in either bridge. The structures of these sulphoxides were readily assignable on the basis of their n.m.r. spectra and [Eu(fod)$_3$]-induced shifts. Sulphoxides (39) and (40) equilibrated thermally in benzene above 110 °C to give a 90 : 10 mixture. The activation parameters ($\Delta G^{\ddagger} = 29.5$ kcal mol^{-1} at 110.7 °C and $\Delta S^{\ddagger} = -4$ e.u.),

[23] C. M. Cimarusti, F. F. Giarusso, P. Grabowich, and S. D. Levine, *Steroids*, 1975, **26**, 359.
[24] J. B. Lambert and S. I. Featherma, *Chem. Rev.*, 1975, **75**, 611.
[25] D. Vedal, O. H. Ellestad, P. Klaboe, and G. Hagen, *Spectrochim. Acta*, 1975, **31A**, 355;
(b) G. Fini and B. Fortuna, *ibid.*, p. 423.
[26] S. Pignataro and C. Distefano, *Chem. Phys. Letters*, 1974, **26**, 356.
[27] C. Mueller, A. Schweig, A. G. Anastassiou, and J. C. Wentzel, *Tetrahedron*, 1974, **30**, 4089.
[28] A. G. Anastassiou, J. C. Wentzel, and B. Y. H. Chao, *J. Amer. Chem. Soc.*, 1975, **97**, 1124.

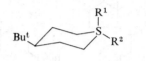

XYZ (43)

(38) $R^1 = R^2 =$ lone pair   (41) R^1 is $=O$, $R^2 =$ lone pair   (43)
(39) $R^1 =$ lone pair, R^2 is $=O$ (42) $R^1 =$ lone pair, R^2 is $=O$
(40) R^1 is $=O$, $R^2 =$ lone pair

when compared with those of sulphoxide epimerizations of known mechanisms,[29] ruled out homolytic C—S bond cleavage, but, coupled with deuterium-labelling experiments, pointed to a suprafacial 1,3-migration of the SO bridge as the pathway for isomerization. The 7,8-dihydro-derivatives of (39) and (40) also isomerized under the influence of heat. In this case a mechanism involving homolytic C—S bond cleavage was indicated.

[13]C n.m.r. spectroscopy is of value in assigning the SO stereochemistry in thian S-oxides.[30] In the model system 4-t-butylthian S-oxide, C-2 and C-3 were shielded by 7.5 and 5.5 p.p.m. in the cis-isomer (41) compared with the trans-isomer (42). These assignments were corroborated by the low-temperature spectrum of thian S-oxide. The shielding of C-3 in (41) is due to a γ-steric effect caused by the SO bond, whereas that at C-2 was shown to be electronic in origin.

Both ^{1}H and [13]C n.m.r. spectra have been obtained for a number of 1-hetero-cyclohexan-4-ones.[31] The data indicated that all systems (heteroatoms = S, SO, SO_2, O, or NMe) exist predominantly in the chair conformation, and that no significant transannular interactions between the heteroatom and the carbonyl group occur. [Eu(fod)₃]-induced shifts are smaller in thiacycloalkanones than in the corresponding carbocycles, due to the decrease in the basicity of the carbonyl oxygen.[32]

The ratio of equatorial to axial imide conformers of thian 1-N-(p-chlorophenyl)-imide is 85 : 15 at −90 °C;[33] for cis-1-thiadecalin 1-β-(N-p-chlorophenyl)imide, only the conformer having the imide function in the equatorial position could be detected. The bicyclic compounds (43; X, Y, Z = O, S, or Se) tend to exist in the double-chair conformation, as determined by [1]H n.m.r. spectroscopy. Appreciable proportions of the chair–boat conformation were detected for those isomers having X and Z = S or Se.[34]

Thian and thiolan S-oxides have a similar basicity in dilute aqueous solutions,[35] but thietan S-oxide is a somewhat weaker base. The axial SO bond of (41) is more readily protonated than the equatorial SO bond of (42). In protonated thian S-oxide the OH group exhibits a preference (by 1.6 kcal mol⁻¹) for the axial position at room temperature.[35] The rates of oxidation of the parent

[29] K. Mislow, *Records Chem. Progr.*, 1967, **28**, 217.
[30] G. W. Buchanan and T. Durst, *Tetrahedron Letters*, 1975, 1683.
[31] J. A. Hirsch and E. Havinga, *J. Org. Chem.*, 1976, **41**, 455.
[32] H. Remane, E. Kleinpeter, and R. Borsdorf, *Z. Chem.*, 1975, **15**, 312.
[33] P. K. Claus, W. Rieder, and F. W. Vierhapper, *Tetrahedron Letters*, 1976, 119.
[34] N. S. Zefirov and S. V. Rogozina, *Tetrahedron*, 1974, **30**, 2345.
[35] R. Curci, F. DiFuria, A. Levi, V. Lucchini, and G. Scorrano, *J.C.S. Perkin II*, 1975, 341.

four-, five-, and six-membered-ring sulphoxides with peroxy-acid vary only slightly in either acidic (1 : 0.7 : 1.9) or basic (1 : 0.7 : 1) media, leading to the suggestion that the geometries of the transition states and the ground states ought to be very similar. Such is easily understandable for the acidic process (nucleophilic attack by the sulphur lone pair on the electrophilic oxygen of the peroxy-acid), but more difficult to comprehend for the basic reaction, which might have been expected to involve attack of the peroxy-acid anion on sulphur, and thus involve rehybridization of sulphur to a trigonal bipyramid. Reduction of the same sulphoxides with KI–HClO₄ [36] showed a much greater rate variation. Under similar reaction conditions, thiolan *S*-oxide is reduced faster than thietan *S*-oxide or DMSO by factors of 5 and 19, respectively, as expected on the basis of a quinquecovalent intermediate. The *trans*-isomer (42) is reduced ten times faster than the *cis*-isomer (41).[36]

Very approximate kinetic measurements showed that the relative rates of reaction of the cyclic sulphilimines (44) with either cyanide ion or tri-n-butyl-phosphine were (44a) > (44c) > (44b).[37] When contrasted with the very small

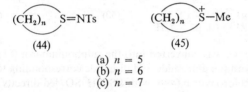

(44)

(45)

(a) *n* = 5
(b) *n* = 6
(c) *n* = 7

variations in the demethylation rates of the sulphonium salts (45) (however, see below), these variations were in agreement with a mechanism involving attack of the nucleophiles at sulphur. The reactions of sulphonium salts such as (45) with a typical nucleophile, N₃⁻, have been studied in some detail.[38] In the parent series the ratios of products of demethylation to those of ring-opening were 47 : 53 and 96 : 4 for (45a) and (45b), respectively. The corresponding ammonium salts behave similarly. The large amount of ring-opening in (45a) was ascribed mainly to ring-strain, reinforced by the absence of eclipsing interactions in the nucleophilic attack on the ring. As expected, α-substitution greatly diminished the amount of ring-opened products.

The equilibrium constant between the sulphonium salts (46) and (47) at 100 °C is 1.45, in favour of the isomer having the equatorial methyl group.[39] Methylation of 4-t-butylthian gives (47) and (46) in an 88 : 12 ratio. The structure of each

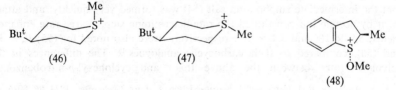

(46)

(47)

(48)

[36] R. Curci, F. DiFuria, A. Levi, and G. Scorrano, *J.C.S. Perkin II*, 1975, 408.
[37] T. Aida, M. Nakajima, T. Inoue, N. Furukawa, and S. Oae, *Bull. Chem. Soc. Japan*, 1975, **48**, 723.
[38] E. L. Eliel, R. O. Hutchins, R. Mebane, and R. L. Willer, *J. Org. Chem.*, 1976, **41**, 1052.
[39] E. L. Eliel, R. L. Willer, A. T. McPhail, and D. K. Onan, *J. Amer. Chem. Soc.*, 1974, **96**, 3021.

isomer was confirmed by an *X*-ray structure determination, which showed that
the ring in (46) is somewhat flattened to relieve the *syn*-1,3 (Me–H)-interactions.
Displacement of the methoxy-group in the alkoxysulphonium salt (48) by either
MeMgBr or Me₂Cd occurred with inversion of configuration.[40] There is, however, considerable loss of stereochemistry, probably due to isomerization of the
starting materials by the basic reagents.

Neighbouring-group participation by sulphur continues to be looked for, and
found, in a variety of solvolytic reactions. Dissolution of the bicyclic sulphide
(49a) in FSO₃H–SO₂ at − 60 °C gave rise to an intermediate which was assigned
the symmetrical sulphurane structure (50) on the basis of its ¹³C n.m.r. spectrum
(four resonances, at δ = 24.2, 29.7, 42.0, and 55.7).[41] This intermediate (50), when

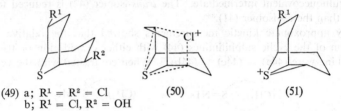

(49) a; R¹ = R² = Cl   (50)   (51)
      b; R¹ = Cl, R² = OH

warmed to − 30 °C, was converted into the sulphonium salt (51). Similar results
were obtained with the dibromide and di-iodide corresponding to (49a), while the
reaction of the chlorohydrin (49b) with FSO₃H–SO₂ led directly to (51).

endo-2-Chloro-7-thia[2,2,1]bicycloheptane (52a) solvolyses at least 5 × 10⁹
times faster than the *exo*-isomer (52b) in acetic acid–sodium acetate solution,
and gives exclusively the *endo*-acetate (52c), whereas the *exo*-isomer gave the

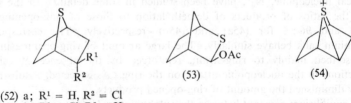

                                    (53)   (54)

(52) a; R¹ = H, R² = Cl
     b; R¹ = Cl, R² = H
     c; R¹ = H, R² = OAc

rearranged acetate (53).[42] The second-order kinetics observed for (52a) showed
that the intermediate sulphonium salt (54) was formed very quickly, and attack
on it by acetate ion constituted the rate-determining step. Large rate enhance-
ments, 5.9 × 10⁶ and 5.4 × 10⁷, respectively, were also noted for (55; R = H)
and (56), compared to their carbocyclic analogues.[43] The differences in the
solvolytic rates between the above thian and cyclohexyl-*p*-nitrobenzoates

[40] K. K. Andersen, R. I. Caret, and I. Karup-Nielsen, *J. Amer. Chem. Soc.*, 1974, **96**, 8026.
[41] J. A. J. M. Vincent, P. Schipper, A. DeGroot, and H. M. Buck, *Tetrahedron Letters*, 1975,
1989.
[42] I. Tabushi, Y. Tamaru, Z. Yoshida, and T. Sugimoto, *J. Amer. Chem. Soc.*, 1975, **97**, 2886.
[43] S. Ikegami, T. Asai, K. Tsuneoka, S. Matsumura, and S. Akaboshi, *Tetrahedron*, 1974, **30**,
2087.

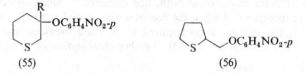

(55)   (56)

decreased to 620 and 91 as the R group was replaced by a methyl and a phenyl group, respectively, indicating less participation by sulphur as the stability of the carbocation increases. No participation by sulphur was observed in the solvolysis of the *p*-nitrobenzoate of 4-hydroxythian.

Addition of Br_2 or ArSCl to 3,3-dichloro-2-thia[2,2,1]bicyclohept-5-ene, and other similarly substituted bicycloheptenes, occurs *via* a sulphonium ion intermediate, and leads in the above instance to 6,7-dibromo-3,3-dichloro-2-thia-[2,2,1]bicycloheptane.[44] The aza-analogue behaves similarly, but when sulphur was replaced by oxygen no participation–rearrangement was observed. The formation of the alcohol (57a) on reduction with $LiBH_4$–BH_3 of the epoxide (58),

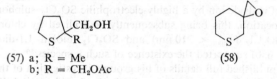

(57) a; R = Me   (58)
b; R = CH_2OAc

and of the hydroxy-acetate (57b) on treatment with acetic acid, indicates participation of sulphur.[45] Surprisingly, no rearrangement products were observed when the sulphur atom was replaced by an NMe group. 2-Bromomethylthiolan and 3-bromothian are equilibrated thermally *via* 1-thionia[3,1,0]bicyclohexyl bromide.[46]

Oxidation of *cis*-2,6-diphenylthian with Br_2–H_2O gave mainly the all-equatorial isomer,[47] while its reaction with Bu^tOCl furnished its epimer. In contrast, oxidation of the 4-hydroxy-thian (59) with Bu^tOCl afforded (60), having the SO

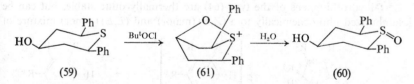

(59)   (61)   (60)

bond equatorially oriented. In this case the cyclic alkoxysulphonium salt (61) is a plausible intermediate. Participation by the 4-hydroxy-group, *via* alkoxyoxo-sulphonium ion intermediates corresponding to (61), has also been observed in the oxidation of 4-hydroxythian *S*-oxides.[48]

The monochlorosulphoxide (62) isomerized in dioxan–HCl at 0 °C to give mixtures of isomers in which either epimerization at sulphur, or at the carbon

[44] M. S. Raasch, *J. Org. Chem.*, 1975, **40**, 161.
[45] S. Ikegami, J. Ohishi, and S. Akaboshi, *Chem. and Pharm. Bull.* (*Japan*), 1975, **23**, 2701.
[46] C. Leroy, M. Martin, and L. Bassery, *Bull. Soc. chim. France*, 1974, 590.
[47] J. Klein and H. Stollar, *Tetrahedron*, 1974, **30**, 2541.
[48] H. Stollar and J. Klein, *J.C.S. Perkin I*, 1974, 1763.

bearing the chlorine atom, or at both, had occurred.[49] Since no deuterium or bromine was incorporated when the reaction was carried out in the presence of DCl or HBr, the epimerization required C—S bond breaking and remaking, presumably *via* the intermediate (63). Further chlorination of (62) gave mainly

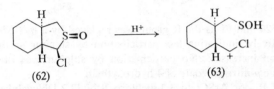

the αα-dichloro-isomer. Chlorination of thiolan 1,1-dioxide with excess SO_2Cl_2 in the dark gives exclusively the 3-chloro-isomer, whereas radical-initiated chlorination gives the 2-chloro- and 3-chloro-isomers, in the ratio 3 : 97. 7-Thia[2,2,1]-bicycloheptane 7,7-dioxide gave 65% of 2-*endo*- and 18% of the 2-*exo*-chloro-isomers. The high *endo/exo* ratio was ascribed to the bulky nature of the SO_2 group preventing *exo* attack. These chlorinations were suggested to involve abstraction of hydride ion by a highly electrophilic SO_2Cl_2–sulphone complex to form a carbocation, this being subsequently neutralized by chloride ion. U.v. spectra of SO_2Cl_2 ($\lambda_{max} < 210$ nm) and SO_2Cl_2–thiolan 1,1-dioxide ($\lambda_{max} = 278$ nm, $\varepsilon = 650$) supported the existence of such a species.[50]

Mock has published full details of his group's investigation of the dissociation of 2,5-dihydrothiophen 1,1-dioxide and 2,7-dihydrothiepin 1,1-dioxide to SO_2 and 1,3-dienes,[51a] and 1,3,5-trienes,[51b, c] respectively. The reactions were concluded to be concerted and suprafacial, with respect to the hydrocarbon portion, in the fragmentation of the five-membered ring, and probably concerted, but antarafacial, with respect to the triene, in the case of the seven-membered ring. The relative rates of reaction of several 1,3-dienes with SO_2 have been reported.[52] Large substituents at position 2 increase the rate of the cycloaddition by favouring the cisoid conformation of the diene. A ρ value of -1.1 was found for a series of 2-aryl-1,3-dienes.

2,5-Dihydrothiophens of the type (64) are thermally quite stable, but can be desulphurized photochemically to a (Z,E) (major) and (E,E) (minor) mixture of

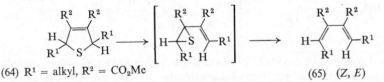

(64) R^1 = alkyl, $R^2 = CO_2Me$ (65) (Z, E)

the dienes (65).[53] The reaction was shown to proceed *via* the intermediacy of a vinyl-thiiran. The sulphoxides corresponding to (64) also yield mixtures of 1,3-dienes, both on photolysis and on thermolysis. The sulphones, on the other

[49] E. Casadevall and M. M. Bouisset, *Tetrahedron Letters*, 1975, 2023.
[50] I. Tabushi, Y. Tamaru, and Z. Yoshida, *Tetrahedron*, 1974, **30**, 1457.
[51] (a) W. L. Mock, *J. Amer. Chem. Soc.*, 1975, **97**, 3666; (b) *ibid.*, p. 3673; (c) W. L. Mock and J. H. McCausland, *J. Org. Chem.*, 1976, **41**, 242.
[52] N. S. Isaacs and A. A. R. Laila, *Tetrahedron Letters*, 1976, 715.
[53] R. M. Kellogg and W. L. Prins, *J. Org. Chem.*, 1974, **39**, 2366.

hand, lost SO_2 thermally in a stereospecific manner, as expected,[51a] but again gave isomeric dienes on photolysis. All of the above reactions, with the exception of the sulphone thermolysis, can be explained by assuming initial formation of a diradical, followed by either cyclization to a three-membered ring or direct loss of the sulphur moiety.

The bicyclic ene-dione (66) underwent a photochemically induced 1,3-shift of the sulphur bridge to give (67), whose structure was established spectroscopically

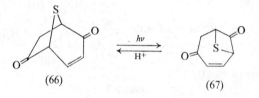

(66)   (67)

and by cleavage with methoxide ion of the cyclobutanone, to give 7-methoxy-carbonyl-Δ^3-thiepin-5-one.[54] Treatment of (67) with acid regenerated (66). Irradiation of the oxothiepan (68) gave the thiolactone (69; R = OAc or $OCOCH_2CCl_3$) in greater than 50% yield.[55] The isolation of several minor products and the possible mechanism of the ring-contraction, for which the authors suggest the intermediacy of (69a), formed by transfer of an electron from sulphur to the excited carbonyl group, have been discussed in detail.[55b]

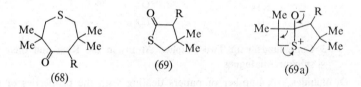

(68)   (69)   (69a)

Ramberg–Bäcklund reactions, discussed in more detail in Chapters 1 and 2, continue to be of great significance in the synthesis of strained alkenes. Bicyclo-[2,2,1]heptenes,[56] [4,2,2]propellanes,[57] 1,4-tetramethylene- and 1,4-penta-methylene-Dewar-benzene,[58] and the ethylene ketal of hydrindane-3,6-dien-1-one[59] have been prepared, utilizing this rearrangement as the final step. 2,5-Dialkyl-sulpholans have been converted, in synthetically useful yields (20—67%), into 1,2-dialkyl-cyclobutenes upon reaction with Bu^nLi in THF followed by reduction with $LiAlH_4$. This most interesting reaction has been applied to the preparation of a wide variety of fused benzocyclobutenes.[60]

[54] H. Tsuruta, M. Ogasawara, and T. Mukai, *Chem. Letters*, 1974, 887.
[55] (a) P. Y. Johnson and M. Berman, *J.C.S. Chem. Comm.*, 1974, 779; (b) *J. Org. Chem.*, 1975, **40**, 3046.
[56] R. G. Carlson and K. D. May, *Tetrahedron Letters*, 1975, 947.
[57] (a) K. Weinges and K. Klessing, *Chem. Ber.*, 1974, **107**, 1925; (b) K. Weinges, K. Klessing, and H. Baake, *ibid.*, 1976, **109**, 796.
[58] K. Weinges and K. Klessing, *Chem. Ber.*, 1974, **107**, 1025; 1976, **109**, 793.
[59] J. Kattenburg, E. R. De Waard, and H. O. Huisman, *Tetrahedron*, 1974, **30**, 463, 3177.
[60] J. M. Photis and L. A. Paquette, *J. Amer. Chem. Soc.*, 1974, **96**, 4715.

A series of reports by Russian authors have appeared, dealing with the reaction of esters[61] and ethers[62] of 3-hydroxysulpholen and its 3-arylamino-[63] and 3,4-epoxy-analogues[64] under basic conditions. Isomerization reactions of sulpholens have been studied.[65] The reaction of the 3-keto-thiolan (70) with $OCH(CH_2)_3CO_2Me$ in pyridine, followed by reduction with $NaBH_4$ (Scheme 3), gave (70a), a compound of potential value in the synthesis of biotin.[66]

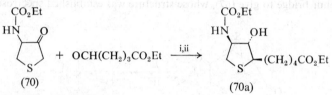

Reagents: i, pyridine; ii, $NaBH_4$.

Scheme 3

Selenium-N-tosylimides, *e.g.* (71), have been prepared by the reaction of selenides with Bu^tOCl and N-sodio-toluene-p-sulphonamides in MeCN. In the presence of H_2O the imide is in equilibrium with the selenoxide (72) and toluene-p-sulphonamide, thus suggesting the successful preparation of (71) from (72) and the amide by azeotropic removal of water.[67]

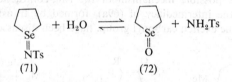

3 Compounds with Two Sulphur Atoms in the Ring and their Oxy-sulphur Analogues

Cyclic Disulphides.—A number of papers dealing with the properties of lipoic acid (73) have appeared. When irradiated in an aldehyde solvent (*e.g.* acet-aldehyde, propionaldehyde), the monoacetylated products (74) of S—S bond fission were obtained in 70—80% yield.[68] The reaction was successfully extended to a series of 3,3-dialkyl-1,2-dithiolans, 1,2-dithians, and acyclic disulphides. Oxidation of (73) with a variety of reagents, including singlet oxygen, t-butyl

[61] (a) T. E. Bezmenova, T. S. Lutsii, A. F. Rekasheva, and Y. N. Usenko, *Khim. geterotsikl. Soedinenii*, 1975, 1060 (*Chem. Abs.*, 1976, **84**, 30 010); (b) S. M. Lukashov and T. E. Bezmenova, *ibid.*, 1974, 625 (*Chem. Abs.*, 1974, **81**, 91 283); (c) Y. N. Usenko and S. M. Lukashov, *Ukrain. khim. Zhur.*, 1974, **40**, 1177 (*Chem. Abs.*, 1974, **82**, 111 879).
[62] T. E. Bezmenova, D. I. Kurlyand, A. F. Rekasheva, and T. S. Lutsii, *Khim. geterotsikl. Soedinenii*, 1975, 1067 (*Chem. Abs.*, 1976, **84**, 16 452).
[63] (a) T. E. Bezmenova and P. G. Dul'nev, *Khim. geterotsikl. Soedinenii*, 1974, 622 (*Chem. Abs.*, 1974, **81**, 63 424); (b) P. G. Dul'nev and T. E. Bezmenova, *ibid.*, p. 1322 (*Chem. Abs.*, 1974, **82**, 97 905).
[64] L. A. Mukhamedova, L. I. Kursheva, and N. P. Anoshina, *Zhur. org. Khim.*, 1975, **11**, 2450.
[65] T. E. Bezmenova, T. E. Lukashov, V. P. Tantsyura, T. S. Lutsii, A. F. Rekasheva, and T. D. Zaika, *Khim. geterotsikl. Soedinenii*, 1975, 1072 (*Chem. Abs.*, 1976, **84**, 4179).
[66] S. D. Mikhno, T. M. Filipova, T. N. Polyanskaya, I. G. Razumova, and V. M. Berezovskii, *Khim. geterotsikl. Soedinenii*, 1974, 1053 (*Chem. Abs.*, 1974, **81**, 169 376).
[67] S. Tamagaki, S. Oae, and K. Sakaki, *Tetrahedron Letters*, 1975, 649.
[68] M. Takagi, S. Goto, and T. Matsuda, *J.C.S. Chem. Comm.*, 1976, 92.

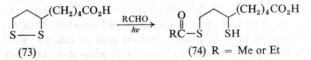

(73)   (74) R = Me or Et

hydroperoxide, or peroxyacetic acid, gave mixtures of thiolsulphinates and thiolsulphonates which have been separated, but, so far, not unambiguously identified.[69] Lipoic acid effectively quenches the oxidation of rubrene by singlet oxygen.[70] The amide of (73) has been prepared in 65% yield by the reaction of Na_2S_2 with 6,8-dichloro-octanoic acid amide.[71] The spectroscopic properties (u.v., i.r., n.m.r.) of nor- and tetranor-lipoic acid have been recorded.[72]

The disulphide (75) and the dihydrothiophen (76) were produced in small (<1%) yield upon photolysis of *O*-ethyl thioacetate in 1,2-dimethylenecyclohexane.[73] The probable pathway to (76) involved rearrangement of the intermediate vinyl episulphide (77). The disulphide (75) could arise from (76) by

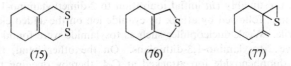

(75)   (76)   (77)

reaction with atomic sulphur or, conceivably, from the reaction of S_2 with the diene. Both (75) and (76) were synthesized unambiguously from the above diene; the former by bromination to 1,2-dibromomethylcyclohexene, conversion into the corresponding dithiol, and then oxidation with $FeCl_3$; the latter by the reaction of SO (generated thermally from ethylene episulphoxide) with 1,2-dimethylenecyclohexane, followed by reduction by $LiAlH_4$. Treatment of $\alpha\omega$-dihalides with Na_2S and sulphur in DMF leads to a mixture of cyclic and polymeric disulphides.[74] This mixture was generally not purified, but reduced with $LiAlH_4$ to give $\alpha\omega$-dithiols cleanly, uncontaminated by cyclic sulphides. The interaction of alkyl-lithiums with 1,2-dithian leads to polymeric disulphides.[75] 4-Hydroxy-1,2-dithiolan [76] has been isolated from *Bruguiera cylindrica*, along with other previously identified derivatives of 1,2-dithiolan.[77] Irradiation of 1,2-dithian with circularly polarized light in a hydrocarbon glass at 77 K produced an excess of one enantiomer.[78]

1,3-Dithiolans and 1,3-Dithians.—Phenylacetonitrile and $(CH_2SCN)_2$ gave 2-cyano-2-phenyl-1,3-dithiolan in 45% yield under phase-transfer conditions.[79]

[69] F. E. Stary, L. S. Jindal, and R. W. Murray, *J. Org. Chem.*, 1975, **40**, 58.
[70] B. Stevens, S. R. Perez, and R. D. Small, *Photochem. and Photobiol.*, 1974, **19**, 315.
[71] V. M. Tursin, L. G. Chebotareva, and A. M. Yurkevich, *Khim. Farm. Zhur.*, 1975, **9**, 26 (*Chem. Abs.*, 1976, **84**, 105 455).
[72] J. C. H. Shih, P. B. Williams, L. D. Wright, and D. G. McCormick, *J. Heterocyclic Chem.*, 1974, **11**, 119.
[73] R. Jahn and U. Schmidt, *Chem. Ber.*, 1975, **108**, 630.
[74] E. L. Eliel, V. S. Rao, S. Smith, and R. O. Hutchins, *J. Org. Chem.*, 1975, **40**, 524.
[75] M. L. Hallensteben, *Makromol. Chem.*, 1974, **175**, 3315.
[76] A. Kato and J. Takahashi, *Phytochemistry*, 1976, **15**, 220.
[77] H. Yanagawa, T. Kato, Y. Kitahara, N. Takahashi, and Y. Kato, *Tetrahedron Letters*, 1972, 2549.
[78] B. Nelander and B. Norden, *Chem. Phys. Letters*, 1974, **28**, 384.
[79] M. Makosza and M. Fedorynski, *Synthesis*, 1974, 274.

The dithiolanylium salts (78; X = Br) were formed in very high yield and with complete exclusion of (79) upon bromination of (80), even when R = Ph.[80] In contrast, cyclization of (80), using a proton acid as catalyst, gave (79; X = ClO_4^-) when R = H, a mixture of (78) and (79) when R = Me, and exclusively

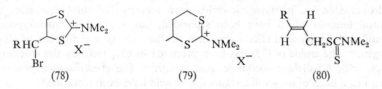

(78)   (79)   (80)

the six-membered-ring species when R = Ph. 2-Cyano-2-dimethylamino-1,3-dithiolan reacts with both electrophilic and nucleophilic reagents.[81] For example, with electrophiles such as n-butyl bromide, benzyl bromide, or benzoyl chloride, the corresponding nitriles were obtained in 22, 41, and 97% yield, respectively, presumably *via* initial ionization to 2-dimethylamino-1,3-dithiolan-2-ylium cyanide, followed by attack by cyanide ion on the added electrophile to give the nitrile. Weak nucleophiles such as tosylamide were found to attack at C-2 and give 2-tosylamino-1,3-dithiolans. On the other hand, the powerful nucleophile thiophenoxide ion attacked at C-4, thereby opening the ring and regenerating a thiocarbamate. Detailed procedures for the preparation of 2,2-(ethylenedithio)- and 2,2-(trimethylenedithio)-cyclohexanone have appeared.[82] The cleavage of a number of 2,2-(trimethylenedithio)cycloalkanones with KOH in Bu^{t}OH at 60 °C afforded dithianyl ω-carboxylic acids (81), generally in >90% yield, unless the carbonyl group is severely hindered.[83] In a related reaction,

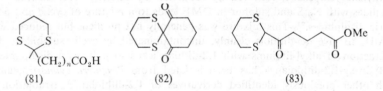

(81)   (82)   (83)

2-acyl-dithians, *e.g.* (83), were prepared from 1,3-diketones and $(TsSCH_2)_2CH_2$ in methanolic base, presumably *via* a diacyl-dithian intermediate (82).[84]

1,3-Dithian-4,6-diones were obtained in about 40—50% yield from *gem*-dithiols and malonyl chlorides;[84] 1,3-dithiolan-4,5-dione can be similarly prepared. The above dithians, which are thio-analogues of Meldrum's acids, were found to be soluble in 5% bicarbonate solution, and can be alkylated at the 5-position, or converted into the 5-benzylidene derivative. Their reaction with 2 equivalents of Bu^{n}Li, followed by methyl iodide, gave, in low yield, products resulting from methylation at the 2-position, *via* the 2,5-dianion.[85]

[80] K. Hiratani, T. Nakai, and M. Okawara, *Chem. Letters*, 1974, 1041.
[81] K. Hiratani, T. Nakai, and M. Okawara, *Bull. Chem. Soc. Japan*, 1974, **47**, 904.
[82] (*a*) R. B. Woodward, I. J. Pachter, and M. L. Scheinbaum, *Org. Synth.*, 1974, **54**, 37; (*b*) *ibid.*, p. 39.
[83] J. A. Marshall and D. E. Seitz, *J. Org. Chem.*, 1974, **39**, 1814.
[84] R. J. Bryant and E. McDonald, *Tetrahedron Letters*, 1975, 3841.
[85] H. J. Schauble, W. A. Van Saun, jun., and J. D. Williams, *J. Org. Chem.*, 1974, **39**, 2946.

The first example of a stable thio-analogue of a Meisenheimer complex has been isolated.[86] The salt (84) (red crystals; m.p. 157 °C) was obtained by treatment of 2-(picrylthio)ethanethiol with NaOMe, followed by careful precipitation with toluene. The existence of the trimethylene analogue of (84) was inferred from spectroscopic measurements, but this material could not be isolated.

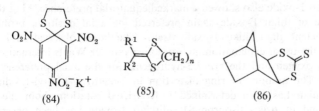

(84) (85) (86)

Keten thioacetals such as (85) have been prepared by two routes. The first gives yields of 50—70% ($n = 2$), and involves the reaction of bis(dimethyl-alumino)propane-1,3-dithiolate with esters;[87a] it is patterned after the reaction earlier described, involving the analogous ethane-1,2-dithiolate.[87b] In the second method, a Grignard reagent is condensed with CS_2, the product is converted into its lithium salt, allowed to react with $Br(CH_2)_nCl$, and then cyclized.[87c]

Ethanedithiol reacts with CO in the presence of selenium, oxygen, and triethylamine to give ethylene dithiocarbonate in 90% yield.[88] Similar conditions, employing the disulphides of mercaptoethanol or mercaptoethylamine, also result in the formation of five-membered-ring heterocycles. Trithiocarbonates are formed from the reaction of epoxides,[89a] and the carbonates (poor yields) [89b] or the thionocarbonates (fair to good yields) [89b] of ethanediols, with potassium *O*-ethyl xanthate. The mechanisms of these interconversions were investigated in detail. The strained alkene norbornene reacted with CS_2 and S under 5 atm pressure, affording the trithiocarbonate (86) in 70% yield;[90] when cyclohexene was used, the expected analogue was formed only to the extent of about 1%.

There is still a strong interest in the stereochemistry of 1,3-dithian and its derivatives, with most studies utilizing n.m.r. spectroscopy to determine the preferred geometry of these molecules. Variable-temperature ^1H n.m.r.[91, 92] and dipole-moment studies [92] have confirmed the earlier reports,[93] that the preferred conformation of 1,3-dithian 1-oxide is the one in which the S=O bond is equatorial (84 : 16 at −81 °C). The barrier to inversion in this compound is 9.8 kcal mol⁻¹ at −70 °C.[91] Oxidation of 2-trimethylsilyl-1,3-dithian with either sodium periodate or *m*-chloroperoxybenzoic acid gave preferentially the thermo-dynamically preferred equatorial S=O isomer.[94] 1,3-Dithian 1,3-dioxide appears

[86] E. Farina, A. C. Veracini, and F. Pietra, *J.C.S. Chem. Comm.*, 1974, 672.

[87] (a) E. J. Corey and A. P. Kozikowski, *Tetrahedron Letters*, 1975, 925; (b) E. J. Corey and D. J. Beames, *J. Amer. Chem. Soc.*, 1973, **95**, 5829; (c) J. Meijer, P. Vermeer, and L. Brandsma, *Rec. Trav. chim.*, 1975, **94**, 83.

[88] P. Koch and E. Perotti, *Tetrahedron Letters*, 1974, 2899.

[89] (a) E. M. Ali, N. G. Kardouche, and N. L. Owen, *J.C.S. Perkin I*, 1975, 748; (b) N. G. Kardouche and N. L. Owen, *ibid.*, p. 754.

[90] J. Petermann and H. Plieninger, *Tetrahedron*, 1975, **31**, 1209.

[91] S. A. Khan, J. B. Lambert, O. Hernandez, and F. A. Carey, *J. Amer. Chem. Soc.*, 1975, **97**, 1468.

[92] M. J. Cook and A. P. Tonge, *J.C.S. Perkin II*, 1974, 762.

[93] M. J. Cook and A. P. Tonge, *Tetrahedron Letters*, 1973, 849.

[94] F. A. Carey, O. Hernandez, I. C. Taylor, jun., and R. F. Bryan, *Preprints Div. Chem., Amer. Chem. Soc.*, 1974, **19**, 261.

9

to exist mainly in the diequatorial form; it gave a temperature-dependent spectrum, which has been suggested to be due to intermolecular association rather than conformational changes.[91] The sulphimide group in 1,3-dithian-1-*p*-tolylsulphonylsulphimide, produced in 60% yield by treatment of 1,3-dithian with Chloramine-T, showed a strong equatorial preference.[95] The 5-keto-derivative of 1,3-dithian 1-oxide also showed a marked equatorial preference,[96] but the 3-keto-derivative of thian 1-oxide again preferred the axial position, probably to a greater extent than thian 1-oxide itself. The latter observation brings into question the usual explanation, an attractive van der Waals interaction between the axial oxygen and the *syn* 3,5-hydrogens, for the axial preference in thian 1-oxide. The amount of ring distortion in 2-methyl-2-t-butyl-1,3-dithian and -1,3-oxathian has been determined.[97] Restricted rotation about the exocyclic C—C bond in *trans*-2-isopropyl-1,3-dithian 1-oxide has been observed. The preferred conformation is one in which the methine hydrogen of the isopropyl group has a 1,3-*syn* relationship with the equatorial SO bond.[98]

Allinger and Kao [99] have completed calculations of the relative energies of the possible chair, twist, and boat conformations of a number of sulphoxides containing six-membered rings, including thian 1-oxide, 1,3-dithian 1-oxide, the monoxide and two dioxides of 1,4-dithian, and the various oxides of 1,3,5-trithian. The calculations generally confirmed known experimental results, but cast doubt upon the originally suggested origin (see above) for the preference for S=O to be axial in thian 1-oxide.

Eliel and co-workers [100] have pointed out some analogies and apparent anomalies in the differences in chemical shifts of protons and carbons that are axially and equatorially orientated, in cyclohexane and in heterocycles that contain six-membered rings, including 1,3-dithian. For example, in cyclohexane the axial H resonates at higher field than the equatorial; for 1,3-dioxan the 5-H_a is downfield from its equatorial partner, but for 1,3-dithian the opposite situation exists. The situation is exactly reversed for the 2-positions in these heterocycles, with H_a being at higher field than H_e in 1,3-dioxan and at lower field in 1,3-dithian, respectively. These variations cannot be explained on the basis of the anisotropy of X—C-4 and X—C-6 since these bonds have the same disposition *vis à vis* the corresponding protons at C-2 and C-5. Although no firm conclusions were drawn, the possibility that these changes are due to a paramagnetic shift, charge-alternation effects, or a stereochemical relationship with the lone pairs on the heteroatoms has been suggested.

The acid-catalysed equilibration of various alkyl-substituted 1,3-dithians has been carried out, and the conformational energies of the groups have been re-estimated.[101] The general conclusions were that alkyl substituents in the 2-position have slightly larger conformational energies in 1,3-dithian than in cyclohexane, while those of substituents in the 5-position are somewhat smaller.

[95] R. B. Greenwald, D. H. Evans, and J. R. DeMember, *Tetrahedron Letters*, 1975, 3885.
[96] K. Bergensen, B. M. Carden, and M. J. Cook, *Tetrahedron Letters*, 1975, 4479.
[97] K. Bergensen, B. M. Carden, and M. J. Cook, *J.C.S. Perkin II*, 1976, 345.
[98] M. Anteunis, L. VanAcker, and D. Daneels, *Bull. Soc. chim. belges*, 1974, **83**, 361.
[99] N. L. Allinger and J. Kao, *Tetrahedron*, 1976, **32**, 529.
[100] E. L. Eliel, V. S. Rao, F. W. Vierhapper, and G. Z. Juaristi, *Tetrahedron Letters*, 1975, 4339.
[101] K. Pihlaja, *J.C.S. Perkin II*, 1974, 890.

trans-2-Alkyl-5-*p*-chlorophenyl-1,3-dithians prefer the expected diequatorial conformation, as determined by studies of dipole moments.[102]

Ring expansion of the 1,3-dithiolans to 2,5-dihydro-1,4-dithiins (88) has been accomplished by two routes. Thus the treatment of (87) with 1 equivalent of ethyl *N*-chlorocarbamate gave (88) in 48% yield.[103] The same type of product

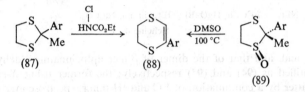

(87) (88) (89)

was obtained in 80—90% yield by heating a number of sulphoxides of the type (89) in DMSO at 100 °C for about 15 h.[104] These compounds, having the SO and Me in the *cis* arrangement that is required for the subsequent elimination, were obtained with high stereoselectivity by oxidation with *m*-chloroperoxy-benzoic acid. Both rearrangements involve ring-opening to an intermediate ω-vinyl-sulphenyl derivative, which recyclizes.

Corey and Hase [105] have found that 1,3-dithiolans and 1,3-dithians can be converted in good yields into dimethyl acetals, cyclic acetals, or thioacetals (or thioketals) upon reaction of the mono-*S*-methylated derivatives with methanol, ethylene glycol, and mercaptoethanol, respectively. These interchanges do not appear to involve the free carbonyl compound. The disulphonium salt prepared by the reaction of triethyloxonium tetrafluoroborate with 2-phenyl-1,3-dithiolan can be used to prepare benzylidene derivatives of 1,2-, 1,3-, and 1,4-diols (30—80%) under basic conditions.[106] This reaction appears to have considerable potential as a route to benzylidenated sugars.

2-Chloro-1,3-dithian, obtained by the reaction of 1,3-dithian and *N*-chloro-succinimide, undergoes dissociation to chloride and dithianium ions in liquid SO_2, as judged by the peak at $\delta = 10.65$ in the n.m.r. spectrum due to the 2-methine hydrogen.[107] Reaction with phenol and *NN*-dimethylaniline produced 2-(*p*-hydroxyphenyl)- and 2-(*p*-dimethylaminophenyl)-1,3-dithian, in 75 and 30% yield, respectively. The dithian derivatives of the *o*-hydroxy-benzaldehydes (90) have been obtained as intermediates in the reaction shown in Scheme 4.[108]

Acid-catalysed cyclization of (91) gave 65% of a mixture of double-bond isomers (92a, b).[109] Electrochemical reduction of ethylene trithiocarbonate resulted in initial fragmentation to ethylene and the $CS_3^{\cdot-}$ radical anion.[110] The structures of the products formed by dissolution of (93) [111] in concentrated

[102] L. Anglioni, I. D. Blackburne, R. A. Y. Jones, and A. R. Katritzky, *Gazzetta*, 1974, **140**, 541.
[103] C. H. Chen, *Tetrahedron Letters*, 1976, 25.
[104] H. Yoshino, Y. Kawazoe, and T. Taguchi, *Synthesis*, 1974, 713.
[105] E. J. Corey and T. Hase, *Tetrahedron Letters*, 1975, 3267.
[106] M. R. Munavu and H. H. Szmant, *Tetrahedron Letters*, 1975, 4543.
[107] K. Arai and M. Oki, *Tetrahedron Letters*, 1975, 2183.
[108] P. G. Gassmann and R. D. Amick, *Tetrahedron Letters*, 1974, 3463.
[109] N. H. Andersen, Y. Yamamoto, and A. D. Dennison, *Tetrahedron Letters*, 1975, 4547.
[110] F. J. Goodman and J. O. Chambers, *J. Org. Chem.*, 1975, **40**, 627.
[111] J. F. Blount, D. L. Coffen, and F. Wong, *J. Org. Chem.*, 1974, **39**, 2374.

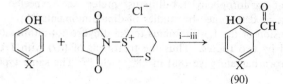

Reagents: i, NEt₃, −70°C; ii, HgO–BF₃,OEt₂; iii, aq. Na₂CO₃.

Scheme 4

sulphuric acid, and that of the dimer of β-mercaptocinnamic aldehyde,[112] have been identified as (94) and (95) respectively; the former using X-ray methods and the latter by a combination of ^{13}C and ^1H n.m.r. spectroscopy. Compound (93) is obtained by the condensation of mercaptoacetic acid and ethanedithiol.

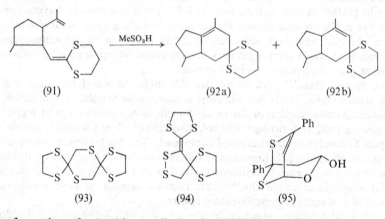

The formation of a transient radical cation intermediate, suggested to have a three-electron bond between the two sulphur atoms, has been detected in the oxidation of 1,4-dithian by hydroxyl radicals.[113]

4 Compounds containing Three or More Sulphur Atoms

1,2,3- and 1,2,4-Trithiolans, 1,2,3- and 1,3,5-Trithians, 1,2,4,5-Tetrathian, and Larger Rings containing Three or More Sulphur Atoms.—4,5-Diphenyl-1,2,3-trithiolan and 3,5-diphenyl-1,2,4-trithiolan have been isolated from *Petiveria alliacea*; benzaldehyde, benzoic acid, and stilbene were also isolated from this species.[114] A series of 5,5-dialkyl-1,2,3-trithians have been prepared by the reaction of appropriately substituted 1,3-dithiols with SCl₂, or from the dimesylate of 1,3-diols with Na₂S₄.[115] In each case the yields were 50—80%, based on a 60—70% conversion, and the trithians were accompanied by various amounts of 1,2-dithiolans or 1,2-dithians. Variable-temperature n.m.r. data showed a barrier to ring inversion of 15.0 and 13.8 kcal mol⁻¹ for the 5,5-dimethyl- and

[112] M. Pulst, M. Weissenfels, and L. Beyer, *Tetrahedron*, 1975, **31**, 3107.
[113] D. Bahnesmann and K. D. Asmus, *J.C.S. Chem. Comm.*, 1975, 238.
[114] E. K. Adesogan, *J.C.S. Chem. Comm.*, 1974, 906.
[115] G. Goor and M. Anteunis, *Synthesis*, 1975, 329.

5,5-diethyl-derivatives.[116] It was concluded that in the 5-methyl-5-alkyl series (alkyl = Et, Pr^n, Pr^i, or Bu^s), the alkyl group prefers the axial position, but with alkyl groups such as isobutyl and neopentyl, and in the 5-methyl-5-phenyl series, the methyl group is mainly axial.[117]

An interesting series of papers on the existence of an anomeric effect in electronegatively substituted 1,3,5-trithians, 1,3-dithians, and thians has appeared. Thus, the copper(I)-catalysed decomposition of benzoyl t-butyl peroxide in the presence of 1,3,5-trithian in benzene gave a mixture of 2-benzoyloxy- and 2,4-dibenzoyloxy-1,3,5-trithian (96) and (97), respectively.[118-121] The dibenzoyloxy-derivative (97), formed as the major product when a 3 : 1 ratio of peroxide to

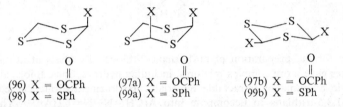

O	O	O
‖	‖	‖
(96) X = OCPh	(97a) X = OCPh	(97b) X = OCPh
(98) X = SPh	(99a) X = SPh	(99b) X = SPh

trithian was employed, was obtained as a single stereoisomer whose n.m.r. spectrum indicated that it existed exclusively in the conformation having both benzoyloxy-groups axial (97a).[118, 119] A similar axial preference was found for the monobenzoylated derivative (96). These assignments were based mainly on the following argument. The reaction of (97) with thiophenol in benzene resulted in the formation of a mixture of *cis*- and *trans*-2,4-di(phenylthio)-1,3,5-trithian. The n.m.r. spectrum of the *trans*-isomer showed an AB pattern at $\delta = 3.70$ and 5.09, $\Delta\nu = 1.3$ p.p.m., indicating that a compound having an axial *S*-phenyl group causes a large $\Delta\nu$ in the remaining methylene group. The spectrum of the *cis*-isomer (99), taken at -83 °C in CDCl$_3$–CS$_2$, showed two AB systems at 3.88 and 4.15 ($\Delta\nu = 0.27$), and 3.66 and 5.22 ($\Delta\nu = 1.6$), in a 3.2 : 1 ratio. The less prominent pair, having the larger chemical-shift difference, was assigned to the diaxial conformation (99a) and the more prominent one, having the smaller chemical-shift difference, to the diequatorial conformation (99b). Since the remaining diastereotopic hydrogens of (97) showed a difference of chemical shift of 1.58 p.p.m., and the spectrum of (97) is temperature-invariant, the above conclusions regarding the preferred conformation, *i.e.* (97a), seem reasonable. In addition, the methine hydrogens of (97a) showed long-range W-coupling.

The axial preference of the 2-SPh group in 1,3,5-trithian (98), taken in CDCl$_3$–CS$_2$ and [^{2}H$_6$]acetone at -80 °C, is 8.2 : 1 and 1.6 : 1, respectively.[120] For the 2-carboxylic acid and 2-methoxycarbonyl group the axial preferences in acetone at -80 °C were 1.5 : 1 and 9 : 1, respectively.[121] In the corresponding 2-substituted 1,3-dithians these groups showed higher axial preferences, namely 5.5 : 1 and about 20 : 1.[121] Solvent changes gave predictable results. In the less

[116] G. Goor and M. Anteunis, *Heterocycles*, 1975, **3**, 363.
[117] M. Anteunis and M. Goor, *Bull. Soc. chim. belges*, 1974, **83**, 463.
[118] T. Sugawara, H. Iwamura, and M. Oki, *Bull. Chem. Soc. Japan*, 1974, **47**, 1496.
[119] T. Sugawara, H. Iwamura, and M. Oki, *Tetrahedron Letters*, 1975, 879.
[120] M. Oki, T. Sugamura, and H. Iwamura, *Bull. Chem. Soc. Japan*, 1974, **47**, 2457.
[121] K. Arai, H. Iwamura, and M. Oki, *Bull. Chem. Soc. Japan*, 1975, **48**, 3319.

polar solvents the electronegative substituents show a greater axial preference, and thus support the various assignments. Decomposition of the t-butylperoxy-ester of 2,4,6-trimethyl-1,3,5-trithian-2-carboxylic acid in toluene gave diastereomeric mixtures of 2-benzyl-2,4,6-trimethyl- and 2,4,6-trimethyl-1,3,5-trithians.[122]

Spiro-trithians, *e.g.* (100), when pyrolysed at 200—300 °C and 10—20 mmHg, afforded mainly a mixture of thiones and enethiols.[123] The method is particularly

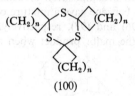

(100)

suitable for the preparation of norbornane-2-thione. Pyrolysis at atmospheric pressures gave more complex mixtures, including hydrocarbons, thiols, and considerable amounts of intractable materials. Hydrazine hydrate caused 2,4,6-triaryl-1,3,5-trithians to decompose into $ArCH=N-N=CHAr$.[124] The mass spectra of $(CH_2S)_n$,[125] where $n = 3$, 4, or 5, and $(ArCHS)_3$ [126] have been recorded.

Dynamic n.m.r. measurements on the 1,2,4,5-tetrathian derivative (101) have revealed the barriers for the twist to chair and chair to twist conformations as

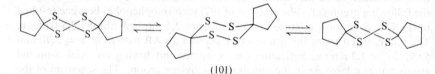

(101)

15.7 and 16.2 kcal mol^{-1}, respectively, at 26.2 °C.[127] These interconversions presumably require rotation about one of the S—S bonds to give a half-chair-type transition state. The angle strain in this transition state, coupled with the known *cis* barrier (up to 9 kcal mol^{-1}) to rotation in acyclic disulphides, could account for these values. The twist to twist barrier is not known, but must be greater than 16 kcal mol^{-1}.

Cycloheptatriene, sulphur, and pyridine, when they react in sulpholane at 70 °C for 72 h, afford 7,8,9-trithiabicyclo[4,3,1]deca-2,4-diene in 21% yield.[128]

Sulphur-containing Macrocyclic Rings.—Two reviews on the synthesis and conformational properties of these molecules have appeared.[129, 130] A host of new sulphur-containing cyclophanes and macrocycles have been prepared during the

[122] K. Arai, H. Iwamura, and M. Oki, *Tetrahedron Letters*, 1975, 1181.
[123] P. S. Fraser, L. V. Robbins, and W. S. Chilton, *J. Org. Chem.*, 1974, **39**, 2509.
[124] J. B. Chattopadhyaya and A. V. R. Rao, *Indian J. Chem.*, 1975, **13**, 632.
[125] M. E. Peach, E. Weissflog, and N. Pelz, *Org. Mass. Spectrometry*, 1975, **10**, 781.
[126] J. B. Chattopadhyaya and A. V. R. Rao, *Org. Mass Spectrometry*, 1974, **9**, 649.
[127] C. Bushweller, G. Bhat, L. J. Letendre, J. A. Brunelle, H. W. Bilofsky, H. Ruben, D. H. Templeton, and A. Zalkin, *J. Amer. Chem. Soc.*, 1975, **97**, 65.
[128] H. Fritz and C. D. Weiss, *Tetrahedron Letters*, 1974, 1859.
[129] J. S. Bradshaw and J. Y. K. Hui, *J. Heterocyclic Chem.*, 1974, **11**, 649.
[130] Z. Yoshida and F. Imashiro, *Kagaku (Kyoto)* 1975, **30**, 167.

past two years, either as intermediates to other cyclophanes[131] *via* various types of sulphur-extrusion reactions, or for studies of conformational effects,[132] electronic effects,[133] studies of chemical interactions between the aromatic rings,[134] for heavy-metal complexing,[135] and for n.m.r. studies.[136] The syntheses were, in the main, variations of the preparation of (102). As stated previously

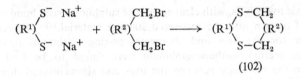

$$(102)$$

(see Vol. 3, p. 177), limitations of space do not permit discussion of these compounds, and the interested reader is referred to the review.[129]

5 Compounds containing Sulphur and Oxygen

Sultines, Sultones, and Related Systems.—The β-sultine (103) has been isolated as a relatively stable crystalline solid.[137] It decomposes quantitatively to 2-methyl-1,1-diphenylpropene, with a half-life of 23 h in CH_2Cl_2 solution at 30 °C. Partial resolution of the cyclic sultines (104) and (105) was accomplished by interrupting

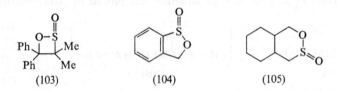

(103)   (104)   (105)

the reaction of these compounds with (S)-2-methyl-1-butylmagnesium chloride or (S)-2-phenyl-1-butylmagnesium chloride when the reaction was about 50—60% complete.[138] The recovered material was enriched to the extent of about 8 and 40%, respectively, in the (S)-isomer of each sultine when the above

[131] (a) V. Boekelheide and R. H. Mitchell, *Jerusalem Symp. Quantum Chem. Biochem.*, 1971, **3**, 150; (b) R. Gray and V. Boekelheide, *Angew. Chem.*, 1975, **87**, 138; (c) J. T. Craig, B. Halton, and S.-F. Siong, *Austral. J. Chem.*, 1975, **28**, 913; (d) K. Galuszko, *Roczniki Chem.*, 1975, **49**, 1597; (e) F. Voegtle and G. Hohner, *Angew. Chem.*, 1975, **87**, 522; (f) F. Voegtle and J. Gruetze, *ibid.*, p. 543; (g) R. H. Mitchell, T. Otsubo, and V. Boekelheide, *Tetrahedron Letters*, 1975, 219; (h) T.-F. Tam, P.-C. Wong, T.-W. Siu, and T.-L. Chan, *J. Org. Chem.*, 1976, **41**, 1289.

[132] (a) F. Voegtle, J. Gruetze, R. Naetscher, W. Wieder, E. Weber, and R. Gruen, *Chem. Ber.*, 1975, **108**, 1694; (b) W. D. Ollis, J. F. Stoddart, and M. J. Nogradi, *Angew. Chem.*, 1975, **87**, 168; (c) W. D. Ollis and J. F. Stoddart, *ibid.*, 1974, **84**, 812, 813; (d) K. Sakamoto and M. Oki, *Chem. Letters*, 1974, 1173; (e) F. Imashiro, M. Oda, T. Iida, Z. Yoshida, and I. Tabushi, *Tetrahedron Letters*, 1976, 371.

[133] R. Danieli, A. Ricci, and J. H. Ridd, *J.C.S. Perkin II*, 1976, 290.

[134] F. Voegtle, W. Wieder, and H. Foerster, *Tetrahedron Letters*, 1974, 4361.

[135] F. Voegtle, E. Weber, W. Wehner, R. Naetscher, and J. Gruetze, *Chem.-Ztg.*, 1974, **98**, 562.

[136] (a) K. Sakamoto and M. Oki, *Chem. Letters*, 1975, 615; (b) A. Ricci, R. Danieli, R. A. Phillips, and J. H. Ridd, *J. Heterocyclic Chem.*, 1974, **11**, 551; (c) R. E. DeSimone, M. J. Albright, W. J. Kennedy, and L. A. Ochrymowycz, *Org. Magn. Resonance*, 1974, **6**, 583.

[137] T. Durst and B. Gimbarszevsky, *J.C.S. Chem. Comm.*, 1975, 724.

[138] J. Nokami, N. Kunieda, and K. Masayoshi, *Tetrahedron Letters*, 1975, 2179.

Grignard reagents were used.[139] In the presence of resolved 1-phenyl-2,2,2-trifluoroethanol, non-identical n.m.r. spectra were obtained due to diastereomeric complexes formed with each enantiomer of (104) and (105). An n.m.r. study of *cis*-4-chloro-4-methyl-1,2-oxathian 2-oxide showed that it existed in the conformation having both the chlorine and oxygen atoms axially oriented.[140] Sulphinate esters, including 1,2-oxathiolan 2-oxide and 1,2-oxathian 2-oxide, undergo basic hydrolysis, with cleavage of the sulphur–oxygen bond and without exchange of oxygen into the recovered starting material.[141] The relative rates of exchange for the five- and six-membered-ring sultines compared to the acyclic ester methyl methanesulphinate were found to be 8 : 1 : 3.3. The difference in the reactivity between the five- and six-membered-ring sultines is much smaller than that found in sultones or sulphites. An unusual sultine, 3,3-dibromo-5-phenyl-1,2-oxathiolan 2-oxide, was formed in unspecified yield upon heating hexabromodimethyl sulphone in the presence of styrene.[142]

The reduction of a number of sultones in the bornyl and camphene series with lithium aluminium hydride can lead to a variety of products, depending on the reaction conditions. Thus 10-isobornyl sultone, when reduced in ether at room temperature, gives the corresponding sultine in 70% yield, whereas reduction in refluxing THF afforded 1-mercaptomethyl-2-*exo*-hydroxy-7,7-dimethyl-bicyclo-[2,2,1]heptane.[143] The propargylic sulphite (106), when heated to 180 °C, gave sultone (107) in 22% yield;[144] its structure was proved by *X*-ray studies. The

$$(Bu^tC{\equiv}CCH_2O)_2SO \xrightarrow{180\,°C} \left[CH_2{=}C{=}\overset{\overset{\displaystyle Bu^t}{|}}{C}SO_2OCH_2C{\equiv}CBu^t \right] \longrightarrow$$

(106)　　　　　　　　　　　　　　　　(108)　　　　　　　　(107)

formation of (107) was suggested to involve an initial sigmatropic rearrangement to the sulphonate ester (108), followed by an intramolecular stepwise [2 + 2]-cycloaddition. 1,2-Oxathiolan 2,2-dioxides were obtained in excellent yield when liquid SO_3 reacted with but-1-enes at −78 °C.[145] 2,6-Dichlorostyrene and SO_3 in dichloroethane gave a 1 : 2 styrene–SO_3 adduct, 6-(2,6-dichlorophenyl)-1,3-dioxa-2,4-dithian 2,2,4,4-tetroxide, in 96% yield.[146]

The first cyclic sulphenate (109) has been reported.[147] This compound was prepared in 85% yield by the reaction of the mercapto-alcohol (110) with Br_2 in CCl_4 containing pyridine at −5 °C. The reaction of (109) with water or moist air gave the corresponding sulphinate, while treatment with 2 equivalents of the potassium salt of hexafluoro-2-phenylpropan-2-ol afforded the sulphurane (111).

[139] W. H. Pirkle and M. S. Hoekstra, *J. Amer. Chem. Soc.*, 1976, **98**, 1832.
[140] K. S. Dhami, *Indian J. Chem.*, 1974, **12**, 278.
[141] A. A. Najam and J. G. Tillet, *J.C.S. Perkin II*, 1975, 858.
[142] C. Y. Kelly and M. Carmack, *Tetrahedron Letters*, 1975, 3605.
[143] J. Wolinsky and R. L. Marhenke, *J. Org. Chem.*, 1975, **40**, 1766.
[144] T. Beetz, R. M. Kellogg, C. T. Kiers, and A. Piepenbrock, *J. Org. Chem.*, 1975, **40**, 3308.
[145] M. D. Robbins and C. D. Broaddus, *J. Org. Chem.*, 1974, **39**, 2459.
[146] J. C. Sheehan and U. Zoller, *J. Org. Chem.*, 1975, **40**, 1179.
[147] G. W. Astrologes and J. C. Martin, *J. Amer. Chem. Soc.*, 1975, **97**, 6909.

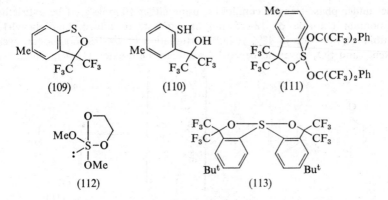

(109)　　　　　(110)　　　　　(111)

(112)　　　　　　　(113)

Variable-temperature n.m.r. studies suggested that (111) is best assigned the structure shown, in which the five-membered ring occupies the diequatorial position. In contrast, the five-membered ring of the sulphurane (112), prepared at $-60\ ^{\circ}\mathrm{C}$ by the interaction of $(\mathrm{MeO})_2\mathrm{S}$ with 3,3-dimethyl-1,2-dioxetan, was assigned an axial-equatorial conformation on the basis of the non-equivalence of the two methoxy-groups in the n.m.r. spectrum.[148] Full details of the study of sulphuranes related to (113) have appeared.[149]

1,3-Oxathiolans, 1,3-Oxathians, 1,4-Oxathians, and Related Compounds.— 1,3-Oxathiolan was obtained by cracking the polymer obtained from formaldehyde and mercaptoethanol.[150] The corresponding 3-oxide and 3,3-dioxide undergo fragmentation in the presence of strong base to give formaldehyde and the potassium salts of ethenesulphenic acid and ethenesulphinic acid, respectively. The interaction of mercaptoethanol with $[\mathrm{Ni(CO)_3(py)}]$ gave 1,3-oxathiolan-2-one.[151] Methyl 3-carboxymethylene-1,4-oxathian-2-one was formed by the reaction of mercaptoethanol with dimethyl acetylenedicarboxylate and subsequent reductive cyclization with sodium hydride.[152] The reaction of mercaptoacetic acid with epoxides such as 2-methyl-2-acetyloxiran afforded 1,4-oxathian-2-ones.[153] 2,2-Dimethyl-1,3-oxathiolan reacted with Cl_2 to give 2-methyl-5,6-dihydro-1,4-oxathiin,[154] but similar treatment of 2,2-diphenyl-, 2,2-di-isopropyl-, and 2,2-hexamethylene-1,3-dithiolans regenerated the precursor ketones as the major reaction product. The mechanism of the formation of 1,3-oxathiolans from ketones and mercaptoethanol has been reinvestigated.[155] The initially formed hemithioketal can either cyclize to the oxathiolan, or add a second mole of mercaptoethanol, to give a dithioketal. Dimethyl sulphone and benzaldehyde

[148] B. S. Campbell, D. B. Denney, D. Z. Denney, and L. Shih, *J. Amer. Chem. Soc.*, 1975, **97**, 3850.
[149] J. C. Martin and E. F. Perozzi, *J. Amer. Chem. Soc.*, 1974, **96**, 3155.
[150] K. Schank, R. Wilmes, and G. Ferdinand, *Internat. J. Sulfur Chem.*, 1973, **8**, 397.
[151] P. Koch and E. Perotti, *J. Organometallic Chem.*, 1974, **81**, 111.
[152] K. Fickenstecher, *Arch. Pharm.*, 1975, **308**, 286.
[153] I. G. Tishchenko, P. M. Malashko, L. I. Lukashik, and A. K. Tuchkovskii, *Vestnik Akad. Navuk. Belorussk. S.S.R.*, Ser. khim. Navuk., 1975, 116 (*Chem. Abs.*, 1975, **83**, 178 659).
[154] G. T. Wilson and M.-G. Huang, *J. Org. Chem.*, 1976, **41**, 966.
[155] L. Fournier, G. Lamaty, A. Natat, and J. P. Roque, *Tetrahedron*, 1975, **31**, 1025.

react under phase-transfer conditions, using either 10 mole% of benzyltriethyl-ammonium chloride or 18-crown-6 as catalyst, to afford 2,6-diphenyl-1,4-oxathian 4,4-dioxide in 48% yield.[156] Irradiation of 1,4-benzoquinones, cyclo-hexene, and SO_2 gave the benzo-1,4-oxathian 4,4-dioxide (114) in about 60%

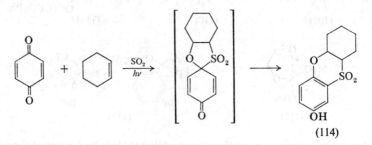

(114)

yield.[157] This product was presumably formed by addition of SO_2 to the 1,4-di-radical formed from the quinone and the alkene, followed by ring-enlargement and aromatization. t-Butylethylene and vinyl acetate were also used, in place of cyclohexene.

2-Acetoxy-1,4-oxathian, when heated with 6-chloropurine and a trace of TsOH, yielded the two derivatives (115) and (116).[158] Interestingly, (115) existed

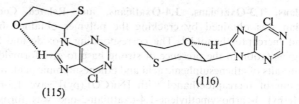

(115)   (116)

mainly in the conformation having the purine group in an axial position, while in (116) this group preferred the equatorial position.[159, 160] Hydrogen-bonding between the ring oxygen and the remaining hydrogen in the five-membered ring was suggested as a possible explanation for this interesting behaviour. The conformational preference of the purine group in a number of other heterocycles has been determined in $CDCl_3$ and CD_3CN.[160] In the former solvent, the per-centages of axial isomer when the substituent is in the 2-position are: tetra-hydropyran (35%), thian (6%), 1,4-dioxan (44%), 1,4-dithian (62%), compound (115) (98%), and compound (116) (24%). In CD_3CN the percentage of axially oriented purine was found to be generally somewhat smaller than in $CDCl_3$.

The difference in energy between the chair and twist conformations for 1,3-oxathian has been found to be 5.5 kcal mol^{-1} at 25 °C, using the equilibrations of 2-t-butyl-2,6-dimethyl-1,3-oxathian as a probe.[161] Activation parameters for

[156] G. W. Gokel, H. M. Gerdes, and N. W. Rebert, *Tetrahedron Letters*, 1976, 653.
[157] M. W. Wilson and S. W. Wunderly, *J. Amer. Chem. Soc.*, 1974, **96**, 7350.
[158] W. A. Szarek, D. M. Vyas, and B. Achmatowicz, *J. Heterocyclic Chem.*, 1975, **12**, 123.
[159] W. A. Szarek, D. M. Vyas, A. M. Sepulchre, S. D. Gero, and G. Lukacs, *Canad. J. Chem.*, 1974, **52**, 2041.
[160] W. A. Szarek, D. M. Vyas, and B. Achmatowicz, *Tetrahedron Letters*, 1975, 1553.
[161] K. Pihlaja and P. Pasanen, *J. Org. Chem.*, 1974, **39**, 1948.

the chair to twist interconversion in 1,4-oxathian, based on n.m.r. analysis, are $\Delta H^{\neq} = 8.78 \pm 0.7$ kcal mol^{-1}, $\Delta S^{\neq} = 0.45 \pm 0.35$ e.u., and $\Delta G^{\neq} = 8.69 \pm 0.3$ kcal mol^{-1} at -96 °C.[162] The small entropy factor suggests that one of the twist conformations is more important than the other possible ones. Another study [163] quotes values of $\Delta G^{\neq} = 7.5$ and 6.7 kcal mol^{-1} for the chair–chair interconversion of 1,4-oxathian and 1,4-oxaselenan. The value for 1,4-oxatelluran was smaller than the values given above, but could not be determined.

The axial : equatorial ratio for oxygen in 1,3-oxathian 3-oxide in [²H₆]acetone is greater than 8 : 1 at -95 °C; however, in CDCl₃ at 38 °C, no preference exists.[164] In 1,4-oxathian 4-oxide the conformation with oxygen axial is preferred by 0.68 kcal mol^{-1} at -80 °C.[165] Equilibrations and 220 MHz n.m.r. studies of various 2-alkyl-4-methyl- and 2-alkyl-4,4-dimethyl-1,3-oxathiolans confirmed earlier conclusions that the 1,3-oxathiolan ring is less flexible than the 1,3-dioxan ring, and somewhat more puckered.[166] The data indicated a preferred envelope structure, in which either the oxygen or C-5 was the 'flap' atom.

Cyclic Sulphites and Related Compounds.—Sulphites of 1,3-diols, in which the diastereomer having the SO group in the less favoured equatorial position is formed to a relatively greater extent, are obtained if the reaction of the diols with SOCl₂ is carried out in the presence of a large excess of pyridine or more than 2 equivalents of triethylamine.[167] When the pyridine/SOCl₂ ratio was less than 2, considerable isomerization to the more stable axial S=O isomer occurred. Similar results have been obtained using steroidal 1,3-diols as substrates.[168] The structures of these sulphites continue to be investigated, both experimentally (by n.m.r. techniques [169]) and by empirical calculations.[170]

Ethylene sulphites have been obtained in good yield, albeit in only about 5% conversion, by irradiation of pentenes in the presence of SO₂ and O₂ at -20 °C for 24 h.[171] The proposed photochemical process involved the triplet state of SO₂ and a charge-transfer complex between SO₂ and the alkene. The electron-diffraction spectra of *trans,trans-* and *trans,cis-*1,2-dimethylethylene sulphite have been obtained.[172] The derived structures agree with those based on n.m.r. data.

162 J. R. Jensen and R. A. Neese, *J. Amer. Chem. Soc.*, 1975, **97**, 4922.
163 J. C. Barnes, G. Hunter, and M. W. Lown, *J.C.S. Perkin II*, 1975, 1354.
164 K. Bergesen, M. J. Cook, and A. P. Tonge, *Org. Magn. Resonance*, 1974, **6**, 127.
165 D. M. Frieze and S. A. Evans, *J. Org. Chem.*, 1975, **40**, 2690.
166 R. Keskinen, A. Nikkilä, K. Pihlaja, and F. G. Riddell, *J.C.S. Perkin II*, 1974, 466.
167 L. Cazaux, G. Chassaing, and P. Maroni, *Tetrahedron Letters*, 1975, 2517.
168 M. D. Beile, C. E. Malmberg, and A. T. Rowland, *Steroids*, 1975, **26**, 29.
169 P. Maroni, L. Cazaux, G. Chassaing, I. Prejzner, and L. T. Tran, *Bull. Soc. chim. France*, 1975, 1258.
170 P. Maroni, L. Cazaux, J. P. Gorrichon, P. Tisnes, and J. G. Wolf, *Bull. Soc. chim. France*, 1975, 1253.
171 P. W. Jones and A. H. Adelman, *Tetrahedron*, 1974, **30**, 2053.
172 H. J. Geise and E. Van Leare, *Bull. Soc. chim. belges*, 1975, **84**, 775.

6
Thiophens and their Selenium and Tellurium Analogues

BY S. GRONOWITZ

1 General

This Report covers the period April 1974—March 1976, and owing to space limitations the subject is treated more summarily than in the previous Reports.

Extensive investigations of [13]C n.m.r. spectra of thiophens and selenophens and [77]Se spectra of selenophens have led to an increased understanding of substituent effects in these systems.

Optically active 3,3'-bithienyls have served as excellent models for the study of conformational and configurational effects on the c.d. curves of atropisomeric compounds.

Spinelli and co-workers have been studying increasingly subtle effects in the nucleophilic substitution of thiophens, and have contributed greatly to an understanding of these reactions.

New cyclization reactions leading to thiophens have been discovered. In particular, the reaction between α-mercapto-ketones and vinylphosphonium salts appears very promising. Many useful ring-closures leading to thiophens have been extended to selenophens and tellurophens.

The interest in fused systems, connected with the search for pharmacologically active compounds, is still increasing. An immense amount of work has been carried out on thienopyrimidine derivatives.

2 Monocyclic Thiophens

Synthesis of Thiophens by Ring-closure Reactions.—A new general method for the preparation of alkylated 2,5-dihydrothiophens (3), through the reaction of α-mercapto-ketones (1) or α-mercapto-aldehydes with vinylphosphonium salts

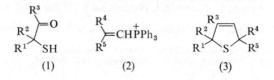

(1) (2) (3)

(2) in anhydrous pyridine, has been reported.[1, 2] The reaction proceeds through a Michael addition of the thiolate ion to the vinylphosphonium salt, followed by a Wittig reaction of the intermediate phosphorous ylide with the elimination of

[1] J. M. McIntosh, H. B. Goodbrand, and G. M. Masse, *J. Org. Chem.*, 1974, **39**, 202.
[2] J. M. McIntosh and R. S. Steevensz, *Canad. J. Chem.*, 1974, **52**, 1934.

triphenylphosphine oxide. Evidence has been found which shows that the *cis*-2,5-dialkyl-2,5-dihydrothiophens are preferentially formed.[3] The 2,5-dihydrothiophens are dehydrogenated to di- or tri-alkylthiophens in high yields by the unorthodox use of chloranil in t-butyl alcohol or pyridine,[4] provided that the 2- and 5-carbons are not fully substituted. This method allows a flexible synthesis of alkylated thiophens from readily available starting materials. When α-mercapto-aldehydes or -ketones were reacted with buta-1,3-dien-1-yltriphenyl-phosphonium salts (4), the expected 2-vinyl-2,5-dihydrothiophen formed by

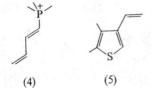

(4)   (5)

β-attack of the thiolate ion on the phosphonium salt or the 2,7-dihydrothiepin formed by δ-attack were not obtained. Instead, 3-vinylthiophens (5) and triphenylphosphine were formed.[5] These results probably depend upon retardation of the Wittig cyclization, and a mechanism for this competing reaction was suggested. The reaction of succinic thioanhydride (6) with the stabilized phosphorus ylide (7) gave diethyl thiophen-2,5-diacetate (8) in 50% yield.[6]

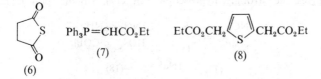

$Ph_3P=CHCO_2Et$ $EtCO_2CH_2$ CH_2CO_2Et

(6)   (7)   (8)

The usefulness of the Gewald reaction for the synthesis of 2-amino-3-carbonyl-substituted thiophens has been amply demonstrated during the past two years (*cf.* ref. 6*a*). These thiophens are very useful starting materials for complex condensed systems of pharmacological interest. I.r. and n.m.r. studies showed some anomalies which were interpreted in terms of electron delocalization in the enaminocarbonyl structure.[6a] From cyanoacetamide, benzylacetone, and sulphur, the thiophen (9) was obtained.[7] 3-Aryl-substituted derivatives such as (10) were obtained from the appropriate aryl methyl ketone, ethyl cyanoacetate, and sulphur.[8] It has been claimed that (11) is obtained from the reaction of acetoacetic acid anilide with malononitrile and sulphur.[9] This hardly seems to be correct, as (12) is the product expected in the Gewald reaction.[10] It was found that 2-amino-3-ethoxycarbonyl-thiophens (13) react with sodium ethoxide in ethanol, pre-

[3] J. M. McIntosh and G. M. Masse, *J. Org. Chem.*, 1975, **40**, 1294.
[4] J. M. McIntosh and H. Khalil, *Canad. J. Chem.*, 1975, **53**, 209.
[5] J. M. McIntosh and F. P. Seguin, *Canad. J. Chem.*, 1975, **53**, 3526.
[6] W. Flitsch, J. Schwiezer, and U. Strunk, *Annalen*, 1975, 1967.
[6a] M. Robba, J. M. Lecomte, and M. Gugnon de Sevricourt, *Bull. Soc. chim. France*, 1974, 2864.
[7] A. Cruceyra, V. Gomez Parra, and R. Madroñero, *Anales de Quím.*, 1975, **71**, 103.
[8] R. F. Koebel, L. L. Needham, and C. DeWitt Blanton, jun., *J. Medicin. Chem.*, 1975, **18**, 192.
[9] L. G. Sharanina and S. N. Baranov, *Khim. geterotsikl. Soedinenii*, 1974, 196.
[10] K. Gewald, *Chem. Ber.*, 1965, **98**, 3571.

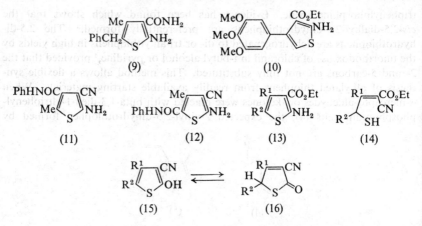

(9)   (10)

(11)   (12)   (13)   (14)

(15)   (16)

sumably to give the intermediate (14), which ring-closes to give 2-hydroxy-3-cyano-thiophens (15). In some cases these compounds exist in the tautomeric form (16).[11] This is a very useful method for the preparation of such compounds, as attempts to prepare them from ethyl 2-cyano-3-methylcinnamate and sulphur or from mercapto-acetone and ethyl cyanoacetate were unsuccessful.

The reaction of malononitrile with carbonyl sulphide and sodium ethoxide in alcohol gives the strongly hygroscopic salt (17), which upon reaction with

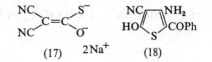

(17) $2Na^+$ (18)

phenacyl bromide yields the tetra-substituted thiophen (18).[12] The reaction of α-halogeno-ketones (19) with the condensation product between nitromethane and aryl isothiocyanates (20) yields 2-anilino-3-nitro-thiophens (21).[13] A somewhat

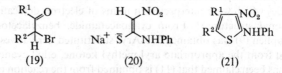

(19)   (20)   (21)

similar route to 3-nitro-thiophens uses the nitro-keten aminal (22), which upon condensation with phenyl isothiocyanates yields (23), which gives (24) with phenacyl bromide.[14] The easily available condensation products from malonic acid derivatives, such as those from malononitrile and ethyl thio- or dithio-acetate (25), can be utilized for the synthesis of 3-amino-2-dimethylsulphonium-substituted thiophens (27).[15] The salt of (25) can be alkylated with chloromethyl

[11] K. Gewald, H. Jablokoff, and M. Hentschel, *J. prakt. Chem.*, 1975, **317**, 861.
[12] H. Schäfer and K. Gewald, *J. prakt. Chem.*, 1975, **317**, 337.
[13] H. Schäfer and K. Gewald, *Z. Chem.*, 1975, **15**, 100.
[14] S. Rajappa, B. G. Advani, and R. Sreenivasan, *Synthesis*, 1974, 656.
[15] G. Gölz and K. Hartke, *Arch. Pharm.*, 1974, **307**, 663.

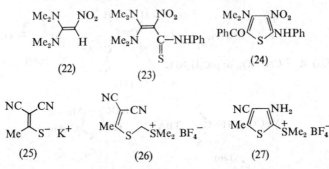

(22) (23) (24)

(25) (26) (27)

methyl sulphide and the thioether alkylated with trimethyloxonium tetrafluoroborate to give (26). The condensation of (26) to give (27), which proceeds *via*
a sulphur ylide, could only be achieved by cyanide ion catalysis. The general use
of this interesting thiophen synthesis is somewhat limited, as the nitrile group
and the sulphonium sulphur have to be *cis*-oriented in intermediate (26).

An alternative to the Gewald reaction for the synthesis of 2-amino-3-thio-
phencarboxylic acids consists in the reaction of the enamine (28) with an aryl
isothiocyanate to give (29), followed by reaction with phenacyl bromide to yield
the thiophen (30).[16] The last step is an example of the 2-acylthiophen synthesis of

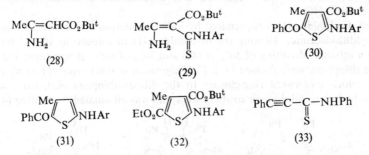

(28) (29) (30)

(31) (32) (33)

Smutny. The t-butyl ester was used in order to achieve hydrolysis of (30), which
required reaction with trifluoroacetic acid at 0 °C for 4—5 h. Treatment of (30)
with warm trifluoroacetic acid for 10 min yielded (31). The reaction of (29) with
ethyl α-chloroacetate gave (32).[16] *N*-Phenyl phenylpropynthioamide (33) reacts
in a similar manner with active bromomethylene derivatives, such as bromo-
acetonitrile, bromonitromethane, and *p*-nitrobenzyl bromide, in the presence of
triethylamine to give the thiophen (33a).[17] Instead of enamine thioesters such as
(29), 2-aminovinyl thioketones (34) react with α-bromo-ketones in the presence
of triethylamine[18] to give 2-acyl-thiophens (35). By the ion-pair extraction
method, tetrabutylammonium salts of acetylacetone were allowed to react with
carbon disulphide to give (36), which upon reaction with chloroacetone gave a
45% yield of the thiol (37), in addition to a thieno[2,3-*b*]thiophen derivative.[19]

[16] S. Rajappa, B. G. Advani, and R. Sreenivasan, *Indian J. Chem.*, 1974, **12**, 4.
[17] W. Ried and L. Kaiser, *Synthesis*, 1976, 120.
[18] J. C. Meslin, Y. T. N'Guessan, H. Quiniou, and F. Tonnard, *Tetrahedron*, 1975, **31**, 2679.
[19] L. Dalgaard, L. Jensen, and S.-O. Lawesson, *Tetrahedron*, 1974, **30**, 93.

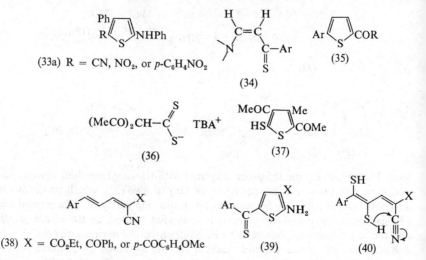

(33a) R = CN, NO$_2$, or *p*-C$_6$H$_4$NO$_2$

(34)

(35)

(36)

(37)

The reaction between sulphur and ethyl 2-cyano-5-arylpenta-2,4-dienoate or the corresponding ketones (38) gave 20—60% yields of 2-amino-5-thioaryl-thiophens (39).[20] It was suggested that the reaction proceeds through the intermediate (40).

The disulphide (41) was prepared by iodine oxidation of the anion of methyl phenyldithioacetate. Heating the disulphide (41) in toluene to 100 °C for 3 h gave a mixture consisting of 30% of (44) and 70% of (45). It was suggested that these thiophens were formed by a [3,3]-sigmatropic rearrangement resulting in (42), which underwent ring-closure to the dihydrothiophen (43), from which either hydrogen sulphide or methyl mercaptan was eliminated.[21] Heating (41) in

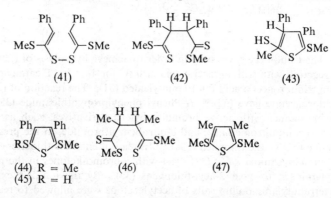

(41)

(42)

(43)

(44) R = Me
(45) R = H

(46)

(47)

the presence of one equivalent of potassium t-butoxide gave only (45). With methyl dithiopropionate, the expected disulphide could not be isolated, but instead the bis(thio) ester (46) was obtained, which upon treatment with potassium t-butoxide and subsequent alkylation with methyl iodide gave the thiophen (47)

[20] N. K. Son, R. Pinel, and Y. Mollier, *Bull. Soc. chim. France*, 1974, 471.
[21] F. C. V. Larsson, L. Brandsma, and S.-O. Lawesson, *Rec. Trav. chim.*, 1974, **93**, 258.

in 74% yield.[21] Reaction of ketones (48) with POCl₃ and DMF (Vilsmeier formylation) gave (49), which upon successive reaction with sodium sulphide and halogen compounds CH₂XR (where R is an electron-withdrawing group such as NO₂, CO₂Et, or CN) gave thiophens (50) in the presence of a basic catalyst.[22] With XCH(R³)CO₂Et (R³ = alkyl or aryl) the intermediate (51) was isolated, and in the course of an attempted distillation it cyclized to the thiophen (52).[22] These reactions are further developments of the Fiesselmann reaction.

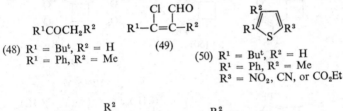

(48) R¹ = Bu^t, R² = H
     R¹ = Ph, R² = Me

(49)

(50) R¹ = Bu^t, R² = H
     R¹ = Ph, R² = Me
     R³ = NO₂, CN, or CO₂Et

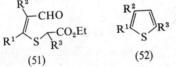

(51)

(52)

Reaction of (53a) with potassium t-butoxide in THF at 20 °C gave a 50% yield of (54). The treatment of (53b) with potassium hydroxide in methanol gave a low yield of (55), while the reaction of (53c) with potassium t-butoxide in t-butanol

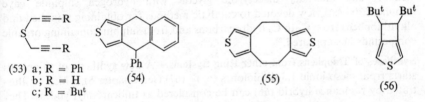

(53) a; R = Ph
     b; R = H
     c; R = Bu^t

(54)

(55)

(56)

gave (56) in 50% yield.[23] These reactions are suggested to proceed by rearrangement of the bis(acetylene) to a bis(allene), which then undergoes an intramolecular allene dimerization to give a heterocyclic bis(methylene) biradical or its equivalent.[23] The formation of (58) in 36% yield upon reaction of (57) with sodium sulphide in acetone most probably proceeds through a similar mechanism.[24] 3-Thiabicyclo[3,2,0]hepta-1,4-diene has been synthesized by a Wittig reaction

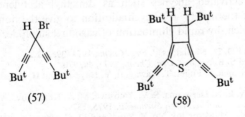

(57)

(58)

[22] P. Cagniant and G. Kirsch, *Compt. rend.*, 1975, **281**, C, 35.
[23] P. J. Garratt and S. B. Neoh, *J. Amer. Chem. Soc.*, 1975, **97**, 3255.
[24] H. Hauptmann, *Tetrahedron Letters*, 1974, 3589.

between cyclobuta-1,2-diene and a bis(ylide) derived from dimethyl sulphide $\alpha\alpha'$-bis(triphenylphosphonium) dichloride, and its reactions have been studied.[24a] Another route to the strained cyclobutane derivatives exploits the classical reaction of 1,4-diketones with phosphorus pentasulphide; (59) gives (60), and the tricyclic derivative (62) is obtained from (61), but in very low yield.[25] Another

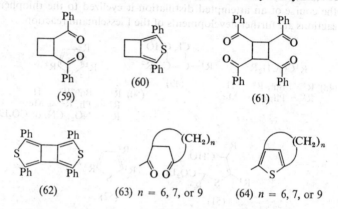

(59)   (60)   (61)

(62)   (63) n = 6, 7, or 9   (64) n = 6, 7, or 9

example of this reaction is the synthesis of the heterophane (64) from (63).[26] The conformations of the aliphatic chain were studied by n.m.r. Upon heating to 560 °C, divinyl sulphide forms thiophen.[27] Reaction of tertiary diacetylene alcohols and ditertiary diacetylene glycols with hydrogen sulphide gave thiophens.[27a] A review devoted to catalytic methods for obtaining thiophen and alkylthiophens from C_2 to C_6 hydrocarbons and from sulphur-containing organic compounds has appeared.[28]

Syntheses of Thiophens from other Ring Systems.—A new synthesis of thiophens starts from meso-ionic 1,3-dithiolones (65)—(67) (see Chapter 8), which as illustrated by resonance hybrid (66) can be considered as thiocarbonyl ylides. They

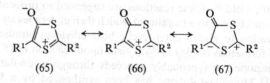

(65)   (66)   (67)

combine with activated alkynes such as dimethyl acetylenedicarboxylate at 90—130 °C in a thermal [3 + 2] cycloaddition to give the non-isolable primary adducts (68) which, by rapid elimination of carbonyl sulphide, give the thiophen

[24a] P. J. Garratt and D. N. Nicolaides, *J. Org. Chem.*, 1974, **39**, 2222.
[25] P. J. Garratt and S. B. Neoh, *J. Org. Chem.*, 1975, **40**, 970.
[26] S. Hirano, T. Hiyama, S. Fujita, T. Kawaguti, Y. Hayashi, and H. Nozaki, *Tetrahedron*, 1974, **30**, 2633.
[27] M. G. Voronkov, E. N. Deryagina, S. V. Amosova, M. A. Kuznetsova, V. V. Kryuchkov, and B. A. Trofimov, *Khim. geterotsikl. Soedinenii*, 1975, 1579.
[27a] A. N. Volkov, Yu. M. Skvortsov, Yu. V. Kind, and M. G. Voronkov, *Zhur. org. Khim.*, 1974, **10**, 174.
[28] M. A. Ryashentseva, Yu. A. Afanas'eva, and Kh. M. Minachev, *Khim. geterotsikl. Soedinenii*, 1971, 1299.

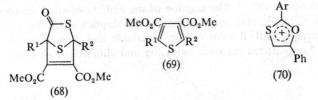

(69) in 67—100% yield.[29] The groups R¹ and R² are combinations of aryl groups. Hirai and Ishiba have increased the usefulness of 2-aryl-1,3-oxathiolium salts (70) for the synthesis of thiophens described by Hartmann *et al.*,[30a] by preparing them through the reaction of potassium thiolbenzoate with phenacyl bromide and cyclization of the resulting thiolester in concentrated sulphuric acid. These salts react with sodium salts of active methylene compounds (71) to give (72),

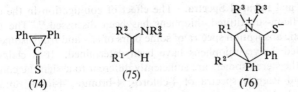

which upon treatment with sodium ethoxide in ethanol gives the thiophens (73) in high yield.[30] Diphenylcyclopropenethione (74) reacts with enamines (75) at 0—20 °C to give 1:1 adducts, most probably with structure (76).

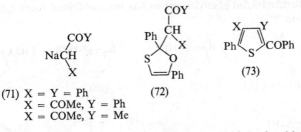

An adduct of this type (77) reacted with dimethyl acetylenedicarboxylate at room temperature by a 1,3-dipolar addition to yield the unstable intermediate (78), which in a very interesting fragmentation gave diphenylcyclopropene (79)

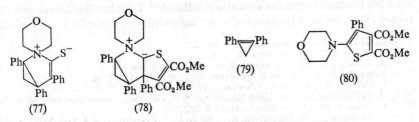

²⁹ H. Gotthardt, M. C. Weisshuhn, and B. Christl, *Chem. Ber.*, 1976, **109**, 753.
³⁰ K. Hirai and T. Ishiba, *Heterocycles*, 1975, **3**, 217.
³⁰ᵃ H. Hartmann, H. Schafer, and K. Gewald, *J. prakt. Chem.*, 1973, **315**, 497.

and the thiophen (80).[31] The reaction of the dianion of ethyl acetoacetate with propylene sulphide (81) gave the tetrahydrothiophen derivative (82),[32] which, upon treatment with lithium di-isopropylamide and phenylselenenyl bromide, followed by hydrogen peroxide oxidation and elimination of benzeneselenenic

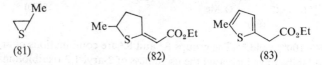

(81) (82) (83)

acid, was transformed into (83).[33] The reaction of (84) with aniline gave the dihydro-derivative (85), which was aromatized by chloranil to give (86).[34] Oxidation of (87) with 40% hydrogen peroxide resulted in the hydroxy-thiophen (88) in 90% yield.[35] The action of copper-bronze on (89) gave the thiophen (90).[36] 2-Benzoyl-3,5-diphenylthiophen has been obtained from 2,4,6-triphenyl-pyrylium ion.[36a]

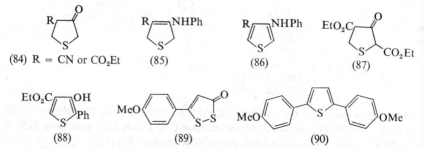

(84) R = CN or CO_2Et (85) (86) (87)

(88) (89) (90)

Ultraviolet and Infrared Spectra.—The effect of conjugation in the excited state of some silicon-substituted thiophens has been discussed.[37] The u.v. and i.r. characteristic absorption spectra of some pairs of *cis*- and *trans*-isomers of 2'- and 3'-substituted 2-styrylthiophens have been determined. It is claimed that the spectra of the stereoisomers are sufficiently different to assign the configuration.[38] The i.r. and Raman spectra of 4-chloro-, 4-bromo-, 4-iodo-, and 4-cyano-2-fluoro-thiophens have been studied in the region 500—540 cm^{-1}. The fundamental frequencies have been assigned, and normal co-ordinate calculations have been performed. The force constants were adjusted to reproduce accurately the fundamental frequencies. Thermodynamic functions and mean amplitudes of vibration for these molecules were also calculated.[39] Thiophen-2-aldehyde and 19 related aldehydes have been studied under high resolution in the i.r. C=O region.

[31] T. Eicher and S. Böhm, *Chem. Ber.*, 1974, **107**, 2238.
[32] T. A. Bryson, *J. Org. Chem.*, 1973, **38**, 3428.
[33] C. A. Wilson, jun. and T. A. Bryson, *J. Org. Chem.*, 1975, **40**, 880.
[34] O. Hromatka, D. Binder, and K. Eichinger, *Monatsh.*, 1974, **105**, 1164.
[35] A. P. Stoll and R. Süess, *Helv. Chim. Acta*, 1974, **57**, 2487.
[36] M. A.-F. Elkaschef, F. M. E. Abdel-Megeid, and A. A. Elbarbary, *Tetrahedron*, 1974, **30**, 4113.
[36a] C. L. Pedersen, *Acta Chem. Scand. (B)*, 1975, **29**, 791.
[37] V. A. Kuznetsov, A. N. Egorochkin, A. I. Burov, E. A. Chernyshev, V. I. Savushkina, V. Z. Anisimova, and O. V. Kuz'min, *Doklady Akad. Nauk S.S.S.R.*, 1974, **216**, 1062.
[38] S. Fisichella, G. Scarlata, and M. Torre, *Gazzetta*, 1974, **104**, 1237.
[39] A. Rogstad, *Spectrochim. Acta*, 1975, **31**, *A*, 1749.

It was established that the multiple absorptions shown by many of the aldehydes are caused by Fermi resonance. While deuteriation of an aldehyde may lead to an increase in the complexity of its spectrum, bromination is often effective in diagnosing the occurrence of Fermi resonance.[40] The i.r. spectra of a large number of organosilicon derivatives of thiophen have been investigated.[41] The shifts of the stretching vibrations of the hydroxy-group of phenol and pentachlorophenol during the formation of intermolecular hydrogen bonds with 2-thienyl phenyl ketones containing substituents in the benzene ring have been measured and correlated with σ^+ constants.[42]

Dipole Moments and Conformational Analyses.—Interest in determining the conformational equilibrium in a variety of thienylcarbonyl derivatives by a variety of physical methods has continued. On the basis of Kerr-effect and dipole-moment data, it was shown that thiophen-2-aldehyde and its 5-chloro- and 5-bromo-derivatives exist exclusively in the O,S-*cis* conformation at 25 °C in CCl$_4$ solution. Under the same conditions, 5-nitrothiophen-2-aldehyde and thiophen-3-aldehyde are mixtures of two conformers.[43] Comparison with the analogous furan and selenophen compounds indicates that the percentage of the OX-*cis* form increases at equilibrium on going from furan to selenophen. It is suggested that this can be explained by increased Coulomb interaction between the carbonyl oxygen and the heteroatom.[44] A ^{13}C n.m.r. study also confirmed that thiophen-2-aldehyde exists in the O,S-*cis* conformation.[45] The occurrence of rotational isomers in 35 thiophen and furan-2-carbonyl fluorides and chlorides has been investigated by ^{19}F n.m.r. and i.r. spectrometry. At room temperature, solutions of the fluorides in trichlorofluoromethane, carbon tetrachloride, and acetonitrile contain appreciable amounts of the *syn* and *anti* isomers.[46] In connection with this work, comparison of methods for resolving overlapping i.r. curves has been carried out.[47] Russian workers have measured i.r. spectra, dipole moments, Kerr constants, and the temperature dependence of dipole moments for some acid chlorides of the thiophen series, and claim that 2-thenoyl chloride exists only in the O,S-*cis* conformation and that the energy barrier towards rotation is higher in acid chlorides than in the corresponding aldehydes.[48] The dipole moment of thiophen-2-carbothioamide has been measured and its conformation discussed.[49] Conclusions about the preferred conformations of 2-styrylthiophens have been made on the basis of dipole moment measurements.[50]

[40] D. J. Chadwick, J. Chambers, G. D. Meakins, and R. L. Snowden, *J.C.S. Perkin II*, 1975, 604.
[41] S. Ya. Khorshev, N. S. Vyazankin, A. N. Egorochkin, E. A. Chernyshev, V. I. Savushkina, O. V. Kuz'min, and V. Z. Anisimova, *Khim. geterotsikl. Soedinenii*, 1974, 477.
[42] N. F. Pedchenko, N. D. Trusevich, and V. F. Lavrushin, *Khim. geterotsikl. Soedinenii*, 1974, 482.
[43] A. S. Kuzharov, V. N. Sheinker, E. G. Derecha, O. A. Osipov, and D. Ya. Movshovich, *Zhur. obshchei Khim.*, 1974, **44**, 2008.
[44] V. N. Sheinker, O. A. Osipov, V. I. Minkin, E. G. Derecha, R. M. Minyaev, V. A. Troilina, A. S. Kuzharov, and N. N. Magdesieva, *Zhur. obshchei Khim.*, 1974, **44**, 1314.
[45] D. J. Chadwick, G. D. Meakins, and E. E. Richards, *Tetrahedron Letters*, 1974, 3183.
[46] D. J. Chadwick, J. Chambers, G. D. Meakins, and R. L. Snowden, *J.C.S. Perkin II*, 1976, 1.
[47] D. J. Chadwick, J. Chambers, and R. L. Snowden, *J.C.S. Perkin II*, 1974, 1181.
[48] V. N. Sheinker, A. S. Kuzharov, D. Ya. Movshovich, Z. N. Nazarova, V. A. Piven', and O. A. Osipov, *Zhur. obshchei Khim.*, 1975, **45**, 884.
[49] G. C. Pappalardo and S. Gruttadauria, *J.C.S. Perkin II*, 1974, 1441.
[50] S. Gruttadauria, G. C. Pappalardo, G. Scarlata, and M. Torre, *J.C.S. Perkin II*, 1974, 1580.

N.M.R. Spectra.—The development of Fourier transform [13]C n.m.r. spectrometry has led to an increased use of this technique in the thiophen field. A large number of 2- and 3-monosubstituted thiophens have been studied in detail. Unequivocal shift assignments were in some cases obtained from deuteriated derivatives. Both direct and long-range coupling constants fall in well-defined intervals. The shifts due to the substituents were compared with those in benzene and also with ^{1}H and ^{19}F shifts of thiophens. Linear correlations with the two-parameter equation of Swain and Lupton were obtained. The *ortho*-shifts of derivatives having substituents with $-I$, $-M$-effects were explained by assuming the existence of an alternating inductive effect, which places negative charge on the *ortho*-carbons.[51] The influence of complex formation between aluminium chloride and carbonyl-substituted thiophens on the electron density has been studied by [13]C n.m.r. spectroscopy.[52]

The [1]H n.m.r. spectra of Grignard reagents in the thiophen series,[53] and of a great number of organosilicon derivatives of thiophen,[54, 55] have been investigated, and in the latter case utilized for the interpretation of the transmission of the electronic effect of organosilicon substituents. The nuclear spin coupling constants of the chalcogen heterocyclics have been calculated by finite perturbation theory in the ZDO approximation.[56] The [1]H n.m.r. spectra of macrocyclic ansa compounds including a thiophen ring have been studied in order to obtain information on the conformation of such rings.[57] The [1]H n.m.r. spectra of some 4-phenylthienopyrimidines, 2-acetyl-3-benzoylthiophen, 5-phenyl-1,4-thienodiazepines, and 5-phenyl-1,4-thienodiazepine 4-oxides have been analysed in detail.[58] Proton–fluorine coupling over six bonds has been observed between the β-hydrogen of the thiophen ring and the fluorine atom in the *o*-fluorophenyl derivatives of the above-mentioned compounds.[59] α-(2-Thienyl)ethylamine was used in an n.m.r. study of lanthanide chelates.[60] Europium and praseodymium thenoyltrifluoroacetonate have been suggested as chemical shift reagents for n.m.r.[61]

Various Physical Properties.—E.s.r. spectra of some 5-substituted 2-thienyl nitroxides, generated by scavenging triethylsilyl radicals with substituted nitro-thiophens, and nitro-anions, have been recorded in order to determine the effective substituent delocalizing power with respect to an unpaired electron in the thiophen series.[62] Triethylsilyloxy nitroxide radicals of thiophen have been produced

[51] S. Gronowitz, I. Johnson, and A.-B. Hörnfeldt, *Chem. Scripta*, 1975, **7**, 76.
[52] L. I. Belen'kii, I. B. Karmanova, Yu. B. Vol'kenshtein, P. V. Petrovskii, L. A. Fedorov, and Ya. L. Gol'dfarb, *Izvest. Akad. Nauk S.S.S.R., Ser. khim.*, 1974, 1725.
[53] G. J. Martin, B. Mechin, Y. Leroux, C. Paulmier, and J. C. Meunier, *J. Organometallic Chem.*, 1974, **67**, 327.
[54] A. N. Egorochkin, N. S. Vyazankin, A. I. Burov, E. A. Chernyshev, V. I. Savushkina, and B. M. Tabenko, *Khim. geterotsikl. Soedinenii*, 1972, 911.
[55] A. N. Egorochkin, N. S. Vyazankin, A. I. Burov, E. A. Chernyshev, V. I. Savushkina, and O. V. Kuz'min, *Khim. geterotsikl. Soedinenii*, 1972, 1483.
[56] V. Galasson, *Chem. Phys. Letters*, 1975, **32**, 108.
[57] F. D. Alashev, A. V. Kessenikh, S. Z. Taits, and Ya. L. Gol'dfarb, *Izvest. Akad. Nauk S.S.S.R., Ser. khim.*, 1974, 2022.
[58] T. Hirohashi, S. Inaba, and H. Yamamoto, *Bull. Chem. Soc. Japan*, 1975, **48**, 147.
[59] T. Hirohashi, S. Inaba, and H. Yamamoto, *Bull. Chem. Soc. Japan*, 1975, **48**, 974, 3373.
[60] K. Ajisaka and M. Kainosho, *J. Amer. Chem. Soc.*, 1975, **97**, 1761.
[61] V. M. Potapov, E. G. Rukhadze, I. G. Il'ina, and V. G. Bakhmut-skaya, *Zhur. obshchei Khim.*, 1974, **44**, 462.
[62] C. M. Camaggi, R. Leardini, and G. Placucci, *J.C.S. Perkin II*, 1974, 1195.

from 2- and 3-nitrothiophen. The e.s.r. spectra showed the presence of rotational isomers at low temperature.[63] Some 2,5-disubstituted thiophens have been used in a study of factors influencing the stabilities of nematic liquid crystals.[64]

Electrophilic Substitution.—Quantitative aspects of the electrophilic substitution of furan, thiophen, pyrrole, and other five-membered heteroaromatic systems are treated in a review by Marino.[65] Gol'dfarb and his school are continuing their extensive work on electrophilic substitution in simple thiophens. The action of benzoyl chloride on 2-acetylthiophen in an excess of aluminium chloride at 110—115 °C without solvent gave a complex mixture of 2-benzoylthiophen, 2,4- and 2,5-diacetylthiophen, 2-acetyl-4(5)-benzoylthiophen, and 2,4- and 2,5-dibenzoylthiophen.[66, 67] The transacylations of 2-acylthiophens were demonstrated by heating with excess AlCl₃ in the absence of an acylating agent, which gave 2,4-diacetylthiophen in 40% yield with no 2,5-isomer as impurity. Transacetylation was also demonstrated by heating 2-acetylthiophen with AlCl₃ and benzene to 110 °C, which gave acetophenone.[66] Acylation of 3-acyl-thiophens, on the other hand, is considerably easier, and not complicated by side-reactions. Comparatively high yields of 2,4-diacyl-thiophens are obtained.[67] MO Calculations (LCAO–SCF) on the protonated and non-protonated thiophen-3-aldehyde appear to confirm these differences in the reaction pattern of 2- and 3-acylthiophens, as they indicate that electrophilic attack should occur at the 5-position, while attack on the carbon atom to which the formyl group is attached is unlikely in this case, in contrast to that of the 2-isomer.[68] MO Calculations of other properties of thiophens having $-I$, $-M$-substituents have previously been reported.[69] It was shown by n.m.r. that in the reaction of 2,5-dimethylthiophen with AlCl₃ and acetyl chloride, or chloroacetyl chloride, in methylene chloride at -40 to $+30$ °C, a stable *C*-protonated σ-complex (91) was formed.[70] This explains some of the

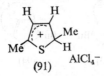

(91)   AlCl₄⁻

anomalous ratios of acylation products previously found. Due to protonation, 2,5-dimethylthiophen is no longer capable of reacting with an acylating agent, which therefore attacks the second aromatic component of the mixture, even if it is the relatively unreactive benzene. This is illustrated by the fact that, on addition of benzene at -30 °C to a mixture of acetyl chloride, 2,5-dimethylthiophen, and an excess of AlCl₃, 3-acetyl-2,5-dimethylthiophen and aceto-

[63] C. M. Camaggi, L. Lunazzi, G. F. Pedulli, G. Placucci, and M. Tiecco, *J.C.S. Perkin II*, 1974, 1226.
[64] M. J. S. Dewar and R. M. Riddle, *J. Amer. Chem. Soc.*, 1975, **97**, 6658.
[65] G. Marino, *Khim. geterotsikl. Soedinenii*, 1973, 579.
[66] A. P. Yakubov, L. I. Belen'kii, and Ya. L. Gol'dfarb, *Zhur. org. Khim.*, 1973, **9**, 2436.
[67] A. P. Yakubov, L. I. Belen'kii, and Ya. L. Gol'dfarb, *Zhur. org. Khim.*, 1973, **9**, 1959.
[68] Ya. L. Gol'dfarb, G. M. Zhidomirov, I. A. Abronin, and L. I. Belen'kii, *Zhur. org. Khim.*, 1974, **10**, 846.
[69] Ya. L. Gol'dfarb, G. M. Zhidomirov, N. D. Chuvylkin, and L. I. Belen'kii, *Khim. geterotsikl. Soedinenii*, 1972, 155.
[70] L. I. Belen'kii, A. P. Yakubov, and Ya. L. Gol'dfarb, *Zhur. org. Khim.*, 1975, **11**, 424.

phenone are formed in the ratio 7 : 3. These results stress the danger of using competing reactions to evaluate relative reactivities of aromatic compounds if large excesses of aromatic compounds are not used and the formation of stable σ-complexes is not taken into account.[70] The observation of stable *C*-protonated species also explains some of the anomalies observed in the alkylation and acylation of alkylthio-thiophens.[70]

The bromination of 2-cyanothiophen with one equivalent of bromine in the presence of an excess of $AlCl_3$ gives 2-cyano-, 4-bromo-2-cyano-, 5-bromo-2-cyano-, and 4,5-dibromo-2-cyano-thiophen in the proportions 16 : 70 : 2 : 12, from which pure 4-bromo-2-cyanothiophen can easily be obtained.[71] Nitration of 2-cyanothiophen in concentrated sulphuric acid led to mixtures of almost equal amounts of 4- and 5-nitrothiophen-2-carboxylic acid.[71] The chloromethylation of 2-acetylthiophen and 2-formylthiophen with αα'-bis(chloromethyl) ether in 60—100% sulphuric acid has been studied. An increase in the acidity of the medium promoted the formation of 4-substituted products.[72] From these products some otherwise difficultly obtainable 2,4-disubstituted thiophens were prepared.[73]

The bromination of 2-acetylthiophen in the 5-position can be effected by *N*-bromosuccinimide in an acetic anhydride–acetic acid mixture (70% yield),[74] or by hypobromous acid generated *in situ* from bromine and silver nitrate in acetic acid–aqueous perchloric acid (31% yield).[75] For the bromination of 3-acetylthiophen, the swamping catalyst method (excess $AlCl_3$) and bromine was used.[76] Treating 3-nitrothiophen with an excess of chlorine in chloroform using catalytic amounts of $AlCl_3$ gave 2,4,5-trichloro-3-nitrothiophen in high yield.[77] When the thiophenophane (92) was treated with an excess of Bu^tCl–$SnCl_4$ in CS_2 at room temperature, a deep-seated rearrangement occurred; t-butyl groups

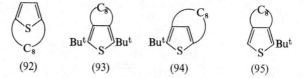

(92)   (93)   (94)   (95)

were introduced in the 2- and 5-position of the thiophen nucleus with a concomitant migration of the octamethylene bridge to the 3- and 4-positions. In addition to (93), (94) and (95) were also formed.[78]

The acetylation of some alkoxy- and acetoxy-thiophens under Friedel–Crafts conditions has been described. Some of the by-products obtained were isolated and their mode of formation was discussed.[79] Friedel–Crafts acylation on thienyl-

[71] Ya. L. Gol'dfarb, G. P. Gromova, and L. I. Belen'kii, *Izvest. Akad. Nauk S.S.S.R., Ser. khim.*, 1974, 2275.
[72] L. I. Belen'kii, É. I. Novikova, and Ya. L. Gol'dfarb, *Khim. geterotsikl. Soedinenii*, 1971, 1353.
[73] I. B. Karmanova, Yu. B. Vol'kenshtein, and L. I. Belen'kii, *Khim. geterotsikl. Soedinenii*, 1973, 490.
[74] V. A. Smirnov and A. E. Lipkin, *Khim. geterotsikl. Soedinenii*, 1973, 185.
[75] T. J. Broxton, L. W. Deady, J. D. McCormack, L. C. Kam, and S. H. Toh, *J.C.S. Perkin I*, 1974, 1769.
[76] B.-P. Roques, M.-C. Fournié-Zaluski, and R. Oberlin, *Bull. Soc. chim. France*, 1975, 2334.
[77] O. Hromatka, D. Binder, and G. Pixner, *Monatsh.*, 1975, **106**, 1103.
[78] R. Helder and H. Wynberg, *Tetrahedron*, 1975, **31**, 2551.
[79] J. F. Bagli and E. Ferdinandi, *Canad. J. Chem.*, 1975, **53**, 2598.

acetic acid derivatives has been studied in connection with work on anti-inflam-matory agents.[80] Mannich reactions have been carried out with 2- and 3-methoxy-thiophen and 3,4-dimethoxythiophen,[81] as well as with 4,5-dialkyl-2-acylamino-thiophens.[82] In connection with work on flavone-type compounds containing a thiophen ring, 3-methoxythiophen was treated with cinnamoyl chlorides under Friedel–Crafts conditions to give (96), which upon addition of bromine and reac-tion with pyridine hydrochloride gave the desired compounds (97).[83] The

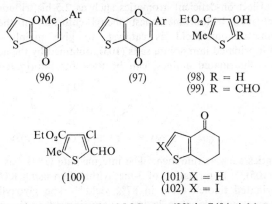

(96)   (97)   (98) R = H
                        (99) R = CHO

(100)   (101) X = H
             (102) X = I

Vilsmeier formylation of (98) at 30—35 °C gave (99) in 74% yield, while at 100 °C (100) was formed in 50% yield.[84] Nitration of (98) gave the 5-nitro-derivative.[85] The ketone (101) was iodinated to (102) with iodine and HgO in benzene.[85a]

2-Hydroxy-3-cyano-thiophens [11] and 2-anilino-3-nitro-thiophens [13] couple with benzenediazonium salts in the 5-position. 2,4-Dinitrobenzenediazonium ions couple normally with 2-t-butylthiophen to give the 5-azo dye, but with 2,3,5-trimethylthiophen and tetramethylthiophen, coupling occurs through methyl groups to give the corresponding 2,4-dinitrophenylhydrazones. 2,5-Dimethyl-thiophen gives both the 3-azo dye and the 2,4-dinitrophenylhydrazone of 5-methyl-thiophen-2-aldehyde.[86] 2,4-Dinitrobenzenediazonium ions react with thiophen, and with both methylthiophens, in acidic media to give the corresponding dinitrophenylthiophens, while anisole under the same conditions gives the corresponding azo-dye.[87] 2-Carboxythiophen-3-diazonium chloride undergoes self-coupling to give substituted azothiophens.[88]

Further evidence for the mechanism suggested for the iodination of thiophens by iodine and nitric acid has been presented. For 2-phenylthiophen the kinetic

[80] F. Clémence, O. Le Martret, R. Fournex, G. Plassard, and M. Dagnaux, *Chim. Thérap.*, 1974, 9, 390.
[81] J. M. Barker, P. R. Huddleston, and M. L. Wood, *Synthetic Comm.*, 1975, 5, 59.
[82] V. I. Shvedov, I. A. Kharizomenova, N. V. Medvedeva, and A. N. Grinev, *Khim. geterotsikl. Soedinenii*, 1975, 918.
[83] G. Henrio and J. Morel, *Tetrahedron Letters*, 1974, 2167.
[84] V. I. Shvedov, V. K. Vasil'eva, and A. N. Grinev, *Khim. geterotsikl. Soedinenii*, 1972, 427.
[85] V. I. Shvedov, V. K. Vasil'eva, O. B. Romanova, and A. N. Grinev, *Khim. geterotsikl. Soedinenii*, 1973, 1024.
[85a] J.-P. Conjat, P. Cagniant, D. Cagniant, and M. Mirjolet, *Tetrahedron Letters*, 1975, 2885.
[86] S. T. Gore, R. K. Mackie, and J. M. Tedder, *J.C.S. Chem. Comm.*, 1974, 272.
[87] M. G. Bartle, R. K. Mackie, and J. M. Tedder, *J.C.S. Chem. Comm.*, 1974, 271.
[88] M. G. Reinecke and R. H. Walter, *J.C.S. Chem. Comm.*, 1974, 1044.

hydrogen isotope effect depends upon substrate concentration, which is consistent with the reaction scheme.[89]

Some more exotic electrophilic reagents have been applied successfully to thiophens. Thus, pentafluorobenzenesulphenyl chloride reacts with thiophen in ether in the presence of catalytic amounts of $SnCl_4$ to give 2-pentafluorophenyl-thio-thiophen.[90] Thiophen and 2-methylthiophen give trifluoromethylthio-derivatives with trifluoromethanesulphenyl chloride in chloroform in the presence of pyridine.[91] Electron-deficient aromatics such as 2,5-bis(trifluoromethylthio)-thiophen react with trifluoromethanesulphenyl chloride, in the presence of trifluoromethanesulphonic acid as catalyst, to give mainly the 3-chloro-derivative.[92] Fluorinated immonium salts (103), obtained by the action of boron trifluoride on α-fluorinated amines, can be used for fluorinated acylation of

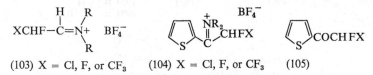

(103) X = Cl, F, or CF_3 (104) X = Cl, F, or CF_3 (105)

reactive aromatics such as thiophen. The intermediate (104) was hydrolysed to the thiophen (105).[93] Fluorination of 3-methylthiophen using $KCoF_4$ at 200 °C gave the fluorinated thiolen (106) in 87% yield.[94] The glyoxylic acid–amide adducts (107) react with thiophen in 10% sulphuric acid–acetic acid to give the

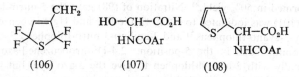

(106)   (107)   (108)

N-acyl-α-amino-acid (108).[95] 2-Bromo-3-methylthiophen has been carboxylated in the 5-position in high yield, using pyrocatechol dichloromethylene acetal.[96] After esterification, the bromine was reduced by zinc dust, providing a route to isomer-free 4-bromothiophen-2-carboxylic acid.[96] In the presence of stannic chloride, thiophen reacts with ethoxycarbonyl isocyanate or isothiocyanate to give *N*-ethoxycarbonylthiophen-2-carboxamide and *N*-ethoxycarbonylthiophen-2-thiocarboxamide, respectively.[96a]

The ionic hydrogenation of thiophens with a mixture of triethylsilane and trifluoroacetic acid, consisting of electrophilic protonation followed by hydride addition, has been used for the synthesis of (109) from (110) and of the tetra-hydrothiophen derivative (111) from the corresponding thiophen.[97] Also, a series of 2-(ω-diethylaminoalkyl)thiophens (112) has been hydrogenated with

[89] A. R. Butler and A. P. Sanderson, *J.C.S. Perkin II*, 1974, 1214.
[90] T. S. Leong and M. E. Peach, *J. Fluorine Chem.*, 1975, **5**, 545.
[91] T. S. Croft and J. J. McBrady, *J. Heterocyclic Chem.*, 1975, **12**, 845.
[92] A. Haas and V. Hellwig, *J. Fluorine Chem.*, 1975, **6**, 521.
[93] C. Wakselman and M. Tordeux, *J.C.S. Chem. Comm.*, 1975, 956.
[94] I. W. Parsons, P. M. Smith, and J. C. Tatlow, *J. Fluorine Chem.*, 1975, **5**, 269.
[95] D. Ben-Ishai, I. Satati, and Z. Berler, *J.C.S. Chem. Comm.*, 1975, 349.
[96] M. Némec, M. Janda, J. Šrogl, and I. Stibor, *Coll. Czech. Chem. Comm.*, 1974, **39**, 3527.
[96a] E. P. Papadopoulos, *J. Org. Chem.*, 1974, **39**, 2540.
[97] D. N. Kursanov, Z. N. Parnes, G. I. Bolestova, and L. I. Belen'kii, *Tetrahedron*, 1975, **31**, 311.

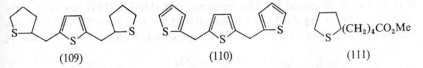

(109)   (110)   (111)

varying yields [98] with this useful method, which, however, is limited to relatively activated thiophens.[97] The cycloaddition of tetracyanoethylene oxide to thiophens, which gives the adducts (113), shows some similarity to electrophilic substitution, as competitive experiments show the following order of reactivity: furan > 2-methylthiophen $\approx$ benzo[*b*]furan > benzo[*b*]thiophen > selenophen > thiophen > 2-chlorothiophen. The mechanism of the cycloaddition has been discussed.[99] In contrast to thiophen and 2-methyl- and 2-chloro-thiophen, 2- and

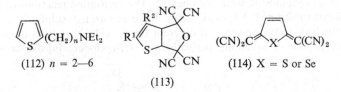

(112) *n* = 2—6

(113)

(114) X = S or Se

3-methoxythiophen and 2-methylthio-thiophen gave no cycloaddition product, but instead the corresponding thenoyl cyanides were obtained in low yields. A probable reaction mechanism for the formation of the thenoyl cyanides was given.[100] The reaction between 2,5-dihalogeno-thiophens or 2,5-dihalogeno-selenophens with tetracyanoethylene oxide gave the interesting derivatives (114).[101] Further structural investigations on the structure of the trimers obtained upon treatment of thiophen and 2-methylthiophen with phosphoric acid have been reported.[102]

Electrophilic Ring-closure Reactions.—The intramolecular acylation of acid chlorides (115) and (116) proceeds at the free 5-position to give (117) and (118).[103] A solid-phase effect (silica gel) gives a 100% increase in yields and the mechanism

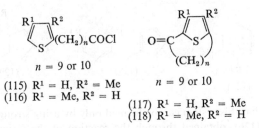

n = 9 or 10

(115) R¹ = H, R² = Me
(116) R¹ = Me, R² = H

n = 9 or 10

(117) R¹ = H, R² = Me
(118) R¹ = Me, R² = H

[98] G. I. Bolestova, E. P. Zakharov, S. P. Dolgova, Z. N. Parnes, and D. N. Kursanov, *Khim. geterotsikl. Soedinenii*, 1975, 1206.
[99] S. Gronowitz and B. Uppström, *Acta Chem. Scand. (B)*, 1975, **29**, 441.
[100] S. Gronowitz and B. Uppström, *Acta Chem. Scand. (B)*, 1974, **28**, 339.
[101] S. Gronowitz and B. Uppström, *Acta Chem. Scand. (B)*, 1974, **28**, 981.
[102] A. Ishigaki and T. Shono, *Bull. Chem. Soc. Japan*, 1975, **48**, 2977.
[103] S. Z. Taits, O. A. Kalinovskii, V. S. Bogdanov, and Ya. L. Gol'dfarb, *Khim. geterotsikl. Soedinenii*, 1972, 170.

of this effect is discussed.[104] The butanoic acid chloride (119) has been ring-closed to (120), using $AlCl_3$ in CS_2 as catalyst.[105] Less than 5% of (121) was formed as by-product. Treatment of the condensation product of thienyl alkyl

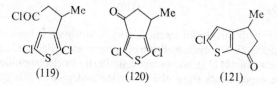

(119)   (120)   (121)

ketones and malononitrile (122; R = Et and Pr^i) with polyphosphoric acid led to (123). When (122; R = Bu^t) was treated in the same way, the expected product (124) was obtained. With (122; R = H or Me), only amide formation (125) and no ring-closure were obtained.[106] The cyclization reactions of several 3-(2'-thienyl)-substituted 3-hydroxy-acids by concentrated sulphuric acid, or acetic anhydride in the presence of different Lewis acids, have been investigated with the purpose of obtaining derivatives of 4H-cyclopenta[b]thiophen-4-one.

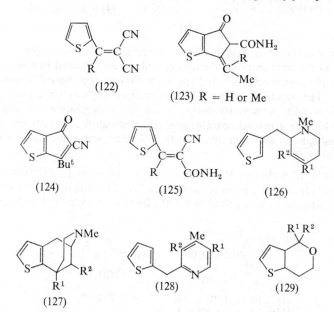

(122)

(123) R = H or Me

(124)   (125)   (126)

(127)   (128)   (129)

It was shown that this system can be formed only by using strong Lewis acids.[107] Treatment of (126), obtained through the reaction of the Grignard reagent of 3-thenyl bromide with a pyridinium compound, followed by sodium borohydride reduction and subsequent cyclization with 48% hydrobromic acid, led to the

[104] S. Z. Taits, O. A. Kalinovskii, B. V. Lopatin, and Ya. L. Gol'dfarb, *Khim. geterotsikl. Soedinenii*, 1973, 624.
[105] J. Skramstad, *Chem. Scripta*, 1975, **7**, 42.
[106] S. W. Schneller and D. R. Moore, *J. Org. Chem.*, 1975, **40**, 1840.
[107] Ch. P. Ivanov and D. M. Mondeshka, *Rev. Chim. (Roumania)*, 1975, **20**, 547.

thieno[2,3-*f*]morphans (127).[108] 2-Thenylmagnesium bromide similarly gave an isomeric thienomorphan.[109] An alternative route to this class of compounds starts with the reaction of 2-thienyl-lithium and cyanopyridines and is followed by reduction of the resulting ketone to give (128). Quaternization and reduction then give (126), which is subsequently transformed into (127) as described previously.[110] 2-(*β*-Hydroxyethyl)thiophen has been reacted with aldehyde acetals such as aminoacetaldehyde diethyl acetals, or ketones such as ethyl acetoacetate, or *N*-methylpiperidone, in the presence of acidic catalysts to give isochroman derivatives (129).[111]

Radical Reactions.—Thiazolyl, pyridyl, and other heteroaryl radicals formed by aprotic diazotization of the corresponding heterocyclic amines substitute homolytically on thiophen with the formation of 2-heteroaryl-thiophens as the main products in 20—50% yield. The results of competitive experiments indicate that the reactivity of thiophen in this reaction at 70—80 °C is slightly higher than that of benzene.[112] The currently accepted mechanism of the decomposition of benzoyl peroxide in thiophen has been criticized on the basis of new experimental results. No free thienyl radicals are involved in the reaction, as demonstrated by scavenging experiments, and the bithienyls formed are probably derived from dimerization of a benzoyloxythiophen radical σ-complex, with subsequent loss of benzoic acid.[113] Nitrene insertion into the thiophen ring has been observed in the thermal decomposition of 2-(2-azidobenzyl)thiophen and similar compounds, leading to thieno[3,2-*b*]quinoline derivatives.[114]

Nucleophilic Substitution.—Spinelli and co-workers are continuing their detailed study of nucleophilic substitution in the thiophen series. Kinetic data from the substitution of piperidine into some 2-substituted 3,5-dinitro-4-methyl- and 3,5-dinitro-thiophens showed the occurrence of a small secondary steric effect, only when the leaving group was SO$_2$Ph.[115] The reactivity of some 2-L-3-nitro-4-R-5-X-thiophens (L = Br or SO$_2$Ph; R = H or Me; X = H or NO$_2$) with various nucleophiles has been measured. The results obtained were discussed with regard to the electronic and steric effects of the nucleophiles.[116] In order to study primary steric effects, the rates of substitution by piperidine and by benzenethiolate ion of some 2-L-5-nitrothiophens and 2-L-3-methyl-5-nitro-thiophens have been measured in methanol. The rate ratios (k_H/k_{Me}) obtained show an absence of primary steric effects in piperidine substitution when L is halogen but not when L = SO$_2$Ph. Benzenethiolate ion substitution data show the presence of a large Reinheimer–Bunnett effect.[117] The logarithms of the rate constants of piperidinodebromination of some 2-bromo-3-X-5-nitro-thiophens gave excellent correlations against σ$_p^-$ and against the logarithm of the piperi-

[108] T. A. Montzka and J. D. Matiskella, *J. Heterocyclic Chem.*, 1974, **11**, 853.
[109] M. Alvarez, J. Bosch, and J. Canals, *Anales de Quím.*, 1975, **71**, 807.
[110] J. Rosch, R. Granados, and F. López, *J. Heterocyclic Chem.*, 1975, **12**, 651.
[111] T. A. Dobson and L. G. Humber, *J. Heterocyclic Chem.*, 1975, **12**, 591.
[112] G. Vernin, J. Metzger, and C. Párkányi, *J. Org. Chem.*, 1975, **40**, 3183.
[113] C.-M. Camaggi, R. Leardini, A. Tundo, and M. Tiecco, *J.C.S. Perkin I*, 1974, 271.
[114] G. R. Cliff, G. Jones, and J. McK. Woollard, *J.C.S. Perkin I*, 1974, 2072.
[115] D. Spinelli, G. Consiglio, R. Noto, and A. Corrao, *J.C.S. Perkin II*, 1974, 1632.
[116] D. Spinelli and G. Consiglio, *J.C.S. Perkin II*, 1975, 1388.
[117] D. Spinelli, G. Consiglio, and T. Monti, *J.C.S. Perkin II*, 1975, 816.

dinodebromination of the corresponding 2-bromo-3-nitro-5-X-thiophens.[118] The reaction rates for piperidino-substitution of some 2-L-3-nitro-5-X-thiophens (L = Cl, Br, I, $OC_6H_4NO_2$-p, or SO_2Ph) have been measured at various temperatures in methanol and have provided information on the influence of the leaving group and the substituent at C-5 on the position of the rate-determining transition state on the reaction co-ordinate.[119]

The n.m.r. spectra of the Meisenheimer complex of 2-nitrothiophen and 2-nitrofuran with methoxide ion have been compared.[120] Evidence has been obtained for the formation of spiro-Meisenheimer compounds (130) and (131)

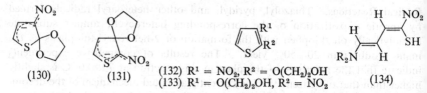

(130)   (131)   (132) $R^1 = NO_2$, $R^2 = O(CH_2)_2OH$
                              (133) $R^1 = O(CH_2)_2OH$, $R^2 = NO_2$ (134)

from (132) and (133), which were prepared from the corresponding bromonitro-thiophens and the sodium salt of ethylene glycol.[121] The reactivity of 2- and 3-nitrothiophen towards some nucleophiles has been studied.[122] Secondary aliphatic amines reacted with 2-nitrothiophen in ethanol to yield bis(4-dialkyl-amino-1-nitro-buta-1,3-dienyl) disulphides *via* the corresponding thiol (134).[123] 2-Chloro-thiophens reacted with hydrogen sulphide at 450—550 °C to give thiophen-2-thiols[124] and with aromatic thiols to give aryl 2-thienyl sulphides.[125] The usefulness of nucleophilic aromatic substitution in the thiophen series is illustrated by Grol's work on dithienothiazines.[126-128] The reaction of the anion of 3-bromothiophen-2-thiol with 2-bromo-3-nitrothiophens in DMSO gave (135) which after reduction to the amine and acylation was ring-closed by a copper-promoted reaction to (136).[126] Alternatively, the nitrogen bridge was first created by copper-promoted reaction between 3-bromothiophen and 3-acetyl-aminothiophen to yield (138), which upon reaction with SCl_2 gave (137).[127]

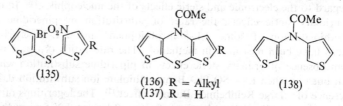

(135)

(136) R = Alkyl   (138)
(137) R = H

[118] D. Spinelli, G. Consiglio, R. Noto, and A. Corrao, *J.C.S. Perkin II*, 1975, 620.
[119] D. Spinelli and G. Consiglio, *J.C.S. Perkin II*, 1975, 989.
[120] G. Doddi, A. Poretti, and F. Stegel, *J. Heterocyclic Chem.*, 1974, **11**, 97.
[121] F. Sancassan, M. Novi, G. Guanti, and C. Dell'Erba, *J. Heterocyclic Chem.*, 1975, **12**, 1083.
[122] C. Dell'Erba, M. Novi, G. Guanti, and D. Spinelli, *J. Heterocyclic Chem.*, 1975, **12**, 327.
[123] G. Guanti, C. Dell'Erba, G. Leandri, and S. Thea, *J.C.S. Perkin I*, 1974, 2357.
[124] M. G. Voronkov, É. N. Deryagina, A. S. Nakhmanovich, and L. G. Klochkova, *Khim. geterotsikl. Soedinenii*, 1974, 712.
[125] M. G. Voronkov, É. N. Deryagina, L. G. Klochkova, E. A. Chernyshev, V. I. Savushkina, and G. A. Kravchenko, *Khim. geterotsikl. Soedinenii*, 1975, 1322.
[126] C. J. Grol, *J. Heterocyclic Chem.*, 1974, **11**, 953.
[127] C. J. Grol, *J.C.S. Perkin I*, 1975, 1234.
[128] C. J. Grol and H. Rollema, *J. Medicin. Chem.*, 1975, **18**, 857.

Applying these principles, a large number of isomeric dithienothiazines and thienobenzothiazines have been prepared.[128] Another route to thienobenzothiazines was based on the electrophilic substitution of 4-methoxy-2-methyl-3-ethoxycarbonylthiophen with nitroarenesulphenyl chlorides.[128a] The reaction of 3-bromo-2-formyl-thiophen and -selenophen with sodium azide in DMSO gave the corresponding 3-azido-2-formyl derivatives, which could be transformed to preparatively useful intermediates such as 3-amino-2-formylthiophen.[129] The reaction between iodo-thiophens and copper acetylides continues to be of great use for the preparation of naturally occurring thiophens.[130, 131] Thus 2,5-di-iodothiophen or 5,5'-di-iodo-2,2'-bithienyl and the copper salt of 2-propynylaldehyde acetal gave (139) and (140) respectively,[130] and 2-acetyl-5-iodo-thiophen and the copper salt of 2-thienylacetylene yielded the naturally occurring

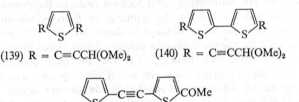

(139) R = C≡CCH(OMe)₂   (140) R = C≡CCH(OMe)₂

(141)

(141).[130] The reaction of 2-thienylacetylene with Cu₂Cl₂ and oxygen gave 1,4-di-(2-thienyl)buta-1,3-diyne.[131] It was demonstrated that the main, and probably radicaloid, reaction patterns of di-(3-thienyl)iodonium chloride in the presence of certain nucleophiles, such as cyanide ion, methoxide ion, and piperidine, can be almost completely altered by the addition of copper salts.[132] Instead of reduction to thiophen and iodothiophen, or polymerization, arylation of the nucleophiles occurred, which led to 3-cyano-3-methoxy- or 3-piperidino-thiophen as the main product in addition to 3-iodothiophen.[132]

Metal-organic Derivatives.—Metallation of thiophens with alkyl-lithiums and halogen–metal exchange between halogeno-thiophens and alkyl-lithiums keep their positions as the most important synthetic routes to substituted thiophens. The directing effect of 3-alkyl groups on the metallation of 3-alkyl-thiophens and halogen–metal exchange of 3-alkyl-2,5-dibromo-thiophens with various organo-lithium and magnesium reagents has been studied. In the metallation reaction as well as in the formation of Grignard reagents, steric hindrance from the alkyl group was observed. On the other hand, in the halogen–metal exchange of 3-alkyl-2,5-dibromothiophen with butyl-lithium or ethylmagnesium bromide, release of steric strain caused an increase in the reactivity of the 2-position over that of the 5-position on going from 2,5-dibromo-3-methyl- to 2,5-dibromo-3-t-butyl-

[128a] V. I. Shvedov, O. B. Romanova, V. K. Vasil'eva, V. P. Pakhomov, and A. N. Grinev, *Khim. geterotsikl. Soedinenii*, 1973, 741.
[129] S. Gronowitz, C. Westerlund, and A.-B. Hörnfeldt, *Acta Chem. Scand. (B)*, 1975, **29**, 224.
[130] F. Bohlmann and J. Kocur, *Chem. Ber.*, 1974, **107**, 2115; 1975, **108**, 2149.
[131] F. Bohlmann and A. Suwita, *Chem. Ber.*, 1975, **108**, 515.
[132] S. Gronowitz and B. Holm, *Chem. Scripta*, 1974, **6**, 133.

thiophen.[133] Metallation of 2,2-di(3-thienyl)-1,3-dioxolan occurs in the 2,2′-positions, and by reaction with sulphur the lithium derivative was converted into dithienothiopyrone derivatives.[134] The dilithiation of di-3-thienylphosphine oxides also occurs in the 2,2′-positions. This compound was then treated with a great variety of reagents.[135] Compound (142), obtained by the reaction of

(142) X = S or Se

4-bromo-3-thienyl-lithium or 4-bromo-3-selenienyl-lithium with DMF, undergoes metallation with butyl-lithium in the 2-position *ortho* to the protected aldehyde group.[136] This was utilized for the synthesis of β-bromo-substituted thieno-[2,3-*b*]thiophen and seleno[2,3-*b*]thiophen derivatives.[136] Thienyl-lithium derivatives were used for the synthesis of a variety of deuteriated thiophen aldehydes needed for conformational studies,[137] and for 'the fixed conformation ester' 4,5-dihydrothieno[2,3-*c*]pyran-7-one, in which the key step was the halogen–metal exchange of 3-bromothiophen-2-aldehyde diethyl acetal followed by reaction with ethylene oxide.[138] The reaction of 2,3,5,6-di-*O*-cyclohexylidene-D-mannolactone with 2-thienyl-lithium has been investigated.[139] Convenient methods for the synthesis of isomer-free 3-nitrothiophen, 3-nitroselenophen, as well as 3-nitrofuran, by the reaction of the corresponding diaryliodonium salts with sodium nitrite have been developed. The iodonium salts are easily obtained from the corresponding lithium derivatives.[140] The reaction of 2,5-dimethylthiophen with alkyl-lithium–*NNN′N′*-tetramethylethylenediamine complexes has been studied.[141, 142] After reaction with carbon dioxide, varying relative amounts of 3-carboxy-5-methyl-3-thienylacetic acid and 2,5-dimethyl-3-thiophencarboxylic acid were formed, depending upon the solvent. A maximum yield of 16% of the former acid was obtained when hexane was used as solvent. The low yield was shown to be due partly to competing ring-opening reactions.[142] The reaction of tetramethylthiophen and 3,4-dichloro-2,5-dimethylthiophen with lithiating reagents was studied.[142] The coupling of a variety of thienyl-lithiums with cupric chloride has been used for the preparation of bithienyls (see p. 274). A review on the use of thienyl-lithium derivatives in synthesis, including their ring-opening reactions, has appeared.[142a]

[133] S. Gronowitz, B. Cederlund, and A.-B. Hörnfeldt, *Chem. Scripta*, 1974, **5**, 217.
[134] C. J. Grol, *Tetrahedron*, 1974, **30**, 3621.
[135] J.-P. Lampin and F. Mathey, *J. Organometallic Chem.*, 1974, **71**, 239.
[136] Ya. L. Gol'dfarb, I. P. Konyaeva, and V. P. Litvinov, *Izvest. Akad. Nauk S.S.S.R., Ser. khim.*, 1974, 1570.
[137] D. J. Chadwick, J. Chambers, P. K. G. Hodgson, G. D. Meakins, and R. L. Snowden, *J.C.S. Perkin I*, 1974, 1141.
[138] D. J. Chadwick, J. Chambers, G. D. Meakins, and R. L. Snowden, *J.C.S. Perkin I*, 1975, 523.
[139] Yu. A. Zhdanov, V. G. Alekseeva, and V. N. Fomina, *Doklady Akad. Nauk S.S.S.R.*, 1974, **219**, 867.
[140] S. Gronowitz and B. Holm, *Synthetic Comm.*, 1974, **4**, 63.
[141] A. J. Clarke, S. McNamara, and O. Meth-Cohn, *Tetrahedron Letters*, 1974, 2373.
[142] S. Gronowitz and T. Frejd, *Acta Chem. Scand. (B)*, 1975, **29**, 818.
[142a] S. Gronowitz, 'Organic Sulphur Chemistry, Structure, Mechanism and Synthesis', ed. C. J. M. Stirling, Butterworth, London, 1975.

The metallation of thiophen and 2- and 3-methylthiophen with butylcaesium and butylpotassium has been studied.[143] Thienylcopper reagents have been used for the synthesis of iodo-thiophens.[144] Thienyltin(iv) compounds have been prepared.[145, 146] The chemistry of thienyl-lead(iv)tricarboxylate has been investigated.[147] The cleavage of the thiophen–silicon bond in tris(trichloro-2-thienyl)-methylsilane with butyl-lithium to give thienyl-lithium derivatives has been achieved.[148] Considerable work on the synthesis and reactions of compounds containing thiophen–phosphorus bonds has appeared,[149–154] and the reactions with αβ-unsaturated acid derivatives have been extensively studied.[150, 152, 154]

Photochemistry of Thiophens.—Increasing interest in the photochemistry of thiophens can be noted. The photochemical reactivity of some five-membered heterocyclic nitro-compounds towards nucleophiles (CN⁻, CNO⁻, OMe⁻, and H₂O) has been investigated. 2-Nitrothiophen, as well as 2-nitrofuran, undergoes smooth photo-substitution of the nitro-group by the nucleophile. 2-Nitrofuran reacts by a first-order process, whereas with 2-nitrothiophen the quantum yield of substitution showed a 'normal' dependence on nucleophile concentration.[155] The u.v. irradiation of 2- and 3-bromothiophen in various solvents produced thiophen as one of the reaction products.[156] The photo-Fries rearrangement of the phenyl ester of thiophen-2-carboxylic acid afforded the corresponding *o*- and *p*-hydroxyphenyl ketones.[157] The photochemical reactivity of 2- and 3-benzoyl-thiophen and their *p*-cyano- and *p*-methoxy-derivatives has been studied. On irradiation in the presence of isobutylene, all six ketones undergo photocyclo-addition at the carbonyl group to yield thermally unstable oxetans such as (143), which eliminate formaldehyde to give (144).[158] In this connection, the u.v. absorption and phosphorescence emission spectra of some methyl-substituted benzoylthiophens were analysed and partial energy diagrams were constructed.[159] The reaction between 2-benzoylthiophen and tetramethylethylene also led to efficient oxetan formation.[160] However, irradiation of 2-acetylthiophen with tetramethylethylene resulted in [4 + 2] addition to the alkene as the major process, yielding 38% of (145), as well as [2 + 2] addition to give (146) in 10% yield, along with oxetan formation in 11% yield. Isobutene gave similar results.[160]

[143] P. Benoit and N. Collignon, *Bull. Soc. chim. France*, 1975, 1302.
[144] M. T. Rahman and H. Gilman, *J. Indian Chem. Soc.*, 1974, **51**, 1018.
[145] S. Gopinathan, C. Gopinathan, and J. Gupta, *Indian J. Chem.*, 1974, **12**, 623.
[146] C. Gopinathan, S. K. Pandit, S. Gopinathan, A. Y. Sonsale, and P. A. Awasarkar, *Indian J. Chem.*, 1975, **13**, 516.
[147] H. C. Bell, J. R. Kalman, J. T. Pinhey, and S. Sternhell, *Tetrahedron Letters*, 1974, 853.
[148] R. D. Howells and H. Gilman, *J. Organometallic Chem.*, 1974, **77**, 177.
[149] R. Z. Aliev, V. K. Khairullin, and S. F. Makhmutova, *Zhur. obshchei Khim.*, 1976, **46**, 58.
[150] V. K. Khairullin and R. Z. Aliev, *Zhur. obshchei Khim.*, 1973, **43**, 2165.
[151] M. A. Vasyanina and V. K. Khairullin, *Zhur. obshchei Khim.*, 1974, **44**, 48
[152] V. K. Khairullin and R. Z. Aliev, *Zhur. obshchei Khim.*, 1974, **44**, 1683.
[153] V. K. Khairullin, L. I. Nesterenko, V. I. Savushkina, and E. A. Chernyshev, *Izvest. Akad. Nauk S.S.S.R., Ser. khim.*, 1974, 1846.
[154] V. K. Khairullin and R. Z. Aliev, *Zhur. obshchei Khim.*, 1974, **44**, 2120.
[155] M. B. Groen and E. Havinga, *Mol. Photochem.*, 1974, **6**, 9.
[156] A. T. Jeffries and C. Párkányi, *Z. Naturforsch.*, 1976, **31b**, 345.
[157] Y. Kanacka and Y. Hatanaka, *Heterocycles*, 1974, **2**, 423.
[158] D. R. Arnold, R. J. Birtwell, and B. M. Clarke, jun., *Canad. J. Chem.*, 1974, **52**, 1681.
[159] D. R. Arnold and B. M. Clarke, jun., *Canad. J. Chem.*, 1975, **53**, 1.
[160] T. S. Cantrell, *J. Org. Chem.*, 1974, **39**, 2242.

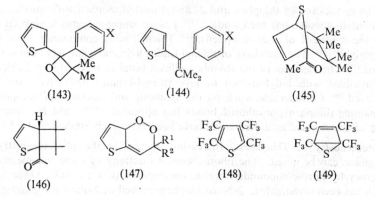

(143) (144) (145)

(146) (147) (148) (149)

Singlet oxygen reacted stereospecifically with 2-vinyl-thiophens in a 1,4-cyclo-addition to give thermally stable 1,4-endoperoxides (147).[161] The previously discovered [2 + 2] photoaddition of benzophenone to 2,5-dimethylthiophen to give oxetans has been extended to other carbonyl derivatives.[162] Mechanisms explaining the formation of pyrroles, obtained by u.v. irradiation of a number of thiophens in the presence of propylamine, have been discussed.[163] Irradiation of (148) led to the 'Dewar' thiophen (149).[164] The structures of dimerization products and Diels–Alder adducts with furans were elucidated.[164]

Electrochemical Reactions.—An electrochemical method for the one-step preparation of 2,5-dihydrothiophen-2-carboxylic acid has been developed.[165] A stable cation radical has been obtained from the electrochemical oxidation of 2-*p*-nitrophenyl-3,4,5-triphenylthiophen.[166] The reduction of 2-acetylthiophen,[167a] 2-benzoylthiophen,[167b] and 2,5-diformylthiophen [168] by electrochemical methods on a mercury electrode has been investigated and compared in the latter case with chemical reductions.[168] Electrochemical reactions have been carried out with some complex thiophens.[168-170]

The Structure and Reactions of Hydroxy-, Mercapto-, and Amino-thiophens.—A new hydroxy-thiophen synthesis consists in the reaction of thiophens with di-isopropylperoxydicarbonate in acetonitrile in the presence of cupric chloride catalysts, which yields thienylisopropylcarbonates in 41—74% yield. Dealkylation and decarboxylation of the esters then give hydroxy-thiophens.[171] Due to

[161] M. Matsumoto, S. Dobashi, and K. Kondo, *Tetrahedron Letters*, 1975, 4471.
[162] C. Rivas and R. A. Bolivar, *J. Heterocyclic Chem.*, 1973, **10**, 967.
[163] A. Couture, A. Delevallee, A. Lablache-Combier, and C. Párkányi, *Tetrahedron*, 1975, **31**, 785.
[164] Y. Kobayashi, I. Kumadaki, A. Ohsawa, Y. Sekine, and H. Mochizuki, *Chem. and Pharm. Bull. (Japan)*, 1975, **23**, 2773.
[165] V. S. Mikhailov, V. P. Gul'tyai, S. G. Mairanovskii, S. Z. Taits, I. V. Proskurovskaya, and Yu. G. Dubovik, *Izvest. Akad. Nauk S.S.S.R., Ser. khim.*, 1975, 888.
[166] M. Libert and C. Caullet, *Compt. rend.*, 1974, **278**, C, 439.
[167] (a) P. Foulatier and C. Caullet, *Compt. rend.*, 1974, **279**, C, 25; (b) P. Foulatier, J.-P. Salaün, and C. Caullet, *Compt. rend.*, 1974, **279**, C, 779.
[168] J.-P. Salaün, M. Salaün-Bouix, and C. Caullet, *Compt. rend.*, 1975, **280**, C, 165.
[169] G. Barbey and C. Caullet, *Tetrahedron Letters*, 1974, 1717.
[170] M. Laćan, I. Tabaković, and Ž. Čeković, *Tetrahedron*, 1974, **30**, 2911.
[171] A. P. Manzara and P. Kovacic, *J. Org. Chem.*, 1974, **39**, 504.

by-product formation, in many cases this method cannot compete with the established methods for hydroxy-thiophen synthesis. Demethylation of substituted methoxy-thiophens with pyridine hydrochloride has been suggested for the synthesis of hydroxy-thiophens.[172] Ionization potentials have been used in the tautomeric analysis of 2-hydroxy-thiophens and 2-hydroxy-selenophens [173] and the corresponding 3-hydroxy-derivatives.[174] Observed values were compared with ionization potentials of compounds derived from either tautomeric form. The methylation of 3-methyl- and 3-t-butyl-2-hydroxythiophen systems with methyl iodide by means of the ion-pair extraction method has been investigated. It was shown that the reactivity of the 3-position became less important in favour of the 5-position when the size of the 3-substituent was increased.[175] With dimethyl sulphate, methylation occurred almost exclusively at the oxygen.[175] Thienyl methacrylates have been synthesized [176] from hydroxy-thiophens and methacrylyl chloride. 4-Ethoxycarbonyl-3-hydroxy-thiophens gave thienylhydrazines with hydrazines.[177] From 2-ethoxycarbonyl-3-hydroxy-thiophens and cyanogen bromide, thienyl cyanates were prepared, which upon reaction with nucleophilic reagents could be transformed to biheterocyclic systems, such as (150).[178] Heating some hydroxy-thiophens with HMPA at 220—230 °C gave dimethylamino-thiophens.[179] Methylenation experiments have been carried out with 3,4-di-hydroxy-2,5-di-(ethoxycarbonyl)-3,4-dihydroxy-thiophens.[180] The diethyl 3,4-methylenedioxythiophen-2,5-dicarboxylates were converted into other symmetrically and unsymmetrically substituted thiophens.[181] The reaction of the

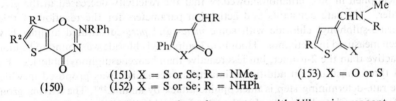

(150)

(151) X = S or Se; R = NMe$_2$
(152) X = S or Se; R = NHPh

(153) X = O or S

5-phenyl-2-hydroxy-thiophen or -selenophen system with Vilsmeier reagent, or *NN'*-diphenylamidine, led to condensation in the 3-position to give (151) and (152), respectively.[182] Hindered rotation in (153) has been studied by the dynamic n.m.r. technique.[183] Some new thiophen-2-thiols have been prepared.[184] Some

[172] G. Henrio, G. Plé, and J. Morel, *Compt. rend.*, 1974, **278**, C, 125.
[173] O. Thorstad, K. Undheim, B. Cederlund, and A.-B. Hörnfeldt, *Acta Chem. Scand. (B)*, 1975, **29**, 647.
[174] O. Thorstad, K. Undheim, R. Lantz, and A.-B. Hörnfeldt, *Acta Chem. Scand. (B)*, 1975, **29**, 652.
[175] B. Cederlund and A.-B. Hörnfeldt, *Chem. Scripta*, 1975, **8**, 140.
[176] R. T. Hawkins, *J. Heterocyclic Chem.*, 1974, **11**, 291.
[177] V. I. Shvedov, Y. I. Trofimkin, V. K. Vasileva, T. F. Vlasova, and A. N. Grinev, *Khim. geterotsikl. Soedinenii*, 1975, 914.
[178] M. Hedayatullah, J. Pailler, and L. Denivelle, *Bull. Soc. chim. France*, 1974, 2161.
[179] E. B. Pedersen and S.-O. Lawesson, *Tetrahedron*, 1974, **30**, 875.
[180] F. Dallacker and V. Mues, *Chem. Ber.*, 1975, **108**, 569.
[181] F. Dallacker and V. Mues, *Chem. Ber.*, 1975, **108**, 576.
[182] L. N. Kurkovskaya, N. N. Shapetko, N. B. Sokolova, and I. Y. Kvitko, *Zhur. org. Khim.*, 1975, **11**, 1091.
[183] V. S. Bogdanov, M. A. Kalik, and Ya. L. Gol'dfarb, *Izvest. Akad. Nauk S.S.S.R., Ser. khim.*, 1974, 598.
[184] K. I. Sadykhov, S. M. Aliev, and M. M. Seidov, *Zhur. org. Khim.*, 1975, **11**, 2157; *Khim. geterotsikl. Soedinenii*, 1975, 344.

3-amino-thiophens and 3-amino-selenophens with electron-attracting groups have been diazotized and transformed into the azido-derivatives.[185, 186] 2-Amino-3-ethoxycarbonyl-thiophens have been diazotized and converted into the 2-hydrazino-derivatives;[187] they have also been tosylated and benzoylated, and the corresponding sodium salts alkylated with ethyl 4-bromobutyrate in reactions connected with the synthesis of thieno[2,3-*b*]azepin-4-ones.[8] The reactivity of the various functional groups of 2-amino-5-(arylthio)thiophens with electron-withdrawing groups in the 3-position has been investigated.[188]

Side-chain Reactivities.—Several papers have appeared in which the influence of the thiophen ring on side-chain reactivity is treated quantitatively. The kinetics of *E*2 elimination from 5-substituted 2-(2-thienyl)ethyl tosylates ($\rho = 2.22$) and bromides ($\rho = 1.89$), and from 2-(3-thienyl)ethyl bromide, have been studied in EtOH–EtONa at 50 °C. Values of σ_α ($+0.26$) and σ_β (-0.05) were calculated for the sulphur hetero-atom.[189] The reactions of triethyl phosphite with chloroacetyl-furans, chloroacetyl-thiophens, and phenacyl chloride yield vinylphosphonates. Kinetic measurements gave the following reactivity order: 2-chloroacetylfuran > phenacyl chloride > 3-chloroacetylfuran $\approx$ 3-chloroacetylthiophen > 2-chloroacetylthiophen.[190] Attempts are made to explain this reactivity order. Bromoacetyl- and iodoacetyl-thiophens give mainly the β-ketophosphonates on reaction with triethyl phosphite.[191] A comparative study of the solvolysis of phenacyl bromide, 2-bromoacetylthiophen, and 2-bromoacetyl-selenophen in 50% ethanol showed [192] that the reactivity decreased in the given order. The rate constants and activation parameters for the reaction of thiophen-3-sulphonyl chloride with some *meta*- and *para*-substituted anilines have been measured in methanol. Thiophen-3-sulphonyl chloride was found to be more reactive than the 2-isomer, but less reactive than benzenesulphonyl chloride. For all three substrates an addition–elimination mechanism was proposed in which the rate-determining step is nucleophilic attack by aniline.[193] The leaving-group effect of F, Cl, and Br in the thiophen-2-sulphonyl halide system has been studied. The results with the sulphonyl chloride and bromide are consistent with the above mechanism, while for the fluoride, S—F bond-breaking was rate-determining.[194] The rates of reaction of 2-thenyl chloride, 2-furfuryl chloride, and benzyl chlorides with different amines by an S_N2 mechanism have been measured. The reactivity order was: furfuryl chloride $\gg$ 2-thenyl chloride $\approx$ benzyl chloride. In S_N1 formolyses in 20% dioxan, 2-chloromethylthiophen was much more reactive than benzyl chloride.[195] It has been found that some halogeno-substituted 2-acetyl-thiophens are not Hammett bases and do not conform with the amide

[185] C. Paulmier, *Compt. rend.*, 1975, **281**, C, 317.
[186] C. Paulmier, G. Ah-Kow, and P. Pastour, *Bull. Soc. chim. France*, 1975, 1437.
[187] M. Hentschel and K. Gewald, *J. prakt. Chem.*, 1974, **316**, 878.
[188] N. K. Son, R. Pinel, and Y. Mollier, *Bull. Soc. chim. France*, 1974, 1359.
[189] E. Baciocchi, V. Mancini, and P. Perucci, *J.C.S. Perkin II*, 1975, 821.
[190] A. Arcoria, S. Fisichella, E. Maccarone, and G. Scarlata, *Gazzetta*, 1975, **105**, 547.
[191] A. Arcoria, S. Fisichella, E. Maccarone, and G. Scarlata, *J. Heterocyclic Chem.*, 1975, **12**, 215.
[192] N. N. Magdesieva and I. V. Leont'eva, *Khim. geterotsikl. Soedinenii*, 1973, 910.
[193] A. Arcoria, E. Maccarone, G. Musumarra, and G. A. Tomaselli, *J. Org. Chem.*, 1974, **39**, 1689.
[194] E. Maccarone, G. Musumarra, and G. A. Tomaselli, *J. Org. Chem.*, 1974, **39**, 3286.
[195] F. Yamamoto, H. Morita, and S. Oae, *Heterocycles*, 1975, **3**, 1.

acidity function, but their protonation is described satisfactorily by the benzo-phenone acidity function.[196]

'Benzylic' Reactivity.—The methyl group in the *o*-methyl-nitro-thiophens showed varying reactivity towards aldehydes under base catalysis. 3-Methyl-2-nitro-thiophen reacted to give a vinyl compound, while 3-methyl-4-nitrothiophen did not react at all. 2-Methyl-3-nitrothiophen with formaldehyde in methanol–water gave, in addition to 3-nitro-2-vinylthiophen, a bis(3-nitro-2-thienyl)cyclobu-tane.[197] The commercial availability of 3-methylthiophen has led to a reinvesti-gation of its usefulness as precursor in the synthesis of 3-substituted aldehydes and ketones.[198] Under liquid-phase conditions in the presence of a catalyst consisting of copper acetate and sodium bromide or 9,10-dibromoanthracene, 2-ethylthiophen is oxidized to give 2-acetylthiophen and 1-(2-thienyl)ethyl acetate.[199] The side-chain bromination with NBS of methyl 2,5-dimethyl-thiophen-3-carboxylate has been investigated.[200] 5-Chloro-2-thenyl cyanide was condensed with carbon disulphide in the presence of sodium hydride to yield (154).[201] The Sommelet and Stevens rearrangements have been studied with

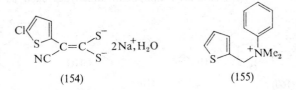

(154)   (155)

(155), and complex product mixtures were obtained.[202, 203] The alkylation of esters of 2-thenylphosphonic acids at the active methylene group has been investigated.[204]

Reactions of Thiophen Aldehydes and Ketones.—The Stobbe condensation of some thienylcarbonyl compounds with dimethyl methylsuccinate in the presence of potassium t-butoxide or sodium hydride gave predominantly the (*E*)-half esters of (156),[205] while condensation with dimethyl homophthalate gave predominantly the (*Z*)-half ester of (157).[206] Thiophen-2-aldehydes were shown to add smoothly to αβ-unsaturated ketones and nitriles, under the catalytic influence of cyanides, to form (158) and (159), respectively.[207] Thiophen-2-aldehydes have been con-densed with aliphatic amines [208] and phenylenediamine [209] to give Schiff bases;

[196] S. V. Tsukerman, L. P. Pivovarevich, L. A. Kutulya, V. G. Gordienko, and V. F. Lavrushin, *Zhur. obshchei Khim.*, 1974, **44**, 683.
[197] S. Gronowitz and I. Ander, *Acta Chem. Scand.* (*B*), 1975, **29**, 513.
[198] J. A. Clarke and O. Meth-Cohn, *Tetrahedron Letters*, 1975, 4705.
[199] T. V. Shchedrinskaya, P. A. Konstantinov, V. P. Litvinov, É. G. Ostapenko, I. V. Zakharov, and M. N. Volkov, *Zhur. obshchei Khim.*, 1974, **44**, 837.
[200] M. Janda, M. Valenta, and P. Holý, *Coll. Czech. Chem. Comm.*, 1974, **39**, 959.
[201] W. O. Foye, J. M. Kauffman, J. J. Lanzillo, and E. F. LaSala, *J. Pharm. Sci.*, 1975, **64**, 1371.
[202] A. G. Giumanini and G. Lercker, *Gazzetta*, 1974, **104**, 415.
[203] A. G. Giumanini and C. Trombini, *J. prakt. Chem.*, 1975, **317**, 897.
[204] V. Lachkova and M. Kirilov, *Annalen*, 1974, 496.
[205] N. R. El-Rayyes and N. A. Al-Salman, *J. prakt. Chem.*, 1975, **317**, 552.
[206] N. R. El-Rayyes and A. H. A. Ali, *J. prakt. Chem.*, 1975, **317**, 1040.
[207] H. Stetter and B. Rajh, *Chem. Ber.*, 1976, **109**, 534.
[208] J. J. Pesek and J. H. Frost, *Synthetic Comm.*, 1974, **4**, 367.
[209] S. Biniecki and F. Herold, *Acta Polon. Pharm.*, 1974, **31**, 417.

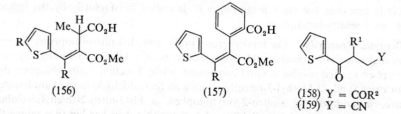

(156)   (157)   (158) Y = COR²
                                                 (159) Y = CN

with 1-acetylindoxyl to give thenylidene indoxyls,[210] with triethylphosphonoacetic acid in the presence of titanium tetrachloride and base to give triethyl thenyl-idenephosphonoacetic acids;[211] and with α-triphenylphosphoranylidene-γ-butyrolactone to give *trans*-α-(2-thenylidene)-λ-γ-butyrolactone.[212] Mono- and di-methine dyes have been prepared from 5-dimethylaminothiophen-2-aldehyde and from the corresponding selenophen aldehyde.[213] 1-(2-Thienyl)alkan-2-ones have been prepared by applying the Darzens glycidic ester synthesis to thiophen-2-aldehyde, using different 2-bromo-esters.[214] The reagent (160) has been used for the preparation of αβ-unsaturated thiophenic aldehydes such as (161) from

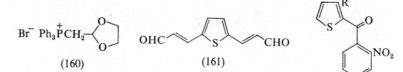

(160)   (161)

(162) R = H or Me

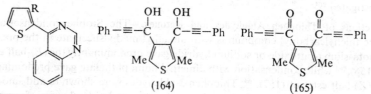

(163) R = H or Me   (164)   (165)

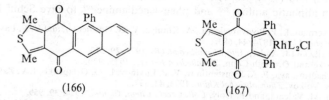

(166)   (167)

[210] J. Štetinová and J. Kováč, *Coll. Czech. Chem. Comm.*, 1975, **40**, 1750.
[211] W. Lehnert, *Tetrahedron*, 1974, **30**, 301.
[212] D. C. Lankin, M. R. Scalise, J. C. Schmidt, and H. Zimmer, *J. Heterocyclic Chem.*, 1974, **11**, 631.
[213] F. A. Mikhailenko, L. I. Shevchuk, and I. T. Rozhdestvenskaya, *Khim. geterotsikl. Soedinenii*, 1975, 316.
[214] J. D. Belcher, jun., D. S. Hunter, D. G. Hutson, R. L. McBroom, and E. H. Sund, *J. Chem. and Eng. Data*, 1975, **20**, 206.

2,5-diformylthiophen.[215] The Leuchart reaction of (162) does not proceed normally, due to nitro-group participation, and yields (163).[216] Phenyl 2-thienyl ketone [217] and 2-acetylthiophen [218] have been reacted with ethynylenemagnesium bromide to give the substituted but-2-ene-1,4-diols. 3,4-Diformyl-2,5-dimethyl-thiophen and phenylethynylmagnesium bromide gave the diol (164), which was oxidized with MnO_2 to the ketone (165). Thermal isomerization of (165) yielded the quinone (166) which, with tris(triphenylphosphine)rhodium(I) chloride, was transformed to the interesting rhodium complex (167).[219] A new method for the synthesis of (168), starting from the condensation of 2-acetylthiophen with diethyl oxalate, followed by reaction with formaldehyde, has been described.[220] Diazo-(phenyl)(2-thienyl)methane and diazo-(2-thienyl)methane were prepared from the corresponding hydrazones by oxidation with peroxyacetic acid in the presence

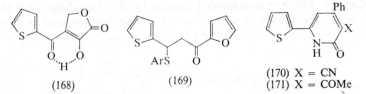

(168)   (169)   (170) X = CN
(171) X = COMe

of iodine and tetramethylguanidine in 1,2-dichloroethane.[221] The geometrical isomers of the 2-benzothiazolylhydrazones of 2-formyl- and 2-acetyl-thiophen have been studied.[222]

Some new chalcones derived from thiophen-2-aldehyde have been obtained [223] and their Michael additions investigated. With arylthiolate anions (169) was

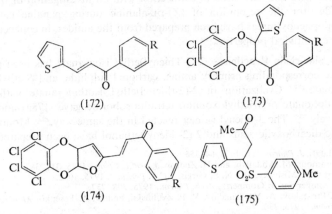

(172)   (173)

(174)   (175)

[215] T. M. Cresp, M. V. Sargent, and P. Vogel, *J.C.S. Perkin I*, 1974, 37.
[216] M. Srinivasan and J. B. Rampal, *Tetrahedron Letters*, 1974, 2883.
[217] J. Krupowicz, K. Sapiecha, and R. Gaszczyk, *Roczniki Chem.*, 1974, **48**, 2067.
[218] I. M. Gverdtsiteli and M. D. Chanturiya, *Zhur. obshchei Khim.*, 1975, **45**, 2349.
[219] E. Müller and W. Winter, *Annalen*, 1975, 605.
[220] J. Paris, M. Payard, and P.-J. Bargnoux, *Compt. rend.*, 1974, **278**, C, 1149.
[221] J. R. Adamson, R. Bywood, D. T. Eastlick, G. Gallagher, D. Walker, and E. M. Wilson, *J.C.S. Perkin I*, 1975, 2030.
[222] S. Kwon, M. Tanaka, and K. Isagawa, *Nippon Kagaku Kaishi*, 1974, 1526 (*Chem. Abs.*, 1975, **82**, 30 858).
[223] B. S. Holla and S. Y. Ambekar, *J. Indian Chem. Soc.*, 1973, **50**, 673.

272   *Organic Compounds of Sulphur, Selenium, and Tellurium*

obtained.[224] Michael condensation of the benzylidine derivative of 2-acetyl-thiophen with ethyl cyanoacetate and acetoacetamide gave (170) and (171), respectively.[225] The chalcones (172) react with tetrachloro-*o*-benzoquinone to give (173), in contrast to the furan analogues, in which the furan ring was claimed to react to give (174).[226] Reaction of 2-thenylideneacetone with toluene-*p*-sulphinic acid gave (175), the Mannich reaction of which was investigated.[227] New β-diketones containing pyridine, thiophen, and furan rings have been synthesized.[228] A new chelating agent, 1,1,1-trifluoro-4-(2-thienyl)-4-selenobut-3-en-2-one, has been prepared.[229]

Reactions of Cyano- and Carboxy-thiophens.—Amidines and imino-ethers [230] as well as triazoles and tetrazines [231] have been prepared from cyano-thiophens. The modified Lossen rearrangement of sodium *NN*-dihydroxythiophen-2,3-di-carboxamide with benzenesulphonyl chloride furnished a mixture of (176) and (177) in 54% yield in a 1 : 3 ratio.[232] Azo dyes have been prepared by diazotization of *N*-(4′-aminophenyl)thiophen-2-carboxamides.[233] 2-Thienylthiocarbonyl

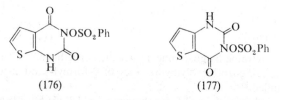

(176)   (177)

chloride has been prepared by a new method.[234] The hydrazide of thiophen-2-carboxylic acid has been used in connection with an investigation of the use of hydrazides for the resolution of (Z)-DL-alanine during papaine catalysis.[235] *N*-(2-Thiophenoyl)ureas have been prepared from the amides, in connection with a study of antiviral properties.[236]

Various Side-chain Reactions.—2-(2-Thienyl)ethyl isocyanate has been prepared from the corresponding primary amine, carbonyl sulphide, and *S*-ethyl chloro-thioformate.[237] Cyclization of 2-(2-thienyl)ethyl isothiocyanate with methyl fluorosulphonate or triethyloxonium tetrafluoroborate gave (178a) and (178b), respectively.[238] The 3-thienyl isomer reacted in the same way.[238] Monoesters of aliphatic dicarboxylic acids and 2-(2-thienyl)ethanol have been prepared in con-

[224] G. Soldati, *J. Pharm. Sci.*, 1975, **64**, 355.
[225] A. Sammour, M. Abdallah, and H. Zoorob, *J. prakt. Chem.*, 1975, **317**, 387.
[226] N. Latif, N. Mishriky, and N. S. Girgis, *J.C.S. Perkin I*, 1975, 1052.
[227] P. Messinger and J. Gompertz, *Arch. Pharm.*, 1975, **308**, 737.
[228] N. S. Prostakov, A. Ya. Ismailov, V. P. Zvolinskii, and D. A. Fesenko, *Khim. geterotsikl. Soedinenii*, 1973, 230.
[229] T. Honjo, *Chem. Letters*, 1974, 481.
[230] B. Decroix and P. Dubus, *Compt. rend.*, 1974, **279**, C, 343.
[231] P. Dubus, B. Decroix, J. Morel, and C. Paulmier, *Compt. rend.*, 1974, **278**, C, 61.
[232] K.-Y. Tserng and L. Bauer, *J. Org. Chem.*, 1975, **40**, 172.
[233] S. Fisichella, M. Longo, G. Scarlata, and M. Torre, *Ann. Chim. (Italy)*, 1974, **64**, 505.
[234] H. Viola and R. Mayer, *Z. Chem.*, 1975, **15**, 348.
[235] J. L. Abernethy, D. Srulevitch, and M. J. Ordway, jun., *J. Org. Chem.*, 1975, **40**, 3445.
[236] D. G. O'Sullivan and A. K. Wallis, *Z. Naturforsch.*, 1975, **30b**, 600.
[237] M. W. Gittos, R. V. Davies, B. Iddon, and H. Suschitzky, *J.C.S. Perkin I*, 1976, 141.
[238] R. V. Davies, B. Iddon, T. McC. Paterson, M. W. Pickering, H. Suschitzky, and M. W. Gittos, *J.C.S. Perkin I*, 1976, 138.

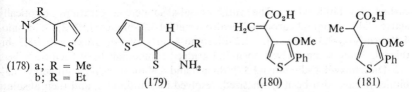

(178) a; R = Me
b; R = Et

(179)

(180)

(181)

nection with work on macrocyclic thiophens.[239] The stable 3-aminoprop-2-enethione (179) has been synthesized.[240] Asymmetric hydrogenation of (180) with Kagan's catalyst gives the (+)-form of (181) in 97% yield, and in 88% optical purity.[35] Thiophen isosteres of phenylethanolamines have been prepared, starting from the bromoacetyl-thiophens.[79] The reactions of dithienyl disulphides and thiocyanato-thiophens with various reagents have been studied.[241, 242] A review on the syntheses, reactions, and properties of some selenides of the thiophen, selenophen, and furan series has appeared.[243] The reactions of thenoyltrifluoroacetone with the dichlorides of organic selenides and selenoxides [244] as well as with tri- and di-organotin(IV) compounds [245] have been studied. Thenoyltrifluoroacetone analogues have been synthesized.[246] The Diels–Alder adduct from 2-vinylthiophen and 4-phenyl-1,2,4-triazoline-3,5-dione has been used in hydrogen-transfer reactions.[247]

Bi- and Poly-heterocycles.—Extensive work on the synthesis, optical resolution, and conformation of 3,3'-bithienyls, as well as detailed studies of their c.d. curves, has appeared.[248-255] Most 3,3'-bithienyls were prepared by the coupling of 3-thienyl-lithium derivatives followed by the introduction of the appropriate substituents by electrophilic substitution [249] or by the modification of substituents by reduction,[248, 250] oxidation,[248, 249, 255] or side-chain bromination.[248] In particular, halogen–metal exchange in bromo-substituted 3,3'-bithienyls was very useful for the modification of substituents.[249, 250, 255b] Ullman coupling of 3-bromo-2,4-dimethoxycarbonylthiophen was used for the synthesis of 2,2',4,4'-tetra-

[239] S. Z. Taits, A. A. Dudinov, F. D. Alashev, and Ya. L. Gol'dfarb, *Izvest. Akad. Nauk S.S.S.R., Ser. khim.*, 1974, 148.
[240] G. Duguay, *Compt. rend.*, 1975, **281**, C, 1077.
[241] Z. V. Todres, F. M. Stoyanovich, Ya. L. Gol'dfarb, and D. N. Kursanov, *Khim. geterotsikl. Soedinenii*, 1973, 632.
[242] V. I. Shvedov, I. A. Kharizomenova, O. B. Romanova, V. K. Vasileva, and A. N. Grinev, *Khim. geterotsikl. Soedinenii*, 1975, 911.
[243] V. P. Litvinov, A. N. Sukiasyan, and Ya. L. Gol'dfarb, *Khim. geterotsikl. Soedinenii*, 1972, 723.
[244] N. N. Magdesieva, R. A. Kyandzhetsian, and V. M. Astafurov, *Zhur. org. Khim.*, 1975, **11**, 508.
[245] B. P. Bachlas and R. R. Jain, *J. Organometallic Chem.*, 1974, **82**, 359.
[246] Y. A. Fialkov, P. A. Yufa, A. G. Goryushko, N. K. Davidenko, and L. M. Yagupolskii, *Zhur. org. Khim.*, 1975, **11**, 1066.
[247] W. A. Pryor, J. H. Coco, W. H. Daly, and K. N. Houk, *J. Amer. Chem. Soc.*, 1974, **96**, 5591.
[248] E. Wiklund and R. Håkansson, *Chem. Scripta*, 1974, **6**, 76.
[249] E. Wiklund and R. Håkansson, *Chem. Scripta*, 1974, **6**, 137.
[250] E. Wiklund and R. Håkansson, *Chem. Scripta*, 1974, **6**, 174.
[251] E. Wiklund and R. Håkansson, *Chem. Scripta*, 1974, **6**, 226.
[252] R. Håkansson and E. Wiklund, *Chem. Scripta*, 1975, **7**, 120.
[253] R. Håkansson, S. Gronowitz, J. Skramstad, and T. Frejd, *Chem. Scripta*, 1975, **7**, 131.
[254] R. Håkansson and E. Wiklund, *Chem. Scripta*, 1975, **7**, 173.
[255] (a) R. Håkansson, B. Nordén, and E. Wiklund, *Acta Chem. Scand.* (B), 1974, **28**, 695; (b) R. Håkansson and A. Svensson, *Chem. Scripta*, 1975, **7**, 186.

carboxy-3,3'-bithienyl.[255b] The syntheses and absolute configurations of optically active 4,4'-dibromo-2-carboxy-2'-hydroxymethyl-3,3'-bithienyl,[248] 2,2'-dibromo-4,4'-dicarboxy-3,3'-bithienyl, 2,2'-dibromo-4-carboxy-4'-hydroxymethyl-3,3'-bithienyl,[249] and some derived open and bridged compounds have been described. 2,2'-Dicarboxy-4,4'-dimethyl-3,3'-bithienyl and 4,4'-dicarboxy-2,2'-dimethyl-3,3'-bithienyl have also been synthesized, resolved into antipodes, and their absolute configurations determined.[250] In the ^{1}H n.m.r. spectra the methylene protons of the bridging lactones and oxepins showed non-equivalence, which allowed determination of the energy barrier separating the enantiomers.[248, 249] The rate of racemization of some optically active 2,2'- and 4,4'-dicarboxy-3,3'-bithienyls has been determined.[251] The absolute configurations, conformations, and c.d. spectra of 4,4'-dicarboxy-2,2',5,5'-tetramethyl-3,3'-bithienyl have been compared with those of the corresponding selenophen and benzene derivatives.[253] The synthesis, resolution, absolute configuration, c.d. spectra, and rates of racemization of 4,4'-dicarboxy-2,2'-diformyl-3,3'-bithienyl and 2,2',4,4'-tetracarboxybithienyl have been studied.[255b] The quasi-racemate method has been successfully applied in the bithienyl series for relative configuration determinations.[249, 253] Comparison of c.d. in liquid solution and in the crystal state has been used for the study of conformations of 3,3'-bithienyls.[255a]

A new promising method for the synthesis of mixed biaryls consists in the reaction of the ethanolamine esters of diarylboronic acids with NBS in dichloromethane/aqueous buffer at pH 9. In this way, 2- and 3-phenylthiophen, 2,2'-bithienyl, and 2-(2-thienyl)furan were obtained.[256] The nitration of 2,3'-bithienyl has been studied.[257] 2-Nitro-2'-amino-3,3'-bithienyl and 3-nitro-3'-amino-2,2'-bithienyl have been found to be stable.[258] Carbonyl derivatives of 2-aryl-thiophens have been prepared,[259] and reactivity constants for 2-aryl-thiophens determined.[260] Some 2,5-bis(4'-n-alkyldiphenyl-4''-yl)thiophens, which are liquid crystals, have been prepared from the corresponding 1,4-diketones and P_4S_{10}.[261] Various naphthyl-thiophens have been synthesized in order to study their anti-inflammatory[262] and antibacterial properties.[263]

A large amount of work on compounds where the thiophen ring is connected to another five-membered aromatic heterocycle has appeared. Reacting thioamides, *e.g.* 2-thiocarbamoyl-thiophens or -selenophens, with α-halogeno-ketones, according to the method of Hantzsch, led to a variety of 2-(thienyl)-thiazoles and 2-(selenienyl)thiazoles.[264] Thiophens attached to two thiazolyl rings,[264] and 2-(5-nitro-2-thienyl)thiazole, needed for a study[265] of antiprotozoal activity, have been prepared similarly. The somewhat exotic thienylthiazole (182)

[256] G. M. Davies, P. S. Davies, W. E. Paget, and J. M. Wardleworth, *Tetrahedron Letters*, 1976, 795.

[257] C. Dell'Erba, G. Guanti, and G. Garbarino, *J. Heterocyclic Chem.*, 1974, **11**, 1017.

[258] R. Guilard, J. C. Nonciaux, E. Laviron, and P. Fournari, *J. Heterocyclic Chem.*, 1975, **12**, 191.

[259] V. K. Polyakov, Z. P. Zaplyuisvechka, and S. V. Tsukerman, *Khim. geterotsikl. Soedinenii*, 1974, 136.

[260] G. I. Matyushecheva, A. I. Tolmachev, A. A. Shulezhko, L. M. Shulezhko, and L. M. Yagupolskii, *Zhur. obshchei Khim.*, 1976, **46**, 162.

[261] H. Schubert, I. Sagitdinov, and J. V. Svetkin, *Z. Chem.*, 1975, **15**, 222.

[262] J. S. Kaltenbronn and T. O. Rhee, *J. Medicin. Chem.*, 1974, **17**, 654.

[263] N. V. Stulin, A. E. Lipkin, D. A. Kulikova, and E. A. Rudzit, *Khim.-farm. Zhur.*, 1975, **9**, 20.

[264] P. Chauvin, J. Morel, P. Pastour, and J. Martinez, *Bull. Soc. chim. France*, 1974, 2079.

[265] J. P. Verge and P. Roffey, *J. Medicin. Chem.*, 1975, **18**, 794.

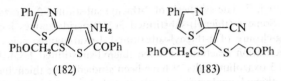

(182)   (183)

was prepared from (183) by ring-closure of the thiophen ring.[266] The decomposition of thienylthiazolidines has been studied.[267] The radical anions derived from 2-(2-thienyl)thiazole have been investigated.[268]

Some 5-nitro-2-thienyl-pyrazoles have been prepared for a study of their antimicrobial activity.[269] The oximes of thienyl- and selenienyl-carboxamides, prepared from the nitriles and hydroxylamine, give 1,2,4-oxadiazoles by reaction with triethyl orthoformate in BF_3 etherate.[270] Starting from cyano-thiophens and cyano-selenophens, thienyl- and selenienyl-triazoles and tetrazines were prepared.[232, 271] From the β-diketone (184) the triheterocyclic (185) was obtained

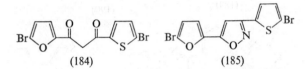

(184)   (185)

by reaction with hydroxylamine.[272] 2-(2-Thienyl)indole has been prepared by the Fischer synthesis.[273] It is formylated in the indolic β-position.[273] 2-(3-Thienyl)-indole derivatives were obtained when methyl indole-3-dithiocarboxylates were reacted with phenacyl bromide in acetone to give (186), which reacted with active methylene compounds to form (187).[274]

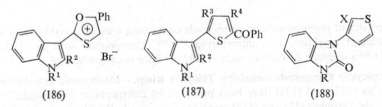

(186)   (187)   (188)

2-(Thienyl)pyrimidines have been synthesized from thiophenic amidines, and their nitration studied.[275] 5-(Thienyl)pyrimidines and 2-(thienyl)pyrazines were prepared from thenyl cyanides and their reactions explored.[276] The synthesis and hydrolytic stability of 2,4-dihalogeno-6-(thienyl)-substituted *sym*-triazines have

[266] H. Schäfer and K. Gewald, *J. prakt. Chem.*, 1974, **316**, 684.
[267] J. J. Pesek and J. H. Frost, *Tetrahedron*, 1975, **31**, 907.
[268] G. F. Pedulli, P. Zanirato, A. Alberti, and M. Tiecco, *J.C.S. Perkin II*, 1975, 293.
[269] L. K. Kulikova and L. V. Cherkesova, *Khim.-farm. Zhur.*, 1974, **8**, 18.
[270] P. Dubus, B. Decroix, J. Morel, and P. Pastour, *Ann. Chim. (France)*, 1975, **10**, 331.
[271] R. K. M. R. Kallury, T. G. S. Nath, and V. R. Srinivasan, *Austral. J. Chem.*, 1975, **28**, 2089.
[272] T. Lesyak and S. Nelek, *Khim. geterotsikl. Soedinenii*, 1975, 162.
[273] B. S. Holla and S. Y. Ambekar, *J. Indian Chem. Soc.*, 1974, **51**, 965.
[274] Y. Tominaga, Y. Matsuda, and G. Kobayashi, *Heterocycles*, 1976, **4**, 9.
[275] J. Pankiewicz, B. Decroix, and J. Morel, *Compt. rend.*, 1975, **281**, C, 39.
[276] J. Bourguignon, J.-M. Boucly, J.-C. Clinet, and G. Queguiner, *Compt. rend.*, 1975, **281**, C, 1019.

been investigated.[277] The synthesis of 2-thienyl-substituted 2-tetrazines has been described.[278] Some 6-(thienyl)-substituted 3(2*H*)-pyridazinones have been prepared by ring-closure of 4-thienyl-substituted 4-oxabutanoic acids with hydrazine.[279] Recyclization reactions of some thienyl-substituted 1,3,4-thiadiazolium salts [280] and 1,3-oxazolium salts [281] have been studied. The thienylbenzimidazole (188) [282] and thienyl analogues of substituted flavonoids (189) [283] have been prepared. Kauffmann has continued his interesting work on the syntheses and

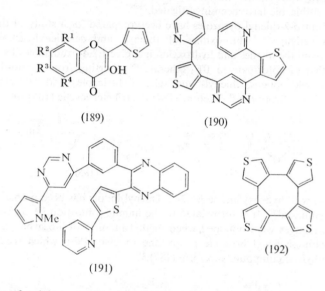

(189)   (190)

(191)

(192)

properties of polyheterocyclic compounds containing π-excessive (thiophen) and π-deficient (pyridine, pyrimidine) rings.[284-288] Compounds such as (190),[285] (191),[286] and cyclic derivatives such as (192) [287] have been synthesized.

Macrocyclic Compounds containing Thiophen Rings.—Macrocyclic keto-lactones such as (193) and (194) have been prepared by electrophilic ring-closure,[289] as were the ketones (117) and (118) mentioned previously.[103] The lactam (195) has

[277] J. K. Chakrabarti, A. F. Cockerill, G. L. O. Davies, T. M. Hotten, D. M. Rackham, and D. E. Tupper, *J.C.S. Perkin II*, 1974, 861.
[278] S. A. Lang, jun., B. D. Johnson, and E. Cohen, *J. Heterocyclic Chem.*, 1975, **12**, 1143.
[279] E. A. Steck, R. P. Brundage, and L. T. Fletcher, *J. Heterocyclic Chem.*, 1974, **11**, 755; W. V. Curran and A. Ross, *J. Medicin. Chem.*, 1974, **17**, 273.
[280] O. P. Shvaika and V. I. Fomenko, *Zhur. org. Khim.*, 1974, **10**, 377.
[281] O. P. Shvaika and V. I. Fomenko, *Zhur. org. Khim.*, 1974, **10**, 2429.
[282] O. Hromatka, D. Binder, and K. Eichinger, *Monatsh.*, 1975, **106**, 555.
[283] K. A. Thaker and P. R. Muley, *J. Indian Chem. Soc.*, 1975, **52**, 243.
[284] B. Greving, A. Woltermann, and T. Kauffmann, *Angew. Chem.*, 1974, **86**, 475.
[285] A. Mitschker, U. Brandl, and T. Kauffmann, *Tetrahedron Letters*, 1974, 2343.
[286] E. Wienhöfer and T. Kauffmann, *Tetrahedron Letters*, 1974, 2347.
[287] T. Kauffmann, B. Greving, J. König, A. Mitschker, and A. Woltermann, *Angew. Chem.*, 1975, **87**, 745.
[288] T. Kauffmann, B. Muke, R. Otter, and D. Tigler, *Angew. Chem.*, 1975, **87**, 746.
[289] Ya. L. Gol'dfarb, S. Z. Taits, F. D. Alasher, A. A. Dudinov, and O. S. Chizhov, *Khim, geterotsikl. Soedinenii*, 1975, 40.

also been obtained.[290] There has been continued interest in layered compounds, and the *anti* (196) and *syn* form (197) have been separated.[291] Furthermore, double-, triple- (198), and quadruple-layered paracyclothiophenophanes have been obtained.[292]

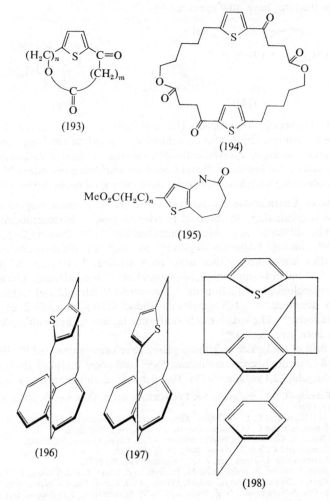

(193)

(194)

(195)

(196) (197)

(198)

Reactions Leading to Destruction of the Thiophen Ring.—Raney-nickel de-sulphurization has been used for the synthesis of aliphatic compounds and for structure determination. From (158) and (159) aliphatic γ-diketones and γ-keto-carboxylic acids have been obtained.[207] ε-Caprolactams and ε-enantholactams such as (199) have been prepared by desulphurization and transformed to amino-

[290] Ya. L. Gol'dfarb, B. P. Fabrichnyi, I. F. Shalavina, and S. M. Kostrova, *Zhur. org. Khim.*, 1975, **11**, 2400.
[291] S. Mizogami, N. Osaka, T. Otsubo, Y. Sakata, and S. Misumi, *Tetrahedron Letters*, 1974, 799.
[292] N. Osaka, S. Mizogami, T. Otsubo, Y. Sakata, and S. Misumi, *Chem. Letters*, 1974, 515.

278   *Organic Compounds of Sulphur, Selenium, and Tellurium*

acids.[293] From (195) ε-amino-dicarboxylic acids were obtained.[290] Aliphatic amino-ketones have been prepared from 2-(ω-dialkylaminoalkyl)thiophens by the action of lithium and t-butanol in liquid ammonia.[294] Further details of the cycloaddition of dimethyl acetylenedicarboxylate to 3-dialkylamino-thiophens, leading to thiepins, have now appeared.[295]

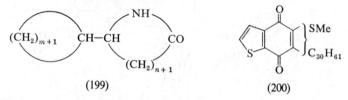

(199)   (200)

Naturally Occurring Thiophens.—The isolation of a terpenoid benzo[*b*]-thiophen-4,7-quinone (200) from an acidophilic bacterium has been reported.[296] New naturally occurring acetylenic thiophen derivatives have been isolated from *Berkheya* and *Cullumia* species [131] and from *Chrysanthemum macrotum*.[297] Several naturally occurring acetylenic thiophens have been synthesized.[130, 131, 132, 297, 298]

Thiophens of Pharmacological Interest.—Interest in pharmacologically active thiophens is continuing. In the field of tricyclic psychopharmaceutically active compounds, dithieno- and thieno-benzothiazines,[126, 128] thieno[3,2-*c*]-2-benzo-thiepins,[299] thieno[2,3-*b*]benzothiepins [300] as well as dithieno-analogues of amitryptyline and nortryptyline have been studied.[301] Interest in thienodi-azepines [77, 302] and in 4-oxothieno[3,2-*b*]pyrimidines [303] is continuing. The synthesis and pharmacological evaluation of 2,3-dihydro-1*H*-thieno[2,3-*e*](1,4)diazepines has been described.[303a] 2-(Thienyl)-substituted 4(3*H*)-quinazolinones showed hypnotic effects.[304] The sedative action of some lactams of aminothienylalkanoic acids has been investigated.[305]

In the field of analgesics, thienomorphans have been synthesized [108-110] as well as some 6-(2-thenoyl)benzoxazolinones [306, 307] and some thiophen derivatives of 3,8-diazabicyclo[3,2,1]octane.[308] In the field of anti-inflammatory agents, a

293 B. P. Fabrichnyi, I. F. Shalavina, Ya. L. Gol'dfarb, and S. M. Kostrova, *Z hur. org. Khim.* 1974, **10**, 1956.
294 Ya. L. Gol'dfarb and E. P. Zakharov, *Khim. geterotsikl. Soedinenii*, 1975, 1499.
295 D. N. Reinhoudt and C. G. Kouwenhoven, *Tetrahedron*, 1974, **30**, 2093.
296 M. De Rosa, A. Gambacorts, and L. Minale, *J.C.S. Chem. Comm.*, 1975, 392.
297 F. Bohlmann and C. Zdero, *Chem. Ber.*, 1975, **108**, 739.
298 T. B. Patrick and J. L. Honegger, *J. Org. Chem.*, 1974, **39**, 3791.
299 M. Rajšner, J. Metyš, B. Kakáč, and M. Protiva, *Coll. Czech. Chem. Comm.*, 1975, **40**, 2905.
300 M. Rajšner, E. Svátek, J. Metyš, and M. Protiva, *Coll. Czech. Chem. Comm.*, 1974, **39**, 1366.
301 B. Yom-Tov, S. Gronowitz, S. B. Ross, and N. E. Stjernström, *Acta Pharm. Suecica*, 1974, **11**, 149.
302 O. Hromatka, D. Binder, and K. Eichinger, *Monatsh.*, 1975, **106**, 375.
303 L. Lorente, R. Madroñero, and S. Vega, *Anales de Quím.*, 1974, **70**, 974.
303a F. J. Tinney, J. P. Sanchez, and J. A. Nogas, *J. Medicin. Chem.*, 1974, **17**, 624.
304 T. Hisano, M. Ichikawa, A. Nakagawa, and M. Tsuji, *Chem. and Pharm. Bull.* (*Japan*), 1975, **23**, 1910.
305 Y. I. Vikhlyaev, T. A. Klygul, E. I. Slynko, Ya. L. Gol'dfarb, I. B. P. Fabrichnyi, I. F. Shalavina, and S. M. Kostrova, *Khim.-farm. Zhur.*, 1974, **8**, 8.
306 J.-P. Bonte, D. Lesieur, C. Lespagnol, M. Plat, J.-C. Cazin, and M. Cazin, *Chim. Thérap.*, 1974, **9**, 491.
307 J.-P. Bonte, D. Lesieur, C. Lespagnol, J.-C. Cazin, and M. Cazin, *Chim. Thérap.*, 1974, **9**, 497.
308 L. Fontanella, E. Occelli, and E. Testa, *Farmaco* (*Pavia*), Ed. Sci., 1975, **30**, 742.

large number of compounds of the type (201)[309] and (202)[80] were shown to be of special interest. Also, some 4- and 5-(2-thienyl)-1-naphthaleneacetic acids showed anti-inflammatory properties.[262] 4-(2-Thenoyl)-2,3-dichlorophenoxyacetic acid showed a notable diuretic activity and also very strong hypuricemic activity.[310]

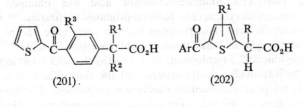

(201).   (202)

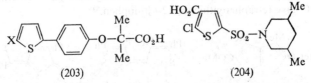

(203)   (204)

Some aminobenzoic acid diuretics containing the thiophen ring have also been studied.[311] The syntheses and lipid-lowering properties of (203)[312] and (204)[313] have been described. 3-β-(2-Thienyl-alanine-8-lysine)vasopressine has been synthesized by the solution technique.[314] A comparison of the bioavailability of [2,5-^{14}C]thiophen after oral and rectal administration in mice has been reported.[315] The nitro-reduction of carcinogenic 5-nitro-thiophens by rat-tissues has been studied.[316] Thiophen-ring-containing compounds with antiradiation,[317] anti-depressant,[318] antiparasitic,[319-322] antibacterial,[323, 324] antiviral,[237] and broncho-dilator activity[325] have been studied. A thia-steroid, 14,15-dehydro-A-nor-3-thiaequilenin, has been synthesized.[325a] In connection with studies on the syn-

[309] P. G. H. Van Daele, J. M. Boey, V. K. Sipido, M. F. L. De Bruyn, and P. A. J. Janssen, *Arzneim.-Forsch.*, 1975, **25**, 1495.
[310] G. Thuillier, J. Laforest, B. Cariou, P. Bessin, J. Bonnet, and J. Thuillier, *Chim. Thérap.*, 1974, **9**, 625.
[311] P. W. Feit, O. B. Tvaermose Nielsen, and H. Bruun, *J. Medicin. Chem.*, 1974, **17**, 572.
[312] S. Gronowitz, R. Svenson, G. Bondesson, O. Magnusson, and N. E. Stjernström, *Acta Pharm. Suecica*, 1974, **11**, 211.
[313] B. Dafgård, S. Gronowitz, G. Bondesson, O. Magnusson, and N. E. Stjernström, *Acta Pharm. Suecica*, 1974, **11**, 309.
[314] C. W. Smith, M. F. Ferger, and W. Y. Chan, *J. Medicin. Chem.*, 1975, **18**, 822.
[315] J.-L. Chanal, M.-T. Calmette, B. Bonnaud, and H. Cousse, *Chim. Thérap.*, 1974, **9**, 641.
[316] C. Y. Wang, C. W. Chiu, and G. T. Bryan, *Biochem. Pharmacol.*, 1975, **24**, 1563.
[317] W. O. Foye, J. M. Kauffman, J. J. Lanzillo, and E. F. LaSala, *J. Pharm. Sci.*, 1975, **64**, 1371.
[318] W. O. Foye and J. P. Speranza, *Chim. Thérap.*, 1974, **9**, 177.
[319] J. K. Chakrabarti and A. Todd, *Chim. Thérap.*, 1974, **9**, 146.
[320] H. R. Wilson, G. R. Revankar, and R. L. Tolman, *J. Medicin. Chem.*, 1974, **17**, 760.
[321] T. R. Herrin, J. M. Pauvlik, E. V. Schuber, and A. O. Geiszler, *J. Medicin. Chem.*, 1975, **18**, 1216.
[322] M. M. El-Kerdawy, A. A. Samour, and A. A. El-Agamey, *Pharmazie*, 1975, **30**, 76.
[323] D. Ducher, J. Couquelet, R. Cluzel, and J. Couquelet, *Chim. Thérap.*, 1973, **8**, 552.
[324] F. Yoneda and T. Nagamatsu, *Chem. and Pharm. Bull. (Japan)*, 1975, **23**, 2001.
[325] G. E. Hardtmann, G. Koletar, O. R. Pfister, J. H. Gogerty, and L. C. Iorio, *J. Medicin. Chem.*, 1975, **18**, 447.
[325a] S. R. Ramadas and P. S. Srinivasan, *Chem. and Ind.*, 1974, 307.

thesis and chemistry of cephalosporin antibiotics, thiophen-2-acetic acid has been used as a side-chain.[326−342]

3 Thienothiophens, their Benzo-derivatives, and Analogous Compounds

Synthesis.—Thieno[2,3-*b*]thiophencarboxylic acid was obtained from 3-(3-thienyl)-2-mercaptoacrylic acid by halogen-promoted cyclization.[343] Chlorination of thiophenacrylic acids and thiophendiacrylic acids gave various chlorinated thiophthencarboxylic acids and thienothiophthencarboxylic acids.[344] 3-Bromo- [136] and 3-methylthieno[2,3-*b*]thiophen, and 3-methylthieno[3,2-*b*]thiophen [345] were prepared by introducing acetyl groups and the thioacetic acid residue in the appropriate thiophen, followed by Dieckmann cyclization. The thiophthen (205) has been obtained in one step from acyclic starting material.[19] Heating (206) to over 200 °C gave tetraphenylthieno[3,2-*b*]thiophen.[346]

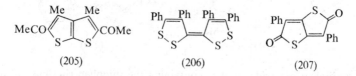

(205)   (206)   (207)

Reactions.—All the mono- and poly-bromo-derivatives,[347] as well as iodo-derivatives,[348] of thieno[2,3-*b*]thiophen have been prepared by direct bromination or iodination, or from the lithium compounds, and characterized by their n.m.r. spectra. Boronic acids were obtained from several alkyl- and aryl-thienothiophens *via* lithiation and were converted by hydrogen peroxide oxidation into the tautomeric hydroxy-derivatives.[349] It was shown by n.m.r. that all thieno[2,3-*b*]thiophen systems exist as thieno[2,3-*b*]thiophen-2(3*H*)-ones, while in the case of

[326] D. A. Berges, *J. Medicin. Chem.*, 1975, **18**, 1264.
[327] R. Bywood, G. Gallagher, G. K. Sharma, and D. Walker, *J.C.S. Perkin I*, 1975, 2019.
[328] Y. Maki and M. Sako, *J. Amer. Chem. Soc.*, 1975, **97**, 7168.
[329] H. Yanagisawa, M. Fukushima, A. Ando, and H. Nakao, *Tetrahedron Letters*, 1976, 259.
[330] R. N. Guthikonda, L. D. Cama, and B. G. Christensen, *J. Amer. Chem. Soc.*, 1974, **96**, 7584.
[331] R. R. Chauvette and P. A. Pennington, *J. Amer. Chem. Soc.*, 1974, **96**, 4986.
[332] A. E. Bird, *J. Pharm. Sci.*, 1975, **64**, 1671.
[333] S. Karady, T. Y. Cheng, S. H. Pines, and M. Sletzinger, *Tetrahedron Letters*, 1974, 2625.
[334] S. Karady, T. Y. Cheng, S. H. Pines, and M. Sletzinger, *Tetrahedron Letters*, 1974, 2629.
[335] N. G. Steinberg, R. W. Ratcliffe, and B. G. Christensen, *Tetrahedron Letters*, 1974, 3567.
[336] M. Ochiai, O. Aki, A. Morimoto, T. Okada, K. Shinozaki, and Y. Asahi, *J.C.S. Perkin I*, 1974, 258.
[337] R. R. Chauvette and P. A. Pennington, *J. Medicin. Chem.*, 1975, **18**, 403.
[338] R. Reiner, U. Weiss, and P. Angehrn, *Chim. Thérap.*, 1975, **10**, 10.
[339] D. O. Spry, *J. Org. Chem.*, 1975, **40**, 2411.
[340] R. A. Firestone, N. S. Maciejewicz. and B. G. Christensen, *J. Org. Chem.*, 1974, **39**, 3384.
[341] D. O. Spry, *J.C.S. Chem. Comm.*, 1974, 1012.
[342] H. Peter, H. Rodriguez, B. Müller, W. Sibral, and H. Bickel, *Helv. Chim. Acta*, 1974, **57**, 2024.
[343] S. W. Schneller and J. D. Petru, *Synthetic Comm.*, 1974, **4**, 29.
[344] B. Capron, C. Paulmier, and P. Pastour, *Bull. Soc. chim. France*, 1975, 2575.
[345] V. P. Litvinov, T. V. Shchedrinskaya, P. A. Konstantinov, and Ya. L. Gol'dfarb, *Khim. geterotsikl. Soedinenii*, 1975, 492.
[346] H. Behringer and E. Meinetsberger, *Tetrahedron Letters*, 1975, 3473.
[347] P. Fournari and P. Meunier, *Bull. Soc. chim. France*, 1974, 583.
[348] P. Meunier and P. Fournari, *Bull. Soc. chim. France*, 1974, 587.
[349] G. Martelli, L. Testaferri, M. Tiecco, and P. Zanirato, *J. Org. Chem.*, 1975, **40**, 3384.

the thieno[3,2-*b*]thiophens, the two isomeric thieno[3,2-*b*]thiophen-2(3*H*)-ones and -2(5*H*)-ones could be identified. The reaction of the potential hydroxy-thieno-thiophens with diazoalkanes and their condensation with aldehydes and ketones were also studied.[350] The compound (207) has been prepared from pulvinic acid lactone and thiolacetic acid.[351]

Physical Properties.—Triethylsilyloxyl nitroxide radicals of thienothiophens [63] and radical anions of three dithienothiophen 7,7-dioxides produced by potassium reduction in DME have been studied.[352]

Non-classical Thienothiophens and Related Systems.—Phosphorus pentasulphide treatment of suitable vicinal dibenzoyl heterocycles has been established as a convenient pathway to tetraphenylthieno[3,4-*c*]thiophen, 5-methyl-1,3,4,6-tetraphenylthieno[3,4-*c*]pyrrole, and hexaphenylthieno[3,4-*f*]isothionaphthene, and their cycloaddition reactions have been studied.[353] Further details on thieno-[3,4-*c*]pyrazole,[354] thieno[3,4-*c*]furan, and thieno[3,4-*c*]pyrrole [355] have appeared (see Volume 3, p. 449).

4 Benzothiophens and their Benzo-fused Systems

Synthesis of Benzothiophens by Ring-closure Reactions.—Several examples of the synthesis of benzo[*b*]thiophens through the reaction of arylthiols with α-halo-geno-ketones followed by electrophilic ring-closure have been published.[356-358] Indanobenzo[*b*]thiophens [357] and the sulphur isostere 5,6-dihydroxytryptamine [358] were obtained in this way. α-Mercaptocinnamic acids, obtained from the condensation products of benzaldehydes with rhodanine, gave benzo[*b*]-thiophen-2-carboxylic acids upon reaction with iodine.[359, 360] The mechanism of the formation of 3-chloro-2-benzo[*b*]thiophencarbonyl chlorides from cinnamic acids and thionyl chloride has been investigated in detail.[361] Reaction of 4-aryl-2-butanones with thionyl chloride usually gave 3-thietanone formation. An exception was 4-(*m*-hydroxyphenyl)butan-2-one, which gave only 2-acetyl-5-hydroxy-benzo[*b*]thiophen.[362]

Heating (208) with hydrobromic acid in acetic acid gave (2-*p*-nitrophenyl)-benzo[*b*]thiophen.[363] The oxidation of 1-thiochromene with selenium dioxide in boiling ethanol led to 2-(2′-benzo[*b*]thienyl)-1-thiochromene.[363a] Substituted 2-benzoylbenzo[*b*]thiophen-3(2*H*)-ones have been synthesized by cyclization of

350   L. Testaferri, M. Tiecco, and P. Zanirato, *J. Org. Chem.*, 1975, **40**, 3392.
351   J. Weinstock, J. E. Blank, and B. M. Sutton, *J. Org. Chem.*, 1974, **39**, 2454.
352   P. B. Koster, M. J. Janssen, and E. A. C. Lucken, *J.C.S. Perkin II*, 1974, 803.
353   K. T. Potts and D. McKeough, *J. Amer. Chem. Soc.*, 1974, **96**, 4268.
354   K. T. Potts and D. McKeough, *J. Amer. Chem. Soc.*, 1974, **96**, 4276.
355   M. P. Cava, M. A. Sprecker, and W. R. Hall, *J. Amer. Chem. Soc.*, 1974, **96**, 1817.
356   O. Dann, H. Fick, B. Pietzner, E. Walkenhorst, R. Fernbach, and D. Zeh, *Annalen*, 1975, 160.
357   M. Pailer and H. Grünhaus, *Monatsh.*, 1974, **105**, 1362.
358   E. Campaigne, R. B. Rogers, A. Donelson, and T. R. Bosin, *J. Heterocyclic Chem.*, 1973, **10**, 979.
359   E. Campaigne and R. B. Rogers, *J. Heterocyclic Chem.*, 1973, **10**, 963.
360   E. Campaigne and Y. Abe, *J. Heterocyclic Chem.*, 1975, **12**, 889.
361   T. Higa and A. J. Krubsack, *J. Org. Chem.*, 1975, **40**, 3037.
362   A. J. Krubsack, R. Sehgal, W.-A. Loong, and W. E. Slack, *J. Org. Chem.*, 1975, **40**, 3179.
363   T. L. Fletcher, H.-L. Pan, C.-A. Cole, and M. J. Namkung, *J. Heterocyclic Chem.*, 1974, **11**, 815.
363a J. Van Coppenolle and M. Renson, *Compt. rend.*, 1975, **280**, C, 283.

the corresponding 2-(phenacylthio)benzoic acids.[364, 365] Similarly 2-mercapto-
benzoic acid was alkylated with bromoacetaldehyde dimethylacetal and ring-
closed by treatment with acetic anhydride to give 2-formyl-3-hydroxybenzo[*b*]-
thiophen.[366] Treatment of phenylpropiolic acid with hydrogen bromide and
sulphur dioxide led to 3-bromo-2-benzo[*b*]thiophencarboxylic acid.[367]

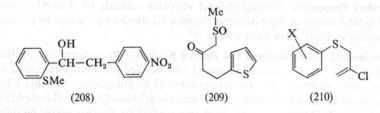

(208)   (209)   (210)

The Stobbe condensation products of some thienylcarbonyl compounds have
been cyclized to benzo[*b*]thiophen derivatives.[205] On acid treatment, β-keto-
sulphoxides of the thiophen series, such as (209), gave 5-hydroxybenzo[*b*]thiophen
as the main product in acetonitrile.[368] When *o*-nitrobenzonitriles reacted with
sodium sulphide in aqueous DMF, the anion of the corresponding *o*-mer-
captobenzonitrile was formed. *In situ* alkylation with chloroacetonitrile, chloroacetone, or phenacyl chloride, and subsequent sulphide-ion-catalysed cyclization,
yielded the corresponding 3-aminobenzo[*b*]thiophen-2-carbonitriles, 2-acetyl-,
or 2-benzoyl-3-aminobenzo[*b*]thiophen, respectively.[369] The chloropropenyl-
thioethers (210) gave 2-methylbenzo[*b*]thiophens when heated at 200 °C in
NN-diethylaniline.[370] Heating of (211), obtained through the condensation of
2-thenyl cyanide with (212), gave (213).[371] Substituted 4-(phenylthio)-4-methyl-

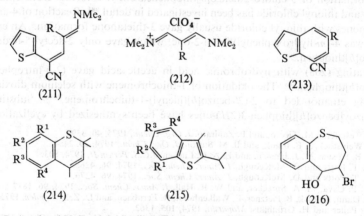

(211)   (212)   (213)

(214)   (215)   (216)

[364] S. B. Awad and N. F. Abdul-Malik, *Austral. J. Chem.*, 1975, **28**, 601.
[365] K. Görlitzer, *Arch. Pharm.*, 1976, **309**, 18.
[366] V. P. Litvinov, L. N. Smirnov, Ya. L. Gol'dfarb, N. N. Peturkhova, and É. G. Ostapenko,
Khim. geterotsikl. Soedinenii, 1975, 480.
[367] I. V. Smirnov-Zamkov and Y. L. Zborovskii, *Zhur. org. Khim.*, 1975, **11**, 1776.
[368] Y. Oikawa, O. Setoyama, and O. Yonemitsu, *Heterocycles*, 1974, **2**, 21.
[369] J. R. Beck and J. A. Yahner, *J. Org. Chem.*, 1974, **39**, 3440.
[370] W. K. Anderson, E. J. LaVoie, and J. C. Bottaro, *J.C.S. Perkin I*, 1976, 1.
[371] C. Jutz, R. M. Wagner, and H.-G. Löbering, *Angew. Chem.*, 1974, **86**, 781.

pentan-2-one underwent cyclodehydration by polyphosphoric acid to give thiochromones (214), which under these conditions rearranged to the benzo[b]-thiophens (215). From 4-naphthylthio-derivatives, 2-isopropyl-3-methylnaphtho-[1,2-b]thiophen was similarly obtained.[372] The ring-contraction of (216) and some other benzothiepin derivatives to benzo[b]thiophens has been reported.[373] Ozonolysis of benzo[b]thiophen followed by reaction with active methylene compounds led to 2-substituted benzo[b]thiophens.[374]

The action of dichloromethyl alkyl ethers and tin(IV) chloride on 2-allyl- or substituted 2-allyl-benzo[b]thiophens yielded the corresponding unsubstituted or 3-substituted derivatives of dibenzothiophen.[375] Dibenzothiophens have been obtained[376] from the photocyclodehydrogenation of diaryl sulphides. The reaction of 1,2,3-benzothiadiazole with phenyl radicals afforded, among other products, dibenzothiophen and 4-phenylthiodibenzothiophen.[377] Dibenzo-thiophen derivatives were also formed in the aprotic diazotization of 2-amino-phenyl 2-methylthiophenyl sulphide.[378] Eighteen substituted 4-naphtho[2,1-b]thio-phenmethanols have been synthesized. The ring system was prepared by photo-cyclization of α-(2-thienyl)-β-phenylacrylic acids.[379] Through reaction with Na₂S in DMF followed by ring-closure, (217) was transformed to (218), which by DDQ

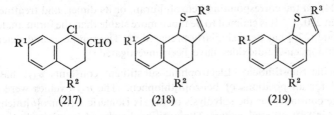

(217)   (218)   (219)

was then aromatized to (219).[380] Several other compounds containing additional fused benzene rings were obtained by this route.[380]

A new route to benzo[c]thiophen *via* the sulphilimine (220) has been found.[381] *trans*-1,4-Dimethoxycarbonyl-2,3-benzodithian, obtained by condensing dimethyl αα'-dibromobenzene-1,2-diacetate with potassium disulphide at low temperature,

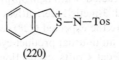

(220)

[372] D. D. MacNicol and J. J. McKendrick, *J.C.S. Perkin I*, 1974, 2493.
[373] A. Chatterjee and B. K. Sen, *J.C.S. Chem. Comm.*, 1974, 626.
[373a] H. Hofmann, H.-J. Haberstroh, B. Appler, B. Meyer, and H. Herterich, *Chem. Ber.*, 1975, **108**, 3596.
[374] K. J. Brown and O. Meth-Cohn, *Tetrahedron Letters*, 1974, 4069.
[375] J. Ashby, M. Ayad, and O. Meth-Cohn, *J.C.S. Perkin I*, 1974, 1744.
[376] K.-P. Zeller and H. Petersen, *Synthesis*, 1975, 532.
[377] L. Benati, P. C. Montevecchi, A. Tundo, and G. Zanardi, *J.C.S. Perkin I*, 1974, 1276.
[378] L. Benati, P. C. Montevecchi, A. Tundo, and G. Zanardi, *J.C.S. Perkin I.*, 1974, 1272.
[379] B. P. Das, M. E. Nuss, and D. W. Boykin, jun., *J. Medicin. Chem.*, 1974, **17**, 516.
[380] P. Cagniant and G. Kirsch, *Compt. rend.*, 1975, **281**, C, 393.
[381] Y. Tamura, H. Matsushima, and M. Ikeda, *Synthesis*, 1974, 277.

was easily decomposed to 1,3-dimethoxycarbonylbenzo[c]thiophen.[382] A zinc-induced intramolecular cyclization of 3,4-bis(bromoacetyl)thiophen led to (221), which upon treatment with DDQ gave the isobenzothiophenquinone (222), and hence the benzo[c]thiophen (223).[383] The benzo[c]thiophen (225) was obtained

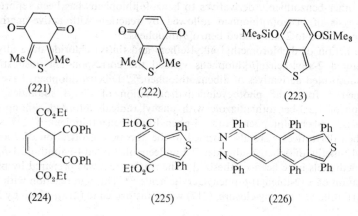

(221)   (222)   (223)

(224)   (225)   (226)

from (224) *via* the corresponding benzo[c]furan, or its dimer, and treatment with sulphur at 270 °C. It is claimed to be much more stable than the furan analogue.[384] Compound (226) has been similarly prepared.[385] The syntheses and reactions of benzo[c]thiophen-2,2-dioxides have been investigated.[386]

Electrophilic Substitution.—Electrophilic substituent constants σ_{Ar}^{+} have been obtained for all positions of benzo[b]thiophen. The σ_{Ar}^{+} values were defined from rate constants for the solvolysis of the six isomeric 1-(benzo[b]thienyl)ethyl chlorides in 80% ethanol–water. The positional order of reactivity in the benzo-[b]thiophen ring was determined to be 3 > 2 > 6 > 5 > 4 > 7. All positions were more reactive than benzene.[387] Ionic hydrogenation has been successfully carried out on benzo[b]thiophens, yielding 2,3-dihydro-derivatives.[388] 2-Piperidinobenzo[b]thiophen, prepared from 2,3-dibromobenzo[b]thiophen, was brominated in the 3-position. Nitration gave the 3,6-dinitro-derivative.[388a] 2,4-Dinitrobenzenediazonium ions reacted with benzo[b]thiophen to give the arylated derivative, while the 2- and 3-methyl derivatives coupled to give azo-dyes.[87] Photocyclization of (227) gave the expected (228), while the isomeric 2-naphthylethylene derivative furnished an unusual product cyclized at the β-position of the naphthalene nucleus.[389] Friedel–Crafts cyclization of the acid chloride (229; X = COCl) gave (230).[390] Nitrene insertion into the 3-position by treatment of

[382] G. Cignarella and G. Cordella, *Gazzetta*, 1974, **104**, 455.
[383] E. Ghera, Y. Gaoni, and D. H. Perry, *J.C.S. Chem. Comm.*, 1974, 1034.
[384] G. Freslon and Y. Lepage, *Compt. rend.*, 1975, **280**, C, 961.
[385] L. Lepage and Y. Lepage, *Compt. rend.*, 1975, **280**, C, 897.
[386] M. P. Cava and J. McGrady, *J. Org. Chem.*, 1975, **40**, 72.
[387] D. S. Noyce and D. A. Forsyth, *J. Org. Chem.*, 1974, **39**, 2828.
[388] G. I. Bolestova, A. N. Korepanov, Z. N. Parnes, and D. N. Kursanov, *Izvest. Akad. Nauk S.S.S.R., Ser. khim.*, 1974, 2547.
[388a] K. E. Chippendale, B. Iddon, H. Suschitzky, and D. S. Taylor, *J.C.S. Perkin I*, 1974, 1168.
[389] A. Croisy, P. Jacquignon, and F. Perin, *J.C.S. Chem. Comm.*, 1975, 106.
[390] B. Iddon, H. Suschitzky, D. S. Taylor, and K. E. Chippendale, *J.C.S. Perkin I*, 1974, 2500.

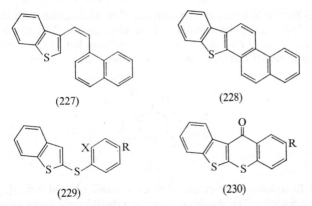

(227)

(228)

(229)

(230)

(229; X = NO₂) with triethyl phosphite or by heating the azide (229; X = N₃) was not successful, but phenothiazine analogues were obtained by the Bamford–Stevens reaction with (229; X = CH=NNHTos).[390]

Metallation and Halogen–Metal Exchange.—3-Formyl-2-benzo[*b*]thiophenthiol has been prepared in a one-pot procedure [391] from 3-bromobenzo[*b*]thiophen by halogen–metal exchange, followed by reaction with DMF, metallation of the intermediate with butyl-lithium, and subsequent reaction with sulphur. Reaction of 2,3,4,5,6,7-hexachlorobenzo[*b*]thiophen with butyl-lithium gave either the 2-lithio- or the 2,6-dilithio-derivative, depending upon conditions.[392] Metallation of benzo[*b*]thiophen with butyl-lithium followed by reaction with benzonitrile and malononitrile gave α-cyano-β-(2-benzo[*b*]thienyl)cinnamonitrile.[392a]

2- and 3-Hydroxybenzo[*b*]thiophens and Related Systems.—During this period, a marked increase in interest in the properties and reactions of these compounds is evident. The composition of the keto–enol equilibrium in some chlorinated 3-hydroxy-benzo[*b*]thiophens has been studied by n.m.r.[393] Russian workers have studied tautomeric equilibria, spectra, and chelating properties of the imines of 3-hydroxy-2-formylbenzo[*b*]thiophen,[366, 394, 395] 2-hydroxy-3-formylbenzo[*b*]thiophen,[396, 397] of 2-formyl-3-mercaptobenzo[*b*]thiophens [398] and of 3-formyl-2-mercaptobenzo[*b*]thiophen.[391] Benzo[*b*]thiophen-3(2*H*)-ones have been transformed into 3-(1-pyrrolidinyl)- and 3-(1-morpholinyl)benzo[*b*]thiophens.[399] These

[391] V. P. Litvinov, Ya. L. Gol'dfarb, V. V. Zelentsov, L. G. Bogdanova, and N. N. Petukhova, *Khim. geterotsikl. Soedinenii*, 1975, 486.

[392] G. M. Brooke and R. King, *Tetrahedron*, 1974, **30**, 857.

[392a] E. Campaigne, D. Mais, and E. M. Yokley, *Synthetic Comm.*, 1974, **4**, 379.

[393] B. Stridsberg and S. Allenmark, *Chem. Scripta*, 1974, **6**, 184.

[394] V. P. Litvinov, Ya. L. Gol'dfarb, and É. G. Ostapenko, *Izvest. Akad. Nauk S.S.S.R.*, *Ser. khim.*, 1974, 2279.

[395] V. I. Usacheva, Z. V. Bren, V. A. Bren, and V. I. Minkin, *Khim. geterotsikl. Soedinenii*, 1975, 623.

[396] Ya. L. Gol'dfarb, V. P. Litvinov, V. S. Bogdanov, G. V. Isagulyants, É. G. Ostapenko, and A. A. Greish, *Izvest. Akad. Nauk S.S.S.R.*, *Ser. khim.*, 1974, 2284.

[397] V. A. Bren, V. I. Usacheva, V. I. Minkin, and M. V. Nekhoroshev, *Khim. geterotsikl. Soedinenii*, 1975, 635.

[398] V. A. Bren, V. I. Usacheva, Zh. V. Bren, B. Ya. Simkin, and V. I. Minkin, *Khim. geterotsikl. Soedinenii*, 1974, 631.

[399] J. Weber and P. Faller, *Bull. Soc. chim. France*, 1975, 783.

compounds behave like enamines in their reactions with activated halides, olefins, acetic anhydride, and methanesulphonyl chloride, giving *C*-substitution.[400] The reaction of thianaphthen-2-one with salicylaldehydes led to (231), which could be aromatized to 6*H*-benzothieno[3,2-*c*][1]benzopyran-6-ones.[401]

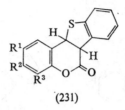

(231)

Side-chain Reactions.—The preparation of a series of azidobenzo[*b*]thiophens has been reported. The 5-azides gave 2-methylthieno[2,3-*g*]benzoxazoles on thermolysis in a mixture of polyphosphoric and acetic acids. 4-Azidobenzo[*b*]-thiophen underwent a Bamberger-type rearrangement under similar conditions to give 4-acetamido-7-acetoxybenzo[*b*]thiophen.[402] Photolysis of 4-azidobenzo[*b*]-thiophen in diethylamine gave a mixture of 4-aminobenzo[*b*]thiophen and 4,4'-azobenzo[*b*]thiophen, while from 5-azidobenzo[*b*]thiophen only 4-amino-5-(diethylamino)benzo[*b*]thiophen was isolated.[403] Photolysis of certain 6-azido-benzo[*b*]thiophens gave a mixture of a 7-amino-6-(diethylamino)benzo[*b*]thiophen and the ring-expanded 6-diethylamino-8*H*-thieno[2,3-*c*]azepine, or either of these compounds, depending on the reaction conditions.[404] Photolysis of 6-azido-2,3-dibromobenzo[*b*]thiophen in an excess of diethylamine for 18 h gave mainly 7-amino-2,3-dibromo-6-(diethylamino)benzo[*b*]thiophen. After 9 h, the major product was 2,3-dibromo-6-diethylamino-8*H*-thieno[2,3-*c*]azepine.[405] The catalytic liquid-phase oxidation of some 2- and 3-alkyl- and 2- and 3-hydroxymethyl-benzo[*b*]thiophens has been studied.[406] 2,3-Diformylbenzo[*b*]thiophen was treated with ethynylmagnesium bromide and the product transformed to rhodium complexes[407] (see p. 271). Thieno[3,2-*c*]phenanthridines were prepared by benzyne-type cyclization of the Schiff base from 4-aminobenzo[*b*]thiophen and *o*-chlorobenzaldehydes.[408] The reaction of 5-aminobenzo[*b*]thiophen with ethyl acetoacetate has been investigated. A number of thienoquinoline derivatives were obtained.[409] Complex styryl derivatives containing benzo[*b*]thiophen rings have been prepared and their spectral properties studied.[410, 411] Cycloaddition reactions to benzo[*b*]thiophens have been studied. The [2 + 2] cycloaddition of 3-pyrroli-dinobenzo[*b*]thiophen to dimethyl acetylenedicarboxylate gave (232) as an

[400] J. Weber and P. Faller, *Compt. rend.*, 1975, **281**, *C*, 389.
[401] R. A. Conley and N. D. Heindel, *J. Org. Chem.*, 1975, **40**, 3169.
[402] B. Iddon, H. Suschitzky, D. S. Taylor, and M. W. Pickering, *J.C.S. Perkin I*, 1974, 575.
[403] B. Iddon, H. Suschitzky, and D. S. Taylor, *J.C.S. Perkin I*, 1974, 579.
[404] B. Iddon, M. W. Pickering, and H. Suschitzky, *J.C.S. Chem. Comm.*, 1974, 759.
[405] B. Iddon, M. W. Pickering, H. Suschitzky, and D. S. Taylor, *J.C.S. Perkin I*, 1975, 1686.
[406] T. V. Shchedrinskaya, V. P. Litvinov, P. A. Konstantinov, Ya. L. Gol'dfarb, and É. G. Ostapenko, *Khim. geterotsikl. Soedinenii*, 1973, 1026.
[407] E. Müller, E. Luppold, and W. Winter, *Chem. Ber.*, 1975, **108**, 237.
[408] S. V. Kessar, P. K. Khullar, and P. Jit, *Indian J. Chem.*, 1973, **11**, 1191.
[409] S. M. A. D. Zayed and A. Emran, *J. prakt. Chem.*, 1974, **316**, 192.
[410] J. Garmatter and A. E. Siegrist, *Helv. Chim. Acta*, 1974, **57**, 945.
[411] A. de Buman and A. E. Siegrist, *Helv. Chim. Acta*, 1974, **57**, 1352.

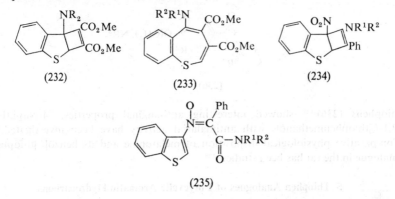

(232) (233) (234)

(235)

intermediate, which rearranged to give (233).[412] 3-Nitrobenzo[*b*]thiophen reacted with 1-dimethylamino-2-phenylacetylene to give a mixture of two 1:1 adducts, (234) and (235).[413] 3-Pyrrolidinobenzo[*b*]thiophen also underwent cycloaddition with 1,3-dipolar reagents such as benzonitrile oxides, diaryl nitrilimines, and electron-deficient dienes.[414] Benzo[*b*]thiophen reacts with tetracyanoethylene oxide.[99]

Reaction at Sulphur.—The oxidation of benzo[*b*]thiophens to sulphoxides by t-butyl hypochlorite has been studied.[415] In the presence of a ruthenium catalyst, dibenzothiophen was oxidized by oxygen to the sulphone.[416] The reactions of benzo[*b*]thiophen 1,1-dioxides with various reagents such as amines,[417] amino-alcohols,[418] 3-aminopropylsilanes,[419] and aromatic sulphonyl chlorides[420] have been studied. Four detailed papers on the reaction of dibenzophenonium salts with aryl-lithiums have appeared in which the mechanism of ligand exchange was elucidated.[421-424]

Pharmacologically Active Compounds.—Benzo[*b*]thiophen-2- and -3-carbox-aldehyde thiosemicarbazones have been prepared and screened for antiviral activity.[425] 3-Benzo[*b*]thiophenoxyaminopropanols[426] and the complex benzo[*b*]-

412 D. N. Reinhoudt and C. G. Kouwenhoven, *Tetrahedron*, 1974, **30**, 2431.
413 D. N. Reinhoudt and C. G. Kouwenhoven, *Tetrahedron Letters*, 1974, 2503.
414 D. N. Reinhoudt and C. G. Kouwenhoven, *Rec. Trav. chim.*, 1974, **93**, 321.
415 P. Geneste, J. Grimaud, J.-L. Olivé, and S. N. Ung, *Tetrahedron Letters*, 1975, 2345.
416 M. A. Ledlie and I. V. Howell, *Tetrahedron Letters*, 1976, 785.
417 F. Sauter and U. Jordis, *Monatsh.*, 1974, **105**, 1252.
418 V. E. Udré and M. G. Voronkov, *Khim. geterotsikl. Soedinenii*, 1972, 1602.
419 V. E. Udré and É, Ya. Lukevits, *Khim. geterotsikl. Soedinenii*, 1973, 493.
420 I. U. Numanov, I. M. Nasyrov, and U. K. Karimov, *Khim. geterotsikl. Soedinenii*, 1973, 425.
421 M. Hori, T. Kataoka, H. Shimizu, and M. Miyagaki, *Chem. and Pharm. Bull. (Japan)*, 1974, **22**, 1711.
422 M. Hori, T. Kataoka, H. Shimizu, and M. Miyagaki, *Chem. and Pharm. Bull. (Japan)*, 1974, **22**, 2004.
423 M. Hori, T. Kataoka, H. Shimizu, M. Miyagaki, and M. Murase, *Chem. and Pharm. Bull. (Japan)*, 1974, **22**, 2014.
424 M. Hori, T. Kataoka, H. Shimizu, and M. Miyagaki, *Chem. and Pharm. Bull. (Japan)*, 1974, **22**, 2020.
425 R. P. Dickinson, B. Iddon, and R. G. Sommerville, *Internat. J. Sulfur Chem.*, 1973, **8**, 233.
426 C. Goldenberg, R. Wandestrick, C. Van Meerbeeck, M. Descamps, J. Bauthier, and R. Charlier, *Chim. Thérap.*, 1974, **9**, 123.

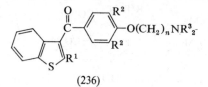

(236)

thiophens (236)[427] showed interesting anti-anginal properties. 4-Naphtho-[2,1-*b*]thiophenmethanols with antimalarial activity have been investigated.[379] Comparative physiological disposition of melatonine and its benzo[*b*]thiophen analogue in the rat has been studied.[428]

5 Thiophen Analogues of Polycyclic Aromatic Hydrocarbons

Thiophen Analogues of Anthracene and Phenanthrene.—The remaining member (237) of the thieno-analogues of anthrone-anthrol has been synthesized from the known (238) by reduction of one of the carbonyl groups to the ketal, followed by

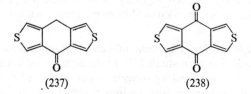

.(237)   (238)

replacement of the hydroxy-group and dechlorination by copper in boiling propionic acid. Only the keto-form was observed by n.m.r.[429] The synthesis[430] and iodination[431] of 2*H*-naphtho[1,8-*bc*]thiophen derivatives has been described. Thienoperylenetetracarboxylic acid and thienoperylene have been synthesized.[432] The gas chromatographic behaviour of condensed heteroaromatic systems including a thiophen ring has been studied.[433]

Thiophen Analogues of Helicenes.—The reaction of seven heterohelicenes with AlCl₃ has been described. The products were compounds in which the two helical termini of a helicene are connected by a σ-bond. The intramolecular ring-closure was limited to hetero-[5]- and -[6]-helicenes.[434] A series of regularly annelated thiaheterohelicenes has been prepared by a new progressive annelation technique based on the key intermediate 2-methylbenzo[1,2-*b*:4,3-*b'*]dithiophen.[435] Two methano-bridged heterohelicenes have been prepared.[436] The synthesis of some

[427]   J. Gubin, N. Claeys, E. Deray, M. Descamps, J. Bauthier, J. Richard, and R. Charlier, *Chim. Thérap.*, 1975, **10**, 418.
[428]   R. P. Maickel, T. R. Bosin, S. D. Harrison, jun., and M. A. Riddle, *Life Sci.*, 1974, **14**, 1735.
[429]   D. W. H. MacDowell and L. F. Ballas, *J. Org. Chem.*, 1974, **39**, 2239.
[430]   U. Folli, D. Iarossi, and F. Taddei, *J.C.S. Perkin II*, 1974, 933.
[431]   M. A. Mostoslavskii, S. I. Saenko, and V. L. Belyaev, *Khim. geterotsikl. Soedinenii*, 1973, 1034.
[432]   V. I. Rogovik, *Zhur. org. Khim.*, 1974, **10**, 1072.
[433]   V. P. Litvinov, V. A. Ferapontov, D. D. Gverdtsiteli, and É. G. Ostapenko, *Khim. geterotsikl. Soedinenii*, 1973, 188.
[434]   J. H. Dopper, D. Oudman, and H. Wynberg, *J. Org. Chem.*, 1975, **40**, 3398.
[435]   P. G. Lehman and H. Wynberg, *Austral. J. Chem.*, 1974, **27**, 315.
[436]   H. Numan and H. Wynberg, *Tetrahedron Letters*, 1975, 1097.

heterocirculenes such as (239) and (240) has been reported. Two classes of circulenes, planar and non-planar, have been recognized based on the study of models. Depending on the ratio of the outer and inner radii, bowl-shaped and

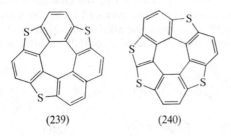

(239)   (240)

corrugated non-planar circulenes might exist. Attempts to prove that [7]-hetero-circulenes belonged to the corrugated type of circulenes were unsuccessful.[437]

Thiophen Analogues of Indene.—Quantitative data for the base-catalysed tautomerization of the 4-methyl- to the 6-methyl-4*H*-cyclopenta[*c*]thiophen have been obtained. The tautomerization was about 10^4 times slower than that of 1-methylindene.[105] 10*H*-Indeno[1,2-*b*][1]benzothiophen has been prepared by treating 2-phenylbenzo[*b*]thiophen-3-carboxaldehyde with sodium methoxide in hot bis(2-methoxyethyl) ether.[438]

Thiophen-fused Tropylium Ions and Related Compounds.—The reaction of some thieno-tropones such as (241) with dimethylsulphoxonium methylide gave cyclo-propane derivatives (242).[439] The reaction of (241) with electrophilic reagents has been studied.[440] The synthesis of compounds of type (241) has been modified

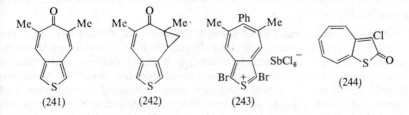

(241)   (242)   (243)   (244)

and the 2-thia-azulenylium salt (243) has been prepared.[441] Reaction of cyclo-heptatrienethione with dichloroacetyl chloride and triethylamine in dichloro-methane gave (244).[442] Thieno[3',2':4,5]cyclohepta[1:2-*b*]naphtho[2,3-*d*]furan and thieno[3',2':4,5]benzo[4,5]bicyclohepta[1,2-*b*:2',1'-*d*]furan have been synthe-sized.[442a]

[437] J. H. Dopper and H. Wynberg, *J. Org. Chem.*, 1975, **40**, 1957.
[438] B. Iddon, H. Suschitzky, and D. S. Taylor, *J.C.S. Perkin I*, 1974, 2505.
[439] R. Guilard and B. Hanquet, *Compt. rend.*, 1974, **278**, C, 295.
[440] M. Hori, T. Kataoka, H. Shimizu, and S. Yoshimura, *Yakugaku Zasshi*, 1974, **94**, 1445.
[441] M. Hori, T. Kataoka, H. Shimizu, and S. Yoshimura, *Yakugaku Zasshi*, 1974, **94**, 1429.
[442] R. Cabrino, C. Biggi, and F. Pietra, *Synthesis*, 1974, 276.
[442a] L. M. Gones, *Compt. rend.*, 1974, **278**, C, 145.

6 Thiophens Fused to Five-membered Aromatic Heterocyclic Rings

Pyrrole- and Furan-fused Thiophens and Related Compounds.—The interest in the chemistry of thienopyrroles is still increasing. Starting from 2-amino-3-carbonyl-substituted thiophens obtained by the Gewald reaction, the *N*-ethoxy-carbonylmethyl derivative (245) was prepared either by direct alkylation with ethyl bromoacetate,[443] or by reductive alkylation with methyl glyoxalate.[444] The compounds (245) were then ring-closed directly,[443] or after acetylation,[444] to the

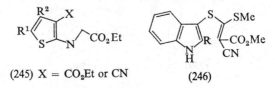

(245) X = CO$_2$Et or CN   (246)

thieno[2,3-*b*]pyrroles, which can be *O*-alkylated or *N*-acetylated.[443] Reaction of ethyl azidoacetate with thiophen-3-carbaldehyde leads to thieno[2,3-*b*]pyrrole. The synthesis of the four formyl derivatives has been described.[445] Another nitrene-insertion reaction is the synthesis of 6-methylthieno[3,2-*b*]pyrrole from 2-isopropenyl-3-nitrothiophen and triethyl phosphite.[197] The starting material can be prepared from bis(2-isopropenyl-3-thienyl)iodonium chloride. Thieno-[3,2-*b*]pyrroles have been obtained through the acid-catalysed reaction of 4-ethoxycarbonyl-5-methyl-3-thienylhydrazines with ketones.[446] Another paper on the synthesis of thieno[3,2-*b*]indoles from (246) has appeared.[447] 4,5-Dialkyl-2-acylamino-thiophens or -3-acylamino-thiophens and oxalyl chloride give the thiophen analogues of the isatin system, 4,5-dioxothieno[2,3-*b*]- and 5,6-dioxo-thieno[3,2-*b*]-pyrroles.[448] *O*-Alkylation of 3-hydroxy-2-methoxycarbonylbenzo-[*b*]thiophen with methyl chloroacetate in the presence of potassium t-butoxide in DMSO, followed by Dieckmann cyclization, led to [1]benzothieno[3,2-*b*]furan derivatives.[449]

Pyrazole-, Thiazole-, and Imidazole-fused Thiophens and Related Systems.—The behaviour of thieno- and selenolo-[-3,2-*d*]pyrazoles in formylation and bromina-tion reactions has been studied. In contrast to benzo[*b*]thiophen, a substituent is directed into the α-position of the condensed thiophen ring.[450] 2-Mercapto-thieno[3,2-*d*]thiazole has been prepared from 2-chloro-3-nitrothiophen and sodium disulphide followed by carbon disulphide. A benzo-fused system was prepared similarly from 3-bromo-2-nitrobenzo[*b*]thiophen.[451] Derivatives of

[443] M. Wierzbicki, D. Cagniant, and P. Cagniant, *Bull. Soc. chim. France*, 1975, 1786.
[444] R. A. Crochet, jun., J. T. Boatright, C. DeWitt Blanton, jun., C. T. Wie, and W. E. Hochholzer, *J. Heterocyclic Chem.*, 1974, **11**, 143.
[445] S. Soth, M. Farnier, and P. Fournari, *Bull. Soc. chim. France*, 1975, 2511.
[446] V. I. Shvedov, Y. I. Trofimkin, V. K. Vasileva, and A. N. Grinev, *Khim. geterotsikl. Soedinenii*, 1975, 1324.
[447] S. Kisaki, Y. Tominaga, Y. Matsuda, and G. Kobayashi, *Chem. and Pharm. Bull. (Japan)*, 1974, **22**. 2246.
[448] V. I. Shvedov, V. K. Vasileva, I. A. Kharizomenova, and A. N. Grinev, *Khim. geterotsikl. Soedinenii*, 1975, 767.
[449] J. R. Beck, *J. Heterocyclic Chem.*, 1975, **12**, 1037.
[450] Yu. N. Koshelev, A. V. Reznichenko, L. S. Éfros, and I. Ya. Kvitko, *Zhur. org. Khim.*, 1973, **9**, 2201.
[451] N. I. Astrakhantseva, V. G. Zhirvakov, and P. I. Abramenko, *Khim. geterotsikl. Soedinenii*, 1975, 1607.

[1]benzothieno[2,3-*c*]pyrazoles[452] and of 1*H*-[1]benzothieno[3,2-*c*]pyrazole[453] were synthesized from benzo[*b*]thiophen 1,1-dioxides by 1,3-dipolar additions of diazoalkanes,[452] or *C*-methyl-*N*-phenyl- and *CN*-diaryl-nitrilimines.[453]

Thieno[2,3-*d*]thiazoles were obtained by thiocyanation of 4,5-dialkyl-2-acylamino-thiophens followed by ring-closure.[454] The synthesis of *meso*-alkyl- and *meso*-alkoxy-thieno[3,2-*d*]thiazolocarbocyanines (247) has been achieved.[455] Reduction of 3-formamido-2-nitrobenzo[*b*]thiophen gave [1]benzothieno[2,3-*d*]-imidazole-2(3*H*)-one in low yield. Efforts to extend these cyclizations to analogous

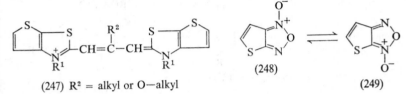

(247) R² = alkyl or O—alkyl   (248)   (249)

thiophens did not succeed.[456] Thieno[2,3-*c*]furazan oxide [(248), (249)] has been synthesized from 3-azido-2-nitrothiophen.[457] The dominating tautomer is (248) and the free energy of activation for the rearrangement has been determined by n.m.r.

7 Thiophens Fused to Six-membered Aromatic Heterocyclic Rings

Thiophen Analogues of Quinoline.—A very convenient method for the synthesis of the parent furano-, thieno- and selenolo-[3,2-*b*]pyridine, and of the 5- and 5,6-substituted derivatives, has been found by applying the Friedländer reaction to 3-amino-2-formyl-furan, -thiophen, and -selenophen. The n.m.r. spectra of these systems have been analysed.[458] 3-Amino-2-benzoyl-4-cyano-5-phenyl-thiophen similarly gives the substituted thieno[3,2-*b*]pyridine (250). Thieno-[2,3-*b*]pyridines are obtained from 2-amino-3-benzoyl-4,5-tetramethylene-thiophen.[459, 459a] 3-Cyanopyridine-2-thiols react with halogen compounds having an electron-attracting group in the α-position, in DMF containing potassium hydroxide, to give 3-aminothieno[2,3-*b*]pyridines. Their hydrolysis and acetylation were investigated.[460] Thieno[2,3-*b*]pyridine has been chlorinated, brominated, and iodinated in the 3-position by means of elemental halogen, silver sulphate, and sulphuric acid. Various transformations of these halogeno-derivatives and their mass-spectral fragmentation have been studied.[461] Treatment of thieno-[2,3-*b*]pyridine with butyl-lithium in hexane–tetramethylethylenediamine at

[452] F. Sauter and G. Büyük, *Monatsh.*, 1974, **105**, 550.
[453] F. Sauter, G. Büyük, and U. Jordis, *Monatsh.*, 1974, **105**, 869.
[454] V. I. Shvedov, I. A. Kharizomenova, and A. N. Grinev, *Khim. geterotsikl. Soedinenii*, 1974, 1204.
[455] V. G. Zhiryakov, N. I. Astrakhantseva, P. I. Abramenko, and I. I. Levkoev, *Doklady Akad. Nauk S.S.S.R.*, 1974, **219**, 870.
[456] P. N. Preston and S. K. Sood, *J.C.S. Perkin I*, 1976, 80.
[457] A. J. Boulton and D. Middleton, *J. Org. Chem.*, 1974, **39**, 2956.
[458] S. Gronowitz, C. Westerlund, and A.-B. Hörnfeldt, *Acta Chem. Scand. (B)*, 1975, **29**, 233.
[459] H. Schäfer, K. Gewald, and M. Hartmann, *J. prakt. Chem.*, 1974, **316**, 169.
[459a] V. I. Shvedov, I. A. Kharitzomenova, and A. N. Grinev, *Khim. geterotsikl. Soedinenii*, 1974, 58.
[460] F. Guerrera, M. A. Siracusa, and B. Tornetta, *Farmaco (Pavia)*, Ed. Sci., 1976, **31**, 21.
[461] L. H. Klemm, R. E. Merrill, F. H. W. Lee, and C. E. Klopfenstein, *J. Heterocyclic Chem.*, 1974, **11**, 205.

292   *Organic Compounds of Sulphur, Selenium, and Tellurium*

− 70 °C led to metallation in the 2-position and not addition to the azomethine bond. Reaction with DMF gave the 2-formyl-derivative. The 3-lithio-derivative was obtained by halogen–metal exchange with 3-bromothieno[2,3-*b*]pyridine.[462] Further studies of the chlorination and *S*-oxidation of thieno[2,3-*b*]pyridine have been reported.[463] The mass-spectral fragmentation patterns of a series of thieno-pyridine *N*-oxides, *S*-oxides, and *SS*-dioxides have been elaborated.[464] The reductive acetylation of nitro-thieno-pyridines has been studied.[465] Polymethine

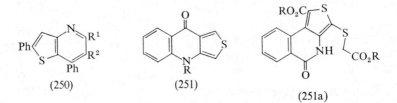

(250)   (251)   (251a)

dyes derived from thieno[2,3-*b*]pyridine have been prepared.[466] Benzothieno- and benzoselenolo-[2,3-*b*]pyridines have been prepared by the reaction of 2-chloro-3-nitro-6-methylpyridine with sodium thiophenoxide or sodium selenophenoxide, reduction of the nitro-groups with stannous chloride in concentrated HCl, diazotization, and thermal decomposition of the diazonium compounds in 25% sulphuric acid.[467, 468] The syntheses of thieno[3,4-*b*]quinoline-9(4*H*)-ones (251)[34] substituted in position C-4, and the similar system (251a),[468a] have been achieved.

Thiophen Analogues of Isoquinoline.—4,7-Dimethylthieno[2,3-*c*]pyridine has been synthesized by hydrolysing the acetal function of (252). This compound was obtained by the reaction of 2-(3-lithio-2-thienyl)-1,3-dioxolan with diacetyl followed by oximation and reduction. 6,7-Dimethylthieno[3,2-*c*]pyridine was obtained in an analogous manner.[469] (2,5-Dimethyl-3-thienyl)acetone gave (253) on acylation. Treatment with perchloric acid gave the pyrylium salt (254), which upon reaction with NH₃ gave (255).[470] The same reaction sequence with (3-benzo-

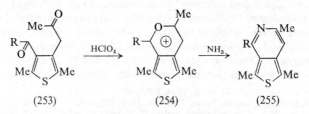

(253)   (254)   (255)

462 L. H. Klemm and R. E. Merrill, *J. Heterocyclic Chem.*, 1974, **11**, 355.
463 L. H. Klemm, R. E. Merrill, and F. H. W. Lee, *J. Heterocyclic Chem.*, 1974, **11**, 535.
464 L. H. Klemm, S. Rottschaefer, and R. E. Merrill, *J. Heterocyclic Chem.*, 1975, **12**, 1265.
465 L. H. Klemm and W. Hsin, *J. Heterocyclic Chem.*, 1975, **12**, 1183.
466 P. I. Abramenko and V. G. Zhiryakov, *Khim. geterotsikl. Soedinenii*, 1975, 475.
467 P. I. Abramenko and V. G. Zhiryakov, *Khim. geterotsikl. Soedinenii*, 1972, 1541.
468 P. I. Abramenko, V. G. Zhiryakov, L. A. Balykova, and T. K. Ponomareva, *Khim. geterotsikl. Soedinenii*, 1974, 796.
468a S. Ueno, Y. Tominaga, R. Natsuki, Y. Matsuda, and G. Kobayashi, *Yakugaku Zasshi*, 1974, **94**, 607.
469 E. Sandberg, *Chem. Scripta*, 1975, **7**, 223.
470 V. I. Dulenko and N. N. Alekseev, *Khim. geterotsikl. Soedinenii*, 1975, 631.

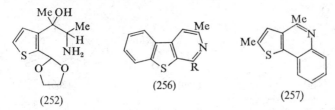

(252)   (256)   (257)

[*b*]thiophen)acetone was used for the synthesis of (256).[471] Decomposition of 3-(*o*-azidobenzyl)-2,5-dimethylthiophen yielded (257).[114] Lanthanide-induced shifts provided a convenient method for the total assignment of the ^{1}H n.m.r. spectra of benzothieno-pyridines and determination of the substitution sites.[472] Some derivatives of 4,5,6,7-tetrahydrothieno[3,2-*c*]pyridines are of pharmacological interest.[473, 474]

Pyrimidine-fused Systems.—There has been considerable interest in this field since the Gewald and Fiesselmann ring-closure reactions, leading to *ortho*-aminocarbonyl derivatives of thiophen and related compounds, opened the route to pyrimidine-fused thiophens. Thieno[3,4-*d*]pyrimidin-4-one has been synthesized from 3-methoxycarbonyl-4-formamido-thiophens. It undergoes electrophilic substitution at the lactam-nitrogen as well as on the thiophen ring.[475] 3,4-Dihydro-4-oxothieno[2,3-*d*]pyrimidines have been synthesized by cyclization of thiophen amino-esters, amino-amides, or amino-nitriles, and their electrophilic substitution reactions studied.[7, 476, 477, 477a] Chloro- and bromo-thieno[2,3-*d*]-pyrimidines have been synthesized by halogenodehydroxylation of thieno-pyrimidones. Nucleophilic substitutions have been studied with these derivatives.[478] The cyclization of 4-hydrazinothieno[2,3-*d*]pyrimidines led to triazolo- and tetrazolo-fused ring systems.[477, 479] Substitution and addition reactions of thieno[2,3-*d*]pyrimidines have been studied.[480] Some 4-amino- and 4-phenyl-thieno[2,3-*d*]pyrimidine 3-oxides have been prepared.[481] 3-Acetylamino-2-acetylthiophen gave thieno[3,2-*d*]pyrimidine. The selenophen analogue was prepared in a similar way.[482] Oxidative photocyclization of the uracil derivative (258) gave the benzothieno-pyrimidine (259).[483] The reaction of 2-amino-3-ethoxycarbonyl-thiophens with 2,3-dibromopropyl isothiocyanate gave (260).[484]

[471] V. I. Dulenko, V. I. Volbushko, L. V. Dulenko, and G. N. Dorofeenko, *Khim. geterotsikl. Soedinenii*, 1974, 273.
[472] G. Gacel, M. C. Fournié-Zaluski, and B. P. Roques, *J. Heterocyclic Chem.*, 1975, **12**, 623.
[473] J. P. Maffrand and F. Eloy, *Chim. Thérap.*, 1974, **9**, 483.
[474] M. Podesta, D. Aubert, and J. C. Ferrand, *Chim. Thérap.*, 1974, **9**, 487.
[475] M. Robba and N. Boutamine, *Bull. Soc. chim. France*, 1974, 1629.
[476] M. Robba, J. M. Lecomte, and M. Cugnon de Sevricourt, *Bull. Soc. chim. France*, 1975, 587.
[477] J. Bourguignon, E. Gougeon, G. Quéguiner, and P. Pastour, *Bull. Soc. chim. France*, 1975, 815; J. Bourguignon, M. Moreau, G. Quéguiner, and P. Pastour, *ibid.*, p. 2483.
[478] M. Robba, J.-M. Lecomte, and M. Cugnon de Sevricourt, *Bull. Soc. chim. France*, 1975, 592.
[479] M. Robba, M. Cugnon de Sevricourt, and J.-M. Lecomte, *J. Heterocyclic Chem.*, 1975, **12**, 525.
[480] M. Robba and M. Cugnon de Sevricourt, *J. Heterocyclic Chem.*, 1975, **12**, 921.
[481] J. Fortea, *J. prakt. Chem.*, 1975, **317**, 705.
[482] G. Ah-Kow, C. Paulmier, and P. Pastour, *Compt. rend.*, 1974, **278**, C, 1513.
[483] S. Senda, K. Hirota, and M. Takahashi, *J.C.S. Perkin I*, 1975, 503.
[484] A. A. Dobosh, I. V. Smolanka, and S. M. Khripak, *Khim. geterotsikl. Soedinenii*, 1974, 134.

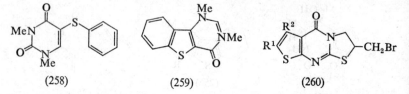

(258)   (259)   (260)

Thieno[2,3-*d*]pyrimidines have been synthesized [459a] from 5-acylamino-2,3-dialkylthiophen-4-aldehydes. 2-Amino-3-cyano-thiophens have been transformed into the 2-cyanamino-3-cyano-thiophens and ring-closed to 4-aminothieno[2,3-*d*]-pyrimidines.[485] Ethyl 3-aminothieno[2,3-*b*]pyridine-2-carboxylate similarly gives (261), whose reactions have been studied.[486] The thieno-pyrimidone (262) has been desulphurized by Raney nickel to the 4-oxo-pyrimidine (263).[487]

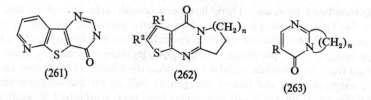

(261)   (262)

(263)

Sauter and co-workers have reported extensive studies on thienopyrimidines to which additional heterocyclic rings, *e.g.* the thiazolo- or triazolo-ring, have been fused.[488-494] Compounds such as (264),[488, 490, 491] (265),[489] (266),[492] (267), and (268) have thus been prepared.

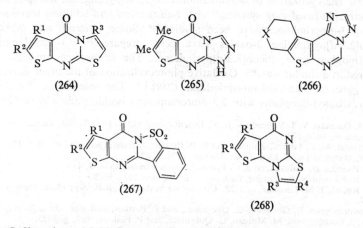

(264)   (265)   (266)

(267)

(268)

[485] L. G. Sharanina and S. N. Baranov, *Khim. geterotsikl. Soedinenii*, 1974, 196.
[486] S. W. Schneller and F. W. Clough, *J. Heterocyclic Chem.*, 1974, **11**, 975.
[487] V. I. Shvedov, I. A. Kharizomenova, and A. I. Grinev, *Khim. geterotsikl. Soedinenii*, 1975, 765.
[488] F. Sauter and W. Deinhammer, *Monatsh.*, 1974, **105**, 452.
[489] F. Sauter and W. Deinhammer, *Monatsh.*, 1974, **105**, 558.
[490] F. Sauter, W. Deinhammer, and K. Danksagmüller, *Monatsh.*, 1974, **105**, 863.
[491] F. Sauter, W. Deinhammer, and K. Danksagmüller, *Monatsh.*, 1974, **105**, 882.
[492] F. Sauter and P. Stanetty, *Monatsh.*, 1975, **106**, 1111.

Miscellaneous Fused Systems.—Ethyl 3-aminothieno[2,3-*b*]pyrazine-2-carboxylate has been prepared from 2-chloro-3-cyanopyrazine and ethyl α-mercaptoacetate, and then converted into (269) by reaction with formamide.[495] Other derivatives of (269) were subsequently prepared. Fused thieno[3,2-*d*]-*vic*-triazin-4-ones

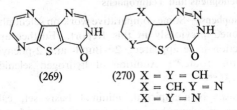

(269)

(270) X = Y = CH
X = CH, Y = N
X = Y = N

(270) have been prepared, starting similarly as above from *ortho*-chlorocyano-pyridine or -pyrazine.[496] The syntheses of some derivatives of thieno[2,3-*c*]-quinoxaline and -quinoline have been described.[497] Long-range proton spin-couplings in thienothiapyrylium cations have been studied.[498]

Studies of the chemistry of the thieno-borazaro-pyridines (271) and (272), prepared by the reaction of *ortho*-formyl- or *ortho*-acetyl-thiophenboronic acids with hydrazines, have continued. Bromination of (271; R = Me) and (272;

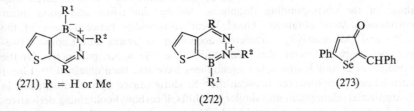

(271) R = H or Me

(272)

(273)

R = Me) with bromine and silver sulphate in conc. sulphuric acid, or with *NN*-dibromoisocyanuric acid in conc. sulphuric acid, gave preparatively useful yields of the 2,3-dibromo-derivatives. Nitration led to mixtures of the 3- (pre-dominantly) and 2-nitro-isomers.[499] Bromination of (271; R = H) or (272; R = H) with bromine in pyridine–CCl₄, or iodination by iodine chloride in pyridine–acetonitrile, led to substitution at the remaining C—H group in the boron-nitrogen-containing ring.[500] Halogen–metal exchange of the iodo-derivative with butyl-lithium, followed by reaction with DMF, gave the formyl derivative. The methoxy derivative could be prepared by nucleophilic substitution from the bromo-derivative.[501] Nitration of (271; R = H) and (272; R = H) with *N*-nitropicolonium tetrafluoroborate in acetonitrile occurs in high yield at the

[493] F. Sauter and W. Deinhammer, *Monatsh.*, 1974, **105**, 1249.
[494] F. Sauter, W. Deinhammer, and P. Stanetty, *Monatsh.*, 1974, **105**, 1258.
[495] S. W. Schneller and F. W. Clough, *J. Heterocyclic Chem.*, 1975, **12**, 513.
[496] S. W. Schneller and F. W. Clough, *Heterocycles*, 1975, **3**, 135.
[497] M. J. Haddadin, N. C. Chelhot, and M. Pieridou, *J. Org. Chem.*, 1974, **39**, 3278.
[498] F. C. Boccuzzi and R. Fochi, *Gazzetta*, 1974, **104**, 671.
[499] S. Gronowitz and C. Roos, *Acta Chem. Scand. (B)*, 1975, **29**, 990.
[500] S. Gronowitz and A. Maltesson, *Acta Chem. Scand. (B)*, 1975, **29**, 461.
[501] S. Gronowitz and A. Maltesson, *Acta Chem. Scand. (B)*, 1975, **29**, 1036.

remaining C–H group in the boron–nitrogen-containing ring.[502] A review of the above-mentioned work has recently appeared.[503] Theoretical studies on borepino-dithiophens have been carried out.[504]

8 Selenophens and Tellurophens

Monocyclic Selenophens.—Some comparative work on selenophens and thiophens has been mentioned previously in this Report. Further applications of the Fiesselmann reaction have appeared. 2-t-Butyl- and 2-phenyl-selenophen were obtained using this route.[505] Addition of hydrogen selenide to diacetylenic ketones led to (273).[506]

Physical Properties. Non-empirical, minimal basis set, calculations of the electronic structure of selenophen, using a combination of scaled molecular and STO-OG atomic functions, have been reported.[507] The infrared intensities of the three ring-stretching modes, V_5, V_6, and V_{15}, of a series of 2- and 3-substituted selenophens have been measured. Using a previously developed valence-bond approach, these intensities were rationalized in terms of the σ_R^0 constants of the substituent groups.[508] Several [13]C n.m.r. studies on selenophens have appeared.[509–511] A systematic study of a series of 2- and 3-substituted derivatives has been undertaken. Both the direct and the long-range couplings fall in well-defined intervals. The shifts caused by the substituents were compared with those of the corresponding thiophens. Strong similarities and good linear correlations were obtained. Good linear correlations between some of the shifts and the reactivity parameters according to Swain and Lupton's two-parameter equation were also observed.[511] The [77]Se n.m.r. parameters for the same set of 2- and 3-substituted selenophens have also been obtained.[512] Linear correlations were observed between the [77]Se shifts caused by the substituents in 2-substituted selenophens and similar ^{13}C shifts, if carbonyl-containing derivatives and 2-nitroselenophen were excluded. The anomalously small downfield shifts, and in some cases upfield shifts (compared with selenophen), in the carbonyl compounds were explained by a through-space binding interaction between Se d-orbitals and the carbonyl oxygen lone-pair in the *cis*-conformation of the 2-carbonyl derivatives. The substituent-caused shifts in the 3-substituted derivatives indicate that electronically the heteroatom and the substituents are '*para*'- and not '*meta*'-related. The substituent-caused [77]Se shifts are about six times larger than the ^{13}C shifts of similarly positioned carbons, indicating a shift

[502] S. Gronowitz and A. Maltesson, *Acta Chem. Scand.* (*B*), 1975, **29**, 457.
[503] S. Gronowitz, *J. Heterocyclic Chem.*, 1975, **12S**, 17.
[504] A. T. Jeffries and C. Párkányi, *J. Phys. Chem.*, 1976, **80**, 287.
[505] P. Cagniant, P. Périn, and G. Kirsch, *Compt. rend.*, 1974, **278**, *C*, 1201; P. Cagniant, G. Kirsch, and P. Périn, *ibid.*, 1974, **279**, *C*, 851.
[506] A. I. Tolmachev and M. A. Kudinova, *Khim. geterotsikl. Soedinenii*, 1974, 274.
[507] R. H. Findlay, *J.C.S. Faraday II*, 1974, **70**, 1397.
[508] G. P. Ford, T. B. Grindley, A. R. Katritzky, M. Shome, J. Morel, C. Paulmier, and R. D. Topsom, *J. Mol. Structure*, 1975, **27**, 195.
[509] F. Fringuelli, S. Gronowitz, A.-B. Hörnfeldt, I. Johnson, and A. Taticchi, *Acta Chem. Scand.* (*B*), 1974, **28**, 175.
[510] M. Garreau, G. J. Martin, M. L. Martin, J. Morel, and C. Paulmier, *Org. Magnetic Resonance*, 1974, **6**, 648.
[511] S. Gronowitz, I. Johnson, and A.-B. Hörnfeldt, *Chem. Scripta*, 1975, **7**, 111.
[512] S. Gronowitz, I. Johnson, and A.-B. Hörnfeldt, *Chem. Scripta*, 1975, **8**, 8.

of 940 p.p.m. per unit charge.[512] [1]H N.m.r. spectra of some magnesium derivatives of selenophen have been obtained.[53]

Reactions. Some 5-substituted 2-hydroxy-selenophens have been synthesized by the acid-catalysed dealkylation of t-butoxy-derivatives.[513, 514] The parent system was obtained by H_2O_2 oxidation of 2-selenophenboronic acid.[514] The equilibrium position between the 3-selenolen-2-one form and the 4-selenolen-2-one form, which favours the former, has been determined, and the kinetics of tautomerization in the 5-methyl system measured and compared with the corresponding furan and thiophen systems. Upon alkylation, using an ion-pair extraction method, the 5-methyl derivative gave mainly *C*-alkylation with methyl iodide and *O*-methylation with dimethyl sulphate.[514] Bis(3-bromo-2-selenienyl) ketones and related ketones have been prepared *via* bromo-selenophenlithium reagents and bromoselenophen aldehydes followed by oxidation of the carbinols.[515] Ullmann coupling reactions then give the selenophen analogues of fluorenone. Substituted selenienyl-lithium derivatives have also been used for the preparation of thieno-selenophens [136] and of selenophen[2,3-*c*]pyrylium salts.[516] (2-Selenienyl)ethanol-amines have been synthesized by the reduction of the corresponding amino-ketones and isonitroso-ketones.[517] The reaction of *N*-acylpyridinium salts with selenophen has been studied.[518] 2,5-Diarylselenophens have been obtained by heating 4-aryl-1,2,3-selenadiazoles.[519]

Selenophens Fused to other Five-membered Aromatics.—4-Bromoselenolo[2,3-*b*]-thiophen has been prepared *via* (274), obtained in a one-pot reaction from 3,4-dibromoselenophen.[136] The metallation of selenolo[3,2-*b*]thiophen with butyl-lithium showed no selectivity, as both α-positions were attacked.[520] Reactions of 2- and 3-selenophenacrylic acids with thionyl chloride in the presence of pyridine led to chlorinated selenolo-thiophens,[344] which were also obtained from thiophenacrylic acids and $SeOCl_2$ in chlorobenzene in the presence of pyridine.[521] Tetrachloroselenolo[3,2-*b*]selenophen was also obtained in this way.[521]

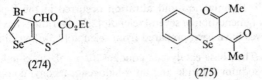

(274)

(275)

Dechlorination was in some cases effected with copper in quinoline.[344, 521] The structure of the selenophthen, m.p. 127.5—128 °C, formed as a by-product in the selenophen synthesis from selenium and acetylene, has been revised to selenolo-

[513] G. Henrio, J. Morel, and P. Pastour, *Ann. Chim. (France)*, 1975, **10**, 37.
[514] B. Cederlund and A.-B. Hörnfeldt, *Acta Chem. Scand. (B)*, 1976, **30**, 101.
[515] C. Maletras, B. Decroix, J. Morel, and P. Pastour, *Bull. Soc. chim. France*, 1974, 1575.
[516] V. I. Dulenko and N. N. Alekseev, *Khim. geterotsikl. Soedinenii*, 1973, 918.
[517] N. N. Magdesieva, T. A. Balashova, and G. M. Dem'yanova, *Khim. geterotsikl. Soedinenii*, 1972, 626.
[518] A. K. Sheinkman, T. V. Stupnikova, and A. A. Deikalo, *Khim. geterotsikl. Soedinenii*, 1973, 1147.
[519] I. Lalezari, A. Shafiee, H. F. Rabet, and M. Yalpani, *J. Heterocyclic Chem.*, 1973, **10**, 953.
[520] V. P. Litvinov, I. P. Konyaeva, and Ya. L. Gol'dfarb, *Izvest. Akad. Nauk S.S.S.R., Ser. khim.*, 1974, 1575.
[521] B. Capron and C. Paulmier, *Compt. rend.*, 1974, **279**, C, 947.

11

[3,2-*b*]selenophen. The structure proof was based on ¹H, ⁷⁷Se, and ¹³C n.m.r. spectra.[522] Condensation of 2- and 3-selenophencarbaldehyde with ethyl azidoacetate gave the vinyl azides, which upon heating yielded selenolo[2,3-*b*]- and selenolo[3,2-*b*]-pyrrole.[523] Selenolo[3,2-*b*][1]benzoselenophen was similarly obtained from 3-chloro-2-formylbenzo[*b*]selenophen.[524]

Benzo[*b*]selenophens and their Benzo-fused Derivatives.—The aryl selenide (275) underwent photocyclization to 2-acetyl-3-methylbenzo[*b*]selenophen.[525] The halogenation of benzo[*b*]selenophen was found to proceed *via* an addition to the selenium followed by reaction at the 2- or 3-position. Bromination in the 3-position is favoured and shows a noticeable isotope effect. Excessive bromination led to 2,3,6-tribromobenzo[*b*]selenophen, which could be debrominated to the 6-bromo-derivative.[526] Some β-(benzo[*b*]selenienyl)propionic acids have been prepared and their cyclizations investigated. The 3-isomer cyclized only to the 2-position, and when this position was blocked, to the 4-position.[527] Electrophilic substitution of 2,3-dihydrobenzo[*b*]selenophen occurred in the 5-position, as proved by aromatization of the product to the benzo[*b*]selenophen derivative.[528] Benzo[*b*]selenophen analogues of gramine, tryptamine, and tryptophan have been prepared.[529] The tautomerism of 3-hydroxybenzo[*b*]selenophen-2-aldehyde anils has been studied.[530] Attempts have been made to prepare benzo[*c*]selenophen.[531] The Rieche–Gross formylation of aryl-(2-benzo[*b*]selenienyl)methane proceeded with simultaneous cyclodehydration and gave selenium analogues of benz[*a*]anthracenes and dibenz[*a,h*]anthracenes.[532]

Selenophens Fused to Six-membered Heterocyclic Aromatic Rings.—Selenolo-[2,3-*c*]-[516] and selenolo[3,2-*c*]-pyrylium salts [533] have been prepared, starting from simple selenophens, and by reaction with ammonia transformed into selenolo-[2,3-*c*]pyridines and selenolo[3,2-*c*]pyridines, respectively. These ring systems have also been obtained from 2- and 3-carbonyl-substituted selenophens by the Pommeranz–Fritsch isoquinoline synthesis employing the acetal of amino-acetaldehyde.[534] Bromination and nitration occurred in the β-position.[534] The mass spectral fragmentation of several selenolo[2,3-*b*]pyridines has been studied.[535] Polymethine dyes have been prepared from selenolo[2,3-*b*]pyridines.[536]

Tellurophens.—The base-catalysed condensation of a β-halogenated vinyl aldehyde with sodium telluride and an α-halogenocarbonyl compound has been

[522] S. Gronowitz, T. Frejd, and A.-B. Hörnfeldt, *Chem. Scripta*, 1974, **5**, 236.
[523] K. N. Java, S. Soth, M. Farnier, and C. Paulmier, *Compt. rend.*, 1975, **281**, C, 793.
[524] P. Cagniant, P. Périn, and G. Kirsch, *Compt. rend.*, 1974, **278**, C, 1011.
[525] A. G. Schultz, *J. Org. Chem.*, 1975, **40**, 3466.
[526] T. Q. Minh, L. Christiaens, and M. Renson, *Bull. Soc. chim. France*, 1974, 2239.
[527] L. Laitem, L. Christiaens, and G. Llabres, *Bull. Soc. chim. France*, 1974, 681.
[528] P. Thibaut, L. Christiaens, and M. Renson, *Compt. rend.*, 1975, **281**, C, 937.
[529] L. Laitem and L. Christiaens, *Bull. Soc. chim. France*, 1975, 2294.
[530] V. I. Minkin, V. A. Bren, and G. D. Palui, *Khim. geterotsikl. Soedinenii*, 1975, 781.
[531] L. E. Saris and M. P. Cava, *J. Amer. Chem. Soc.*, 1976, **98**, 867.
[532] F. Girardin, P. Faller, L. Christiaens, and D. Cagniant, *Bull. Soc. chim. France*, 1974, 2095.
[533] V. I. Dulenko and N. N. Alekseev, *Khim. geterotsikl. Soedinenii*, 1973, 1212.
[534] F. Outurquin, C. Paulmier, and P. Pastour, *Bull. Soc. chim. France*, 1974, 3039.
[535] A. Croisy, P. Jacquignon, P. Pirson, and L. Christiaens, *Org. Mass Spectrometry*, 1974, **9**, 970.
[536] P. I. Abramenko, V. G. Zhiryakov, and T. K. Ponomareva, *Khim. geterotsikl. Soedinenii*, 1976, 56.

used for the synthesis of a variety of 2,5-substituted tellurophens.[537] [13]C N.m.r. spectra of some 2-substituted tellurophens have been compared with those of the corresponding thiophens and selenophens. [125]Te–H Couplings have been determined and compared with [77]Se–H couplings.[509] The conformational preferences and electronic effects in selenophen and tellurophen carbonyl derivatives have been studied by lanthanide-induced shifts.[538] The direction of the dipole moments of selenophen and tellurophen and their tetrahydro-derivatives has been studied by n.m.r.[539] Some complex tellurophenquinones have been prepared by the reaction of rhodium complexes such as (167) with tellurium.[540]

[537] P. Cagniant, R. Close, G. Kirsch, and D. Cagniant, *Compt. rend.*, 1975, **281**, *C*, 187.

[538] S. Caccamese, G. Montaudo, A. Recca, F. Fringuelli, and A. Taticchi, *Tetrahedron*, 1974, **30**, 4129.

[539] F. Fringuelli, S. Gronowitz, A.-B. Hörnfeldt, and A. Taticchi, *J. Heterocyclic Chem.*, 1974, **11**, 827.

[540] E. Müller, E. Luppold, and W. Winter, *Synthesis*, 1975, 265.

7
6a-Thiathiophthens and Related Compounds

BY R. J. S. BEER

1 6a-Thiathiophthens

Full details have been published of X-ray crystallographic studies on the thia-thiophthens (1),[1] (2),[2] (3),[3] and (4),[4] previously reported briefly.[5a] Points of interest in these structures include the shortening of the total S—S—S distance (4.617 Å) in (1), resulting from the presence of the trimethylene bridge, and the interaction of the substituent amino- and methylthio-groups with the thiathiophthen nucleus

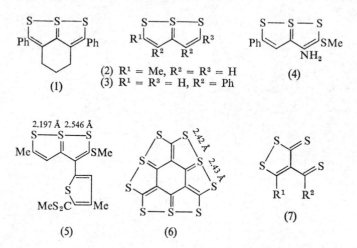

(2) R^1 = Me, R^2 = R^3 = H
(3) R^1 = R^3 = H, R^2 = Ph

(1)   (4)

(5)   (6)   (7)

in (4), reflected in shortened N—C and S—C bond lengths. In another 2-methyl-thiothiathiophthen (5),[6] the methylthio-group is twisted out of the plane of the central bicyclic system, and the S—S distances are markedly unequal.

J. P. Brown's 'sulphocarbon' (6), a sulphur analogue of coronene, contains three fused thiathiophthen systems, with linear S—S—S sequences, in which the S—S distances (2.42, 2.43 Å) are longer than in normal thiathiophthens.[7]

[1] B. Birknes, A. Hordvik, and L. J. Saethre, *Acta Chem. Scand. (A)*, 1975, **29**, 195.
[2] L. J. Saethre and A. Hordvik, *Acta Chem. Scand. (A)*, 1975, **29**, 136.
[3] P. L. Johnson, E. C. Llaguno, and I. C. Paul, *J.C.S. Perkin II*, 1976, 234.
[4] E. C. Llaguno and I. C. Paul, *J.C.S. Perkin II*, 1976, 228.
[5] (*a*) See 'Organic Compounds of Sulphur, Selenium, and Tellurium', ed. D. H. Reid (Specialist Periodical Reports), The Chemical Society, London, 1975, Vol. 3, Ch. 9; (*b*) *ibid*., 1973, Vol. 2, Ch. 9.
[6] E. C. Llaguno, C. T. Mabuni, and I. C. Paul, *J.C.S. Perkin II*, 1976, 239.
[7] L. K. Hansen and A. Hordvik, *J.C.S. Chem. Comm.*, 1974, 800.

[13]C N.m.r. spectra for representative thiathiophthens (with varying degrees of substitution by methyl, phenyl, and methylthio-groups) have been analysed in two papers,[8] which are in good agreement where they overlap. The results indicate symmetry within the n.m.r. time-scale for symmetrically substituted compounds. Substituent chemical-shift effects are similar to those for thiophen derivatives.

Non-empirical calculations of the electronic structure of 6a-thiathiophthen and related 1-oxa-6,6a-dithia-, 1,6-dioxa-6a-thia-, and 6,6a-dithia-1-azapentalenes (see Sections 3 and 5) appear to indicate that these molecules have little resonance energy, and are unlikely to be aromatic in the ordinary sense.[9a] Calculations of dipole moment for the same series of compounds give reasonable values when compared with the limited experimental data available.[9b]

Low-energy (He 584 Å) photoelectron spectra for various members of the heteropentalene family have been reported.[10] The difficulties in assigning the various bands in the spectra to particular molecular orbitals are emphasized. Another theoretical paper discusses the differences between the thiathiophthen system, with an electron-rich three-centre bond favoured by the linear arrangement of the three sulphur atoms, and 1,2-dithiole-3-thione derivatives of type (7), in which tautomerism between isomers is observable.[11]

Further details have been given of the rearrangement, under basic conditions, of 2-alkylthiathiophthens to thiophen derivatives [*e.g.* (8) → (9)],[12] and a method

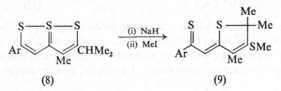

(8)   (9)

has been reported[13] for the preparation of 2-methoxy-5-methylthio-6a-thiathiophthens.

2 Multisulphur Systems

In continuing studies of molecules containing approximately linear arrays of four and five sulphur atoms, it has been found that the symmetrically substituted compound (10) (which crystallizes with 0.5 mole of carbon disulphide) shows a marked departure from symmetry in the distribution of bond lengths.[14] The thiathiophthen (11) exists in two independent forms in the crystal, with marked differences in the S—S distances,[15] but analogies for each pair of values can be found in the literature.[5b]

[8] R. D. Lapper and A. J. Poole, *Tetrahedron Letters*, 1974, 2783; C. Th. Pedersen and K. Schaumburg, *Org. Magn. Resonance*, 1974, **6**, 586.
[9] (a) M. H. Palmer and R. H. Findlay, *J.C.S. Perkin II*, 1974, 1885; (b) M. H. Palmer, R. H. Findlay, and A. J. Gaskell, *ibid.*, p. 420.
[10] R. Gleiter, R. Gygax, and D. H. Reid, *Helv. Chim. Acta*, 1975, **58**, 1591.
[11] G. Calzaferri and R. Gleiter, *J.C.S. Perkin II*, 1975, 559.
[12] A. Josse and M. Stavaux, *Bull. Soc. chim. France*, 1974, 1727; A. Josse, M. Stavaux, and N. Lozach, *ibid.*, p. 1723.
[13] J. L. Burgot and J. Vialle, *Compt. rend.*, 1974, **278**, C, 793 (*Chem. Abs.*, 1974, **80**, 146 058).
[14] J. Sletten, *Acta Chem. Scand.* (A), 1975, **29**, 436.
[15] J. Sletten, *Acta Chem. Scand.* (A), 1974, **28**, 499.

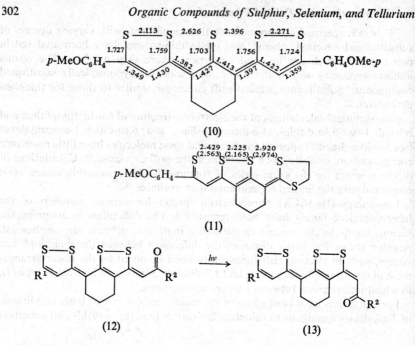

(10)

(11)

(12)   (13)

Extending earlier work on the photo-isomerization of 1,2-dithiolylidene ketones,[5a] systems of type (12) have been shown to isomerize, on irradiation, to relatively stable products, which are probably the geometrical isomers (13).[16] The thermal reversion reactions show first-order kinetics and are accelerated by acids.

3 1,2-Dithiolylidene Aldehydes and Ketones

Full structural details for the dithiolylidene ketone (14) [5a] have been published,[17] and the use of the shift reagent Eu(dpm)$_3$ in the analysis of n.m.r. spectra in this series has been discussed.[18]

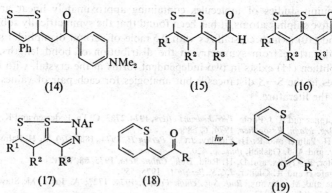

(14)   (15)   (16)

(17)   (18)   (19)

[16] C. Th. Pedersen, C. Lohse, and M. Stavaux, *J.C.S. Perkin I*, 1974, 2722.
[17] L. J. Saethre and A. Hordvik, *Acta Cryst.*, 1975, **B 31**, 30.
[18] A. Josse, M. Stavaux, and N. Lozac'h, *Bull. Soc. chim. France*, 1975, 1873.

Aldehydes of type (15), regarded as 1-oxa-6,6a-dithiapentalenes (16), readily couple with diazonium fluoroborates.[19] When R^3 = H, the products are 6,6a-dithia-1,2-diazapentalene-3-aldehydes (17; R^3 = CHO), but, when the starting material carries an alkyl substituent in the 3-position, the formyl group is lost, and the coupling reaction leads to 3-alkyl-6,6a-dithia-1,2-diazapentalenes (17; R^3 = alkyl).

Of interest in connection with earlier discussions of the structure of 1,2-di-thiolylidene ketones [5] are three papers dealing with analogous thiopyran deriva-tives of type (18). The absence of typical carbonyl-stretching bands from the i.r. spectra is taken to indicate interaction between the sulphur atom and the adjacent oxygen atom.[20] Irradiation of a series of compounds of this structure yields photo-products that are believed to be the geometrical isomers (19).[21]

4 1,3-Dithiole Derivatives

Experimental details have been published for addition reactions between acetylenes (21; R^1, R^2 = CO$_2$Me, CO$_2$Et, or H) and 1,2-dithiole-3-thiones (20), and for the conversion of the resulting 'isothiathiophthens' (22) into thiathio-phthens with thioacetamide or thiobenzamide.[22] The similar reaction between 1,2-dithiole-3-thiones and benzyne, variously generated, has been described.[23]

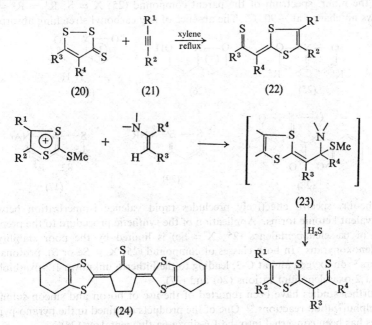

[19] R. M. Christie and D. H. Reid, *J.C.S. Perkin I*, 1976, 880.
[20] (a) J.-P. Sauvé and N. Lozac'h, *Bull. Soc. chim. France*, 1974, 1196; (b) R. Pinel, K. S. Nguyen, and Y. Mollier, *Compt. rend.*, 1974, **278**, C, 729 (*Chem. Abs.*, 1974, **80**, 144 906).
[21] C. Th. Pedersen, C. Lohse, N. Lozac'h, and J.-P. Sauvé, *J.C.S. Perkin I*, 1976, 166.
[22] H. Davy and J. Vialle, *Bull. Soc. chim. France*, 1975, 1435; H. Davy and J.-M. Decrouen, *ibid.*, 1976, 115.
[23] D. Paquer and R. Pou, *Bull. Soc. chim. France*, 1976, 120.

The major products are benzisothiathiophthens (22; $R^1R^2 = CH=CH\cdot CH= CH$), sometimes accompanied by the corresponding benzothiathiophthens.

The product obtained by the action of thioacetic acid on phenylacetylene has now been shown to be (22; $R^1 = Ph$, $R^2 = H$, $R^3 = Me$, $R^4 = H$), which can be synthesized from 5-methyl-1,2-dithiole-3-thione and phenylacetylene.[24]

A new route to isothiathiophthens has been reported,[25] involving the reaction of 2-methylthio-1,3-dithiolium salts with enamines, and treatment of the inter-mediate products (23) with hydrogen sulphide. More complex examples, *e.g.* (24), have also been prepared.

A study of the mass spectra of isothiathiophthens shows that these compounds behave quite differently from the thiathiophthens, which characteristically lose hydrogen and substituent groups. The iso-compounds generally fragment by loss of the appropriate acetylene residue.[26]

5 Other Analogues of 6a-Thiathiophthens

The preparation of 1,6-dioxa-6a-thiapentalenes (25; $X = S$; $R^1 = R^2 = H$; $R^1 = R^2 = Me$; $R^1 = R^2 = CO_2Et$; $R^1 = Ph$, $R^2 = H$; *etc.*) by the thallium trifluoroacetate route[5a] has been described in detail.[27] The n.m.r. data for symmetrically substituted compounds indicate real or time-averaged symmetry, and the n.m.r. spectrum of the parent compound (25; $X = S$; $R^1 = R^2 = H$) shows no change at $-90\ ^\circ C$. The absence of any carbonyl stretching absorption

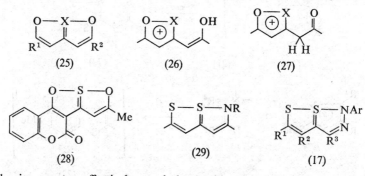

in the i.r. spectra effectively precludes rapid valence isomerization between equivalent ketonic forms. Application of the synthetic procedure to the prepara-tion of 6a-selenapentalenes (25; $X = Se$) is limited by the poor stability of 4-selenoxopyrans. In both classes of compound (25; $X = Se$ or S), protonation occurs[28] on oxygen and at C-3, leading to the hitherto unknown 1,2-oxathiolium and 1,2-oxaselenolium cations (26) and (27).

Further studies have been reported of the use of boron and silicon sulphides in sulphurization reactions.[29] One of the products obtained in the pyrano-pyrone series has been converted into the 1,6-dioxa-6a-thiapentalene (28).

[24] C. Th. Pedersen, *Acta Chem. Scand.* (*B*), 1974, **28**, 367.
[25] E. Fanghänel, *J. prakt. Chem.*, 1975, **317**, 137.
[26] C. Th. Pedersen, H. Davy, J. Møller, and J. Vialle, *Acta Chem. Scand.* (*B*), 1974, **28**, 964.
[27] D. H. Reid and R. G. Webster, *J.C.S. Perkin I*, 1975, 775.
[28] D. H. Reid and R. G. Webster, *J.C.S. Perkin I*, 1975, 2097.
[29] F. M. Dean, J. Goodchild, A. W. Hill, S. Murray, and A. Zahman, *J.C.S. Perkin I*, 1975, 1335.

Diazo-coupling reactions of thiathiophthens, 1,2-dithiolylidene aldehydes (see Section 3), and 6,6a-dithia-1-azapentalenes (29) all lead to 6,6a-dithia-1,2-diazapentalene derivatives (17), which can also be obtained from 2-methyl- (or methylene)-1,2-dithiolium salts by coupling with diazonium fluoroborates.[30] When R^3 = H in structure (17), further coupling can take place at this position. 1,2-Diselenolium salts give similar products but in poor yields. Preliminary *X*-ray data on the 5-t-butyl-1-phenyl compound (17; R^1 = Bu^t, R^2 = R^3 = H, Ar = Ph) clearly support the bicyclic formulation. The bond lengths are S—S 2.426 Å, and S—N 1.841 Å.

A study of the mass spectra of 1,6-dioxa-6a-thia-2,5-diazapentalenes [5] emphasizes the stability of this ring system, and of the corresponding 6a-selena- and 6a-tellura-compounds.[31]

3-Acylimino-1,2,4-dithiazoles (30a) do not show carbonyl stretching frequencies above *ca.* 1600 cm⁻¹, suggesting the possibility that these compounds are bicyclic, as in structure (30b).[32]

An *X*-ray study [33] on the trithiadiazapentalene derivative (31) confirms the structure previously assigned.[5a] The S—S distances are in the range typical of

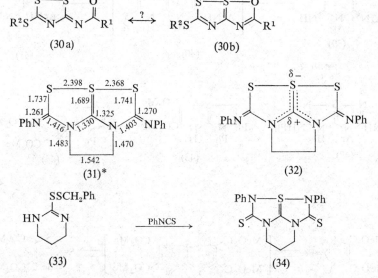

thiathiophthens but, significantly, the central C—S bond is shorter than the outer C—S bonds. Inspection of the bond lengths in the N—C—N—C—N—C—N sequence suggests that the molecule is best represented as (32), with the exocyclic C—N links as virtually pure double bonds, and with relatively little π-overlap between C-2 and N-3, and between N-4 and C-5. This stable, well-defined compound can hardly be regarded as aromatic, since it is effectively a

[30] R. M. Christie and D. H. Reid, *J.C.S. Perkin I*, 1976, 228.
[31] M. Perrier, R. Pinel, and J. Vialle, *J. Heterocyclic Chem.*, 1975, **12**, 639.
[32] D. Wobig, *Annalen*, 1975, 1018.
[33] K.-T. Wei, I. C. Paul, R. J. S. Beer, and A. Naylor, *J.C.S. Chem. Comm.*, 1975, 264.

* Only one of two similar but crystallographically independent forms is shown

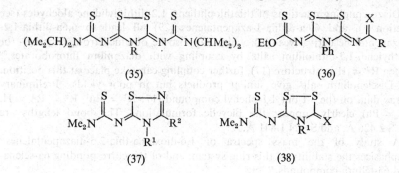

(35)   (36)

(37)   (38)

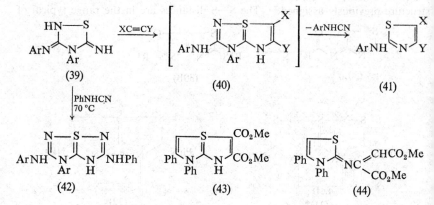

(39)   (40)   (41)

(42)   (43)   (44)

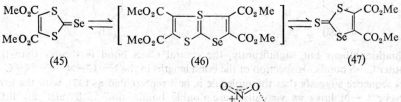

(45)   (46)   (47)

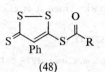

(48)

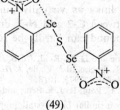

(49)

derivative of a hypothetical tetrahydro-1,6,6a-trithia-3,4-diazapentalene. A possible relationship to the electron-rich three-centre-bond system of the triselenocyanate anion is suggested.

When the method used for the preparation of compound (31) is applied to the disulphide (33), the product is the 6a-thia-1,3,4,6-tetra-azapentalene derivative (34).[33]

Structural data for dithiazole derivatives (35; R = Me)[34] and (35; R = p-NO$_2$-C$_6$H$_4$)[35] indicate molecular dimensions similar to those of systems studied previously.[5a] Some closely related structures (36; X = O or NR) have been described,[36] and an interesting series of papers[37] deals with various compounds derived from 2-iminodithiazoles which could be formulated as bicyclic systems, for example (37) and (38; X = S or NR). In the last of the four papers, the methylation reactions of these and other compounds are discussed, in relation to earlier work on the methylation of thiathiophthens.

Hector's bases (39) undergo addition reactions with phenylcyanamide to give products for which bicyclic structures (42) are favoured,[38] although the alternative monocyclic tautomeric structures cannot be excluded in the absence of X-ray data. With activated acetylenes, Hector's bases are converted into 2-arylaminothiazoles (41), presumably *via* bicyclic intermediates (40). An attempt to prepare the related structure (43) yielded the 2-iminothiazole (44).

The bicyclic system (46) has been invoked as a transient species in the interconversion of the 1,3-dithiole-2-selone (45) and the 1-selena-3-thiole-2-thione (47) brought about by dimethyl acetylenedicarboxylate in refluxing toluene.[39]

5-Acylthio-1,2-dithiole-3-thiones (48) show normal carbonyl-stretching frequencies in the i.r., indicating that there is no significant interaction between the carbonyl oxygen atom and the adjacent S—S bond.[40]

A crystallographic study[41] of bis-(*o*-nitrophenylselenenyl) sulphide (49) reveals that the atoms O···Se—S—Se···O are approximately linear. Reference is made to related compounds with similar linear sequences.

[34] J. Sletten, *Acta Chem. Scand. (A)*, 1974, **28**, 989.
[35] J. Sletten, *Acta Chem. Scand. (A)*, 1975, **29**, 317.
[36] J. Goerdeler, W. Kunnes, and F. M. Panshiri, *Chem. Ber.*, 1976, **109**, 848.
[37] J. E. Oliver and A. B. DeMilo, *J. Org. Chem.*, 1974, **39**, 2225; J. E. Oliver and R. T. Brown, *ibid.*, p. 2228; J. E. Oliver and J. L. Flippen, *ibid.*, p. 2233; J. E. Oliver, *ibid.*, p. 2235.
[38] K. Akiba, T. Tsuchiya, M. Ochiumi, and N. Inamoto, *Tetrahedron Letters*, 1975, 455, 459.
[39] M. V. Lakshmikantham and M. P. Cava, *J. Org. Chem.*, 1976, **41**, 879.
[40] N. Loayza and C. Th. Pedersen, *Acta Chem. Scand. (B)*, 1976, **30**, 88.
[41] R. Eriksen, *Acta Chem. Scand. (A)*, 1975, **29**, 517.

8
1,2- and 1,3-Dithioles

BY R. J. S. BEER

Limitations of space restrict detailed discussion of the many interesting topics which come within the scope of this Report. Of these, the most important is undoubtedly the preparation and study of the so-called 'organic metals' (see Section 7). An attempt has been made to provide a comprehensive list of references, but it has not been possible to include references to the patent literature.

1 1,2-Dithioles

1,2-Dithiole-3-thiones have been isolated in low yields from high-temperature reactions of various organic compounds with sulphur,[1, 2] and a new preparation of 1,2-dithiole-3-ones (2) from disulphides of type (1) has been described.[3] Oxidation with bromine of the dithiomalonamides (3) gives 3,5-diarylamino-1,2-dithiolium bromides (4), which yield imines (5) on treatment with base.[4] 3-Arylimines (6; R^1 = Ph, R^2 = Cl, R^3 = Ar) are obtainable from the corre-

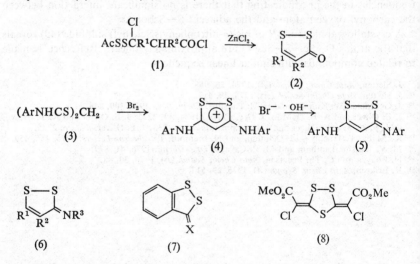

[1] J. Perregaard, I. Thomsen, and S.-O. Lawesson, *Acta Chem. Scand.* (*B*), 1975, **29**, 538.
[2] I. R. Gelling and M. Porter, *Tetrahedron Letters*, 1975, 3089.
[3] T. P. Vasil'eva, M. G. Lin'kova, O. V. Kil'disheva, and I. L. Knunyants, *Izvest. Akad. Nauk S.S.S.R., Ser. khim.*, 1974, 700 (*Chem. Abs.*, 1974, **81**, 3836).
[4] A. D. Grabenko, L. N. Kulaeva, and P. S. Pel'kis, *Khim. geterotsikl. Soedinenii*, 1974, 924 (*Chem. Abs.*, 1974, **81**, 120 520).

sponding 3-methylthiodithiolium salts [5] by reaction with the appropriate amine, and a new route to imines of type (6; R^3 = SO_2Ar or CO_2R) involving reaction of a dithiolium salt with R^3NCl_2 has appeared.[6] Several papers report syntheses of benzo-1,2-dithiole-3-thione (7; X = S) and some derivatives.[7]

The products obtained by the action of bases and of Grignard reagents on 4,5-dichloro-1,2-dithiole-3-one include the 1,2,4-trithiolan (8),[8] and 1,3-dithioles (9; R = CO_2Me or COSR).[9] 1,2-Dithiole-3-one is reported to give ring-opened trisulphides, *e.g.* RSSSCH=CHCOCl, with sulphenyl chlorides.[10] Reactions of 1,2-dithiole-3-ones with diphenyldiazomethane have also been investigated.[11]

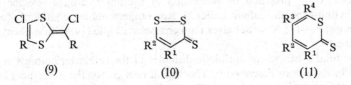

(9)   (10)   (11)

In a study of the oxidation of various heterocyclic thiones with peroxyacetic acid, 1,2-dithiole-3-thiones (10) carrying electron-withdrawing substituents have been found to give 1,2-dithiole-3-ones rather than dithiolium salts.[12] Experimental details have been published for addition reactions [13a] of 1,2-dithiole-3-thiones with acetylenes,[14] and with benzyne.[15] Reactions with enamines, leading to thiopyran derivatives (11), have been described.[16]

A full account has been given of photoadditions of 1,2-dithiole-3-thiones with olefins.[13a, 17] Similar reactions with benzodithiole-3-thione (7; X = S) lead to blue quinonoid compounds (12), which in solution, at room temperature, are in equilibrium with dimeric forms.[18]

Other reactions of benzodithiole (7; X = S) recently reported include those with ethylene- and trimethylene-diamines [19] and with diphenyldiazomethane.[20]

[5] Y. N'Guessan and J. Bignebat, *Compt. rend.*, 1975, **280**, C, 1323.
[6] F. Boberg, G.-J. Wentrup, and M. Koepke, *Synthesis*, 1975, 502.
[7] J. P. Brown and M. Thompson, *J.C.S. Perkin I*, 1974, 863; M. G. Voronkov and L. N. Khokhlova, *Zhur. org. Khim.*, 1974, **10**, 811 (*Chem. Abs.*, 1974, **81**, 25 590; see also *Chem. Abs.*, 1974, **81**, 120 599).
[8] F. Boberg and M. Ghoudikian, *Annalen*, 1975, 1513.
[9] F. Boberg, M. Ghoudikian, and M. H. Khorgami, *Annalen*, 1974, 1261.
[10] T. P. Vasil'eva, M. G. Lin'kova, O. V. Kil'disheva, and I. L. Knunyants, *Izvest. Akad. Nauk S.S.S.R., Ser. khim.*, 1975, 2610 (*Chem. Abs.*, 1976, **84**, 90 085).
[11] M. A.-F. Elkaschef, F. M. E. Abdel-Megeid, and A. A. Elbarbary, *Tetrahedron*, 1974, **30**, 4113.
[12] J. L. Charlton, S. M. Loosmore, and D. M. McKinnon, *Canad. J. Chem.*, 1974, **52**, 3021.
[13] (a) See 'Organic Compounds of Sulphur, Selenium, and Tellurium,' ed. D. H. Reid (Specialist Periodical Reports), The Chemical Society, London, 1975, Vol. 3, Ch. 10; (b) *ibid.*, 1973, Vol. 2, Ch. 9.
[14] H. Davy and J. Vialle, *Bull. Soc. chim. France*, 1975, 1435; H. Davy and J.-M. Decrouen, *ibid.*, 1976, 115.
[15] J.-M. Decrouen, D. Paquer, and R. Pou, *Compt. rend.*, 1974, **279**, C, 259; D. Paquer and R. Pou, *Bull. Soc. chim. France*, 1976, 120.
[16] F. Ishii, M. Stavaux, and N. Lozac'h, *Tetrahedron Letters*, 1975, 1473.
[17] R. Okazaki, F. Ishii, K. Okawa, K. Ozawa, and N. Inamoto, *J.C.S. Perkin I*, 1975, 270.
[18] P. de Mayo and H. Y. Ng, *J.C.S. Chem. Comm.*, 1974, 877.
[19] J. P. Brown, *J.C.S. Perkin I*, 1974, 869.
[20] S. Tamagaki, R. Ichihara, and S. Oae, *Bull. Chem. Soc. Japan*, 1975, **48**, 355.

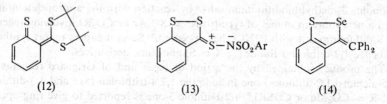

(12)   (13)   (14)

Reactions of thiocarbonyl ylides of type (13) [13b] have been further explored,[21] and several papers deal with the chemistry of benzo-1,2-dithiole-3-selone (7; X = Se),[20, 22, 23] prepared by the action of sodium hydrogen selenide on the 3-methylthiobenzodithiolium iodide.[22] Rearrangements are commonly found in this series; *e.g.* (7; X = Se) gives the benzoselenathiole (14) with diphenyldiazomethane.[20]

New total syntheses of antibiotic dithioles of the thiolutin, holomycin group, *e.g.* (15), have been described.[24] Theoretical reasons for the differences between 6a-thiathiophthens (see Chapter 7) and the tautomeric system (16) have been

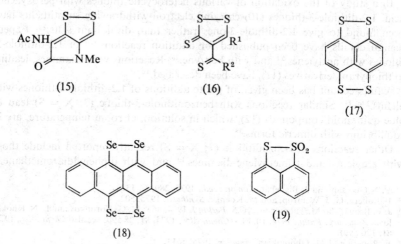

(15)   (16)   (17)

(18)   (19)

discussed.[25] The bis-dithiole derivative (17), prepared from naphthalene-1,5-dithiol [26] in low yield with sulphur dichloride and aluminium chloride, and the related bis-diselenole (18),[27] form complexes with tetracyanoquinodimethane (see Section 7). The chemistry of the dioxide (19) has also been studied.[28]

[21] S. Tamagaki, K. Sakaki, and S. Oae, *Bull. Chem. Soc. Japan*, 1975, **48**, 2983, 2985, 2987.
[22] S. Tamagaki, K. Sakaki, and S. Oae, *Heterocycles*, 1974, **2**, 39, 45 (*Chem. Abs.*, 1974, **80**, 120 820, 120 823).
[23] S. Tamagaki, K. Sakaki, and S. Oae, *Tetrahedron Letters*, 1974, 1059.
[24] K. Hagio and N. Yoneda, *Bull. Chem. Soc. Japan*, 1974, **47**, 1484.
[25] G. Calzaferri and R. Gleiter, *J.C.S. Perkin II*, 1975, 559.
[26] F. Wudl, D. E. Schafer, and B. Miller, *J. Amer. Chem. Soc.*, 1976, **98**, 252.
[27] M. L. Khidekel and E. B. Yagukskii, *Izvest. Akad. Nauk S.S.S.R., Ser. khim.*, 1975, 1213 (*Chem. Abs.*, 1975, **83**, 58 726).
[28] S. Tamagaki, H. Hirota, and S. Oae, *Bull. Chem. Soc. Japan*, 1974, **47**, 2075.

2 1,2-Dithiolium Salts

Oxidation of dithiomalonamides with hydrogen peroxide in the presence of hydrochloric acid leads to 3,5-diamino-4-chloro-1,2-dithiolium chlorides (20).[29] The formation and deacetylation (or debenzoylation) of 3,5-diaryl-4-acyloxy-1,2-dithiolium salts [13a] have been described,[30] and an earlier investigation of the mass spectra of the thermolysis products of dithiolium salts has been extended to include 3-alkylthio- and 3-arylthio-derivatives.[31] Studies on the photochemical formation of 1,2-dithiolyl radicals and dithioketonate anions from dithiolium salts [13a] have been continued.[32]

Dimeric products (21), previously obtained electrochemically,[13b] have been prepared by reduction of 1,2-dithiolium salts with zinc.[33] These products oxidize, either in air or with selenium dioxide, to highly coloured products of type (22).

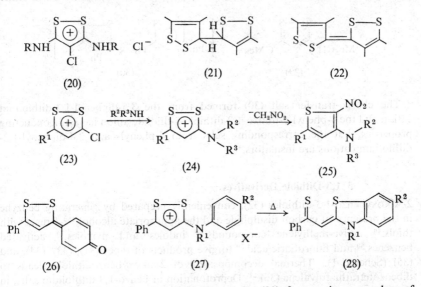

(20)   (21)   (22)

(23)   (24)   (25)

(26)   (27)   (28)

The reactions of 3-chloro-1,2-dithiolium salts (23) feature in a number of papers. With secondary amines, the products are 3-amino-1,2-dithiolium salts (24), which undergo an interesting reaction with the carbanion of nitromethane to give thiophen derivatives (25).[34] Aromatic amines,[35] phenols, and phenol ethers [36] are substituted at the *para*-position by 3-chloro-5-phenyl-1,2-dithiolium cations. The product obtained with phenol can be deprotonated to the quinonoid compound (26), and 3-arylamino-1,2-dithiolium salts (27; R^1 = aryl or $PhCH_2$) are converted thermally into benzothiazole derivatives (28).[37] Other studies deal

[29] L. Menabue and G. C. Pellacani, *J.C.S. Dalton*, 1976, 455.
[30] D. Barrillier, P. Rioult, and J. Vialle, *Bull. Soc. chim. France*, 1976, 444.
[31] C. Th. Pedersen, N. L. Huaman, and J. Møller, *Acta Chem. Scand.* (*B*), 1974, **28**, 1185.
[32] C. Th. Pedersen and C. Lohse, *Acta Chem. Scand.* (*B*), 1975, **29**, 831.
[33] H. Behringer and E. Meinetsberger, *Tetrahedron Letters*, 1975, 3473.
[34] B. Bartho, J. Faust, R. Pohl, and R. Mayer, *J. prakt. Chem.*, 1976, **318**, 221.
[35] B. Bartho, J. Faust, and R. Mayer, *Z. Chem.*, 1975, **15**, 440.
[36] J. Faust, B. Bartho, and R. Mayer, *Z. Chem.*, 1975, **15**, 395.
[37] B. Bartho, J. Faust, and R. Mayer, *Tetrahedron Letters*, 1975, 2683.

with the conversion of 3-chloro-1,2-dithiolium salts into 1,2-dithiole-3-thiones with thioacetic acid [38] and with the preparation of labelled 3,4,5-trichloro-1,2-dithiolium chlorides.[39]

Ring-opening reactions of dithiolium salts with amines [13a] continue to receive attention. The behaviour of 3-styryl derivatives deuteriated in various positions,[40] and of 3-ethoxy-1,2-dithiolium salts,[41] has been studied. Further examples have been provided of the formation of isothiazoles and pyrazoles by the action of ammonia and hydrazine, respectively.[42] 3,5-Di-isopropyl-1,2-dithiolium cations are deprotonated by aqueous ethanolic ammonia to give the dithiole (29).[42]

The reactivity of 3-methyl (or -methylene) groups in 1,2-dithiolium salts [13a] has been exploited in a new synthesis of 6,6a-dithia-1,2-diazapentalenes [43] (see Chapter 7).

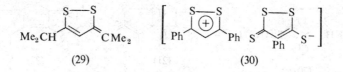

(29)   (30)

The charge-transfer salt (30) formed from the 3,5-diphenyl-1,2-dithiolium cation and the 4-phenyl-3-thioxo-1,2-dithiole-5-thiolate anion has semiconducting properties, but the corresponding salts from 4-phenyl- and 3,4-diphenyl-1,2-dithiolium cations are insulators.[44]

3 1,3-Dithiole Derivatives

2-Alkoxybenzo-1,3-dithioles (32), conveniently prepared by generating benzyne in the presence of carbon disulphide and the appropriate alcohol,[45a, b] react with thiols,[46] active-methylene compounds,[47] indoles and pyrroles,[48] activated benzenes,[49] and fluoroboric acid [50] to give products of types (31), (33), (34), and (35) (Scheme 1). Thermal decomposition of 2-alkoxybenzodithioles leads to dibenzotetrathiafulvalene (36).[51] Deprotonation of benzo-1,3-dithiolium salts, in the presence of methanol, acetone, and other reactants, produces compounds of types (32) and (34), as well as dimers of type (36).[52]

[38] G.-J. Wentrup, M. Koepke, and F. Boberg, *Synthesis*, 1975, 525.
[39] F. Boberg, R. Wiedermann, and J. Kresse, *J. Labelled Compounds*, 1974, **10**, 297 (*Chem. Abs.*, 1974, **81**, 136 025).
[40] G. Duguay, C. Métayer, and H. Quiniou, *Bull. Soc. chim. France*, 1974, 2853.
[41] J. Faust, *Z. Chem.*, 1975, **15**, 478.
[42] S. Coen, J.-C. Poite, and J.-P. Roggero, *Bull. Soc. chim. France*, 1975, 611.
[43] R. M. Christie and D. H. Reid, *J.C.S. Perkin I*, 1976, 228.
[44] N. Loayza and C. Th. Pedersen, *J.C.S. Chem. Comm.*, 1975, 496.
[45] (a) J. Nakayama, *Synthesis*, 1975, 38; (b) *J.C.S. Perkin I*, 1975, 525.
[46] J. Nakayama, *Synthesis*, 1975, 436.
[47] J. Nakayama, *J.C.S. Perkin I*, 1976, 540.
[48] J. Nakayama, M. Imura, and M. Hoshino, *Chem. Letters*, 1975, 1319.
[49] J. Nakayama, *Synthesis*, 1975, 170.
[50] J. Nakayama, K. Fujiwara, and M. Hoshino, *Chem. Letters*, 1975, 1099.
[51] J. Nakayama, *Synthesis*, 1975, 168.
[52] G. Scherowsky and J. Weiland, *Annalen*, 1974, 403.

Various spiro-derivatives of benzo-1,3-dithiole have been described,[53-55] including a series of photochromic compounds (37).[53]

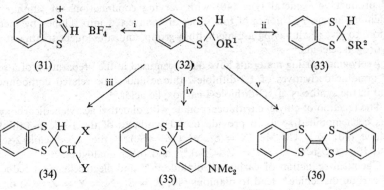

(31)   (32)   (33)

(34)   (35)   (36)

Reagents: i, HBF₄; ii, R²SH–HOAc; iii, HOAc–XCH₂Y; iv, PhNMe₂; v, Δ

Scheme 1

4 1,3-Dithiole-2-thiones and Selenium-containing Analogues

The reaction of dithiocarbamates and diselenocarbamates with α-halogenated carbonyl compounds has proved to be a useful first step in the synthesis of 1,3-dithiole-2-thiones and their analogues. Thus 2-dimethylamino-1,3-dithiolium salts (38)[56a, b] and diselenolium salts (39)[57a-c] are readily obtainable by this route, and these salts are converted into thiones (40; X = Y = S or Se, Z = S) or selones (40; X = Y = S or Se, Z = Se) with hydrogen sulphide or hydrogen selenide. The dihydroxybenzo-1,3-dithiole-2-thione (41) has been prepared from *p*-benzoquinone and dithiocarbamic acids.[58]

The Mayer–Gebhardt procedure (involving sodium acetylide, sulphur, and carbon disulphide) has been extended to the synthesis of 1,3-diselenole-2-selone

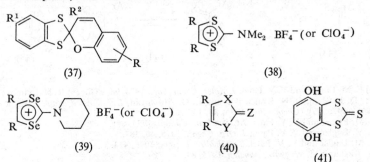

(37)   (38)

(39)   (40)   (41)

[53] P. Appriou and R. Guglielmetti, *Bull. Soc. chim. France*, 1974, 510.
[54] L. Dalgaard and S.-O. Lawesson, *Tetrahedron Letters*, 1973, 4319.
[55] W. Schroth, H. Bahn, and G. Huck, *Z. Chem.*, 1973, **13**, 465.
[56] (*a*) Y. Ueno, A. Nakayama, and M. Okawara, *Synthesis*, 1975, 277; (*b*) P. Calas, J. M. Fabre, M. Khalife-El-Saleh, A. Mas, E. Torreilles, and L. Giral, *Compt. rend.*, 1975, **281**, C, 1037.
[57] (*a*) K. Bechgaard, D. O. Cowan, and A. N. Bloch, *J.C.S. Chem. Comm.*, 1974, 937; (*b*) K. Bechgaard, D. O. Cowan, A. N. Bloch, and L. Henriksen, *J. Org. Chem.*, 1975, **40**, 746; (*c*) J. R. Andersen and K. Bechgaard, *ibid.*, 1975, **40**, 2016.
[58] R. L. N. Harris and L. T. Oswald, *Austral. J. Chem.*, 1974, **27**, 1309.

(40; R = H, X = Y = Z = Se),[59a, b] and of 1-selena-3-thiole-2-selone,[60] but an attempt to use this route to obtain 1-selena-3-thiole-2-thione gave a mixture of products of general type (40) with varying combinations of sulphur and selenium atoms.[61] Thiones of this series are converted into selones by methylation, to give a salt having a 2-methylthio-group, and subsequent treatment with hydrogen selenide.[62]

Acetylenic starting materials have also been used in the preparation of a series of amidine derivatives of 1,3-dithiole-2-thione and some related compounds,[63] and in the synthesis of 1,3-dithiole-4-carboxylic acids.[64]

The reaction of ethylene trithiocarbonate with dimethyl acetylenedicarboxylate has been modified so as to provide useful syntheses of the 1-selena-3-thiole-2-selone (40; R = CO$_2$Me, X = S, Y = Z = Se),[65a] the 1,3-diselole-2-selone (40; R = CO$_2$Me, X = Y = Z = Se),[65b] and other analogues.[65c]

Chemical reduction of carbon disulphide,[66, 67] and electrochemical reduction of carbon diselenide[68] lead to dianions (40; R = S$^-$, X = Y = Z = S) or (40; R = Se$^-$, X = Y = Z = Se), which on methylation yield the corresponding bis(methylthio)- or bis(methylseleno)-compounds.

Annelated 1,3-dithiole-2-thiones[69] and selenium analogues[70] have been prepared by thermal decomposition of thiadiazoles and selenadiazoles in the presence of carbon disulphide, and 1,3-dithiole-2-thiones are also obtainable from enamines and thiuram disulphides, *via* intermediates of type (42).[71]

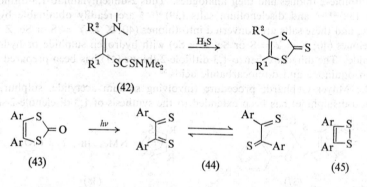

[59] (a) E. M. Engler and V. V. Patel, *J. Amer. Chem. Soc.*, 1974, **96**, 7376; (b) R. N. Lyubovskaya, Ya. D. Lipshan, O. N. Krasochka, and L. O. Atovmyan, *Izvest. Akad. Nauk S.S.S.R., Ser. khim.*, 1976, 179 (*Chem. Abs.*, 1976, **84**, 135 562).
[60] E. M. Engler and V. V. Patel, *J.C.S. Chem. Comm.*, 1975, 671.
[61] E. M. Engler and V. V. Patel, *J. Org. Chem.*, 1975, **40**, 387.
[62] E. M. Engler and V. V. Patel, *Tetrahedron Letters*, 1976, 423.
[63] W. Ried and M. Wegwitz, *Annalen*, 1975, 89.
[64] C. U. Pittman and M. Narita, *J.C.S. Chem. Comm.*, 1975, 960.
[65] (a) M. V. Lakshmikantham, M. P. Cava, and A. F. Garito, *J.C.S. Chem. Comm.*, 1975, 383; (b) M. V. Lakshmikantham and M. P. Cava, *J. Org. Chem.*, 1976, **40**, 882; (c) *ibid.*, p. 879.
[66] U. Reuter and G. Gattow, *Z. anorg. Chem.*, 1976, **421**, 143.
[67] G. Steimecke, R. Kirmse, and E. Hoyer, *Z. Chem.*, 1975, **15**, 28.
[68] E. M. Engler, D. C. Green, and J. Q. Chambers, *J.C.S. Chem. Comm.*, 1976, 148.
[69] H. K. Spencer, P. M. Cava, F. G. Yamagishi, and A. F. Garito, *J. Org. Chem.*, 1976, **41**, 730.
[70] H. K. Spencer, M. V. Lakshmikantham, M. P. Cava, and A. F. Garito, *J.C.S. Chem. Comm.*, 1975, 867.
[71] E. Fanghänel, *J. prakt. Chem.*, 1975, **317**, 123.

The 1,3-dithiole-2-one (43; Ar = p-Me$_2$NC$_6$H$_4$), on photolysis at room temperature, gives the dark-red crystalline dithione (44; Ar = p-Me$_2$NC$_6$H$_4$), from which, by irradiation in methylene chloride at $-50\,°$C, a yellow product, believed to be the dithiet (45), is obtained.[72]

5 1,3-Dithiolium Salts

Several references to 1,3-dithiolium salts have already been cited (Sections 3 and 4) and others are to be found in Sections 6 and 7.

Reduction of the 1,3-dithiolium cation with hexacarbonylvanadate anion yields the colourless dimer (46), which can be oxidized to tetrathiafulvalene with manganese dioxide.[73] Reduction of 2-methylthio-1,3-dithiolium salts with sodium borohydride and treatment of the resulting 2-methylthio-1,3-dithiole with fluoroboric acid provides a convenient small-scale preparation of 1,3-dithiolium fluoroborate.[74] The method has been extended to the preparation of 1-selena-3-thiolium and 1,3-diselenolium fluoroborates.[75] For large-scale preparation of the 1,3-dithiolium cation, the route involving peroxy-acid oxidation of 1,3-dithiole-2-thione and isolation of the hexafluorophosphate is recommended.[76] This method fails with selenium-containing compounds.[75]

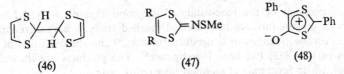

(46)   (47)   (48)

Reactions of 2-methylthio-1,3-dithiolium salts with sodium azide have been investigated;[77] the products include 3-methylthio-1,4,2-dithiazines and sulphenimides (47), from which 2-amino-1,3-dithiolium salts and other related compounds can be prepared.

Reactions of the meso-ionic dithiolium salt (48) have been reported with diphenylcyclopropenone,[78] benzyne,[79] dimethyl acetylenedicarboxylate,[80] and other compounds[81] (see Chapter 6).

6 Dithiafulvene Derivatives

2-Alkynyl alkanedithioates (49) yield dithiafulvenes (50) on treatment with bases, *via* dialkylidene-1,3-dithioles.[82] The indene derivative (51)[83] and other dithiafulvenes,[84] including isothiathiophthens (see Chapter 7), have been prepared by condensations involving 2-methylthio-1,3-dithiolium salts.

[72] W. Kusters and P. de Mayo, *J. Amer. Chem. Soc.*, 1974, **96**, 3502.
[73] A. R. Siedle and R. B. Johannesen, *J. Org. Chem.*, 1975, **40**, 2002.
[74] F. Wudl and M. L. Kaplan, *J. Org. Chem.*, 1974, **39**, 3608.
[75] E. M. Engler and V. V. Patel, *Tetrahedron Letters*, 1975, 1259.
[76] L. R. Melby, H. D. Hartzler, and W. A. Sheppard, *J. Org. Chem.*, 1974, **39**, 2456.
[77] E. Fanghänel, *J. prakt. Chem.*, 1976, **318**, 127.
[78] H. Matsukubo and H. Kato, *J.C.S. Chem. Comm.*, 1974, 412.
[79] S. Nakazawa, T. Kiyosawa, K. Hirakawa, and H. Kato, *J.C.S. Chem. Comm.*, 1974, 621.
[80] H. Gotthardt, M. C. Weisshuhn, and B. Christl, *Chem. Ber.*, 1976, **109**, 753, 740.
[81] H. Matsukubo and H. Kato, *J.C.S. Perkin I*, 1975, 632; *J.C.S. Chem. Comm.*, 1975, 840.
[82] P. Vermeer, J. Meijer, H. J. T. Bos, and L. Brandsma, *Rec. Trav. chim.*, 1974, **93**, 51.
[83] G. Seitz and H. G. Lehmann, *Arch. Pharm.* (*Weinheim*), 1974, **307**, 853 (*Chem. Abs.*, 1975, **82**, 72 737).
[84] E. Fanghänel, *J. prakt. Chem.*, 1975, **317**, 137.

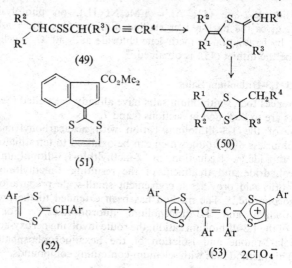

(49)

(50)

(51)

(52)

(53) $2ClO_4^-$

The mechanism of the base-catalysed decomposition of 4-aryl-1,2,3-selena-diazoles to diselenafulvenes has received detailed study.[85] Several reports have appeared on the synthesis of benzodithiafulvenes,[86] and the oxidation of dithia-fulvenes of type (52) has been investigated.[87] The products are the red-violet salts (53), which form part of an interesting redox system.

7 Tetrathiafulvalenes and Selenium-containing Analogues

The discovery that the charge-transfer salt derived from tetrathiafulvalene (54; R = H) and tetracyanoquinodimethane (55) has metal-like behaviour,[13a] and shows a remarkable electrical conductivity over a wide temperature range, has stimulated a large volume of work on the design and synthesis of 'organic metals'.[88]

Details have been given[74, 76] for the preparation of tetrathiafulvalene (TTF) by deprotonation of 1,3-dithiolium fluoroborate and hexafluorophosphate. Improvements in the synthesis have been suggested,[89-91] and modifications described[76, 89] for deuteriated TTF. Oxidation of the dihydro-compound (46) with manganese dioxide to give TTF has been reported.[73]

The dithiolium cation-deprotonation route has also been applied to the syn-thesis of various tetrasubstituted derivatives of TTF (54; R = Me, Et, Ph,

[85] M. H. Ghandehari, D. Davalian, M. Yalpani, and M. H. Partovi, *J. Org. Chem.*, 1974, **39**, 3906; F. Malek-Yazdi and M. Yalpani, *ibid.*, 1976, **41**, 729.

[86] G. Purrello and P. Fiandaca, *J.C.S. Perkin I*, 1976, 692; T. Takeshima, N. Fukada, E. Okabe, F. Mineshima, and M. Muraoka, *J.C.S. Perkin I*, 1975, 1277; N. S. Loseva, L. E. Nivorozhkin, N. I. Borisenko, and V. I. Minkin, *Khim. geterotsikl. Soedinenii*, 1974, 1485 (*Chem. Abs.*, 1975, **82**, 57 586).

[87] R. Mayer and H. Kröber, *J. prakt. Chem.*, 1974, **316**, 907.

[88] A. F. Garito and A. J. Heeger, *Accounts Chem. Res.*, 1974, **7**, 232; A. D. Yoffe, *Chem. Soc. Rev.*, 1976, **5**, 51.

[89] D. Dolphin, W. Pegg, and P. Wirz, *Canad. J. Chem.*, 1974, **52**, 4078.

[90] H. Anzai, *Denshi Gijutsu Sogo Kenkyujo Iho*, 1975, **39**, 667 (*Chem. Abs.*, 1976, **84**, 164 654).

[91] F. Wudl, *J. Amer. Chem. Soc.*, 1975, **97**, 1962.

etc.),[56b, 92] including bis(cyclopenteno)- and bis(cyclohexeno)-derivatives [54; RR = (CH₂)₃] and [54; RR = (CH₂)₄],[56, 69] which, when combined with tetracyanoquinodimethane (TCNQ), give salts with remarkably different electrical properties.[69]

A second route to TTF derivatives involves desulphurization of 1,3-dithiole-2-thiones (or, better, deselenization of 2-selones; see below) with phosphines or phosphites. For the preparation of tetracyano-TTF (54; R = CN), the use of triphenyl phosphite is recommended;[93] alternatively, 4,5-dicyano-1,3-dithiole-2-one is reported to be deoxygenated in high yield by trimethyl phosphite.[94]

Symmetrical bis(benzo)-derivatives of TTF have been prepared by the deprotonation route,[52] by desulphurization with triethyl phosphite,[95] by generating

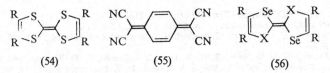

(54)　　　　　　　(55)　　　　　　　(56)

benzyne in the presence of carbon disulphide,[45b] and by thermal decomposition of 2-alkoxybenzodithioles.[51] A new route, involving the reaction of aromatic *ortho*-dithiols with tetrachloroethylene, has provided an unsymmetrical bis-(benzo)-TTF.[96] An *X*-ray study of a bis(benzo)-TTF–TCNQ adduct has been published.[97]

Replacement of the four sulphur atoms of TTF by four selenium atoms leads, in the TCNQ complex, to enhanced metallic behaviour.[59a] The required tetraselenafulvalene (56; R = H, X = Se) was prepared by deselenization of 1,3-diselenole-2-selone; a convenient alternative route has been described.[65b] Similar methods have been used to prepare tetramethyltetraselenafulvalene (56; R = Me, X = Se)[57a] (which forms two salts with TCNQ, one a red insulator and the other a black conductor), a series of similar derivatives,[57b] and a tetrakis-methylselenotetraselenafulvalene (56; R = MeSe, X = Se).[68]

Diselenadithiafulvalene (56; R = H, X = S) has been prepared as a mixture of *cis*- and *trans*-isomers;[65a, 60] the TCNQ salt shows high conductivity. The conversion of 1-selena-3-thiole-2-thiones into 2-selones is a useful practical device in this area, since the selones are more readily converted into fulvalenes with phosphorus compounds.[62] Various annelated diselenadithiafulvalenes have been synthesized and their complexes with TCNQ studied.[70]

Tetraselenafulvalene and diselenadithiafulvalene have been compared (by cyclic voltammetry and measurement of gas-phase ionization potentials) with TTF and are found to be weaker electron donors.[98] The TCNQ complex of

[92] P. Calas, J. M. Fabre, E. Torreilles, and L. Giral, *Compt. rend.*, 1975, **280**, C, 901.
[93] Z. Yoshida, T. Kawase, and S. Yoneda, *Tetrahedron Letters*, 1975, 331.
[94] M. G. Miles, J. D. Wilson, D. J. Dahm, and J. H. Wagenknecht, *J.C.S. Chem. Comm.*, 1974, 751.
[95] G. Scherowsky and J. Weiland, *Chem. Ber.*, 1974, **107**, 3155.
[96] G. S. Bajwa, K. D. Berlin, and H. A. Pohl, *J. Org. Chem.*, 1976, **41**, 145.
[97] R. P. Shibaeva and O. V. Yarochkina, *Doklady Akad. Nauk S.S.S.R.*, 1975, **222**, 91 (*Chem. Abs.*, 1975, **83**, 58 695).
[98] E. M. Engler, F. B. Kaufman, D. C. Green, C. E. Klots, and R. N. Compton, *J. Amer. Chem. Soc.*, 1975, **97**, 2921.

hexamethylenetetraselenafulvalene [56; RR = $(CH_2)_3$, X = Se] remains 'metallic' to very low temperatures (0.045 K), and a study of its crystal structure indicates that this is the most two-dimensional member of the TTF–TCNQ family of organic metals.[99]

The possibility of combining the electron-rich heterofulvalenes with electron acceptors other than TCNQ has received some attention. Acceptor systems studied include 2,3-dichloro-5,6-dicyano-*p*-benzoquinone,[100] tetracyanomucononitrile,[101] dimethyl-TCNQ (which forms a highly conducting salt with tetramethyltetraselenafulvalene, but an insulating salt with hexamethylenetetraselenafulvalene),[102] and complex nickel compounds.[103]

[99] T. E. Phillips, T. J. Kistenmacher, A. N. Bloch, and D. O. Cowan, *J.C.S. Chem. Comm.*, 1976, 334.

[100] Y. Ueno and M. Okawara, *Chem. Letters*, 1974, 1135.

[101] F. Wudl and E. W. Southwick, *J.C.S. Chem. Comm.*, 1974, 254.

[102] J. R. Andersen, C. S. Jacobsen, G. Rindorf, H. Soling, and K. Bechgaard, *J.C.S. Chem. Comm.*, 1975, 883.

[103] L. V. Interrante, K. W. Browall, H. R. Hart, I. S. Jacobs, G. D. Watkins, and S. H. Lee, *J. Amer. Chem. Soc.*, 1975, 97, 889; J. S. Kasper, L. V. Interrante, and C. H. Secaur, *ibid.*, p. 890; P. Calas, J. M. Fabre, M. Khalife-El-Saleh, A. Mas, E. Torreilles, and L. Giral, *Tetrahedron Letters*, 1975, 4475.

9
Thiopyrans and Related Compounds

BY U. EISNER

1 Introduction

This chapter follows essentially the same plan as in the previous Volume [1b] except that all benzothiopyrans (thiochromans, thiochromenes, thioisochromans, etc.) have been covered in Section 7 and all benzothiopyrones (thiocoumarins, thiochromones, etc.) in Section 8. Thiopyrylium salts are included in Section 6, regardless of whether or not they are monocyclic, and similarly all thiopyran 1,1-dioxides are treated in Section 5.

Reviews have appeared on thiopyrans,[2a] thiopyrylium salts,[2b] and thiochromanones and related compounds.[3] Biological aspects of some thioxanthone drugs and other thiopyran derivatives have been reviewed.[4] Space limitations preclude citation of the patent literature, but it should be noted that many patents have appeared that are concerned with juvenile-hormone intermediates [1b] and with biologically active thioxanthene derivatives.

2 Dihydrothiopyrans

Trithiocarbonate SS-dioxides $ArSO_2CS_2Ar$ undergo [4 + 2] cycloaddition with 1,3-dienes to afford the unstable Δ^3-dihydrothiopyrans (1), which readily lose $ArSO_2H$ to give the corresponding $2H$-thiopyrans.[5] Addition of $ArSO_2H$ to these $2H$-thiopyrans gives the 4-arylsulphinyl-2-arylthio-Δ^2-dihydrothiopyrans isomeric with (1). Evidence for the existence of the unstable 4-thioxotetrahydrofuran-3-ones was obtained [6] by trapping them with 2,3-dimethylbuta-1,3-diene, which yielded the spiro-compounds (2); these underwent unusual mass spectral

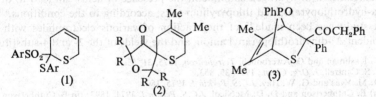

[1] 'Organic Compounds of Sulphur, Selenium, and Tellurium', ed. D. H. Reid (Specialist Periodical Reports), The Chemical Society, London, (a) Vol. 2; (b) Vol. 3.
[2] V. G. Kharchenko, S. N. Chalaya, and T. M. Konovalova, (a) Khim. geterotsikl. Soedinenii, 1974, 1155; (b) ibid., 1975, 147.
[3] S. W. Schneller, Adv. Heterocyclic Chem., 1975, 18, 59.
[4] E. Hirschberg, Antibiotics, 1975, 3, 274; P. E. Hartman and P. B. Hulbert, J. Toxicol. Environ. Health, 1975, 1, 243.
[5] J. A. Boerma, N. H. Nilsson, and A. Senning, Tetrahedron, 1974, 30, 2735.
[6] G. Hoehne, F. Marschner, K. Praefcke, and P. Weyerstahl, Chem. Ber., 1975, 108, 673.

fragmentations. Initial 1,3-cycloaddition of the oxyallyl cation $Ph\overset{+}{C}HC(\overset{-}{O})=$ CHPh to the P=S group of 3,4-dimethyl-1-phenylphosphole-1-sulphide followed by a complex rearrangement was postulated [7] to account for the final product (3). Oxidation of (3) with peroxy-acid removed the phosphorus bridge and formed a 2H-thiopyran. Irradiation of thiobenzophenone in cyclo-octatetraene led to a bridged Δ^3-dihydrothiopyran by [4 + 2] cycloaddition. Other dienes added in a [2 + 2] manner.[8] Details [1a] of the Diels–Alder reactions of MeSC(=S)CN have been published.[9]

The bridged heterocycles (4; X = S, Se, or Te) have been prepared [10] from the dibromides. Compound (4; X = Te), which extruded tellurium on heating, did not give the expected bicyclo[3,3,0]octa-2,6-diene, but instead gave the fluxional bicyclo[5,1,0]octa-2,5-diene.[10b]

Addition of SCl_2 to cyclohepta-3,5-dienone gave (5), which readily rearranged to 8-thia[3,2,1]octa-3,6-dien-2-one with loss of HCl.[11]

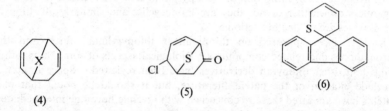

(4)   (5)   (6)

Pyrolysis of the spiro-compound (6) and its congeners did not result in the expected retro-Diels–Alder reaction but furnished thiophens, fluorene, and 2-(9-fluorenyl)thiophens.[12] The anhydride (7) gave [13] the retro-Diels–Alder product on heating, together with a cycloheptatriene formed by a rearrangement for which there is an earlier precedent. As these reactions are not given [1b] by the corresponding diester, subtle structural differences in the systems (7) clearly have a profound influence on the course of the thermolysis.

Dihydrothiopyrans have been shown [14a] to disproportionate to tetrahydro-thiopyrans and thiopyrylium salts with acid, thus explaining why reaction [1] of 1,5-diketones with hydrogen sulphide in the presence of acids can lead to di- or tetra-hydrothiopyrans, and thiopyrylium salts, according to the conditions.[14b, c]

A study has been made [15] of the reaction of various electrophiles with the ambident Δ^3-dihydrothiopyranyl anion, and the yields of the 2- and 4-substituted

[7] Y. Kashman and O. Awerbouch, *Tetrahedron*, 1975, **31**, 53.
[8] T. S. Cantrell, *J. Org. Chem.*, 1974, **39**, 853.
[9] D. M. Vyas and G. W. Hay, *J.C.S. Perkin I*, 1975, 180.
[10] (a) E. Cuthbertson and D. D. McNicol, *J.C.S. Perkin I*, 1974, 1893; (b) E. Cuthbertson and D. D. McNicol, *J.C.S. Chem. Comm.*, 1974, 498.
[11] P. H. McCabe and W. Routledge, *Tetrahedron Letters*, 1976, 85.
[12] B. Koenig, J. Martens, K. Praefcke, A. Schoenberg, H. Schwarz, and L. Zeisberg, *Chem. Ber.*, 1974, **107**, 2931.
[13] J. F. King, R. M. Emanoza, and E. G. Lewars, *Canad. J. Chem.*, 1974, **52**, 2409.
[14] (a) V. G. Kharchenko, S. N. Chalaya, T. V. Stolbova, and S. K. Klimenko, *Zhur. org. Khim.*, 1975, **11**, 2447; (b) S. K. Klimenko, M. N. Berezhnaya, and V. G. Kharchenko, *ibid.*, 1974, **10**, 2425; (c) V. G. Kharchenko, S. N. Chalaya, and L. G. Chichenkova, *Khim. geterotsikl. Soedinenii*, 1975, 643 (*Chem. Abs.*, 1976, **84**, 74 046; 1975, **82**, 86 264; 1975, **83**, 114 151).
[15] S. Torii, H. Tanaka, and Y. Tomotaki, *Chem. Letters*, 1974, 1541.

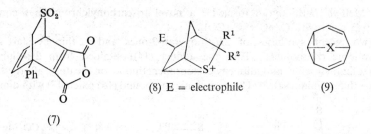

(8) E = electrophile   (9)

(7)

products have been determined. The rearrangement which occurs on bromination and other electrophilic addition reactions of 2-thianorbornenes has been shown [16] to involve the intermediate ion (8).

The effect of the heteroatom on the fluxional behaviour of the heterobarbaralanes (9; X = NR, S, or SO$_2$) has been examined.[17] Photoelectron spectroscopy of the distorted vinyl sulphide 9-thiabicyclo[3,3,1]non-1-ene [1b] and its oxygen and carbon analogues was carried out [18] in order to examine the consequences of steric inhibition of conjugation. It was concluded that participation of *d*-orbitals need not be invoked for the sulphide. The mass spectra of some dihydrothiopyrans have been recorded.[19]

3 2*H*-Thiopyrans and Related Compounds

2*H*-Thiopyrans may occasionally be isolated [14b] from the reaction of 1,5-diketones, hydrogen sulphide, and acids, although they usually disproportionate under these conditions. Dienones give 2*H*-thiopyrans with the same reagents.[20] [4 + 2] Cycloaddition of 3-dialkylaminovinyl aryl thiones with acrylic acid derivatives and related compounds afforded [21] the 2*H*-thiopyrans (10; R^1 = H, OEt; R^2 = CN, CONH$_2$, COMe, *etc.*). The intermediate 4-dialkylamino-Δ^2-dihydrothiopyrans were sometimes isolated. Factors affecting the addition were studied, and some reactions of (10) and the derived thiopyrylium salts were described. The preparation of the azido-thiopyrans (11; R = H or Ph) by the action of sodium azide on the thiopyrylium salts has been reported.[22] Their thermolysis afforded substituted pyridines, thiophens, and sulphur, which were the decomposition products of the expected thiazepine intermediates.

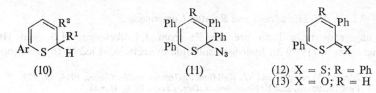

(10)   (11)   (12) X = S; R = Ph
                                                      (13) X = O; R = H

[16] M. Raasch, *J. Org. Chem.*, 1975, **40**, 161.
[17] A. G. Anastassiou, E. Reichmanis, and J. C. Wetzel, *Tetrahedron Letters*, 1975, 1651.
[18] C. Batich, E. Heilbronner, C. B. Quinn, and J. R. Wiseman, *Helv. Chim. Acta*, 1976, **59**, 512.
[19] H. Remane, *Z. Chem.*, 1974, **14**, 313; J.-P. Pradère, G. Duguay, and H. Quiniou, *Org. Mass Spectrometry*, 1976, **11**, 293.
[20] C. Fournier, D. Paquer, and M. Vazeux, *Bull. Soc. chim. France*, 1975, 2753.
[21] J.-P. Pradère, Y. T. N'Guessan, and H. Quiniou, *Tetrahedron*, 1975, **31**, 3059.
[22] J. P. Le Roux, J. C. Cherton, and P. L. Desbene, *Compt. rend.*, 1975, **280**, C, 37.

6-Methyl-2*H*-thiopyran formed [23] a novel triscarbonylchromium π-complex on treatment with $(MeCN)_3Cr(CO)_3$.

Two recent syntheses of 2*H*-thiopyran-2-ones and -2-thiones start from meso-ionic compounds. Thus the adduct (14) obtained from 2,3-diphenyl-thiazolium 4-oxide and diphenylcyclopropenethione on heating afforded (12) and a thiazocinium salt.[24] The meso-ionic compound (15) reacted [25] with dimethyl

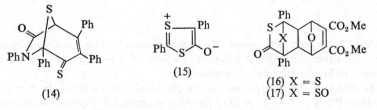

(14)

(15)

(16) X = S
(17) X = SO

8-oxabicyclo[2,2,1]hepta-2,5-diene-2,3-dicarboxylate to yield the adduct (16), which on thermolysis underwent double fragmentation with loss of COS to give 2,5-diphenylthiophen. The corresponding sulphoxide (17), however, on heating, extruded SO to give the thiopyrone (13).

The reaction of 1,2-dithiole-3-thiones with enamines gave [26] substituted thio-pyranthiones such as (12). Another synthesis of (12) employed [27a] the reaction of 1,2-dithiolium salts with $CNCH_2CSNH_2$. Various syntheses of substituted thiopyranylidenes (18) have been reported [27] and their photoisomerization about the exocyclic double bond has been studied.[28] The thermal isomerization of

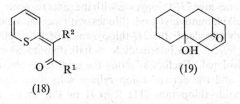

(18)

(19)

3*H*-pyran-2-thiones to 2*H*-thiopyran-2-ones, which involves 1,5-sigmatropic hydrogen shifts, has been investigated,[29a] and the mass spectral fragmentations of the isomers have been compared.[29b]

4 4*H*-Thiopyrans and Related Compounds

4*H*-Thiopyrans have been prepared [30a] from 1,5-diketones, H_2S, and HCl. The reaction of (19) with hydrogen sulphide in acetic acid led [30b] to C—C bond

[23] K. Oefele, A. Wurzinger, and W. Kalbfuss, *J. Organometallic Chem.*, 1974, **69**, 279.
[24] K. T. Potts, J. Baum, and E. Houghton, *J. Org. Chem.*, 1976, **41**, 818.
[25] H. Matsukubo and H. Kato, *J.C.S. Chem. Comm.*, 1975, 840.
[26] F. Ishii, M. Stavaux, and N. Lozac'h, *Tetrahedron Letters*, 1975, 1473.
[27] (a) K. S. N'Guyen and R. Pinel, *Bull. Soc. chim. France*, 1974, 1356; (b) J. P. Sauvé and N. Lozac'h, *ibid.*, p. 1196.
[28] C. T. Pedersen, C. Lohse, N. Lozac'h, and J. P. Sauvé, *J.C..S Perkin I*, 1976, 166.
[29] W. H. Pirkle and W. V. Turner, (a) *J. Org. Chem.*, 1975, **40**, 1617; (b) *ibid.*, p. 1644.
[30] (a) L. I. Lelyukh and V. G. Kharchenko, *Zhur. org. Khim.*, 1974, **10**, 1547; (b) V. G. Kharchenko and A. F. Blinokhvatov, *ibid.*, p. 2462; (c) V. G. Kharchenko and S. N. Chalaya, *ibid.*, 1975, **11**, 1540 (*Chem. Abs.*, 1974, **81**, 120 393; 1975, **82**, 125 232; 1975, **83**, 114 153).

cleavage, with the formation of 2,3:5,6-bis(tetramethylene)-4*H*-thiopyran. 4*H*-Pyrans gave 4*H*-thiopyrans under similar conditions.[30c] 4*H*-Thiopyran-3,5-dialdehydes were prepared [31] from the corresponding 4*H*-pyrans by base-induced ring opening followed by ring closure with P_4S_{10}.

2,6-Disubstituted 4*H*-thiopyran-4-ones and selenopyran-4-ones have been prepared by the reaction of the acetylenic ketones $(RC{\equiv}C)_2CO$ with thiourea [32] and hydrogen selenide,[33] respectively.

There are several reports [34] of the [4 + 2] cycloaddition of dimethyl acetylene-dicarboxylate (DMAD) to thioacylketen thioacetals $R^1C({=}S)\overset{\underset{|}{R^2}}{C}{=}\overline{CSCR^3R^4}$-$\underline{CR^3R^4S}$, which yielded the thioketals (20). Cycloaddition of DMAD to some heterocycles containing the grouping $>C{=}CHC({=}S)SMe$ was reported [35] to yield a variety of products, the nature of which depended on the starting material.

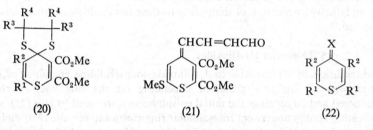

(20)   (21)   (22)

These products included some compounds believed to be complex spiro-4*H*-thiopyrans.[35a] The thiopyranylidene (21) isolated from the reaction of DMAD with the ester of 2-dithiocarboxymethylene-1-methyl-1,2-dihydropyridine was postulated to arise from a spiro-dihydropyridine intermediate.[35b]

The reaction of (15) with the diphenylcyclopropenes $\overline{CPh{=}CPhC}{=}X$ [X = O, $C(CN)_2$, or $C(CN)CO_2Et$] gave [36] the thiopyranylidenes [22; X = O, $C(CN)_2$, or $C(CN)CO_2Et$; $R^1 = R^2 = Ph$] in low yields. The nature of the products obtained on reaction of meso-ionic compounds with cyclopropenes is clearly dependent on subtle structural variations.[24, 36]

Interest in sesquifulvalene analogues and related compounds continues.[1] A number of 2,6-disubstituted derivatives [22; R^1 = Me or SMe; R^2 = H; X = substituted cyclopentadienyl,[37, 38] $C(COR)_2$,[38] CYZ [39] where Y, Z = NO_2,

[31] J. M. Brown and F. Sondheimer, *Angew. Chem. Internat. Edn.*, 1974, **13**, 337.
[32] K. G. Migliorese and S. I. Miller, *J. Org. Chem.*, 1974, **39**, 843.
[33] A. I. Tolmachev and M. A. Kudinova, *Khim. geterotsikl. Soedinenii*, 1974, 274 (*Chem. Abs.*, 1974, **81**, 13 365).
[34] P. de Mayo and H. Y. Ng, *J.C.S. Chem. Comm.*, 1974, 877; R. Okazaki, A. Kitamura, and N. Inamoto, *ibid.*, 1975, 257; Y. Tominaga, Y. Morita, Y. Matsuda, and G. Kobayashi, *Chem. and Pharm. Bull. (Japan)*, 1975, **23**, 2390.
[35] (a) G. Kobayashi, Y. Matsuda, Y. Tominaga, and K. Mizuyama, *Chem. and Pharm. Bull. (Japan)*, 1975, **23**, 2749; (b) Y. Tominaga, K. Mizuyama, and G. Kobayashi, *ibid.*, 1974, **22**, 1670.
[36] H. Matsukubo and H. Kato, *J.C.S. Perkin I*, 1975, 632.
[37] G. Seitz and H. G. Lehmann, *Annalen*, 1975, 331.
[38] B. Eistert and T. J. Arackal, *Chem. Ber.*, 1975, **108**, 2397.
[39] I. Belsky, H. Dodink, and Y. Shvo, *J. Org. Chem.*, 1974, **39**, 989.

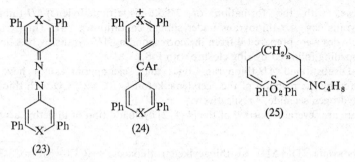

(23)   (24)   (25)

CN, CO_2Me, *etc.*] has been prepared. The redox systems (23; X = S or NMe) [40] and the pyrylocyanines (24; X = O, S, or Se) [41] have also been described. Synthetic routes to (22), (23), and (24) employ either the appropriate thiopyranones [37, 39] or a 4-substituted thiopyrylium salt. [38, 40, 41]

Catalytic hydrogenation of thiopyrans to their tetrahydro-derivatives has been reported. [42]

5 Thiopyran 1,1-Dioxides

Cycloaddition of diphenylthiiren 1,1-dioxide to pyrrolidine enamines of cyclic ketones gave [43] different products, depending on the ring size. With six-membered and larger rings the initial episulphone rearranged to give (25), which on heating readily underwent transannular ring closure to bicyclic [2,6]-diphenyl-3(N-pyrrolidinyl)-4,5-polymethylene-5,6-dihydrothiopyran 1,1-dioxides. Bicyclic dihydrothiophen 1,1-dioxides were also formed.

Photolysis of the diones o-$RCH_2C_6H_4COCOMe$ (R = H or Me) in benzene–sulphur dioxide afforded [44] the base-sensitive compound (26).

A number of heterocyclic derivatives (27; X = S, O, or NR) and their benzologues have been synthesized [45] in order to examine the involvement of the SO_2 group in the conjugation of the molecule. Investigation of their spectral and chemical properties led to the conclusion that these compounds (27) could not be considered as pseudo-azulenes, and that there was little interaction of the S—O bond with the conjugated system.

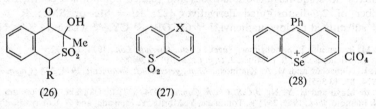

(26)   (27)   (28)

[40] S. Huenig and G. Ruider, *Annalen*, 1974, 1415.
[41] S. V. Krivun, A. I. Buryak, S. V. Sayapina, O. F. Voziyanova, and S. N. Baranov, *Khim. geterotsikl. Soedinenii*, 1973, 1004 (*Chem. Abs.*, 1974, **80**, 49 227).
[42] V. G. Kharchenko, N. S. Smirnova, S. N. Chalaya, A. S. Tatarinov, and L. G. Chichenkova, *Zhur. org. Khim.*, 1975, **11**, 1543 (*Chem. Abs.*, 1975, **83**, 178 726) and earlier papers.
[43] M. H. Rosen and G. Bonet, *J. Org. Chem.*, 1974, **39**, 3805.
[44] N. K. Hamer, *J.C.S. Chem. Comm.*, 1975, 557.
[45] G. A. Pagani, *J.C.S. Perkin II*, 1974, 1389; *ibid.*, p. 1392.

The Diels–Alder reaction was used as a probe for aromaticity in the ketones $\overline{CH=CHXCH=CHCO}$ (X = O, S, SO$_2$, or NR). Thiopyran-4-one 1,1-dioxide was the only compound to react with a diene.[46]

Evidence has been provided[47] that photolysis of 2H-thiopyran 1,1-dioxides and their benzo-derivatives does involve sulphene intermediates. Details of the synthesis[1b] of a thiabenzene oxide have appeared.[48]

6 Thiopyrylium Salts

The synthesis of 2,4,6-trisubstituted thiopyrylium salts has been modified.[49] Four new selenaphenanthrenium salts have been prepared.[50] 3,5-Diphenylthiopyrylium salts have been synthesized.[51]

Reduction of thiopyrylium salts to 2H- and 4H-thiopyrans by trichlorosilane[51] and by lithium aluminium hydride[52] has been reported. Addition of organometallic reagents to thiopyrylium and selenopyrylium salts, which was (see Volume 2, p. 387) believed to yield thia- and selena-benzenes, has now been shown[53] to give oligomers. Some authentic thia-[51, 53] and selena-benzenes[53] have been prepared. Nucleophilic addition[54a] to, and nitration[54b] of, 9-phenylselenoxanthylium salts (28) has been studied.

4-Chlorothiopyrylium salts react with nucleophiles to give either 4-substituted thiopyrylium salts by addition–elimination, or products arising from ring opening. The course of the reaction depends on whether the attacking nucleophile is a hard or a soft base.[55] Reactions of 4-chlorothiopyrylium salts with active methylene compounds, which result in (22)[38] or (24),[41] and of 4-(methylthio)-thiopyrylium salts with hydrazine, which give (23),[40] were mentioned in Section 4. 2,4,6-Triarylthiopyrylium salts react with hydrazines,[56] yielding 1,2-diazepines and pyrazolines.

A series of interesting papers[57] deals with the dimers of thiopyrylium 3-oxides (29) and their benzologues. Their structures and stereochemistry are highly dependent on the substitution pattern of the monomers (29). Dimers linked through the 2,6- and 2′,6′-carbons, and the 2,6- and 2′,4′-carbons, have been isolated, and their formation may be envisaged as the coupling of the appropriate resonance hybrids (30)—(32).

[46] J. A. Hirsch, R. W. Kosley, R. P. Morin, G. Schwarzkopf, and R. D. Brown, *J. Heterocyclic Chem.*, 1975, **12**, 785.
[47] J. F. King, E. G. Lewars, D. R. K. Harding, and R. M. Enanoza, *Canad. J. Chem.*, 1975, **53**, 3656; C. R. Hall and D. J. H. Smith, *Tetrahedron Letters*, 1974, 3633.
[48] Y. Tamura, H. Taniguchi, T. Miyamoto, M. Tsunekawa, and M. Ikeda, *J. Org. Chem.*, 1974, **39**, 3519.
[49] G. A. Reynolds, *Synthesis*, 1975, 638.
[50] I. Degani and R. Fochi, *Ann. Chim. (Italy)*, 1973, **63**, 319 (*Chem. Abs.*, 1974, **81**, 63 466).
[51] A. G. Hortmann, R. L. Harris, and J. A. Miles, *J. Amer. Chem. Soc.*, 1974, **96**, 6119.
[52] S. K. Klimenko, M. N. Berezhnaya, T. V. Stolbova, I. Ya. Evtushenko, and V. G. Kharchenko, *Zhur. org. Khim.*, 1975, **11**, 2173 (*Chem. Abs.*, 1976, **84**, 30 820).
[53] J. Stackhouse, G. H. Senkler, B. E. Maryanoff, and K. Mislow, *J. Amer. Chem. Soc.*, 1974, **96**, 7835; *ibid.*, 1975, **97**, 2718.
[54] M. Hori, T. Kataoka, C. Hsu, Y. Ashahi, and E. Mizuta, (*a*) *Chem. and Pharm. Bull. (Japan)*, 1974, **22**, 27; (*b*) *ibid.*, p. 21.
[55] S. Yoneda, T. Sugimoto, O. Tanaka, Y. Moriya, and Z. Yoshida, *Tetrahedron*, 1975, **31**, 2669.
[56] D. J. Harris, G. Y.-P. Kan, V. Snieckus, and E. Klingsberg, *Canad. J. Chem.*, 1974, **52**, 2798.
[57] S. Baklien, P. Groth, and K. Undheim, *J.C.S. Perkin I*, 1975, 2099; *Acta Chem. Scand. (B)*, 1976, **30**, 24, and earlier papers.

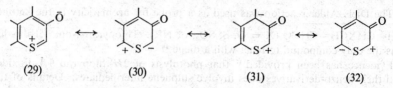

(29)   (30)   (31)   (32)

Charge-transfer complexes of thiopyrylium salts with aromatic hydrocarbons and olefins have been investigated.[58] The n.m.r. spectra of thio- and seleno-pyrylium salts [59a] and their benzo- [59a, b] and thieno-derivatives [59c] have been examined. The results of further MO calculations have been advanced [60] to support the earlier [1b] contention that *d*-orbitals are not significantly involved in bonding in thiopyrylium salts, despite the apparently contradictory evidence based on n.m.r. shifts.

7 Benzothiopyrans and Related Compounds

Geranyl phenyl sulphides or sulphones, $Me_2C=CHCH_2CMe=CHCHXY$ (33; X = H; Y = SPh or SO_2Ph), cyclized to cyclocitral derivatives under acid conditions, whereas the corresponding dithioacetal (33; X = Y = SPh) gives [61] the thiochroman (34). The novel heterocycles (35) have been prepared by standard methods from the corresponding tellurochromanones.[62] The critical step in the synthesis of the latter consists of reduction of the appropriate diaryl ditelluride with borohydride and subsequent reaction with 3-chloropropionic acid under carefully controlled conditions.

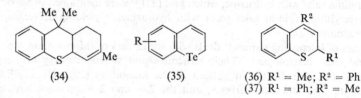

(34)   (35)   (36) R^1 = Me; R^2 = Ph
                                            (37) R^1 = Ph; R^2 = Me

Cyclization of the sulphide $PhSCHMeCH_2COPh$ under acidic conditions gave the expected thiochromene (36), together with the rearranged product (37). Both compounds disproportionated under the experimental conditions, and the thiopyrylium salts and the thiochromans corresponding to (36) and (37) were isolated. The same products were obtained from either of the isomeric salts $\overline{CH=CPhSPhCHMe}^+$ and $\overline{HC=CMeSPhCHPh}^+$, which were shown to be inter-convertible *via* a thiacyclobutadiene.[63]

[58] Z. Yoshida, T. Sugimoto, and S. Yoneda, *Bull. Chem. Soc. Japan*, 1975, **48**, 1519.
[59] (a) A. I. Tolmachev, L. M. Shulezko, and M. Yu. Kornilov, *Ukrain. khim. Zhur.*, 1974, **40**, 287 (*Chem. Abs.*, 1974, **81**, 3730); (b) I. Degani, R. Fochi, and G. Spunta, *Ann. Chim. (Italy)*, 1973, **63**, 527 (*Chem. Abs.*, 1975, **82**, 42 797); (c) F. Catti Boccuzzi and R. Fochi, *Gazzetta*, 1974, **104**, 671 (*Chem. Abs.*, 1975, **82**, 72 286).
[60] M. H. Palmer, R. H. Findlay, W. Moyes, and A. J. Gaskell, *J.C.S. Perkin II*, 1975, 841.
[61] S. Torii, K. Uneyama, and M. Isihara, *Chem. Letters*, 1975, 479.
[62] N. Deren, J.-L. Piette, J. Van Coppenolle, and M. Renson, *J. Heterocyclic Chem.*, 1975, **12**, 423.
[63] R. S. Devhard, V. N. Gogte, and B. D. Tilak, *Tetrahedron Letters*, 1974, 3911.

Desulphurization of 3,4-diarylthiochromenes was investigated as a possible route to 1,1,2-triarylprop-1-enes of known configuration, but although some stereoselectivity was obtained, the results were not promising enough for its recommendation as a general method.[64]

Further details[1b] of the rearrangement–ring contraction of the ketone ArSCMe$_2$CH$_2$COMe, which gives benzothiophens with polyphosphoric acid, have been reported.[65] Additional[1b] analogues of thiochroman clathrate hosts have been prepared in order to test their inclusion properties.[66] The thermolysis of 2,2-dimethyl-1-(toluene-*p*-sulphonylimino)thiochroman has been investigated.[67] In benzene it gives *N*-[2-(3-methylbut-3-enyl)phenylthio]-*N*-(2'-methyl-thiochroman-2'-ylmethyl)toluene-*p*-sulphonamide.

The thioisochroman (38) was produced together with a thietan when a mixture of thiobenzophenone and methyl acrylate was irradiated at 589 nm. Since (38) was also produced from the same reagents in a dark reaction, the present

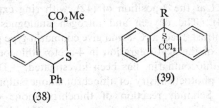

(38)

(39)

findings[68] contradict earlier results (see Vol. 3, p. 115) regarding the formation of the thietan. Thiophosgene undergoes[69] regiospecific [4 + 2] cycloaddition with 9-substituted anthracenes to give (39).

The preparation and reactions of 1-azidothioisochroman and its *SS*-dioxide have been recorded.[70] Photochemical ring-opening of 4-phenylthioisochromenes yielded 3-phenylindene as the principal product; an *ortho*-quinonoid intermediate was postulated.[71]

Conformational studies of 4-hydroxythiochroman and related compounds,[72] and mass spectral fragmentations of thiochromans,[73] have been reported.

8 Benzothiopyrones

Intramolecular cyclization, catalysed by aluminium chloride, of *p*-RC$_6$H$_4$X-COCH=CHPh (X = S, O, or NR') to the thiocoumarins (40) or their oxygen and nitrogen analogues occurs[74] with loss of the terminal phenyl group. When R = Bu^t the alkyl group is also lost. Oxidation of thiochromenes with selenium dioxide in ethanol gives 2-(2'-benzo[*b*]thienyl)-1-thiochromones by a complex

[64] D. J. Collins and J. J. Hobbs, *Austral. J. Chem.*, 1974, **27**, 1545.
[65] D. D. MacNicol and J. J. McKendrick, *J.C.S. Perkin I*, 1974, 2493.
[66] A. D. U. Hardy, J. J. McKendrick, and D. D. MacNicol, *J.C.S. Chem. Comm.*, 1974, 972.
[67] M. Kise, M. Murase, M. Kitano, T. Tomita, and H. Murai, *Tetrahedron Letters*, 1976, 691.
[68] H. Gotthardt, *Chem. Ber.*, 1974, **107**, 1856.
[69] H. Allgeier and T. Winkler, *Tetrahedron Letters*, 1976, 215.
[70] H. Boehme and F. Ziegler, *Annalen*, 1974, 734.
[71] A. Padwa, A. Au, G. A. Lee, and W. Owens, *J. Org. Chem.*, 1975, **40**, 1142.
[72] H. Hanaya, S. Onodera, and H. Kudo, *Bull. Chem. Soc. Japan*, 1974, **47**, 2607.
[73] H. Budzikiewicz and U. Lenz, *Org. Mass Spectrometry*, 1975, **10**, 992.
[74] T. Manimaran, T. K. Thiruvengadam, and V. T. Ramakrishnan, *Synthesis*, 1975, 739.

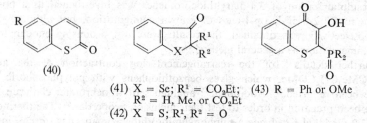

(40)

(41) X = Se; R¹ = CO₂Et;
R² = H, Me, or CO₂Et
(42) X = S; R¹, R² = O

(43) R = Ph or OMe

reaction. Selenochromenes give a different reaction.[75] Under different conditions another type of product results (see p. 335). Reduction of (41) with hypophosphorous acid yielded 3-substituted 4-hydroxy-selenocoumarins *via* the intermediates $[o\text{-EtO}_2\text{CCR}^1\text{R}^2\text{CH(OH)C}_6\text{H}_4\text{Se}\!-\!]_2$, but the selenochromanone was reduced to the selenochromene by the same reagent.[76] Phosphoryldiazoalkanes $R_2\text{P(O)CHN}_2$ react at the 2-position of (42), with ring expansion to give the thiochromone (43). The oxygen and nitrogen analogues (42; X = O, NR; R^1R^2 = O), in contrast, react at C-3 and give coumarin-type products.[77]

The mechanism of the acyl migration in 4-acetoxythiocoumarin, which gives 3-acetyl-4-hydroxythiocoumarin, has been investigated.[78] Details of the extensive work on the photochemistry of thiochromanone sulphoxides[1] have been published.[79] The Schmidt reaction of thiochromanone oxides and related compounds has been investigated.[80] A number of 2*H*-thiopyran-2-thiones and their benzologues has been prepared in order to study their behaviour on oxidation (H_2O_2–AcOH) which gives either a thiopyrylium salt or a 2*H*-thiopyrone.[81]

The ¹³C n.m.r. spectra of a wide variety of thiopyrones, thiopyranthiones, their oxides, and their benzo-derivatives have been recorded.[82]

9 Thioxanthenes and Thioxanthones

A high-temperature synthesis of thioxanthene from thiophenol and *o*-chloro- or *o*-bromo-toluene has been reported.[83] Irradiation of the thiol esters (44; X = Cl, Br, I, or SO₂Me; Y = H) afforded the thioxanthone (45; Y = H) and *p*-tolyl disulphide. This rearrangement is believed to involve a 1,3-shift of the $\text{SC}_6\text{H}_4\text{Me}$ group.[84] When the dichloro-compound (44; X = Y = Cl) was irradiated, photosubstitution occurred,[84c] giving the sulphide (45; Y = *p*-Me-$\text{C}_6\text{H}_4\text{S}$) instead of the expected chloride (45; Y = Cl). The same product was

[75] J. Van Coppenolle and M. Renson, *Compt. rend.*, 1975, **280**, C, 283.
[76] R. Weber, L. Christiaens, P. Thibaut, M. Renson, A. Croisy, and P. Jacquignon, *Tetrahedron*, 1974, **30**, 3865.
[77] W. Disteldorf and M. Regitz, *Annalen*, 1976, 225.
[78] J. Lehmann and H. Wamhoff, *Annalen*, 1974, 1287.
[79] I. W. J. Still, P. C. Arora, M. S. Chauhan, M.-H. Kwan, and M. T. Thomas, *Canad. J. Chem.*, 1976, **54**, 455.
[80] I. W. J. Still, T. M. Thomas, and A. M. Clish, *Canad. J. Chem.*, 1975, **53**, 276.
[81] J. L. Charlton, S. M. Loosmore, and D. M. McKinnon, *Canad. J. Chem.*, 1974, **52**, 3021.
[82] I. W. J. Still, N. Plavac, D. M. McKinnon, and M. S. Chauhan, *Canad. J. Chem.*, 1976, **54**, 280; *ibid.*, 1975, **53**, 2880.
[83] M. G. Voronkov, E. N. Deryagina, A. S. Nakhmanovich, L. G. Klochkova, and G. M. Ivanova, *Khim. geterotsikl. Soedinenii*, 1974, 429 (*Chem. Abs.*, 1974, **81**, 2550).
[84] G. Buchholz, J. Martens, and K. Praefcke, (*a*) *Synthesis*, 1974, 666; (*b*) *Tetrahedron*, 1974, **30**, 2565; (*c*) *Tetrahedron Letters*, 1975, 3213.

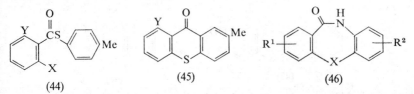

(44)   (45)   (46)

obtained when the fluoride (45; Y = F) was irradiated in the presence of *p*-thiocresol.

Benzyne intermediates are probably involved in the ring-closure of 2-(or 3-) chloro-2'-(methylthio)benzophenones to thioxanthones.[85] Xanthones were prepared analogously. Syntheses of several pharmacologically useful thioxanthene derivatives have been described.[86] Ceric ammonium nitrate oxidizes thioxanthene to thioxanthone.[87]

Although alkalis react [1b] with thioxanthone *SS*-dioxides to give the hydroxy-benzophenonesulphonic acids, sodamide in liquid ammonia gives thiazepinones (46; X = SO$_2$), presumably *via* an aryne intermediate.[88a] Thioxanthones, on the other hand, are decarbonylated with sodamide in toluene to give diaryl sulphides with only traces of (46; X = S).[88b] Thioxanthone did not undergo the Schmidt reaction, but the corresponding sulphone and sulphoxide afforded [89] the expected thiazepinones (46; X = SO$_2$ and SO, respectively). In the latter case some thioxanthone sulphoximide was also formed. Beckmann rearrangement of thioxanthone oximes proceeded normally to yield (46; X = S).[90]

The conformational preferences of 9-alkylthioxanthenes and their *S*-oxides have been investigated by n.m.r. techniques, including solvent shifts.[91] The magnitude of the chemical shifts of the 9-protons in the xanthene and thioxanthene anions indicates that their central rings are paratropic.[92]

Solutions of 9,9-dialkyl-thioxanthene cation radicals have been prepared and their e.s.r. and n.m.r. spectra studied.[93] The conformations of anion radicals derived from thioxanthene *SS*-dioxide and its 9-substituted derivatives have been investigated by e.s.r. techniques.[94]

10 Complex Thiopyran Derivatives

A novel carbene-insertion reaction at the 3-position of the benzothiophen group was postulated for the formation of (47) from the tosylhydrazone of 2-(2'-benzo-thienylthio)-5-nitrobenzaldehyde under Bamford–Stevens conditions. The new

[85] M. S. Gibson, S. M. Vines, and J. W. Walthew, *J.C.S. Perkin I*, 1975, 155.
[86] N. Latif, N. Mishriki, and K. A. Mohsen, *J.C.S. Perkin I*, 1974, 875; C. Kaiser, P. J. Fowler, D. H. Tedeschi, B. M. Lester, E. Garvey, C. L. Zirkle, A. J. Saggiomo, and E. A. Nodiff, *J. Medicin. Chem.*, 1974, **17**, 57; I. Nabih and A. Zayed, *J. Pharm. Sci.*, 1974, **63**, 1806.
[87] B. Rindone and C. Scolastico, *J.C.S. Perkin I*, 1975, 1398.
[88] (*a*) O. F. Bennett, J. Johnson, and S. Galletto, *J. Heterocyclic Chem.*, 1975, **12**, 1211; (*b*) O. F. Bennett and J. Collins, *Tetrahedron Letters*, 1976, 965.
[89] S. Palazzo, L. I. Ciannola, and S. Caronna, *J. Heterocyclic Chem.*, 1974, **11**, 839.
[90] K. Nagarajan, C. L. Kulkarni, and A. Venkateswarlu, *Indian J. Chem.*, 1974, **12**, 247.
[91] A. L. Ternay and S. A. Evans, *J. Org. Chem.*, 1974, **39**, 2941; *ibid.*, 1975, **40**, 2993.
[92] H. W. Vos, Y. N. Baker, C. MacLean, and N. H. Velthorst, *Org. Magn. Resonance*, 1974, **6**, 245.
[93] D. Deavenport, J. T. Edwards, A. L. Ternay, E. T. Strom, and S. A. Evans, *J. Org. Chem.*, 1975, **40**, 103.
[94] P. Lambelet and E. A. C. Lucken, *J.C.S. Perkin II*, 1976, 164.

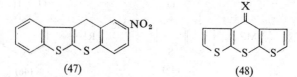

(47)   (48)

heterocycle (47) was also synthesized by other methods.[95] The related compounds (48; X = O, S, or OCH$_2$CH$_2$O) were obtained unexpectedly when the ethylenethioketal of 3,3-dithienyl ketone was treated with butyl-lithium and sulphur.[96] Conventional methods were used to prepare various heterocycles having benzofuran annelated to a thiopyran, *e.g.*, 1*H*-[1]benzothiopyrano-[3,2-*b*]benzofuran[97] and 1*H*-thiopyrano[3,4-*b*]benzofuran[98] derivatives. Syntheses of thieno[4,3,2-*de*]thiochromans[99] and 4*H*-selenolo[3,2-*c*]benzo-thiopyran and -selenin derivatives[100] have also been reported.

When the phthalides (49) or (50) were heated with concentrated sulphuric acid, rearrangement to (52) took place.[101a] Mixtures of (49), (51), and (52) were

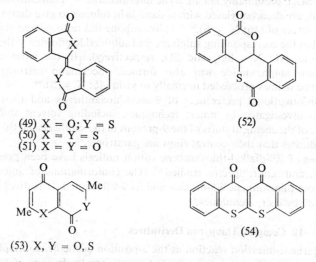

(49) X = O; Y = S   (52)
(50) X = Y = S
(51) X = Y = O

(53) X, Y = O, S   (54)

also produced when phthalic or thiophthalic anhydride was heated with phthalide or its thio-analogues.[101b] Preparations of the systems (53) and derived thiones have been described.[102] When the keten derivative (MeO$_2$C)$_2$C=C(SPh)$_2$ was heated with polyphosphoric acid the pharmacologically useful (54) was formed.[103]

[95] B. Iddon, H. Suschitzky, D. S. Taylor, and K. E. Chippendale, *J.C.S. Perkin I*, 1974, 2500.
[96] C. J. Grol, *Tetrahedron*, 1974, **30**, 3621.
[97] K. Goerlitzer, *Arch. Pharm.*, 1975, **308**, 272.
[98] P. Cagniant and G. Kirsch, *Compt. rend.*, 1974, **279**, C, 829.
[99] A. Ricci, D. Balucani, and G. Grandolini, *J. Heterocyclic Chem.*, 1974, **11**, 515.
[100] P. Cagniant, G. Kirsch, and P. Perin, *Compt. rend.*, 1974, **279**, C, 851.
[101] (*a*) C. W. Bird and D. Y. Wong, *Tetrahedron*, 1975, **31**, 31; (*b*) S. Yada and K. Itabashi, *Yuki Gosei Kagaku Kyokai Shi*, 1974, **32**, 188 (*Chem. Abs.*, 1974, **81**, 136 021).
[102] O. Caputo, L. Cattel, A. Ceruti, and G. Rua, *Ann. Chim. (Italy)*, 1974, **64**, 685 (*Chem. Abs.*, 1976, **84**, 59 265).
[103] F. Eiden and H.-D. Schweiger, *Synthesis*, 1974, 511; G.P. 2 412 582 (*Chem. Abs.*, 1976, **84**, 31 034).

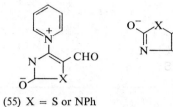

(55) X = S or NPh

(56)

Compounds in which other heterocyclic rings (pyrazoles,[104] isoxazoles,[104] imidazoles[105]) are fused to thiopyrans have been described. The interesting thiazolium salts (56) were prepared[106] by allowing the betaines (55) to react with 4-thioxothiazolidin-2-ones. Application of the Fischer indole synthesis and the Friedlander reaction to tetrahydrothiopyran-3-one led to the expected indolo- and quinolino-derivatives.[107] Ring closure of (57) under acidic conditions afforded the rearranged product (58).[108] Syntheses of benzothiopyrano[4,3,2-*de*]-quinazolines (59) from 1-amino-thioxanthones and of their *S*- and *N*-oxides have been described.[109]

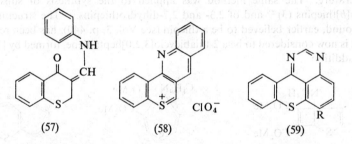

(57)   (58)   (59)

[104] A. V. El'tsov, A. I. Grigor'eva, and I. Ya. Kvitko, *Zhur. org. Khim.*, 1974, **10**, 1128; A. Fravolini, G. Grandolini, and A. Martani, *Gazzetta*, 1973, **103**, 1073; I. Ito, T. Ueda, and N. Oda, *J. Pharm. Soc. Japan*, 1975, **95**, 879 (*Chem. Abs.*, 1974, **81**, 63 552, 136 043; *ibid.*, 1976, **84**, 4845).
[105] H. J. J. Loozen, B. J. van der Beek, E. F. Godefroi, and H. M. Brick, *J. Heterocyclic Chem.*, 1975, **12**, 1039.
[106] S. N. Baranov, R. O. Kochkanyan, G. I. Belova, and A. N. Zaritovskii, *Doklady. Akad. Nauk S.S.S.R.*, 1975, **222**, 101 (*Chem. Abs.*, 1975, **83**, 114 273).
[107] A. Croisy, A. Ricci, N. Jančevska, P. Jacquignon, and D. Balucan, *Chem. Letters*, 1976, 5, and earlier papers.
[108] V. N. Gogte, K. A. R. Sastry, and B. D. Tilak, *Indian J. Chem.*, 1974, **12**, 1147.
[109] F. Eiden and J. Dusemund, *Arch. Pharm.*, 1974, **307**, 701; J. Dusemund, *ibid.*, 1975, **308**, 230.

10
Thiepins and Dithiins

BY U. EISNER

1 Thiepins

Since this topic was last reviewed,[1b] a book on seven-membered heterocycles containing oxygen and sulphur has been published.[2]

A number of successful thiepin syntheses has become available. The first monocyclic thiepin has been prepared[3a] by cycloaddition at $-30\,°C$ of dimethyl acetylenedicarboxylate to enamines derived from 2,3-dihydrothiophen-3-ones. The adduct (1) rearranged to (2), which rapidly extruded sulphur at room temperature. The same method was applied to the synthesis of substituted benzo[b]thiepins (3)[3b] and of 2,3- and 2,7-dihydrothiepins.[3c] The structure of a compound, earlier believed to be a thiepin (see Vol. 3, p. 440), has been revised,[4] and it is now considered to be a 2-thiabicyclo[3,2,0]heptadiene, formed by [2 + 2] cycloaddition.

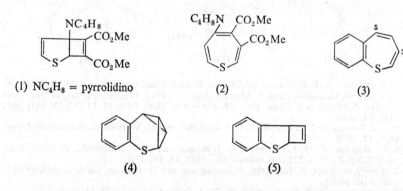

(1) NC_4H_8 = pyrrolidino   (2)   (3)

(4)   (5)

Benzo[b]thiepin (3), and some of its derivatives, have been prepared[5] by the rhodium-catalysed isomerization of the valence isomers (4). The effect of substituents in the 3-, 4-, and 5-positions on the thermal stability of (3) has been

[1] 'Organic Compounds of Sulphur, Selenium, and Tellurium', ed. D. H. Reid (Specialist Periodical Reports), The Chemical Society, London, (a) Vol. 1; (b) Vol. 2; (c) Vol. 3.
[2] 'Chemistry of Heterocyclic Compounds', ed. A. Weissberger and E. C. Taylor, Wiley–Interscience, New York, 1972, Vol. 26.
[3] D. N. Reinhoudt and C. G. Kouwenhoven, (a) Tetrahedron, 1974, 30, 2093; (b) ibid., p. 2431; (c) Rec. Trav. chim., 1973, 92, 865.
[4] D. N. Reinhoudt, H. C. Volger, C. G. Kouwenhoven, H. Wynberg, and R. Helder, Tetrahedron Letters, 1972, 5269.
[5] (a) I. Murata, T. Tatsuoka, and Y. Sugihara, Angew. Chem. Internat. Edn., 1974, 13, 142; (b) I. Murata and T. Tatsuoka, Tetrahedron Letters, 1975, 2697.

investigated.[5b, 6] Improved methods for the preparation of enol ethers and esters under mild conditions have permitted [6] the synthesis of further [1a] 3,4,5-trisubstituted benzo[*b*]thiepins from 4-phenyl-2,3,4,5-tetrahydrobenzo[*b*]thiepin-3,5-dione. Another synthesis of (3) involves dehydrohalogenation of 2-chloro-2,3-dihydrobenzo[*b*]thiepin as the last step.[7] Photoisomerization of (3) to (5) has been reported.[5a, 8]

Attempted synthesis of [*d*]annelated thiepins from heterocyclic dialdehydes and $X(CH_2COR)_2$ ($X = S, SO, or SO_2$; $R = OMe, Ph, etc.$) (*cf.* ref. 1*b*) has met with only limited success,[9] sulphur-free products being isolated in most cases.

Interest in the powerfully neuroleptic 10,11-dihydrodibenzo[*b*, *f*]thiepins (6) continues, and some 30 papers and many patents have been published by Protiva's group alone [10] since the last Report.[1b] An improvement in the original synthesis of (6) consists [11] of the use of thallium nitrate for the ring expansion

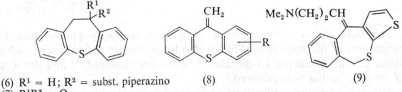

(6) R¹ = H; R² = subst. piperazino   (8)   (9)
(7) R¹R² = O

of the 9-methylenethioxanthene (8) to the key intermediate (7). Antihistamines such as (9),[12] anti-inflammatory dibenzo[2,3:6,7]thiepin[4,5-*d*]imidazoles,[13] and some tetrahydrobenzo[*b*]thiepins [14] have also been prepared.

Substituted benzo[*b*]thiepin 1-oxides have been synthesized [15] by oxidation of the corresponding benzothiepins (3). The thermal stability of the oxides is lower than that of the parent compounds. The diyne (10) has been isomerized to tribenzo- and dibenzoazuleno-thiepin *S*-oxide derivatives with rhodium and palladium complexes, respectively.[16]

3,4,5,6-Tetrachloro-2,7-dihydrothiepin was obtained as a by-product in the Glaser coupling, using cuprous chloride, of dipropargyl sulphide;[17] its ring inversion has been studied by n.m.r. spectroscopy.[17]

[6] H. Hofmann, H.-J. Haberstroh, B. Appler, B. Meyer, and H. Herterich, *Chem. Ber.*, 1975, **108**, 3596.
[7] V. J. Traynelis, Y. Yoshikawa, J. C. Sih, L. J. Miller, and J. R. Livingston, *J. Org. Chem.*, 1973, **38**, 3978.
[8] H. Hofmann and B. Meyer, *Tetrahedron Letters*, 1972, 4597.
[9] A. Corvers, A. De Groot, and E. F. Godefroi, *Rec. Trav. chim.*, 1973, **92**, 1368; C. V. Greco, F. C. Pellegrini, and M. A. Pesce, *J. Heterocyclic Chem.*, 1972, **9**, 967.
[10] *e.g.* K. Šindelář, J. O. Jílek, V. Bártl, J. Metyšová, B. Kakáč, J. Holubek, E. Svátek, J. Pomykáček, and M. Protiva, *Coll. Czech. Chem. Comm.*, 1976, **41**, 910, and earlier papers.
[11] K. Šindelář, B. Kakáč, E. Svátek, J. Holubek, M. Rajsner, J. Metyšová, and M. Protiva, *Coll. Czech. Chem. Comm.*, 1974, **39**, 333.
[12] M. Rajsner, J. Metyš, B. Kakáč, and M. Protiva, *Coll. Czech. Chem. Comm.*, 1975, **40**, 2905.
[13] J. G. Lombardino, *J. Heterocyclic Chem.*, 1974, **11**, 17.
[14] F. Kvis, E. Svátek, R. Zahradník, and M. Protiva, *Coll. Czech. Chem. Comm.*, 1972, **37**, 3808.
[15] H. Hofmann and H. Gaube, *Angew. Chem. Internat. Edn.*, 1975, **14**, 812.
[16] E. Mueller and G. Zountas, *Chem.-Ztg.*, 1974, **98**, 41.
[17] K. G. Ornell, W. Carruthers, and M. G. Pellatt, *Spectrochim. Acta*, 1972, **A28**, 753.

The bridged 2,5-dihydrothiepin (11) has been prepared [18] by photochemical or thermal isomerization of 2,3-epithionorbornene; a nonconcerted mechanism has been postulated for the reaction.

Details of the previously reported [1b] synthesis and reactions of the bridged 2,7-dihydrothiepin (12) and its S-oxide have been published.[19] The corresponding sulphone and its homologues have been converted into homotropylium sulphinate complexes.[20]

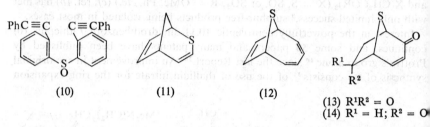

(10)   (11)   (12)

(13) $R^1R^2 = O$
(14) $R^1 = H$; $R^2 = O$

The stereochemical consequences of orbital symmetry control and the mechanism of the reversible addition of sulphur dioxide to hexatrienes, which yields 2,7-dihydrothiepin 1,1-dioxides,[1a] have been examined,[21a] and the scope of the reaction has been explored.[21b]

The photochemical synthesis of a thiepin 1,1-dioxide from a 4H-thiopyran-4-one 1,1-dioxide has been reported earlier (see Vol. 3, p. 528). The β-keto-sulphides (13) and (14) underwent photochemical rearrangement, involving a 1,3-sulphur shift, which yielded complex 2,3,4,5-tetrahydrothiepin-3-one derivatives.[22]

3,3,6,6-Tetramethyl-1-thiacyclohept-4-yne dimerized [23a] on heating with $(PhCN)_2PdCl_2$ in THF to give the sterically stabilized cyclobutadiene (15). The structure of (15) and of its iron tricarbonyl complex have been determined [23b] by X-ray analysis.

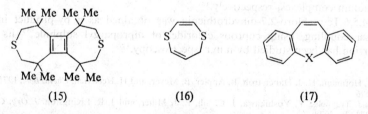

(15)   (16)   (17)

Base-induced ring contraction of benzo- [24a] and dibenzo-α-bromothiepinones [24b] to thiopyran and thiophen derivatives has been described.

[18] T. Fujisawa and R. Kobori, *J.C.S. Chem. Comm.*, 1972, 1298.
[19] A. G. Anastassiou, J. C. Wetzel, and B. Y.-H. Chao, *J. Amer. Chem. Soc.*, 1975, **97**, 1124.
[20] L. A. Paquette, U. Jacobssen, and S. V. Ley, *J. Amer. Chem. Soc.*, 1976, **98**, 152.
[21] (a) W. L. Mock, *J. Amer. Chem. Soc.*, 1975, **97**, 3666, 3673; (b) W. L. Mock and J. H. McCausland, *J. Org. Chem.*, 1976, **41**, 242.
[22] H. Tsuruta, M. Ogasawara, and T. Mukai, *Chem. Letters*, 1974, 887.
[23] (a) H. Kimling and A. Krebs, *Angew. Chem. Internat. Edn.*, 1972, **11**, 932; (b) H. Irngartinger, H. Kimling, A. Krebs, and R. Maeusbacher, *Tetrahedron Letters*, 1975, 2571.
[24] (a) P. M. Weintraub and A. D. Still, *J.C.S. Chem. Comm.*, 1975, 784; (b) I. Ueda, *Bull. Chem. Soc. Japan*, 1975, **48**, 2306.

5*H*-1,4-Dithiepin (16) has been synthesized and the derived anion shown not to be aromatic.[25] A study of the electronic spectra of the tricyclic compounds (17) (X = CH₂, NH, O, CO, S, or Se) appears to indicate a non-planar (boat) conformation for these systems.[26]

Theoretical aspects of thiepins were reported earlier (see Vol. 3, p. 755). A general method for evaluating resonance energies has been successfully applied to thiepin.[27]

2 Dithiins

The chemistry of 1,2- and 1,3-dithiins has been reviewed.[28]

1,2-Dithiins.—The cyclic structure of 1,2-dithiins [1b] has been further substantiated by a determination of the heat of formation of 3,6-diphenyl-1,2-dithiin.[29]

The reaction of the disulphide ClC(E)=C(E)SSC(E)=C(E)SCN (E = CO_2Me) with sodium thiophenolate unexpectedly yielded [30] the dithiin (18). Addition of sulphene or phenylsulphene to 3-aminovinyl-thiones afforded [31] 1,2-dithiin 1,1-dioxides (see Vol. 3, p. 238) or their dihydro-derivatives. Oxidation of 1-thiochromene with selenium dioxide yielded [32] the novel heterocycle (19).

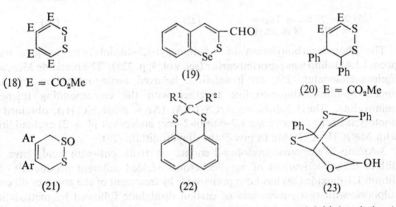

(18) E = CO_2Me

(19)

(20) E = CO_2Me

(21)

(22)

(23)

Photochemically generated diatomic sulphur added to 1,2-bis(methylene)-cyclohexane to afford a 3,6-dihydro-1,2-dithiin.[33] The dihydrodithiin (20) was formed by thermal rearrangement of a 1,3-dithiin.[34] Addition of disulphur monoxide to 2,3-disubstituted buta-1,3-dienes yielded [35] the rather unstable dihydrodithiin 1-oxides (21).

[25] I. Murata and K. Nakasuji, *Tetrahedron Letters*, 1975, 1895.
[26] T. Toth and L. Klasinc, *Z. Naturforsch.*, 1974, **29a**, 1371.
[27] J. Aihara, *Bull. Chem. Soc. Japan*, 1975, **48**, 1501.
[28] U. Eisner and T. Krishnamurthy, *Internat. J. Sulfur Chem.* (B), 1972, 7, 101.
[29] G. Geiseler and J. Sawistowsky, *Z. phys. Chem.* (Leipzig), 1973, **253**, 333 (*Chem. Abs.*, 1974, **80**, 26 643).
[30] W. Ried and W. Ochs, *Chem. Ber.*, 1972, **105**, 1093.
[31] J. C. Meslin, Y. T. N'Guessan, and H. Quiniou, *Tetrahedron*, 1975, **31**, 2679.
[32] J. Van Coppenolle and M. Renson, *Tetrahedron*, 1975, **31**, 2099.
[33] R. Jahn and U. Schmidt, *Chem. Ber.*, 1975, **108**, 630.
[34] U. Eisner and T. Krishnamurthy, *Tetrahedron*, 1971, **27**, 5753.
[35] R. M. Dodson, K. S. Srinisavan, K. S. Sharma, and R. F. Sauers, *J. Org. Chem.*, 1972, **37**, 2367.

Further details of the ring contraction of 1,4-dimethoxycarbonyl-2,3-benzo-dithian to a benzo[*c*]thiophen [36] have been reported (see Vol. 3, p. 452).

1,3-Dithiins.—A new synthesis of 1,3-dithiins from aromatic aldehydes, acetylenes, and H_2S has been described.[34] Carbenoids react with 1,2-dithia-acenaphthene to afford [37] the insertion product (22; R = Ph or tosyl). Pyrolysis of the ditosyl derivative yields the ketone (22; RR = O).

The dimer of β-mercaptocinnamaldehyde has been shown [38] to have the structure (23).

1,4-Dithiins and Related Compounds.—Diacetylenic sulphides react [39] with Na_2X (X = S, Se, or Te) to give the mono- and di-substituted heterocycles (24). Other syntheses of 1,4-dithiins from acetylenes are the reactions of amidines of phenylpropiolic acid with sulphur (or selenium) in DMF,[40] and of dimethyl acetylenedicarboxylate with hydrogen bromide in liquid sulphur dioxide.[41] In the latter reaction, sulphuric acid is formed by disproportionation.

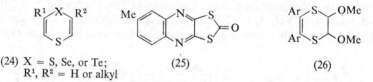

(24) X = S, Se, or Te;   (25)   (26)
     R^1, R^2 = H or alkyl

The photodecarbonylation of 4,5-diphenyl-1,3-dithiol-2-one to give tetra-phenyl-1,4-dithiin was reported earlier (see Vol. 3, p. 228). The pesticide Morestan (quinomethionate) (25), on irradiation, behaved analogously [42] in affording dimethyl-*p*-dithiinodiquinoxaline together with the corresponding thienodi-quinoxaline. The 1,2-dithione $ArCS \cdot CSAr$ (Ar = *p*-$Me_2NC_6H_4$), obtained by photolysis of the appropriate 1,3-dithiol-2-one, underwent [4 + 2] cycloaddition with $MeOCH=CHOMe$ to give [43] the dihydrodithiin (26).

3-Amino-2-thiocyanatocyclohex-2-enone, on treatment with acid, gave the dithiin (27). Derivatives of acyclic ketones yielded different products.[44] The dithiin 1,1-dioxide (28) has been prepared [45] by treatment of the anion of dibenzyl sulphone with two equivalents of carbon disulphide followed by methylation.

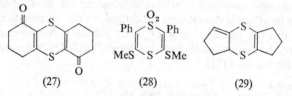

(27)   (28)   (29)

[36] G. Cignarella and G. Cordella, *Gazzetta*, 1974, **104**, 455.
[37] S. Tamagaki and S. Oae, *Bull. Chem. Soc. Japan*, 1972, **45**, 960.
[38] M. Pulst, M. Weissenfels, E. Kleinpeter, and L. Beyer, *Tetrahedron*, 1975, **31**, 3107.
[39] A. Zilverschoon, J. Meijer, P. Vermeer, and L. Brandsma, *Rec. Trav. chim.*, 1975, **94**, 163.
[40] W. Ried and M. Wegwitz, *Annalen*, 1975, 89.
[41] I. V. Smirnov-Zamkov and Yu. L. Zborovskii, *Zhur. org. Khim.*, 1973, **9**, 1591 (*Chem. Abs.*, 1973, **79**, 115 535).
[42] W. F. Gray, I. H. Pomerantz, and R. D. Ross, *J. Heterocyclic Chem.*, 1972, **9**, 707.
[43] W. Kusters and P. de Mayo, *J. Amer. Chem. Soc.*, 1974, **96**, 3502.
[44] E. Schmitz and H. Striegler, *J. prakt. Chem.*, 1971, **313**, 1125.
[45] D. Ladurée, P. Rioult, and J. Vialle, *Bull. Soc. chim. France*, 1973, 637.

The structure of an alleged stereoisomer of 2,3:5,6-bis(trimethylene)-1,4-dithiin has been revised [46] to that of the positional isomer (29), thus disproving the existence of stereoisomers due to ring inversion in doubly co-ordinated sulphur compounds.

Several more [1a] syntheses of dihydro-1,4-dithiins by ring expansion of ethylene-thioketals [47a, b] or the derived *S*-monoxides [47c] have been reported. One of these [47b] involves extensive rearrangement of a steroid.

Numerous derivatives of (30) have been prepared [1a] and shown to display a wide spectrum of biological activity, *e.g.* acaricidal,[48] sedative,[49] *etc.* Further examples may be found in the patent literature.

(30)

The reaction of 2,5-diaryl-1,4-dithiin *SS*-tetroxides with selenium afforded diarylselenophens instead of the expected diselenins.[50]

Benzo- and Dibenzo-1,4-dithiins.—An improved synthesis [51] of benzo-1,4-dithiin and -oxathiin is based on a novel intramolecular sulphenylation of the thiol-sulphonates $PhY(CH_2)_2SSO_2Me$ (Y = S or O) in the presence of aluminium chloride. Phenyl-lithium reacts with bromobenzene and carbon disulphide to give 2,3-diphenyl-1,4-benzodithiin *via* a benzyne intermediate.[52]

A new, versatile synthesis of thianthrene and its derivatives by intramolecular ring closure of the radical (31) (derived from the corresponding diazonium salt) has been developed.[53] Thianthrene may be prepared [54] from benzene and PhSSCl, while 2,3,7,8-tetra-alkoxythianthrenes are produced [54] in low yield in a one-step synthesis from 1,2-dialkoxybenzenes and S_2Cl_2. The corresponding tetramethoxy-selenanthrenes have been prepared [55] by the reaction of 1,2-dimethoxybenzene with SeO_2. Fusion of phenanthrene with sulphur afforded tetrabenzo[*a,c,h,j*]-thianthrene.[56] Fusion of halogenated aromatic or heterocyclic compounds with sulphur has been similarly used to prepare more complex thianthrenes for use as herbicides and fungicides.[57]

[46] R. M. Moriarty, C. C. Chien, and C. W. Jefford, *Tetrahedron Letters*, 1973, 4429.
[47] (a) H. Yoshino, Y. Kawazoe, and T. Taguchi, *Synthesis*, 1974, 713; (b) J. R. Williams and G. M. Sarkisian, *Tetrahedron Letters*, 1974, 1109; (c) C. H. Chen, *ibid.*, 1976, 25.
[48] F. Gregan, V. Konečný, and P. Hrnciar, *Chem. Zvesti*, 1975, **29**, 250 (*Chem. Abs.*, 1975, **83**, 97 164).
[49] Z. Kleinrok, E. Przegalinski, Z. Borzecki, I. Zebrowska-Lupina, M. Wielosz, and W. Goral, *Pol. J. Pharmacol. Pharm.*, 1974, **26**, 419 (*Chem. Abs.*, 1975, **82**, 118 866).
[50] I. Lalezari, A. Shafiee, and A. Rashidbaigi, *J. Heterocyclic Chem.*, 1976, **13**, 57.
[51] J. H. Verheijen and H. Kloosterziel, *Synthesis*, 1975, 451.
[52] M. Yokoyama, T. Kondo, N. Miyasa, and M. Torri, *J.C.S. Perkin I*, 1975, 160.
[53] L. Benati, P. C. Montevecchi, A. Tundo, and G. Zanardi, *J.C.S. Perkin I*, 1974, 1272.
[54] Z. S. Ariyan and R. L. Martin, *J.C.S. Perkin I*, 1972, 1687.
[55] T. Weiss, W. Nitsche, F. Boehnke, and G. Klar, *Annalen*, 1973, 1418.
[56] M. Zander and W. H. Franke, *Chem. Ber.*, 1973, **106**, 2752.
[57] G.P. 2 224 834; 2 229 163 (*Chem. Abs.*, 1974, **80**, 48 009; 83 076).

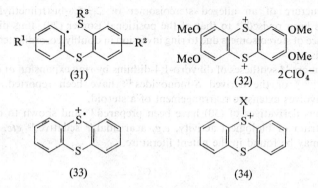

(31)   (32)   2ClO₄⁻

(33)   (34)

The first thianthrene dication (32) has been isolated.[58] Work on the thianthrene cation radical (33) continues.[1b] Its reaction with aliphatic amines [59a] yielded the sulphimides (34; X = N̄R); phenols and aromatic amines were attacked [59b] in the *para*-position to afford (34; X = *p*-HOC₆H₄ or *p*-R₂NC₆H₄), and enolizable ketones yielded [59c] the β-keto-sulphonium salts (34; X = CHR¹COR²). The latter compounds reacted with nucleophiles Y⁻ to give the α-substituted ketones YCHR¹COR²,[21] and might prove to be useful intermediates for their synthesis.[59c] Reinvestigation of the kinetics of the reaction of (33) with phenol and with anisole has provided evidence [60] which appears to disprove the disproportionation mechanism postulated earlier [1b, 59b] for these reactions.

[58] R. S. Glass, W. J. Britt, W. N. Miller, and G. S. Wilson, *J. Amer. Chem. Soc.*, 1973, **95**, 2375.
[59] (a) K. Kim and H. J. Shine, *J. Org. Chem.*, 1974, **39**, 2537; (b) K. Kim, V. J. Hull, and H. J. Shine, *ibid.*, p. 2534; (c) K. Kim, S. R. Mani, and H. J. Shine, *ibid.*, 1975, **40**, 3857.
[60] U. Svanholm and V. D. Parker, *J. Amer. Chem. Soc.*, 1976, **98**, 997.

11
Isothiazoles and Related Compounds

BY F. KURZER

1 Introduction

In view of the fundamental importance of the monocyclic isothiazole structure, it is not surprising that isothiazoles have rapidly passed from being laboratory curiosities to the status of a class of closely studied and well-documented compounds. The period under review has witnessed continued progress in their synthesis, by new and by the extension of established methods, and in their increasingly systematic investigation. Special attention may be drawn to current work concerned with the conversion of penicillin derivatives into isothiazoles, to theoretical and physical studies aimed at quantifying the chemical behaviour of this ring-system, and to kinetic studies designed to elucidate reaction mechanisms. Condensed structures incorporating isothiazole continue to receive their share of attention. In presenting the new material, the arrangement adopted in the previous volumes of these Reports has been retained.

2 Isothiazoles

Synthesis.—As before (see Vol. 3, p. 541), the synthetic routes to isothiazoles have been classified according to the nature of the fragments from which the heterocyclic ring is ultimately constructed.

From 1,2-Dithiolans (Type C). The general synthesis of isothiazoles from 1,2-dithiolans (see Vol. 2, p. 563; Vol. 3, p. 543) by reaction with ammonia or amines continues to be exploited. 3-Ethoxy-1,2-dithiolium salts (1) are cleaved by primary amines to the linear thioesters (2), which are in turn convertible into

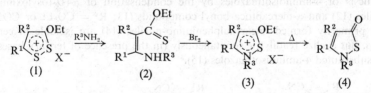

the appropriate isothiazole-derivatives (3) and (4). Reaction with ammonia gives the expected 3-ethoxyisothiazoles.[1] 3-Bromothio-1,2-dithiolium bromides react similarly with aromatic amines to give isothiazoline-5-thiones, which undergo scission with Grignard reagents[2] to give 3-alkylthio-1-arylamino-3-thioxo-2-arylprop-1-enes, ArNHCH=CArCSSR.

[1] J. Faust, *Z. Chem.*, 1975, **15**, 478.
[2] F. Boberg and W. von Gentzkow, *J. prakt. Chem.*, 1973, **315**, 965.

339

From β-Aminocrotonic Esters (*Type C*). The synthesis of isothiazoles from β-aminocrotonic esters (see Vol. 1, p. 370) has been exemplified by the condensation of *p*-nitrobenzoyl and 2-furoyl isothiocyanates (6) with ethyl β-methylaminocrotonate (7), which produces pyrimidine derivatives (5) in the absence, and substituted isothiazoles (8) in the presence, of oxidizing agents such as bromine.

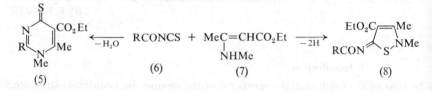

Under standard conditions 5-nitrofuroyl isothiocyanate reacts with the enamine to give the isothiazoles (8) directly.[3]

From β-Chlorovinylmethine-immonium and Bunte Salts (*Types C and E*). β-Chlorovinylmethine-immonium salts (9) react with alkali-metal thiocyanates in the presence of aromatic amines to give isothiazoles (10). The reaction is considered to proceed by a nucleophilic displacement to give eventually the β-thiocyanovinyl aldehyde anils $ArC(SCN)=CHCH=NR^1$, which cyclize *in situ*. The reaction of (9) with sodium thiosulphate gives Bunte salts, which react with amines (R^1R^2NH) to give the stable β-aminovinyl thioketones (11), and hence, by oxidative cyclization, the isothiazolium salts (10; $R^2 = H$)[4] (Scheme 1).

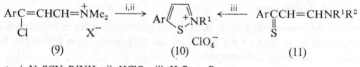

Reagents: i, $NaSCN–R^1NH_2$; ii, $HClO_4$; iii, H_2O_2 or Br_2

Scheme 1

From Sulphenimino-nitriles (*Type E*). A preliminary report has described a new synthesis of 4-aminoisothiazoles by the condensation of α-(*O*-tosyloximino)-nitriles (12) and α-mercaptocarbonyl compounds (13; $R^2 = CO_2Et$ or $COMe$). The primarily formed linear sulphenimino-nitriles (14) are isolable in certain cases, but cyclize readily or spontaneously (in the presence of tertiary bases) to the substituted 4-aminoisothiazoles (15).[5]

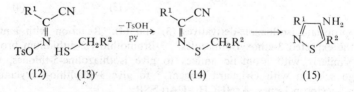

[3] S. Mishio, J. Matsumoto, and S. Minami, *J. Pharm. Soc. Japan*, 1975, **95**, 1342.
[4] J. Liebscher and H. Hartmann, *Z. Chem.*, 1974, **14**, 189.
[5] K. Gewald, P. Bellmann, and H. J. Jänsch, *Z. Chem.*, 1975, **15**, 18.

From Thioxadiazepines (*Type G*). Sulphur di-imines react with 2 moles of phenylketen below $-50\,^{\circ}C$ to yield 1-substituted 3,4-diphenylpyrroline-2,5-diones and 2-substituted 4,5-diphenyl-5-(substituted)-carbamoylisothiazolidin-3-ones (17) (18—39%) as the main products.[6] The reaction is likely to involve

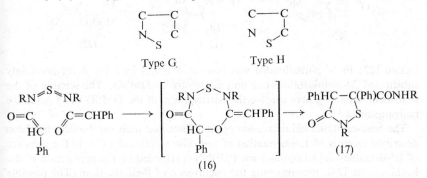

Type G   Type H

unstable intermediate thioxadiazepines (16), an example of which has been isolated.[7] Oxidation of (17) with *m*-chloroperoxybenzoic acid gives the sulphoxide.[6]

From Thioacylcyanomethanephosphonate Esters (*Type H*). The condensation of diethyl thioacylcyanomethanephosphonate *S*-esters (18) with amino-compounds (hydrazine, hydroxylamine, *etc.*) yields a variety of heterylphosphonate esters.

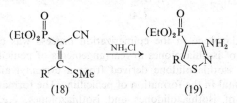

The use of chloramine in this reaction produces 3-amino-4-diethylphosphono-isothiazole (19) in low yield.[8]

From Penicillins (*Type C*). The important problem of the relationship between penicillin, cephalosporin, and isothiazole derivatives has recently interested several research groups. This topic is also considered in Chapter 16. The anion of azetidinesulphenic acid (20) undergoes spontaneous fragmentation and re-cyclization, giving the isothiazolone derivative (21) in high yield.[9] Direct treatment of penicillin sulphoxides with bases has previously given isothiazoles similarly,[10] and further examples continue to be reported.[11] The 4-mercapto-

[6] T. Minami and T. Agawa, *J. Org. Chem.*, 1974, **39**, 1210.
[7] T. Minami, K. Yamataka, Y. Oshiro, T. Agawa, N. Yasuoka, and N. Kasai, *J. Org. Chem.*, 1972, **37**, 3810.
[8] O. Günther and K. Hartke, *Arch. Pharm.*, 1975, **308**, 693.
[9] T. S. Chou, J. R. Burgtorf, A. L. Ellis, S. R. Lammert, and S. Kukolja, *J. Amer. Chem. Soc.*, 1974, **96**, 1609; G. A. Koppel and S. Kukolja, *J.C.S. Chem. Comm.*, 1975, 57; T. Kamiya, T. Teraji, M. Hashimoto, O. Nakaguchi, and T. Oku, *J. Amer. Chem. Soc.*, 1975, **97**, 5020.
[10] R. B. Morin, B. G. Jackson, R. A. Mueller, E. R. Lavagnino, W. B. Scanlon, and S. L. Andrews, *J. Amer. Chem. Soc.*, 1963, **85**, 1896; *ibid.*, 1969, **91**, 1401.
[11] Y. Tachibana, K. Nakagawa, and H. Fukawa, *J. Chem. Soc. Japan, Ind. Chem. Sect.*, 1976, 148.

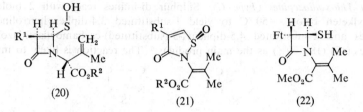

(20)   (21)   (22)

lactam (22; Ft = phthalimido) was converted (40%) into the correspondingly substituted Δ^4-isothiazolin-3-one on treatment with DMSO. The scission of the four-membered ring may involve the participation of the DMSO in an oxidative rearrangement.[12]

The base-catalysed epimerization of penicillins and their sulphoxides has been described recently.[13] Epimerization of penicillin sulphoxides (23) in the presence of 1,5-diazabicyclo[4,3,0]non-5-ene ('DBN') is attended by the production of the isothiazolone (24), presupposing the occurrence of β-elimination. The possible

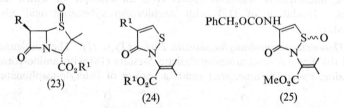

(23)   (24)   (25)

role of the isothiazolone in the epimerization process was considered.[14] The ready formation of isothiazolones in rearrangements of penicillin sulphoxides[10] suggests that an isothiazolidone derived from a cysteinylvalinesulphenic acid may be instrumental in the formation of penicillin. The formation and chemical transformations of isothiazolidones and isothiazolones, *e.g.* (25) from acyl cysteinylvaline dipeptides, have been described, and attempts to obtain penicillins from certain of such isothiazole-precursors are reported to be under way.[15]

Meso-ionic Isothiazoles.—The synthesis and chemical properties of novel meso-ionic compounds of the dehydrodithizone (2,3-diphenyltetrazolium 5-sulphide) family derived from pyrazole, isothiazole, and isoxazole have been described.[16] The member (28) of the isothiazole series is accessible from 3-methylisothiazole

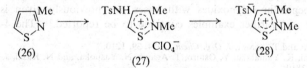

(26)   (27)   (28)

[12] M. D. Bachi and O. Goldberg, *J.C.S. Perkin I*, 1974, 1184.
[13] P. Claes, A. Vlietinck, E. Roets, H. Vanderhaege, and S. Toppet, *J.C.S. Perkin I*, 1973, 932; A. Vlietinck, E. Roets, P. Claes, G. Janssen, and H. Vanderhaege, *J.C.S. Perkin I*, 1973, 937.
[14] M. Fukumura, N. Hamma, and T. Nakagome, *Tetrahedron Letters*, 1975, 4123.
[15] R. B. Morin, E. M. Gordon, and J. R. Lake, *Tetrahedron Letters*, 1973, 5213.
[16] G. V. Boyd and T. Norris, *J.C.S. Perkin I*, 1974, 1028.

(26) by successive nitration (in the 4-position), reduction, tosylation, and methylation. Final deprotonation with aqueous alkali converts (27) into the meso-ionic *N*-(2,3-dimethyl-4-isothiazolium) toluene-*p*-sulphonamidate (28). Attempts to effect electrophilic substitution at its free ring position were unsuccessful.[16]

Physical Properties.—I.r. and Raman spectra of 5-monosubstituted isothiazoles (5-CH$_3$, -CD$_3$, -Cl, -Br, or -I) in their liquid and solid states have been correlated with the fundamental modes of vibration of the ring system.[17] Detailed experimental and theoretical studies of the ^{13}C and ^{1}H n.m.r. spectra of isothiazoles, including long-range couplings, have been reported.[18, 19] A comparison with theoretical results based on a CNDO model revealed some deficiencies.

Chemical Properties.—*Photolytic and Free-radical Reactions.* The photoisomerization reactions of twelve phenylmethyl-thiazoles and -isothiazoles have been compared, and the following order of decreasing reactivity was observed: 2-phenylthiazole or 5-phenylisothiazole > 5-phenylthiazole or 3-phenylisothiazole > 4-phenylthiazole or 4-phenylisothiazole. Most of the rearranged compounds are formed either by a valence-bond isomerization mechanism, by way of bicyclic intermediates, where the phenyl group is conjugated to the cycle, or by the Kellogg mechanism involving a 180° rotation about the bonds adjacent to the sulphur atom. The original paper must be consulted for details of the extensive data and their detailed discussion [20] (see Vol. 3, p. 544). Further details have been reported [21] of the photoisomerization of 5- to 3-phenylisothiazole in an ether–deuterium oxide system, which is considered to involve tricyclic sulphonium cations as intermediates (see Vol. 2, p. 563).

A preliminary report [22] has described the photolysis of methyl α-isopropylidene-4-phenylacetamido-3-isothiazolone-2-acetate (29), a degradation product of penicillin G sulphoxide methyl ester, to the 2-benzylthiazole-4-carboxamide derivative (30; 29%), the corresponding oxazole (28%), and the linear dipeptide

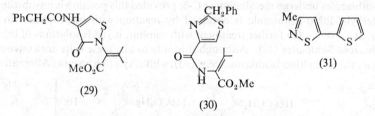

(29)   (30)   (31)

(18%). Recyclization of (29) to the thiazole (30) is accountable in terms of intramolecular trapping of the intermediate thioaldehyde, itself a possible biogenetic precursor of penicillins, by an amide side-chain.

3-Methylisothiazol-5-ylthiophen (31) was formed (30%) by aprotic diazotization of 5-amino-3-methylisothiazole by amyl nitrite in the presence of thiophen.[23]

[17]   G. Mille, J. C. Poite, J. Chouteau, and J. Metzger, *Canad. J. Chem.*, 1975, **53**, 1642.
[18]   R. E. Wasylishen, T. R. Clem, and E. D. Becker, *Canad. J. Chem.*, 1975, **53**, 596.
[19]   R. Faure, J. R. Llinas, E. Vincent, and M. Rajzmann, *Canad. J. Chem.*, 1975, **53**, 1677.
[20]   C. Riou, J. C. Poite, G. Vernin, and J. Metzger, *Tetrahedron*, 1974, **30**, 879.
[21]   M. Maeda, A. Kawahara, M. Kai, and M. Kojima, *Heterocycles*, 1975, **3**, 389.
[22]   Y. Maki and M. Sako, *Tetrahedron Letters*, 1976, 375.
[23]   G. Vernin, J. Metzger, and C. Parkanyi, *J. Org. Chem.*, 1975, **40**, 3183.

Alkylation. Rates of quaternization of isothiazole and its 2,1-, and 1,2-benzologues have been examined as a part of a wider study on azoles. The isothiazoles react with methyl iodide in DMSO, and with dimethyl sulphate, affording novel *N*-methyl derivatives. Benzo-fusion exerts a rate-diminishing effect, but this influence is very small in 1,2-benzisothiazole. The stability of the heterocycles apparently increases in the order 2,1-benzisoxazole < N^2-methylindazole < 2,1-benzisothiazole, which parallels the well-established order of stability and aromaticity of furan < pyrrole < thiophene.[24]

Electrophilic Reactions. Further studies aimed at the correlation of the reactivity of five-membered heteroaromatic rings towards electrophiles have been reported. Acid-catalysed hydrogen exchange rates were measured (n.m.r.) for isoxazole, isothiazole, and their 3- and 5-methyl derivatives. Extrapolated values for the rate constants at 100 °C and pH 0 were obtained and compared with those for other ring systems. Quantitative effects of the methyl groups reflect the relative degree of bond fixation, and indicate that the aromaticity increases in the series isoxazole, pyrazole, and isothiazole.[25]

The rates of nitration of isothiazole (and its 3- and 5-methyl derivatives) have been measured.[26] While isothiazole and its 3- and 5-methyl derivatives are nitrated as free bases, the 3,5-dimethyl-compound is nitrated as the conjugate acid. Isothiazoles are nitrated in the 4 position ~ 10^4 times slower than benzene. The order of reactivity for nitration is pyrazole > isothiazole > isoxazole, in contrast to that of five-membered heterocycles containing *one* heteroatom, which for trifluoroacetylation is pyrrole > furan > thiophen.[26]

Nucleophilic Reactions. Rates of solvolysis in 80% ethanol of the 2-[3-, 4-, and 5-isothiazolyl]-2-chloropropanes have been determined. From these rates, and the rates of solvolysis of cumyl chlorides bearing electron-withdrawing substituents, σ^+ values appropriate for the replacement of a benzene by an isothiazolyl ring have been determined.[27]

Isothiazoles undergo metallation at C-5, provided this position is unsubstituted. 4-Methyl-5-lithio-isothiazole is obtained by reaction with butyl-lithium in dry THF at − 70 °C. On further treatment with sulphur, it yields solutions of lithium isothiazole-5-thiolates (32). Although unstable to air, these salts are convertible *in situ* into 5-alkylthio-isothiazoles, *e.g.* (33), with alkylating agents. Alternatively,

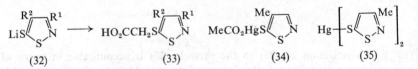

(32)   (33)   (34)   (35)

they may be transformed, in low yield, into stable mercuric thiolates (34) and (35). 3-Methylisothiazole undergoes the same series of reactions. The method promises to be useful for the synthesis of isothiazol-5-yl alkyl sulphides, which have so far been relatively inaccessible.[28]

[24] M. Davis, L. W. Deady, and E. Homfeld, *Austral. J. Chem.*, 1974, **27**, 1221.
[25] S. Clementi, P. P. Forsythe, C. D. Johnson, A. R. Katritzky, and B. Terem, *J.C.S. Perkin II*, 1974, 399.
[26] A. R. Katritzky, H. O. Tarhan, and B. Terem, *J.C.S. Perkin II*, 1975, 1620.
[27] D. S. Noyce and B. B. Sandel, *J. Org. Chem.*, 1975, **40**, 3381.
[28] D. E. Horning and J. M. Muchowski, *Canad. J. Chem.*, 1974, **52**, 2950.

The scission of isothiazolium salts (36) to amine-thiones (37) is influenced by the nature of the substituent, R, and the anion, X^-. When R = H, an additional product arises, for which several possible structures, generally representable by (38), are being considered.[29]

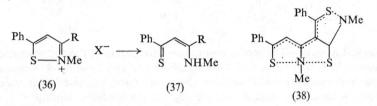

(36)   (37)   (38)

Action of Peroxyacetic Acid. The action of hydrogen peroxide in acetic acid on heterocyclic thiones eliminates the exocyclic sulphur as sulphate, and produces either heteroaromatic cations or oxo-compounds, depending on structural and electronic features.[30] Benzo-substitution, or the presence of strongly electron-withdrawing groups in the heterocycle, favours the latter product, while the presence of a tervalent nitrogen atom favours the former. Accordingly *N*-alkyl-isothiazoline-5-thiones, *N*-arylisothiazoline-3-thiones, and benzisothiazoline-5-thiones gave the corresponding isothiazolium salts.[30] The results are in accord with the findings of earlier studies.[31]

Action of Acetylenic Reagents. Isothiazoline-5-thiones (39) react rapidly with acetylenic reagents (*e.g.* dibenzoylacetylene) to form mono-adducts of type (40). *N*-Phenylmaleimide gives an adduct, but diethyl azodicarboxylate gave only

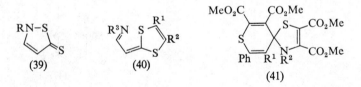

(39)   (40)   (41)

decomposition products.[32] In contrast, the 3-thiones react only slowly with dimethyl acetylenedicarboxylate and related acetylenes, to give mixtures of products from which diadducts formulated as the spiranes (41) may be isolated in moderate yield.[32]

The electron-deficient heterocycle 4-nitroisothiazole (42) reacts with the electron-rich compound 1-dimethylamino-2-phenylacetylene (43; $R^1 = R^2 = Me$) to give a mixture of the [2 + 2]-cycloadduct (44) and the nitrone (45), their ratio depending on the polarity of the solvent. The regio-selective character of the cycloaddition, as shown by the formation of only one of the two possible isomers of (44), is consistent with the polarization of the two π-systems.[33]

[29] H. Ullah and P. Sykes, *Chem. and Ind.* 1973, 1163.
[30] J. L. Charlton, S. M. Loosmore, and D. M. McKinnon, *Canad. J. Chem.*, 1974, **52**, 3021.
[31] M. E. Hassan and D. M. McKinnon, *Canad. J. Chem.*, 1973, **51**, 3081; G. E. Bachers, D. M. McKinnon, and J. M. Buchshriber, *ibid.*, 1972, **50**, 2568.
[32] M. S. Chauhan, M. E. Hassan, and D. M. McKinnon, *Canad. J. Chem.*, 1974, **52**, 1738.
[33] D. N. Reinhoudt and C. G. Kouwenhoven, *Tetrahedron Letters*, 1974, 2503.

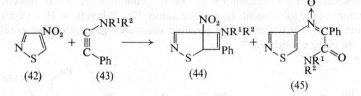

(42)   (43)   (44)

(45)

Biological Aspects.—The modes and rates of dissipation of the two microbicidal compounds 5-chloro-2-methyl-4-isothiazolin-3-one and 2-methyl-4-isothiazolin-3-one (both calcium chloride addition compounds) over a range of environmental conditions have been studied, and the degradation products identified.[34]

3 1,2-Benzisothiazoles

Synthesis.—Several new methods of preparing 1,2-benzisothiazoles have been reported. Ultimately, they all involve the cyclization of an arylsulphenamide incorporating an *ortho*-carbonyl function.

Thus, the novel method of producing sulphenimines from diaryl disulphides, silver nitrate, and ammonia, applied to bis(2-acetyl-4-methylphenyl) disulphide (46), affords 3,5-dimethyl-1,2-benzisothiazole (47) directly, in 30% yield.[35]

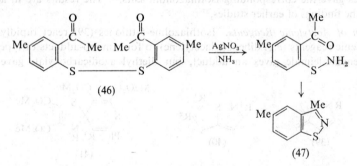

(46)

(47)

A general synthesis of 2-substituted 1,2-benzisothiazolin-3-ones from 2-mercaptobenzoic acid involves esterification, halogenation to the sulphenyl halide, conversion into the sulphenamide, and cyclization with a strong base.[36] The synthesis of 3-oxo-3*H*-1,2-benzisothiazole 1-oxides by the action of hydrazoic acid on 2-sulphinylbenzoic acids has been extended (see Vol. 2, p. 579). 2-Sulphinylbenzamides (48) react with hydrazoic acid in polyphosphoric acid to give 3-imino-3*H*-1,2-benzisothiazole 1-oxides (49) together with the related *SS*-dioxide and 1,2-benzisothiazolin-3-one. The preparation of a number of functional derivatives of (49), and an unusual alkaline ring-scission to (50), were also described.[37]

[34] S. F. Krzeminski, S. K. Brackett, and J. D. Fisher, *J. Agric. Food Chem.*, 1975, **23**, 1060; S. F. Krzeminski, C. K. Brackett, J. D. Fisher, and J. F. Spinnler, *ibid.*, 1975, **23**, 1068.
[35] F. A. Davis, W. A. R. Slegeir, S. Evans, A. Schwartz, D. L. Goff, and R. Palmer, *J. Org. Chem.*, 1973, **38**, 2809.
[36] J. C. Grivas, *J. Org. Chem.*, 1975, **40**, 2029.
[37] P. Stoss and G. Satzinger, *Chem. Ber.*, 1975, **108**, 3855.

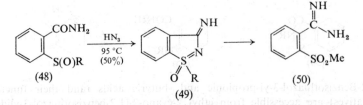

(48)   (49)   (50)

Mass Spectra.—Detailed fragmentation patterns have been proposed for 1,2-benzisothiazoles and the mass spectra compared with those of 1,2-benziso-selenazoles and the corresponding benzothiazoles.[38-40] The parent compound, 1,2-benzisothiazole, eliminates HCN after partial inter-ring hydrogen scrambling.[39] Some 3-substituted derivatives show unusual fragmentation patterns.[39] The kinetic energy spectra and relative abundances of metastable ions of 1,2-benzisothiazole and benzothiazole ions are accountable, on comparison, in terms of a primary isomerization of either molecular ion to common structures before these undergo metastable loss of HCN or CS.[40]

Chemical Reactions.—Photolysis of 1,2-benzisothiazole cleaves the S—N bond, yielding bis(2-cyanophenyl) disulphide (12%). In contrast to the behaviour of the oxygen and nitrogen analogues (benzo[*d*]isoxazole and 1*H*-indazole), no transposition product (*i.e.* benzothiazole) is formed in this reaction.[41]

For a comparative study[24] of the rates of *N*-methylation, see p. 344. Nucleophilic replacement of the halogeno-substituent in 2-ethyl-3-chloro-1,2-benzisothiazolium chloride (52) by (thio)phenols readily yields quaternary isothiazolium

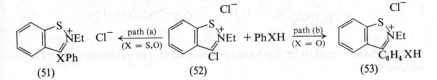

(51)   (52)   (53)

salts (51), which are dehydrohalogenated at 140 °C to 3-phenoxy- or 3-phenylthio-1,2-benzisothiazoles. In an alternative reaction [path (b)], phenol undergoes electrophilic aromatic substitution by the hetero-salt, yielding successively the quaternary salt (53) and 3-(4-hydroxyphenyl)-1,2-benzisothiazole. In favourable cases, the quaternary salts (51) and (53) are isolable as perchlorates.[42]

1,2-Benzisothiazolinone (54; R = H) dissolves in aqueous sodium bisulphite to yield, instead of the expected Bunte salt (55; R = SO$_3$Na), an addition-compound of structure (56; R = SO$_3$Na). Cyanide ion reacts similarly to give (56; R = CN). In spite of the acidity of its NH group (54; R = H) reacts with phenyl isocyanate and sodium cyanate to yield the ureido-derivatives (54; R = CONHPh and CONH$_2$).[43]

[38] A. Croisy, P. Jacquignon, R. Weber, and M. Renson, *Org. Mass. Spectrometry*, 1972, **6**, 1321.
[39] A. Selva and E. Gaetani, *Org. Mass Spectrometry*, 1973, **7**, 327.
[40] A. Selva, U. Vettori, and E. Gaetani, *Org. Mass Spectrometry*, 1974, **9**, 1161.
[41] M. Ohashi, A. Ezaki, and T. Yonezawa, *J.C.S. Chem. Comm.*, 1974, 617.
[42] H. Böshagen and W. Geiger, *Chem. Ber.*, 1974, **107**, 1667.
[43] W. V. Farrar, *Z. Naturforsch.*, 1974, **29b**, 693.

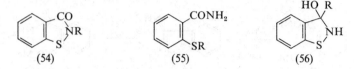

(54)   (55)   (56)

1,2-Benzisothiazol-3-yl-propionic and -butyric acids (and their functional derivatives) are accessible from ethyl 2-cyano-2-(1,2-benzisothiazol-3-ylidene)-acetate by reaction with ethyl chloroacetate or acrylonitrile, respectively.[44]

Biological Activity.—1,2-Benzisothiazoles bearing acidic 3-substituents (CO_2H, CH_2CO_2H, CONHOH, or $CH_2CONHOH$ groups) possess varying degrees of anti-inflammatory, analgesic, and antipyretic activity.[45] 3-Carboxamides exert antihistaminic and anticholinergic effects.[46] 1,2-Benzisothiazol-3-yl-acetic, -propionic and -butyric acids, and their ethyl esters, amides, and nitriles, display auxin-like properties, approaching and exceeding (in some instances) those of indol-3-ylacetic acid.[44] The action of 1,2-benzisothiazole, its derivatives, and selenium analogues on lysosomes and mitochondria *in vitro* has been studied.[47]

4 1,2-Benzisothiazole 1,1-Dioxides

Synthesis.—A series of 4-substituted 6-carboxy-3-phenyl-1,2-benzisothiazole 1,1-dioxides (58) has been produced in almost quantitative yields by the thermolysis of 3-substituted 4-benzoyl-5-sulphamoylbenzoic acids (57). The diuretic

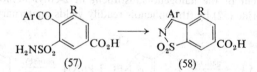

(57)   (58)

effect of (58) is found to be due to the presence of (57) in equilibrium with (58) in plasma.[48]

Chemical Properties.—The action of Grignard reagents on saccharin has been reinvestigated.[49] It affords 3-alkyl- (or aryl-)substituted 1,2-benzisothiazole 1,1-dioxide derivatives (59), together with the tertiary alcohols (60), the relative amounts depending on the proportions of the reagents employed and the experi-

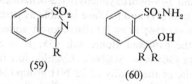

(59)   (60)

[44] C. Branca, V. Plazzi, M. Vitto, F. Bordi, D. S. Fracassini, and N. Bagni, *Phytochemtisry*, 1975, **14**, 2545.
[45] T. Vitali and G. Bertaccini, *Farmaco, Ed. Sci.*, 1974, **29**, 109.
[46] E. Molina, L. Zappia, L. Amoretti, and P. L. Catellani, *Ateneo Parmense, Acta Bio-med.*, 1974, **45**, 183 (*Chem. Abs.*, 1975, **83**, 254).
[47] P. van Caneghem, *Biochem. Pharmacol.*, 1974, **23**, 3491.
[48] O. B. T. Nielsen, H. Bruun, C. Bretting, and P. W. Feit, *J. Medicin. Chem.*, 1975, **18**, 41.
[49] R. A. Abramovitch, E. M. Smith, M. Humber, B. Purtschert, P. C. Srinivasan, and G. M. Singer, *J.C.S. Perkin I*, 1974, 2589.

mental conditions. The alcohol (60) arises undoubtedly by nucleophilic displacement at C-3 in the intermediate formed by addition of 1 mole of reagent. In no case was this intermediate isolable. The reaction of saccharin with organolithium compounds at − 78 °C affords the 3-substituted derivatives (59) exclusively, thus providing a general method for their preparation.[49]

An alternative approach to 3-substituted 1,2-benzisothiazole 1,1-dioxides bearing substituents in the benzene ring is to allow the appropriately substituted benzenesulphonamide to react successively with an acyl chloride and n-butyllithium.[49]

The action of nucleophiles (MNuc; M = H or Na) on 3-alkoxy-1,2-benzisothiazole 1,1-dioxides (62; R = Me or Pr^i) yields either 3-substituted 1,2-benzisothiazole 1,1-dioxides (63) or saccharin (61) as main products. The nature of the nucleophile determines which pathway predominates.[50]

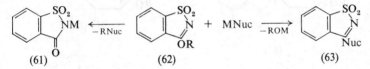

(61)   (62)   (63)

The action of sodium alkyl carbonates on 3-chloro-1,2-benzisothiazole 1,1-dioxide (64) produces mixtures of *N*-alkoxycarbonylsaccharins (65) and pseudosaccharin anhydride (66), together with dialkyl pyrocarbonates. The product ratio of (65) to (66) decreases in the order (R =) Me > Et > Pr^i > Bu^t. The initial step in the reaction is considered to be a nucleophilic displacement of chlorine by ROCOO$^-$ ion.[51]

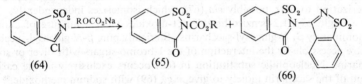

(64)   (65)   (66)

The interaction of heterocyclic hydroxy- and potential hydroxy-compounds with hexamethylphosphoric triamide (HMPA) at 220—230 °C replaces the hydroxy- with a dimethylamino-group.[52] The numerous examples include the conversion of saccharin into 3-dimethylaminobenzisothiazole *SS*-dioxide. In the presence of pyrrolidine, the 3-(1′-pyrrolidyl)-compound is obtained.[53] Improved yields of *N*-carboxymethyl-1,2-benzisothiazolin-3-one 1,1-dioxide have been obtained from sodium saccharin by the action of sodium bromoacetate, or chloroacetonitrile, followed by hydrolysis.[54]

A technique of reducing carboxylic acids in the form of their *N*-acyl-saccharins to the corresponding aldehydes, using sodium bis(2-methoxyethoxy)aluminium hydride (SDA), has been described. The *N*-acyl-saccharins are easily prepared

[50] N. Matsumura, Y. Otsuji, and E. Imoto, *Nippon Kagaku Kaishi* (*J. Chem. Soc. Japan, Ind. Chem. Sect.*), 1974, 1532 (*Chem. Abs.*, 1975, **82**, 4168).
[51] N. Matsumura, Y. Otsuji, and E. Imoto, *Nippon Kagaku Kaishi* (*J. Chem. Soc. Japan, Ind. Chem. Sect.*) 1974, 1539 (*Chem. Abs.*, 1975, **82**, 57 595).
[52] N. O. Vesterager, E. B. Pedersen, and S. O. Lawesson, *Tetrahedron*, 1973, **29**, 321; H. Vorbrüggen, *Synthesis*, 1973, 301.
[53] E. B. Pedersen and S. O. Lawesson, *Tetrahedron*, 1974, **30**, 875.
[54] C. S. Giam and R. A. Lockhart, *Org. Prep. Proced. Internat.*, 1974, **6**, 1.

by the reaction of the acids with saccharin chloride, and can be used in the reduction step without isolation.[55]

Saccharin Glucosides.—3-Chloro-1,2-benzisothiazole 1,1-dioxide (64) reacts with nucleophiles (ROH, RSH, or RNH_2) to yield products derived from the imide form of saccharin (63). Its reaction in pyridine with 2,3,4,6-tetra-*O*-methyl(or -acetyl)-$\alpha\beta$-D-glucopyranose similarly gives the readily crystallizable 3-(α-D-glucosyl)-1,2-benzisothiazole 1,1-dioxides (67), irrespective of the configuration of the monosaccharide used.[56] In the absence of base, the reaction

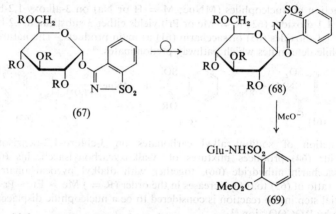

(67)

(68)

Glu-NHSO2

MeO2C

(69)

requires higher temperatures, but then proceeds with simultaneous O → N isomerization to (68), possibly *via* (67), which isomerizes independently. Thus, pure α-*O*-glucosyl-1,2-benzisothiazole 1,1-dioxides yield a mixture of anomeric (predominantly β) *N*-glucosyl-saccharins. Sterically pure β-*N*-glucosyl-saccharins (68) are accessible by the interaction of the 1-bromo-sugars with silver or sodium saccharin. Nucleophilic substitution in (68) occurs exclusively at the carbonyl carbon of the saccharin moiety to give, *e.g.*, (69) with sodium methoxide.[56]

The reaction can be applied to other carbohydrates and two or three saccharin-residues may be introduced. It is useful for characterizing monosaccharides which may be recovered by acid hydrolysis.[56]

Analytical Procedures.—Methods have been described for determining saccharin by chromatographic,[57-61] u.v.,[61-63] and *X*-ray fluorescence[64] spectrometric

[55] N. S. Ramegowda, M. N. Modi, A. K. Koul, J. M. Bora, C. K. Narang, and N. K. Mathur, *Tetrahedron*, 1973, **29**, 3986.
[56] A. Klemer, G. Uhlemann, S. Chahin, and M. N. Diab, *Annalen*, 1973, 1943.
[57] Y. Hoshino and T. Suzuki, *Shokuhin Eiseigaku Zasshi*, 1975, **16**, 182 (*Chem. Abs.*, 1975, **83**, 204 877).
[58] I. Nagai, H. Oka, M. Tasaka, and A. Oka, *Eisei Kagaku*, 1975, **21**, 261 (*Chem. Abs.*, 1976, **84**, 42 107).
[59] B. Unterhalt, *Z. Lebensm.-Unters. Forsch.*, 1975, **159**, 161 (*Chem. Abs.*, 1976, **84**, 42 087).
[60] Y. Tanaka, K. Ikebe, R. Tanaka, and N. Kunita, *Shokuhin Eiseigaku Zasshi*, 1975, **16**, 295 (*Chem. Abs.*, 1976, **84**, 88 059).
[61] H. Jacin, *Dtsch. Lebensm.-Rundschau*, 1975, **71**, 428 (*Chem. Abs.*, 1976, **84**, 88 062).
[62] M. Nakamura, K. Watabe, T. Kirigaya, Y. Yazawa, A. Watabe, Y. Suzuki, and T. Kawamura, *Shokuhin Eiseigaku Zasshi*, 1975, **16**, 264 (*Chem. Abs.*, 1976, **84**, 88 058).
[63] M. M. Hussein, H. Jacin, and F. B. Rodriguez, *J. Agric. Food Chem.*, 1976, **24**, 36.
[64] T. Yamada and T. Nakaoka, *Shokuhin Eiseigaku Zasshi*, 1975, **16**, 7 (*Chem. Abs.*, 1975, **83**, 162 252).

techniques. An ion-selective electrode that is responsive to saccharin has been developed, capable of measuring saccharin concentration in the presence of other sweetening substances such as sucrose, glucose, sodium cyclamate, and sorbitol over a concentration range of 10^{-1} to 10^{-5} mol l^{-1}.[65]

Biological Properties.—Only a selection of this voluminous part of the literature can be given. A bibliography of publications dealing with the biological effects of saccharin has been provided,[66] and its metabolism, toxicity, and carcinogenicity have been reviewed.[67, 68] Prolonged administration of diets containing sodium saccharin to rats did not lead to an increase in the incidence of tumours, but high doses were associated with reduced body weight in both sexes, and decreased longevity in male rats.[69] The excretion pattern of saccharin in the rat and other mammals, after oral ingestion of ^{14}C-labelled sodium saccharin, has been studied.[70]

5 1,2-Benzisoselenazoles

Information is available [38, 39] concerning the mass spectra of 1,2-benzisoselenazoles (see p. 347).

Electrophilic and nucleophilic substitution reactions in the 1,2-benziso-selenazole nucleus [71] have been studied in some detail. Nitration and bromination occur at C-5 and C-7, giving mixtures of the appropriate derivatives. Nucleophilic amination by potassamide produces 3-amino-1,2-benzisoselenazole (70;

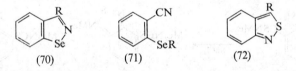

(70)   (71)   (72)

$R = NH_2$). The synthesis of the 3-carboxylic acid (70; $R = CO_2H$) and its functional derivatives, starting with ozonolysis of the 3-styryl-analogue (70; $R = CH:CHPh$), has also been described. Butyl-lithium cleaves the hetero-ring bonds of 1,2-benzisoselenazole (70; $R = H$) with the formation of products such as (71) and the corresponding aldehyde.[72]

6 2,1-Benzisothiazoles

Synthesis.—Synthetic work in this field has been mainly confined to the elaboration of existing methods, most of which involve cyclizations of suitable *ortho*-disubstituted benzenes.

[65] N. Hazemoto, N. Kamo, and Y. Kobatake, *J. Assoc. Offic. Analyt. Chemists*, 1974, **57**, 1205.
[66] H. S. Warren, 'Biological Effects of Saccharin', NTIS, Springfield, Va., 1973 (24 pp.).
[67] I. C. Munro, B. Stavric, and R. Lacombe, *Toxicol. Ann.*, 1974, 71, (*Chem. Abs.*, 1976, **84**, 57 428).
[68] B. A. Becker and G. R. Thompson, *Proc. West. Pharmacol. Soc.*, 1975, **18**, 306 (*Chem. Abs.*, 1975, **83**, 162 221).
[69] I. C. Munro, C. A. Moodie, D. Krewski, and H. C. Grice, *Toxicol. Appl. Pharmacol.*, 1975, **32**, 513.
[70] E. J. Lethco and W. C. Wallace, *Toxicology*, 1975, **3**, 287.
[71] R. Weber and M. Renson, *J. Heterocyclic Chem.*, 1973, **10**, 267.
[72] R. Weber and M. Renson, *J. Heterocyclic Chem.*, 1975, **12**, 1091.

Thus, the synthesis of 2,1-benzisothiazoles by the action of thionyl chloride on *o*-toluidines (see Vol. 2, p. 576) has found further application.[73] *o*-Benzylaniline affords 3-phenyl-2,1-benzisothiazole (72; R = Ph) nearly quantitatively, but *o*-toluidines functionally substituted in their methyl group yield only tars or *N*-sulphinyl-derivatives. *o*-Ethylaniline gives several products: 3-methyl-2,1-benzisothiazole (72; R = Me) may first be formed, but is rapidly chlorinated at its methyl group, to produce, after hydrolysis, 2,1-benzisothiazole-3-carboxylic acid (30%) (72; R = CO$_2$H). Another product, previously erroneously regarded as 3-methyl-2,1-benzisothiazole (72; R = Me), appears to be 2-(2,1-benziso-thiazol-3-yl)-7-ethylbenzothiazole. More highly substituted *o*-alkyl-anilines produce, apart from traces of 2,1-benzisothiazoles, merely *N*-sulphinyl-amines.[73]

The production of chlorinated by-products is largely avoided when *N*-sul-phinylmethanesulphonamide (MeSO$_2$NSO) is used as the cyclization agent for *o*-toluidines or *N*-sulphinyl-*o*-toluidines, in boiling pyridine. The procedure has been successful in the production of numerous 2,1-benzisothiazoles (including naphth[1,2-*c*]isothiazole), but failed to yield the hydroxy- or amino-substituted heterocycles from the corresponding substituted *o*-toluidines.[74]

Further examples of the reduction of *o*-nitrothiocarbonyl- and the oxidation of *o*-aminothiocarbonyl-compounds to 2,1-benzisothiazoles are provided by the reduction of *NNN'N'*-tetramethyl-2-nitroisophthalthioamide (73) and the oxidation of the amine (74) with peroxyacetic acid. The amine (74) is the major product (80%) of the reduction of (73).[75]

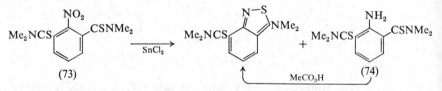

(73)   (74)

The method of converting 2,1-benzisoxazoles into the corresponding 2,1-benzisothiazoles by the action of phosphorus pentasulphide (see Vol. 1, p. 369; Vol. 2, p. 576) has also been further applied.[75]

2,1-Benzisothiazoline 2,2-dioxides incorporating a 1-amino-alkyl group (76; R^1 = R^2 = Me, R^1 = Me, R^2 = CH$_2$Ph; *n* = 2 or 3) have, after several unsuccessful attempts, been synthesized from 2-chloro-sulphonanilides (75) by condensation with the appropriate amino-alkyl chloride in DMF and subsequent ring-closure with potassamide in liquid ammonia.[76]

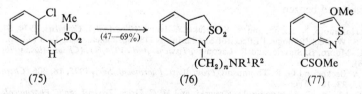

(75)   (76)   (77)

[73] M. Davis, T. G. Paproth, and L. J. Stephens, *J.C.S. Perkin I*, 1973, 2057.
[74] G. M. Singerman, *J. Heterocyclic Chem.*, 1975, **12**, 877.
[75] M. S. Chauhan and D. M. McKinnon, *Canad. J. Chem.*, 1975, **53**, 1336.
[76] J. A. Skorcz, J. T. Suh, and R. L. Germershausen, *J. Heterocyclic Chem.*, 1974, **11**, 73.

Chemical Properties.—A study of the n.m.r. spectra, in DMSO, over a temperature range, has shown that methyl 3-methoxy-2,1-benzisothiazole-7-thiono-carboxylate (77) undergoes valence tautomerism above 200 °C.[75]

The photolytic decomposition of the substituted 2,1-benzisothiazole (78) gives the benzophenone (79) (31%) or the dimeric dibenzodiazocine (80) (39%) as main products, depending on the experimental conditions.[41]

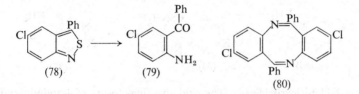

(78)   (79)   (80)

The formulation of 2,1-benzisothiazolin-3-one as ketonic tautomers (see Vol. 3, p. 558) has been confirmed by the preparation of its *N*- and *O*-alkylated derivatives, by unambiguous routes, and examination of their u.v. spectra. The same conclusions have been reached about 5-chloro-2,1-benzisothiazolin-3-one.[77] Other nucleophilic displacements of the 3-halogen in 3-chloro- and 3,5-dichloro-2,1-benzisothiazole have been reported: the auxin analogue (72; R = CH₂CO₂H) was obtained in the course of this work.[78] The results of a comparative study [24] of the rates of *N*-methylation are included under isothiazoles (p. 344).

7 Other Condensed Ring Systems incorporating Isothiazole

A number of condensed hetero-structures incorporating isothiazole, some of them novel, have been described, and are dealt with in this section in ascending order of ring complexity.

Isothiazolo[5,1-*e*]isothiazoles [6a-Thia-1,6-diazapentalenes].—Full details have now been provided of a synthesis of compounds of this novel class of hypervalent heterocyclic compounds, which are structurally analogous to 6a-thiathiophthens (see Vol. 2, p. 585), from 6-methyl-1,6a-dithia-6-azapentalenes (81) by successive alkylation and reaction with methylamine.[79]

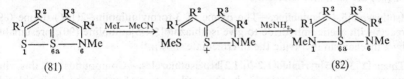

(81)   (82)

The ¹H n.m.r. spectra of the isothiazolo[5,1-*e*]isothiazoles (82) show magnetic equivalence of ring protons or substituents at the pairs of sites N-1/N-6, C-2/C-5, and C-3/C-4, thus demonstrating that, in solution, the compounds possess real or time-averaged C_{2v} symmetry. Their chemical properties resemble those of enamines and pyrroles. Their electron-rich nature, suggested by n.m.r. chemical

[77] M. Davis, L. W. Deady, E. Homfeld, and S. Pogany, *Austral. J. Chem.*, 1975, **28**, 129.
[78] M. Davis, E. Homfeld, J. McVicars, and S. Pogany, *Austral. J. Chem.*, 1975, **28**, 2051.
[79] A. S. Ingram, D. H. Reid, and J. D. Symon, *J.C.S. Perkin I*, 1974, 242.

shift data, is reflected in their power of forming stable black 1 : 1 charge-transfer complexes with 1,3,5-trinitrobenzene.[79]

Isothiazolo[5,4-*b*]pyridine.—The synthesis of 2-aminoisothiazoles from *o*-mercaptonitriles and hydroxylamine-*O*-sulphonic acid in basic media has been successfully extended to the production of 3-aminoisothiazolo[5,4-*b*]pyridines (83) from 3-cyanopyridine-2-thiones.[80]

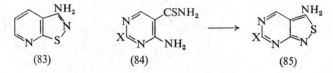

(83)   (84)   (85)

Isothiazolo[3,4-*d*]pyrimidines.—The oxidative ring-closure of 4-aminopyrimidine-5-thioamides (84) provides 3-aminoisothiazolo[3,4-*d*]pyrimidines (85) in excellent yields. Similarly substituted uracils, which are accessible by the interaction of 4-amino-uracils with isothiocyanates (including acyl and sulphonyl isothiocyanates), readily afford a series of 3-amino-4,5,6,7-tetrahydroisothiazolo-[3,4-*d*]pyrimidine-4,6-diones by the same cyclization.[81]

(86)   (87)   (88)

Isothiazolo[5,4-*d*]pyrimidines.—Successive reaction of 5-amino-3-methylisothiazole-4-carbonitrile (86; X = CN) with orthoesters and primary amines produces 4-amino-3-methylisothiazolo[5,4-*d*]pyrimidines, apparently by rearrangement of the primarily produced isomers, which were not isolated.[82] Condensation of (86; X = CN) with amidines gave the same products in one stage, albeit in very low yields. The 4-carboxamide (86; X = NH_2CO) may in turn be cyclized with orthoesters to 5*H*-isothiazolo[5,4-*d*]pyrimidin-4-ones (88), and thiono-analogues of (88) are similarly accessible.[83]

Isothiazolo[5,4-*b*]quinoline.—The oxime of 3-formylquinoline-2(1*H*)-thione (89) reacts with acetic anhydride to give isothiazolo[5,4-*b*]quinoline (90), presumably by a mechanism involving the intermediate shown.[84]

Thieno[2′,3′: 4,5]pyrimido[1,2-*b*][1,2]benzisothiazoles.—Compounds of this ring system, *e.g.* (91), are obtainable by the condensation of pseudosaccharin chloride and 2-aminothiophen-3-amide, followed by cyclization of the resulting intermediate in boiling glacial acetic acid. The corresponding 13*H*[1]benzothieno-(five membered) ring system is similarly accessible.[85]

[80] K. Gewald, U. Schlegel, and H. Schäfer, *J. prakt. Chem.*, 1975, **317**, 959.
[81] R. Niess and H. Eilingsfeld, *Annalen*, 1974, 2019.
[82] J. A. Montgomery and H. J. Thomas, *J. Org. Chem.*, 1963, **28**, 2304.
[83] R. C. Anderson and Y. Y. Hsiao, *J. Heterocyclic Chem.*, 1975, **12**, 883.
[84] R. Hull, *J.C.S. Perkin I*, 1973, 2911.
[85] F. Sauter and W. Deinhammer, *Monatsh.*, 1974, **105**, 1249.

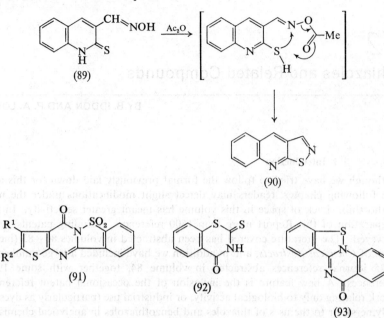

(89)

(90)

(91)

(92)

(93)

1,2-Benzisothiazolo[2,3-a]quinazoline.—The interaction of 2H-1,3-benzothiazine-2-thione-4(3H)-one (92) with anthranilamide yields a complex mixture, from which 5H[1,2]benzisothiazolo[2,3-a]quinazolin-5-one (93) is isolable. This appears to arise in fact from an intermediate, 3H[1,2]benzodithiole-3-thione, whose presence in the mixture was demonstrated, and which was separately condensed with anthranilamide to give (93).[86]

[86] S. Palazzo, L. I. Giannola, and M. Neri, *J. Heterocyclic Chem.*, 1975, **12**, 1077.

12
Thiazoles and Related Compounds

BY B. IDDON AND P. A. LOWE

1 Introduction

Although we have tried to follow the format previously laid down for this and the following chapter, readers may detect slight modifications under the new authorship. Lack of space in this volume has meant greater selectivity. In the preparation of this Report more than 1500 references, including patents, were surveyed. The literature covered has been abstracted in volumes 80—83 (inclusive) of *Chemical Abstracts*, and in addition we have included a large number of 1975 journal references abstracted in volume 84, together with some 1976 references. A new feature is the inclusion of the occasional patent reference. Work relating only to biological activity, or industrial use (particularly as dyes or polymers), or to the uses of thiazoles and benzothiazoles in analytical chemistry has been excluded. It would be impossible, for example, to cover all the references to tetramisole (the anthelmintic) or those describing the uses of benzothiazole-2-thiol and its derivatives as rubber additives.

β-Lactams are covered in Chapter 16 of this volume, and have been reviewed recently.[1, 2] The role of thiazoles in food flavours[3] and in analytical chemistry,[4] the diazotization of 2-amino-thiazoles and the uses of the derived diazonium salts,[5] the reactions of carbonyl compounds (especially ketones) with sulphur and ammonia or an amine to give Δ³-thiazolines,[6a] synthetic uses of Δ²-thiazoline-2-thiol,[6b] and polychloro-thiazoles and -benzothiazoles[7] have been reviewed also. The proceedings[8] of the VIth International Conference on Organic Sulphur Chemistry, held at Bangor in 1974, are worth consulting. We have not reviewed the chemistry contained therein, since much of it has been published elsewhere. Noteworthy, however, is Metzger's chapter on the 'gear effect' in thiazole and related systems.

The most interesting developments in thiazole chemistry during the period under review have been those concerned with meso-ionic compounds, their

[1] A. K. Mukerjee and A. K. Singh, *Synthesis*, 1975, 547.
[2] P. G. Sammes, *Chem. Rev.*, 1976, **76**, 113.
[3] J. A. Maga, *C.R.C. Crit. Rev. Food Sci. Nutr.*, 1975, **6**, 153 (*Chem. Abs.*, 1976, **84**, 3351).
[4] H. R. Hovind, *Analyst*, 1975, **100**, 769.
[5] R. N. Butler, *Chem. Rev.*, 1975, **75**, 241.
[6] (a) F. Asinger, W. Leuchtenberger, and H. Offermanns, *Chem.-Ztg.*, 1974, **98**, 610; (b) K. Hirai and Y. Kishida, *Heterocycles*, 1974, **2**, 185; *Yuki Gosei Kagaku Kyokai Shi*, 1974, **32**, 20 (*Chem. Abs.*, 1974, **80**, 132 700).
[7] B. Iddon and H. Suschitzky, in 'Polychloroaromatic Compounds', ed. H. Suschitzky, Plenum Press, London, 1974, Ch. 2, p. 309.
[8] 'Organic Sulphur Chemistry; Structure, Mechanism, and Synthesis', ed. C. J. M. Stirling, Butterworths, London, 1975.

synthesis and reactions. The first meso-ionic selenium compound has been synthesized, the structure of the sulphur-containing antibiotic althiomycin has been established, several Meisenheimer complexes derived from thiazoles have been obtained, the first detailed kinetic studies of electrophilic substitution of thiazoles have been carried out, the adducts formed from thiazoles and dimethyl acetylenedicarboxylate have been re-examined and their structures resolved beyond doubt, whilst studies on the synthesis and reactions of thiazolium salts are producing many interesting results. Organolithium derivatives of Δ^2-thiazolines continue to be of interest in synthesis, whilst amminium salts derived from 2-imino-3-hydroxy-Δ^4-thiazolines are excellent alkoxycarbonylation agents. Metzger's group has observed restricted rotation of ethyl groups in Δ^4-thiazolines, caused by a 'gear effect'.

Compounds with the general formula (1) have been shown to be carcinogenic in animal experiments.[9]

2 Synthesis of Thiazoles [10]

Hantzsch's Synthesis (Type A; S—C—N + C—C).—The majority of thiazoles continue to be prepared by this method. The intermediate formation of 4-hydroxy-Δ^2-thiazolines (see Vol. 3, p. 567 and Vol. 1, p. 379) has been demonstrated again,[11, 12] whilst Metzger's group [13] has detected thioimidate intermediates in Hantzsch reactions by the use of ^{1}H n.m.r. spectroscopy.

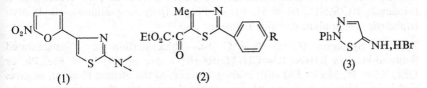

(1)   (2)   (3)

Aromatic thioamides react with $MeC(O)CHClC(O)CO_2Et$ to give a 3 : 1 ratio of ethyl 5-acetyl-2-aryl-thiazole-4-carboxylates and the isomeric products (2).[14] Modifications of the Hantzsch synthesis continue to appear. Thus, 2-aryl-5-aroyl-thiazoles may be prepared from the N'-thioaroyl-NN-dimethylformamidines $ArC(S)N=CHNMe_2$ and α-bromo-ketones [15] [strictly speaking, this is a Type F (C—N—C—S + C) synthesis], whilst 5-amino-2-phenyl-1,2,3-thiadiazolium salts [*e.g.* (3)] and compounds with the general structure RCH_2CN (*e.g.* R = CN or CO_2Et) yield 4-amino-thiazoles.[16]

[9] C.-Y. Wang, C.-W. Chiu, B. Kaiman, and G. T. Bryan, *Biochem. Pharmacol.*, 1975, **24**, 291; C.-Y. Wang, C.-W. Chiu, and G. T. Bryan, *Drug Metab. Dispos.*, 1975, **3**, 89; C.-Y. Wang, K. Muraoka, and G. T. Bryan, *Cancer Res.*, 1975, **35**, 3611; E. Kunze, A. Schauer, and G. Kruesmann, *Z. Krebsforsch. Klin. Onkol.*, 1975, **84**, 143; E. Erturk, *Ankara Univ.*, *Vet. Fak. Derg.*, 1974, **21**, 296 (*Chem. Abs.*, 1976, **84**, 84 979).

[10] The classification is that used in previous volumes; the syntheses of meso-ionic thiazoles are dealt with separately.

[11] P. Chauvin, J. Morel, P. Pastour, and J. Martinez, *Bull. Soc. chim. France*, 1974, 2079.

[12] R. J. C. Kleipool and A. C. Tas, *Riechst. Aromen, Koerperpflegem.*, 1973, **23**, 326, 329 (*Chem. Abs.*, 1974, **80**, 27 158).

[13] A. Babadjamian, J. Metzger, and M. Chanon, *J. Heterocyclic Chem.*, 1975, **12**, 643.

[14] B. Tornetta, F. Guerrera, and A. Ronsisvalle, *Ann. Chim.* (*Italy*), 1974, **64**, 477 (*Chem. Abs.*, 1976, **84**, 17 210).

[15] J. C. Meslin and H. Quiniou, *Tetrahedron*, 1975, **31**, 3055; *Synthesis*, 1974, 298.

[16] K. Gewald and U. Hain, *J. prakt. Chem.*, 1975, **317**, 329.

Other Type A Syntheses.—Successive treatment of lactone (4) with hydrochloric acid and thiourea yields 2-amino-5-hydroxyethyl-4-methylthiazole, an intermediate in the synthesis of vitamin B_1.[17] $\beta\beta$-Dichloro-α-amino-acrylonitrile, $Cl_2C=C(CN)NH_2$, prepared from hydrogen cyanide and dichloroacetonitrile, reacts with thioformamide in the necessary presence of toluene-*p*-sulphonic acid to give thiazole-4-carbonitrile.[18] 2-Methylthiazole-4-carbonitrile may be prepared similarly. 4-Aryl-3-arylamino-5-imino-1,2,4-thiadiazolidines (Hector's

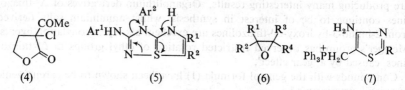

(4)   (5)   (6)   (7)

bases) (see p. 389) undergo 1,3-dipolar cycloaddition reactions with alkynes $R^1C{\equiv}CR^2$ in chloroform to give 2-arylamino-thiazoles *via* intermediates (5; Ar^2 = Ph or *p*-MeC$_6$H$_4$; $R^1 = R^2 = CO_2Me$ or PhCO; R^1 = Ph, $R^2 = CO_2Me$), which eliminate Ar^1NHCN.[19] 2,2,2-Trichloro-1-phenylethanol reacts with base to give an epoxide (6; R^1 = Ph, $R^2 = H$, $R^3 = R^4 = Cl$), which condenses with thioacetamide to yield 4-hydroxy-2-methyl-5-phenylthiazole.[20] The 4-amino-thiazoles (7; R = H, Et, or Ph) may be prepared by allowing β-cyano-vinyltriphenylphosphonium bromide, $Ph_3\overset{+}{P}CH=CHCN\ Br^-$, to react with a thioamide $RC(S)NH_2$ (R = H, Et, or Ph); a (β-cyano-γ-amino-γ-alkyl)-allyl-triphenylphosphonium bromide is formed also.[21]

Type B Syntheses (C—C—N + C—S).—Thiocyanation of $\alpha\beta$-unsaturated β-amino-ketones $R^3HNCR^1=CHC(O)R^2$ (R^1 = Me or Ph; R^2 = Me, Ph, or OEt; R^3 = H, Me, or Ph) with cyanogen occurs at the olefinic H-atom, or gives a 2-imino-Δ^4-thiazoline, or its rearranged isomer (8), depending upon the structure of the starting material and reaction temperature.[22] The thiocyanated alkenes are cyclized to mixtures of the cyclized products in acid or alkali, or else thermally.

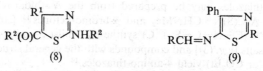

(8)   (9)

Type C Syntheses (C—C—N—C + S).—The reaction of α-aminobenzyl cyanide, $PhCH(NH_2)CN$, with aldehydes and sulphur in the presence of an organic base results in good yields of anils (9), which are hydrolysed by acids to 2-substituted 5-amino-4-phenylthiazoles.[23]

[17] B. V. Passet, G. N. Kul'bitskii, V. Ya. Samarenko, and L. I. Vekshina, *Khim.-Farm. Zhur.*, 1974, **8**, 48 (*Chem. Abs.*, 1975, **82**, 57 596).
[18] G. D. Hartman, M. Sletzinger, and L. M. Weinstock, *J. Heterocyclic Chem.*, 1975, **12**, 1081.
[19] K. Akiba, M. Ochiumi, T. Tsuchiya, and N. Inamoto, *Tetrahedron Letters*, 1975, 459.
[20] W. Reeve and E. R. Barron, *J. Org. Chem.*, 1975, **40**, 1917.
[21] C. Ivancsics and E Zbiral, *Monatsh.*, 1975, **106**, 839.
[22] T. Tokumitsu and T. Hayashi, *Yuki Gosei Kagaku Kyokai Shi*, 1975, **33**, 478 (*Chem. Abs.*, 1976, **84**, 17 205).
[23] K. Gewald, H. Schoenfelder, and U. Hain, *J. prakt. Chem.*, 1974, **316**, 299.

Type D Syntheses (C—C—S + C—N).—α-Mercapto-ketones $R^2C(O)CHR^1SH$ (*e.g.* R^1 = Me, R^2 = Et) react with nitriles R^3CN (R^3 = Me, Ph, or $PhCH_2$) or thiocyanates R^3SCN [R^3 = Me, Bu^n, CH_2Ph, CH_2CO_2Et, $CHMeC(O)Et$, *etc.*] in the presence of sodium cyanide to give thiazoles, and with liquid hydrogen cyanide to give the corresponding 2,2′-bithiazolyl.[24]

Type F Syntheses (C—N—C—S + C).—A synthesis of 2-amino-thiazoles based on a corresponding thiophen synthesis (where NH is equivalent to CH) involves the reaction of an alkyl halide R^1CH_2X with an *N*-acyl-thiourea in the presence of base, and proceeds *via* intermediate (10).[25]

(10)   (11)

Type K Syntheses (C—C—N—C—S).—1,2-Dichloroethyl isothiocyanate reacts with thioureas to give thiadiazines, at room temperature, or thiazoles (11) at 80—100 °C.[26] The former rearrange to the latter on heating or in the presence of base. With ureas, the only ureido-derivative isolated corresponded to (11; R^1 = H, R^2 = R^3 = Me).[26]

Synthesis of Meso-ionic Thiazoles.—The reactions of meso-ionic thiazoles are currently of great interest (see p. 368), but only a few syntheses are known. A new synthesis (Type A) of *anhydro*-4-hydroxythiazolium hydroxides (12; R^1 = Me or Ph, R^2 = Ph or $PhCH_2$, R^3 = Ph, p-ClC_6H_4, or p-$O_2NC_6H_4$) involves the reaction of a thioamide $R^1C(S)NHR^2$ with a *gem*-dicyano-epoxide (6; R^1 = R^2 = CN, R^3 = Ar, R^4 = H).[27] *anhydro*-2-Aryl-5-hydroxy-3-methylthiazolium hydroxides (14; R^1 = Ar, R^2 = Me, R^3 = H) may be prepared by the reaction

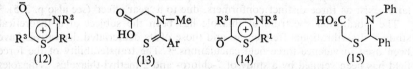

(12)   (13)   (14)   (15)

of compounds (13) with acetic anhydride in the presence of triethylamine; if the reactants are not kept cold, acetylation of the product occurs in the 4-position.[28] In a similar manner, *anhydro*-2,3-diphenyl-4-hydroxythiazolium hydroxide (12; R^1 = R^2 = Ph, R^3 = H) may be prepared by cyclization of the condensation product (15) of bromoacetic acid and thiobenzanilide (a Type A synthesis).[29] Other meso-ionic thiazoles may be prepared from bromoacetic acid or α-bromophenylacetic acid and a thiobenzanilide.[29]

[24] F. Asinger, K. Fabian, H. Vossen, and K. Hentschel, *Annalen*, 1975, 410.

[25] J. Liebscher and H. Hartmann, *Z. Chem.*, 1974, **14**, 470.

[26] R. Lantzsch and D. Arlt, *Synthesis*, 1975, 675.

[27] M. Baudy and A. Robert, *J.C.S. Chem. Comm.*, 1976, 23.

[28] K. T. Potts, J. Baum, E. Houghton, D. N. Roy, and U. P. Singh, *J. Org. Chem.*, 1974, **39**, 3619.

[29] K. T. Potts, E. Houghton, and U. P. Singh, *J. Org. Chem.*, 1974, **39**, 3627.

3 Physical Properties of Thiazoles

Althiomycin is a sulphur-containing antibiotic that was first isolated from *Streptomyces althioticus* in 1957, whose structure has now been shown [30] to be (16) by chemical and spectroscopic methods and by the unequivocal formulation of a derivative, bisanhydroalthiomycin, by *X*-ray crystallographic analysis.[31]

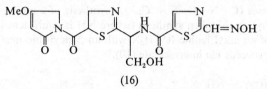

(16)

2-Amino-4-aryl-thiazoles give Meisenheimer complexes with 2,4,6-trinitro-chlorobenzene.[32] The tautomeric properties of thiazoles continue to attract attention. Unlike most 4-hydroxy-thiazoles, which prefer to exist in their keto-forms, 4-hydroxy-2-methyl-5-phenylthiazole exists as a stable enol.[20] The spectroscopic properties of the meso-ionic thiazole (12; R^1 = Me, R^2 = Ph, R^3 = *p*-ClC$_6$H$_4$), particularly its [1]H n.m.r. spectrum, support the postulate that it exists in solution (*e.g.* in chloroform) as a mixture of tautomers, with some contribution by (17).[27] [1]H N.m.r. spectroscopy has been used to study the tautomerism of various 2,4-diamino-thiazoles.[33] 2-Heteroarylamino-thiazoles exist predominantly in

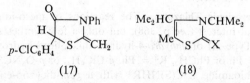

(17)   (18)

the amino-form; association by hydrogen-bonding between the NH groups and the N-atom of the thiazole ring occurs in solution.[34] [1]H N.m.r. spectroscopy shows that compounds (18; X = O or S) and the *S*-methyl derivative of the latter exist as three distinct conformers, due to a 'gear effect' (see also p. 375).[35]

The vibrational spectrum of thiazole has been the subject of a theoretical study.[36] Its vibrational frequencies and those of its deuteriated derivatives have been used in valence force field calculations.[37] The transferability of the force field has been verified by a study of 2-chloro- and 2-methyl-thiazoles. Thiazoles have also been the subjects for detailed i.r. and Raman spectral studies,[38] magnetic rotatory dispersion studies,[39] and mass spectrometric analysis.[40]

[30] H. Sakakibara, H. Naganawa, M. Ohno, K. Maeda, and H. Umezawa, *J. Antibiotics*, 1974, **27**, 897.
[31] N. Nakamura, Y. Iitaka, H. Sakakibara, and H. Umezawa, *J. Antibiotics*, 1974, **27**, 894.
[32] A. Sharma, B. K. Sinha, and G. B. Behera, *Indian J. Chem.*, 1975, **13**, 509.
[33] O. Ceder and B. Beijer, *Tetrahedron*, 1975, **31**, 963.
[34] J. Boedeker, H. Pries, D. Roesch, and G. Malewski, *J. prakt. Chem.*, 1975, **317**, 953.
[35] A. Lidén, C. Roussel, M. Chanon, J. Metzger, and J. Sandström, *Tetrahedron Letters*, 1974, 3629.
[36] D. Ya. Movshovich, V. N. Sheinker, A. D. Garnovskii, and O. A. Osipov, *Zhur. org. Khim.*, 1975, **11**, 1740 (*Chem. Abs.*, 1975, **83**, 170 181).
[37] T. Avignon, E. J. Vincent, J. Raymond, and M. Chaillet, *J. Mol. Structure*, 1974, **21**, 319.
[38] G. Davidovics, C. Garrigou-Lagrange, J. C. Caffarel, and J. Chouteau, *J. Chim. phys.*, 1975, **72**, 527.
[39] M.-F. Bruniquel, D. Bouin, and J.-F. Labarre, *J. Chim. phys.*, 1973, **70**, 1369.
[40] R. Tabacchi, *Helv. Chim. Acta*, 1974, **57**, 324.

4 Chemical Properties of Thiazoles

Electrophilic Substitution.—2-Amino-thiazoles and their *N-*[41] or 4-substituted derivatives [41-44] and 2-aryl-thiazoles and their 4-substituted derivatives [45] undergo electrophilic substitution predominantly or exclusively in the 5-position. 2-Aryl-thiazoles may undergo substitution in the non-heteroaromatic ring, particularly (as in the case of 5-ethyl-2-phenylthiazole [46]) if the 5-position is blocked or if the non-heteroaromatic ring is activated.[45] Mononitration of 2-(2-thienyl)thiazole occurs exclusively in the vacant α-position of the thiophen ring, and 2-(3-thienyl)-thiazole similarly gives the mononitro-compound (19).[47] The situation is not

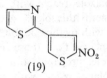

(19)

always straightforward, however. Thus, when the bromination of 4-chloro-methyl-2-phenyl- or 4-chloromethyl-2-(*p*-tolyl)-thiazole is carried out in the presence of silver sulphate, complex formation preferentially activates the aryl ring, but in the absence of the silver salt, substitution occurs predominantly in the 5-position of the thiazole ring.[45] Nitration of this *p*-tolyl compound with acetyl nitrate at ambient temperature gives the 5-substituted derivative, whilst substitution occurs preferentially in the non-heteroaromatic ring at 60 °C with the same reagent.[45] The first detailed kinetic studies of electrophilic substitution of thiazoles show [48, 49] that 2,4- and 2,5-dialkyl-thiazoles are nitrated *via* their protonated forms at all acidities. The 5-position is more reactive than the 4-position by a factor of two. A conjugate-acid mechanism is followed also with phenyl-substituted thiazoles and 2-methoxy-4-methylthiazole, but 4-methyl- and 3,4-dimethyl-2-thiazolone are nitrated *via* their free bases.[48]

It is noteworthy that perchlorothiazole may be prepared by chlorination of 2,4-dichlorothiazole with chlorine in the presence of antimony trichloride.[50]

Nucleophilic Substitution.—The reactivities of 2-halogeno-thiazoles with arenethiols are explained by the existence of an acid–base equilibrium preceding the substitution step; the formation of an ion-pair or of thiazolium and arene-thiolate ions is the rate-determining step.[51] Some 4- or 5-substituted 2-chloro-thiazoles apparently react with methoxide ion in a normal aza-activated nucleo-philic substitution, the reaction being influenced by the substituent. In the case

[41] J. Boedeker, S. Hauser, U. Selle, and H. Koeppel, *J. prakt. Chem.*, 1974, **316**, 881.
[42] G. Vernin, M. A. Leberton, H. J. M. Dou, J. Metzger, and G. Vernin, *Bull. Soc. chim. France*, 1974, 1085.
[43] F. Gagiu, Gh. Csavassy, and Al. Cacoveanu, *Farmacia (Bucharest)*, 1974, **22**, 305 (*Chem. Abs.*, 1975, **82**, 43 234).
[44] T. S. Croft and J. J. McBrady, *J. Heterocyclic Chem.*, 1975, **12**, 845.
[45] I. Simiti and M. Farkas, *Acta Chim. Acad. Sci. Hung.*, 1974, **83**, 381.
[46] A. Benkó and I. Rotaru, *Monatsh.*, 1975, **106**, 1027.
[47] P. Chauvin, J. Morel, P. Pastour, and J. Martinez, *Bull. Soc. chim. France*, 1974, 2099.
[48] A. R. Katritzky, C. Ögretir, H. O. Tarhan, H. M. Dou, and J. V. Metzger, *J.C.S. Perkin II*, 1975, 1614.
[49] S. Ilkay and H. O. Tarhan, *Chim. Acta Turc.*, 1973, **1**, 123 (*Chem. Abs.*, 1975, **82**, 15 880).
[50] F. E. Herkes, U.S.P. 3 907 819/1975 (*Chem. Abs.*, 1975, **83**, 206 248).
[51] M. Bosco, V. Litturi, L. Troisi, L. Forlani, and P. E. Todesco, *J.C.S. Perkin II*, 1974, 508.

of 2-chloro-5-nitrothiazole, however, a good yield of product is obtained only if methoxide ion is not used in excess.[52] Otherwise the two Meisenheimer complexes (20) and (21) are formed, which decompose to give mainly unidentified

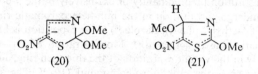

(20)   (21)

products. 2-Chloro-5-nitrothiazole reacts with methoxide ion *ca.* 1.6×10^7 times faster than 2-chloro-5-methylthiazole.[52] A close study of the reactions of 2,4- and 2,5-dichlorothiazoles with methoxide ion has revealed that the 2-chloro-substituent is more easily displaced than those in the 4- or 5-position in these compounds.[52] Perfluorothiazole, prepared from the perchloro-compound and potassium fluoride in sulpholane, reacts with sodium hydrogen sulphide by displacement of the 2-fluorine atom.[53] The presence of a 2-amino-group in a 5-bromo-thiazole increases the reactivity towards nucleophiles compared with the compound having a free 2-position by a factor of *ca.* 10^4.[54] Surprisingly, however, such activation is absent when a 2-ethylamino-group is present. Enhanced reactivity in the former class of compounds has been attributed[54] to the higher reactivity of a tautomer.

Nucleophiles attack 2-alkylthio-thiazolium salts (22) either at C-2, with elimination of the alkylthio-group [*e.g.* (22; R = Me) → (23; X = S) with HS⁻, or → (23; X = NCH₂Ph) with PhCH₂NH₂], or at the C-atom adjacent to

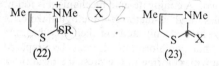

(22)   (23)

sulphur in the alkylthio-group [*e.g.* (22; R = Me) → (23; X = S) with aniline or phenylhydrazine].[55] 2-Acylthio-3,4-dimethylthiazolium fluorosulphonates react similarly with nucleophiles at C-2 (with smaller nucleophiles, *e.g.* HO⁻), but also at the carbonyl group of the side-chain, resulting in deacylation to give the thione [*e.g.* (22; R = COPh, X = SO₃F) → (23; X = S) with aniline or benzylamine].[55]

Thiazolyl Radicals and their Reactions.—The reactions have been reported of a large number of 4- and 5-mono- and 4,5-di-substituted thiazol-2-yl radicals (generated by aprotic diazotization of 2-amino-thiazoles) with substrates such as benzene,[42, 47, 56] pyridine,[56, 57] and thiophen[47] (see Vol. 3, p. 573). The radical

[52] G. Bartoli, O. Sciacovelli, M. Bosco, L. Forlani, and P. E. Todesco, *J. Org. Chem.*, 1975, **40**, 1275.
[53] K. Sasse, G. Beck, and L. Eue, Ger. Offen. 2 344 134/1975 (Chem, Abs., 1975, **83**, 28 217).
[54] L. Forlani, A. Medici, and P. E. Todesco, *Tetrahedron Letters*, 1976, 201.
[55] Y. Gelernt and P. Sykes, *J.C.S Perkin I*, 1974, 2610.
[56] G. Vernin, H. J. M. Dou, J. Metzger, and G. Vernin, *Bull. Soc. chim. France*, 1974, 1079.
[57] G. Vernin, M. A. Lebreton, H. J. M. Dou, J. Metzger, and G. Vernin, *Tetrahedron*, 1974, **30**, 4171.

(24) may be prepared by photolysis of a mixture of 2-nitrothiazole and triethyl-silane.[58, 59] Rotation about its C-2—N bond has been studied. The nitroxide N-atom is pyramidal in the transition state, and the radical exhibits a low barrier to rotation.[58] Radical anions generated from various 2-substituted thiazoles with

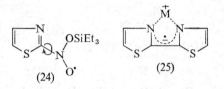

(24) (25)

alkali metals in THF or dimethoxyethane exist as ion pairs.[59] In cases in which the anion contains two adjacent binding sites, formation of a chelate ring [*e.g.* (25)] occurs.

Oxidation and Reduction.—Treatment of 2-mercapto(or methylthio)-4-methyl-thiazole with sodium in liquid ammonia leads to a ring-opening reaction, giving substituted propenethiolates, which are then alkylated with EtBr to give, *e.g.*, EtSCH=CMeN=CHSEt from 2-mercapto-4-methylthiazole. With other 2-substituents (*e.g.* OH, NH₂, NHMe, or NMe₂), ring-opening does not occur. The explanation suggested [60] is that the anion of the mercaptothiazole, having the lowest electron density, is the only member of this series that is capable of giving a radical anion through which ring-opening can occur. The other sub-stituents increase the charge density in the nucleus, so as to render addition of an electron impossible.

Ring Expansion to 1,4-Thiazines.—Takamizawa's ring-expansion reaction of thiazolium salts to 1,4-thiazines by the use of dialkyl acylphosphonates (see Vol. 2, p. 605 and Vol. 3, p. 577) has been extended further.[61]

Alkylation.—3-Alkyl-2-alkylthio-4-methylthiazolium salts may be prepared either by *N*-alkylation of a 2-alkylthio-4-methylthiazole or by *S*-alkylation of the 3-alkyl-4-methyl-Δ⁴-thiazoline-2-thione.[55] In the former case, alkyl exchange in the alkylthio-group can occur. Another problem arises owing to the fact that 2-methylthio-thiazolium salts are dealkylated when heated in ethanol. *N*-Alkyl-ation of 2-alkylthio-4-methylthiazoles with methyl fluorosulphonate overcomes these difficulties.[55] 2-Acylthio-4-methylthiazoles are alkylated normally with methyl fluorosulphonate to give the corresponding *N*-methylthiazolium salt, but successive deacylation and alkylation occur with methyl iodide to give 3,4-di-methyl-2-methylthiothiazolium iodide.[55]

Compound (26) is *O*-methylated by diazomethane in DMF on the carbonyl oxygen atom.[62] Compound (27; X = O) reacts similarly with the same reagent in dioxan, but, in DMF, compound (28; X = O) is formed also. Likewise, compound (28; X = NH) is formed in DMF from compound (27; X = NH).[62]

[58] A. Alberti, M. Guerra, G. F. Pedulli, and M. Tiecco, *Gazzetta*, 1974, **104**, 1301.
[59] G. F. Pedulli, P. Zanirato, A. Alberti, and M. Tiecco, *J.C.S. Perkin II*, 1975, 293.
[60] S. Hoff and A. P. Blok, *Rec. Trav. chim.*, 1974, **93**, 18.
[61] A. Takamizawa and H. Harada, *Chem. and Pharm. Bull. (Japan)*, 1974, **22**, 2818.
[62] H. Schaefer and K. Gewald, *J. prakt. Chem.*, 1975, **317**, 771.

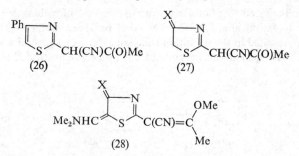

(26)   (27)

(28)

Compounds (28; X = O or NH) are believed to arise by sequential O-methylation and electrophilic attack by the species $Me_2\overset{+}{N}=CHOMe$.

Reactions of Thiazoles with Dimethyl Acetylenedicarboxylate.—A re-examination [63] of the adducts formed between thiazoles and dimethyl acetylenedicarboxylate,[64, 65] using X-ray and [13]C n.m.r. techniques, has been reported. The primary adduct (29; R^1, R^2, R^3 = H, R^4 = CO_2Me) formed from thiazole apparently undergoes rearrangement, either by a [1,5]-sigmatropic shift or *via* the vinyl sulphide, to give compound (29; R^1 = CO_2Me, R^2, R^3, R^4 = H).[63] The compounds 2-, 4-, and 5-mono- and 2,5-di-methylthiazole yield similar products (29; R^1 = CO_2Me,

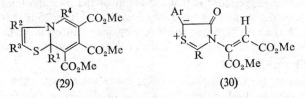

(29)   (30)

R^2, R^3, R^4 = H or Me).[63] By contrast, the adduct formed from 2,4-dimethylthiazole has structure (29; R^1 = R^2 = Me, R^3 = H, R^4 = CO_2Me).

5-Aryl-4-hydroxy-thiazoles react with dimethyl acetylenedicarboxylate to give ylides (30; R = H or Ph).[66] These react further with the ester to give intermediates (31; R = H), which lose sulphur to yield a pyridone, or intermediates

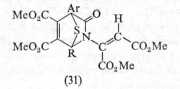

(31)

(31; R = Ph), which fragment to give thiophens. 5-(p-Chlorophenyl)-4-hydroxy-thiazole yields 1 : 1 adducts with dimethyl maleate and dimethyl fumarate.[66]

Organometallic Reagents.—2-Thiazolyl-lithium, prepared from 2-bromothiazole and n-butyl-lithium, reacts with NN-diethylthiazole-2-carboxamide to give

[63] P. J. Abbott, R. M. Acheson, U. Eisner, D. J. Watkin, and J. R. Carruthers, *J.C.S. Chem. Comm.*, 1975, 155; *J.C.S. Perkin I*, 1976, 1269.
[64] D. H. Reid, F. S. Skelton, and W. Bonthrone, *Tetrahedron Letters*, 1964, 1797.
[65] R. M. Acheson, M. W. Foxton, and G. R. Miller, *J. Chem. Soc.*, 1965, 3200.
[66] A. Robert, M. Ferrey, and A. Foucaud, *Tetrahedron Letters*, 1975, 1377.

bis(thiazol-2-yl) ketone,[59] and with 2-nitro-benzaldehydes to give the corresponding secondary alcohols.[67] Metallation of 2-methyl-4-aryl-thiazoles proceeds predominantly in the 5-position, whereas metallation of 2,4-dimethylthiazole occurs in the 2-methyl group [to give (32)].[68] The anions generated at −78 °C reflect the respective kinetic acidities of these positions. At elevated temperatures the thermodynamic acidities prevail, producing the lithiomethyl anions regardless of the nature of the 4-substituent. An apparent primary kinetic isotope effect for the C-5 ring position was determined, which agrees well with the isotope effect for other heterocyclic positions.[68] The lithium compound (32), formed by metallation at −78 °C, dimerizes at ambient temperature (see Vol. 3, p. 578).[69] 2-(2-Thienyl)thiazole is metallated with n-butyl-lithium in the 5-position of the thiazole ring, as shown by incorporation of deuterium at this position on addition of deuterium oxide, and by its other reactions.[47]

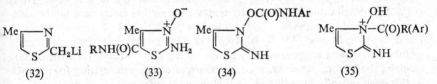

(32)   (33)   (34)   (35)

N-Oxides.—In agreement with calculations, aliphatic isocyanates react with tautomeric 2-amino-4-methylthiazole 3-oxide in the 5-position, to give the amides (33), while aromatic isocyanates react at the *N*-oxide oxygen atom, to give imines (34).[70] This *N*-oxide reacts also with acid chlorides to give the *N*-acyl-ammonium salts (35), which are rearranged by bases. When R is aliphatic, rearrangement results in acylation on the exocyclic N-atom, but when R is aromatic, rearrangement occurs with substitution in the 5-position.[70]

Reactions of 2-Amino-thiazoles.—A number of 2-amino-thiazoles have been diazotized and the resulting diazonium compounds used in synthesis.[71–73] 2-Amino-4-(1- or 2-naphthyl)-thiazoles undergo the Mannich reaction with paraformaldehyde and acetophenone,[74] whilst dehydro-*N*-Mannich bases are formed on heating 2-amino-thiazoles with 1,3,5-triazine in the presence of secondary amines; for example, 2-amino-4-methylthiazole gives compound (36).[75]

Reactions of 2-Hydrazino-thiazoles.—A report in 1971 [76] that 2-hydrazino-5-nitrothiazole reacts with 2-chloroethyl isocyanate at the terminal N-atom has

[67] R. Kalish, E. Broger, G. F. Field, T. Anton, T. V. Steppe, and L. H. Sternbach, *J. Heterocyclic Chem.*, 1975, **12**, 49.
[68] G. Knaus and A. I. Meyers, *J. Org. Chem.*, 1974, **39**, 1192.
[69] G. Knaus and A. I. Meyers, *J. Org. Chem.*, 1974, **39**, 1189.
[70] G. Entenmann, *Tetrahedron*, 1975, **31**, 3131.
[71] E. Gudriniece, I. Gaile, and K. Ziemelis, *Khim. Seraorg. Soedinenii, Soderzh. Neftyakh Nefteprod.*, 1972, **9**, 273 (*Chem. Abs.*, 1975, **82**, 139 990).
[72] W. Knauf, P. Strehlke, and E. Wölk, *European J. Medicin. Chem., Chim. Therap.*, 1975, **10**, 533.
[73] J. Millers and J. Putnins, *Latv. P.S.R. Zinat. Akad. Vestis, Kim. Ser.*, 1975, 499 (*Chem. Abs.*, 1975, **83**, 193 159).
[74] M. Apparao, A. Nayak, and M. K. Rout, *J. Indian Chem. Soc.*, 1975, **52**, 168.
[75] A. Kreutzberger, B. Meyer, and A. Guersoy, *Chem.-Ztg.*, 1974, **98**, 160.
[76] L. M. Werbel and J. R. Battaglia, *J. Medicin. Chem.*, 1971, **14**, 10.

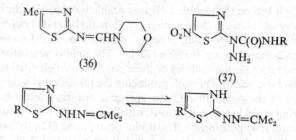

(36)

(37)

(38)

now been shown to be incorrect; in fact the product is the isomeric compound (37; R = CH_2CH_2Cl).[77] Other isocyanates react with 2-hydrazino-5-nitro-thiazole similarly at the exocyclic N-atom next to the ring to give compounds (37). 2-Hydrazinothiazole, however, reacts with methyl isocyanate at the side-chain terminal C-atom.[78] The tautomeric compound 1-isopropylidene-2-(thiazol-2-yl)hydrazine (38; R = H) reacts with methyl isocyanate in methyl cyanide at ambient temperatures by attack on the ring N-atom.[78] At the boiling point of the solvent, however, a 2 : 1 mixture of compounds arising from attack at this position and at the side-chain NH position is formed.

Reactions of Thiazolium Salts.—Thiazolium salts catalyse the addition of aliphatic aldehydes to activated double bonds,[79] whilst asymmetric induction of the benzoin condensation by means of asymmetric thiazolium salts gives products having optical purities as high as 51%.[80] Both reactions envisage the intermediate formation of a secondary alcohol (39) from the aldehyde R^4CHO. A series of papers describe the reactions of thiazolium salts (including thiamine) and thiazolium ylides with heterocumulenes.[81-84] The salts (40; R = Me or CH_2Ph, X = Cl, Br, or I) add to carbodi-imides ArN=C=NAr in the presence of base to give the 1 : 1 adducts (41) and the 1 : 2 adducts (42; X = NAr), the product distribution being largely affected by the substituents in the carbodi-imides and on the thiazolium salt N-atom.[82] By contrast, thiamine reacts with carbodi-imides to give complex products.[83] Addition of the salts (40) to isothiocyanates ArNCS similarly yields either a zwitterionic 1 : 1 adduct (43) or a 1 : 2 adduct (42; X = S), depending on the presence of an electron-withdrawing or electron-donating group, respectively, in the isothiocyanate.[81] 2-Amino-3-hydroxy-thiazolium chlorides [44; R^1 = R^2 = H, alkyl, or aryl; or R^1R^2 = $(CH_2)_4$] undergo reversible adduct formation with alkyl or aryl isocyanates to give lactams (45).[85]

Miscellaneous Reactions.—2-Aryl-thiazoles, obtained by the reaction of thiazoles with dibenzoyl peroxide in the presence of copper, undergo photolysis in the

[77] S. S. Berg and M. P. Toft, *European J. Medicin. Chem., Chim. Therap.*, 1975, **10**, 268.
[78] S. S. Berg, B. J. Peart, and M. P. Toft, *J.C.S. Perkin I*, 1975, 1040.
[79] H. Stetter and H. Kuhlmann, *Angew. Chem. Internat. Edn.*, 1974, **13**, 539.
[80] J. C. Sheehan and T. Hara, *J. Org. Chem.*, 1974, **39**, 1196.
[81] A. Takamizawa, S. Matsumoto, and S. Sakai, *Chem. and Pharm. Bull. (Japan)*, 1974, **22**, 293.
[82] A. Takamizawa, S. Matsumoto, and S. Sakai, *Chem. and Pharm. Bull. (Japan)*, 1974, **22**, 299.
[83] A. Takamizawa and S. Matsumoto, *Chem. and Pharm. Bull. (Japan)*, 1974, **22**, 305.
[84] A. Takamizawa, S. Matsumoto, and I. Makino, *Chem. and Pharm. Bull. (Japan)*, 1974, **22**, 311.
[85] G. Entenmann, *Tetrahedron Letters*, 1974, 3279.

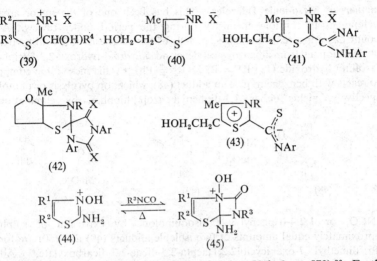

(39)   (40)   (41)

(42)   (43)

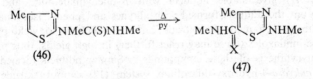

(44)   (45)

presence of iodine to yield 3-aryl-isothiazoles (see Vol. 3, p. 573).[86] Further experiments on these photoisomerizations have been reported [87, 88] involving marking the 4- and 5-positions of 2-phenylthiazole with deuterium or a methyl group. Both groups of workers suggest the intermediacy of bicyclic or tricyclic compounds formed *via* valence-bond isomerizations.

Chloroalkyl-thiazoles have been converted into aldehydes by Sommelet reactions,[89] and a number of condensation and coupling reactions involving the active methylene group in some derivatives of 2-[90] and 4-thiazolylacetonitrile [91] have been reported.

Although 4- and 5-thiazolylglycines have been reported, attempts to prepare 2-thiazolylglycine proved unsuccessful, owing to its instability.[92] 2-Thiazolylacetic acid similarly is unstable, unlike its 4- and 5-isomers.

Thermal isomerization of the thiourea (46) in the presence of pyridine gives thioamide (47; X = S), whose structure was confirmed by oxidation with hydrogen peroxide to the corresponding amide (47; X = O), which was synthesized unambiguously.[93]

(46)   (47)

[86] H. J. M. Dou, G. Vernin, and J. Metzger, *Chim. Acta Turc*, 1974, **2**, 82 (*Chem. Abs.*, 1974, **81**, 152 078).
[87] M. Maeda and M. Kojima, *Tetrahedron Letters*, 1973, 3523.
[88] C. Riou, J. C. Poite, G. Vernin, and J. Metzger, *Tetrahedron*, 1974, **30**, 879.
[89] I. Simiti and G. Hintz, *Pharmazie*, 1974, **29**, 443.
[90] H. Schaefer and K. Gewald, *J. prakt. Chem.*, 1974, **316**, 684.
[91] V. P. Khilya, V. Szabo, L. G. Grishko, D. V. Vikhman, and F. S. Babichev, *Zhur. org. Khim.*, 1973, **9**, 2561 (*Chem. Abs.*, 1974, **80**, 70 747).
[92] M. Hatanaka and T. Ishimaru, *Bull. Chem. Soc. Japan*, 1973, **46**, 3600.
[93] Y. Yamamato, R. Yoda, and C. Tamura, *Chem. Letters*, 1975, 1147.

Reactions of Meso-ionic Thiazoles.—This has been one of the major areas of development in thiazole chemistry during the period under review. A full account has appeared (see Vol. 3, p. 449) of the synthesis of 3,4-dibenzoyl-2,5-diphenylthiophen from dibenzoylacetylene and *anhydro*-4-hydroxy-2,3,5-triphenyl-thiazolium hydroxide (12; $R^1 = R^2 = R^3 = Ph$).[94] This meso-ionic compound also reacts with benzyne to give an adduct (48), which, on pyrolysis or photolysis, respectively, yields either 1,3-diphenylbenzo[c]thiophen by elimination of

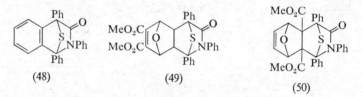

(48)   (49)   (50)

PhNCO, or 1,2,4-triphenyl-3(2*H*)-isoquinolone by extrusion of sulphur.[95] Approximately equal amounts of two isolable adducts (49) and (50) are formed with dimethyl 7-oxabicyclo[2,2,1]hepta-2,5-diene-2,3-dicarboxylate.[96] Adduct (50) may be used to prepare thiophens or pyridones. Meso-ionic thiazole (14; $R^1 = R^3 = Ph$, $R^2 = Me$) similarly reacts with this bicyclic diene to give non-isolable adducts, which decompose with elimination of a furan and carbonyl sulphide to give pyrroles.[96] With various 2,3-diphenylcyclopropenylidenes it gives adducts [51; X = O, S, NTs, $C(CN)_2$, or $C(CN)CO_2Et$], which also extrude carbonyl sulphide with concomitant ring-expansion and formation of a pyridine.[97]

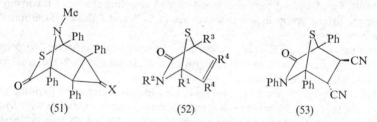

(51)   (52)   (53)

anhydro-2-Aryl-5-hydroxy-3-methylthiazolium hydroxides (14; $R^1 = Ar$, $R^2 = Me$, $R^3 = H$) give stable adducts with *N*-phenylmaleimide and dimethyl fumarate, but the adducts formed with alkynes and phenyl isocyanate or iso-thiocyanate can be made to extrude carbonyl sulphide to yield pyrroles or meso-ionic imidazoles (these may react further) in high yields, thus providing a convenient synthesis of these compounds.[28] Similar adducts formed from the isomeric *anhydro*-4-hydroxy-thiazolium system (12) cannot extrude carbonyl sulphide. Instead they must either lose sulphur or undergo a retro-Diels–Alder reaction with elimination of an isocyanate.[29] For example, with alkynes $R^4C≡CR^4$, the unstable adducts (52) are formed, which extrude isocyanate R^2NCO under remarkably mild conditions to yield thiophens, provided that the

[94] K. T. Potts and D. McKeough, *J. Amer. Chem. Soc.*, 1974, **96**, 4268.
[95] S. Nakazawa, T. Kiyosawa, K. Hirakawa, and H. Kato, *J.C.S. Chem. Comm.*, 1974, 621.
[96] H. Matsukubo and H. Kato, *J.C.S. Chem. Comm.*, 1975, 840.
[97] H. Matsukubo and H. Kato, *J.C.S. Perkin I*, 1975, 632.

starting material (12) contains a 5-phenyl substituent. In the absence of a 5-phenyl group, a 2-pyridone is formed preferentially, by extrusion of sulphur, which suggests that thermal decomposition of the primary adduct (52) is controlled more by steric than by electronic factors. Several stable 1 : 1 adducts [*e.g.* (53), formed with fumaronitrile] obtained from the meso-ionic thiazoles (12) and electron-deficient dipolarophiles can be made to lose hydrogen sulphide in the presence of strong base with the generation of a pyridone.[98]

Irradiation of the meso-ionic thiazole (12; $R^1 = SMe$, $R^2 = Me$, $R^3 = Ph$) results in a novel type of rearrangement for a meso-ionic system, giving rise to the isomeric system 3-methyl-4-methylthio-5-phenyl-2-thiazolone.[99] The kinetics of the reactions of the meso-ionic systems (54) and (55) with chloroacetic acid have been studied.[100]

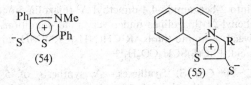

(54)　　　　　　　(55)

5 Synthesis of Δ²-Thiazolines

Type A (S—C—N + C—C) Syntheses.—4-Hydroxy-Δ²-thiazolines are isolable as intermediates in certain Hantzsch reactions (see p. 357). Condensation of oxalyl chloride with hydrazones $R^1R^2C=NNPhC(S)NH_2$ [$R^1 = R^2 = Me$ or $R^1R^2 = (CH_2)_4$] yields the Δ²-thiazoline-4,5-diones (56),[101] whilst the *gem*-dicyano-epoxides (6; $R^1 = Ar$, $R^2 = H$, $R^3 = R^4 = CN$) react with thiourea

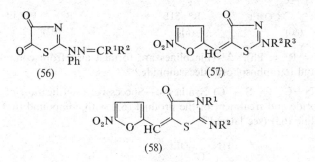

(56)　　　　　　　(57)

(58)

to give 2-amino-5-aryl-Δ²-thiazolin-4-ones.[102] Adducts formed between thiourea or its *N*-methyl and *NN*- or *NN'*-dimethyl derivatives and ethyl (5-nitro-2-furyl)propiolate cyclize readily, to give Δ²-thiazolin-4-ones (57) or 2-imino-thiazolidin-4-ones (58), or a mixture of both.[103]

[98] K. T. Potts, J. Baum, and E. Houghton, *J. Org. Chem.*, 1974, **39**, 3631.
[99] O. Buchardt, J. Domanus, N. Harrit, A. Holm, G. Isaksson, and J. Sandström, *J.C.S. Chem. Comm.*, 1974, 376.
[100] P. B. Talukdar and A. Chakraborty, *Indian J. Chem.*, 1975, **13**, 661.
[101] R. Neidlein and H.-G. Hege, *Chem.-Ztg.*, 1974, **98**, 512.
[102] M. Ferrey, A. Robert, and A. Foucaud, *Compt. rend.*, 1973, **277**, C, 1153.
[103] E. Åkerblom, *Chemica Scripta*, 1974, **6**, 35.

Type B (C—C—N + C—S) Syntheses.—Benzoyl isothiocyanate reacts with
2,2-dimethyl-3-dimethylamino-2*H*-azirine to give the interesting dipolar compound 4,4-dimethyl-Δ^2-thiazoline-5-dimethylimminium-2-benzocarboxamidate
(59), whose structure was established by *X*-ray analysis.[104] The dipolar compound

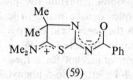

(59)

(59) is stable in organic solvents such as DMSO or chloroform, but the addition
of water converts it into the thiourea PhC(O)NHC(S)NHCMe$_2$C(O)NMe$_2$, while
acid converts it into 2-benzamido-4,4-dimethyl-Δ^2-thiazolin-5-one. A convenient
synthesis of 2-phenyl-Δ^2-thiazolines under very mild conditions consists of the
interaction between α-amino-thiols R^{1}CH(NH$_2$)CR^{2}R^{3}SH and thiobenzoyl-
mercaptoacetic acid, PhC(S)SCH$_2$CO$_2$H.[105]

Type C (C—C—N—C + S) Syntheses.—A synthesis of 5-substituted Δ^2-
thiazoline-4-carboxylates that is useful for the preparation of penicillin derivatives involves the interaction of isocyano-acrylates (60) with hydrogen sulphide.[106]
Depending on the nature of R^1 and R^2, the reaction proceeds *via* either intermediate (61) (as shown by i.r. spectroscopy) or intermediate (62; X = S; isolable

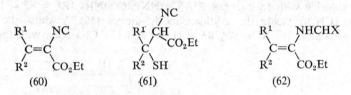

when R^1 = R^2 = Ph). Δ^2-Thiazolines are formed also from compounds (62;
X = O) and tetraphosphorus decasulphide.[106]

Type E (N—C—C—S + C) Syntheses.—Successive esterification of L-cysteine
hydrochloride and treatment of the product (63) with compound (64) yields the
Δ^2-thiazoline (65) (see later).[107]

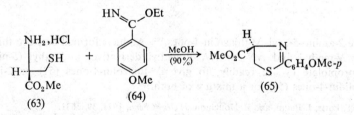

[104] U. Schmid, H. Heimgartner, H. Schmid, P. Schönholzer, H. Link, and K. Bernauer, *Helv.
Chim. Acta*, 1975, **58**, 2222.
[105] N. Suzuki and Y. Izawa, *Tetrahedron Letters*, 1974, 1863.
[106] U. Schöllkopf and D. Hoppe, *Angew. Chem. Internat. Edn.*, 1973, **12**, 1006.
[107] S. Nakatsuka, H. Tanino, and Y. Kishi, *J. Amer. Chem. Soc.*, 1975, **97**, 5010.

Type K (C—C—N—C—S) Syntheses.—A novel synthesis of 2-amino-5-aryl-Δ^2-thiazolines, which occurs in preference to isoquinoline formation, takes place on cyclization of NN'-disubstituted thioureas $ArCH(OMe)CH_2NHC(S)NHR$ with mercuric chloride, phosphorus oxychloride, or polyphosphoric acid.[108] β-Halogenoethyl isothiocyanates react with compounds containing an active hydrogen atom (*e.g.* secondary amines, thiols, alcohols) in the presence of sodium hydride to give Δ^2-thiazolines.[109] Thus, 2-chloro(or bromo)ethyl isothiocyanates and 5-substituted indoles give the Δ^2-thiazolines (66; R = H, OMe, or NO_2).[110] Analogous benzimidazoles may be prepared similarly.[111] A biogenetic-type synthesis of the bicyclic penicillin–cephalosporin antibiotics from an acyclic

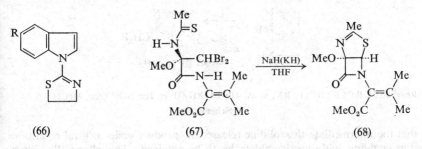

(66)   (67)   (68)

tripeptide equivalent has been reported which involves a double cyclization of compound (67) to form both the thiazoline and β-lactam rings of the β-lactam thiazoline system (68) in one step.[112] We refer elsewhere to the loss of acetonitrile from thiazolidines to yield Δ^2-thiazolines (see p. 378).

6 Chemical Properties of Δ^2-Thiazolines

Photolysis at 253.7 nm in acetonitrile of 2,4- or 2,5-dimethyl-Δ^2-thiazoline gives the same products, namely mixtures of the *N*-alkenyl-thioamides $MeC(S)NH-CH_2CMe{=}CH_2$ and $MeC(S)NHCH{=}CHMe$, apparently *via* the same intermediate, namely the aziridine valence-bond isomer (69) (see Vol. 3, p. 587).[113]

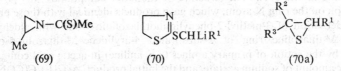

(69)   (70)   (70a)

Other substituted Δ^2-thiazolines behave similarly, although compounds with a 2-substituent (*e.g.* Ph, CH=CHPh, SH, or 3-pyridyl) conjugated with the double bond in the ring are inert.[113] These results explain why indoline, dihydrobenzofurans, and dihydroanthracenes are capable of being photoaromatized by excited

108 A. Mohsen, M. E. Omar, M. S. Ragab, and A. A. B. Hazzaa, *Pharmazie*, 1974, 29, 445.
109 R. E. Hackler and T. W. Balko, *Synthetic Comm.*, 1975, 5, 143.
110 V. L. Narayanan and R. D. Haugwitz, S. African P. 73 01 198/1973 (*Chem. Abs.*, 1975, 83, 58 796).
111 R. D. Haugwitz and V. L. Narayanan, Ger. Offen. 2 446 259/1975 (*Chem. Abs.*, 1975, 83, 79 248).
112 S. Nakatsuka, H. Tanino, and Y. Kishi, *J. Amer. Chem. Soc.*, 1975, 97, 5008.
113 T. Matsuura and Y. Ito, *Tetrahedron*, 1975, 31, 1245.

acetone molecules, but not Δ^2-thiazolines, which are either recovered unchanged or undergo other photochemical reactions, for example conversion into N-vinyl-thioamides, as described above.[114]

Lithium derivatives of 2-substituted Δ^2-thiazolines continue to be of interest in synthesis. Thus the organolithium compounds (70; R^1 = H or Ph), prepared by metallation of 2-methylthio- or 2-benzylthio-Δ^2-thiazoline, react with aldehydes or ketones R^2R^3CO to yield the corresponding carbinols, which are useful because they decompose on being heated, or in the presence of base, to give episulphides (70a) in high yields.[115] Δ^2-Thiazolines have been employed to prepare aldehydes, as shown in Scheme 1.[116] A major feature of this process is

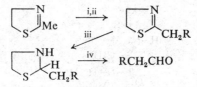

Reagents: i, Bu^{n}Li–THF; ii, RX; iii, Al–Hg–Et$_2$O(H$_2$O); iv, HgCl$_2$–80% aq. MeCN

Scheme 1

that the intermediate thiazolidine releases the product under neutral conditions, thus enabling acid-sensitive aldehydes to be obtained. This allows the hitherto elusive primary adducts of aldol condensations, β-hydroxy-aldehydes R^1R^2C-(OH)CH$_2$CHO, to be synthesized.[117] Extensions of the process allow the synthesis of aldehydes R^1R^2CHCHO and $R^1R^2R^3CCHO$ from 2-methyl-Δ^2-thiazoline by stepwise introduction of two or three alkyl groups,[116] and the synthesis of cyclic aldehydes, such as indane-2-carbaldehyde,[116] and aldehydes such as PhC(O)NHCMeRCHO (R = H or CO$_2$Me).[118] 2-Lithiomethyl-Δ^2-thiazolines react with epoxides to form two isomeric secondary alcohols.[119]

5-Arylidene-2-ethylthio-Δ^2-thiazolin-4-ones react with N-methylaniline by displacement of the ethylthio-group.[120] With aniline, however, addition occurs across the endocyclic double bond, and is followed by elimination of ethanethiol to give the imines (71). The structure of the latter compounds was proved by alkylation on the ring N-atom, which gave products identical with those prepared by condensation of 3-methyl-2-phenyliminothiazolidin-4-one with aldehydes.[120]

The Δ^2-thiazoline ring of 2-substituted 4-arylidene-Δ^2-thiazolin-5-ones is opened by the action of primary amines (*e.g.* aniline) in acetic acid containing a catalytic amount of sodium acetate, and the initial product, ArCH=C(CONHPh)-NHC(S)R, cyclizes with the formation of imidazoles (72) and (73), respectively.[121] The Δ^2-thiazoline-4,5-diones [56; $R^1R^2 = (CH_2)_{4-7}$] rearrange on

[114] T. Matsuura and Y. Ito, *Bull. Chem. Soc. Japan*, 1974, **47**, 1724.
[115] C. R. Johnson, A. Nakanishi, N. Nakanishi, and K. Tanaka, *Tetrahedron Letters*, 1975, 2865.
[116] A. I. Meyers and J. L. Durandetta, *J. Org. Chem.*, 1975, **40**, 2021.
[117] A. I. Meyers, J. L. Durandetta, and R. Munavu, *J. Org. Chem.*, 1975, **40**, 2025.
[118] T. Okutome, Y. Sakurai, M. Kurumi, H. Kawamura, S. Sato, and K. Yamaguchi, *Chem. and Pharm. Bull. (Japan)*, 1975, **23**, 48.
[119] A. I. Meyers and E. D. Mihelich, *Heterocycles*, 1974, **2**, 181.
[120] S. M. Ramsh, G. S. Antonova, A. I. Ginak, and E. G. Sochilin, *Zhur. org. Khim.*, 1975, **11**, 1755 (*Chem. Abs.*, 1975, **83**, 193 156).
[121] A. M. Khalil, I. I. Abd El-Gawad, and M. Hammouda, *Austral. J. Chem.*, 1974, **27**, 2035.

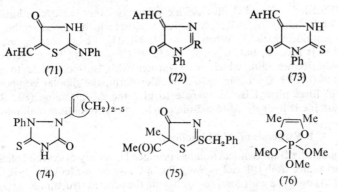

(71)   (72)   (73)

(74)   (75)   (76)

heating to give the triazolidines (74).[101] Conversion of 2-benzylthio-Δ^2-thiazoline-4,5-dione into the 5,5-disubstituted compound (75) may be effected with the biacetyl–trimethyl phosphite adduct (76).[122]

7 Δ^3-Thiazolines

These compounds are relatively unknown. Ketones react with a mixture of sulphur and ammonia to give a mixture of products, depending on the temperature of reaction and the ratio of ketone to sulphur. With a ketone : sulphur ratio of 1 : 8 at $-70\,°C$, the main product from ethyl methyl ketone is an imidazole, with the Δ^3-thiazoline (77) as a by-product, whereas a ketone : sulphur ratio of 2 : 1 produces compound (77) in good yield.[123] 2-Alkyl-substituted Δ^3-thiazolines may be prepared by allowing an aldehyde to react with ammonia (or an ammonium salt) and 2-mercaptoacetaldehyde, or its disulphide,[124] or dimer[125] (78), and they are aromatized to thiazoles with DDQ.[125] The insecticidal and herbicidal Δ^3-thiazoline (79) arises on heating either $CCl_3CCl_2CCl=NCCl_2CCl_3$ or 4-chloro-2,5-bis(trichloromethyl)thiazole with sulphur at $220—240\,°C$.[126]

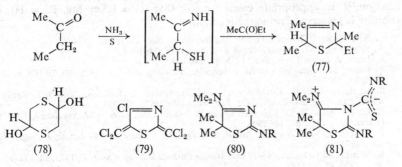

(78)   (79)   (80)   (81)

[122] F. Ramirez, C. D. Telefus, and V. A. V. Prasad, *Tetrahedron*, 1975, **31**, 2007.
[123] F. Asinger and W. Leuchtenberger, *Annalen*, 1974, 1183.
[124] P. Dubs and M. Pesaro, Swiss P. 563 379/1975 (*Chem. Abs.*, 1975, **83**, 179 043); Swiss P. 565 516/1973 (*Chem. Abs.*, 1975, **83**, 206 244); Swiss P. 565 515/1975 (*Chem. Abs.*, 1975, **83**, 206 243).
[125] P. Dubs and M. Pesaro, *Synthesis*, 1974, 294.
[126] G. Beck and H. Holtschmidt, Ger. Offen. 2 331 795/1975 (*Chem. Abs.*, 1975, **82**, 170 883).

2,2-Dimethyl-3-dimethylaminoazirine undergoes thermal cycloaddition reactions with isothiocyanates RNCS to give products which depend on the structure of the isothiocyanate.[127]   When R = p-$O_2NC_6H_4$, PhCO, or Ts, an isolable zwitterion is formed, but, when R = Me, $PhCH_2$, or Ph, these cyclize to give the Δ^3-thiazolines (80), which react further with isothiocyanate to give thiazolidines (81).[127] On being heated in chloroform, the dipolar compound (81; R = Ph) loses phenyl isothiocyanate to give the Δ^3-thiazoline (80; R = Ph). This enamine is hydrolysed to 5,5-dimethyl-2-phenyliminothiazolidin-4-one.[127]

8 Synthesis of Δ^4-Thiazolines

Once again, the Hantzsch reaction has produced a variety of Δ^4-thiazolines.[128-134] Isothiocyanates R^1NCS (R^1 = Me, Ph, o-, m-, or p-MeC_6H_4, p-BrC_6H_4, or p-ClC_6H_4) condense with compounds having the general structure $R^2C(O)CH_2CN$ [*e.g.* ethyl cyanoacetate (R^2 = OEt), *N*-substituted cyanoacetamides (R^2 = NH_2, NHPh, morpholino, or piperidino)] in DMF in the presence of base, to give the corresponding 4-amino-Δ^4-thiazoline-2-thione (82).[135] Thiazolidine-2,4-dione is converted into 4-chloro-2-oxo-Δ^4-thiazoline-5-carbaldehyde (83)

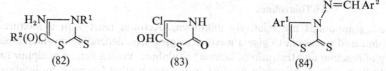

(82)   (83)   (84)

with phosphorus oxychloride in DMF.[131]   4,5-Diphenyl-Δ^4-thiazolin-2-one is obtained from PhC(O)CH(SCN)Ph on treatment with sulphuric acid.[136] 4-Amino-5-cyanomethyl-Δ^4-thiazoline-2-thiones may be prepared by addition of RNHC-(S)SH to fumaronitrile,[137] whilst cyclization of the dithiocarbazic esters $Ar^1C(O)$-$CH_2SC(S)NHN=CHAr^2$ gives the Δ^4-thiazolines (84).[138]

Thiazino-oxazolidinones (85; $R^1 = R^2$ = H or Me, R^3 = Me) are converted, in the presence of strong bases, into the Δ^4-thiazolines (86) *via* an initial elimination to give an enethiolate ion, which cyclizes, probably by an internal S_N2 reaction.[139] In appropriate cases, *e.g.* for (85; $R^1 = R^2$ = Me, R^3 = H), the chirality is preserved during the reaction.

[127] E. Schaumann, E. Kausch, and W. Walter, *Chem. Ber.*, 1974, **107**, 3574.
[128] A. Singh and A. S. Uppal, *Austral. J. Chem.*, 1975, **28**, 1049.
[129] V. Jiram, *J. Indian Chem. Soc.*, 1975, **52**, 240.
[130] F. Russo, M. Santagati, and M. Alberghina, *Farmaco, Ed. Sci.*, 1975, **30**, 70 (*Chem. Abs.*, 1975, **82**, 139 995).
[131] S. N. Baranov, R. O. Kochkanyan, A. N. Zaritovskii, G. I. Belova, and S. S. Radkova, *Khim. geterotsikl. Soedinenii*, 1975, 85 (*Chem. Abs.*, 1975, **83**, 9899).
[132] R. G. Dubenko, V. D. Konysheva, and P. Pelkis, *Khim. geterotsikl. Soedinenii*, 1975, 650 (*Chem. Abs.*, 1975, **83**, 114 299).
[133] G. L'Abbé, E. Van Loock, R. Albert, S. Toppet, G. Verhelst, and G. Smets, *J. Amer. Chem. Soc.*, 1974, **96**, 3973.
[134] S. Bilinski and L. Bielak, *Ann. Univ. Mariae Curie-Sklodowska, Sect. D*, 1973, **28**, 171 (*Chem. Abs.*, 1975, **82**, 57 618).
[135] M. B. Devani, C. J. Shishoo, S. D. Patel, B. Mukherji, and A. C. Padhya, *Indian J. Chem.*, 1975, **13**, 532.
[136] T. I. Temnikova, V. V. Kashina, and A. S. Atavin, *Zhur. org. Khim.*, 1973, **9**, 2140 (*Chem. Abs.*, 1974, **80**, 27 155).
[137] H. Nagase, *Chem. and Pharm. Bull.* (*Japan*), 1974, **22**, 505.
[138] Y. Usui, T. Iwatani, and I. Aoki, Japan. Kokai 74 20 309/1974 (*Chem. Abs.*, 1975, **82**, 155 838).
[139] A. G. W. Baxter and R. J. Stoodley, *J.C.S. Chem. Comm.*, 1975, 251.

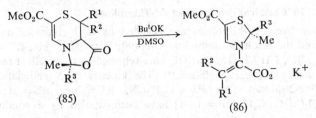

(85)   (86)

4-Alkyl-5-sulphonylimino-1,2,3,4-thiatriazolines react by cycloaddition with ynamines, with subsequent loss of nitrogen, to give Δ^4-thiazolines, *e.g.* (87) → (88; R^1 = Me, R^2 = NEt_2).[133] The reaction is regiospecific, in that the amine

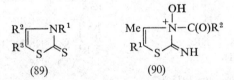

(87)   (88)

function is placed in the 4-position (see also p. 376). Keto-stabilized phosphorus ylides $[R^2C(O)CH=PPh_3 \leftrightarrow R^2C(O^-)=CH\overset{+}{P}Ph_3]$ react similarly, giving intermediates which undergo a Wittig-type elimination to give the Δ^4-thiazolines (88; R^1 = H, R^2 = Me, Ph, or p-$O_2NC_6H_4$).[133] A detailed discussion of the mechanism of these reactions is given in the original paper.

9 Physical Properties of Δ^4-Thiazolines

Hindered rotation ('gear effect') (see also p. 360) in a number of Δ^4-thiazoline-2-thiones [*e.g.* (89; R^1 = Et, R^2 = CMe_3, R^3 = Me)] has been studied, using a ^1H n.m.r. spectroscopic technique (this is the first evidence for restricted rotation of ethyl groups),[140] and a complete n.m.r. bandshape analysis has been carried out on 3,4-di-isopropyl-Δ^4-thiazoline-2-thione (89; R^1 = R^2 = Pr^i, R^3 = H) in

                  OH
                  |
R^2⌐—NR^1 Me⌐—$\overset{+}{N}$—$C(O)R^2$

R^3⌐—S—S R^1⌐—S—NH

(89)   (90)

order to study the interaction of the two intermeshing groups.[141] A study of the long-range ^{31}P–^1H coupling constants in twenty-one 2-imino-3-methyl-Δ^4-thiazolines that are phosphorylated on the imino-group suggests that coupling is associated with an extended planar zig-zag configuration.[142] Photoelectron spectra of Δ^4-thiazoline-2-thione and several of its derivatives have been recorded and analysed.[143]

[140] R. Gallo, A. Liden, C. Roussel, J. Sandström, and J. Metzger, *Tetrahedron Letters*, 1975, 1985.
[141] R. E. Carter, T. Drakenberg, and C. Roussel, *J.C.S. Perkin II*, 1975, 1690.
[142] C. K. Tseng and A. Mihailovski, *Org. Magn. Resonance*, 1974, 6, 494.
[143] C. Guimon, G. Pfister-Guillouzo, M. Arbelot, and M. Chanon, *Tetrahedron*, 1974, 30, 3831.

10 Chemical Properties of Δ⁴-Thiazolines

2-Imino-3-hydroxy-Δ^4-thiazolines are acylated by acyl chlorides on the ring N-atom, and the resulting amminium salts (90; R^1 = H or Me, R^2 = Me, Ph, CH=CH₂, Buᵗ, CH₂Cl, OMe, or OEt) are reported to be excellent reagents for acylation and alkoxycarbonylation.[144] The kinetics of S-methylation of the Δ^4-thiazoline-2-thiones [89; R^1 = $(CH_2)_nX$, R^2 = Me, R^3 = H; X = H, Ph, or 3,4-(MeO)₂C₆H₃; n = 1—4] have been studied by a conductometric technique.[145]

11 Synthesis of Thiazolidines

Type A (S—C—N + C—C) Syntheses.—A further variety of 2-iminothiazolidin-4-ones has been prepared, mainly by Indian workers, by the well-established procedure of allowing an α-chloro-carboxylic acid (usually chloroacetic acid) to react with a N-mono- or NN'-di-substituted thiourea.[129, 146—155] Condensation of dithiobiuret with 1,2-dibromoethane in boiling ethanol gives 2-imino-3-thiocarbamoylthiazolidine.[156] 2,2,2-Trichloro-1-phenylethanol reacts with thiosemicarbazide in the presence of base to give a mixture of dihydro-2-imino-6-phenyl-2H-1,3,4-thiadiazin-5(6H)-one (18%) and 5-phenylthiazolidine-2,4-dione-2-hydrazone (91; Ar = Ph, R = NH₂) (10%), together with sixteen unidentified compounds.[20] The reaction presumably proceeds *via* the epoxide (6; R^1 = Ph, R^2 = H, R^3 = R^4 = Cl) and the intermediate PhCH(COCl)SC(=NH)-NHNH₂, which can cyclize at one of three N-atoms (see also p. 358). Acetone and benzaldehyde thiosemicarbazones react similarly with 2,2,2-trichloro-1-aryl-ethanols to give thiazolidin-4-ones (91; Ar = Ph, p-ClC₆H₄, or p-MeOC₆H₄; R = N=CR¹R²) (*ca.* 65%), which prefer to exist as thiazolidines rather than the tautomeric Δ^2-thiazolines.[20] Cyclization to six-membered rings is excluded in these cases. 4-Alkyl-5-sulphonylimino-1,2,3,4-thiatriazolines react with enamines R¹₂NCR²=CR³R⁴ to give thiazolidines, *e.g.* (87) → (92).[133] The reaction is regiospecific in that the amino-group is always placed in the 4-position

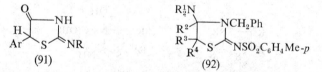

(91)   (92)

[144] G. Entenmann, *Synthesis*, 1975, 377.
[145] K. Datta, C. Roussel, and J. Metzger, *Bull. Soc. chim. France*, 1974, 2135.
[146] R. P. Rao and S. R. Singh, *J. Indian Chem. Soc.*, 1973, **50**, 492.
[147] K. F. Modi and J. J. Trivedi, *J. Indian Chem. Soc.*, 1973, **50**, 564.
[148] P. N. Dhal and A. Nayak, *J. Inst. Chem., Calcutta*, 1973, **45**, 127 (*Chem. Abs.*, 1974, **80**, 82 771).
[149] P. N. Dhal, T. E. Achary, A. Nayak, and M. K. Rout, *J. Indian Chem. Soc.*, 1973, **50**, 680.
[150] S. K. Chaudhari, M. Verma, A. K. Chaturvedi, and S. S. Parmar, *J. Pharm. Sci.*, 1975, **64**, 614.
[151] P. N. Dhal, T. E. Achary, and A. Nayak, *Indian J. Chem.*, 1975, **13**, 753.
[152] A. A. Shaikh, *J. Inst. Chem., Calcutta*, 1974, **46**, 135 (*Chem. Abs.*, 1975, **82**, 125 312).
[153] R. Shyam and I. C. Tiwari, *Agric. and Biol. Chem. (Japan)*, 1975, **39**, 715.
[154] P. M. Pawar and R. A. Pawar, *J. Univ. Poona, Sci. Technol.*, 1974, **46**, 57.
[155] G. L'abbé, G. Verhelst, C.-C. Yu, and S. Toppet, *J. Org. Chem.*, 1975, **40**, 1728.
[156] Y. Yamamoto, R. Yoda, and S. Kouda, *Kyoritsu Yakka Daigaku Kenkyu Nempo (Japan)*, 1973, 46 (*Chem. Abs.*, 1974, **81**, 120 524).

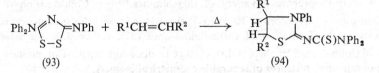

(93)   (94)

of the product. Compound (87) reacts similarly with ketens to give 5-mono- and 5,5-di-substituted thiazolidin-4-ones.[155] Imino-dithiazoles react with alkenes to give a corresponding imino-thiazolidine, *e.g.* (93) → (94; $R^1 = R^2 = CO_2Et$; $R^1 = H$, $R^2 = CN$ or CO_2Me).[157]

Type B (C—C—N + C—S) Syntheses.—The product obtained in 1889 by Gabriel when 2-bromoethylamine reacted with an excess of methyl isothiocyanate, believed to be *NN'*-dimethyl-*N'*-(Δ^2-thiazolin-2-yl)thiourea, has now been shown[158] to be 3-*N*-methylthiocarbamoyl-2-methyliminothiazolidine (95; $R^1 = R^2 = Me$). It is hydrolysed by acid to 3-*N*-methylthiocarbamoylthiazolidin-2-one, which may be prepared from thiazolidin-2-one and methyl isothiocyanate. 3-*N*-Alkylthiocarbamoyl-2-alkylimino-thiazolidines (95; $R^1 = Me$, $R^2 = Et$) and (95; $R^1 = Et$, $R^2 = Me$) may be prepared by stepwise reaction of 2-bromo-ethylamine with two different isothiocyanates.[158] Also contrary to previous

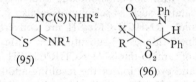

(95)   (96)

reports, methyl isothiocyanate has been shown to react with 2-aminoethyl hydrogen sulphate, $H_2N(CH_2)_2OSO_3H$, to give compound (95; $R^1 = R^2 = Me$); 2-methylthio-Δ^2-thiazoline reacts sequentially with methylamine and methyl isothiocyanate to give the same compound, *via* 2-methylamino-Δ^2-thiazoline.[158] The reaction of 1-arylsulphonyl-aziridines with alkyl, alkenyl, or aryl isothiocyanates in the presence of sodium iodide gives 2-alkyl- or 2-aryl-imino-3-arylsulphonyl-thiazolidines.[159]

Type C (C—C—N—C + S) Syntheses.—Benzylideneaniline reacts with α-halogeno- or α-alkylthio-acyl chlorides RCHXC(O)Cl, in the presence of sulphur dioxide and base, to give mixtures of a thiazolidine [96: X and R = Cl, Cl; Br, Br; Cl, H; Cl, Bu^n; Cl, Ph; S(CH$_2$)$_2$S; S(CH$_2$)$_3$S; SEt, SEt; or SMe, H] and an azetidin-2-one.[160] The reaction is considered to proceed *via* the generation of a keten and the formation of betaines, which either cyclize or react further with sulphur dioxide.

Type D (C—C—S + C—N) Syntheses.—Extensions of the well-known reactions between Schiff bases and 2-mercaptoacetic acid (see Vol. 3, pp. 602, 603) have

[157] J. Goerdeler and H. W. Linden, *Tetrahedron Letters*, 1975, 3387.
[158] Y. Yamamoto, R. Yoda, and M. Matsumura, *Chem. and Pharm. Bull. (Japan)*, 1975, **23**, 2134.
[159] V. I. Markov and D. A. Danileiko, *Zhur. org. Khim.*, 1974, **10**, 1262 (*Chem. Abs.*, 1974, **81**, 105 381).
[160] D. Belluš, *Helv. Chim. Acta*, 1975, **58**, 2509.

been used to prepare a variety of thiazolidines.[161-164] Condensation of the imidic ester hydrochlorides $RCH_2C(OEt)=\overset{+}{N}H_2\ Cl^-$ with 2-mercaptoacetamide results in the formation of thiazolidin-4-ones (97), rather than the isomeric Δ^2-thiazolin-4-ones.[165] $MeC(OEt)=\overset{+}{N}H_2\ Cl^-$ does not react. Compounds (97) are obtained as mixtures of separable geometrical isomers.

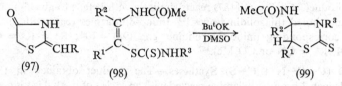

(97)   (98)   (99)

2-Acetamido-alkenethiolates $MeC(O)NHCR^2=CR^1S^-$ condense with iso-thiocyanates R^3NCS to give dithiocarbamates (98).[166] In the presence of a strong base these are converted into thiazolidine-2-thiones (99). When compound (99; R^1 = H, R^2 = R^3 = Me) is treated with trifluoroacetic acid, elimination of acetamide occurs, with formation of 3,4-dimethyl-Δ^4-thiazoline-2-thione.[166]

Type E (N—C—C—S + C) Syntheses.—High yields of 4-substituted thiazolidin-2-ones are formed when the amino-thiols $H_2NCHRCH_2SH$ or their disulphides are treated successively with carbon monoxide and oxygen in the presence of triethylamine and a catalyst such as selenium (see also p. 386).[167] In a similar manner the amino-thiols $RNHCH_2CMe_2SH$ are converted into 3-substituted 5,5-dimethylthiazolidine-2-thiones (3-substituent = p-Me$_2$NC$_6$H$_4$ or p-PhNH-C$_6$H$_4$).[168] Addition of amino-thiols $H_2NCHR^3CH_2SH$ to the cyano-allenes $R^1R^2C=C=CHCN$ and cyclization of the resulting adducts in the presence of base yields thiazolidines (100).[169] Distillation of the products at atmospheric pressure results in loss of acetonitrile, with formation of 2,4-disubstituted Δ^2-thiazolines.

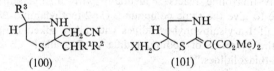

(100)   (101)

Type K (C—C—N—C—S) Syntheses.—The adduct of allyl isothiocyanate and dimethyl malonate cyclizes in the presence of halogen (Cl$_2$, Br$_2$, or I$_2$) to give

161 H. Lettau, H. Matschiner, R. Matuschke, and J. Hauschild, East Ger. P. 107 047/1974 (*Chem. Abs.*, 1975, **82**, 72 976).
162 R. R. Astik, G. B. Joshi, and K. A. Thaker, *J. Indian Chem. Soc.*, 1975, **52**, 1071.
163 M. Susse and H. Dehne, *Z. Chem.*, 1975, **15**, 268.
164 M. S. Raasch, *J. Heterocyclic Chem.*, 1974, **11**, 587; U.S.P. 3 853 902/1974 (*Chem. Abs.*, 1975, **82**, 112 059).
165 O. Ceder, U. Stenhede, K.-I. Dahlquist, J. M. Waisvisz, and M. G. van der Hoeven, *Acta Chem. Scand.*, 1973, **27**, 1914.
166 S. Hoff and A. P. Blok, *Rec. Trav. chim.*, 1974, **93**, 75.
167 N. Sonoda, G. Yamamoto, K. Natsukawa, K. Kondo, and S. Murai, *Tetrahedron Letters*, 1975, 1969; P. Koch and E. Perrotti, Ger. Offen. 2 437 132/1975 (*Chem. Abs.*, 1975, **82**, 156 282); P. Koch and E. Perrotti, *Tetrahedron Letters*, 1974, 2899.
168 J. J. D'Amico and W. E. Dahl, *J. Org. Chem.*, 1975, **40**, 1224.
169 Z. T. Fomum, P. D. Landor, S. R. Landor, and G. B. Mpango, *Tetrahedron Letters*, 1975, 1101.

the corresponding thiazolidine (101),[170] and not a 1,3-thiadiazine derivative as claimed previously.[171]

12 Physical Properties of Thiazolidines

Good agreement has been found between the calculated and experimentally determined ^{13}C n.m.r. chemical shifts for thiazolidine and several of its derivatives.[172] The configurations (*erythro* or *threo*) of the diastereomeric 2-amino-3-mercaptobutyric acids have been determined unambiguously from the derived *N*-formyl-thiazolidines (102), using ^{1}H and ^{13}C n.m.r. spectroscopy.[173] The spectra of the thiazolidines, recorded in $[^{2}H_{6}]DMSO$, reveal the presence of two conformers, arising from restricted rotation of the formyl group, with a low rate of interconversion. From a comparison of the ^{1}H n.m.r. coupling constants of some 2-substituted thiazolidines and of 2-aminoethanethiol derivatives, the former

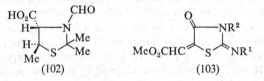

(102)   (103)

are considered to exist in conformations close to envelopes, with either C-4 or C-5 as the 'flap' atom and the 2-substituent *anti* to the flap.[174] Raman and i.r. spectroscopic studies of thiazolidine and its *N*-deuterio-derivative are compatible with the existence of only one half-chair conformation,[175] whilst similar studies of the three monomethyl-thiazolidines suggest the existence of two conformations at ambient temperatures.[176] The structures of the products derived from the reaction of thioureas or *N*-substituted thioureas with dimethyl acetylenedicarboxylate have been the subject of detailed debate; three ring structures have received support. The reaction between *N*-methylthiourea and the ester has been shown to give compounds (103; R^{1} = H, R^{2} = Me) and (103; R^{1} = Me, R^{2} = H). The structure of the former compound has been established by an *X*-ray analysis.[177, 178]

In an attempt to account for the ability of thiazolidine-2,4-dione to inhibit metallic corrosion, its crystal and molecular structures have been studied.[179] A study has been made of the distribution of electronic charge in various thiazolidine-4-carboxylic acids in relationship to their chelating power,[180] and the photoelectron spectra of thiazolidine-2-thione and its *N*-methyl and 4,4-dimethyl

[170] K. S. Dhami, *Indian J. Chem.*, 1975, **13**, 989.
[171] G. Just and P. Rossy, *J. Org. Chem.*, 1972, **37**, 318.
[172] R. Faure, J. R. Llinas, E. J. Vincent, and J. L. Larice, *Compt. rend.*, 1974, **279**, C, 717.
[173] S. Toppet, P. Claes, and J. Hoogmartens, *Org. Magn. Resonance*, 1974, **6**, 48.
[174] G. E. Wilson and T. J. Bazzone, *J. Amer. Chem. Soc.*, 1974, **96**, 1465.
[175] M. Guiliano, G. Davidovics, J. Chouteau, J. L. Larice, and J. P. Roggero, *J. Mol. Structure*, 1975, **25**, 329.
[176] M. Guiliano, G. Davidovics, J. Chouteau, J. L. Larice, and J. P. Roggero, *J. Mol. Structure*, 1975, **25**, 343.
[177] J. F. B. Mercer, G. M. Priestley, R. N. Warrener, E. Adman, and L. H. Jensen, *Synthetic Comm.*, 1972, **2**, 35.
[178] E. Adman, L. H. Jensen, and R. N. Warrener, *Acta Cryst.*, 1975, **31B**, 1915.
[179] G. R. Form, E. S. Raper, and T. C. Downie, *Acta Cryst.*, 1975, **31B**, 2181.
[180] E. Catrina, Z. Simon, and Gh. Catrina, *Rev. Roumaine Chim.*, 1975, **20**, 651.

derivatives have been recorded and analysed.[143] In the mass spectra of thiazolidines, both 1,4- and 2,5-ring-cleavage is apparently common, and a 2-substituent is readily lost, with formation of a stabilized carbonium ion.[181]

13 Chemical Reactions of Thiazolidines

Ring-cleavage Reactions.—The reduction of Δ^2-thiazolines to thiazolidines and the hydrolytic ring-opening of these compounds as a practical route to various aldehydes has been mentioned (p. 372) (see also Vol. 3, pp. 607, 608). The ring-cleavage reactions of 5-arylidene-thiazolidine-2,4-diones and related compounds are complex. The reaction of the parent compound, and its 3-phenyl derivative, with piperidine involves cleavage of the 1,2-bond to give compounds (104; $R^1 = H$ or Ph, $R^2 = NC_5H_{10}$).[182] It now appears that these compounds react similarly with benzylamine (see Vol. 3, p. 607), *i.e.* by cleavage of the 1,2-bond. In the presence of an excess of benzylamine, and after longer reaction times, arylacetic acids are the major isolable products.[182] These were previously thought to arise by an initial cleavage of the 4,5-bond. Hydrazine also reacts with 5-arylidene-thiazolidine-2,4-diones by an initial cleavage of the 1,2-bond, but further reaction of the products with hydrazine is reported[183] to yield triazinones. It has, however, been claimed[184] that the products of similar reactions are 3-aryl-5-pyrazolones. 5-Arylidene-3-phenylthiazolidine-2,4-diones are reported[183] to react with hydrazine to yield the corresponding biuret

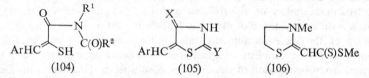

(104)   (105)   (106)

$H_2NNHC(O)NPhC(O)NHN=CHCH_2Ar$. Treatment of the substituted 5-arylidene-thiazolidin-2-ones (105; X = NH or S, Y = O) with hydrazines has been reported[184] to give 1,2,4-triazolin-3-one derivatives. With compounds (105; X = O, Y = S) and hydrazine, however, 2-amino-5-arylidene-Δ^2-thiazolin-4-ones are formed.[184]

With acetic anhydride or dicyclohexylcarbodi-imide, 5-hydroxy-thiazolidine-2-thiones give 2-keto-isothiocyanates $R^1C(O)CR^2R^3NCS$, which are useful starting materials for the synthesis of imidazolines, oxazolidines, and other heterocyclic systems.[185]

Other Reactions.—Further examples of condensation of a 5-methylene group in thiazolidines with aldehydes have been reported.[151, 154, 186, 187] 3-Aryl-2-arylimino-

[181] M. M. Vestling and R. L. Ogren, *J. Heterocyclic Chem.*, 1975, **12**, 243.
[182] A. R. A. Raouf, M. T. Omar, and M. M. El-Attal, *Acta Chim. Acad. Sci. Hung.*, 1974, **83**, 367.
[183] A. R. A. Raouf, M. T. Omar, and M. M. El-Attal, *Acta Chim. Acad. Sci. Hung.*, 1975, **87**, 187.
[184] V. N. Artemov, S. N. Baranov, N. A. Kovach, and O. P. Shvaika, *Doklady Akad. Nauk S.S.S.R.*, 1973, **211**, 1369 (*Chem. Abs.*, 1974, **80**, 3422).
[185] J. C. Jochims and A. Abu-Taha, *Chem. Ber.*, 1976, **109**, 154.
[186] E. V. Vladzimirskaya and B. M. Kirichenko, *Farm. Zhur. (Kiev)*, 1975, **30**, 41 (*Chem. Abs.*, 1975, **83**, 9879).
[187] R. P. Rao and S. R. Singh, *J. Indian Chem. Soc.*, 1973, **50**, 600.

thiazolidin-4-ones react with diazonium salts, diazonium salts plus copper(II) chloride, or mercury(I) acetate, to give the corresponding 5-arylazo-derivatives, 5-aryl derivatives, or derivatives that are mercurated in the aryl groups, respectively.[187] The (*E*)- and (*Z*)-configurations of the 5-arylidene derivatives derived from thiazolidine-2,4-dione [183] and 2-iminothiazolidin-4-ones [108] have been isolated. 3-Substituted 2-iminothiazolidin-4-ones undergo ring-expansion with carbon disulphide, to give thiazines.[188]

Methyl 3-methyl-2-methylenethiazolidine-α-dithiocarboxylate (106) reacts with dimethyl acetylenedicarboxylate to give the spiro(thiazolidine cyclopentadiene) (107),[189] whilst *o*-aminobenzenethiol reacts with 3-substituted 5-methoxycarbonyl-methylene-2-oxo-(or thioxo-)thiazolidin-4-ones to give compounds (108; X = O

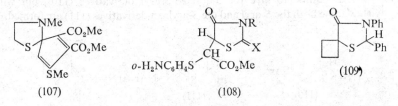

(107)   (108)   (109)

or S, R = Me or CH₂Ph), which cyclize on heating, with loss of methanol, to the benzo-1,4-thiazinone.[190] Thiazolidines are oxidized by hydrogen peroxide in a mixture of acetic acid and acetic anhydride to the corresponding sulphoxides.[191]

Reductive dehalogenation of compound (96; X = Cl, R = H) or treatment of compound (96; X = H, R = SMe) with Raney nickel and alkylation of the product (96; X = R = H) with 1,3-dibromopropane in the presence of sodium hydride yields the spiro-compound (109).[160] Compound (96; X = Cl, R = H) is alkylated similarly by methyl iodide, and reductive dehalogenation of the product yields compound (96; X = H, R = Me). The 5,5-dichloro-compound (96; X = R = Cl) undergoes extrusion of sulphur dioxide on being heated, to give an azetidin-2-one.

14 Rhodanines, Isorhodanines, and Thiorhodanines

The reactions of isothiocyanates with α-mercaptoacetic acid continue to be the major method for the synthesis of rhodanines.[192-195] A facile and uniform procedure for the large-scale preparation of some *N*-alkyl-rhodanines has been reported, which involves allowing a primary amine RNH₂ (R = Me, Et, allyl, or PhCH₂) to react with carbon disulphide and ammonia, followed by treatment of the resulting ammonium alkyldithiocarbamate RNHC(S)S⁻ NH₄⁺ with chloro-

[188] E. Fischer, I. Hartmann, and H. Priebs, East Ger. P. 108 991/1974 (*Chem. Abs.*, 1975, **83**, 58 854).
[189] G. Kobayashi, Y. Matsuda, Y. Tominaga, and K. Mizuyama, *Chem. and Pharm. Bull. (Japan)*, 1975, **23**, 2749.
[190] H. Nagase, *Chem. and Pharm. Bull. (Japan)*, 1974, **22**, 42.
[191] M. Susse and H. Dehne, *Z. Chem.*, 1975, **15**, 303.
[192] M. Dzurilla and E. Demjanova, *Chem. Zvesti*, 1974, **28**, 693.
[193] V. Knoppová, M. Uher, and M. Hroboňová, *Zb. Pr. Chemickotechnol. Fak. SVST*, 1972, 61 (publ. 1974) (*Chem. Abs.*, 1975, **82**, 72 868).
[194] M. Uher, V. Knoppová, and P. Zálupský, *Chem. Zvesti*, 1975, **29**, 240.
[195] V. Knoppová and L. Drobnica, *Coll. Czech. Chem. Comm.*, 1974, **39**, 1589.

acetic acid.[196] A similar method has been used for the synthesis of the 3,5-disub-
stituted rhodanines.[197]

The i.r. spectra of rhodanine, isorhodanine, and thiorhodanine, together with
those of several derivatives, have been studied in detail.[198-202]

N-Aryl-rhodanines are readily brominated in the 5-position with bromine in
acetic acid.[203] Rhodanine is arylated by 2,4-dinitrochlorobenzene, in the
presence of base, exclusively on nitrogen,[204] and it condenses, as do its *N*-methyl
and *N*-phenyl derivatives, with various aldehydes to give 5-arylidene derivatives
as mixtures of their separable (*E*)- and (*Z*)-isomers.[183] The reaction of phthalic
anhydride with 3-aryl-rhodanines in acetic anhydride in the presence of sodium
acetate gives the 3-aryl-5-phthalylidene-rhodanines.[205] Arenediazonium salts
react with rhodanine to give the 5-arylazo-*S*-aryl derivatives (110), but with
5-ethylrhodanine both the *S*-aryl and the *S*-arylazo-derivatives (111) are formed.[206]

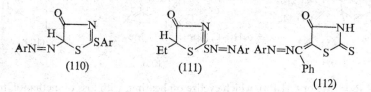

         (110)   (111)   (112)

N-Aryl-rhodanines couple normally in the 5-position.[207] Whereas the sodium
salt of 5-benzylidene-rhodanine reacts with diazonium salts to give the *S*-arylated
derivatives, 5-benzylidene-rhodanine itself reacts at the methine proton to give
the azo-derivatives (112).[208] Compounds (112) react further, to give *S*-arylated
derivatives.

5-Arylidene-rhodanines react with morpholine or piperidine to give eventually
the Δ²-thiazolin-4-ones (113).[183, 209] 5-Arylidene-3-phenyl-rhodanines react with

[196] F. E. Condon, D. Shapiro, P. Sulewski, I. Vasi, and R. Waldman, *Org. Prep. Proced. Internat.*, 1974, **6**, 37.

[197] M. Augustin and W.-D. Rudorf, *J. prakt. Chem.*, 1974, **316**, 520.

[198] R. S. Lebedev, V. N. Yukhimets, V. I. Yakimenko, A. N. Gromnitskii, A. E. Kovalev, R. P. Chumakova, M. I. Starchenko, and N. N. Golokhvasthov, *Voprosy Mol. Spektroskopii*, 1974, 63 (*Chem. Abs.*, 1975, **82**, 162 260).

[199] A. V. Korshunov, V. E. Volkov, R. S. Lebedev, and V. I. Yakimenko, *Voprosy Mol. Spektroskopii*, 1974, 57 (*Chem. Abs.*, 1975, **82**, 162 258).

[200] R. S. Lebedev and V. I. Yakimenko, *Voprosy Mol. Spektroskopii*, 1974, 60 (*Chem. Abs.*, 1975, **82**, 162 259).

[201] A. V. Korshunov, V. E. Volkov, R. S. Lebedev, and V. N. Yukhimets, *Voprosy Mol. Spektroskopii*, 1974, 308 (*Chem. Abs.*, 1975, **82**, 85 967).

[202] K. A. V'yunov, A. I. Ginak, and E. G. Sochilin, *Zhur. priklad. Spektroskopii*, 1975, **23**, 917 (*Chem. Abs.*, 1976, **84**, 51 662).

[203] H. K. Gakhar, S. H. Chander, and H. C. Khurana, *J. Indian Chem. Soc.*, 1975, **52**, 155.

[204] S. P. Gupta and P. Dureja, *J. Indian Chem. Soc.*, 1974, **51**, 936.

[205] A. M. Khalil, A. A. Fadda, and M. M. Yousef, *J. Indian Chem. Soc.*, 1973, **50**, 734.

[206] E. P. Nesynov, M. M. Besprozvannaya, and P. S. Pelkis, *Khim. geterotsikl. Soedinenii*, 1973, 1487 (*Chem. Abs.*, 1974, **80**, 82 785).

[207] Md. I. Husain and S. K. Agarwal, *Indian J. Pharm.*, 1975, **37**, 89.

[208] E. P. Nesynov and M. M. Besprozvannaya, *Ukrain. khim. Zhur.*, 1975, **41**, 842 (*Chem. Abs.*, 1975, **83**, 178 904).

[209] A. R. A. Raouf, M. T. Omar, S. M. A. Omran, and K. E. El-Bayoumy, *Acta Chim. Acad. Sci. Hung.*, 1974, **83**, 359.

an excess of these amines by ring-cleavage to give the corresponding thiourea, *e.g.* OC$_4$H$_8$NC(S)NHPh.[209] This is analogous to the behaviour of 5-arylidene-3-phenylthiazolidine-2,4-diones (see p. 380). 5-Arylidene-3-methyl-rhodanines react with benzylamine to give compounds (114).[183]

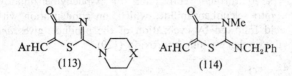

N-Allyl-rhodanine is cyclized by strong acids on the oxygen atom, rather than on the more nucleophilic sulphur atom, to give compound (115).[210] ^{1}H N.m.r. studies suggest that this is due to initial protonation on the thiocarbonyl group. Addition of 2,3-dimethylbuta-1,3-diene to 3-substituted 5-methoxy-carbonyl-methylidene-rhodanines gives the spiro-compounds (116).[211] 3-Ethyl-rhodanine is nitrosated on heating with 9-nitrosocarbazole and hydrochloric acid or on treatment with nitrous acid, to give the 5-nitroso-compound, whereas

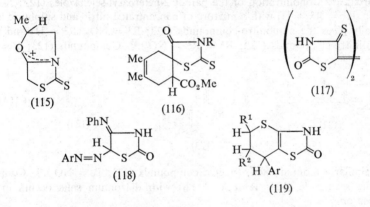

similar treatment of isorhodanine gives the dimer (117).[212] 5-Arylidene-isorhodanines react with diazomethane in ether–chloroform to give both the *N*- and *S*-methylated derivatives, but only the latter is formed in ether.[213] Ethyl bromoacetate gives only an *S*-alkylated product. 5-Arylazo-isorhodanine reacts with diazomethane in ether to give both the *N*- and *S*-methylated products.[213] Aniline displaces the methylthio-group from the latter compounds, with formation of the anils (118). 5-Arylidene-rhodanines react with activated alkenes R^{1}CH=CHR^2 to give adducts (119; R^1 = H, R^2 = CN or CO$_2$Et; R^1 = Ph, R^2 = NO$_2$ or CO$_2$Et).[213, 214]

[210] S. P. McManus, K. Y. Lee, and C. U. Pittman, *J. Org. Chem.*, 1974, **39**, 3041.
[211] H. Nagase, *Chem. and Pharm. Bull. (Japan)*, 1974, **22**, 1661.
[212] B. E. Zhitar, V. N. Melikova, and S. N. Baranov, *Zhur. org. Khim.*, 1973, **9**, 2434 (*Chem. Abs.*, 1974, **80**, 70 744).
[213] N. A. El L. Kassab and N. A. Messeha, *J. prakt. Chem.*, 1973, **315**, 1017.
[214] N. A. El L. Kassab, S. O. Abdallah, and N. A. Messeha, *J. prakt. Chem.*, 1974, **316**, 209.

15 Selenazoles

A number of selenazoles have been prepared by an extension of the Hantzsch synthesis.[11, 215–217] As in the Hantzsch synthesis (see p. 357), intermediate Δ^2-selenazolines have been isolated.[11] The first meso-ionic selenium compound has been prepared through conversion of *N*-phenylbenzimidoyl chloride, PhCCl=NPh, into selenobenzanilide, which, on condensation with α-bromophenylacetic acid followed by cyclization of the product, gives *anhydro*-2,3,5-triphenyl-4-hydroxyselenazolium hydroxide (120).[218]

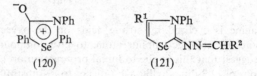

(120)   (121)

The u.v. spectra of compounds (121) have been compared with those of the corresponding thiazoles.[217]

Various reactions of ethyl 2-aminoselenazole-4-carboxylate have been reported.[215] Mononitration of the parent 2-heteroaryl-selenazoles (122; X = S or Se, $R^1 = R^2 = H$) with a mixture of concentrated nitric and sulphuric acids initially gives the mononitro-compounds (122; $R^1 = NO_2$, $R^2 = H$) and then the dinitro-compounds (122; $R^1 = R^2 = NO_2$).[47] Compounds (123; R = H)

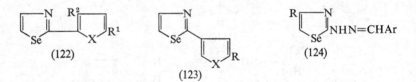

(122)   (123)   (124)

are similarly mononitrated, to give compounds (123; $R = NO_2$).[47] Coupling of the selenazole (124; R = Ar = Ph) with diazonium salts occurs in the 5-position.[216]

2-Iminoselenazolidin-4-ones react with isothiocyanates at the ring N-atom to give thioureas (125), but with isocyanates they react with the exocyclic N-atom to give ureas (126).[219] New merocyanine dyes containing selenazolidine rings

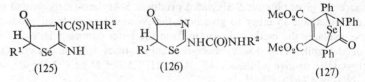

(125)   (126)   (127)

[215] A. Shafiee and I. Lalezari, *J. Heterocyclic Chem.*, 1975, **12**, 675.
[216] A. A. Tsurkan and Z. I. Popova, *Sbornik Nauchn. Tr., Ryazan. Med. Inst.*, 1975, **50**, 64 (*Chem. Abs.*, 1976, **84**, 59 326).
[217] S. Bilinski and T. Urban, *Acta Polon. Pharm.*, 1975, **32**, 159 (*Chem. Abs.*, 1976, **84**, 30 970).
[218] M. P. Cava and L. E. Saris, *J.C.S. Chem. Comm.*, 1975, 617.
[219] H. Siaglo, S. Andrzejewski, E. Kleczek, and D. Prelicz, *Pol. J. Pharmacol. Pharm.*, 1975, **27**, 57.

have been obtained.[220] Compared with the analogous dyes of the isorhodanine series, the long-wavelength maxima of their i.r. spectra have bathochromic shifts. The kinetics of condensation of *p*-dimethylaminobenzaldehyde with selenazolidine-2,4-dione have been studied in n-butanol in the presence of piperidine.[221]

The meso-ionic selenazole (120) is an extremely reluctant dipolarophile, but when heated under reflux for one week with dimethyl acetylenedicarboxylate in benzene it gives a pyridone, presumably by loss of selenium from the intermediate (127).[218]

[220] V. E. Kononenko, B. E. Zhital, and S. N. Baranov, *Khim. geterotsikl. Soedinenii*, 1973, 1493 (*Chem. Abs.*, 1974, **80**, 82 817).
[221] S. N. Baranov, B. E. Zhitar, and V. E. Kononenko, *Ukrain. khim. Zhur.*, 1974, **40**, 1168 (*Chem. Abs.*, 1975, **82**, 85 651).

13
Condensed Ring Systems Incorporating Thiazole

BY B. IDDON AND P. A. LOWE

1 Introduction [1]

Apart from the considerable industrial interest shown in the benzothiazole system, new methods of synthesis continue to attract the most interest. Noteworthy among those now reviewed is a new condensation of *o*-aminobenzenethiol with acetylenes, which results in cleavage of the acetylenic bond.

The 85-year-old controversy concerning the structures of Hector's, Dost's, and Hugershoff's bases has been partially resolved by establishing the structures of several Dost's bases and of two Hugershoff's bases. Confusion over the structures of the products obtained from benzothiazole and dimethyl acetylenedicarboxylate has finally been resolved.

Other noteworthy advances in benzothiazole chemistry include the use of 3-methylbenzothiazoline-2-thione as a sulphur-transfer reagent (oxirans → thiirans), the isolation of a Meisenheimer-type adduct when 6-nitrobenzothiazole reacts in the 2-position with methoxide ion, Wasserman's work on the reaction of benzothiazoles with singlet oxygen, and the proof that *p*-benzoquinone is incorporated into the luciferin of live Japanese fireflies.

Among other condensed thiazole ring systems, imidazo[2,1-*b*]thiazoles (tetramisole is a member of this class), thiazolo-pyrimidines, and thiazolo-pyridines continue to attract the most interest. Attention is concentrated on synthesis and biological activity. Unexpectedly, dihydrothiazolo[3,2-*a*]pyridinium bromide is readily brominated when the pyridine ring has a 3-nitro-substituent but not when it has a 3-amino-substituent.

2 Synthesis of Benzothiazoles

From *o*-Aminobenzenethiols (Type A; S—C$_6$H$_4$—N + C).—An improvement on the standard procedures (milder conditions) is the use of DMF dimethyl acetal instead of formic acid or alkyl orthoformates.[2] A 64% yield of benzothiazolin-2-one is obtained when *o*-aminobenzenethiol is treated sequentially with carbon monoxide and oxygen in the presence of triethylamine and selenium in DMF.[3] *o*-Aminobenzenethiol forms substituted vinylamine adducts with acetylenic nitriles RC≡CCN. These cyclize in the presence of sodium ethoxide, and the products lose acetonitrile on distillation at atmospheric pressure to give 2-sub-

---

[1] See general comments in Introduction to Chapter 12.
[2] B. Stanovnik and M. Tišler, *Synthesis*, 1974, 120.
[3] N. Sonoda, G. Yamamoto, K. Natsukawa, K. Kondo, and S. Murai, *Tetrahedron Letters*, 1975, 1969.

stituted benzothiazoles (1).[4] This constitutes a further example of cleavage of a triple bond, furnishing benzothiazoles (see Vol. 3, p. 619). Allenic nitriles $R^1R^2C=C=CHCN$ react similarly to give (1; $R = R^1R^2CH$).[4] The adduct formed between *o*-aminobenzenethiol and triphenyl(prop-2-ynyl)phosphonium bromide eliminates methyltriphenylphosphonium bromide with simultaneous cyclization, to give (1; $R = Me$).[5]

Diethyl monothiono-oxalate, $EtOC(S)CO_2Et$, reacts with *o*-aminobenzenethiol to give ethyl benzothiazole-2-carboxylate (1; $R = CO_2Et$),[6] while esters of

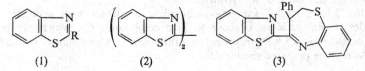

(1)　　　　　　(2)　　　　　　(3)

dithiono-oxalic acid in the presence of toluene-*p*-sulphonic acid condense to give 2,2'-bibenzothiazolyls (2).[7] 2-Hydroxy-3,4-diphenylcyclobut-2-en-1-one condenses to give the benzothiazole [1; $R = C(O)CHPhCH_2Ph$].[8] When the 4-phenyl group is replaced with a toluene-*p*-sulphonyl group, a more complex reaction occurs, and a mixture of the benzothiazole [1; $R = C(O)CHPhCH_2SO_2C_6H_4Me$-*p*] and compound (3) is obtained.[8]

An intermediate has been isolated from the reaction between *o*-aminobenzenethiol and dimethyl cyanimidodithioate (see Vol. 3, pp. 617, 618); this yields bis(benzothiazol-2-yl)amine, thus further elucidating the mechanism.[9]

Jacobson–Hugershoff Synthesis (Type B; C_6H_5—N—C—S).—Examples of compounds prepared by this well-established method include (oxidizing agent in parentheses) the imine (4) (Br_2 in $CHCl_3$),[10] the 2-substituted benzothiazoles [1; $R = CH(CN)C(O)Ph$] (potassium ferricyanide),[11] and the benzothiazoles (1; $R = NH_2$, NHPh, *etc.*) variously substituted in the benzene ring with methyl, bromo-, chloro-, or nitro-groups (plumbophosphates).[12]

Other Type B Syntheses.—The interaction of 2-picoline with 3-substituted anilines in the presence of sulphur under Willgerodt–Kindler conditions, reported

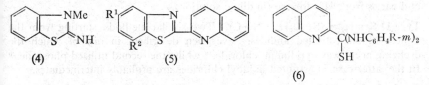

(4)　　　　(5)　　　　　　　　(6)

[4] Z. T. Fomum, P. D. Landor, S. R. Landor, and G. B. Mpango, *Tetrahedron Letters*, 1975, 1101.
[5] E. E. Schweizer and S. V. De Voe, *J. Org. Chem.*, 1975, **40**, 144.
[6] K. Thimm and J. Voss, *Z. Naturforsch.*, 1975, **30b**, 292.
[7] H. Hoppe and K. Hartke, *Arch. Pharm.*, 1975, **308**, 526.
[8] W. Ried and H. Knorr, *Chem. Ber.*, 1975, **108**, 2750.
[9] L. S. Wittenbrook, *J. Heterocyclic Chem.*, 1975, **12**, 37.
[10] S. W. Horgan, J. K. Woodward, A. Richardson, and A. A. Carr, *J. Medicin. Chem.*, 1975, **18**, 315.
[11] R. G. Dubenko and E. F. Gorbenko, *Khim. geterotsikl. Soedinenii*, 1974, 500 (*Chem. Abs.*, 1974, **81**, 49 612).
[12] M. S. A. El-Meligy and S. A. Mohamed, *J. prakt. Chem.*, 1974, **316**, 154.

earlier (see Vol. 3, p. 621), has been extended to 4-picoline [13] and 2-methyl-quinoline,[14] with similar results. In the latter case, 5-substituted benzothiazoles (5; R^2 = H) are formed exclusively (similar results are obtained in DMF), while oxidative cyclization of the thiocarbanilides, also formed in the reaction, with potassium ferricyanide in a modified Jacobsen procedure gives both 5- and 7-substituted benzothiazoles (5; R^2 = H and R^1 = H, respectively).[14] The α-(2-quinolyl)-$\alpha\alpha$-NN'-diarylamino-methanethiols (6) are now proposed as intermediates. These presumably cyclize by dehydrogenation, with simultaneous elimination of aniline or a derivative.

Ethyl 2-bromocyclohexanone-6-carboxylate condenses with aromatic thio-anilides in the presence of base to give compounds (7), which are readily dehydrated to 4,5,6,7-tetrahydrobenzothiazoles.[15] Compounds with Ar = NHAr may be prepared similarly.[15]

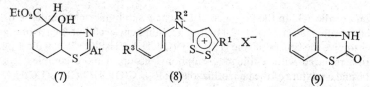

(7)   (8)   (9)

3-Amino-1,2-dithiolium salts (8; R^1 = aryl, R^2 = aryl or arylalkyl, R^3 = H or Me) undergo thermal rearrangement to give the corresponding vinylogous benzothiazoline-2-thiones [=CHC(S)R^1 in the 2-position].[16]

Type C Syntheses (N—C$_6$H$_4$—S—C).—Conversion of *o*-nitroarylthio-acetic acids, prepared from α-mercaptoacetic acid and the appropriate derivative of *o*-nitro-chlorobenzene, into the corresponding benzothiazolin-2-one (9) can be brought about by treatment with acetic anhydride followed by hydrolysis with base or acid [17] (see Vol. 3, p. 622).

Thermal decomposition of 2-(cyanoethylthio)benzenediazonium tetrafluoro-borate in acetonitrile gives 2-methylbenzothiazole (1; R = Me) together with products in which the diazonium group has been replaced with fluoro- or acetamido-groups.[18] The yield of the benzothiazole is concentration-dependent and varies from 41% to zero (in dilute solution).

Type D Syntheses (S—C$_6$H$_4$—N—C).—Two similar methods, starting from the 2-methylthio-substituted anilides, have been described, in one of which the cyclizing agent was pyridinium chloride,[19] whilst the second utilized phosgene.[20] In the latter case, substituted imidoyl chlorides are probably intermediates.

Synthesis from Thiazoles.—2-Aryl-4-cyanomethyl-thiazoles condense with the cation Me$_2$NCH=CPhCH=NMe$_2$+ in the presence of base, and the products

[13] T. Hisano and M. Tamayama, *Yakugaku Zasshi*, 1973, **93**, 1356 (*Chem. Abs.*, 1974, **80**, 82 777).
[14] T. Hisano and M. Ichikawa, *Chem. and Pharm. Bull. (Japan)*, 1974, **22**, 2051.
[15] P. H. L. Wei, U.S.P. 3 859 280/1975 (*Chem. Abs.*, 1975, **82**, 140 120).
[16] B. Bartho, J. Faust, and R. Mayer, *Tetrahedron Letters*, 1975, 2683.
[17] A. F. Aboulezz, S. M. A. Zayed, W. S. El-Hamouly, and M. I. El-Sheikh, *Egyptian J. Chem.*, 1973, **16**, 355 (*Chem. Abs.*, 1974, **81**, 152 075).
[18] D. C. Lankin, R. C. Petterson, and R. A. Velazquez, *J. Org. Chem.*, 1974, **39**, 2801.
[19] R. Royer, J. P. Lechartier, and P. Demerseman, *Bull. Soc. chim. France*, 1973, 3017.
[20] K. Pilgrim and R. D. Skiles, *J. Org. Chem.*, 1974, **39**, 3277.

lose dimethylamine at 160—220 °C to give 2-aryl-4-cyano-6-phenyl-benzo-thiazoles.[21]

Synthesis from 3,5-Bis(arylimino)-1,2,4-thiadiazolidines (Dost's Bases); Structure of Hugershoff's Bases.—Oxidation of the thioureas RNHC(S)NHPh with nitrous acid in aqueous ethanol gives Dost's bases (10; R = Me or Et), which rearrange with acid to give benzothiazoles [1; R = NMeC(:NPh)NHMe or NEtC(:NPh)-NHEt] (Hugershoff's bases).[22] The factors governing benzothiazole formation need clarification, since not all 2,4-disubstituted-3,5-bis(arylimino)-1,2,4-thia-diazolidines (10) are rearranged under these conditions. The structure of compound (10; R = Me) has now been established conclusively by *X*-ray crystallo-graphic analysis, while the structures of the Hugershoff's bases have been estab-lished, using [13]C n.m.r. spectroscopy.[22] The product of oxidation of the thiourea MeNHC(S)NHPh with benzoyl peroxide in methylene chloride has been shown to be (10; R = Me) and not 4-methyl-3-methylimino-2-phenyl-5-phenylimino-1,2,4-thiadiazolidine, as previously suggested.[23] This work casts some doubt on

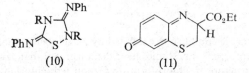

(10)   (11)

the validity of the hypothesis that Hector's bases (Dost's bases in this case) are formed only in acidic polar solvents, while non-polar neutral solvents favour the formation of benzothiazoles by radical processes.

Synthesis from 1,4-Benzothiazines.—(See Vol. 2, p. 658.) A new synthesis of benzothiazoles, prompted by the suggestion that firefly luciferin arises from *p*-benzoquinone and cysteine, involves an initial condensation of the former with cysteine ethyl ester hydrochloride, to give [24] the isolable intermediate (11), which on acetylation isomerizes to give the 7-acetoxy-1,4-benzothiazine ester. This can be converted into the corresponding amide, which is oxidized in the presence of base to give 2-ethoxycarbonyl-6-hydroxybenzothiazole. It is sug-gested [24] that this ring-contraction proceeds *via* the rearrangement of an intermediate peroxide.

Experiments [25] showing positive incorporation of [^{14}C]-*p*-benzoquinone into live Japanese fireflies, *Luciola cruciata*, indicate that the benzothiazole nucleus in firefly luciferin does arise in the manner previously suggested.

Other Syntheses.—A high yield of 2-phenylbenzothiazole is obtained when benzaldehyde, benzyl alcohol, benzyl benzoate, or thiobenzanilide is heated (~185 °C) with aniline and sulphur in HMPA.[26]

[21] C. Jutz, R. M. Wagner, and H.-G. Löbering, *Angew. Chem. Internat. Edn.*, 1974, **13**, 737.
[22] C. Christophersen, T. Øttersen, K. Seff, and S. Treppendahl, *J. Amer. Chem. Soc.*, 1975, **97**, 5237.
[23] J. M. Sprague and A. H. Land, in 'The Chemistry of Heterocyclic Compounds', ed. R. C. Elderfield, Wiley, New York, 1957, Vol. 5, p. 511; F. Kurzer and P. M. Sanderson, *J. Chem. Soc.*, 1963, 3333.
[24] F. McCapra and Z. Razavi, *J.C.S. Chem. Comm.*, 1975, 42.
[25] K. Okada, H. Iio, and T. Goto, *J.C.S. Chem. Comm.*, 1976, 32.
[26] J. Perregaard and S.-O. Lawesson, *Acta Chem. Scand.* (*B*), 1975, **29**, 604.

3 Physical Properties of Benzothiazoles

Tautomerism of benzothiazoles continues to attract attention. 2-Phenacyl-benzothiazole (13), for example, has been shown [27] by i.r., n.m.r., and chemical evidence to exist in three tautomeric forms. The spectroscopic properties (u.v., i.r.) of compounds (12; Ar = Ph, p-MeC$_6$H$_4$, or p-ClC$_6$H$_4$) indicate that they exist as chelate ring structures.[28] ^{1}H N.m.r. spectroscopy has been used to study

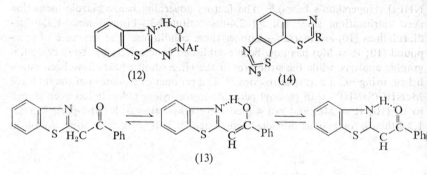

(12)   (14)

(13)

the azide $\rightleftharpoons$ tetrazole equilibrium exhibited by 2-azidobenzothiazole (1; R = N$_3$) in deuteriochloroform.[29] I.r. studies have demonstrated that azide (14; R = Me) has the azide structure in chloroform or carbon tetrachloride, while its isomer with the 2-azidothiazole ring fused in the 5,4- instead of the 6,7-position exists in chloroform as an azide $\rightleftharpoons$ tetrazole tautomeric mixture.[30] In potassium bromide discs, compound (14; R = N$_3$) and 2,6- and 2,7-di-azidobenzothiazoles exist as azido-tetrazoles, while in chloroform and carbon tetrachloride they exist as equilibrium mixtures of azido-tetrazoles and di-azides.[30]

Chemical shifts and coupling constants have been reported for a series of 6-substituted 2-fluorobenzothiazoles.[31] A large cross-ring coupling constant ($J_{2,6}$ 10.4 Hz) was found for the 2,6-difluoro-compound. Examination of the ^{1}H n.m.r. spectra of some 2-alkylamino-4-phenylbenzothiazoles indicates that the alkyl groups are abnormally shielded, the least abnormality existing in the case of the 2-t-butylamino-compound.[32] It is suggested that this phenomenon is due to the existence of dimeric hydrogen-bonded species. The extent of the shielding is a function of the solvent, the concentration, and the size of the alkyl group.

The mass spectrometric breakdown of benzothiazole has been studied [33, 34] and compared with that of 1,2-benzisothiazole. Comparisons have been made between the mass spectra of 3-methylbenzothiazolin-2-one hydrazone and 2-hydrazinobenzothiazole, and between those of their 4-, 5-, 6-, and 7-chloro-

[27] D. Nardi, A. Tajana, and R. Pennini, *J. Heterocyclic Chem.*, 1975, **12**, 139.
[28] S. K. Mukerji and N. C. Sogani, *Bull. Acad. polon. Sci., Sér. Sci. chim.*, 1975, **23**, 463 (*Chem. Abs.*, 1975, **83**, 205 377).
[29] R. Faure, J. P. Galy, G. Guisti, and E. J. Vincent, *Org. Magn. Resonance*, 1974, **6**, 485.
[30] T. F. Grigorenko, L. F. Avramenko, and V. Ya. Pochinok, *Ukrain. khim. Zhur.*, 1975, **41**, 1222 (*Chem. Abs.*, 1976, **84**, 73 205).
[31] M. Calafell, J. Elguero, and A. Fruchier, *Org. Magn. Resonance*, 1975, 7, 84.
[32] K. Nagarajan and R. K. Shah, *Indian J. Chem.*, 1973, **11**, 879.
[33] A. Maquestiau, Y. Van Haverbeke, R. Flammang, and J. Pierard, *Bull. Soc. chim. belges*, 1975, **84**, 213 (*Chem. Abs.*, 1975, **83**, 77 890).
[34] A. Selva, U. Vettori, and E. Gaetani, *Org. Mass Spectrometry*, 1974, **9**, 1161.

derivatives.[35] The mass spectra of 2-aminobenzothiazole 3-oxide and its 4-alkyl and 4-aryl derivatives have also been studied.[36] Comparisons have been made between the behaviour of 3-phenylbenzothiazolin-2-one and the corresponding thione,[37] and between benzothiazolin-2-one, benzoxazolin-2-one, and benzimidazolin-2-one [38] in the mass spectrometer and on pyrolysis.

The (*E*)- and (*Z*)-configurations of compounds (15; R = H, Me, or Ph; Ar = Ph, 2-furyl, 2-thienyl, *etc.*) have been prepared by condensation of 2-hydrazinobenzothiazole with *o*-aminobenzophenones,[39] or the appropriate heterocyclic aldehydes or ketones,[40] the isomers then being separated. Their

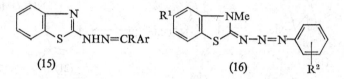

(15)   (16)

structures have been established by spectroscopic methods. Photochemical isomerization of the geometrical isomers of the triazenes (16; R^1 = H, Cl, or OMe; R^2 = *e.g.*, H, *m*-NO$_2$, *p*-Cl, or *p*-OMe) has been studied at 320 and 436 nm [41] and their mass spectrometric fragmentation has been investigated.[42]

4 Chemical Properties of Benzothiazoles

Oxidation and Reduction.—The electrochemical oxidation of benzothiazole-2-thiol to its disulphide has been further studied.[43] This conversion may be accomplished with ozone in the presence of potassium iodide to prevent ozonization.[44, 45] Desulphurization of benzothiazoles as a preparative procedure does not appear to have been exploited to the extent that it has in the field of thiophen chemistry. Noteworthy, however, is the desulphurization of sugars (17; *e.g.*, *n* = 3, R = H) to 1-deoxy-1-(*N*-methylanilino)-alditals.[46]

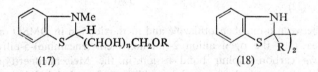

(17)   (18)

[35] J.-C. Richer, P. Lapointe, M. Beljean, and M. Pays, *Canad. J. Chem.*, 1975, **53**, 3677.

[36] G. Entenmann, *Org. Mass. Spectrometry*, 1975, **10**, 831.

[37] D. C. K. Lin, M. L. Thomson, and D. C. DeJongh, *Canad. J. Chem.*, 1975, **53**, 2293.

[38] M. L. Thomson and D. C. DeJongh, *Canad. J. Chem.*, 1973, **51**, 3313.

[39] S. Kwon and K. Isagawa, *Nippon Kagaku Zasshi*, 1974, 524 (*Chem. Abs.*, 1975, **82**, 3686).

[40] S. Kwon, M. Tanaka, and K. Isagawa, *Nippon Kagaku Zasshi*, 1974, 1526 (*Chem. Abs.*, 1975, **82**, 30 858).

[41] E. Fanghaenel, R. Haensel, W. Ortmann, and J. Hohlfeld, *J. prakt. Chem.*, 1975, **317**, 631.

[42] E. Fanghaenel, B. Adler, and R. Haensel, *J. prakt. Chem.*, 1975, **317**, 712.

[43] H. Berge, H. Millat, and B. Strübing, *Z. Chem.*, 1975, **15**, 37.

[44] V. A. Yakobi, V. P. Tsipenyuk, V. A. Ignatov, B. A. Ponomarev, and P. P. Karpukhin, *Zhur. priklad. Khim.*, 1974, **47**, 2417 (*Chem. Abs.*, 1975, **82**, 42 645).

[45] L. P. Petrenko, V. A. Yakobi, M. V. Lyashchenko, and T. P. Dubinskaya, *Izvest. V.U.Z.*, *Khim. i khim. Tekhnol.*, 1975, **18**, 304 (*Chem. Abs.*, 1975, **82**, 156 166).

[46] R. Bognar and P. Herczegh, *Org. Prep. Proced. Internat.*, 1975, **7**, 111.

Treatment of bibenzothiazolin-2-yls (18) with zinc acetate yields zinc complexes,[47] the free azomethines of which are unknown. Reduction of the coloured compounds with sodium borohydride gives colourless complexes, which are decomposed by alkali to give $[o\text{-}(HS)C_6H_4NHCHR\text{-}]_2$.

Reactions of Benzothiazoles with Singlet Oxygen.—Singlet oxygen reacts with 4,5,6,7-tetrahydro-2-methylbenzothiazole in methylene chloride to yield the transannular peroxide (19), which undergoes successive Baeyer–Villiger rearrangement and tautomerism *via* (20) to the isolable *N*-acetyl-thio-isoimide

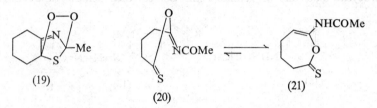

(21).[48] If irradiation is carried out in methanol, the disulphide $[MeO_2C(CH_2)_4\text{-}C(O)S\text{-}]_2$ is formed, presumably *via* hydrolysis of (21), or its precursor (20), and successive photolytic conversion and oxidation of the intermediate thiol.

Nucleophilic Substitution.—Compound (22) is formed when 2-chlorobenzothiazole reacts with 2,6-dimethylaniline at 170 °C.[10] The ring N-atom in the initially formed 2-(2,6-dimethylanilino)benzothiazole reacts in preference to the 2,6-dimethylaniline with any remaining 2-chlorobenzothiazole. The reversible

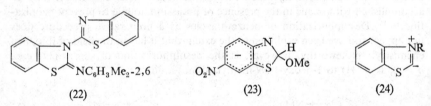

reaction between 6-nitrobenzothiazole and methoxide ion in DMSO or DMSO–MeOH leads to the open anion 2-(methoxymethyleneamino)-5-nitrobenzenethiolate *via* carbon–sulphur bond fission in the Meisenheimer-type adduct (23).[49] The anion can be trapped by addition of methyl iodide or bromine. [1]H N.m.r. spectroscopy shows that rapid addition of methanol to the C=N bond of the anion occurs in DMSO–MeOH to give the substituted dimethyl acetal.[49]

N-Alkyl-2-azidobenzothiazolium tetrafluoroborates exhibit ambident reactivity with nucleophiles. With hard nucleophiles, *e.g.* hydroxide ion, methoxide ion, or dimethylamine, reaction occurs[50] at the 2-position with displacement of azide ion. In contrast, softer nucleophiles, *e.g.* azide ion, cyanide ion, toluene-*p*-

[47] J. L. Corbin and D. E. Work, *Canad. J. Chem.*, 1974, **52**, 1054.
[48] H. H. Wasserman and G. R. Lenz, *Tetrahedron Letters*, 1974, 3947.
[49] G. Bartoli, F. Ciminale, and P. E. Todesco, *J.C.S. Perkin II*, 1975, 1472; G. Bartoli, M. Fiorentino, F. Ciminale, and P. E. Todesco, *J.C.S. Chem. Comm.*, 1974, 732.
[50] H. Balli, *Helv. Chim. Acta*, 1974, **57**, 1912.

sulphinate ion, or triphenylphosphine, react initially at the terminal nitrogen atom of the azido-group. The product of reaction with azide ion is unstable, and loses nitrogen to give a 2-tetrazo-3-alkyl-benzothiazoline. This can be made to lose a further two molecules of nitrogen with generation of the ylide (24; R = Et), which reacts [50] at the terminal nitrogen atom of the *N*-alkyl-2-azido-benzothiazolium tetrafluoroborate.

Reactions of Benzothiazoles with Dimethyl Acetylenedicarboxylate.—Earlier confusion [51, 52] over the structures of the products obtained by the reaction of benzothiazole with dimethyl acetylenedicarboxylate in methanol has finally been clarified.[53-55] Initial reaction at nitrogen yields the dipolar intermediate, which reacts further either by protonation, nucleophilic attack by water at the 2-position, and sequential ring-opening and ring-closure to give the 1 : 1 adduct (25), or by further reaction with dimethyl acetylenedicarboxylate to give a tricyclic 1 : 2 adduct. This either undergoes a [1,5]-sigmatropic rearrangement to yield the 1 : 2 adduct (26) or rearranges, possibly *via* a vinyl sulphide, to give an isomeric 1 : 2 adduct (27). The structures of the three adducts (25)—(27) have been

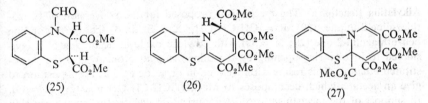

(25)　　　　　　　　(26)　　　　　　　　(27)

established conclusively by ^{1}H and [13]C n.m.r. spectroscopy, mass spectrometry, and *X*-ray analysis. Compound (27) is obtained when benzothiazole reacts with the ester in DMF at room temperature.[55] The products obtained in methanol appear to depend on the reaction conditions and on the work-up technique. McKillop and Sayer [56] have found that the benzothiazine (25) is formed exclusively in the presence of an excess of water, but that in anhydrous methanol the reaction follows the alternative routes. Methyl 3-benzyl-2-methylene-3*H*-benzothiazoline-α-dithiocarboxylate undergoes cycloaddition with dimethyl acetylenedicarboxylate with monodesulphurization, to give the corresponding spiro(benzothiazoline-2,1'-cyclopenta-2,4-diene) derivative.[57] The reactions of 2-aminobenzothiazole with alkynes, including dimethyl acetylenedicarboxylate, are discussed elsewhere (p. 412).

Organometallic Derivatives.—A novel, one-step asymmetric synthesis of a benzothiazole nucleoside analogue has been reported [58] which involves a reaction

[51] D. H. Reid, F. S. Skelton, and W. Bonthrone, *Tetrahedron Letters*, 1964, 1797.
[52] R. M. Acheson, M. W. Foxton, and G. R. Miller, *J. Chem. Soc.*, 1965, 3200.
[53] H. Ogura, H. Takayanagi, K. Furuhata, and Y. Iitaka, *J.C.S. Chem. Comm.*, 1974, 759.
[54] P. J. Abbott, R. M. Acheson, U. Eisner, D. J. Watkin, and J. R. Carruthers, *J.C.S. Chem. Comm.*, 1975, 155; *J.C.S. Perkin I*, 1976, 1269.
[55] H. Ogura, K. Kikuchi, H. Takayanagi, K. Furuhata, Y. Iitaka, and R. M. Acheson, *J.C.S. Perkin I*, 1975, 2316.
[56] A. McKillop and T. S. B. Sayer, *Tetrahedron Letters*, 1975, 3081.
[57] G. Kobayashi, Y. Matsuda, Y. Tominaga, and K. Mizuyama, *Chem. and Pharm. Bull.* (*Japan*), 1975, **23**, 2749; *Heterocycles*, 1974, **2**, 309.
[58] H. Ogura and H. Takahashi, *J. Org. Chem.*, 1974, **39**, 1374.

between benzothiazol-2-yl-lithium (1; R = Li) and either di-isopropylidene L-gulofuranolactone or D-ribofuranolactone. 2-Trimethylsilylbenzothiazole (1; R = SiMe₃) reacts with trichlorogermane to yield 2-trimethylsilylbenzothiazolium trichlorogermane, which is converted on pyrolysis at 91 °C into compound (28).[59] The N—Ge bond in this compound is highly reactive,[59] and is cleaved by boron trichloride or hydrogen chloride. Exchange reactions occur with aromatic

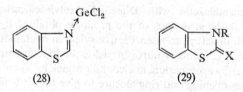

(28)   (29)

nitrogen bases.[59] Photolysis of (28) with metal hexacarbonyls of chromium, molybdenum, or tungsten yields metal complexes whose structures have been studied spectroscopically.[60] 2-Trimethylsilylbenzothiazole and related compounds form Pd^{II} chloride complexes with PdCl₂,(PhCN)₂.[61]

Alkylation Reactions.—The mechanism proposed for the conversion of benzothiazoline-2-thione (29; R = CH₂CH₂OH, X = S) into the corresponding benzothiazolin-2-one (29; X = O) with ethylene oxide in acetic acid [62] is considered to involve ring-opening by the thione sulphur atom to give the β-substituted ethanol. This reacts with acetate ion or acetic acid at the 2-position to give an acetate, which decomposes to MeC(O)SCH₂CH₂OH and (29; X = O). In support of the mechanism, no desulphurization occurred when the reaction was attempted in acetone or methanol.

The 2-amino-benzothiazoles (1; R = NH₂, NHMe, or NHPh) react with acrylic acid to give the 2-imino-compounds (29; R = CH₂CH₂CO₂H; X = NH, NMe, or NPh), whilst the 2-imino-compounds (29; R = Me, Ph, or CH₂CH₂-CO₂H; X = NH) yield compounds (29; R = Me, Ph, or CH₂CH₂CO₂H; X = NCH₂CH₂CO₂H).[63]

The reaction of benzothiazole with *p*-iodonitrobenzene at 200 °C in the presence of cuprous oxide gives a 72% yield of 2-(*p*-nitrophenyl)benzothiazole. 2-(*p*-Methoxyphenyl)thiazole and 2,2′-bibenzothiazolyl may be prepared similarly.[64] Whilst suggesting the organo-copper compound (1; R = Cu) as a probable intermediate, the authors have also considered the possibility of a copper-complexed carbene as an intermediate.

Reactions of Benzothiazolium Salts.—*N*-Methyl quaternary salts react [65] in an unexpected manner with phenacyl chloride in the presence of base to give 2-benzoyl-4-methyl-4*H*-1,4-benzothiazine (30) (see also Chapter 15, p. 460).

[59] P. Jutzi, H.-J. Hoffmann, and K.-H. Wyes, *J. Organometallic Chem.*, 1974, **81**, 341.
[60] P. Jutzi and H.-J. Hoffmann, *Chem. Ber.*, 1974, **107**, 3616.
[61] P. Jutzi and H. Heusler, *J. Organometallic Chem.*, 1975, **102**, 145.
[62] P. Sohar and G. H. Denny, *J. Heterocyclic Chem.*, 1973, **10**, 1015.
[63] N. K. Rozhkova and V. A. Saprykina, *Uzbek. khim. Zhur.*, 1974, **18**, 60 (*Chem. Abs.*, 1974, **81**, 120 526).
[64] J. Chodowska-Palicka and M. Nilsson, *Synthesis*, 1974, 128.
[65] J. A. Van Allan, J. D. Mee, C. A. Maggiulli, and R. S. Henion, *J. Heterocyclic Chem.*, 1975, **12**, 1005.

Chloroacetone reacts similarly. Bromomalonic ester gives [29; R = Me; X = C(CO$_2$Et)$_2$]; the intermediate in this case cannot lose a proton, but loses hydrogen bromide instead. Secondary α-halogeno-ketones CHBr(COR)$_2$ react analogously.[65]

3-Methylbenzothiazolium iodide reacts with phenylmagnesium bromide in nitrogen to give 3-methyl-2-phenylbenzothiazoline. In air, however, oxidative ring-cleavage occurs *via* a homolytic mechanism to give the disulphide (31).[66] By contrast, 3-methyl-2-phenylbenzothiazolium iodide and phenylmagnesium bromide give only 2,2-diphenyl-3-methylbenzothiazoline in air or nitrogen. With

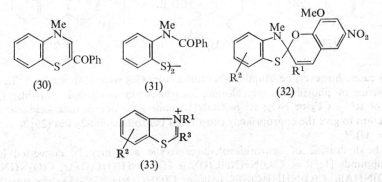

(30)   (31)   (32)

(33)

lithium aluminium hydride or sodium borohydride under nitrogen, 3-methyl-2-phenylbenzothiazolium iodide is reduced to 3-methyl-2-phenylbenzothiazoline, but in air these reactions yield only the disulphide (31).[66] Studies of photochromic benzothiazolinic spiropyrans (32; R^1 = alkoxy or aryloxy), prepared [67] by condensation of (33; R^1 = Me; R^3 = CH$_2$Me, *etc.*) with substituted *o*-hydroxy-benzaldehydes, continue. The synthesis and properties of photochromic 6-methacryloylamino-benzothiazolino-spiro-pyrans {*e.g.* [32; R^1 = Me, R^2 = 6-NHC(O)CMe=CH$_2$]} have been investigated, and a copolymer with methyl methacrylate has been prepared.[68]

Reactions of Other Derivatives.—Condensation of 2-methyl-benzothiazoles with aromatic aldehydes may be carried out in 50% aqueous sodium hydroxide at room temperature in the presence of triethylbenzylammonium chloride.[69] In some cases the intermediate carbinols are isolable under these conditions. 2-Methylbenzothiazoles undergo the Vilsmeier–Haack reaction to give products [1; R = C(:CHOH)CHO] that can be converted into other heterocycles; *e.g.*, alkyl- or aryl-hydrazines furnish pyrazoles.[70]

Photodehydrodimerization of benzothiazole to give 2,2'-bibenzothiazolyl (30% yield) has been carried out in acetonitrile with air, but was unsuccessful in other

[66] M. Hori, T. Kataoka, H. Shimizu, Y. Imai, and H. Fujimura, *Yakugaku Zasshi*, 1975, **95**, 634 (*Chem. Abs.*, 1975, **83**, 193 149).
[67] J. Kister, A. Blanc, E. Davin, and J. Metzger, *Bull. Soc. chim. France*, 1975, 2297; A. Samat, J. Kister, F. Garnier, J. Metzger, and R. Guglielmetti, *ibid.*, p. 2627; H. Pommier, A. Samat, J. Metzger, and R. Guglielmetti, *J. Chim. phys.*, 1975, **72**, 589.
[68] R. M. Gitina, A. L. Prokhoda, I. P. Yudina, E. L. Zaitseva, and V. A. Krongauz, *Khim. geterotsikl. Soedinenii*, 1973, 1639 (*Chem. Abs.*, 1974, **81**, 50 073).
[69] V. Dryanska and C. Ivanov, *Tetrahedron Letters*, 1975, 3519.
[70] M. R. Jayanth, H. A. Naik, D. R. Tatke, and S. Seshadri, *Indian J. Chem.*, 1973, **11**, 1112.

solvents.[71] The mechanism of this reaction appears to be complex; quantum yields depend on the solvent composition, pH, water content, and concentration. Studies [72] of the highly reactive 2-alkoxycarbonyl-benzothiazole *N*-oxides (34; $R^3 = CO_2Et$) (see Vol. 3, pp. 622, 629) have continued. On treatment with ArNCO or ArNSO, the interesting ethyl 1-(arylimino)(S^{IV})benzothiazole-2-carboxylates (35) are produced, with evolution of CO_2 or SO_2, respectively. In

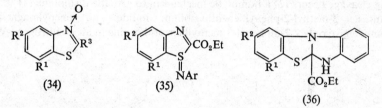

(34)   (35)   (36)

two cases, however, benzimidazo-benzothiazoles (36) were also obtained. In the presence of phosphorus oxychloride, 2-carbamoylbenzothiazole *N*-oxides (34; R^1 or R^2 = CF$_3$ or NO$_2$; R^3 = CONH$_2$) undergo a new intramolecular redox reaction to give the appropriately substituted benzothiazolin-2-ones (29; R = H, X = O).[73]

The hydrazide of benzothiazole-2-carboxylic acid may be converted into compounds [1; R = C(O)NHNHC(O)Alk, C(O)NHNHC(O)Ar, C(O)NHNHC(O)NHAlk, C(O)NHNHC(S)NHAlk, or C(O)NHNHC(S)NHAr] by standard procedures. Cyclization of the oxygen compounds with polyphosphoric acid [74] or phosphorus oxychloride,[75] or of the sulphur compounds with red lead [75] or iodine in sodium hydroxide,[75] yields the corresponding 1,3,4-oxadiazole (37; X = O). The 1,3,4-thiadiazoles (37; X = S) may be prepared by treatment of the oxygen compounds with phosphorus pentasulphide [74] or by cyclization of the sulphur compounds with acid.[75] Compounds (38; X = O or S; Y = N) may be prepared similarly from benzothiazole-6-carboxylic acid.[74] When compounds [1; R = C(O)NHNHC(S)NHAlk or C(O)NHNHC(S)NHAr] are cyclized with 5% aqueous sodium hydroxide or polyphosphoric acid they yield compounds (37;

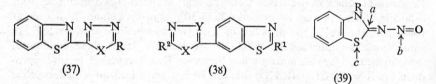

(37)   (38)   (39)

X = NAlk; R = SH).[75] The 6-(2-aminothiazol-2-yl)benzothiazoles (38; X = S; Y = CH; R^2 = NHAlk) have been prepared [76] *via* Hantzsch reactions.

Sulphur-transfer reagents are currently of interest. 3-Methylbenzothiazoline-2-thione is reported [77] to convert oxirans into thiirans in trifluoroacetic acid

[71] K. H. Grellmann and E. Tauer, *Tetrahedron Letters*, 1974, 375.
[72] K. Wagner, K. Ley, and L. Oehlmann, *Chem. Ber.*, 1974, **107**, 414.
[73] K. Wagner and L. Oehlmann, *Chem. Ber.*, 1974, **107**, 305.
[74] S. N. Sawhney, J. Singh, and O. P. Bansal, *J. Indian Chem. Soc.*, 1974, **51**, 886.
[75] S. N. Sawhney, J. Singh, and O. P. Bansal, *Indian J. Chem.*, 1975, **13**, 804.
[76] S. N. Sawhney, J. Singh, and O. P. Bansal, *J. Indian Chem. Soc.*, 1975, **52**, 561.
[77] V. Calò, L. Lopez, L. Marchese, and G. Pesce, *J.C.S. Chem. Comm.*, 1975, 621.

stereospecifically, and in high yield, by a mechanism involving nucleophilic ring-opening by the thione sulphur and transfer of the benzothiazolium moiety to the oxiran oxygen atom.

Diazotization of 2-amino-benzothiazoles can be made to yield the diazonium tetrafluoroborates.[78] Addition of alkali to chilled solutions of these salts gives diazotates (1; R = N_2O^- Na^+), but nitroso-amines (1; R = NHNO) are obtained from 50% aqueous acetic acid.[78]

The ambident reactivity of 3-substituted 2-nitrosoimino-benzothiazolines (39) has been explored further.[79] Nucleophiles can attack at three centres (a, b, and c) (see Vol. 3, p. 630). Grignard reagents react at the 2-position, as well as at the N-atom of the nitroso-group, to give a variety of products, depending on the structure (and hence the nucleophilic reactivity) of the Grignard compound. Organolithium compounds can also react at these positions, as well as at the S-atom. The differences in behaviour of Grignard and organolithium compounds have been attributed to their different co-ordination positions with (39).

5 Benzoselenazoles and Naphthoselenazoles

Oxidation of *N*-methyl-*N′*-phenyl- or *NN′*-diphenyl-selenourea with hydrogen peroxide gives the benzoselenazolyl-guanidines [40; R = N(Me)C(=NPh)-NHMe or N(Ph)C(=NPh)NHPh], analogues of Hugershoff's bases (see p. 389).[80] These compounds may be prepared also from the corresponding 2-amino-benzoselenazoles (40; R = NHMe or NHPh) and PhN=C=NPh or PhN=C=NMe. Their structures have been established by a comparison of their ^{13}C n.m.r. spectra with those of 2-methylaminobenzoselenazole and 2-imino-3-methylbenzoselenazoline.[80]

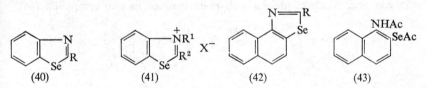

(40)   (41)   (42)   (43)

The chemical shifts of the protons of the methyl groups of *N*-methyl-benzoselenazolium salts have been compared with those for the protons of the methyl groups in similar compounds.[81] Benzoselenazole, tritiated in the 2-position, has been prepared, and the triton chemical shift has been shown to be the same as that of the corresponding proton chemical shift.[82] Benzoselenazol-2-yl 2-formyl-fur-5-yl sulphide has been synthesized from potassium benzoselenazole-2-

[78] V. V. Shaburov, O. V. Vasil'eva and A. V. El'tsov, *Khim. geterotsikl. Soedinenii*, 1974, 367 (*Chem. Abs.*, 1974, **81**, 25 595).

[79] K. Akiba, T. Kawamura, M. Ochiumi, and N. Inamoto, *Heterocycles*, 1973, **1**, 35; K. Akiba, K. Ishikawa, and N. Inamoto, *ibid.*, 1974, **2**, 555; K. Akiba, T. Kawamura, M. Hisaoko, and N. Inamoto, *Bull. Chem. Soc. Japan*, 1975, **48**, 3262; M. Hisaoka, K. Akiba, and N. Inamoto, *ibid.*, p. 3266; K. Akiba, M. Hisaoka, T. Kawamura, and N. Inamoto, *ibid.*, p. 3270; M. Hisaoka, K. Akiba, and N. Inamoto, *ibid.*, p. 3274.

[80] S. Treppendahl, *Acta Chem. Scand.* (*B*), 1975, **29**, 385.

[81] M. Davis, L. W. Deady, and E. Homfeld, *J. Heterocyclic Chem.*, 1974, **11**, 1011.

[82] J. M. A. Al-Rawi, J. P. Boxsidge, C. O'Brien, D. E. Caddy, J. A. Elvidge, J. R. Jones, and E. A. Evans, *J.C.S. Perkin II*, 1974, 1635.

thiolate and 5-bromo(or iodo)-2-furaldehyde, and its u.v. spectrum has been compared with those of similar compounds.[83]

The kinetics of the reactions of 2-azido-3-ethylbenzoselenazolium tetra-fluoroborate (41; R^1 = Et; R^2 = N_3; X = BF_4) with pyrazoles, in which a diazo-group is transferred to the pyrazole and 2-imino-3-ethylbenzoselenazoline is generated, have been studied.[84]

The naphthoselenazole derivative (42; R = NH_2) may be prepared by a Hantzsch-type synthesis from selenourea and 2-bromo-3,4-dihydronaphthalen-1($2H$)-one-3-sulphonic acid followed by fusion of the product with sodium hydroxide.[85] This amine (42; R = NH_2) is ring-opened by sequential treatment with sodium borohydride and acetic anhydride to give compound (43), which may be cyclized to give the naphthoselenazole (42; R = Me).

Selenium analogues of other condensed thiazole ring systems are dealt with in the appropriate Section.

6 Structures Comprising Two Five-membered Rings (5,5)

Thiazolo-[2,3-c]- and -[3,2-b]-[1,2,4]thiadiazoles [C_2N_2S-C_3NS].—Condensations of α-amino-N-heterocycles with chlorocarbonylsulphenyl chloride in the presence of an organic tertiary base can give rise either to a 2,3- or to a 3,4-fused [1,2,4]-thiadiazolone. Thus, 2-aminothiazole yields $2H$-thiazolo[3,2-b][1,2,4]thiadiazol-2-one in THF (see Vol. 3, p. 684).[86] The isomeric 2,3-fused system is formed in ethanol-free chloroform.[87] 2-Amino-Δ^2-thiazoline similarly yields 5,6-dihydro-thiazolo[2,3-c][1,2,4]thiadiazol-3-one, the structure of which has been confirmed by X-ray analysis.[87]

2-Aminothiazole reacts with trichloromethanesulphenyl chloride to give a sulphenamide, which condenses with *m*-nitroaniline to give compound (44).[88]

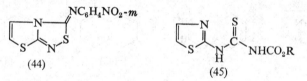

(44)   (45)

p-Nitroaniline yields an analogous product. Oxidation of thioureas (45) with a variety of oxidants leads to ring-closure and good yields of 2-alkoxycarbamyl-imino-thiazolo[3,2-b][1,2,4]thiadiazolines.[89]

Thiazolo-[2,3-c]- and -[3,2-b]-[1,2,4]triazoles [C_2N_3-C_3NS].—4-Allyl-5-phenyl-1,2,4-triazole-3-thione cyclizes in the presence of bromine to give a hydro-bromide of the thiazolo[2,3-c][1,2,4]triazole (46).[90]

[83] R. Kada and J. Kováč, *Chem. Zvesti*, 1975, **29**, 402.
[84] H. Balli, B. Hellrung, and A. Kneubühler-Hof, *Helv. Chim. Acta*, 1974, **57**, 1178.
[85] A. D. Ezekiel, Ger. Offen. 2 449 533/1975 (*Chem. Abs.*, 1975, **83**, 97 306).
[86] K. Pilgram and R. D. Skiles, *J. Org. Chem.*, 1973, **38**, 1575.
[87] D. Baldwin and P. van den Broek. *J.C.S. Perkin I*, 1975, 375.
[88] K. T. Potts and J. Kane, *J. Org. Chem.*, 1975, **40**, 2600.
[89] M. Nagano, M. Oshige, T. Kinoshita, T. Matsui, J. Tobitsuka, and K. Oyamada, *Chem. and Pharm. Bull. (Japan)*, 1973, **21**, 2408.
[90] M. M. Tsitsika, S. M. Khripak, and I. V. Smolanka, *Khim. geterotsikl. Soedinenii*, 1974, 1425 (*Chem. Abs.*, 1975, **82**, 43 276).

Cyclization of the triazoles (47) in 48% hydrobromic acid yields the corresponding thiazolo[3,2-*b*][1,2,4]triazole, isolable as the perchlorate salts.[91] In polyphosphoric acid, the triazole (48) gives the thiazolo[3,2-*b*][1,2,4]triazole (49; $R^1 = R^2 = Ph$), and not the isomer, as previously reported.[92]

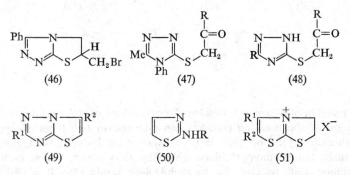

(46)　　　　　(47)　　　　　(48)

(49)　　　　　(50)　　　　　(51)

3-Amination of the thiazoles (50; R = CHO or COMe) with *O*-(mesitylene-sulphonyl)hydroxylamine gives the corresponding salts, which cyclize on heating or on treatment with polyphosphoric acid to give a high yield of the thiazolo-[3,2-*b*][1,2,4]triazoles (49; R^1 = H or Me; R^2 = H).[93] Treatment of these salts with alkali yields 3-amino-2-imino-Δ^4-thiazoline. In a similar manner, 2,3-diaminothiazolium arenesulphonates condense with acid chlorides to give the corresponding thiazolo[3,2-*b*][1,2,4]triazole (49).[94] The intermediate 2-*N*-acyl compounds may be isolated and cyclized in polyphosphoric acid.

Thiazolo[2,3-*b*]thiazoles [C_3NS-C_3NS].—The 5,6-dihydrothiazolo[2,3-*b*]thiazolium salts (51; R^1 = Me or Ph; R^2 = H, COMe, or CO_2Et) may be prepared [95] by allowing a 2-mercapto-Δ^2-thiazoline to react with an α-halogeno-ketone $R^1C(O)CHR^2X$. Their reactions with various nucleophiles have been studied extensively.[95] Nucleophilic attack occurs on the polarized $C=N^+$ bond to give an adduct. This is followed by an elimination, which depends on the basicity, polarizability, and other properties of the reagent to induce cleavage of the S-7—C-7a bond, cleavage of both the S-7—C-7a and N-4—C-5 bonds with elimination of thiiran, or further attack of the reagent on C-6 or on S-7.

Pyrazolo-[3,4-*d*]- and -[5,1-*b*]-thiazoles [$C_3NS-C_3N_2$].—Cyclization of 3-amino-2-methylthiazolium hydrogen sulphate with acetic anhydride yields the pyrazolo-[5,1-*b*]thiazole (52),[96] while condensation of the 3-mercapto-1,2-dihydro-pyrazoles (53) with chloroacetic acid gives compounds (54).[97]

[91] V. A. Kovtunenko, V. N. Bubnovskaya, and F. S. Babichev, *Khim. geterotsikl. Soedinenii*, 1975, 138 (*Chem. Abs.*, 1975, **82**, 140 028).

[92] K. S. Dhaka, J. Mohan, V. K. Chadha, and H. K. Pujari, *Indian J. Chem.*, 1974, **12**, 485.

[93] Y. Tamura, H. Hayashi, J.-H. Kim, and M. Ikeda, *J. Heterocyclic Chem.*, 1973, **10**, 947; Y. Tamura, H. Hayashi, E. Saeki, J.-H. Kim, and M. Ikeda, *ibid.*, 1974, **11**, 459.

[94] Y. Tamura, M. Ikeda, and H. Hayashi, Japan. Kokai 75 30 897/1975 (*Chem. Abs.*, 1975, **83**, 206 279); Japan. Kokai 75 30 898/1975 (*Chem. Abs.*, 1975, **83**, 206 280).

[95] H. Ohtsuka, H. Toyofuku, T. Miyasaka, and K. Arakawa, *Chem. and Pharm. Bull. (Japan)*, 1975, **23**, 3234; H. Ohtsuka, T. Miyasaka, and K. Arakawa, *ibid.*, 1975, **23**, 3243, 3254.

[96] H. Koga, M. Hirobe, and T. Okamoto, *Chem. and Pharm. Bull. (Japan)*, 1974, **22**, 482.

[97] K. Peseke, East Ger. P. 103 006/1974 (*Chem. Abs.*, 1974, **81**, 25 662).

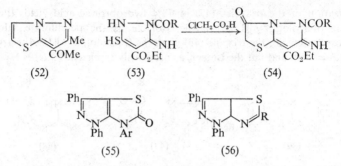

(52)   (53)   (54)

(55)   (56)

The reaction of 4-bromo-1,3-diphenyl-5-pyrazolone with *N*-aryl-thioureas and cyclization of the resulting products can give rise to two types of pyrazolo-[3,4-*d*]thiazoles, (55) or (56; R = NHAr), respectively.[98] With potassium thiocyanate the bromo-pyrazolone gives the thiocyanate, which cyclizes in hydrochloric acid to give the pyrazolo[3,4-*d*]thiazole (56; R = Cl).[98] The pyrazolo[3,4-*d*]thiazole ring system is also formed by the reaction of acid hydrazides with 4-imino-thiazolidin-2-ones.[99]

Imidazo[2,1-*b*]thiazoles [C$_3$NS-C$_3$N$_2$].—Tetramisole is a member of this ring system. Consequently, much of the work published in this area during the period under review is in the patent literature or in biological journals, which the reader should consult for complete coverage.

Further compounds of this class have been prepared from 2-amino-thiazoles or -thiazolines by established procedures.[100-102] Compounds (57; X = ArCH) react with aldehydes at the remaining methylene group, while compounds (57; X = H, H) react first at the methylene group that is adjacent to nitrogen, to give arylidene derivatives.[103]

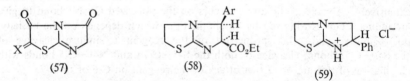

(57)   (58)   (59)

cis-Isomers of α-bromo-cinnamates react with 2-amino-Δ²-thiazoline to give the 2,3,5,6-tetrahydroimidazo[2,1-*b*]thiazoles (58). These reactions are highly regiospecific, and only small amounts of the alternative products are formed.[104] The *trans*-isomers of α-bromo-cinnamates give analogous *trans*-products. The

[98] H. K. Gakhar, V. Parkash, and K. Bhushan, *J. Indian Chem. Soc.*, 1974, **51**, 941.
[99] P. G. Sekachev, *Khim. geterotsikl. Soedinenii*, 1973, 1351 (*Chem. Abs.*, 1974, **80**, 27 156).
[100] J.-F. Robert, A. Xicluna, and J. J. Panouse, *European J. Medicin. Chem., Chim. Therap.*, 1975, **10**, 59.
[101] M. Patra, S. K. Mahapatra, and B. Dash, *J. Indian Chem. Soc.*, 1974, **51**, 1031.
[102] B. S. Drach, I. Yu. Dolgushina, and A. D. Sinitsa, *Khim. geterotsikl. Soedinenii*, 1974, 928 (*Chem. Abs.*, 1974, **81**, 105 382).
[103] A. H. Harhash, M. H. Elnagdi, M. E. Sobhy, and K. M. Foda, *Indian J. Chem.*, 1975, **13**, 238.
[104] R. Bayles, P. W. R. Caulkett, T. P. Seden, and R. W. Turner, *Tetrahedron Letters*, 1975, 4587.

postulated mechanism involves a *cis*-Michael addition, followed by cyclization of the resulting adduct.[104]

A 95% yield of tetramisole hydrochloride (59) is obtained by condensation of thiazolidin-2-one or -2-thione with 2-phenylaziridine in the presence of hydrogen chloride.[105] Optically active aziridine yields optically active tetramisole.

2-Mercapto-Δ^2-imidazoline condenses with ethyl α-chloroacetoacetate to give 5,6-dihydro-4H-imidazo[2,1-b]thiazole (60; R^1 = Me; R^2 = CO_2Et).[106] Compound (60; R^1 = p-ClC_6H_4; R^2 = CH_2CO_2H) may be prepared similarly.[107] 2-Mercapto-Δ^2-imidazoline also reacts with aromatic isothiocyanates in the presence of mercury bis(phenylacetylide) to form the betaines (62), probably *via*

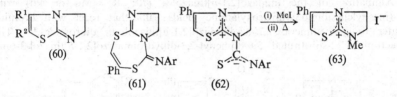

(60) (61) (62) (63)

the corresponding dithiazepine (61).[108] The betaines (62) are formed also by the reaction of 3-phenyl-5,6-dihydroimidazo[2,1-b]thiazole with the appropriate isothiocyanate. Aryl isocyanates react similarly. The betaines (62) are alkylated by methyl iodide on the exocyclic S-atom, and the initial products decompose on pyrolysis under reduced pressure to give the methiodides (63).[108]

Various thiohydantoins (64) react with bromine to give the corresponding 2-bromomethylimidazo[2,1-b]thiazole (65).[109] With glycine, 4-arylidene-2-benzylthio-Δ^2-thiazolin-5-ones are converted into the 4-arylidene-2-thiohydantoin-1-acetic acids (66), which cyclize in acetic anhydride to yield imidazo-[2,1-b]thiazole-2,5-diones (67).[103] A route to 6-acetyl-4-methyl-imidazo[2,1-b]-thiazoles involves hydrogenolysis of compounds (68) followed by cyclization of the ring-cleavage products.[110]

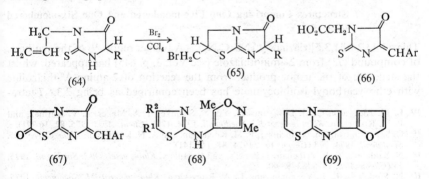

(64) (65) (66)

(67) (68) (69)

[105] R. Fellous, Fr. Demande 2 224 472/1974 (*Chem. Abs.*, 1975, **82**, 156 317).
[106] J.-F. Robert and J. J. Panouse, *Compt. rend.*, 1974, **278**, C, 1289.
[107] P. H. L. Wei and C. Bell, U.S.P. 3 853 872/1974 (*Chem. Abs.*, 1975, **82**, 140 202).
[108] W. Ried, W. Merkel, and S.-W. Park, *Annalen*, 1975, 79.
[109] E. G. Delegan, I. V. Smolanka, and Yu. V. Melika, *Khim. geterotsikl. Soedinenii*, 1974, 1572 (*Chem. Abs.*, 1975, **82**, 72 870).
[110] V. Sprio, O. Migliara, and E. Ajello, *J. Heterocyclic Chem.*, 1974, **11**, 91.

Imidazo[2,1-*b*]thiazole has been shown by ¹H n.m.r. spectroscopy to protonate on the N-atom of the imidazole ring,[111] whilst the mass spectra of some 2-alkyl-substituted imidazo[2,1-*b*]thiazoles have been correlated with their structures.[112]

Compound (69) is brominated in the vacant imidazole ring position by bromine in acetic acid or with sodium hypobromite, but bromination with bromine in chloroform occurs preferentially in the vacant α-position of the furan ring, and in this case a mixture of mono- and di-bromo-compounds is obtained.[113] Compound (69) undergoes the Mannich reaction preferentially in the vacant imidazole ring position; the products undergo further substitution (*e.g.* nitration) in the α-position of the furan ring.[114]

Amination of the imidazo[2,1-*b*]thiazoles (70; R = Me or Ph) with *O*-(mesitylenesulphonyl)hydroxylamine yields salts that react with bromine water to give dimeric 7,7'-azoimidazo[2,1-*b*]thiazolium dibromides.[115] The reaction of 2-substituted 5,6-diphenyl-2,3-dihydroimidazo[2,1-*b*]thiazol-3-ones

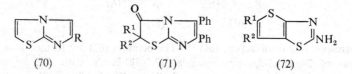

(70)   (71)   (72)

(71; R² = H) with primary amines results in cleavage of the thiazole ring and gives substituted imidazoles, whereas treatment of these compounds with benzene-diazonium chloride yields the 2-phenylazo-derivatives (71; R² = PhN=N).[116]

Pyrrolo[2,1-*b*]thiazoles [C₃NS-C₄N].—An *X*-ray crystallographic study has been made of 3,6-dimethyl-5-thioformylpyrrolo[2,1-*b*]thiazole [117] (see Vol. 3, p. 643).

Thieno[2,3-*d*]thiazoles [C₃NS-C₄S].—The thieno[2,3-*d*]thiazoles [72; R¹ = R² = Me; R¹R² = (CH₂)₄] may be prepared by heating 2-acylamino-3-thiocyanato-thiophens in ethyl benzoate.[118]

7 Structures Comprising One Five-membered and One Six-membered Ring (5,6)

Thiazolo[3,2-*a*][1,3,5]triazines [C₃NS-C₃N₃].—A further report [119] of the synthesis of compound (73) from 2-aminothiazole (see Vol. 3, p. 645) has appeared, whilst the structure of the major product from the reaction of 2-amino-Δ²-thiazoline with ethoxycarbonyl isothiocyanate has been confirmed as being 2,3,6,7-tetra-

[111] L. M. Alekseeva, G. G. Dvoryantseva, Yu. N. Sheinker, I. A. Mazur, B. V. Kurmaz, and P. M. Kochergin, *Khim. geterotsikl. Soedinenii*, 1974, 1206 (*Chem. Abs.*, 1975, **82**, 16 791).
[112] O. S. Anisimova, Yu. N. Sheinker, P. M. Kochergin, and A. N. Krasovskii, *Khim. geterotsikl. Soedinenii*, 1974, 778 (*Chem. Abs.*, 1974, **81**, 151 141).
[113] N. Saldabols, L. L. Zeligman, J. Popelis, and S. Hillers, *Khim. geterotsikl. Soedinenii*, 1975, 55 (*Chem. Abs.*, 1975, **83**, 9898).
[114] N. Saldabols, L. L. Zeligman, and L. A. Ritevskaya, *Khim. geterotsikl. Soedinenii*, 1975, 1208 (*Chem. Abs.*, 1976, **84**, 30 960).
[115] E. E. Glover and K. D. Vaughan, *J.C.S. Perkin I*, 1974, 1137.
[116] M. I. Ali, M. A. Abou-State, and A. F. Ibrahim, *J. prakt. Chem.*, 1974, **316**, 147.
[117] A. Sharma and R. C. G. Killean, *Acta Cryst.*, 1974, **30B**, 2869.
[118] V. I. Shvedov, I. A. Kharizomenova, and A. N. Grinev, *Khim. geterotsikl. Soedinenii*, 1974, 1204 (*Chem. Abs.*, 1975, **82**, 16 733).
[119] G. Barkinow and H. Ebeling, *Z. Chem.*, 1974, **14**, 356.

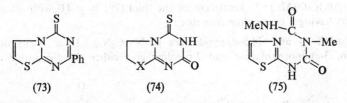

(73)   (74)   (75)

hydro-4*H*-thiazolo[3,2-*a*][1,3,5]triazin-2-one-4-thione (74; X = S) by means of
X-ray crystallographic analysis.[120] This product is analogous to that similarly
obtained from 2-aminothiazole. In this reaction the behaviour of 2-amino-
thiazole has been compared with that of other heterocyclic amines.[121] Compound
(74; X = S) reacts with diazomethane at both the N- and S-atoms of the
triazine ring.[120] The biuret (75) cyclizes in aqueous acetic acid with loss of
methylamine to give 3-methyl-2,3-dihydro-4*H*-thiazolo[2,3-*a*][1,3,5]triazine-2,4-
dione.[122]

2-Amino-Δ²-selenazoline condenses similarly with ethoxycarbonyl isothiocyanate to give 2,3,6,7-tetrahydro-4*H*-selenazolo[3,2-*a*][1,3,5]triazin-2-one-4-
thione (74; X = Se).[120]

Thiazolo[2,3-*c*][1,2,4]triazines [C₃NS-C₃N₃].—Condensation of triazine (76;
R = SEt) with chloroacetic acid gives 5,6-diphenyl-5*H*-thiazolo[2,3-*c*][1,2,4]-
triazin-3(2*H*)-one (77), which condenses with benzaldehyde to give the arylidene

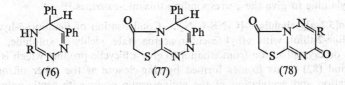

(76)   (77)   (78)

derivative, and reacts with aromatic amines to give the 3-arylamino-compounds
(76; R = NHAr) (see next Section also).[123] These arylidene derivatives may be
prepared similarly in a 'one-pot' reaction from 5,6-diaryl-4,5-dihydro[1,2,4]-
triazine-3(2*H*)-thiones.[123]

Thiazolo[3,2-*b*][1,2,4]triazines [C₃NS-C₃N₃].—Investigations into the reactivity
of 6-substituted 2,3-dihydro-7*H*-thiazolo[3,2-*b*][1,2,4]triazine-3,7-diones (78) have
shown that, whilst condensation with aromatic aldehydes gives the 2-arylidene
derivatives, the reaction with primary aromatic amines yields the corresponding
2,5-dihydro-3-arylamino[1,2,4]triazin-5-one, and not the expected anilides.[124]

Thiazolo[3,2-*b*]pyridazines [C₃NS-C₄N₂].—α-Chloro-ketones interact with 3,6-
dimercaptopyridazines to give the thiazolo[3,2-*b*]pyridazinium salts [79; R = H

[120] D. L. Klayman and T. S. Woods, *J. Org. Chem.*, 1974, **39**, 1819.
[121] T. Matsui, M. Nagano, J. Tobitsuka, and K. Oyamada, *Chem. and Pharm. Bull. (Japan)*, 1974, **22**, 2118.
[122] A. Étienne and B. Bonte, *Bull. Soc. chim. France*, 1975, 1419.
[123] M. I. Ali, A. M. Abd-Elfattah, H. A. Hammouda, and S. M. Hussein, *Indian J. Chem.*, 1975, **13**, 109.
[124] M. I. Ali, A. A. El-Sayed, and H. A. Hammouda, *J. prakt. Chem.*, 1974, **316**, 163.

or CH(Alk)C(O)Alk].[125] Treatment of the thiol (79; R = H) with alkali gives products having a novel ylide structure.

Thiazolo-[4,5-*c*]- and -[4,5-*d*]-pyridazines [$C_3NS-C_4N_2$].—In a Willgerodt-type reaction, 3-aminopyridazine and 2-picoline give either the thioamide at low

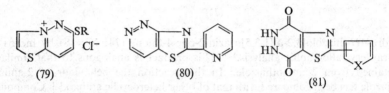

(79)   (80)   (81)

temperatures or 2-(2-pyridyl)thiazolo[4,5-*c*]pyridazine (80) at higher temperatures.[126] Some 3-aminopyridazine-5(2*H*)-thione is also produced in the latter case. Oxidative cyclization of the thioamide with potassium ferricyanide also gives the thiazolo-pyridazine (80). Condensation of 4,5-diethoxycarbonyl-2-heteroaryl-thiazoles with hydrazine yields the thiazolo[4,5-*d*]pyridazines (81; X = O, S, or Se).[127] The corresponding selenazolo[4,5-*d*]pyridazines [127] and compounds derived from substituted hydrazines and ethyl 2-aryl-5-acetyl-thiazole-4-carboxylates may be prepared similarly.[128]

Thiazolo[5,4-*c*]pyridazines [$C_3NS-C_4N_2$].—5-Alkoxy-4-acylmethyl-2-aminothiazoles react similarly with hydrazine to give thiazolo[5,4-*c*]pyridazines and with hydroxylamine to give the corresponding thiazolo-oxazines.[129]

Thiazolo[3,2-*a*]pyrimidines [$C_3NS-C_4N_2$].—Condensation of 6-amino-4-hydroxy-pyrimidine-2-thiol with ethyl α-chloroacetoacetate yields a sulphide, which cyclizes on crystallization from ethanol to give a bicyclic product which is either compound (82) or its isomer formed by ring-closure at the other nitrogen.[130] Dehydration and acetylation of the amino-group occur with acetic anhydride.

Cyclization of 5-arylidene-2-(β-ethoxycarbonylethylimino)thiazolidin-4-ones (prepared from ethyl acrylate and 5-arylidene-2-iminothiazolidin-4-ones) with acetic anhydride gives the thiazolo[3,2-*a*]pyrimidines (83).[103]

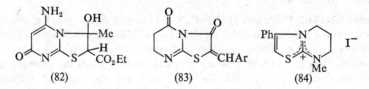

(82)   (83)   (84)

[125] K. Arakawa, Japan. Kokai 73 92 393/1973 (*Chem. Abs.*, 1974, **80**, 95 998) and 73 91 092/1973 (*Chem. Abs.*, 1974, **81**, 120 669).
[126] L. Kramberger, P. Lorenčak, S. Polanc, B. Verček, B. Stanovnik, and M. Tišler, *J. Heterocyclic Chem.*, 1975, **12**, 337.
[127] P. Chauvin, J. Morel, P. Pastour, and J. Martinez, *Bull. Soc. chim. France*, 1974, 2099.
[128] B. Tornetta, F. Guerrera, and G. Ronsisvalle, *Ann. chim. (Italy)*, 1974, **64**, 477 (*Chem. Abs.*, 1976, **84**, 17 210).
[129] A. Sammour, M. I. B. Selim, and E. A. Soliman, *Egyptian J. Chem.*, 1972, **15**, 311 (*Chem. Abs.*, 1974, **80**, 108 473).
[130] S. C. Bell, C. Gochman, and P. H. L. Wei, *J. Heterocyclic Chem.*, 1975, **12**, 1207.

Cyclic thioureas containing six-membered rings condense with aromatic isothiocyanates in the presence of mercury bis(phenylacetylide), in a similar manner to that described previously (see p. 401) for 2-mercaptoimidazoline, to give, eventually, the methiodide (84).[108] Thiazolo[3,2-*a*]pyrimidine-5,7-diones, on alcoholysis or aminolysis, give the corresponding 2-substituted thiazoles [50; R = C(O)CH$_2$CO$_2$Alk or C(O)CH$_2$C(O)NHAlk].[131]

2-Amino-selenazoles, substituted in the 4- or 5-positions, react with dimethyl acetylenedicarboxylate or ethyl propiolate to give the corresponding 7*H*-selenazolo[3,2-*a*]pyrimidin-7-ones (85; R = H or CO$_2$Me).[132] 2-Amino-1,3,4-selenadiazoles react similarly with ethyl propiolate, but the products react further with

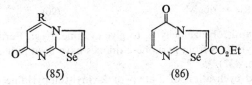

(85)   (86)

the ester to yield a betaine, which extrudes a nitrile or HCN to give the selenazolo[3,2-*a*]pyrimidinone (86).[132]

Thiazolo[3,2-*c*]pyrimidines [C$_3$NS-C$_4$N$_2$].—2,3-Dihydro-6-(ethoxycarbonyl)thiazolo[3,2-*a*]pyrimidin-5-one undergoes hydrolysis, rearrangement, and decarboxylation to give 2,3-dihydrothiazolo[3,2-*c*]pyrimidin-5-one on treatment with hydrochloric acid.[133]

Thiazolo[4,5-*d*]pyrimidines [C$_3$NS-C$_4$N$_2$].—4-(β-D-Ribofuranosyl)-4,5,6,7-tetrahydrothiazolo[4,5-*d*]pyrimidine-5,7-dione (88), the 1-thio-analogue of 3-isoxanthosine, has been synthesized by allowing the uracil derivative (87) to react with

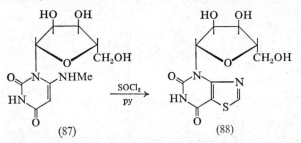

(87)   (88)

thionyl chloride in pyridine.[134] Thiazolo[4,5-*d*]pyrimidines have also been obtained by oxidative cyclization of thioureas.[135]

Thiazolo[5,4-*d*]pyrimidines [C$_3$NS-C$_4$N$_2$].—Further details have been published of the preparation of thiazolo[5,4-*d*]pyrimidine nucleosides [*e.g.* (89)] from thiazolo[5,4-*d*]pyrimidine-5,7-diones (see Vol. 3, p. 653).[136]

[131] L. B. Dashkevich, B. P. Tarasov, Yu. Khodzhibaev, and M. M. Samoletov, *Zhur. org. Khim.*, 1975, **11**, 2200 (*Chem. Abs.*, 1976, **84**, 59 368).
[132] A. Shafiee and I. Lalezari, *J. Heterocyclic Chem.*, 1975, **12**, 675.
[133] G. R. Brown, *J.C.S. Perkin I*, 1973, 2022.
[134] Y. Mizuno, Y. Watanabe, and K. Ikeda, *Chem. and Pharm. Bull.* (*Japan*), 1974, **22**, 1198.
[135] A. Berger and E. E. Borgaes, U.S.P. 3 772 290/1973 (*Chem. Abs.*, 1974, **80**, 48 027).
[136] C. L. Schmidt and C. B. Townsend, *J. Org. Chem.*, 1975, **40**, 2476.

Thiazolo[3,2-*a*]pyridines [C₃NS-C₅N].—3-Ethoxy-2,3-dihydrothiazolo[3,2-*a*]pyridinium bromides [*e.g.* (90)] are formed by the reaction of the appropriate pyridinethiol with bromoacetal.[137] On heating with acid they eliminate ethanol.[138] Both types of compounds, (90)[137] and its unsaturated analogue,[138] react with

(89) Rib = ribosyl

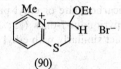

(90)

aldehydes, *e.g.* *p*-nitrobenzaldehyde, to give cyanine dyes. Further syntheses of thiazolo[3,2-*a*]pyridines from pyridine-2-thiones have been published (see Vol. 2, p. 698; Vol. 1, p. 432).[139, 140]

Acid-catalysed cyclization of 2-allyl- or 2-vinyl-thiopyridine or 2-β-hydroxyalkylthio-pyridines (91) proceeds without rearrangement, to give the corresponding dihydrothiazolo[3,2-*a*]pyridinium derivative (93).[141] By contrast, 2-β-hydroxyalkylthio-pyridines (91) react with cold thionyl chloride to give a mixture of isomeric dihydrothiazolo[3,2-*a*]pyridinium derivatives and isomeric 2-β-hydroxyalkylthio-pyridines.[141] These results are in part rationalized by reversible formation of the episulphonium ion (92).

The electron-releasing properties of the amino- or acetoamido-groups in the thiazolo[3,2-*a*]pyridines (93; R¹ = 6- or 8-NH₂, 6- or 8-NHCOMe, R²—R⁵ = H) are insufficient to allow bromination under mild conditions (see Vol. 1, p. 433; Vol. 2, p. 698). Contrary to expectations, the corresponding nitro-

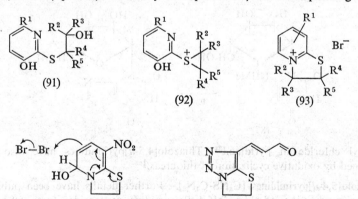

(91)   (92)   (93)

(94)   (95)

[137] E. D. Sych, V. N. Bubnovskaya, L. T. Gorb, and M. Yu. Kornilov, *Khim. geterotsikl. Soedinenii*, 1973, 1254 (*Chem. Abs.*, 1974, **80**, 38 345).
[138] E. D. Sych, L. T. Gorb, V. N. Bubnovskaya, and F. S. Babichev, *Khim. geterotsikl. Soedinenii*, 1974, 1335 (*Chem. Abs.*, 1975, **82**, 100 051).
[139] L. A. Riege and K. Undheim, *Acta Chem. Scand.* (*B*), 1975, **29**, 582.
[140] S. Hagen, G. A. Ulsaker, and K. Undheim, *Acta Chem. Scand.* (*B*), 1974, **28**, 523.
[141] G. A. Ulsaker and K. Undheim, *Acta Chem. Scand.* (*B*), 1975, **29**, 853.

compounds are readily brominated in aqueous methanol.[140] This observation has been rationalized by postulating the formation of the pseudo-base, which is then brominated [*e.g.* (94)]. The electron-withdrawing effect of the nitro-group increases the electron-deficiency of the pyridinium ring such that addition of hydroxide ion occurs.

Diazotization of the aminothiazolo[3,2-*a*]pyridinium bromide (93; R¹ = 8-NH₂, R²—R⁵ = H) gives a mixture of *trans*-3-(7,8-dihydrothiazolo[2,3-*e*]-[1,2,3]triazolyl)propenal (95), whose structure has been confirmed by *X*-ray crystallographic analysis,[142] and the corresponding *cis*-isomer.[140] This reaction is also considered to involve pseudo-base formation followed by ring-opening and ring-closure with rearrangement.[140]

Thiazolo[4,5-*b*(and -*c*)]pyridines [C₃NS-C₅N].—Oxidative cyclization of the thioamide (96) gives 6-methyl-2-(2-pyridyl)thiazolo[4,5-*b*]pyridine.[126] The isomeric thiazolo[4,5-*c*]pyridine and 2-(2-pyridyl)thiazolo[5,4-*b*]pyridine may be prepared similarly.

Thiazolo[5,4-*b*]pyridines [C₃NS-C₅N].—Condensation of 3-amino-2-chloropyridine with carbon disulphide in DMF yields thiazolo[5,4-*b*]pyridine-2-thione (97), which exists predominantly as this tautomer in neutral media.[143] 2-Chloro-3-isocyanatopyridine reacts with various *N*-substituted piperazines to give the

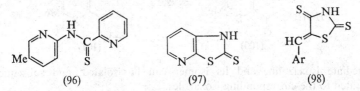

(96)   (97)   (98)

corresponding 2-amino-thiazolo[5,4-*b*]pyridine (see Vol. 3, p. 656).[144] 6-Substituted 2-amino-thiazolo[5,4-*b*]pyridines are formed from 5-amino-2-chloro(or methoxy)pyridine in up to 97% yield on treatment with bromine and potassium thiocyanate.[145] The reaction presumably involves the intermediate formation of the corresponding 3-amino-2-thiocyanatopyridine, which then cyclizes.

Thiopyrano-7*H*-[2,3-*d*]thiazoles [C₃NS-C₅S].—The 5-arylidene-1,3-thiazolidine-2,4-dithiones (98; Ar = Ph, *o*-ClC₆H₄, or *o*-, *p*-MeOC₆H₄) undergo 1,4-cyclo-addition with various alkenes R¹CH=CHR² (*e.g.* R¹ = CN or CO₂Et, R² = H) and related compounds to give tetrahydrothiopyrano-7*H*-[2,3-*d*]thiazole-2-thiones.[146]

8 Structures Comprising Two Five-membered Rings and One Six-membered Ring (5,5,6)

[1,2,4]Triazolo-[3,2-*b*]- and -[3,4-*b*]-benzothiazoles [C₂N₃-C₃NS-C₆].—Cyclization of 4-(2-halogeno-aryl)-1-acyl-3-thiosemicarbazides in aqueous potassium

[142] C. Roemming, *Acta Chem. Scand.* (*A*), 1975, **29**, 282.
[143] W. O. Foye, J. M. Kauffman, J. J. Lanzillo, and E. F. LaSala, *J. Pharm. Sci.*, 1975, **64**, 1371.
[144] B. G. Khadse, M. H. Shah, and C. V. Deliwala, *Bull. Haffkine Inst.*, 1975, 3, 27 (*Chem. Abs.*, 1975, **83**, 193 225).
[145] C. O. Okafor, *J. Org. Chem.*, 1973, **38**, 4383.
[146] N. A. El L. Kassab, S. O. A. Allah, and H. A. El R. Ead, *Z. Naturforsch.*, 1975, **30b**, 441.

hydroxide yields 3-mercapto-triazoles.[147] Treatment of these or the starting
thiosemicarbazides with sodium hydride in DMF gives the [1,2,4]triazolo-
[3,4-*b*]benzothiazoles (99).

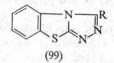

(99)

2-Methyl- and 2-phenyl-[1,2,4]triazolo[3,2-*b*]benzothiazoles are formed when
2,3-diaminobenzothiazolium arenesulphonates are treated with acetyl or benzoyl
chloride.[94b] The intermediate 2-*N*-acylated compounds are isolable and may be
cyclized with polyphosphoric acid. Analysis of the ^{1}H n.m.r. spectrum of
[6-^{2}H][1,2,4]triazolo[3,4-*b*]benzothiazole allows all the ^{1}H chemical shifts in the
parent compound to be assigned.[148]

Thiazolo[4,5-*g*]benzisoxazoles [$C_3NO-C_3NS-C_6$].—The reaction of the α-hydroxymethylene ketone (100) with hydroxylamine gives the novel heterocycle 7-methyl-
4,5-dihydrothiazolo[4,5-*g*]benzisoxazole (101).[149] Under suitable conditions the

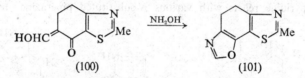

(100)   (101)

intermediate isoxazoline and its isomer can be isolated, and subsequently
dehydrated to the corresponding isoxazoles.

Bis(thiazolo)[4,5-*b*:5,4-*e*]thiopyrylium Salts [$C_3NS-C_3NS-C_5S$].—Compounds
(104) may be prepared from the betaine (102) and compounds (103).[150] When
R = Ph, an intermediate is isolable, and it cyclizes, either in acid or on heating,
to yield the product.

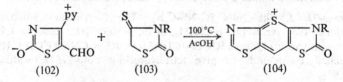

(102)   (103)   (104)

Thiazolo[2,3-*b*]benzothiazoles [$C_3NS-C_3NS-C_6$].—Addition of bromine to
2-allylthiobenzothiazole yields 3-bromomethyl-2,3-dihydrothiazolo[2,3-*b*]benzo-
thiazol-4-ium bromide.[151]

[147] J. H. Wikel and C. J. Paget, *J. Org. Chem.*, 1974, **39**, 3506.
[148] R. Faure, G. Guisti, J. P. Galy, E. J. Vincent, and J. Elguerro, *Bull. Soc. chim. France*, 1974,
      2967.
[149] A. Fravolini, G. Grandolini, and A. Martani, *Gazzetta*, 1973, **103**, 755.
[150] S. N. Baranov, R. O. Kochkanyan, G. I. Belova, and A. N. Zaritovskii, *Doklady Akad. Nauk
      S.S.S.R.*, 1975, **222**, 101 (*Chem. Abs.*, 1975, **83**, 114 273).
[151] M. Kocevar, B. Stanovnik, and M. Tišler, *Croat. Chem. Acta*, 1973, **45**, 457 (*Chem. Abs.*,
      1974, **80**, 37 043).

Thiazolo-[2,3-*f*]- and -[3,2-*e*]-purines [C₃NS-C₃N₂-C₄N₂].—A thiazole ring can be annelated to the purine ring system by standard procedures[152] to give, *e.g.*, compounds (105) obtained from the corresponding theophylline derivatives.[153]

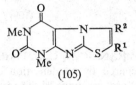

(105)

Cyclization of 6-amino-8-(2-hydroxyethylthio)purine with polyphosphoric acid gives a mixture of 4-amino-6,7-dihydrothiazolo[2,3-*f*]- and 4-amino-7,8-dihydro-thiazolo[3,2-*e*]-purines.[154] The 8-propargylthio-derivatives are cyclized on treatment with hydrogen halides to give the corresponding thiazolo[3,2-*e*]-purines.[155]

Deuteriation of thiazolo[2,3-*f*]purine in deuteriosulphuric acid has been shown, by the use of n.m.r. spectroscopy, to occur first on a N-atom of the pyrimidine ring and then on a N-atom of the imidazole ring.[111]

Thiazolo-[3,2-*a*]- and -[3,4-*a*]-benzimidazoles [C₃NS-C₃N₂-C₆].—Members of this ring system have been prepared from 2-mercaptobenzimidazoles by conventional methods.[156] Cyclization of the sulphide, prepared from benzimidazole-2-thiol and ethyl α-chloroacetoacetate, in acetic anhydride containing pyridine has been shown previously[157] to give a thiazolo[3,2-*a*]benzimidazole. In the absence of pyridine, compound (106) is also isolable.[130] The ester group in the latter is readily hydrolysed. A novel depropargylative cyclization of 1-pro-pargyl-2-propargylthiobenzimidazole on reaction with HMPA leads to 3-methyl-

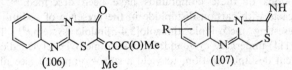

(106)   (107)

thiazolo[3,2-*a*]benzimidazole.[158] 2-Chloromethyl-benzimidazoles react with ammonium thiocyanate to give the thiazolo[3,4-*a*]benzimidazoles (107), the imino-group of which readily reacts with electrophiles.[159]

Pyrazolo[4,3-*g*]benzothiazoles [C₃NS-C₃N₂-C₆].—2*H*-7-Methyl-4,5-dihydropyra-zolo[4,3-*g*]benzothiazole (108) is obtained when compound (100) is treated with hydrazine or semicarbazide.[149, 160]

[152] H. Uno, A. Irie, and K. Hino, *Chem. and Pharm. Bull.* (*Japan*), 1975, **23**, 450; Japan. Kokai 73 72 195/1973 (*Chem. Abs.*, 1974, **80**, 48 038).
[153] M. I. Yurchenko, P. M. Kochergin, and A. N. Krasovskii, *Khim. geterotsikl. Soedinenii*, 1974, 693 (*Chem. Abs.*, 1974, **81**, 63 585).
[154] K. Hino, A. Irie, and H. Uno, *Chem. and Pharm. Bull.* (*Japan*), 1975, **23**, 1696.
[155] H. Uno and A. Irie, Japan. Kokai 73 80 598/1973 (*Chem. Abs.*, 1974, **80**, 121 006).
[156] J. Mohan, V. K. Chadha, and H. K. Pujari, *Indian J. Chem.*, 1973, **11**, 1119.
[157] J. J. D'Amico, R. H. Campbell, and E. C. Guinn, *J. Org. Chem.*, 1964, **29**, 865.
[158] K. K. Balasubramanian and B. Venugopalan, *Tetrahedron Letters*, 1974, 2645.
[159] R. D. Haugwitz, B. V. Maurer, and V. L. Narayanan, *J. Org. Chem.*, 1974, **39**, 1359; R. Haugwitz and V. L. Narayanan, U.S.P. 3 819 618/1974 (*Chem. Abs.*, 1974, **81**, 105 520).
[160] A. Fravolini, G. Grandolini, and A. Martani, *Gazzetta*, 1973, **103**, 1057.

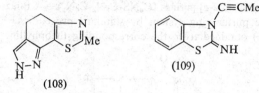

(108)   (109)

Imidazo[2,1-*b*]benzothiazoles [C₃NS-C₃N₂-C₆].—Further members of this class of compounds have been prepared by condensation of a 2-aminobenzothiazole with a substituted ω-bromo-acetophenone.[101] The imine (109) is converted into 2-methylimidazo[2,1-*b*]benzothiazole in dilute sulphuric acid in the presence of mercuric sulphate.[161]

Thiazolo-[3,2-*a*]- and -[3,4-*a*]-indoles [C₃NS-C₄N-C₆].—A convenient synthesis of the thiazolo[3,2-*a*]indolones (110; R¹ = H or Me; R² = H, NH₂, NHMe, NHCO₂CH₂Ph, or CONHPh), analogues of anhydrogliotoxin and mono-desthiosecogliotoxin, involves the reaction of α-mercaptoacetic acids (or their

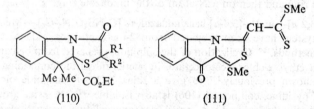

(110)   (111)

esters) R¹R²C(SH)CO₂H with ethyl 3,3-dimethyl-3*H*-indole-2-carboxylate.[162, 163] Various reactions of these compounds have been described.[163] 1-Acetyl-3-indolinone reacts with carbon disulphide in the presence of sodium hydride to give an interesting meso-ionic thiazolo[3,4-*a*]indole system.[164] The dithio-carboxylate (111) undergoes cycloaddition with dimethyl acetylenedicarboxylate, with concurrent desulphurization, to yield a spiro-compound (see also pp. 381 and 393).[57]

Thiazolo-[4,5-*b*]- and -[5,4-*b*]-indoles [C₃NS-C₄N-C₆].—Syntheses of compound (112) and its [4,5-*b*]-isomer have been reported, the former from 1-methylisatin and the latter starting from 1-methyl-2-acetamidoindole.[165]

Thiazoloisoindoles [C₃NS-C₄N-C₆].—Photolysis of *N*-(benzylthiomethyl)phthali-mide gave a 27% yield of the cyclized product (113). With the corresponding methylthiomethyl compound, cyclization also occurred, but in lower yield.[166]

[161] P. M. Kochergin, A. N. Krasovskii, and N. P. Grin, U.S.S.R. P. 443 039/1974 (*Chem. Abs.*, 1975, **82**, 31 325).
[162] H. C. J. Ottenheijm, N. P. E. Vermeulen, and L. F. J. M. Breuer, *Annalen*, 1974, 206.
[163] H. C. J. Ottenheijm, J. J. M. L. Hoffmann, P. T. M. Biessels, and A. D. Potman, *Rec. Trav. chim.*, 1975, **94**, 138.
[164] Y. Tominaga, Y. Matsuda, and G. Kobayashi, *Yakugaku Zasshi*, 1975, **95**, 980 (*Chem. Abs.*, 1976, **84**, 31 011).
[165] P. I. Abramenko, *Zhur. Vsesoyuz. Khim. obshch. im D. I. Mendeleeva*, 1973, **18**, 714 (*Chem. Abs.*, 1974, **80**, 95 808).
[166] Y. Sato, H. Nakai, H. Ogiwara, and T. Mizoguchi, *Tetrahedron Letters*, 1973, 4565.

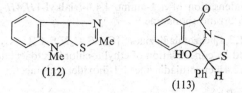

(112)   (113)

Furo[3,2-*e*]benzothiazoles [$C_3NS-C_4O-C_6$].—2-Aminofuro[3,2-*e*]benzothiazoles may be prepared [167] by a Hantzsch-type synthesis involving the appropriately substituted 4,5,6,7-tetrahydrobenzo[*b*]furan.

Thiazolo[3,2-*a*]thieno[2,3-*b*(and -*d*)]pyrimidines [$C_3NS-C_4S-C_4N_2$].—An attempt to prepare thiazolo[3,2-*a*]thieno[3,2-*d*]pyrimidines containing basic side-chains by the reaction of a 5-allyl-6-mercapto-thienopyrimidine with bromine followed by base led to the formation of the 2-methyl compound (114).[168, 169] Subsequent

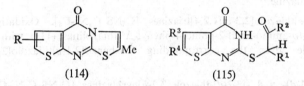

(114)   (115)

bromination and reaction with morpholine gave the desired side-chain.[170-172] Usually the 2-(β-oxo-alkylthio)-thieno[3,2-*d*]pyrimidin-4(3*H*)-ones (115; R^3 = R^4 = Me) cyclize in sulphuric acid to give the thiazolo[3,2-*a*]thieno[2,3-*d*]-pyrimidines, together with a small amount of the [2,3-*b*]-isomer.[173] Compound (115; R^1 = H; R^2—R^4 = Me), prepared from chloroacetone and the corresponding 2-mercaptothieno[2,3-*d*]pyrimidin-4(3*H*)-one, spontaneously cyclizes to give only the [2,3-*d*]-isomer.[173]

Indeno[1,2-*d*]thiazoles [$C_3NS-C_5-C_6$].—2-Nitroindane-1,3-dione reacts with various thioureas to give 2-amino-indeno[1,2-*d*]thiazol-8-ones.[174] The primary amine may be diazotized and converted into the 2-bromo-compound by standard procedures.

9 Structures Comprising One Five-membered Ring and Two Six-membered Rings (5,6,6)

Pyrimido[2,1-*b*]thiazolo[4,5-*e*][1,3,4]thiadiazines [$C_3NS-C_3N_2S-C_4N_2$].—The title compounds (116; R^1 = Me or Et; R^2 = H, Me, Cl, OMe, or OEt) may be

[167] W. A. Remers and G. S. Jones, *J. Heterocyclic Chem.*, 1975, **12**, 421.
[168] F. Sauter and W. Deinhammer, *Monatsh.*, 1974, **105**, 452.
[169] F. Sauter, W. Deinhammer, and K. Danksagmueller, *Monatsh.*, 1974, **105**, 863.
[170] F. Sauter, W. Deinhammer, and K. Danksagmueller, *Monatsh.*, 1974, **105**, 882.
[171] I. V. Smolanka, A. A. Dobosh, and S. M. Khripak, *Khim. geterotsikl. Soedinenii*, 1973, 1289 (*Chem. Abs.*, 1974, **80**, 3465).
[172] A. A. Dobosh, I. V. Smolanka, and S. M. Khripak, *Khim. geterotsikl. Soedinenii*, 1974, 134 (*Chem. Abs.*, 1974, **80**, 95 876).
[173] F. Sauter, W. Deinhammer, and P. Stanetty, *Monatsh.*, 1974, **105**, 1258.
[174] V. Barkane, E. Gudriniece, and D. Skerite, *Latv. P.S.R. Zinat. Akad. Vestis., Khim. Ser.*, 1975, 345 (*Chem. Abs.*, 1975, **83**, 178 899).

prepared by condensation of a 1-amino-4,4,6-trialkyl-1*H*,4*H*-pyrimidine-2-thiol with an *N*-aryl-5-bromo-rhodanine.[175]

Thiazolo[2,3-c][1,2,4]benzothiadiazines [C₃NS-C₃N₂S-C₆].—Cyclization of the sulphide prepared by condensation of ethyl α-chloroacetoacetate with 6-chloro-3-mercapto-4*H*-[1,2,4]benzothiadiazine 1,1-dioxide with acetic anhydride in the

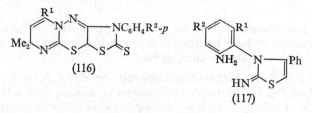

(116)   (117)

presence of pyridine yields the appropriately substituted thiazolo[2,3-c][1,2,4]-benzothiadiazine.[130]

Benzo[2,1-e]thiazolo[2,3-c][1,2,4]triazines [C₃NS-C₃N₃-C₆].—Oxidative cyclization of 3-(*o*-amino-aryl)-2-imino-4-phenyl-Δ⁴-thiazolines (117) with *N*-bromosuccinimide gives the corresponding 9*H*-benzo[2,1-c]thiazolo[2,3-c][1,2,4]-triazine.[176]

Pyrido-[2,3-d]- and -[3,2-d]-thiazolo[3,2-b]pyridazines [C₃NS-C₄N₂-C₅N].—Two of the compounds investigated by mass spectrometry for which fragmentation patterns have been proposed are salts of 3-phenylpyrido[2,3-d]thiazolo[3,2-b]-pyridazine and the isomeric 3-phenylpyrido[3,2-d]thiazolo[3,2-b]]pyridazine.[177]

Pyrido[2,3-d]thiazolo[3,2-a]pyrimidines [C₃NS-C₄N₂-C₅N].—Pyrido[2,3-d]thiazolo[3,2-a]pyrimidin-5-ones are formed by condensation of a 2-amino-thiazole with 2-chloro-5-nitronicotinic acids.[178]

Pyrido[1,2-a]thiazolo[5,4-e]pyrimidines [C₃NS-C₄N₂-C₅N].—Condensation of the thiazole (118) with ammonium carbonate yields compound (119).[179] The same

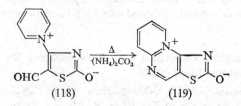

(118)   (119)

compound arises by elimination of hydrogen chloride from the product of condensation of 2-aminopyridine with 4-chloro-5-formyl-Δ⁴-thiazolin-2-one.

Pyrimido[2,1-b]benzothiazoles [C₃NS-C₄N₂-C₆].—2-Aminobenzothiazole reacts with methyl propiolate in THF to give largely 2*H*-pyrimido[2,1-b]benzothiazol-

[175] H. K. Gakhar, S. H. Chander, and H. C. Khurana, *J. Indian Chem. Soc.*, 1975, **52**, 155.
[176] S. K. Vasudeva, M. P. Mahajan, and N. K. Ralhan, *Indian J. Chem.*, 1973, **11**, 1204.
[177] V. Kramer, M. Medved, B. Stanovnik, and M. Tišler, *Org. Mass Spectrometry*, 1974, **8**, 31.
[178] S. Singh, *J. Indian Chem. Soc.*, 1973, **50**, 358.
[179] R. O. Kochkanyan, G. I. Belova, V. S. Garkusha-Bozhko, A. N. Zaritovskii, and S. N. Baranov, *Khim. geterotsikl. Soedinenii*, 1975, 1426 (*Chem. Abs.*, 1976, **84**, 43 968).

2-one (120; R = H) (45% yield).[180] With dimethyl acetylenedicarboxylate at room temperature, a complex mixture is obtained, containing methyl 2-oxo-2*H*-pyrimido[2,1-*b*]benzothiazole-4-carboxylate (120; R = CO_2Me) (4%) and its isomer, methyl 4-oxo-4*H*-pyrimido[2,1-*b*]benzothiazole-2-carboxylate (2%). In boiling ethanol, only compound (120; R = CO_2Me) is obtained.[180] Compound (120; R = H) is hydrolysed with sodium methoxide in methanol to give an imino-ester resulting from cleavage of the pyrimidine ring.[180]

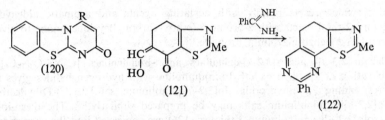

(120)

(121)

(122)

Pyrimido[5,4-*g*]benzothiazoles [$C_3NS-C_4N_2-C_6$].—The α-hydroxymethylene ketone (121) reacts with benzamidine to give 8-methyl-5,6-dihydro-2-phenylpyrimido-[5,4-*g*]benzothiazole (122).[181]

Thiazolo[2,3-*b*]quinazolines [$C_3NS-C_4N_2-C_6$].—Annelation of a thiazole ring to the quinazoline ring system may be carried out by standard procedures.[182] Ester (123) reacts normally with hydrazine or phenylmagnesium bromide, but undergoes a complex rearrangement with piperidine or *N*-methylpiperazine.[182]

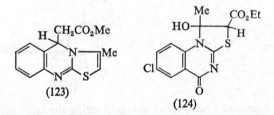

(123)

(124)

Thiazolo[3,2-*a*]quinazolines [$C_3NS-C_4N_2-C_6$].—Condensation of 6-chloro-4(1*H*)-quinazolinone-2-thiol with ethyl α-chloroacetoacetate and recrystallization of the resulting sulphide from ethanol yields compound (124).[130] This is dehydrated in acetic anhydride. 4-Allylthiazolidino[3,2-*a*]quinazolin-5(4*H*)-ones have been prepared from 3-allyl-2-thionoquinazolin-4(3*H*)-one.[183]

Pyrido[2,1-*b*]benzothiazoles [$C_3NS-C_5N-C_6$].—Members of this class of compound are formed by the reaction of benzothiazoles with dimethyl acetylene-dicarboxylate (see p. 393). The pyrido[2,1-*b*]benzothiazolium perchlorates (125) may be prepared by heating a benzothiazolium perchlorate with acetylacetone at 150 °C.[184] Compound (125; R = NO_2) may be reduced to the corresponding

[180] H. Reimlinger, M. A. Peiren, and R. Merenyi, *Chem. Ber.*, 1975, **108**, 3894.
[181] A. Fravolini, G. Grandolini, and A. Martani, *Gazzetta*, 1973, **103**, 1063.
[182] R. Hull and M. L. Swain, *J.C.S. Perkin I*, 1976, 653.
[183] S. K. P. Sinha and M. P. Thakur, *J. Indian Chem. Soc.*, 1974, **51**, 457.
[184] F. S. Babichev, V. N. Bubnovskaya, A. G. Goncharenko, and M. Yu. Kornilov, *Ukrain. khim. Zhur.*, 1975, **41**, 1281 (*Chem. Abs.*, 1976, **84**, 74 164).

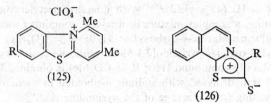

(125)

(126)

amine, which reacts readily with acylating agents and aromatic aldehydes. Pyridine-2-thiol reacts with 2-bromo-6-(ethoxycarbonyl)cyclohexanone to give a pyrido[2,1-*b*]benzothiazolium bromide.[185]

Thiazolo-[2,3-*a*]- and -[3,2-*a*]-quinolines and -isoquinolines [C₃NS-C₅N-C₆].— Cyclization of 2-(β-hydroxyethylthio)quinoline with hydrogen halides gives the corresponding 2,3-dihydrothiazolo[3,2-*a*]quinolinium salt.[186] 2,3-Dihydrothiazolo[2,3-*a*]isoquinolinium salts may be prepared similarly.[186] The meso-ionic thiazolo[2,3-*a*]isoquinolinium-2-thiones (126) are converted into the corresponding imidazo-compounds by successive treatment with methyl iodide and a primary aliphatic amine.[187] Certain thiazolo[2,3-*a*]isoquinolinium salts undergo ring-opening and a rearrangement reaction with active-hydrogen compounds in the presence of base to give pyrrolo-isoquinolines.[188]

Thiazolo[5,4-*c*]isoquinoline [C₃NS-C₅N-C₆].—The urea (127; R = CONHPh) eliminates aniline at 270—280 °C to give the thiazolo[5,4-*c*]isoquinoline (128).[189]

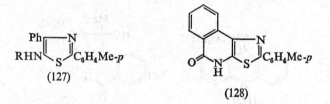

(127)

(128)

Thermal ring-closure can be effected also by heating the ethyl carbamate (127; R = CO₂Et).[189]

Naphtho[2,1-*d*]thiazoles [C₃NS-C₆-C₆].—The mass spectra of 3-methylnaphtho-[2,1-*d*]benzothiazolin-2-one hydrazone and 2-hydrazinonaphtho[2,1-*d*]thiazole have been compared.[35] Compounds (129), which have been studied as potential

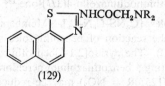

(129)

[185] P. H. L. Wei, U.S.P. 3 823 153/1974 (*Chem. Abs.*, 1974, **81**, 105 491).
[186] V. N. Gogte, R. N. Sathe, and B. D. Tilak, *Indian J. Chem.*, 1973, **11**, 1115.
[187] P. B. Talukdar, S. K. Sengupta, and A. K. Datta, *Indian J. Chem.*, 1973, **11**, 1257.
[188] K. Mizuyama, Y. Matsuo, Y. Tominaga, Y. Matsuda, and G. Kobayashi, *Heterocycles*, 1975, **3**, 533 (*Chem. Abs.*, 1975, **83**, 164 056).
[189] G. Winters and N. Di Mola, *Tetrahedron Letters*, 1975, 3877.

anaesthetics, may by synthesized from 2-aminonaphtho[2,1-*d*]thiazole by sequential treatment with chloroacetyl chloride and the appropriate secondary amine.[190] (See also following Sections.)

10 Structures Comprising Two Five-membered and Two Six-membered Rings (5,5,6,6)

Tetrazolo- and [1,2,4]Triazolo-[5,4-*b*]naphtho[2,1-*d*]thiazoles [CN$_4$-C$_3$NS-C$_6$-C$_6$].— The novel tetrazolo[5,4-*b*]naphtho[2,1-*d*]thiazole (130) and the closely related [1,2,4]triazolo[5,4-*b*]naphtho[2,1-*d*]thiazole are obtained from 2-hydrazino-naphtho[2,1-*d*]thiazole by standard procedures.[191]

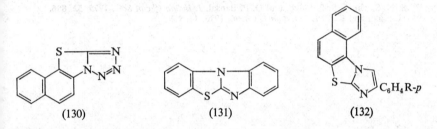

(130)   (131)   (132)

Benzimidazo[2,1-*b*]benzothiazoles [C$_3$NS-C$_3$N$_2$-C$_6$-C$_6$].—The parent compound (131) is formed in 46% yield on pyrolysis of 1-(benzothiazol-2-yl)benzotriazole.[192] This yield is higher than that obtained on photolysis.

Imidazo[2,1-*b*]naphtho-[1,2-*d*]- and -[2,1-*d*]-thiazoles [C$_3$NS-C$_3$N$_2$-C$_6$-C$_6$].—Compounds (132; R = Me, Cl, OMe, or OEt) are obtained on condensation of the appropriate 2-amino-naphthothiazole with a *para*-substituted ω-bromo-aceto-phenone.[101, 193] The isomer is obtained similarly.

11 Structures Comprising One Five-membered Ring and Three Six-membered Rings (5,6,6,6)

Thiazolo[4,5-*a*]phenothiazines [C$_3$NS-C$_4$NS-C$_6$-C$_6$].—The reaction of 1-amino-2-bromophenothiazine with carbon disulphide yields thiazolo[4,5-*a*]pheno-thiazine-2-thiol (133).[194]

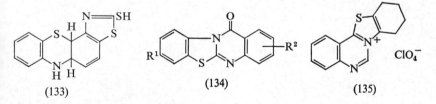

(133)   (134)   (135)

[190] P. N. Bhargava and I. C. Tiwari, *Indian J. Pharm.*, 1974, **36**, 93.
[191] J. V. Singh, *J. Indian Chem. Soc.*, 1974, **51**, 443.
[192] D. C. K. Lin and D. C. DeJongh, *J. Org. Chem.*, 1974, **39**, 1780.
[193] J. V. Singh, *J. Indian Chem. Soc.*, 1974, **51**, 559.
[194] Z. I. Ermakova, A. N. Gritsenko, and S. V. Zhuravlev, *Khim. geterotsikl. Soedinenii*, 1974, 202 (*Chem. Abs.*, 1974, **80**, 133 370).

Benzothiazolo-[2,3-*b*]- and -[3,2-*c*]-quinazolines [C_3NS-C_4N_2-C_6-C_6].—When various derivatives of anthranilic acid are condensed with 6-substituted 2-chlorobenzothiazoles they give the 12*H*-benzothiazolo[2,3-*b*]quinazolin-12-ones (134).[195] These are hydrolysed by potassium hydroxide to the anthranilic acid derivatives, which revert to the tetracyclic compounds with acidic ethanol. A synthesis of 8,9,10,11-tetrahydrobenzothiazolo[3,2-*c*]quinazolin-7-ium perchlorate (135) (having a thiaza-steroidal skeleton) involves cyclization of 4-[(α-oxocyclohexyl)-thio]quinazoline.[196]

Acknowledgements: The Reporters thank Miss M. Cradden, Mrs. P. Cassell, and Miss J. A. Fearnley for typing the manuscripts of Chapters 12 and 13.

[195]　S. N. Sawhney, S. P. Singh, and O. P. Bansal, *J. Indian Chem. Soc.*, 1975, **52**, 886.
[196]　H. Singh and K. B. Lal, *Indian J. Chem.*, 1973, **11**, 959.

14
Thiadiazoles and Selenadiazoles

BY F. KURZER

1 Introduction

The chemistry of the four isomeric series of thiadiazoles and their selenium analogues continues to be a field of fruitful research. In the period under review, much attention has been devoted to photolytic reactions and their mechanistic interpretation. Flash thermolysis of 1,2,3-thiadiazoles has made the highly active thioketens available for further synthetic use. The investigation of the interesting properties of the Δ^3-1,3,4-thiadiazoline system continues, and interest in different kinds of meso-ionic compounds has not diminished. Butler's review [1] of the diazotization of heterocyclic primary amines contains sections devoted to isothiazoles and thiadiazoles. Of the numerous condensed ring systems, special mention may be made of the 1,2,5-thiadiazolo[3,4-c][1,2,5]thiadiazole, on account of its exceptionally pronounced aromatic nature, and of benzo[1,2-c,3,4-c',5,6-c"]tris[1,2,5]thiadiazole, because of its highly symmetrical structure. Many thiadiazole derivatives possess biological activities, of varying promise, and their potential usefulness for diverse medical, agricultural, and industrial purposes has been claimed.

2 1,2,3-Thiadiazoles

Synthesis.—*Pechmann's Synthesis.* The scope of Pechmann's synthesis, the 1,3-cycloaddition of aryl isothiocyanates to diazomethane, has been extended, using methyl, phenyl, and 1-alkenyl isothiocyanates. The resulting 5-(substituted amino)-1,2,3-thiadiazoles are rearranged, under the influence of an excess of diazomethane, to 1-substituted-5-methylthio-1H-1,2,3-triazoles (1) [2] (see Vol. 2, p. 717).

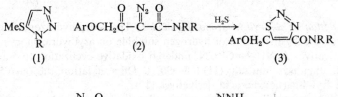

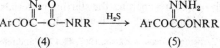

[1] R. N. Butler, *Chem. Rev.*, 1975, **75**, 241.
[2] S. Hoff and A. P. Blok, *Rec. Trav. chim.*, 1974, **93**, 317.

From Diazo-keto-amides. A classical synthesis of 1,2,3-thiadiazoles (3) involves the cyclization of α-diazo-β-keto-amides (2) [3] with hydrogen sulphide and catalytic amounts of ammonia. Under similar conditions, diazo-compounds of structure [4; R = Me or Et, RR = (CH₂)₅] gave the hydrazones (5) rather than the desired 1,2,3-thiadiazoles (3).[4]

From Hydrazones. The synthesis of 1,2,3-thiadiazole-4-carboxylic acids by the action of thionyl chloride on ethoxycarbonylhydrazones of α-keto-acids has been applied to β-phenylpyruvic acid. The reaction produces a mixture of 5-phenyl-1,2,3-thiadiazole-4-carboxylic acid (6; X = OH), its acid chloride, and the oxadiazine (7), which is converted into (6; X = Cl) in boiling thionyl chloride.[4]

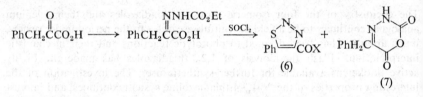

Tosylhydrazones of cycloalkanones (8) undergo cyclization with thionyl chloride to give cycloalkeno-1,2,3-thiadiazoles (9), probably by way of the stages shown. Further action of peroxy-acids successively furnishes the 2-oxides and 1,1,2-trioxides.[5]

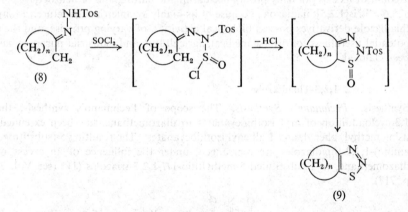

By Oxidation of α-Arylhydrazono-thioamides. The α-arylhydrazono-thioamides (10), prepared by the action of hydrogen sulphide on arylhydrazonocyanoacetic acid derivatives ArNHN=CRCN, undergo oxidative cyclization to 5-amino-2-aryl-1,2,3-thiadiazolium salts (11) (65—90%). On acetylation, they give 5-acetyl-imino-2-aryl-4-carboxylic acid derivatives (12).[6]

The action of alkalis on (11) regenerates the starting materials (10). Treatment with acetoacetic ester produces, by S—N bond fission and recyclization, good

[3] J. H. Looker and J. W. Carpenter, *Canad. J. Chem.*, 1967, **45**, 1727, and references given therein.
[4] N. P. Peet and S. Sunder, *J. Heterocyclic Chem.*, 1975, **12**, 1191.
[5] H. P. Braun and H. Meier, *Tetrahedron*, 1975, **31**, 637.
[6] K. Gewald and U. Hahn, *J. prakt. Chem.*, 1975, **317**, 329.

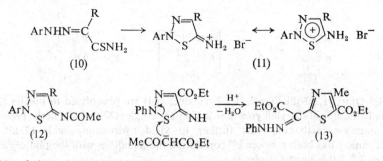

(10)   (11)   (12)   (13)

yields of the substituted thiazoles (13); malonic acid dinitrile and cyanoacetic ester react analogously.[6]

Meso-ionic 1,2,3-Thiadiazoles.—The action of thionyl choride on the meso-ionic 3-aryl-4-carboxymethylsydnones (14) unexpectedly produces high yields of 3-aryl-4,5-dichloro-1,2,3-thiadiazolium salts (15), also obtainable from (16) with phosphorus oxychloride. On oxidation with hydrogen peroxide, the salts (15) are converted into novel meso-ionic 1,2,3-thiadiazoles of structure (17).[7]

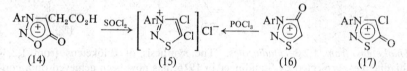

(14)   (15)   (16)   (17)

Chemical Properties.—*Photolytic Reactions.* The mechanism of the photolysis of 1,2,3-thiadiazole involves the intermediate formation of the thioketen (20) and thiiren (19), which decompose to the observed photoproducts, carbon disulphide, and ethynethiol (18), respectively. The growth rates of bands of comparable intensity in the i.r. spectra of (20) and (18) indicate that these species are increasing at a similar rate. The photolysis of 4- and 5-deuteriothiadiazole was similarly studied.[8]

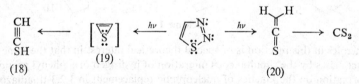

(18)   (19)   (20)

Photolysis of 1,2,3-thiadiazole 2-oxides (21) yields the isomeric 3-oxides (23), probably by way of intermediate oxadiaziridines (22). As with other *N*-oxides, competitive deoxygenation results in the simultaneous formation of the parent 1,2,3-thiadiazoles, which may become the exclusive products, but, being photo-labile, are liable to be further cleaved (see Vol. 3, p. 672). Information has been provided concerning dipole moments and i.r. and ¹H and ¹³C n.m.r. spectra of (21) and (23), and their behaviour on electron impact.[9]

[7] S. I. Burmistrov and V. A. Kozinskii, *Zhur. Org. Khim.*, 1974, **10**, 891.
[8] A. Krantz and J. Laureni, *J. Amer. Chem. Soc.*, 1974, **96**, 6768.
[9] H. P. Braun, K. P. Zeller, and H. Meier, *Annalen*, 1975, 1257.

(21)   (22)   (23)

4,5-Diaryl-1,2,3-thiadiazole 1,1,2-trioxides (24) are photolysed to nitriles (25), small amounts of diarylalkynes (26), and intermediates (27), which, in the presence of aqueous methanol, react further to yield hydroxamic acids (28).[10] A mechanism has been proposed [10] correlating these findings with the photolysis of other 1,2,3-thiadiazole oxides.[9]

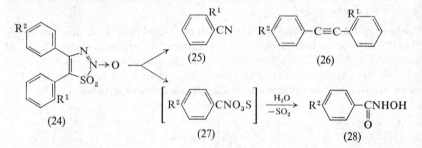

(24)   (25)   (26)

(27)   (28)

Thioketens from 1,2,3-*Thiadiazoles.* The synthesis of thioketens from 1,2,3-thiadiazoles, projected by Staudinger in 1916, has now been achieved on a convenient preparative scale (1—10 g), in good yield, by the flash thermolysis of 1,2,3-thiadiazoles at 580 °C and 10^{-4} Torr in a specially designed apparatus.[11] The thermolysis of 1,2,3-thiadiazoles, preferably in boiling diglycol, also provides a practical source of thioketens for immediate further synthetic use, *e.g.* in the production of thioesters, as shown in Scheme 1. The use of 4- and 5-phenyl-1,2,3-

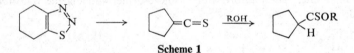

Scheme 1

thiadiazoles in this reaction is of special theoretical interest in that the same final product arises by the simultaneous migration of hydrogen or a phenyl group.[12]

Information on the kinetics of nucleophilic replacement in 1,2,3-thiadiazoles [80] is included in the appropriate section on 1,3,4-thiadiazoles (see p. 436).

Biochemical Properties.—1-Aryl-3-(1,2,3-thiadiazol-5-yl)ureas are inhibitors of energy conservation in respiration and photosynthesis. The most effective example (Ar = 3,4-dichlorophenyl) uncouples ATP formation in isolated chloroplasts or mitochondria, at a concentration of *ca.* 2 or 9 μmol l^{-1}, respectively.[13]

[10] H. Meier, G. Trickes, and H. P. Braun, *Tetrahedron Letters*, 1976, 171.
[11] G. Seybold and C. Heibl, *Angew. Chem. Internat. Edn.*, 1975, **14**, 248.
[12] H. Meier and H. Bühl, *J. Heterocyclic Chem.*, 1975, **12**, 605.
[13] G. Hauska, A. Trebst, C. Koetter, and H. Schulz, *Z. Naturforsch.*, 1975, **30c**, 505.

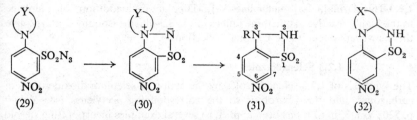

(29)   (30)   (31)   (32)

1,2,3-Benzothiadiazoles.—Further details have been reported of the thermolysis of *o*-dialkylaminobenzenesulphonyl azides (29) (see Vol. 2, p. 718). The meso-ionic benzothiadiazoles (30) are produced initially, and can be isolated from the dimethylamino-azide (Y = Me$_2$) or when Y is a part of a six-membered ring. In suitable cases, thermolysis of (29) gives (31) directly by a Cope-type elimination of an alkyl group from (30). The azides (29) and the derived benzothiazoles [30; Y = (CH$_2$)$_5$] decompose smoothly in the presence of copper, to afford the fused thiadiazine (32) in 48 and 73% yield, respectively.[14] The azide [29; Y = (CH$_2$)$_5$] gives [31; R = (CH$_2$)$_5$Cl] with 4M-HCl.

Phenyl radicals react with 1,2,3-benzothiadiazole to give diphenyl sulphide, dibenzothiophen, thianthrene, 4-phenylthiodibenzothiophen, and 1,2-bis(phenyl-thio)benzene. It is suggested that the reaction involves attack of phenyl radicals on the sulphur atom of the reactant, to produce an intermediate *o*-(phenylthio)-phenyl radical, from which each of the products may be derived. Similarly, methyl radicals yield a mixture of thioanisole, thianthrene, 4-methylthiodi-benzothiophen, and 1-methylthio-2-phenylthiobenzene. The experiments are claimed to be the first examples of the attack of a radical on a hetero-aromatic sulphur atom [15] (see Vol. 3, p. 678).

1,2,3-Thiadiazolo[5,4-*b*]indoles.—Arenesulphonyl azides react vigorously with indoline-2-thione in pyridine to afford a product that has been formulated as the condensed thiadiazole (34), arising presumably *via* the intermediate (33). The bis-(indoline) (35), formed as a by-product, becomes the main product with 1-methyl-indoline-2-thione, in accord with the suggested mechanism.[16]

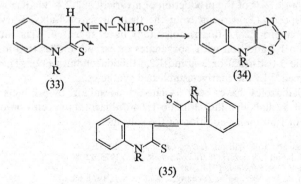

(33)   (34)

(35)

[14] J. Martin, O. Meth-Cohn, and H. Suschitzky, *J.C.S. Perkin I*, 1974, 2451; *J.C.S. Chem. Comm.*, 1971, 1319.
[15] L. Benati, P. C. Montevecchi, A. Tundo, and G. Zanardi, *J.C.S. Perkin I*, 1974, 1276.
[16] A. S. Bailey, J. F. Seager, and Z. Rashid, *J.C.S. Perkin I*, 1974, 2384.

1,2,3-Thiadiazolo[3,4-*a*]isoquinolines.—1,3-Dipolar cycloaddition of sulphenes to the highly reactive azomethine imines of the 3,4-dihydroisoquinoline type gives this new ring system [17] (see Chapter 3, Section 2).

3 1,2,3-Selenadiazoles

The synthesis of 1,2,3-selenadiazoles by the action of selenium dioxide on semi-carbazones, and their pyrolysis to the corresponding acetylenes (see Vol. 2, p. 720; Vol. 3, p. 673) has been applied to several examples incorporating steroid, phenanthrenyl, fluorenyl, and naphthyl residues. The method may prove useful as a route to pharmacologically active 17-ethynyl-steroids.[18] A large number of aryl-1,2,3-selenadiazoles [19] as well as the three isomeric 4-[(2-, 3-, and 4-)pyridyl]-1,2,3-selenadiazoles [20] have been similarly obtained. The n.m.r. spectra of aryl-1,2,3-selenadiazoles have been recorded and discussed.[19]

4 1,2,4-Thiadiazoles

Synthesis.—As in the past, the majority of the synthetic routes to 1,2,4-thiadiazoles are ultimately based on ring-closures involving suitably placed imino- and mercapto-groups (Type C). Additional types of syntheses, classified according to the fragments from which the hetero-ring is constructed (see Vol. 3, p. 679) are listed as methods **F** and **G**.

From Perthiocyanic Acid. The use of the dianion of 3,5-dimercapto-1,2,4-thiadiazole ('perthiocyanic acid') (see Vol. 1, p. 448; Vol. 2, p. 722) as a convenient source of other derivatives has been applied to the production of 3,5-bis-(2,4-dinitrophenylthio)-1,2,4-thiadiazole.[21]

Type A Syntheses [N—C—S + C—N]. *From 1,2,3,4-thiatriazoles.* Two independent reports [22, 23] of the production of a variety of 1,2,4-thiadiazolidines from a reactive species (37) arising from the thermolysis of 4-alkyl-5-arylsulphonyl-imino-4,5-dihydro-1,2,3,4-thiatriazoles (36) have appeared. The intermediates (37) undergo 1,2-addition with isocyanates or carbodi-imides to afford excellent yields of the 3-oxo- (38) or 3-imino-1,2,4-thiadiazolidines (39). Their structure was confirmed [23] by alternative established syntheses.

1,2,4-Thiadiazoles have been isolated [24] as minor by-products in certain reactions of 5-(dialkylamino)-3-imino-1,2,4-dithiazoles in aceto- or benzo-nitrile as a result of 1,3-dipolar addition of the solvent.

[17] W. E. Truce and J. R. Allison, *J. Org. Chem.*, 1975, **40**, 2260.
[18] H. Golgolab and I. Lalezari, *J. Heterocyclic Chem.*, 1975, **12**, 801.
[19] A. Caplin, *J.C.S. Perkin I*, 1974, 30.
[20] I. Lalezari, A. Shafiee, and S. Yazdany, *J. Pharm. Sci.*, 1974, **63**, 628.
[21] R. Seltzer, U.S.P. 3 816 441/1974 (*Chem. Abs.*, 1974, **81**, 105 523).
[22] R. Neidlein and K. Salzmann, *Synthesis*, 1975, 52.
[23] G. L'Abbé, G. Verhelst, C. C. Yu, and S. Toppet, *J. Org. Chem.*, 1975, **40**, 1728.
[24] J. E. Oliver and A. B. DeMilo, *J. Org. Chem.*, 1974, **39**, 2225.

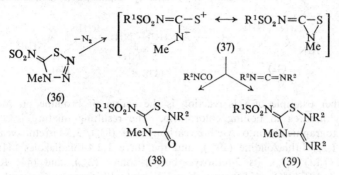

(36)   (37)   (38)   (39)

Type B Syntheses [N—C—N + C—S]. *From amidines.* 5-Chloro-3-(2,2-dichloro-1,1-difluoroethyl)-1,2,4-thiadiazole has been produced from the appropriate amidinium salt and trichloromethanesulphenyl chloride, and has been converted into a number of functional derivatives.[25]

Type C Syntheses [N—C—N—C—S]. *From amidino-thiono-compounds.* The general synthesis of 1,2,4-thiadiazoles by the oxidative cyclization of compounds incorporating the amidino-thiono-grouping [—C(=NH)NHCS—] has been extended [26] by the synthesis of 1-acyl(and -sulphonyl)-3-thioacyl-guanidines (41) from acyl- or sulphonyl-guanidines (40) with thioacid O-esters in the presence of sodium hydride, followed by their oxidation with hydrogen peroxide in pyridine.

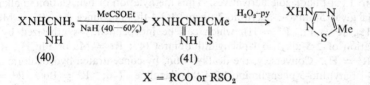

(40)   (41)

$$X = RCO \text{ or } RSO_2$$

Oxidation of thioureas. The well-established synthesis of 1,2,4-thiadiazole derivatives by the oxidation of arylthioureas [27] has been further studied by an examination of the oxidation products of mixtures of equimolar quantities of *two* thioureas.[28-31] Oxidation of a mixture of *sym*-diarylthiourea and thiourea by hydrogen peroxide in acidified ethanol yields 3-amino-4-aryl-5-arylimino-Δ^2-1,2,4-thiadiazolines (42; R = H). These compounds are considered to be formed by the cyclization of amidino-thioureas (43), which can be isolated from the oxidation of mixtures of 1-alkyl-3-arylthioureas and thiourea.[29] The overall mechanism resembles that postulated for the formation of 'Hector's Bases.' [27] The oxidation of binary mixtures of *sym*-diaryl- and N-alkyl-thioureas similarly furnishes the trisubstituted thiadiazolines (42; R = alkyl).[31]

[25] H. Röchling and G. Hörlein, *Annalen*, 1974, 504.
[26] B. Junge, *Annalen*, 1975, 1961.
[27] F. Kurzer and P. M. Sanderson, *J. Chem. Soc.*, 1963, 3336, and references therein.
[28] C. P. Joshua and P. N. K. Nambisan, *Indian J. Chem.*, 1973, **11**, 1272.
[29] C. P. Joshua and P. N. K. Nambisan, *Indian J. Chem.*, 1974, **12**, 962.
[30] P. N. K. Nambisan, *Tetrahedron Letters*, 1974, 2907.
[31] C. P. Joshua and P. N. K. Nambisan, *Indian J. Chem.*, 1975, **13**, 241.

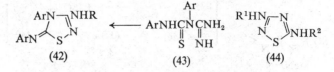

(42)   (43)   (44)

Another example of this reaction is the action of bromine on *N*-ethoxy-carbonylthiourea in boiling chloroform. The resulting mixture is separable chromatographically into *N*-ethoxycarbonylurea (8%), 3,5-bis(ethoxycarbonyl-imino)-1,2,4-dithiazolidine (29%), and the three 1,2,4-thiadiazoles (44; R^1 = R^2 = CO$_2$Et) (8%), its 2-(ethoxycarbonyl)-isomer (2%), and (44; R^1 = H, R^2 = CO$_2$Et) (3%). *N*-Ethoxycarbonyl-*N'*-methylthiourea reacts analogously. In view of the complexity of the mixture of products, and the low proportion of 1,2,4-thiadiazoles formed, the utility of the reaction is limited.[32]

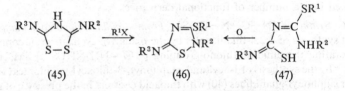

(45)   (46)   (47)

From dithiazoles. The alkylation of 'phenylthiuret' (45) is attended by isomeriza-tion to 1,2,4-thiadiazole derivatives. Thus methylation or benzylation in alkaline media gives 5-imino-3-alkylthio-2-phenyl-1,2,4-thiadiazolines (46; R^1 = Me or PhCH$_2$, R^2 = Ph, R^3 = H), which can be independently synthesized by the oxidation of 2-*S*-alkyliso-1-phenyldithiobiuret (47; R^1 = Me or PhCH$_2$, R^2 = Ph, R^3 = H). Conversely, the dealkylation, by concentrated hydrochloric acid, of 3-t-butylthio-5-phenylimino-1,2,4-thiadiazoline (46; R^1 = Bu^t, R^2 = H, R^3 = Ph), obtained from (47; R^1 = Bu^t, R^2 = H, R^3 = Ph), proceeds with the opposite isomerization, yielding (45) instead of the expected (46; R^1 = R^2 = H, R^3 = Ph). Possible mechanisms of these reactions, and their significance for the structure of phenylthiuret, have been discussed. However, the issue is compli-cated by the fact that different alkyl halides produce different structural types of the 1,2,4-thiadiazole system.[33]

The thermal decomposition of 5-(dialkylamino)-3-alkylimino-1,2,4-dithiazoles (48) gives products derived from 2 molecules of the reactant, with loss of one atom of sulphur. They are formulated as the substituted 3,5-bis(thioureido)-1,2,4-thiadiazolidines (49) and are thought to arise by a mechanism involving inter-mediate spiranes. The 3-phenylimino-analogues of (48) decompose differently, yielding benzothiazole derivatives.[34] The same class of 1,2,4-thiadiazolines (49) arise also by the successive alkylation and aminolysis of the 1,2,4-dithiazoles (50).[35]

[32] M. Nagano, M. Oshige, T. Matsui, J. Tobitsuka, and K. Oyamada, *Chem. and Pharm. Bull. (Japan)*, 1973, **21**, 2396.
[33] G. Bhaskaraiah, *Indian J. Chem.*, 1974, **12**, 134.
[34] J. E. Oliver and J. L. Flippen, *J. Org. Chem.*, 1974, **39**, 2233.
[35] J. E. Oliver, *J. Org. Chem.*, 1974, **39**, 2235.

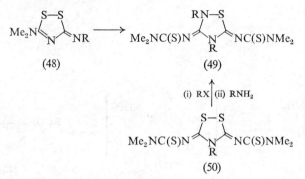

From oxaziridines. The reaction of oxaziridines (51) with phenyl isothiocyanate yields carbodi-imides, but, under mild conditions, produces considerable amounts of the 1,2,4-thiadiazolidines (53a) and/or (53b). The reactions are likely to proceed by reaction paths outlined in Scheme 2, involving the intermediate

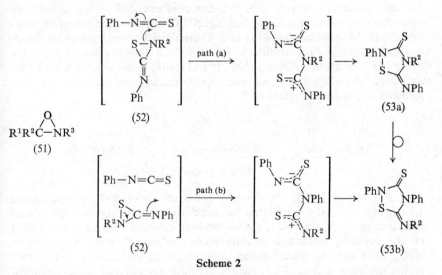

Scheme 2

thiaziridine-imines (53a), formed by condensation of the oxaziridines with phenyl isothiocyanate. Members of series (53a) are isomerized to (53b) in good yield in boiling benzene in the presence of catalytic amounts of triethyl phosphite. 2-t-Butyloxaziridine behaves exceptionally in producing an oxadiazolidinethione, probably for steric reasons.[36]

From 1,2,4-oxadiazoles. The reaction of 5-substituted 3-amino-1,2,4-oxadiazoles (56) with phenyl isothiocyanate produces substituted 3,5-diamino-1,2,4-thia-diazoles (58), presumably by the thermal rearrangement of the intermediates (57).[37] The reaction is an extension of the heterocyclic rearrangements of the

[36] M. Komatsu, Y. Ohshiro, K. Yasuda, S. Ichijima, and T. Agawa, *J. Org. Chem.*, 1974, **39**, 957.
[37] M. Ruccia, N. Vivona, and G. Cusmano, *J.C.S. Chem. Comm.*, 1974, 358.

15

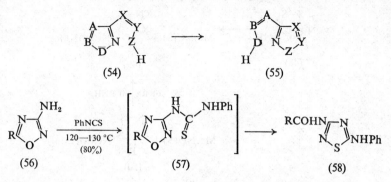

(54)   (55)

(56)   (57)   (58)

general type (54) → (55), being the first example in which sulphur forms part of the X–Y–Z chain.

Type D Syntheses [C—N—C—S—N]. The synthesis of Δ^4-1,2,4-thiadiazolines from potassium methyl cyanoiminodithiocarbonate, MeS C(=NCN)SK, and *N*-chloro-amidines (see Vol. 3, p. 682) has been extended by the inclusion of other *N*-chloro-cyanamide derivatives, *e.g.* NH$_2$C(=NCl)OR, as one of the reactants.[38] In this case, condensation at 5 °C produces 2-alkoxyimidoyl-3-imino-5-methylthio-Δ^4-1,2,4-thiadiazolines in 41—79% yield. They are dealkylated to the amides by methanolic hydrogen chloride and readily undergo aminolysis to the amidines. *N*-Chloro-guanidines and *N*-chloroguanylurethane may be used similarly in this general synthesis.[38]

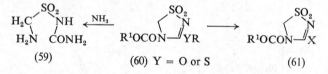

(59)   (60) Y = O or S   (61)

Type F Syntheses [N—C—N—S—C]. The synthesis of Δ^2-1,2,4-thiadiazoline 1,1-dioxides[39] by the cyclization of *N*-chloromethyl sulphonylamidines and related compounds (see Vol. 1, p. 446) has been extended[40] by a re-investigation of the acylating cyclization of chloromethyl sulphonylisoureas ROC(NH$_2$)= NSO$_2$CH$_2$Cl and the corresponding thioureas, using chloroformates ROCOCl as the acylating agents. The reaction provides a variety of 3,4-disubstituted Δ^2-1,2,4-thiadiazoline 1,1-dioxides (60). The 3-alkylthio-group is displaced by ammonia, furnishing the corresponding 3-amines (61; X = NH$_2$), but ring-opening to aminomethyl sulphonylureas (59) occurs on more vigorous treatment. In the chlorination of (60; Y = O), attack is at the 3-position, to give (61; X = Cl), which undergoes nucleophilic replacement at the 3-chloro-atom; a route to a variety of functional derivatives of this ring system it thus made available. Readiness in facile ring-opening occasionally diverts the course of the reaction in this sense.[40]

[38] T. Fuchigami and K. Odo, *Bull. Chem. Soc. Japan*, 1975, **48**, 310.
[39] A. Lawson and R. B. Tinkler, *Chem. Rev.*, 1970, **70**, 593.
[40] A. Etienne, A. Le Berre, G. Longchambon, G. Lochey, and B. Cucumel, *Bull. Soc. chim. France*, 1974, 1580.

Type G Syntheses [C—N—S + C—N]. Nitrile sulphides R—C≡N→S have recently become available [41] as reactive intermediates in the thermolysis (at *ca.* 190 °C) of 5-substituted 1,3,4-oxathiazol-2-ones; they may be trapped by 1,3-dipolar cycloadditions, *e.g.* with acetylenes, resulting in S,N-heterocycles. Cycloaddition of nitrile sulphides R¹CNS to nitriles R²CN provides a new general synthesis of 3,5-disubstituted 1,2,4-thiadiazoles (62).[42] Yields are moderate, but are satisfactory when electrophilic nitriles are used in conjunction with aromatic nitrile sulphides. Minor amounts of (63) are formed as by-products,

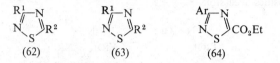

(62) (63) (64)

possibly by sulphur-transfer to the nitrile. Ethyl cyanoformate reacts particularly smoothly, yielding (64), which may be hydrolysed and decarboxylated to the parent thiadiazole.[42] The method is useful in providing 3,5-non-identically disubstituted 1,2,4-thiadiazoles in which there is no uncertainty about the position of the substituents.

Physical Properties.—The spectral properties of a series of 3-amino-4-aryl-5-aryl(or alkyl)-imino-4,5-dihydro-1,2,4-thiadiazoles [28—31] are in accord with the proposed structures.[43] The n.m.r. spectra indicate the preferred existence of the dihydro- rather than the tautomeric tetrahydro-thiadiazole structure. Mass spectral fragmentation patterns were proposed for these,[43] and for compound (70).[47]

An n.m.r. investigation of the energy barrier to internal rotation of the dimethylamino-group in five-membered heterocyclic compounds has included a 1,2,4-thiadiazole (65) and variously substituted 1,3,4-thiadiazoles (66).[44] Compound (65) is iso-π-electronic with 4-dimethylaminopyrimidine, and is comparable

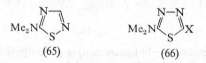

(65) (66)

in the positions of its nitrogen atoms. The closeness of their energy barriers reflects the analogies between cyclic compounds in which a sulphur atom is replaced by a CH=CH group. In the 1,3,4-thiadiazoles (66), the rotational barriers are 0.4—0.7 kcal mol⁻¹ higher than in the corresponding 1,3,4-oxadiazoles, due to a difference in bonding interaction between the dimethylamino-group and the heterocyclic ring system. A variety of other aspects opened up by these measurements have been discussed.[44]

An *X*-ray analysis of 2,4-dimethyl-1,2,4-thiadiazolidine-3,5-dithione has shown its hetero-ring to be planar, due presumably to conjugation, which does

[41] J. E. Franz and L. L. Black, *Tetrahedron Letters*, 1970, 1381 (see also Vol. 3, p. 542).
[42] R. K. Howe and J. E. Franz, *J. Org. Chem.*, 1974, **39**, 962.
[43] C. P. Joshua and K. N. Rajasekharan, *Austral. J. Chem.*, 1975, **28**, 591.
[44] T. Liljefors, *Org. Magn. Resonance*, 1974, **6**, 144.

not include the N—S bond.[45] Considerable repulsion distorts the geometry of the peripheral sulphur and methyl groups.[45] Crystallographic data and an *X*-ray analysis have also been reported for (49; R = Me), which is found to be a non-planar structure.[34]

Chemical Properties.—*Alkylation.* Treatment of 3,5-diethoxy-1,2,4-thiadiazole (67) with benzyl bromide in boiling acetonitrile ('Hilbert–Johnson reaction') slowly yields a monoalkylation product (68). Chloromethyl benzyl ether effects the alkylation rapidly in good yield. The use of ribosyl halides should make novel nucleosides accessible.[46]

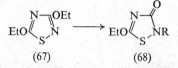

$$(67) \longrightarrow (68)$$

Cycloaddition. 4-Aryl-3-arylimino-5-imino-1,2,4-thiadiazolidines (69) ('Hector's Bases') undergo 1,3-dipolar cycloaddition with arylcyanamides to yield 1 : 1 adducts, for which the condensed 1,2,4-thiadiazolo-[5,1-*e*][1,2,4]thiadiazole structure (70) has been proposed. The alternative guanidino-thiadiazolidine

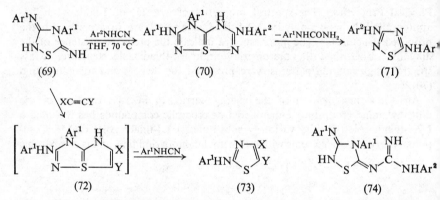

structure (74), with strong S—N interaction, is not entirely ruled out. On treatment with sodium ethoxide, (70) forms 1,3-bis(arylamino)-1,2,4-thiadiazoles (71).[47] The thermal decomposition of (69; Ar¹ = Ar² = Ph or *p*-MeC₆H₄) likewise produces (70) (*ca.* 50%), together with (71), as a by-product.[48] Similarly, the 1,3-dipolar cycloaddition of acetylenes to Hector's bases (69) gives 2-arylamino-thiazoles (73), probably by way of the intermediate 1,2,4-thiadiazolo-[5,1-*a*][1,3]thiazoles (72).[48]

The action of aldehydes and ketones on 1,2,4-thiadiazole-3,5-disulphenamide [49] yields successively mono- and di-alkylidene-sulphenamides.[50]

[45] C. L. Raston, A. H. White, A. C. Willis, and J. N. Varghese, *J.C.S. Perkin II*, 1974, 1096.
[46] G. J. Koomen and U. K. Pandit, *Heterocycles*, 1975, **3**, 539.
[47] K. Akiba, T. Tsuchiya, M. Ochiumi, and N. Inamoto, *Tetrahedron Letters*, 1975, 455.
[48] K. Akiba, M. Ochiumi, T. Tsuchiya, and N. Inamoto, *Tetrahedron Letters*, 1975, 549.
[49] H. Garzia-Fernandez, *Compt. rend.*, 1967, **265**, C, 88.
[50] V. A. Ignatov, N. V. Zhorkin, G. A. Blokh, R. A. Akchurina, and L. M. Agafonova, *Zhur. obshchei Khim.*, 1974, **44**, 2553.

Biological Activity.—5-Ethoxy-3-(trichloromethyl)-1,2,4-thiadiazole controls root rot (caused by *Phytophthora cinnamomi*) in plants such as *Chamaecyparis lawsoniana* and *Eucalyptus marginata*.[51]

5 Condensed 1,2,4-Thiadiazoles

Isoxazolo[3,2-c][1,2,4]thiadiazoles.—The condensation of heterocyclic amines (the amino-group of which forms part of a hetero-amidine moiety) with tri-chloromethanesulphenyl chloride is a general route to condensed ring-systems incorporating the 1,2,4-thiadiazole ring (see Vol. 2, p. 724). The use of 5-sub-stituted 3-amino-isoxazoles (75) in this reaction produces 3*H*-isoxazolo[3,2-c]-[1,2,4]thiadiazoles (76) from the intermediate sulphenamides, which are isolable and separately cyclizable. The products are remarkably light-sensitive, (76; R = m-NO$_2$C$_6$H$_4$) turning from greenish-gold to orange in less than 1 hour by a

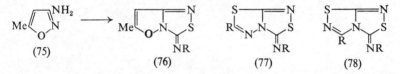

(75)   (76)   (77)   (78)

process as yet not elucidated.[52] Similarly prepared from the appropriate amino-heterocycle were 3*H*-1,2,4-thiadiazolo[3,4-b][1,3,4]oxadiazoles, 3*H*-thiazolo-[2,3-c][1,2,4]thiadiazoles, 3*H*-1,3,4-thiadiazolo[2,3-c][1,2,4]thiadiazoles (77), and 3*H*-1,2,4-thiadiazolo[4,3-d][1,2,4]thiadiazoles (78).[52]

Thiazolo[2,3-c][1,2,4]thiadiazoles.—The condensation of α-amino-N-heterocycles with chlorocarbonylsulphenyl chloride (chlorothioformyl chloride, ClCOSCl) can give rise to either a 3,4- or a 2,3-fused 1,2,4-thiadiazolone (see Vol. 3, p. 684). The former reaction occurs when 2-aminothiazole or 2-amino-Δ²-thiazoline is condensed with this reagent in chloroform, affording thiazolo[2,3-c][1,2,4]-thiadiazol-3-one (79) or the 5,6-dihydro-analogue, the structure of which has been confirmed by *X*-ray analysis.[53] For another synthesis of this class see above.[52]

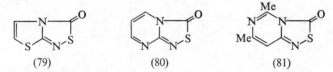

(79)   (80)   (81)

The use of 2-amino-(and 4-amino-2,6-dimethyl-)pyrimidine in this reaction similarly afforded 1,2,4-thiadiazolo[4,3-a]pyrimidin-3-one (80) (34%) and 5,7-dimethyl-1,2,4-thiadiazolo[4,3-c]pyrimidin-3-one (81) (39%), respectively. Very dilute ethanolic hydrochloric acid cleaves the thiadiazole ring of (81), yielding ethyl 2,6-dimethylpyrimidin-4-yl carbamate. Possible mechanisms of the reactions have been discussed.[53]

[51]   A. L. Bertus, *Plant Dis. Rep.*, 1974, **58**, 437 (*Chem. Abs.*, 1974, **81**, 100 526).
[52]   K. T. Potts and J. Kane, *J. Org. Chem.*, 1975, **40**, 2600.
[53]   D. Baldwin and P. van den Broek, *J.C.S. Perkin I*, 1975, 375.

3*H*-1,2,4-Thiadiazolo[3,4-*b*][1,3,4]oxadiazole.—See ref. 52.

3*H*-1,2,4-Thiadiazolo[4,3-*d*][1,2,4]thiadiazole (78).—See ref. 52.

3*H*-1,3,4-Thiadiazolo[2,3-*c*][1,2,4]thiadiazole (77).—See ref. 52.

1,2,4-Thiadiazolo[2,3-*a*]pyridines.—A series of derivatives of this ring system has been obtained by the oxidative cyclization of *N*-ethoxycarbonyl-*N'*-pyridyl-thioureas (82). The action of alkalis on (83) ruptures the thiadiazole ring, affording, *e.g.*, 2-cyanaminopyridine. Catalytic reduction also cleaves the S—N bond, with regeneration of the initial thiourea derivative (82). The synthesis is applicable to the production of 2*H*-1,2,4-thiadiazolo[2,3-*b*]pyridazines and the novel 2*H*-1,2,4-thiadiazolo[2,3-*a*]pyrazines (84).[54]

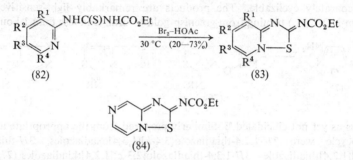

(82)   (83)

(84)

1,2,4-Thiadiazolo-[4,3-*a*]- and -[4,3-*c*]-pyrimidines.—The 3-substituted 3*H*-1,2,4-thiadiazolo[4,3-*a*]pyrimidine (85; R = 2-pyridyl) undergoes a Dimroth-type rearrangement in 10% ethanolic hydrochloric acid or sodium hydroxide, forming the 2-substituted 2*H*-1,2,4-thiadiazolo[2,3-*a*]pyrimidine (86), the structure of which was confirmed by its independent synthesis from the appropriate disubstituted thiourea by oxidation with sulphuryl chloride. In contrast, the substituted 1,2,4-thiadiazolo[4,3-*c*]pyrimidine (87) does not give rearranged products, but is cleaved to the 3-acetyl derivative on treatment with 10% acid, and to (88) with alkali. The latter corresponds to the addition of the elements of water to the starting material, and is a rare example of the isolation of a Dimroth inter-

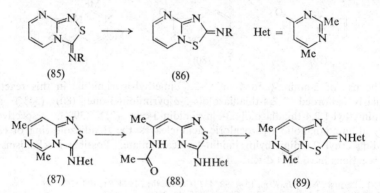

(85)   (86)

(87)   (88)   (89)

⁵⁴ B. Koren, B. Stanovnik, and M. Tisler, *Org. Prep. Proced. Internat.*, 1975, **7**, 55.

mediate. It is cyclized to (89) by boiling phosphorus oxychloride.[55] For another synthesis of this ring-system (80), see ref. 53.

1,2,4-Thiadiazolo[2,3-b]pyridazines.—See ref. 54.

1,2,4-Thiadiazolo[4,3-c]pyrimidine (81).—See ref. 53.

1,2,4-Thiadiazolo[2,3-a]pyrazine (84).—See ref. 54.

6 1,3,4-Thiadiazoles

Synthesis.—A variety of 1,3,4-thiadiazoles continue to be produced by well-established methods, generally in connection with biological or medical screening programmes. The methods of preparation include well-tried cyclizations,[56] replacement reactions in the pre-formed heteronucleus, as well as the production of simple functional derivatives.[57] With very few exceptions, the compounds concerned are derived from 2-amino-1,3,4-thiadiazole.

The standard synthesis of 5-substituted 2-amino-1,3,4-thiadiazoles from thiosemicarbazides and carboxylic acids has been improved by the use of a mixture (3 : 1) of polyphosphoric acid and concentrated sulphuric acid as the condensing agent; the use of either acid alone produces lower yields.[58]

[55] K. T. Potts and J. Kane, *J. Org. Chem.*, 1974, **39**, 3783.

[56] S. Yoshina and K. Yamamoto, *Yakugaku Zasshi*, 1974, **94**, 1312 (*Chem. Abs.*, 1975, **82**, 43 108) [2-subst. 5-(3,4-diaryl-2-furyl)-1,3,4-thiadiazoles]; S. N. Sawhney, J. Singh, and O. P. Bansal, *J. Indian Chem. Soc.*, 1974, **51**, 886 [2-, 6-(1,3,4-thiadiazol-2-yl)benzothiazoles]; G. S. Goldin, M. V. Maksakova, V. G. Poddubnyi, and A. N. Koltsova, *Zhur. obshchei Khim.*, 1975, **45**, 2661 [2-amino-5-(ω-trimethylsilyl)ethyl-1,3,4-thiadiazole]; H. Röchling and G. Hörlein, *Annalen*, 1974, 504 [2-amino-5-(2,2-dichloro-1,1-difluoroethyl)-1,3,4-thiadiazole]; I. Saramet, *Farmacia (Bucharest)*, 1975, **23**, 35 (*Chem. Abs.*, 1976, **84**, 70 208) [4-(2-amino-1,3,4-thiadiazol-5-yl) diphenyl sulphones]; I. Simiti and H. Demian, *Farmacia (Bucharest)*, 1973, **21**, 713 (*Chem. Abs.*, 1974, **81**, 63 506) [2-amino-5-(2-arylthiazol-4-yl)-1,3,4-thiadiazoles]; F. Russo and M. Santagati, *Farmaco, Ed. Sci.*, 1976, **31**, 41 [2-amino-5-(benzothiazol-2-yl)-1,3,4-thiadiazoles]; R. Bognar, C. Jaszberenyi, and A. Tokes, *Magyar Kém. Folyóirat*, 1974, **80**, 114 [2-aryl-5-(flavan-4-yl) amino-1,3,4-thiadiazoles]; E. V. P. Tao, *Synth. Comm.*, 1974, **4**, 249 [1-(5-*t*-butyl-1,3,4-thiadiazol-2-yl)-1,3-dimethylurea]; G. Werber, F. Buccheri, and M. L. Marino, *J. Heterocyclic Chem.*, 1975, **12**, 581 [2-amino-5-benzoyl-1,3,4-thiadiazoles and Δ²-1,3,4-thiadiazolines].

[57] I. Lalezari, A. Shafiee, A. Badali, M. M. Salimi, M. A. Khoyi, F. Abtahi, and M. R. Zarrindast, *J. Pharm. Sci.*, 1975, **64**, 1250 [5-subst. 2-(*NN*-dialkylaminoethyl)amino-1,3,4-thiadiazoles]; J. J. Lukes and K. A. Nieforth, *J. Medicin. Chem.*, 1975, **18**, 351 [2-amino-5-mercapto-, 2-amino-5-sulphonamido-1,3,4-thiadiazoles, *S*-benzyl and *N*-alkyl derivatives]; D. Oteleanu and M. A. Dimitriu, *Farmacia (Bucharest)*, 1975, **23**, 25 (*Chem. Abs.*, 1976, **84**, 74 145) [2-amino-5-alkylthio-1,3,4-thiadiazoles, 2-tosyl derivatives]; D. Oteleanu and M. A. Dimitriu, *Farmacia (Bucharest)*, 1974, **22**, 461 (*Chem. Abs.*, 1976, **84**, 74 186) [as previously, 2,3-disulphonyl derivatives]; J. Heindl, E. Schröder, and H. W. Kelm, *European J. Med. Chem.-Chim. Thér.*, 1975, **10**, 121 [2-(*NN*-dialkylaminomethyleneamino)-5-nitro-1,3,4-thiadiazoles, 2-heteroarylthio-5-nitro-1,3,4-thiadiazoles]; S. A. S. El-Dine and O. Clauder, *Acta Chim. Acad. Sci. Hung.*, 1975, **84**, 85 (*Chem. Abs.*, 1975, **82**, 125 322) [2-subst. 2-amino-5-nitro-1,3,4-thiadiazoles]; A. K. Sengupta and K. Avasthi, *J. Indian Chem. Soc.*, 1975, **52**, 750 [5-subst. 2-phenoxyacetylamino-1,3,4-thiadiazoles]; A. K. Sengupta and A. K. Ramrakhyani, *J. Indian Chem. Soc.*, 1973, **50**, 742 [*S*-(2-arylamino-1,3,4-thiadiazol-5-yl) mercaptoacetic acids]; S. A. S. El-Dine and A. A. B. Hazzaa, *Pharmazie*, 1974, **29**, 761 [*N*-(2-alkylthio-1,3,4-thiadiazol-5-yl)dithiocarbamic acid esters]; A. K. Sengupta and K. Avasthi, *J. Indian Chem. Soc.*, 1975, **52**, 433 [5-subst. 1,3,4-thiadiazol-2-yl)acetylaminodithiocarbamic acids]; F. Russo, M. Santagati, and M. Alberghina, *Farmaco, Ed. Sci.*, 1975, **30**, 1031 [1-(5-alkyl-1,3,4-thiadiazol-2-yl)-ureas, -3-methoxycarbonylureas, and -thioureas]; A. R. El Nasser, A. R. N. Ossman, and M. Khalifa, *Pharmazie*, 1975, **30**, 540 [3-(2-hydroxy-1,3,4-thiadiazol-5-yl)-quinazol-4-ones].

[58] E. V. P. Tao and C. F. Christie, *Org. Prep. Proced. Internat.*, 1975, **7**, 179.

The successful synthesis of stable 4-hydroxy-3-thiosemicarbazides (90) [59] has opened the way to novel, substituted, 2-hydroxyamino-1,3,4-thiadiazoles (91), by their conventional condensation with imino-esters [60] or with acid halides R^2COCl and subsequent dehydrative cyclization.[61]

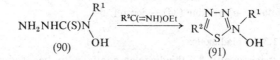

The interaction of 1,2-dihydro-4H-benzo[d][1,3]thiazine-2,4-dithione (92) with acid hydrazides provides very satisfactory yields of 5-substituted 3-(o-aminophenyl)-1,3,4-thiadiazoles (94), undoubtedly by way of the intermediate N-(2-aminothiobenzoyl)-N'-acyl-hydrazines (93).[62]

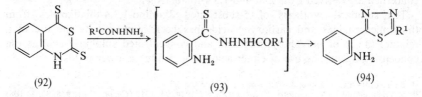

From (Thio)carbonohydrazides. 1,3,4-Thiadiazoles arise in the acidic cyclization of thioacylated (thio)carbonohydrazides. Thioaroylthioacetic acids [63] convert carbonohydrazide into 1-thioaroylcarbonohydrazides (95; R = H, X = O). 1-Phenyl- and 1-benzylidene-carbonohydrazide react to give similar compounds. All are readily cyclized by mineral acids to 2-aryl-5-hydroxy-1,3,4-thiadiazoles (96), with loss of hydrazine or its appropriate derivative. The action of acetic anhydride produces 2-acetoxy-5-phenyl-1,3,4-thiadiazoles directly, in one stage.[64]

1-Thiobenzoylthiocarbonohydrazide (95; Ar = Ph, R = H, X = S) tends to cyclize to a 1,2,4-triazole rather than to a 1,3,4-thiadiazole derivative. However, its interaction with acetylacetone gives 2-mercapto-5-phenyl-1,3,4-thiadiazole

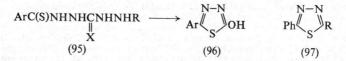

(97; R = SH), probably by the scission and cyclization of an intermediate pyrazolyl derivative. The labile 1-phenyl-5-thiobenzoylthiocarbono-hydrazide (95; Ar = R = Ph, X = S) is cyclized by acid to 2-phenyl-5-phenylhydrazino-1,3,4-thiadiazole (97; R = NHNHPh) (25—40%). 2-Aryl-5-phenylazo-1,3,4-thiadiazoles, *e.g.* (97; R = N=NPh), arising as by-products in the corresponding

[59] P. Gröbner and E. Müller, *Monatsh.*, 1974, **105**, 969.
[60] H. Weidinger and J. Kranz, 'Festschrift Carl Wurster' (1960), Badische Anilin- & Soda-Fabrik AG., p. 119.
[61] E. Müller and P. Gröbner, *Monatsh.*, 1975, **106**, 27.
[62] S. Leistner and G. Wagner, *Z. Chem.*, 1974, **14**, 305.
[63] F. Kurzer, *Chem. and Ind.*, 1961, 1333; K. M. Doyle and F. Kurzer, *ibid.*, 1974, 803.
[64] R. Esmail and F. Kurzer, *J.C.S. Perkin I*, 1975, 1781.

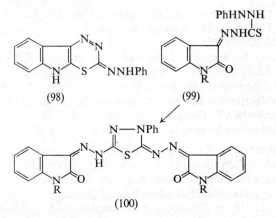

(98)   (99)

(100)

alkaline cyclization, are the principal products of the oxidation of the 5-phenylhydrazino-analogues (97; R = NHNHPh).[65]

The product of the interaction of isatin with phenylthiocarbonohydrazide in glacial acetic acid was originally incorrectly formulated [66] as the thiadiazine (98), but has now been identified as the substituted 1,3,4-thiadiazoline (100).[67] It arises in moderate yield from the primarily formed phenylthiocarbonohydrazone (99) with loss of hydrogen sulphide and phenylhydrazine. In strongly acidic media, the reaction proceeds differently, affording excellent yields of spiro-thiadiazolines. Acenaphthenequinone gives rise to an analogous heterocyclic spirane.[67]

Formation from 4-Hydrazino-quinazoline. The interaction of 4-hydrazoquinazoline (101) and carbon disulphide in the presence of base unexpectedly yields 2-(2-aminophenyl)-5-mercapto-1,3,4-thiadiazole (103). Its formation has been accounted for by a mechanism involving the intermediate spiro-compound (102).[68]

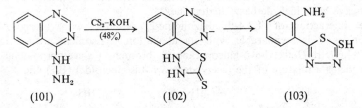

(101)   (102)   (103)

Physical Properties.—A study of the u.v. spectra and acid–base equilibria of α-acylamino-N-heterocycles has included measurements for 2-acylamino-1,3,4-thiadiazoles. The formation of their sodium salts is accompanied by a strong bathochromic displacement in the near-u.v. band. The ionization constants indicate that the stability of the anions depends on both the electron-attracting

[65] R. Esmail and F. Kurzer, *J.C.S. Perkin I*, 1975, 1787.
[66] P. C. Guha and S. K. Roy-Choudhury, *J. Indian Chem. Soc.*, 1928, **5**, 49.
[67] A. B. Tomchin and V. S. Dmitrukha, *Zhur. org. Khim.*, 1975, **11**, 184.
[68] G. M. Coppola and G. E. Hardtmann, *J. Org. Chem.*, 1974, **39**, 2467.

properties of the acyl group and the resonance stabilization of the exocyclic N=C bond. The latter effect increases in the series pyridines < pyrimidines < thiazoles < thiadiazoles.[69]

In an extensive investigation of the π-electron structure of N- and S-heterocycles, the application of fundamental PPP MO calculations and [13]C n.m.r. and u.v. spectral measurements have provided information concerning the changes in the π-electron distribution due to 2-amino- and 2-formyl-substituents in 1,3,4-thiadiazole derivatives.[70] Energy barriers to rotation of the dimethylamino-group in 1,3,4-thiadiazoles have been determined [44] (see p. 427).

Chemical Properties.—*Alkylation*. Methylation of 2-amino-5-benzoyl-1,3,4-thiadiazole and its derivatives has been systematically examined. The direction of methylation, which occurs at the N-3 atom of the ring and the exocyclic amino-group, is influenced by the structure of the substrate, the nature of the reagent (MeI, Me$_2$SO$_4$, or CH$_2$N$_2$), and the reaction conditions.[71] 2-(Substituted amino)-1,3,4-thiadiazoline-5-thiones (104) are *S*-methylated to (105) by one equivalent of methyl iodide, in the presence or absence of alkali. Further methylation occurs at N-3 or N-4, but may lead to mixtures, the composition of which depends on the ratio of the reactants and on the degree of substitution of the 2-amino-group in (105). At 190 °C, (105; R^1 = R^2 = Me) isomerizes to (106);

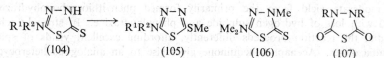

this rearrangement appears to precede the formation of the *N*-4-methylated products. On thermolysis, the *N*-3- or *N*-4-dimethylated products are demethylated to the corresponding thiones.[72]

1,3,4-Thiadiazolidine-2,5-dione yields a mono- and a di-potassium salt, which are *N*-alkylated with alkyl halides in DMF to give (107). The structure was confirmed for (107; R = PhCH$_2$) by the action of alkaline hydrogen peroxide, which gave benzaldehyde benzylhydrazone.[73]

The interaction of 1,3,4-thiadiazolidine-2,5-dithione with bromo-monosaccharides[74a] is comparable with its alkylation. Treatment with 2 moles of 2,3,4,6-tetra-*O*-acetyl-α-D-glucopyranosyl bromide ('α-acetobromoglucose', ABG) yields a mixture of the (acetylated) *SS'*-bis(glucoside) (108) and *SN*-bis-

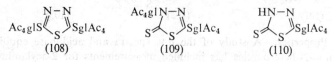

gl = β-D-glucopyranosyl

[69] D. Suciu and Z. Györfi, *Rev. Chim.* (*Roumania*), 1974, **19**, 671.
[70] H. Hartmann and R. Radeglia, *J. prakt. Chem.*, 1975, **317**, 657.
[71] G. Werber, F. Buccheri, and N. Vivona, *J. Heterocyclic Chem.*, 1975, **12**, 841.
[72] M. B. Kolesova, L. I. Maksimova, and A. V. Eltsov, *Zhur. org. Khim.*, 1973, **9**, 2613.
[73] S. W. Mojé and P. Beak, *J. Org. Chem.*, 1974, **39**, 2951.
[74] (*a*) G. Wagner and B. Dietzsch, *J. prakt. Chem.*, 1973, **315**, 915; (*b*) G. Wagner, B. Dietzsch, and U. Krake, *Pharmazie*, 1975, **30**, 694.

glucoside (109). The latter is formed exclusively by the action of ABG on the monomercury(II) salt of the 2,5-dithione.[74a] The reaction may be terminated at the monoglucosidation stage (110) by the use of an excess of the aglycone. Both (108) and (110) undergo transglycosidation to (109) nearly quantitatively in anhydrous solvents in the presence of mercury(II) bromide. Comparable results are obtained by the use of the monomethylated derivatives [73] or 2-phenyl-1,3,4-thiadiazoline-5-thione [74b] as aglycones. The u.v. spectra of the numerous glucosides isolated confirm the assigned structures, and show that examples incorporating a thioamide group capable of tautomerizing, *e.g.* (110), occur predominantly in the lactam form.[74]

Ring Fission. The kinetics of the hydrolytic ring-scission [75] of 2-methylthio-4,5-diphenyl-1,3,4-thiadiazolium iodide (111) to *N*-benzoyl-*N*-phenyldithiocarbamate (113) have been studied spectrophotometrically.[76] Above pH 4, the reaction is of the first order and is almost independent of pH, in accord with the rate-determining formation of a neutral intermediate, *e.g.* (112), *via* a cationic transition state.

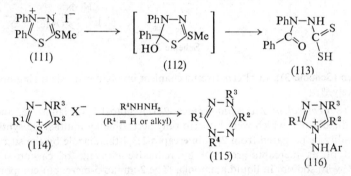

The partial inhibition of the hydrolysis observed at pH < 4, and the negative salt effect, may be attributed to a specific cation–anion interaction. The proposed mechanism receives further support from the observed effects on the hydrolysis rates of increased viscosity of the medium and of added inorganic nucleophiles.[77]

Hydrazine or alkylhydrazines cleave and recyclize 1,3,4-thiadiazolium salts (114) to 1,2- or 1,4-dihydro-1,2,4,5-tetrazines (115) in high yield. The action of arylhydrazine results in the alternative recyclization of the probable intermediate ArNHN=CR^{2}NR^{3}N=CR^{1}SH, to 4-amino-1,2,4-triazolium salts (116).[78]

5-Amino-2-imino-3-phenacyl-1,3,4-thiadiazolines (117) isomerize in boiling ethanol to 5-amino-3-mercapto-1-phenacyl-1,2,4-triazoles (118).[79] This example

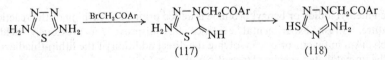

[75] P. B. Talukdar, S. K. Sengupta, and A. K. Datta, *Indian J. Chem.*, 1972, **10**, 1070; P. B. Talukdar, S. Banerjee, and A. Chakraborty, *Bull. Chem. Soc. Japan*, 1970, **43**, 125; *Indian J. Chem.*, 1972, **10**, 610.
[76] P. B. Talukdar and A. Chakraborty, *J. Indian Chem. Soc.*, 1974, **51**, 600.
[77] P. B. Talukdar and A. Chakraborty, *J. Indian Chem. Soc.*, 1975, **52**, 893.
[78] O. P. Shvaika and V. I. Fomenko, *Zhur. org. Khim.*, 1974, **10**, 377.
[79] A. Sitte, R. Wessel, and H. Paul, *Monatsh.*, 1975, **106**, 1291.

of the well-known Dimroth rearrangement, usually performed in basic media, is noteworthy by the readiness with which it occurs under the mild neutral conditions.

Nucleophilic Substitution. A study of the kinetics of the nucleophilic replacement of halogen by methoxide in four halogeno-1,3,4-thiadiazoles and two 1,2,3-isomers has provided values for the pseudo-first-order rate constants and activation energies of these reactions. A mechanism consistent with the observations

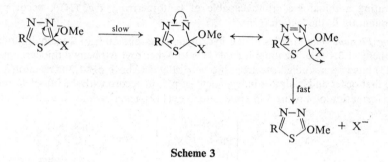

Scheme 3

is shown (Scheme 3); an alternative mechanism involving transient ring-opening is unlikely.[80]

Reductive Removal of Mercapto-groups. 2-Mercapto-1,3,4-thiadiazole, and the parent thiadiazole, which were hitherto only accessible by multi-stage syntheses, may readily be prepared from 2,5-dimercapto-1,3,4-thiadiazole by the successive removal of its mercapto-groups,[81] by reductive cleavage of carbon–sulphur bonds[82] with sodium in liquid ammonia. The 2-amino-5-mercapto-compound is similarly convertible into 2-amino-1,3,4-thiadiazole (up to 65%).[81]

Dimerization. The lithio-1,3,4-thiadiazole (120), formed by lithiation at −78 °C, is methylated at low temperatures to give the 2-ethyl derivative, but it dimerizes at ambient temperatures to give (121). The open-chain thioimine (119) is isolable

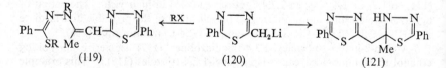

when methyl iodide or benzyl bromide is added prior to quenching of the reaction mixture. Fresh experimental evidence has disproved a previously held[83] mechanism in favour of one involving the direct addition of the lithio-thiadiazole to the unmetallated species[84] (see also Vol. 3, p. 578).

[80] B. Modarai, M. H. Ghandehari, H. Massoumi, A. Shafiee, I. Lalezari, and A. Badali, *J. Heterocyclic Chem.*, 1974, **11**, 343.
[81] S. Hoff, *Tetrahedron Letters*, 1974, 843.
[82] S. Hoff, A. P. Blok, and E. Zwanenburg, *Rec. Trav. chim.*, 1973, **92**, 879.
[83] A. I. Meyers and G. N. Knaus, *J. Amer. Chem. Soc.*, 1973, **95**, 3408.
[84] G. N. Knaus and A. I. Meyers, *J. Org. Chem.*, 1974, **39**, 1189.

Formation of Metal Complexes. 1,3,4-Thiadiazole-2,5-dithiol, a potential analytical reagent for the detection and determination of cations, forms complex salts with Cu^I, Zn^{II}, Ag^I, Cd^{II}, Tl^I, Pb^{II}, Pd^o, and Pt^o. The Cu^I, Ag^I, and Tl^I complexes are linear; those of the other ions are tetrahedral. Bonding occurs through the thione- or thiol-sulphur, except in the Pd^o and Pt^o complexes, where it involves the nitrogen.[85] The complexes of Ru^{III}, Ru^{II}, Rh^{III}, Pd^{IV}, Ir^{III}, and Pt^{IV} were similarly studied.[86]

Polymeric 1,3,4-Thiadiazoles. Knoevenagel condensation of 2,5-dimethyl-1,3,4-thiadiazole with mono- and di-aldehydes of the benzene and thiophen series has given a number of polymers of type (123), as well as their oligomers (122); their

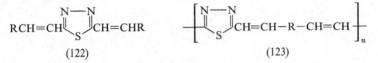

$$\text{(122)} \qquad\qquad\qquad \text{(123)}$$

physical properties, including detailed spectral characteristics, have been recorded.[87]

Analytical. Details have been described of a method of estimating 5-acetamido-1,3,4-thiadiazole-2-sulphonamide (Acetazolamide) by high-pressure liquid chromatography, sensitive to 25 ng ml^{-1} of the drug in plasma. The compound is a carbonic anhydrase inhibitor, used to lower intraocular pressure by a reduced production of aqueous humour.[88] 2-Methyl-5-sulphanilamido-1,3,4-thiadiazole (Sulfamethizole) may be estimated by a coulometric titration.[89]

Biochemical Aspects. The zinc-containing metalloenzyme carbonic anhydrase (carbonic hydro-lyase, EC 4.2.1.1) is specifically inhibited by univalent anions and by aromatic sulphonamides. ΔH^o (25 °C) for the binding reactions between bovine carbonic anhydrase and 3-methyl-2-acetylimino-1,3,4-thiadiazoline-5-sulphonamide (Methazolamide) is -59.0 kJ mol^{-1}, and that for 5-phenyl-sulphonamido-1,3,4-thiadiazole-2-sulphonamide (CL 11366) is -57.7 kJ mol^{-1}. The thermodynamic quantities did not correlate with the assumed structural features of the binding process.[90]

5-Substituted 2-anilino-1,3,4-thiadiazoles inhibit oxidative phosphorylation (in rat liver mitochondria) and photosynthetic phosphorylation (in spinach chloroplasts).[91]

Biological Activity.—5-Substituted 2-(NN-dialkylaminoethyl)amino- and 2-(N-methylpiperazinyl)-1,3,4-thiadiazoles display antihistamine, anticholinergic, and norepinephrine-potentiating activities.[57a]

2,2'-Methylenedi-imino-bis(1,3,4-thiadiazole) is effective against various tumours at non-toxic doses in mice, and significantly suppresses the production

[85] M. R. Gajendragad and U. Agarwala, *Indian J. Chem.*, 1975, **13**, 697.
[86] M. R. Gajendragad and U. Agarwala, *Austral. J. Chem.*, 1975, **28**, 763.
[87] G. Kossmehl, I. Dornacher, and G. Manecke, *Makromol. Chem.*, 1974, **175**, 1359.
[88] W. F. Bayne, G. Rogers, and N. Crisologo, *J. Pharm. Sci.*, 1975, **64**, 402.
[89] M. Ishibashi, J. Tokusawa, S. Kojima, A. Ejima, and T. Inoue, *Yakugaku Zasshi*, 1973, **93**, 711 (*Chem. Abs.*, 1973, **79**, 57 718).
[90] J. S. Binford, S. Lindskog, and I. Wadsö, *Biochim. Biophys. Acta*, 1974, **341**, 345.
[91] G. Schaefer, A. Trebst, and K. H. Buechel, *Z. Naturforsch.*, 1975, **30c**, 183 (*Chem. Abs.*, 1975, **83**, 91 476).

of serum antibody against sheep erythrocytes. The activities are *ca.* 5 times those of the parent compound, 2-amino-1,3,4-thiadiazole.[92] The mechanism of its action has been discussed.[93] Toxicity data are available concerning 2-amino-1,3,4-thiadiazole.[94]

1-Methyl-6-(1-methylallyl)bithiourea (Metallibur, ICI 33828), MeNHC(S)-NHNHC(S)NHCHMeCH=CH$_2$, possessing gonadotropin-inhibiting properties, is being used increasingly in animal husbandry. Its stability, its metabolic fate at varying pH, and its thermal decomposition to a variety of 1,2,4-triazole derivatives and 1,3,4-thiadiazoles, *e.g.* (124), have been studied. The reported isolation of (125) is noteworthy.[95]

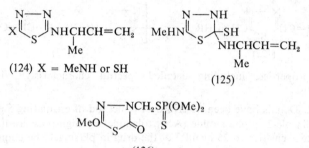

(124) X = MeNH or SH

(125)

(126)

A study of the photodecomposition of the insecticide *OO*-dimethyl [(2-methoxy-Δ²-1,3,4-thiadiazolin-5-on-4-yl)methyl]phosphorodithioate (Methidathion) (126) has provided data that have allowed a detailed photochemical breakdown scheme to be devised.[96]

Meso-ionic 1,3,4-Thiadiazoles.—The synthesis and properties of isomeric pairs of 1,3,4-thiadiazolium-2-olates (128) and of 1,3,4-oxadiazolium-2-thiolates (127) have been described,[97] thus establishing the existence of isomeric meso-ionic compounds, the interconversion of which [(127) → (128)] involves an exchange between endo- and exo-cyclic hetero-atoms.[97] The synthesis of the 1,3,4-thiadiazolium-2-olates (128) is based on an analogous route to the comparable isosydnones, (129) and involves the reaction of carbonyl chloride with substituted

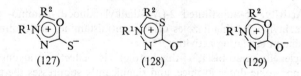

(127)   (128)   (129)

[92] T. Matsumoto, K. Ootsu, and Y. Okada, *Cancer Chemother. Rep., Part I*, 1974, **58**, 331 (*Chem. Abs.*, 1975, **83**, 53 359).

[93] K. Tsukamoto and M. Suno, *Cancer Res.*, 1975, **35**, 2631 (*Chem. Abs.*, 1976, **84**, 154).

[94] N. Rakietan, D. A. Cooney, and R. D. Davis, U.S. Nat. Tech. Inform. Service, AD. Rep. 1973, No. 226610/4GA.

[95] U Olthoff and K. Matthey, *Pharmazie*, 1974, **29**, 20.

[96] W. P. Dejonckheere and R. H. Kips, *J. Agric. Food Chem.*, 1974, **22**, 959.

[97] A. R. McCarthy, W. D. Ollis, and C. A. Ramsden, *J.C.S. Perkin I*, 1974, 627; *J.C.S. Chem. Comm.*, 1968, 499.

N-thioacyl-hydrazines $R^2C(S)NHR^1NH_2$.[98] Unlike isosydnones (129), the meso-ionic 2-olates (128) react with neither alcohols nor amines, and are resistant to hydrolysis by acid.[97]

Further details have been reported[99] of the formation of meso-ionic 1,3,4-thiadiazoles (*e.g.* anhydro-4-methyl-5-phenyl-2-phenylamino-1,3,4-thiadiazolium hydroxide) from isocyanide chlorides and N-methyl-N-thiobenzoylhydrazine followed by treatment with ammonia (see Vol. 2, p. 746). These red, viscous oils give methiodides and are reduced by LiAlH$_4$ to thiosemicarbazides.

The synthesis of meso-ionic 1,3,4-thiadiazole-imines described by Grashey *et al.* (see Vol. 3, p. 698) has been independently described by a Russian group of workers.[100] The conversion of meso-ionic 4,5-diphenyl-1,3,4-thiadiazole-2-thiones into meso-ionic 1,2,4-triazole-3-thiones under the influence of amines[101] has also been further exemplified.[102]

7 Δ²-1,3,4-Thiadiazolines

Although tautomerism introduces a measure of ambiguity into the structural assignment of certain 1,3,4-thiadiazoles, those carrying substituents on the nitrogen atoms in the ring, and some others, are clearly Δ²-1,3,4-thiadiazolines, and are dealt with in this section.

Synthesis.—*From Hydrazonyl Halides.* The interaction of equimolar proportions of hydrazonyl halides PhC(Cl)=NNHAr with thioureas or thiosemicarbazides in ethanol yields 1,3,4-thiadiazolines, generally in excellent yields. The observations are explicable by postulating the initial formation of the hydrazonylated adduct (130), which successively undergoes ring-closure, elimination of hydrazine, and deprotonation. The rate of the elimination step follows the sequence NH$_3$ > NH$_2$NH$_2$ > PhNHNH$_2$ > PhNH$_2$, *i.e.* following the decreasing order of base strength. In the presence of excess of triethylamine, the main products are 1,2,4-triazole derivatives.[103] The use of β-keto-thioacid anilides RCOCH= C(SH)NHAr′ in this reaction produces the Δ²-1,3,4-thiadiazolines (131) in good

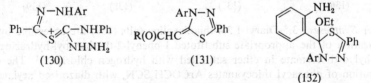

(130)   (131)   (132)

yield, with loss of Ar′NH$_2$. Under more strongly basic conditions, thiazole derivatives arise by an alternative pathway. The formulation of (131) is supported by its conversion into a thiadiazole of established structure.[104] The

[98] A. R. McCarthy, W. D. Ollis, A. N. M. Barnes, L. E. Sutton, and C. Ainsworth, *J. Chem. Soc. (B)*, 1969, 1185.
[99] W. D. Ollis and C. A. Ramsden, *J.C.S. Perkin I*, 1974, 633; *J.C.S. Chem. Comm.*, 1971, 1222.
[100] A. Y. Lazaris, S. M. Shmuilovich, and A. N. Egorochkin, *Khim. geterotsikl. Soedinenii*, 1973, 1345; A. Y. Lazaris and A. N. Egorochkin, *Khim. geterotsikl. Soedinenii*, 1975, 648.
[101] M. Ohta, H. Kato, and T. Kaneko, *Bull. Chem. Soc. Japan*, 1967, **40**, 578; A. Y. Lazaris and A. N. Egorochkin, *Zhur. org. Khim.*, 1970, **6**, 2342.
[102] P. B. Talukdar, S. K. Sengupta, and A. K. Datta, *Indian J. Chem.*, 1972, **10**, 1070.
[103] P. Wolkoff, S. T. Nemeth, and M. S. Gibson, *Canad. J. Chem.*, 1975, **53**, 3211.
[104] D. Pocar, L. M. Rossi, and P. Trimarco, *J. Heterocyclic Chem.*, 1975, **12**, 401.

interaction of 4-mercapto-1,2,3-benzotriazine and *N*-(2,4-dibromophenyl)benzo-hydrazonyl bromide in the presence of triethylamine also yields a substituted Δ^2-1,3,4-thiadiazoline (132) (53%).[105]

The Δ^2-1,3,4-thiadiazoline (133) and the dithiadiazine (134) appear amongst the products of the reaction of bis(α-phenylhydrazonobenzyl) disulphide [PhNHN=C(Ar)S$+_2$ with *N*-phenylbenzohydrazonyl chloride or with oxidizing agents. The dithiadiazine (134) undergoes ring-contraction to (133) in boiling acetone. The structure of (133) is supported by its alternative synthesis from the hydrazonyl halide PhCCl=NNHPh, by the action of sodium sulphide, followed by oxidation with potassium ferricyanide.[106]

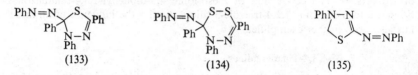

(133) (134) (135)

From Dithizone. Dithizone, PhNHN=C(SH)N=NPh, undergoes *SN*-dialkyl-ation on treatment with di-iodomethane, with simultaneous cyclization, to give the substituted 1,3,4-thiadiazoline (135),[107] a reaction that occurs also under the influence of formaldehyde.[108]

The action of iron pentacarbonyl on the meso-ionic dehydrodithizone (136), expected to yield (137) by carbonyl-insertion,[109] produces the 1,3,4-thiadiazoline derivative (138) (54%), the structure of which has been established by *X*-ray analysis. The mechanism of its formation is not yet clear.[110]

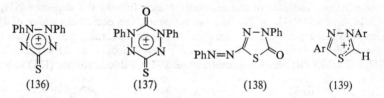

(136) (137) (138) (139)

Other Syntheses. 3,5-Diaryl-1,3,4-thiadiazoline salts (139) are accessible by the cyclization of the appropriate substituted 1-phenyl-2-thiobenzoyl-hydrazines by triethyl orthoformate in ether saturated with hydrogen chloride.[111] The con-densation of phenacyl thiocyanates ArCOCH$_2$SCN, with diazotized arylamines provides a convenient one-step synthesis of 4-aryl-2-benzoyl-5-imino-Δ^2-1,3,4-thiadiazolines (141).[112] Their alternative synthesis from (140) and potassium thiocyanate confirms the assigned structure.[112]

A study of *N*-acetyl-heterocycles as acylating agents has included the prepara-tion and application of 4-acetyl-2-phenyl-1,3,4-thiadiazolin-5-one (and -thione)

[105] P. D. Callaghan, A. J. Elliott, and M. S. Gibson, *J. Org. Chem.*, 1975, **40**, 2131.
[106] D. H. R. Barton, J. W. Ducker, W. A. Lord, and P. D. Magnus, *J.C.S. Perkin I*, 1976, 38.
[107] F. A. Neugebauer and H. Fischer, *Chem. Ber.*, 1974, **107**, 717.
[108] H. M. Irving and M. S. Mahnot, *Talanta*, 1968, **15**, 811.
[109] Y. Becker, A. Eisenstadt, and Y. Shoo, *Tetrahedron*, 1974, **30**, 839.
[110] P. N. Preston, N. J. Robinson, K. Turnbull, and T. J. King, *J.C.S. Chem. Comm.*, 1974, 998.
[111] G. Scherowsky, *Chem. Ber.*, 1974, **107**, 1092; *Tetrahedron Letters*, 1971, 4985.
[112] A. S. Shawali and A. O. Abdelhamid, *Tetrahedron Letters*, 1975, 163.

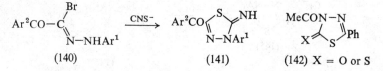

(142). They react 10^4 to 10^6 times as fast as *p*-nitrophenyl acetate with cyclo-hexylamine; they are, moreover, 4—50 times as reactive as the corresponding oxadiazolines. Their use in peptide synthesis is under investigation.[113]

An early claim of the synthesis of 1,3,4-thiadiazolidine-2,5-dione by the action of boiling hydrochloric acid on the substituted bithiourea PhNHNHC(S)-NHNHC(S)NHPh has been disproved, and the product identified as 4-(*p*-amino-phenyl)-1,2,4-triazolidine-3,5-dithione.[73]

Properties.—The mass spectra of di-, tri-, and tetra-aryl-substituted 1,3,4-thiadiazoles show that the major fragmentation reactions occur by cleavage of ring bonds, or by loss of substituents from the hetero-ring, rather than by cleavage within the substituents. They reveal the formation of: (i) thiadiazolium ions by elimination of a C-5 substituent [forming (143)]; (ii) ions generated by the elimination of PhS· arising by transannular phenyl migration [forming (144)], and (iii) ions formed by elimination of thiocarbonyl neutrals in retro-1,3-dipolar cycloadditions. The suggested predominating patterns are outlined in Scheme 4.[114]

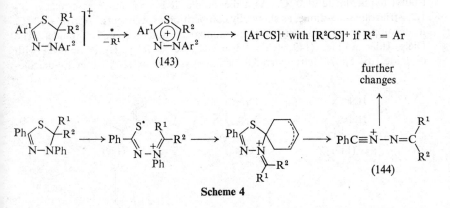

Scheme 4

3,5-Diaryl-1,3,4-thiadiazolines form strongly acidic onium salts (145), which are suitable for examining (by deuterium exchange) the influence of substituents (in Ar) on the kinetic C—H acidity. With one exception, the results conform to the Hammett equation. The salts (145) are deprotonated in boiling chloroform or benzene, and the resulting carbene is stabilized by dimerization to the coloured solids (146). The process is reversed by hydrogen chloride.[111]

---

[113] H. Fukuda, T. Endo, and M. Okawara, *Chem. Letters*, 1973, 1181.
[114] P. Wolkoff and S. Hammerum, *Org. Mass Spectrometry*, 1974, **9**, 181.

(145)   (146)

8 Δ³-1,3,4-Thiadiazolines

The production and synthetic applications of Δ³-1,3,4-thiadiazolines continue to provide novel results that are of exceptional interest.

Formation.—The use of the cycloaddition of azo-compounds to sulphines as a synthesis of Δ³-thiadiazolines (see Vol. 3, pp. 251, 695) has been extended to arylsulphonyl-sulphines.[115]

Alkene Synthesis.—The alkene synthesis by two-fold extrusion processes (see Vol. 3, pp. 246, 693) is discussed in Chapter 2 of this Volume (p. 89). The addition of diazo-compounds to suitable thiones provides an alternative route to Δ³-1,3,4-thiadiazolines, which give episulphides on pyrolysis and hence alkenes on further reaction with triphenylphosphine.[116] The symmetrical 1,3,4-thiadiazoline represents the normal direction of cycloaddition,[116] but the orientation of addition of diazomethane to adamantanethione is governed by the nature of the solvent, and the unsymmetrical adduct predominates in solvents of higher polarity.[117] Δ³-1,3,4-Thiadiazoline-2,5-dione (148) is generated [73] by oxidation of 1,3,4-thiadiazolidine-2,5-dione (147) with cupric chloride, lead tetra-acetate, or t-butyl hypochlorite at 0 °C. As a dienophile, it is as reactive as 4-phenyl-Δ¹-1,2,4-triazoline-3,5-dione, as shown by competition experiments. Above −35 °C, it decomposes to nitrogen, carbon monoxide, and carbonyl sulphide. The Diels–Alder adduct (149; R¹ = R⁴ = CH₂, R² = R³ = H) is reduced with di-imide to 5,6,7,8-tetrahydro-5,8-methanopyridazino[1,2-c]-2-thia-4,9-diazole-

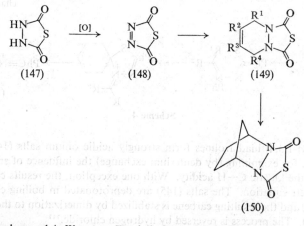

(147)   (148)   (149)

(150)

[115] B. Zwanenburg and A. Wagenaar, *Tetrahedron Letters*, 1973, 5009.
[116] D. H. R. Barton, F. S. Guziec, and I. Shahak, *J.C.S. Perkin I*, 1974, 1794.
[117] A. P. Krapcho, M. P. Silvon, I. Goldberg, and E. G. E. Jahngen, *J. Org. Chem.*, 1974, **39**, 860.

1,3-dione (150), which is pyrolysed at 300—350 °C under nitrogen to a 4.5 : 1 mixture of bicyclopentane and cyclopentene in 18% yield [73] (see also Vol. 3, p. 696).

The thermolysis of Δ^3-1,3,4-thiadiazolines continues to serve as a source of thiocarbonyl ylides.[118] For an account of these highly reactive species see Chapter 2, p. 89.

9 Condensed Ring Systems incorporating 1,3,4-Thiadiazole

Imidazo[2,1-*b*][1,3,4]thiadiazoles.—6-Substituted 2-(benzothiazol-2-yl)imidazo-[2,1-*b*][1,3,4]thiadiazoles (152; Bt = benzothiazol-2-yl) are accessible by the condensation of amino-thiadiazoles and halogeno-ketones in DMF, followed by cyclization of the intermediates (151) in boiling acetic acid. Numerous further applications of this synthesis have been described.[119]

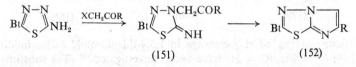

1,3,4-Thiadiazolo[3,2-*a*]pyrimidines.—The synthesis of 1,3,4-thiadiazolo[3,2-*a*]-pyrimidines by the condensation of 2-amino-1,3,4-thiadiazoles and β-di-ketones [120] has been extended [121] by employing β-chlorovinyl ketones, β-chloro-vinyl aldehydes, and the acetal of malondialdehyde. The reaction generally yields only one of the possible isomers, but occasionally leads to inseparable mixtures [121] (see Vol. 3, p. 701).

In a preliminary report, Tsuji [122] has described the Vilsmeier–Haack reaction of 3-amino-6-methyl-4(3*H*)-oxo-2(1*H*)-pyrimidinethione (153). The action of dimethylformamidine and phosphorus oxychloride affords the two formamidines

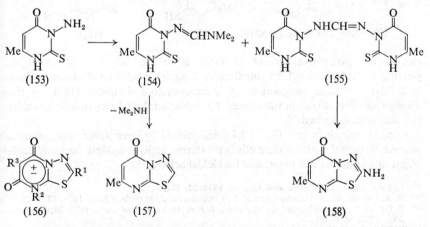

[118] J. Buter, P. W. Reynolds, and R. M. Kellog, *Tetrahedron Letters*, 1974, 2901; R. M. Kellog, M. Noteboom, and J. M. Kaiser, *J. Org. Chem.*, 1975, **40**, 2573.
[119] L. Pentimalli, G. Milani, and F. Biavati, *Gazzetta*, 1975, **105**, 777.
[120] M. K. Pordeli and V. A. Chuiguk, *Ukrain. khim. Zhur.*, 1972, **38**, 1045.
[121] M. K. Pordeli, V. V. Oksanich, and V. A. Chuiguk, *Khim. geterotsikl. Soedinenii*, 1973, 1285.

(154) and (155), which are separately convertible into 7-methyl-5H-1,3,4-thia-diazolo[3,2-a]pyrimidin-5-one (157) and its 2-amino-derivative (158).[122]

Further examples of meso-ionic 1,3,4-thiadiazolo[3,2-a]pyrimidine-5,7-diones (156) have been prepared [123] by previously described methods (see Vol. 3, p. 702). Certain of these compounds possess *in vitro* antibacterial activity.[123]

The u.v. spectra of 5-oxothiadiazolopyrimidines (159) have λ_{max} at *ca.* 300 nm, while those of the 7-oxo-isomerides (161) have λ_{max} at 260—270 nm, suggesting that π-electrons are more delocalized in the former ring system. I.r., n.m.r., and mass spectral data are in accord with this view.[124]

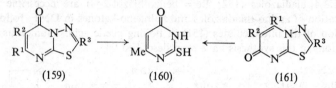

(159)   (160)   (161)

Reactions leading to ring-cleavage in 1,3,4-thiadiazolo[3,2-a]pyrimidines [125] (159; 161; R^1 = Me, R^2 = H) have been reinvestigated.[126] The substituent R^3 influences the choice of the site at which ring-scission occurs; in general, the thiadiazole ring is destroyed preferentially. Thus the isomers (159; 160; R^1 = Me, R^2 = R^3 = H) react with 5% sodium hydroxide to give quantitative yields of methyl(thiouracil) (160) by N—N bond cleavage. When R^3 = alkyl, however, the S—C bond is ruptured, resulting in the corresponding acylamino-pyrimidine derivatives (162) and pyrimidinethiones (163). In contrast, chlorine, at 0 °C,

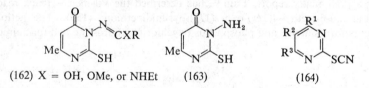

(162) X = OH, OMe, or NHEt   (163)   (164)

cleaves the pyrimidine moiety of (159; R^1 = Me, R^2 = H, R^3 = alkyl), yielding 5-alkyl-2-amino-1,3,4-thiadiazole. Numerous additional examples have been given.[126] The production of 2-thiocyanato-pyrimidines (164) by the alkaline scission of certain substituted 1,3,4-thiadiazolo[3,2-a]pyrimidinium salts has also been described.[127]

Selected members of the 1,3,4-thiadiazolo[3,2-a]pyrimidine ring system possess larvicidal activity (housefly), protective activity against bacterial leaf-blight disease of the rice plant, and herbicidal activity.[128]

[122] T. Tsuji, *Chem. and Pharm. Bull. (Japan)*, 1974, **22**, 471.
[123] R. A. Coburn, R. A. Glennon, and Z. F. Chmielewicz, *J. Medicin. Chem.*, 1974, **17**, 1025.
[124] T. Okabe, E. Taniguchi, and K. Maekawa, *J. Fac. Agric., Kyushu Univ.*, 1975, **20**, 7 (*Chem. Abs.*, 1976, **84**, 4010).
[125] T. Tsuji and Y. Kamo, *Chem. Letters*, 1972, 641.
[126] T. Okabe, E. Taniguchi, and K. Maekawa, *Bull. Chem. Soc. Japan*, 1974, **47**, 2813.
[127] V. A. Chuiguk and Y. M. Volovenko, *Khim. geterotsikl. Soedinenii*, 1974, 1660.
[128] T. Okabe, E. Taniguchi, and K. Maekawa, *J. Fac. Agric., Kyushu Univ.*, 1975, **19**, 91 (*Chem. Abs.*, 1975, **83**, 127 285).

1,3,4-Thiadiazolo[3,2-*a*][1,3,5]triazines.—Members of this ring system (166) have been obtained,[129] within the framework of a more general synthesis of hetero-condensed *sym*-triazines, by converting 2-amino-1,3,4-thiadiazoles, by the action of benzonitrile in the presence of aluminium chloride, into the corresponding amidines (165). Subsequent cyclization with carbonyl chloride produces (166) as yellow crystalline solids in moderate yields.[129] 1,3,4-Thiadiazolo[3,2-*a*][1,3,5]-triazines of type (167) are formed by the alkaline or thermal cyclization of the appropriate *N*-methoxycarbonyl-*S*-methylisothioureas.[130]

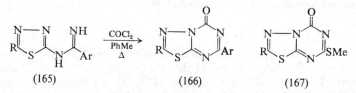

(165) (166) (167)

1,3,4-Thiadiazolo[3,2-*b*]isoquinolines.—Homophthalic anhydride reacts with acyl-hydrazines to yield the hydrazides (168) or the corresponding cyclized tetra-hydroisoquinolines, both of which are cyclized on treatment with phosphorus pentasulphide, to yield the novel 2-substituted 1,3,4-thiadiazolo[3,2-*b*]iso-quinoline-5(5*H*)-thiones (169). Methanolic alkali cleaves the condensed ring system, yielding (170).[131]

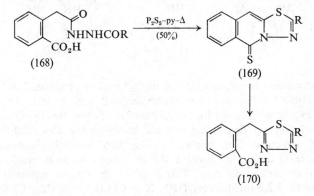

(168)

(169)

(170)

10 Condensed Ring Systems incorporating 1,3,4-Selenadiazole

1,3,4-Selenadiazolo[3,2-*a*]pyrimidines.—Substituted 1,3,4-selenadiazolo[3,2-*a*]-pyrimidinium perchlorates (171) are accessible in 27—64% yield by the interaction of 2-amino-5-phenyl-1,3,4-selenadiazole and dicarbonyl compounds or their derivatives, *e.g.* acetylacetone and malonic dialdehyde acetal.[132]

The reaction of 2-amino-1,3,4-selenadiazoles with dimethyl acetylenedi-carboxylate gives 2-substituted 5-methoxycarbonyl-7*H*-1,3,4-selenadiazolo[3,2-*a*]-pyrimidin-7-ones (172), which may be decarboxylated to the 2-substituted

[129] T. George and R. Tahilramani, *Synthesis*, 1974, 346.
[130] F. Russo, M. Santagati, and M. Alberghina, *Farmaco. Ed. Sci.*, 1975, **30**, 1031.
[131] M. Takahashi, T. Shinoda, H. Osada, and T. Nakajima, *Bull. Chem. Soc. Japan*, 1975, **48**, 2915.
[132] V. A. Chuiguk, D. I. Sheiko, and V. G. Glushakov, *Khim. geterotsikl. Soedinenii*, 1974, 1435.

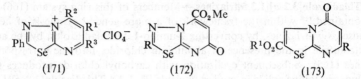

(171)   (172)   (173)

analogues. Isomeric products have not been detected. Ethyl propiolate gave a product to which structure (173) was assigned, but the possibility that it is the 3-substituted isomer could not be excluded.[133]

11 1,2,5-Thiadiazoles

Synthesis.—Another report has appeared of the condensation of sulphur di-imides ArN=S=NAr with oxalyl chloride to give, in this case, 2,5-diaryl-1-arylimino-1,2,5-thiadiazolidine-3,4-diones (174).[134] 3-Chloro-4,4-dimethyl-Δ^2-1,2,5-thiadiazoline *S*-oxide (175) and the dioxide (176) arise in good yield from the interaction of 2-amino-2-cyanopropane with thionyl or sulphuryl chloride, respectively. The halogen is replaced by basic groups.[135]

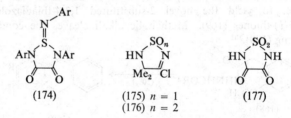

(174)   (175) $n = 1$ (177)
                   (176) $n = 2$

The condensation of sulphamide and diethyl oxalate produces 3,4-dihydroxy-1,2,5-thiadiazole 1,1-dioxide as the dipotassium salt in high yield. The free acid (177) is an excellent precursor for the preparation of other functional derivatives, and for building up fused ring systems incorporating the 1,2,5-thiadiazole 1,1-dioxide structure (see below).[136] The two isomeric benzothiadiazole acetoximes (178) and (179) undergo the Beckmann reaction, resulting in the fission of their carbocyclic ring. This provides an effective synthesis of novel $\alpha\beta$-unsaturated 1,2,5-thiadiazoles (180; X = CO$_2$H, Y = CR=CHCN; X =

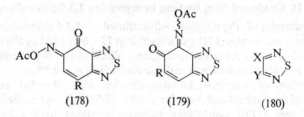

(178)   (179)   (180)

[133] A. Shafiee and I. Lalezari, *J. Heterocyclic Chem.*, 1975, **12**, 675.
[134] R. Neidlein and P. Leinberger, *Angew. Chem. Internat. Edn.*, 1975, **14**, 762.
[135] V. V. Dovlatyan and R. S. Mirzoyan, *Armyan. khim. Zhur.*, 1975, **28**, 412 (*Chem. Abs.*, 1975, **83**, 206 174).
[136] R. Y. Wen, A. P. Komin, R. W. Street, and M. Carmack, *J. Org. Chem.*, 1975, **40**, 2743.

CN, Y = CH=CRCO$_2$H) that are otherwise not readily available.[137] The geometrical configuration in some cases was established by n.m.r. spectroscopy. Thermal decarboxylation of (180; X = CO$_2$H, Y = CMe=CHCN) and (180; X = CO$_2$H, Y = CH=CHCO$_2$H) gave 3-methyl-3-(1,2,5-thiadiazol-3-yl)acrylonitrile and 3-(1,2,5-thiadiazol-3-yl)acrylic acid, respectively, in contrast to that of the oxadiazole analogues, which results in violent decomposition.[137]

Physical Properties.—The ^{1}H n.m.r. spectrum of 2,5-dimethyl-1,2,5-thiadiazolidine 1-oxide has been fully analysed on the basis of an [AB]$_2$ spin system, and compared with those of 3-phenyl-1,2,3-oxathiazolidine 2-oxide, 1,3,2-oxadithiolan 2-oxide, and ethylene sulphites. The effect of ring heteroatoms other than oxygen in the ethylene sulphite structure [138] has thus been elucidated.

The redox behaviour has been determined, in acetonitrile solution at a mercury and platinum electrode, of 3,4-disubstituted and fused 1,2,5-thiadiazoles, including 2,1,3-benzothiadiazoles. All ring systems and their derivatives are reversibly reduced initially in a one-electron step to their radical anion, but nitro- and bromo-derivatives are reduced preferentially at the substituent group.[139]

Chemical Properties.—The rates of quaternization of 1,2,5-thiadiazole, 2,1,3-benzothiadiazole, and the corresponding selenium heterocycle have been examined by a competition method, using dimethyl sulphate in sulpholane.[140]

1,2,5-Thiadiazole, previously thought to be resistant to electrophilic substitution, is in fact as aromatic as thiophen, and undergoes chloromethylation to the bis(chloromethyl)-derivative. This is also obtained as the principal product of the chlorination of 3,4-dimethyl-1,2,5-thiadiazole with *N*-chlorosuccinimide. NBS gives comparable results.[141]

The ready availability of 1,2,5-thiadiazole 1,1-dioxides (177) (see above) has provided data for their comparison with the parent aromatic ring system.[136] The *NO*-, *NN*-, and *OO*-dimethylated derivatives are obtained by suitable alkylation procedures. The action of an excess of phosphorus pentachloride gives the 3,4-dichloro-compound (181; X = Cl), which is rapidly hydrolysed back to the dione (177), and which undergoes ammonolysis or aminolysis to give a variety of 3,4-di(substituted amino)-1,2,5-thiadiazole 1,1-dioxides [*e.g.* (181; X = NR^{1}R^2)]. The amide-like character of the 3,4-diamino-derivative (181; X = NH$_2$) has been noted, as has a pronounced tendency of the 3,4-di-alkoxy-derivatives (181; X = OMe) to transfer their *O*-alkyl groups to the ring nitrogen on thermolysis.[136]

1,2,5-Thiadiazoles and 1,2,5-selenadiazoles (182; Z = S or Se) react with Grignard reagents or lithium alkyls at −70 °C, to yield, after hydrolysis, sulphides

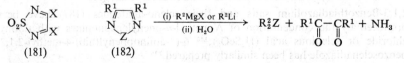

(181)   (182)

[137] V. Cere, D. Dalmonte, S. Pollicino, and E. Sandri, *Gazzetta*, 1975, **105**, 723.
[138] P. Albriktsen and M. Bjoroy, *Acta Chem. Scand. (A)*, 1975, **29**, 414.
[139] E. O. Sherman, S. M. Lambert, and K. Pilgram, *J. Heterocyclic Chem.*, 1974, **11**, 763.
[140] M. Davis, L. W. Deady, and E. Homfeld, *Austral. J. Chem.*, 1974, **27**, 1917.
[141] R. F. Cookson and A. C. Richards, *J.C.S. Chem. Comm.*, 1974, 585.

or selenides, ammonia, and 1,2-dicarbonyl compounds. The evidence suggests that nucleophilic attack occurs at the sulphur (or selenium) atoms, and a possible mechanism of the ring cleavage of the heterocycles has been proposed.[142] In a competitive reaction involving 3,4-diphenyl-1,2,5-thiadiazole and -selenadiazole with methylmagnesium iodide, the selenium ring was cleaved more rapidly.[142] The observation is in accord with the fact that the 1,2,5-selenadiazole ring is also the more susceptible to reducing and oxidizing agents, to heat, and to sunlight.

12 1,2,5-Selenadiazoles

The interaction of 1,2-diketone dioximes with an excess of diselenium dichloride in DMF yields 1,2,5-selenadiazole *N*-oxides, a reaction that is comparable with the corresponding synthesis of 1,2,5-thiadiazole *N*-oxides (see Vol. 2, p. 753). 2,1,3-Benzoselenadiazole *N*-oxide (184), similarly obtained from the dioxime (183), undergoes thermolysis in decalin to give 2,1,3-benzoselenadiazole (185)

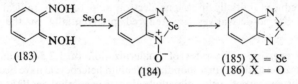

(51%) and benzofurazan (186) (32%). Two competing mechanisms are likely, *viz.* reduction of the *N*-oxide to the parent amine (185), and ring-opening, followed by recyclization, with loss of selenium, to give (186).[143]

A detailed study of the mass spectra of a number of 1,2,5-selenadiazoles (including 2,1,3-benzoselenadiazole and three sulphur analogues) has provided data for a discussion of their fragmentation, as outlined in Scheme 5.[144]

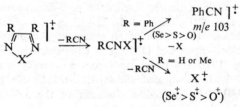

Scheme 5

For the action of Grignard reagents and lithium alkyls on 1,2,5-selenadiazole,[142] see above.

13 2,1,3-Benzothiadiazoles and Selenium Analogues

2,1,3-Benzothiadiazolium salts and their seleno-analogues (187) have been produced by the condensation of *N*-methyl-*o*-phenylenediamines with thionyl chloride or selenious acid (H_2SeO_3);[145] 6-(2-aminophenylthio)-4-methyl-2,1,3-benzoselenadiazole has been similarly prepared.[146]

[142] V. Bertini, A. de Munno, A. Menconi, and A. Fissi, *J. Org. Chem.*, 1974, **39**, 2294.
[143] C. L. Pedersen, *J.C.S. Chem. Comm.*, 1974, 704.
[144] C. L. Pedersen and J. Møller, *Acta Chem. Scand.* (*B*), 1975, **29**, 483.
[145] G. I. Eremeeva, B. K. Strelets, and L. S. Efros, *Khim. geterotsikl. Soedinenii*, 1975, 276.
[146] D. P. Sevbo and T. F. Stepanova, *Zhur. org. Khim.*, 1974, **10**, 1318.

The ring system has been the subject of quantum-chemical calculations.[147] The outstanding feature of the mass spectra of the 2,1,3-benzothiadiazoles (35 examples) is the occurrence of molecular ions of very great abundance.[148] Fragmentations analogous to the benzyne formation postulated in the case of the 2,1,3-benzoselenadiazoles are not encountered. The polarographic behaviour of a large number of 2,1,3-benzothiadiazoles, of 2,1,3-benzoselenadiazole, and of several fused 1,2,5-thiadiazoles has been examined.[139]

The n.m.r. spectra and dissociation constants of a series of monoximes, dioximes, and mono-O-methyldioximes of 2,1,3-benzothiazole-4,5- and -4,7-diones (188), (189) have been measured. The 4,5- and 4,7-di-O-methyldioximinobenzothiadiazoles were also prepared and investigated. Four and two geometrical isomers of the 4,5- and 4,7- derivatives, respectively, have been separated and described.[149]

(187) X = S or Se

(188) X, Y = O, NOH or NOMe

(189) X, Y = O, NOH or NOMe

A number of useful replacement reactions in the 2,1,3-benzothiadiazole series involve the production and reactions of their nitriles.[150] 4,7-Dicyano-2,1,3-benzothiadiazole is of particular interest because of its potent herbicidal[151] and defoliating activity.[152] It is produced by the action of a pyridine–copper(II) cyanide complex on the readily available 4,7-dibromo-compound. Other nitriles of this series are similarly obtained. A second method, of more limited applicability, is the Sandmeyer reaction, which, in the present instance, gives improved results when the amino-2,1,3-benzothiadiazoles are diazotized with nitrosyl-sulphuric acid.[150] The nitriles are convertible by conventional methods into other functional derivatives.[153] Successive coupling with diazotized aniline, and

[147] A. S. Bylina and V. S. Korobkov, *Sintez. Analiz i Struktur Org. Soedinenii*, 1974, 86 (*Chem. Abs.*, 1976, **84**, 4310) (from *Ref. Zhur. Khim.*, 1975, Abstr. No. 16–B–54).
[148] F. C. V. Larsson, S. O. Lawesson, I. Jardine, and R. I. Reed, *Acta Chem. Scand.* (*B*), 1975, **29**, 622.
[149] A. S. Angeloni, D. Dalmonte, S. Pollicino, E. Sandri, and G. Scapini, *Tetrahedron*, 1974, **30**, 3839.
[150] K. Pilgram and R. D. Skiles, *J. Heterocyclic Chem.*, 1974, **11**, 777.
[151] R. H. Schieferstein and K. Pilgram, *J. Agric. Food Chem.*, 1975, **23**, 392.
[152] R. S. Slott, E. R. Bell, and K. Pilgram, U.S.P. 3 478 044 (*Chem. Abs.*, 1968, **69**, 106 711), R. W. Baldwin, U.S.P. 3 501 285 (*Chem. Abs.*, 1970, **72**, 111 477).
[153] K. Pilgram, *J. Heterocyclic Chem.*, 1974, **11**, 835.

reduction, converts 4-hydroxy-2,1,3-benzothiadiazole into the 7-phenylazo- and 7-amino-4-hydroxy-2,1,3-benzothiadiazoles. Several other functional derivatives were similarly obtained by conventional routes.[154] For a comparative study of the rates of N-methylation of 2,1,3-benzothiadiazoles and the selenium analogues, see above.[140]

14 Condensed Ring Systems incorporating 1,2,5-Thiadiazole

1,2,3-Triazolo[4,5-c][1,2,5]thiadiazoles and Selena-analogues.—5-Phenyl-5H-[1,2,3]triazolo[4,5-c][1,2,5]thiadiazole (191; X = S) is obtained (38%) on treatment of 4,5-diamino-2-phenyl-v-triazole (190) with sulphur monochloride.

$$\text{(190)} \longrightarrow \text{(191)}$$

Selenous acid produces the corresponding selenadiazole (191; X = Se). The condensed thiadiazole (191; X = S) is resistant to oxidation, but is reconverted into (190) (90%) by reduction with lithium aluminium hydride. The spectral properties (n.m.r., u.v.) and half-wave potentials of the compounds (191; X = S or Se) point to coplanarity of the phenyl substituent with the bicyclic heteronucleus.[155]

1,2,5-Thiadiazolo[3,4-c][1,2,5]thiadiazoles.—1,2,5-Thiadiazolo[3,4-c][1,2,5]thiadiazole (192) is a novel, exceptionally interesting, compound, the ring system of which is derived by the isoelectronic replacement of all peripheral atoms of naphthalene by hetero-atoms. The resulting symmetrical structure is aromatic, as manifested by its planarity, shortened bond lengths, and stability, which is considerably greater than that of the thieno[3,4-c]thiophens.[156]

The ring system has been synthesized by the action of excess of sulphur dichloride on 3,4-diamino-1,2,5-thiadiazole or its 1,1-dioxide in DMF, or by the condensation of oxamide dioxime [NH$_2$C(=NOH)C(=NOH)NH$_2$] and sulphur dichloride in the same solvent. Experiments aimed at elucidating the mechanism of each of these cyclizations have been reported.[156]

Pure (192) is a white crystalline solid resembling naphthalene in its physical properties. Its high degree of symmetry is reflected in the simplicity of its i.r. spectrum. Hydrolysis yields successively 3,4-diamino-1,2,5-thiadiazole, oxamide, and sulphur.[156]

1,2,5-Thiadiazolo-[3,4-b]- and -[3,4-c]-pyridines.—A study of the specificity of purine bases as substrates for *trans*-N-deoxyribosylase has included the examination of several 1,2,5-thiadiazolo-[3,4-b]- and -[3,4-c]-pyridines (193), (194), and their selenium analogues from this viewpoint.[157]

[154] I. A. Belenkaya, N. S. Tsepova, Y. L. Kostyukovskii, and V. G. Pesin, *Khim. geterotsikl. Soedinenii*, 1973, 926.
[155] A. Matsumoto, M. Yoshida, and O. Simamura, *Bull. Chem. Soc. Japan*, 1974, **47**, 1493.
[156] A. P. Komin, R. W. Street, and M. Carmack, *J. Org. Chem.*, 1975, **40**, 2749.
[157] J. Holguin, R. Cardinaud, and C. A. Salemink, *European J. Biochem.*, 1975, **54**, 575.

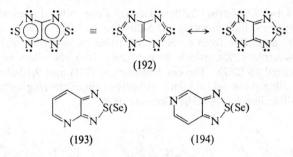

(192)

(193)   (194)

1,2,5-Thiadiazolo[3,4-d]pyridazine.—Successive *S*-benzylation and hydrazinolysis of 4-mercapto-1,2,5-thiadiazolo[3,4-*d*]pyridazine-7(6)-thione yields (195), which is further converted by nitrous acid or boiling formic acid into the tricyclic structures (196; X = N or CH).[158]

(195)   (196)   (197)

1,2,5-Thiadiazolo[3,4-d]pyrimidines.—7-Amino-1,2,5-thiadiazolo[3,4-*d*]pyrimidine (197)[159] displays numerous analogies in its properties with the comparable 4-aminopteridine, including its enzymatic oxidation[159, 160] by xanthine oxidase to the corresponding 5-hydroxy-compound.[160]

1,2,5-Thiadiazolo[3,4-b]pyrazines.—A number of members of this ring system have been prepared by the interaction of 2,3-diamino-pyrazines with thionyl chloride, and their properties studied.[161]

Condensation of 3,4-dimethoxy-1,2,5-thiadiazole 1,1-dioxide with ethylene-diamine gives 4,5,6,7-tetrahydro-1,2,5-thiadiazolo[3,4-*b*]pyrazine 2,2-dioxide (198) in excellent yield.[136]

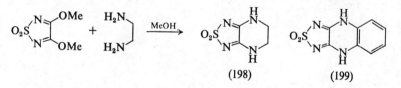

(198)   (199)

1,2,5-Thiadiazolo[3,4-b]quinoxaline.—The condensation of 3,4-dimethoxy-1,2,5-thiadiazole 1,1-dioxide with *o*-phenylenediamine similarly produces 1,3-dihydro-1,2,5-thiadiazolo[3,4-*b*]quinoxaline 2,2-dioxide (199).[136]

[158] A. Majcen, B. Stanovnik, and M. Tisler, *Vestnik. Slovensk. kem. Drustva*, 1974, **21**, 23.
[159] T. A. Krenitzky, S. M. Neill, G. B. Elion, and G. H. Hitchings, *Arch. Biochem. Biophys.*, 1972, **150**, 585.
[160] J. J. McCormack and E. C. Taylor, *Biochem. Pharmacol.*, 1975, **24**, 1636.
[161] Y. C. Tong, *J. Heterocyclic Chem.*, 1975, **12**, 451.

Benzo[1,2-c:3,4-c':5,6-c'']tris[1,2,5]thiadiazole.—Four related condensed tetra-cyclic ring systems incorporating 1,2,5-thiadiazole (202; X = S, Se, CH, or CH=CH) are accessible from a common starting material, benzo[1,2-c:3,4-c']-bis[1,2,5]thiadiazole (200), which is converted into the diamine (201) and variously cyclized to (202). The condensation of (201) and 9,10-phenanthrene-quinone in boiling acetic acid similarly yields the heptacyclic ring system dibenzo-[a,c]bis[1,2,5]thiadiazolo[3,4-h:3',4'-j]phenazine.[162]

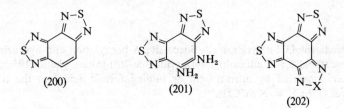

(200)   (201)   (202)

[162] A. P. Komin and M. Carmack, *J. Heterocyclic Chem.*, 1975, **12**, 829.

15
Thiazines

BY G. PROTA

1 Introduction

Certain trends in the investigation of thiazine systems, noted in the previous Volume, have been sustained, with an increasing interest especially in the chemistry of 1,4-benzothiazines, which occur as structural units in pigments of considerable biological significance such as the trichochromes and phaeomelanins. The possible involvement of a 1,4-benzothiazine intermediate in the biosynthesis of firefly luciferin [1] will certainly stimulate further activity in this area. Two reviews,[2, 3] surveying various aspects of the chemistry and biology of the 1,4-benzothiazine pigments found in mammals and birds, have appeared.

The layout of the Chapter adopted in the previous Volumes of this Series is retained, except that further restrictions on space have precluded a fuller presentation of activity in some areas, e.g. fused-ring systems incorporating the thiazine structure. Moreover, the section dealing with simple 1,2-thiazines has been omitted, as no significant work in this field has recently appeared.

2 1,2-Thiazines

Benzo-1,2-thiazines.—As usual, work in this area is limited, and is mainly occasioned by the search for biologically or chemotherapeutically active compounds. In addition to several patents, which, for reasons of space, are not considered, two reports [4, 5] have appeared describing various alkylation and amination reactions of some benzothiazine dioxides, e.g. (2), available by base-induced rearrangement of appropriately N-β-keto-substituted saccharins (1).

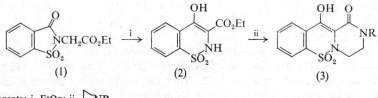

Reagents: i, EtO⁻; ii, ⟨NR

Scheme 1

[1] F. McCapra and Z. Razavi, *J.C.S. Chem. Comm.*, 1975, 42, 492; 1976, 153; K. Okada, H. Iio, and T. Goto, *ibid.*, 1976, 32.

[2] R. H. Thomson, *Angew. Chem. Internat. Edn.*, 1974, **13**, 305.

[3] G. Prota and R. H. Thomson, *Endeavour*, 1976, **35**, 32.

[4] C. R. Rasmussen, *J. Org. Chem.*, 1974, **39**, 1554.

[5] C. R. Rasmussen and D. L. Shaw, *J. Org. Chem.*, 1974, **39**, 1560.

454   *Organic Compounds of Sulphur, Selenium, and Tellurium*

Among the reactions reported, the most interesting is perhaps that leading [5] to the novel piperazines (3) by treatment of (2) with aziridines in DMF or boiling ethanol (Scheme 1).

Some new synthetic approaches to 3-carboxamides of 4-hydroxy-2H-1,2-benzothiazine 1,1-dioxide have also been reported.[6]

3 1,3-Thiazines

Simple 1,3-Thiazines.—The stereospecific and regiospecific synthesis of 5,6-dihydro-4H-1,3-thiazines by polar cycloaddition of thioamido-alkyl ions, generated *in situ* from aldehydes and thioamides, to olefins was reported in the previous Volume (see p. 712). A similar reaction with arylacetylenes PhC≡CR^1 has now been found [7] to give mixtures of isomeric 4H-1,3-thiazines (4) and (5) in a ratio that depends upon the nature of substituents.

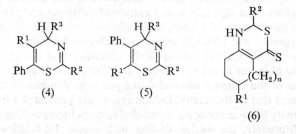

(4)   (5)   (6)

The coupling of carbonyl compounds with β-iminodithiocarboxylic acids (see Vol. 2, p. 762), formed by the action of carbon disulphide and ammonia on cyclic ketones, has been extended [8] to the preparation of some thiazinethiones of type (6), and a modified procedure [9] involves the reaction of (7) with an acyl chloride to give the 1,3-thiazinethiones (8).

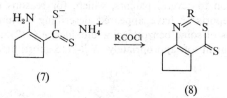

(7)   (8)

A new route to 1,3-thiazinethiones with a wide variety of substituents is provided [10] by the reaction of ββ-dichlorovinyl aryl ketones with primary thioamides in a molar ratio of 1 : 2. Thus (9) reacts with thiobenzamide in boiling acetic acid to give (10), which, by sequential treatments with mercuric acetate and oxalyl chloride, can be converted into the thiazinium salt (11)

[6] J. G. Lombardino and H. A. Watson, jun., *J. Heterocyclic Chem.*, 1976, **13**, 333.

[7] C. Giordano, *Gazzetta*, 1974, **104**, 849.

[8] T. Miyauchi, Y. Kadokura, N. Fukada, and T. Takoshima, *Bull. Chem. Soc. Japan*, 1974, **47**, 1678.

[9] W. Schroth and U. Becker, *Z. Chem.*, 1974, **14**, 51.

[10] W. Schroth, G. Dill, N. T. K. Dung, N. T. M. Khoi, P. T. Binh, H. J. Waskiewicz, and A. Hildebrandt, *Z. Chem.*, 1974, **14**, 52.

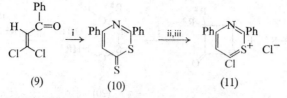

Reagents: i, PhC(S)NH₂; ii, Hg(OAc)₂; iii, (COCl)₂

Scheme 2

(Scheme 2). Treatment[11] of dimethylformamide diethyl acetal with a primary aromatic thioamide leads to the formamidine (12), from which some 6-oxo-6*H*-1,3-thiazines (13) have been obtained by reaction with appropriate ketens.

$$ Me_2N-CH=N-CS-R^1 \quad \xrightarrow[(-Me_2NH)]{R^2CH=C=O} $$

(12)

(13)

4*H*-1,3-Thiazinium betaines, *e.g.* (15), are formed[12] in excellent yields by condensation of monosubstituted arylthioamides or trisubstituted thioureas with chlorocarbonylphenylketen, a new versatile 1,3-dielectrophilic species. Although (15) contains a masked 1,4-dipolar system (16), it does not undergo cycloaddition reactions with acetylenic and olefinic dipolarophiles, its preferred reaction being

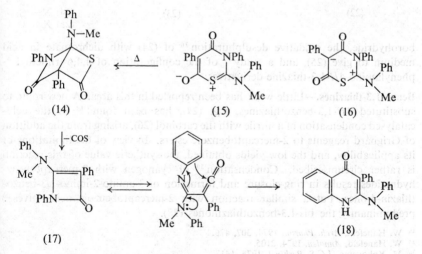

(14) (15) (16)

(17) (18)

a ready thermal elimination of carbonyl sulphide to give the 4-quinolone (18), probably *via* an intermediate or transition state similar to (14), and hence the azetidone (17).

[11] J. C. Meslin and H. Quiniou, *Synthesis*, 1974, 298.

[12] K. T. Potts, R. Ehlinger, and W. M. Nichols, *J. Org. Chem.*, 1975, **40**, 2596.

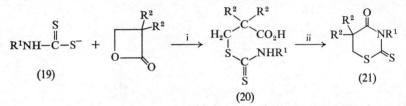

(19)   (21)

(20)

Reagents: i, HCl; ii, Ac₂O

Scheme 3

An improved procedure [13] for preparing 2-thioxo-4-oxotetrahydro-1,3-thiazines is based upon the reaction (Scheme 3) of a propiolactone with ammonium (or sodium) *N*-substituted dithiocarbamates, formed *in situ* from an equimolar amount of a primary amine and carbon disulphide in the presence of ammonium (or sodium) hydroxide. Subsequent cyclization of the resulting adduct (20) with acetic anhydride gives (21), usually in good yields. The method is quite general, and has been adapted [14] to the preparation of tetrahydro-1,3-thiazine-2,4-diones, using RNHC(S)OEt in place of (19). Other aspects of the chemistry of 1,3-thiazines that have been examined recently include a novel hydrogenolysis [15] of the S—C-2 bond, *e.g.* (22) → (23), with LiAlH₄ or sodium

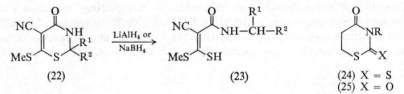

(22)   (23)   (24) X = S
                                                               (25) X = O

borohydride, the oxidative desulphuration [16] of (24) with dichromate in acid medium to give (25), and a study [17] of the configuration of 4,4,6-trimethyl-2-phenylamino-4*H*-1,3-thiazine derivatives.

Benzo-1,3-thiazines.—Little work has been reported in this area. A new route to substituted 4*H*-1,3-benzothiazines, *e.g.* (27), has been found [18] in the acid-catalysed condensation of a nitrile with the carbinol (26), arising from the addition of Grignard reagents to 2-mercaptobenzoic esters. In view of the limitations of its applicability, and the low yields obtained, the synthetic value of this reaction is rather circumscribed. Condensation [19] of cyanogen with 2-mercaptobenz-hydrazides results in ring-closure and formation of 3-amino-2-imino-1,3-benzo-thiazin-4-ones, but a similar reaction with 2-mercaptobenzoic acids gives [20] predominantly the bis-1,3-benzothiazinones (28).

[13] W. Hanefeld, *Arch. Pharm.*, 1974, **307**, 476.
[14] W. Hanefeld, *Annalen*, 1974, 2105.
[15] M. Yokoyama, *J.C.S. Perkin I*, 1975, 1417.
[16] W. Hanefeld, *Annalen*, 1974, 1789.
[17] A. P. Engoyan, Yu. N. Sheinker, J. F. Vlasova, L. A. Ignatova, and A. E. Gekhman, *Khim. geterotsikl. Soedinenii*, 1975, 921 (*Chem. Abs.*, 1975, **83**, 163 508).
[18] V. A. Zagorevskii, K. I. Lopatina, T. V. Sokolova, and S. M. Klynev, *Khim. geterotsikl. Soedinenii*, 1975, 1437 (*Chem. Abs.*, 1976, **84**, 90 098).
[19] N. D. Heindel and L. A. Schaeffer, *J. Pharm. Sci.*, 1975, **74**, 1425.
[20] N. D. Heindel and L. A. Schaeffer, *J. Heterocyclic Chem.*, 1975, **12**, 783.

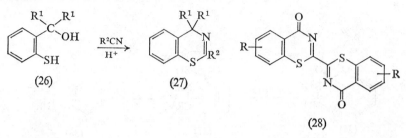

(26)   (27)   (28)

A study of the reaction of some 4*H*-1,3-benzothiazinones, *e.g.* (29), with Grignard reagents has also appeared.[21] Typically, alkylmagnesium bromides, *e.g.* EtMgBr, react additively with (29) to give the dihydrobenzothiazinones (30), while in the case of phenylmagnesium bromide, addition takes place preferentially at the carbonyl function, affording the carbinols (31).

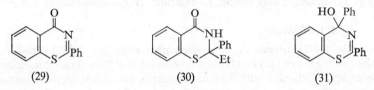

(29)   (30)   (31)

In the area of 4*H*-3,1-benzothiazines, a series of 2-substituted 4-formyl-methylene derivatives (33) have been synthesized [22] by allowing various nucleophiles, especially amines, to react with (32), which may be obtained by ring-opening of 4,7-dichloroquinoline with thiophosgene and barium carbonate. The benzothiazine-1-thiones (34) have been reported [23] to react readily with tetrachloro- or tetrabromo-*o*-benzoquinone, giving the hitherto unknown

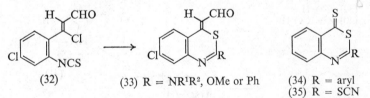

(32)   (33) R = NR¹R², OMe or Ph   (34) R = aryl
                                                    (35) R = SCN

spiranes (36), which, on treatment with anilines or phenylhydrazine, undergo an unusual cleavage reaction, leading to the quinazoline-4-thiones (38) and the tetrahalogenocatechol. This transformation is considered to proceed through the initial attack of the nucleophile on the spiro-carbon atom and the formation of the intermediate (37) by a ring-opening and re-cyclization process which has already been observed.[24]

The conversion of the 2-thiocyanate (35) into bis(4-thioxo-3,1-benzothiazin-2-yl) sulphide by treatment with triethylamine has been described,[25] and further

[21] I. Varga, J. Szabó, and P. Sohar, *Chem. Ber.*, 1975, **108**, 2523.
[22] R. Hull, P. J. Van der Broek, and M. L. Swain, *J.C.S. Perkin I*, 1975, 922.
[23] N. Latif, I. F. Zeid, N. Mishriky, and F. M. Assad, *Tetrahedron Letters*, 1974, 1355.
[24] L. Legrand and N. Lozach, *Bull. Soc. chim. France*, 1961, 1400.
[25] S. Leistner and G. Wagner, *Pharmazie*, 1975, **30**, 542.

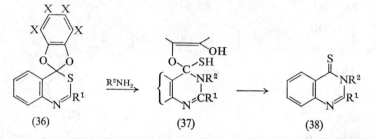

(36)   (37)   (38)

studies on ring-opening reactions of 1,2-dihydro-3,1-benzothiazine-4-thiones with aliphatic amines and ethylenediamine have also appeared.[26] Other papers describe the synthesis and properties of various fused-ring systems incorporating the 1,3-thiazine structure, mainly pyrrolo[2,3-*d*][1,3]thiazine-2-thiones,[27] pyrido-[2,3-*b*][1,3]thiazines,[28] and certain 1,3-thiazino[2,3-*a*]isoquinolinium salts.[29]

4 1,4-Thiazines

Monocyclic 1,4-Thiazines.—A new alternative route to 3,4-dihydro-2*H*-1,4-thiazines of the types (42) or (43) has been developed[30] by direct condensation of 2-mercaptoethylamine with 2,3-dibromoacrylic esters (39) or dibromomaleic hemiesters (40) in the presence of triethylamine. The reaction (Scheme 4) is

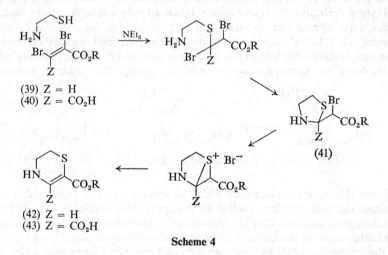

(39) Z = H
(40) Z = CO₂H

(41)

(42) Z = H
(43) Z = CO₂H

Scheme 4

considered to proceed through the initial attack of the thiolate ion at the β-carbon atom of the esters, followed by cyclization and ring-expansion of the resulting thiazolidine intermediates (41) to give, ultimately, (42) or (43).

[26] L. Legrand and N. Lozach, *Bull. Soc. chim. France*, 1975, 1873, 2118.
[27] T. D. Duffy and D. G. Wibberley, *J.C.S. Perkin I*, 1974, 1921.
[28] T. Zawisza and S. Respond, *Roczniki Chem.*, 1975, 49, 743.
[29] H. Singh, V. K. Vij, and K. Lal, *Indian J. Chem.*, 1974, 12, 1242.
[30] J. Alexander, G. Lowe, N. K. McCullum, and G. K. Ruffles, *J.C.S. Perkin I*, 1974, 2092.

A further example of 1,4-thiazine formation by ring-enlargement of 2-substituted thiazolium salts (see Vol. 3, p. 635) is provided [31] by the reaction of some 2-amidino-derivatives (44) with aqueous sodium carbonate, leading to the 2-iminothiazinones (45) in good yields.

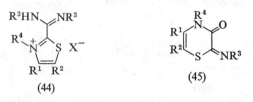

(44)   (45)

When treated with 1 equivalent of potassium t-butoxide, thiazino-oxazolidones of type (46) are reported [32] to undergo a β-elimination to give an enethiolate, *e.g.* (47), which is rapidly converted into the Δ⁴-thiazoline (48), probably by an S_N2-like pathway. This molecular rearrangement finds some analogy in the triethylamine-induced conversion of 1,4-thiazepines into 1,3-thiazines (see Vol. 1, pp. 457, 459).

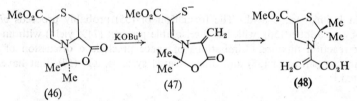

(46)   (47)   (48)

Benzo-1,4-thiazines and Related Compounds.—Although a substantial volume of work has been published on ring homologues of 1,4-benzothiazine, little is known of the chemistry of the parent compound, which seems to be rather elusive. A recent attempt [33] to obtain (50) by an unambiguous route, involving treatment of the amino-acetal (49) with mineral acids in methanolic solution, has resulted in the isolation of the photochromic conjugated dimer (51), which evidently arises by a ready oxidative coupling at C-2 of the parent monomer by the action of atmospheric oxygen. A different course of reaction is observed [34] in the treatment of the same amino-acetal (49) with trifluoroacetic acid, which gives, within a few minutes, the expected 2*H*-1,4-benzothiazine (detected by n.m.r. spectroscopy). This compound gradually reacts to give, after work-up in air, the novel cyanine dye (52). The oxygen analogue of (52) is formed [34] similarly from 1-(*o*-aminophenoxy)-2,2-diethoxyethane.

In connection with a wider study on new indigoid dyes, the trichochrome-like chromophore (55) (see Vol. 2, pp. 772—776) has been obtained [35] by a novel one-step 'inverse indigo synthesis' involving the reaction of 1 mole of 2,3-dichloromaleic anhydride (53) with 2 moles of 2-aminothiophenol or its zinc salt

[31] A. Takamizawa, S. Matsumoto, and I. Makino, *Chem. and Pharm. Bull. (Japan)*, 1974, **22**, 311.
[32] A. G. W. Baxter and R. J. Stoodley, *J.C.S. Chem. Comm.*, 1975, 251; *J.C.S. Perkin I*, 1976, 584.
[33] G. Prota, E. Ponsiglione, and R. Ruggiero, *Tetrahedron*, 1974, **30**, 2781.
[34] F. Chioccara, G. Prota, and R. H. Thomson, *Tetrahedron Letters*, 1975, 811.
[35] B. L. Kaul, *Helv. Chim. Acta*, 1974, **57**, 2665.

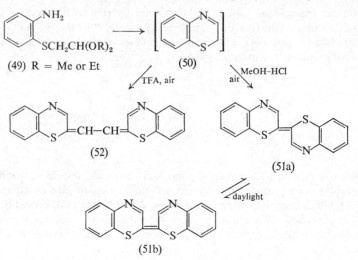

(49) R = Me or Et   (50)

(52)

(51a)

(51b)

(54) in boiling acetic acid. The formation of (55) probably proceeds *via* the monocyclic adduct (56), which is the isolable product (74% yield) with an equimolar reaction mixture. Limitations of space preclude a discussion of some relevant properties of (55) and other related systems, *e.g.* (57), that have been reported.[35]

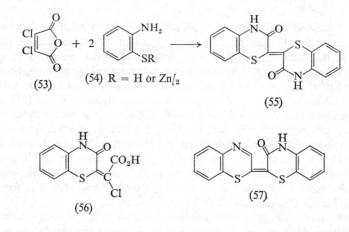

(53)   (54) R = H or Zn/₂

(55)

(56)

(57)

A simple one-step synthesis of 1,4-benzothiazines (59) is found [36] in the ring-expansion of the chlorosulphonium salts of the readily available benzothiazolines (58) (Scheme 5). Treatment [37] of 3-methylbenzothiazolium ion with phenacyl bromide in the presence of triethylamine results in ring-expansion and formation of the 1,4-benzothiazine (61), possibly *via* the thermally unstable intermediate

[36] F. Chioccara, R. A. Nicolaus, E. Novellino, and G. Prota, *Chimica e Industria*, 1976, **58**, 546.
[37] J. A. Van Allan, J. D. Mee, C. A. Maggiulli, and R. S. Henion, *J. Heterocyclic Chem.*, 1975, **12**, 1005.

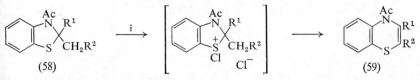

(58)

Reagents: i, SO₂Cl₂, CH₂Cl₂, 25 °C

Scheme 5

(60). Another paper [38] describes the reaction of some chlorobenzothiazinones (62) with triethyl phosphite that leads to the 2-phosphonates (63), from which a series of 2-alkylidene derivatives have been obtained by treatment with various carbonyl compounds, especially aromatic aldehydes.

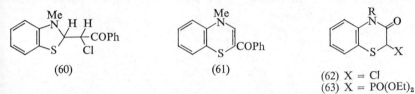

(60)   (61)   (62) X = Cl
                                               (63) X = PO(OEt)₂

The synthesis [39] of 1,4-benzothiazines by acid-catalysed condensation of bis(*o*-aminophenyl) disulphide with ketones has been supplemented by a variant [40] (Scheme 6) in which 1,3-diketones, *e.g.* (64), are condensed with *o*-aminothio-

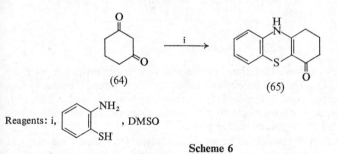

(64)   (65)

Reagents: i, , DMSO

Scheme 6

phenol in DMSO, a solvent capable of converting the thiol into the required disulphide *in situ*. The use of cyclic β-keto-esters as reagents in the synthesis of 1,4-benzothiazines has also been examined recently.[41] In this form, the course of reaction is somewhat different, as illustrated by the condensation between bis(*o*-aminophenyl) disulphide and ethyl 2-oxo-cycloheptane-1-carboxylate,

[38] J. W. Worley, K. W. Ratts, and K. L. Cammack, *J. Org. Chem.*, 1975, **40**, 1731.
[39] V. Carelli, P. Marchini, M. Cardellini, F. Micheletti Moracci, G. Liso, and M. G. Lucarelli, *Tetrahedron Letters*, 1969, 4619; *Ann. Chim.* (*Italy*), 1969, **59**, 1050; V. Carelli, P. Marchini, G. Liso, F. Micheletti Moracci, M. Cardellini, F. Liberatore, and M. G. Lucarelli, *Internat. J. Sulfur Chem.* (*A*), 1971, **1**, 251.
[40] S. Miyano, N. Abe, and K. Sumoto, *J.C.S. Chem. Comm.*, 1975, 760; S. Miyano, N. Abe, K. Sumoto, and K. Teramoto, *J.C.S. Perkin I*, 1976, 1146.
[41] P. Marchini, G. Trapani, G. Liso, V. Berardi, F. Liberatore, and F. Micheletti Moracci, *Internat. J. Sulfur. Chem.*, 1976, in the press.

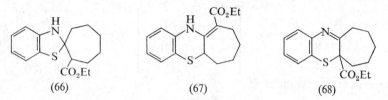

(66)   (67)   (68)

which gives the spiro-benzothiazoline (66) and a mixture of two isomeric benzo-thiazines (67) and (68). The isomer (67) probably arises from (68) by an acid-catalysed rearrangement involving a [1,3]-sulphur migration.

Continuing studies on the autoxidation [42] of 1,4-benzothiazines of the general structure (69) and (70) have confirmed the free-radical chain character of the reaction (Scheme 7). In addition to the sulphoxides (71), the unstable hemi-

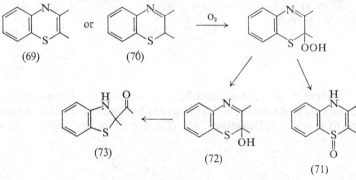

Scheme 7

thioketals (72) are formed, and they readily undergo ring-contraction into the benzothiazolines (73). The new indolobenzothiazine (75), obtained [43] by deoxygen-ation of the *o*-nitrophenyl sulphide (74) by triethyl phosphite under nitrogen, behaves similarly, and, on exposure to air, is rapidly converted into the spiro-oxindole (76) by an analogous oxidative process.

Interest in the field of condensed heterothiazines has also been noticeable, and further examples include some pyrazino[2,3-*b*][1,4]thiazin-6-ones;[44] 6-amino-

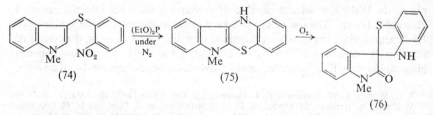

(74)   (75)   (76)

[42] V. Carelli, F. Micheletti Moracci, F. Liberatore, M. Cardellini, M. G. Lucarelli, P. Marchini, G. Liso, and A. Reho, *Internat. J. Sulfur Chem.*, 1973, **8**, 267; G. Liso, P. Marchini, A. Reho, and F. Micheletti Moracci, *ibid.*, 1976, in the press.
[43] A. H. Jackson, D. N. Johnston, and P. V. R. Shannon, *J.C.S. Chem. Comm.*, 1975, 911.
[44] L. A. Myshkina and T. S. Safonova, *Khim. geterotsikl. Soedinenii*, 1975, 66 (*Chem. Abs.*, 1975, **83**, 9982).

pyrimido[4,5-*b*][1,4]thiazines;[45] and 5*H*-pyrimido[4,5-*b*][1,4]thiazine-6-ones, -4,6-diones, and -2,4,6-triones.[46] A study[47] dealing with the characterization of sulphur isosteres of dihydropteridines and their cation radicals by electronic and e.s.r. spectra also merits attention.

Phenothiazines and Related Compounds.—A paper dealing with the laser photolysis (347.1 nm) of phenothiazine (pth) in methanolic and aqueous sodium lauryl sulphate has appeared.[48] An interesting aspect of this work is the study of a new type of redox reaction involving triplet states of metal ions that have suitable reduction potentials, *e.g.* Cu^{2+} or Eu^{3+}, as the reducing agents:

$$pth + M^{n+} \longrightarrow pth^+ + M^{(n-1)+}$$

New evidence,[49] obtained by e.s.r. spectroscopy, suggests that the dye-sensitized or direct light-induced photoxidation of phenothiazine is initiated by an electrophilic attack of singlet oxygen on the unshared electron pair of nitrogen, to give the hydroperoxide (77). Subsequently, depending upon the reaction conditions, the O—O or N—O bonds in (77) are cleaved, to yield phenothiazine nitroxide or a neutral phenothiazinyl radical.

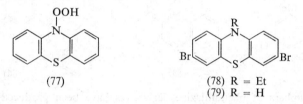

(77)

(78) R = Et
(79) R = H

Bromination of 3,7-dibromo-10-alkyl-phenothiazines, *e.g.* (78), in acetic acid gives[50] purple radical-cations, *e.g.* (80). On subsequent heating in the same solvent, these radical cations undergo electron transfer with bromide ions to form the parent phenothiazines, *e.g.* (78), or are irreversibly dealkylated to yield (79) and, hence, 1,3,7,9-tetrabromophenothiazine, by the action of molecular bromine produced in the previous redox reaction.

In a continued study[51] of *N*-alkyl- and *NN*-dialkyl-sulphilimines, various compounds of type (81) have been prepared (see Vol. 3, p. 725), and the reactions

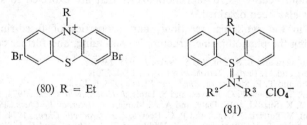

(80) R = Et

(81)

[45] M. P. Nemeryuk and T. S. Safonova, *Khim. geterotsikl. Soedinenii*, 1975, 192.
[46] J. Clark and I. W. Southon, *J.C.S. Perkin I*, 1974, 1805, 1814.
[47] H. Fenner, H. Motschall, S. Ghisla, and P. Hemmerich, *Annalen*, 1974, 1973.
[48] S. A. Alkaitis, S. Beck, and M. Gratzel, *J. Amer. Chem. Soc.*, 1975, **97**, 5723.
[49] I. Rosenthal and R. Poupko, *Tetrahedron*, 1975, **31**, 2103.
[50] H. Chion, P. C. Reeves, and E. R. Biehl, *J. Heterocyclic Chem.*, 1976, **13**, 77.
[51] B. K. Bandlish, A. G. Padilla, and H. J. Shine, *J. Org. Chem.*, 1975, **40**, 2590.

464   *Organic Compounds of Sulphur, Selenium, and Tellurium*

of some members of the series with acids in acetonitrile solution have been examined. With perchloric acid (81; $R^1 = Me$, $R^2 = H$, $R^3 = Bu^t$), for example, is converted quantitatively into the 10-methylphenothiazine cation-radical, whereas with hydrochloric acid the initial formation of the same radical is followed by reduction and chlorination to give mainly 10-methylpheno-thiazine and 3-chloro-10-methylphenothiazine.

Additional evidence [52] has been presented in support of hydro-aromatic species, *e.g.* (82), as intermediates in the thermolysis of aryl 2-azidophenyl sulphides, leading to phenothiazines (Vol. 2, p. 780). A successful synthesis of 1*H*-phenothiazin-1-one (83), a member of a novel class of heterocyclic *o*-quinone-imines, by oxidation of 1-hydroxyphenothiazine in an inert solvent under carefully controlled conditions has been reported.[53] A dominant feature of the chemical behaviour of (83) is its tendency to undergo dimerization to give a rather intractable product, identified tentatively as (84). A multi-step synthesis of 1-methyl-2-aminophenothiazin-3-one has also been described.[54]

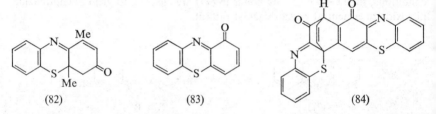

(82)   (83)   (84)

Large numbers of phenothiazine derivatives have been produced from the pre-formed ring system by replacement reactions or by appropriate modifications of existing substituents. They include 3-bromo-10-alkyl-phenothiazines,[55] 7,8-dioxochlorpromazine,[56] which is a possible metabolite of chlorpromazine, certain 2-*N*-alkylamino- and 2-*NN*-dialkylamino-10-methylphenothiazines, obtained [57] *via* phenothiazyne, various 2- and 3-dimethylsulphamidopheno-thiazines [58] of potential pharmacological interest, and a miscellany of 10-sub-stituted derivatives,[59-61] including a ribofuranosyl derivative of phenothiazine [62] and some phenothiazinyl pyrylium salts.[63] A group of 10-alkyl- and 10-amino-2,2'-bis(trifluoromethyl)-3,10'-biphenothiazines that are expected to show CNS activity has also been obtained.[64]

Interest in the synthesis, reactions, and properties of fused-ring systems incorporating the phenothiazine structure is continuing, and further compounds

[52] J. I. G. Cadogan and B. S. Tait, *J.C.S. Perkin I*, 1975, 2396.
[53] I. R. Silberg and M. Bartha, *Tetrahedron Letters*, 1974, 3801.
[54] D. P. Sevbo and T. F. Stepanova, *Zhur. org. Khim.*, 1974, 6, 1318.
[55] E. R. Biehl, T. Daniel, P. C. Reeves, and S. Lapis, *J. Heterocyclic Chem.*, 1974, 11, 247.
[56] A. Zirnis, J. K. Suzuki, J. W. Daly, and A. A. Manian, *J. Heterocyclic Chem.*, 1975, 12, 739.
[57] F. R. Biehl, V. Patrizi, S. Lapis, and P. C. Reeves, *J. Heterocyclic Chem.*, 1974, 11, 965.
[58] V. V. Shavyrina, S. V. Zhuravlev, Yu. I. Vikhlynev, T. A. Klygul, and E. I. Slyn'ko, *Khim. Farm. Zhur.*, 1974, 8, 23.
[59] N. M. Shrivastava, V. N. Sharma, and S. P. Garg, *J. Indian Chem. Soc.*, 1975, 52, 743.
[60] M. M. El-Kerdawy, H. A. Moharram, and M. N. Tolba, *J. prakt. Chem.*, 1974, 316, 511.
[61] D. Lambrou and G. Tsatsas, *Ann. Pharm. France*, 1974, 31, 295.
[62] G. Tapiero and J. L. Imbach, *J. Heterocyclic Chem.*, 1975, 12, 439.
[63] J. Van Allan and G. Reynolds, *J. Heterocyclic Chem.*, 1976, 13, 73.
[64] K. Nagarajan, J. David, and A. J. Nagana Gound, *J. Medicin. Chem.*, 1974, 17, 652.

that have been described include quaternary salts of imidazo[4,5,1-*ki*]phenothiazines [65] and 2-methylmercaptothiazolo[4,5-*b*]phenothiazines,[66] some dimethylpyridophenothiazines,[67] and the pyrimido-dione (86), derived [68] from (85) by thermal cyclization and subsequent hydrolysis.

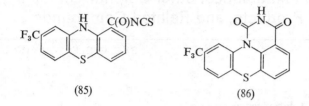

(85)　　　　　　　　　(86)

In the field of phenothiazine analogues, Grol [69] has described the synthesis of some isomeric dithienothiazines and thienobenzothiazines by applying an Ullmann-type cyclization to the appropriate dithienyl and thienyl aryl sulphides. A more convenient route [70] to the dithienothiazine ring system is illustrated by the direct sulphuration of the dithienylamide (87) with sulphur dichloride,

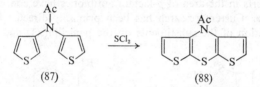

(87)　　　　　　　　　(88)

leading to (88) in good yield. Another interesting paper [71] deals with the synthesis and characterization of several novel tetrazaphenothiazines in which both benzenoid rings of the parent phenothiazine are replaced by pyridazine rings.

[65] Z. I. Ermakova, A. N. Gritsenko, and S. V. Zhuravlev, *Khim. geterotsikl. Soedinenii*, 1974, 372·
[66] V. V. Shavyrina and S. V. Zhoravlev, *Khim. geterotsikl. Soedinenii*, 1974, 643.
[67] A. N. Gritsenko, Z. I. Ermakova, T. Ya. Morhaeva, V. S. Troitskaya, and S. V. Zhuravlev, *Khim. geterotsikl. Soedinenii*, 1975, 50.
[68] J. Weinstock, D. E. Gaitanopoulos, and B. M. Sutton, *J. Org. Chem.*, 1975, **40**, 1914.
[69] C. J. Grol, *J. Heterocyclic Chem.*, 1974, **11**, 953.
[70] C. J. Grol, *J.C.S. Perkin I*, 1975, 1234.
[71] D. S. Wise, jun., and R. N. Castle, *J. Heterocyclic Chem.*, 1974, **11**, 1001.

16

β-Lactam Antibiotics, other Sulphur-containing Natural Products, and Related Compounds

BY J. G. GLEASON AND G. L. DUNN

1 Introduction

The purpose of this Report is to provide a review of the major developments in the chemistry of the β-lactam antibiotics and other sulphur-containing natural products during the period March 1974—March 1976. An excellent monograph[1] and several articles[2] provide a comprehensive review of the pre-1974 literature, and complement this Report.

Research efforts in the area of β-lactam antibiotics have continued at a high level of activity. Current research has been primarily directed in three major areas; modification of the substituents on the penicillin (1) and cephalosporin

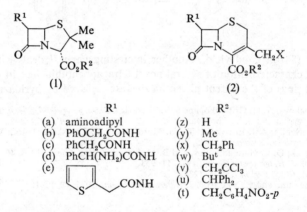

	R^1		R^2
(a)	aminoadipyl	(z)	H
(b)	PhOCH$_2$CONH	(y)	Me
(c)	PhCH$_2$CONH	(x)	CH$_2$Ph
(d)	PhCH(NH$_2$)CONH	(w)	Bu^t
(e)	[thiophene]CONH	(v)	CH$_2$CCl$_3$
		(u)	CHPh$_2$
		(t)	CH$_2$C$_6$H$_4$NO$_2$-p

(2; X = H or OAc) ring systems, penicillin–cephalosporin interconversions, and total synthesis of new β-lactam systems related to these known β-lactam antibiotics. In addition, there have been further efforts to isolate novel β-lactam antibiotics from fermentation processes, while studies on the biosynthesis of these important antibiotics have also continued.

[1] E. H. Flynn, 'Cephalosporins and Penicillins; Chemistry and Biology', Academic Press, New York, 1972.

[2] R. J. Stoodley, *Progr. Org. Chem.*, 1973, **8**, 102; J. H. C. Nayler, *Adv. Drug Res.*, 1973, **7**, 1; R. D. G. Cooper, *MTP. Internat. Rev. Sci.: Biochem.*, Ser. 1, 1973, **6**, 247; J. G. Gleason, in 'Organic Compounds of Sulphur, Selenium, and Tellurium', ed. D. H. Reid (Specialist Periodical Reports), The Chemical Society, London, 1975, Vol. 3, Ch. 4, p. 190.

Several review articles have appeared during the past two years.[3-6] The rearrangements of penicillanic acid derivatives have been the subject of one review,[3] while the total synthesis of nuclear analogues of penicillin and cephalosporins [4] and recent developments in the chemistry of the β-lactam antibiotics have also been reviewed.[5-7]

2 Fermentation and Biosynthesis

The cephamycins, a family of naturally occurring β-lactam antibiotics first described in 1971,[8] differ structurally from the cephalosporins in that they possess an α-methoxy-group at the 7-position. A fourth cephamycin, C-2801*X* (3;a),* has been isolated from two species of *Streptomyces*,[9] along with its close analogues cephamycins A and B, which have been previously described.

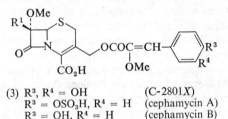

(3) R³, R⁴ = OH   (C-2801*X*)
    R³ = OSO₃H, R⁴ = H   (cephamycin A)
    R³ = OH, R⁴ = H   (cephamycin B)

The screening of fermentation broths for cephalosporin compounds has led to the isolation of several new metabolites. Deacetoxycephalosporin C (4;a) was shown [10, 11] to be produced by no less than 22 strains of fungi and 2 Streptomycetes. The well-known ability of mutagenic agents to produce mutations of micro-organisms has led to the isolation of *N*-acetyl-deacetoxycephalosporin C (5) [12] and the 3-methylthiomethyl derivative (6;a) [13] from mutant strains of *Cephalosporium* species. The novel metabolite (7;a), produced by a *Cephalosporium* mutant, is the first example of a stable naturally occurring dihydrothiazine containing an opened β-lactam ring.[14] This usually unstable structure is

[3] R. J. Stoodley, *Tetrahedron*, 1975, **31**, 2321.
[4] G. Lowe, *Chem. and Ind.*, 1975, 459.
[5] P. G. Sammes, *Chem. Rev.*, 1976, **76**, 113.
[6] A. K. Mukerjee and A. K. Singh, *Synthesis*, 1975, 547.
[7] D. R. Owens, D. K. Luscombe, A. D. Russell, and P. J. Nichols, *Adv. Pharmacol. Chemotherapy*, 1975, **13**, 83–172.
[8] R. Nagarajan, L. D. Boeck, M. Gorman, R. L. Hamill, C. E. Higgens, M. M. Hoehn, W. M. Stark, and J. G. Whitney, *J. Amer. Chem. Soc.*, 1971, **93**, 2308.
[9] H. Fukase, T. Hasegawa, K. Hatano, H. Iwasaki, and M. Yoneda, *J. Antibiotics*, 1976, **29**, 113.
[10] R. Nagarajan, L. D. Boeck, R. L. Hamill, C. E. Higgens, and K. S. Yang, *J.C.S. Chem. Comm.*, 1974, 321.
[11] C. E. Higgens, R. L. Hamill, T. H. Sands, M. M. Hoehn, N. E. Davis, R. Nagarajan, and L. D. Boeck, *J. Antibiotics*, 1974, **27**, 298.
[12] P. Traxler, H. J. Treichler, and J. Nuesch, *J. Antibiotics*, 1975, **28**, 605.
[13] T. Kanzaki, T. Fukita, H. Shirafuji, Y. Fujisawa, and K. Kitano, *J. Antibiotics*, 1974, **27**, 361.
[14] Y. Fujisawa and T. Kanzak, *J. Antibiotics*, 1975, **28**, 372.

* The general code for the substituents R¹ and R², as defined for formulae (1) and (2), will be used throughout this Report. For publications in which several substituents R¹ and R² have been employed, or where a general class of compounds is implied, R¹ and R² will not be specified. Other substituents will be labelled R³, R⁴, *etc.*, but their numbering is to avoid ambiguity with R¹ and R², and does not imply similarity of position of substituents or of their nature.

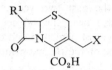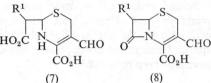

(4) X = H   (7)   (8)
(5) X = H, R¹ = HO₂CCH(NHAc)(CH₂)₃CONH
(6) X = SMe

retained in (7) due to the stabilizing effect of the aldehyde moiety. Compound (7) is believed to arise directly from (8), which is unstable at neutral pH.

The quest to unravel the mysteries of penicillin and cephalosporin biosynthetic pathways continues to attract attention. A brief summary has appeared recently [15] which outlines some of the early biosynthetic hypotheses. A number of reports [15-21] have appeared describing results obtained *via* incorporation of labelled precursors into penicillins and cephalosporins. In summary, the data indicate that the methyl groups of valine are incorporated stereospecifically into both penicillins and cephalosporins, the hydrogen atoms of the two valine methyl groups and the 3-hydrogen of valine are retained during biosynthesis, and both the 2- and 3-hydrogen atoms of cysteine are incorporated into penicillin with retention of configuration. Thus, tripeptide precursor (9;a) would be converted into penicillin (10;a).

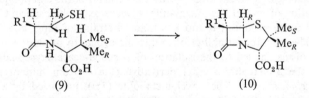

(9)   (10)

Evidence obtained from incorporation studies using a *Cephalosporium acremonium* mutant with (2S,3S)-[3-*Me*-²H₃]methyl[¹⁵N]valine suggests that both penicillin N and cephalosporin C originate from a common tripeptide intermediate, such as δ-(1-α-aminoadipyl)-L-cysteinyl-D-valine (11).[22] It has also been shown that the hydrogen atom on C-3 of L-valine is retained during its incorporation into (11). This result, coupled with the observation that the α-proton of the D-valinyl moiety of (11) is retained when (11) is incorporated into penicillin N by protoplast lysates, provides strong evidence against the intermediacy of free αβ-dehydrovalinyl-tripeptides (12) and related intermediates (13)

[15] D. J. Aberhart and L. J. Lin, *J.C.S. Perkin I*, 1974, 2320.
[16] D. J. Aberhart, L. J. Lin, and J. Y. Chu, *J.C.S. Perkin I*, 1975, 2517.
[17] D. J. Aberhart, J. Y. R. Chu, N. Neuss, C. H. Nash, J. Occolowitz, L. L. Huckstep, and N. Delahiguera, *J.C.S. Chem. Comm.*, 1974, 564.
[18] B. W. Bycroft, C. M. Wels, K. Corbett, A. P. Maloney, and D. A. Lowe, *J.C.S. Chem. Comm.*, 1975, 923.
[19] B. W. Bycroft, C. M. Wels, K. Corbett, and D. A. Lowe, *J.C.S. Chem. Comm.*, 1975, 123.
[20] D. J. Morecombe and D. W. Young, *J.C.S. Chem. Comm.*, 1975, 198.
[21] J. Cheney, C. J. Morres, J. A. Raleigh, A. I. Scott, and D. W. Young, *J.C.S. Perkin I*, 1974, 986.
[22] H. Kluender, F. C. Huang, A. Fritzberg, H. Schnoes, C. J. Sih, P. Fawcett, and E. P. Abraham, *J. Amer. Chem. Soc.*, 1974, **96**, 4054; F. C. Huang, J. A. Chan, C. J. Sih, P. Fawcett, and E. P. Abraham, *J. Amer. Chem Soc.*, 1975, **97**, 3858.

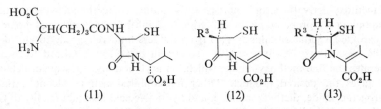

(11) (12) (13)

in the formation of the penam nucleus. This type of structure has long been thought to be a likely intermediate in the biosynthesis of β-lactam antibiotics.[23]

3 Modification of the β-Lactam Ring System

Modification at C-6(7).—A prediction [24] that alkylation at C-6 of penicillin might enhance antibacterial activity has been found generally not to be valid. With the exception of the 7-α-methoxycephalosporins (cephamycins), the activities of these C-6(7) derivatives have been disappointingly poor. However, spurred on by discovery of the potent cephamycin antibiotics, a number of practical synthetic techniques have been developed for replacing the C-6(7) hydrogen of penicillins and cephalosporins with a variety of functional groups.[25-27] Recent investigations have concentrated on stereospecific methoxylation of cephalosporins at C-7. Several general methods have been reported; addition of methanol to 7-acylamino-cephalosporins (14),[28] conversion of 7-methylthio- (15) [29] and 7-halogeno-cephalosporins (16) [30] into the corresponding

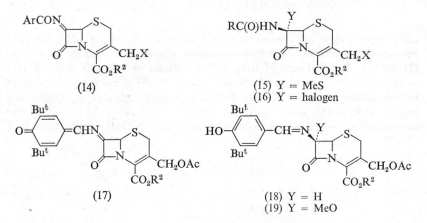

(14)

(15) Y = MeS
(16) Y = halogen

(17)

(18) Y = H
(19) Y = MeO

[23] J. E. Baldwin, S. B. Haba, and J. Kitchin, *J.C.S. Chem. Comm.*, 1973, 790.
[24] J. L. Strominger and D. J. Tipper, *Amer. J. Medicin.*, 1965, **39**, 708.
[25] E. H. W. Bohme, H. E. Applegate, J. B. Ewing, P. T. Funke, M. S. Puar, and J. E. Dolfini, *J. Org. Chem.*, 1973, **38**, 320.
[26] D. B. R. Johnston, S. M. Schmitt, R. A. Firestone, and B. G. Christensen, *Tetrahedron Letters*, 1972, 4917.
[27] W. A. Spitzer, T. Goodson, R. J. Smithey, and I. G. Wright, *J.C.S. Chem. Comm.*, 1972, 1138.
[28] W. H. W. Lunn and E. V. Mason, *Tetrahedron Letters*, 1974, 1311.
[29] H. E. Applegate, J. E. Dolfini, M. S. Puar, W. A. Slusarchyk, B. Toeplitz, and J. Z. Gougoutas, *J. Org. Chem.*, 1974, **39**, 2794.
[30] L. D. Cama and B. G. Christensen, *Tetrahedron Letters*, 1973, 3505.

7-α-methoxy-derivatives, and, more recently, addition of methanol to the quinoidal compound (17), prepared by the oxidation of Schiff-base (18) with lead dioxide.[31] As expected, methanol adds to (17) from the less hindered α-side to afford the stable 7-α-methoxy-Schiff-base (19). Treatment of (19) with Girard Reagent T generates the free methoxy-amine, which, on acylation and subsequent deblocking, gives the desired product. Other nucleophiles such as alcohols, thiols, hydrogen cyanide, diethyl malonate, and hydrazoic acid also add readily to (17), to give good yields of (18) where $Y = OR$, SMe, CN, N_3, or $CH(CO_2Et)_2$.[32]

An alternative approach for the introduction of substituents at C-6 of penicillins utilizes the isocyano-compound (20;x), generated by treating the 7-formylamino-penicillin (21;z, $R^1 = HCONH$) with phosgene.[33] Subsequent reaction of a 1 : 1 epimeric mixture of (20) with reactive halides under alkaline conditions gave predominantly a single epimer of compounds (22)—(24). Similarly, benzyl acrylate, acetone, and methyl methoxycarbonyl disulphide gave (25), (26), and

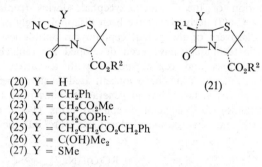

(20) Y = H
(22) Y = CH$_2$Ph
(23) Y = CH$_2$CO$_2$Me
(24) Y = CH$_2$COPh·
(25) Y = CH$_2$CH$_2$CO$_2$CH$_2$Ph
(26) Y = C(OH)Me$_2$
(27) Y = SMe

(21)

(27), respectively. Mild acid hydrolysis converted the isocyanides into the corresponding amines. The 6-α-methoxy-derivatives (21;b,z,Y = OMe) and (21;d,z,Y = OMe), prepared from (27) by published methods, showed very weak antibacterial activity. Acylimine (28;x), resulting from the reaction of 6-oxopenicillanic acid (29;x) with *N*-phenoxyacetyliminotriphenylphosphorane, readily adds methanol or hydrogen cyanide stereospecifically to afford the corresponding 6-α-substituted penicillins (21;b,x,Y = OMe) and (21;b,x,Y = CN).[34]

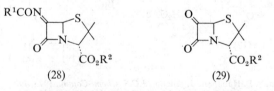

(28)   (29)

A simple and direct preparation of novel 6-α-acylamino-penicillins has been reported.[35] The reaction of a penicillin ester (1;b,v) with *N*-chloro-*N*-sodio-urethane gave (30;b,v) in 80—90% yield. Oxidation of (30) to its sulphoxide and

[31] H. Yanayisawa, M. Fukushima, A. Ando, and H. Nakao, *Tetrahedron Letters*, 1975, 2705.
[32] H. Yanayisawa, M. Fukushima, A. Ando, and H. Nakao, *Tetrahedron Letters*, 1976, 259.
[33] P. H. Bentley and J. P. Clayton, *J.C.S. Chem. Comm.*, 1974, 278.
[34] Y. S. Lo and J. C. Sheehan, *J. Org. Chem.*, 1975, **40**, 191.
[35] M. M. Campbell and G. Johnson, *J.C.S. Chem. Comm.*, 1975, 479.

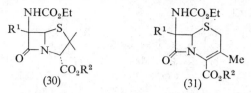

(30)   (31)

rearrangement in DMF–acetic anhydride at 130 °C led to the corresponding deacetoxycephalosporin (31;b,v). However, the stereochemistry of the substituents at C-6(7) has yet to be determined.

Diazopenicillin (32;x) has been shown[36] to be a useful intermediate for the synthesis of other 6-substituted penicillins, but its method of preparation has suffered from low yields. An improved procedure[37] in which *N*-nitroso-amide (33;v) is treated with pyridine in refluxing methylene chloride afforded (32;v) as a crystalline solid in 72% yield. The corresponding 7-diazo-cephalosporins (34)

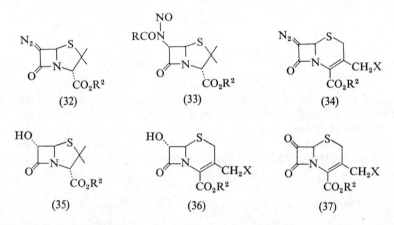

(32)   (33)   (34)

(35)   (36)   (37)

were prepared similarly, but in much lower yields. Hydrolysis of (32) or (34) in aqueous acid gave good yields of the 6(7)-α-hydroxy-derivatives (35) and (36), which serve as useful intermediates for the preparation of (29) and the analogous cephalosporin (37).

An unsuccessful attempt[38] to synthesize a 7-α-fluorocephalosporin from the imino-chloride (38) gave only the α-hydroxy-compound (39). In contrast, penicillin imino-chloride (40), under similar hydrolysis conditions, gave predominantly the ring-opened rearrangement product (41), and none of the desired fluoro-penicillin.

Epimerization.—Though a variety of procedures have been devised to epimerize the C-6(7) position in penicillins and cephalosporins, additional ones continue to appear. Treatment of Schiff-base (42;u) with di-isopropylethylamine in THF or Et₂O gave a 1 : 1 mixture of (42) and (43), with no isomerization of the double

[36] D. Hauser and H. P. Sigg, *Helv. Chim. Acta*, 1967, **50**, 1327.
[37] J. C. Sheehan, Y. S. Lo, J. Loliger, and C. C. Podewell, *J. Org. Chem.*, 1974, **39**, 1444.
[38] W. A. Spitzer, T. Goodson, jun., M. O. Chaney, and N. D. Jones, *Tetrahedron Letters*, 1974, 4311.

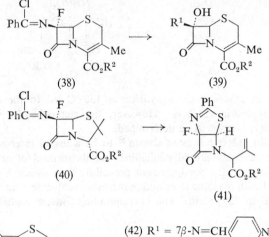

(38)   (39)

(40)   (41)

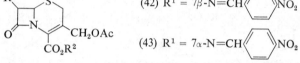

(42) $R^1 = 7\beta\text{-N}=CH\langle\rangle NO_2$

(43) $R^1 = 7\alpha\text{-N}=CH\langle\rangle NO_2$

bond of the dihydrothiazine ring to the Δ^2-position.[39] This contrasts with an earlier finding [40] that, when DMF is used as solvent, up to 30% of the product is a mixture of Δ^2-isomers.

The first step in the base-catalysed conversion of penicillin imino-chloride (44; pom = pivaloyloxymethyl) into the corresponding ketenimine (45) was

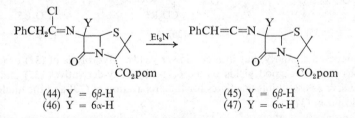

(44) Y = 6β-H   (45) Y = 6β-H
(46) Y = 6α-H   (47) Y = 6α-H

rapid equilibration to a 1 : 9 mixture of (44) and (46), followed by slower dehydro-chlorination to the ketenimines (45) and (47).[41] Hydrolysis of the ketenimines affords a mixture of the epimeric penicillins. The imino-chlorides of other penicillin and cephalosporin esters also epimerize under these conditions.

Epimerization of the penicillin sulphoxide (48;c,x) with diazabicyclononene leads rapidly to a 2 : 3 mixture of (48) and 6-epi-(48).[42] However, *p*-nitrobenzyl ester (48;c,$R^2 = CH_2C_6H_4NO_2$-4) under identical conditions gave a 6 : 4 : 1 mixture

[39] C. U. Kim and D. N. McGregor, *J. Antibiotics*, 1974, **27**, 881.
[40] R. A. Firestone, N. S. Maciejewicz, R. W. Ratcliffe, and B. G. Christensen, *J. Org. Chem.*, 1974, **39**, 437.
[41] R. D. Carroll, E. S. Hamanaka, D. K. Pirie, and W. M. Welch, *Tetrahedron Letters*, 1974, 1515.
[42] M. Fukumura, N. Hamma, and T. Nakagone, *Tetrahedron Letters*, 1975, 4123.

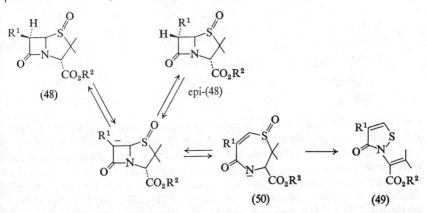

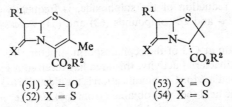

of (48), 6-epi-(48), and isothiazolone (49), while (48;c,u) gave only (48) and (49). An intermediate such as (50) was suggested as a possible precursor for (49), β-elimination being facilitated by increasing the electron-withdrawing ability of the ester group attached to C-4.

Modification at C-7(8).—The β-lactam antibiotics are believed to exert their inhibitory action by irreversible acylation of the transpeptidase enzyme responsible for cross-linking the peptide chains of the peptidoglycan in the final stage of cell-wall synthesis. This irreversible acylation step involves rupture of the β-lactam bond; hence chemical modification near the atoms involved in this bond might have a significant effect on the degree of antibacterial activity. The first modifications reported [43] for the C-7(8) position involved replacement of the β-lactam carbonyl oxygen by sulphur, giving β-thiono-lactams (52) and (54).

(51) X = O
(52) X = S

(53) X = O
(54) X = S

The reaction of (51;b,v) with boron sulphide afforded β-thiono-lactam (52) in 10% yield. Penicillin (53;b,v) similarly was converted into its thiono-derivative (54) in 1% yield. Deblocking of the ester functions produced the free acids, which showed substantially reduced antibacterial activity compared to their oxo counterparts.

4 Modifications in the Thiazine Ring

The sulphur atom of penicillins and cephalosporins is normally presumed to be unreactive toward common alkylating agents. However, treatment of (55;f,y) with methyl fluorosulphonate produced the stable, optically active sulphonium

[43] P. W. Wojtkowski, J. E. Dolfini, D. Kocy, and C. M. Cimarusti, *J. Amer. Chem. Soc.*, 1975, 97, 5628.

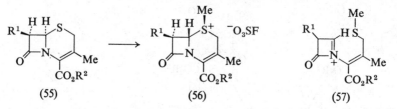

(55)   (56)   (57)

salt (56;f,y) in 30% yield.[44] Inversion at C-6 occurred during alkylation. Onium salt (57) has been postulated as a likely intermediate in this transformation.

Modification at C-2 of the cephalosporin skeleton has attracted limited interest, since those transformations which have been reported have not led to products with significantly improved antibiotic properties. Low-temperature methyl-thiolation of cephem sulphoxide (58) *via* its C-2 anion afforded thiomethyl derivative (59).[45] The reaction proceeded stereospecifically from the least hindered side to give the α-isomer exclusively. When the reaction was carried out at a somewhat higher temperature (-23 °C), the bis(methylthio)-compound (60) was obtained. Methylation of (59) with methyl iodide gave a mixture of

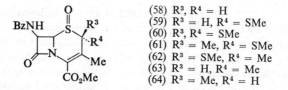

(58) R^3, R^4 = H
(59) R^3 = H, R^4 = SMe
(60) R^3, R^4 = SMe
(61) R^3 = Me, R^4 = SMe
(62) R^3 = SMe, R^4 = Me
(63) R^3 = H, R^4 = Me
(64) R^3 = Me, R^4 = H

(61) and (62). While the methylthio group(s) can be removed reductively with zinc in acetic acid, a mixture of products, resulting from migration of the double bond and partial reduction of the sulphoxide, is frequently obtained. None-theless, both the 2α- and 2β-compounds (63) and (64) were synthesized by this route.

Since modification at C-3 of the cephalosporin structure has led to derivatives with improved antibacterial activity, this area has received a great deal of atten-tion. A key intermediate for C-3 modification is the aldehyde (65;e). For example, the 3,4-dicarboxylic acid (66) was obtained from (65) *via* a seven-step synthesis.[46] The diacid in turn was converted into a number of C-3-modified derivatives, *e.g.* (67), (68). The dipolar *N*-methylnitrone (69), resulting from the reaction of (65) with *N*-methylhydroxylamine, served as an intermediate for the generation of several novel 3-substituted cephalosporins.[47] For example, oxidation of (69) to its sulphoxide, cycloaddition with dimethyl acetylenedicarboxylate, reduction of the sulphoxide, and removal of the protecting groups gave the interesting heterocyclic analogue (70;e). Several similar derivatives were prepared, but all showed weak activity against Gram-negative organisms. Aldehydes (65), (71), and (72) have been utilized in a number of transformations, *e.g.* reaction with

[44] D. K. Herron, *Tetrahedron Letters*, 1975, 2145.
[45] A. Yoshida, S. Oida, and E. Okki, *Chem. and Pharm. Bull. (Japan)*, 1975, **23**, 2507.
[46] D. O. Spry, *J.C.S. Chem. Comm.*, 1974, 1012.
[47] D. O. Spry, *J. Org. Chem.*, 1975, **40**, 2411.

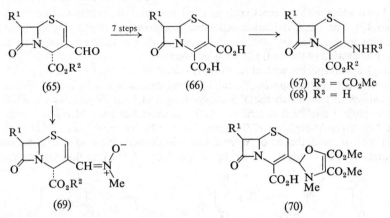

(65)   (66)   (67) $R^3 = CO_2Me$
(68) $R^3 = H$

(69)   (70)

Wittig reagents,[48] decarbonylation,[49] oxidation,[50] oxime formation,[50] and fluorination,[51] to afford a variety of novel cephalosporins. Acid (73;c), also prepared from (65), in which the carboxy-group has been translocated to the abnormal C-3 position, has been reported to have only weak antibacterial activity.[50] Recently, a more direct synthesis of 3-halogenomethyl-cephalosporins (74) and (75) has been described [52] in which a boron trihalide effected the replacement of the 3-acetoxy-group by a halogen atom. Similarly, (75; X = OAc or

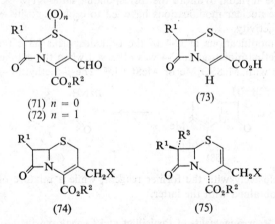

(71) $n = 0$
(72) $n = 1$

(73)

(74)   (75)

OCONH$_2$) can be converted into the corresponding 3-halogeno-compound in good yield when treated with HCl, HBr, or HI.[53] This transformation has been carried out in both the Δ^2-cephalosporin (75; R^3 = H) and the Δ^2-cephamycin (75; R^3 = MeO) series, but fails for the Δ^3- isomers. In addition, the carbamate (75; X = OCONH$_2$) undergoes a variety of replacement reactions in the presence

[48] J. A. Webber, J. L. Ott, and R. T. Vasileff, *J. Medicin. Chem.*, 1975, **18**, 986.
[49] H. Peter and H. Bickel, *Helv. Chim. Acta*, 1974, **57**, 2044.
[50] H. Peter, B. Muller, and H. Bickel, *Helv. Chim. Acta*, 1975, **58**, 2450.
[51] B. Muller, H. Peter, P. Schneider, and H. Bickel, *Helv. Chim. Acta*, 1975, **58**, 2469.
[52] H. Yazawa, H. Nakamura, K. Tanaka, and K. Kariyone, *Tetrahedron Letters*, 1974, 3991.
[53] S. Karady, T. Y. Cheng, S. H. Pines, and M. Sletzinger, *Tetrahedron Letters*, 1974, 2625.

of Lewis acids with reagents such as ethyl acetoacetate, azide, acetonitrile, and phenol.[54] The usual oxidation–reduction isomerization of the double bond of the dihydrothiazine ring afforded the corresponding modified Δ^3-cephalosporin. The preparation of the useful 3-methylene-cephams (76) by reduction of 3-acetoxy-methylcephalosporins with chromium(II) salts,[55] by electrochemical reduction of 3-substituted cephalosporins,[56] and by desulphurization with Raney nickel of cephalosporins in which the 3′-acetoxy-group has been displaced by a sulphur nucleophile [57] has been described. Active interest has focused on reactions of the 3-hydroxy-Δ^3-cephems (77) obtained from (76) by ozonolysis. Treatment of (77) with diazomethane, or with chlorinating/brominating agents, gives compounds (78; X = MeO, Cl, or Br, respectively).[58-60] The fluorinating agent

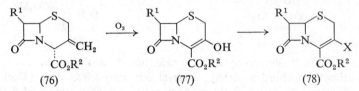

(76)   (77)   (78)

piperidino-sulphur trifluoride afforded (78; X = F).[51] Additionally, the reaction of the tosylate (78; X = Ts) with sulphur nucleophiles has produced a variety of derivatives (78), where X = S-alkyl, S-aryl, S-heterocyclyl, SO-alkyl, or SO$_2$-alkyl.[61] The same tosylate derivative reacts with amines to give (78; X = NHR), which could be acylated to afford the corresponding amides (78; X = NHCOR). Some of these nuclear modifications have led to cephalosporins with improved antibacterial activity.

Structural modifications at C-4 of the cephalosporins have received limited attention. Methylthiolation at C-4 was observed when the sulphoxide (58) was treated either with MeSSO$_2$Me or MeSOCl.[45] The β-methylthio-isomer (79) is

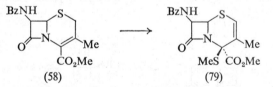

(58)   (79)

produced exclusively with the former reagent, while a mixture of both α- and β-isomers is obtained with the latter.

5 Rearrangements of Penicillins and Cephalosporins

The search for new antibacterial agents has resulted in intensive chemical study and the manipulation of virtually every carbon, oxygen, and sulphur atom of

[54] S. Karady, T. Y. Cheng, S. H. Pines, and M. Sletzinger, *Tetrahedron Letters*, 1974, 2629.
[55] M. Ochiai, O. Aki, A. Morimoto, T. Okada, and K. Morita, *Tetrahedron*, 1975, 31, 115.
[56] M. Ochiai, O. Aki, A. Morimoto, T. Okada, K. Shinozaki, and Y. Asaki, *J.C.S. Perkin I*, 1974, 258.
[57] R. R. Chauvette and P. A. Pennington, *J. Org. Chem.*, 1973, 38, 2994.
[58] R. R. Chauvette and P. A. Pennington, *J. Amer. Chem. Soc.*, 1974, 96, 4986.
[59] R. R. Chauvette and P. A. Pennington, *J. Medicin. Chem.*, 1975, 18, 403.
[60] R. Scartazzini and H. Bickel, *Helv. Chim. Acta*, 1974, 57, 1919.
[61] R. Scartazzini, P. Schneider, and H. Bickel, *Helv. Chim. Acta*, 1975, 58, 2437.

both the penicillin and cephalosporin nucleus. As a result, a surprising number of molecular rearrangements have been observed, at least one of which has formed the basis for a commercially feasible synthesis of cephalosporins from the much more readily available penicillins. Penicillin and cephalosporin rearrangements have been the subject of two recent reviews, and only the most recent developments will be discussed here.

1,2-Bond Cleavage.—The thermal or acid-catalysed rearrangement of penicillin S-oxides to cephalosporins has been shown to proceed through a reactive sulphenic acid (80;y), which may be isolated in crystalline form in low yield.[62] This intermediate slowly reverts on standing to the starting sulphoxide, while under acid

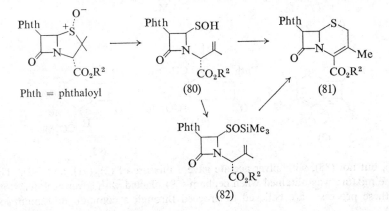

Phth = phthaloyl   (80)   (81)

(82)

catalysis it re-closes to the cephem (81;y). The sulphenic acid intermediate may, alternatively, be trapped by silylation;[63] the silyl ester (82;y) rearranges under acidic catalysis (toluene-*p*-sulphonic acid, DMA) to the cephem (81;y) in 70% yield.

At very low temperature ($-126\ ^{\circ}C$), the lithio-anion of sulphenic acid (83;t) has been trapped by alkylation (methyl fluorosulphonate); at higher temperature,

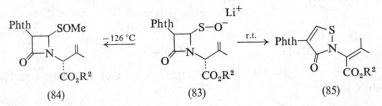

(84)   (83)   (85)

rearrangement to the thiazolone (85;t) takes place.[64] The sulphenyl bromide (86) has been suggested as a transient intermediate formed on reaction of the sulphenic acid (80;y) with phosphorus tribromide; cyclization of (86) at room temperature gave the penicillins (87) and (88), in a 1:1 ratio.[65] Treatment of

[62] T. S. Chou, J. R. Burgtorf, A. L. Ellis, S. R. Lammert, and S. P. Kukolja, *J. Amer. Chem. Soc.*, 1974, **96**, 1609.
[63] T. S. Chou, *Tetrahedron Letters*, 1974, 725.
[64] G. A. Koppel and S. Kukolja, *J.C.S. Chem. Comm.*, 1975, 57.
[65] S. Kukolja, S. R. Lammert, M. R. Gleissner, and A. L. Ellis, *J. Amer. Chem. Soc.*, 1975, **97**, 3192.

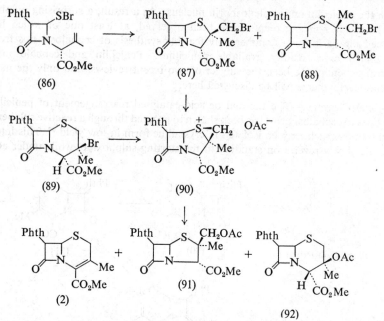

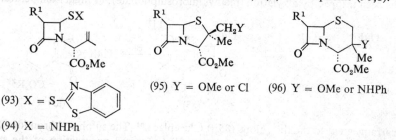

(87), but not (88), with silver acetate gave a mixture of (2), (91), and (92). The same mixture was obtained when cepham (89) reacted with silver acetate. Both of these processes are believed to proceed through a common thiiranium ion (90).

A similar penicillin sulphoxide–cephalosporin transformation, initiated with acetyl chloride–pyridine, has also been described.[66] The disulphide (93;b) and sulphenanilide (94;b) have been prepared from the sulphenic acid (80).[67] These sulphenyl derivatives readily cyclize to penicillins (95;b) and cephems (96;b).

(95) Y = OMe or Cl   (96) Y = OMe or NHPh

(93) X = S—
(94) X = NHPh

A novel cyclization of the 2β-bromomethyl sulphide (87;v) to the tricyclic β-lactam (98;v) in low yield has been reported.[67] Cyclization of the corresponding sulphoxide (97;v) (DBU, DMF, 1 h, −30 °C) gave the tricyclic sulphoxide (99;v) in high yield.

[66] H. Tanida, T. Tsuji, T. Tsushima, H. Ishitolu, T. Irie, T. Yano, H. Matsumura, and K. Tori, *Tetrahedron Letters*, 1975, 3303.
[67] T. Kamiya, T. Teraji, M. Hashimoto, O. Nakaguchi, and T. Oku, *J. Amer. Chem. Soc.*, 1975, **97**, 5020.

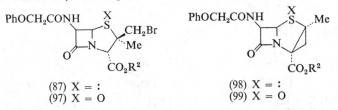

(87) X = :
(97) X = O

(98) X = :
(99) X = O

Penicillin-derived sulphenic acids have been intercepted intramolecularly,[68, 69] as in the case of (100;t),[68] or intermolecularly, by esterification [70] or by addition to acetylenic esters.[71] Reduction of the sulphenic acid [P(OMe)₃, Ac₂O] gave azetidinone thioacetate (102;v, R⁴ = Ac), from which the mercapto-azetidinone (102;v,R⁴ = H) was obtained by silver- or mercury-catalysed cleavage.[72]

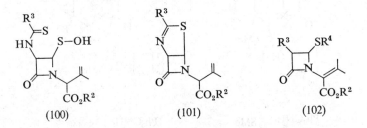

(100)   (101)   (102)

Cleavage of the 1,2-bond has also been achieved by chlorination or bromination of penicillin sulphoxides.[73, 74] The resulting sulphinyl bromide (103;y) or chloride (104;t) cyclizes readily to the Δ³-cephalosporin sulphoxide (2; X = H).[73] The reaction of (103;t) with diazomethane, however, gives a mixture of three novel cepham sulphoxides (105;t), (106;t), and (107;t) as well as the chloromethyl sulphoxide (109;t).[74] These products are believed to result from a diazo-sulphoxide (108;t), which undergoes a rapid intramolecular cyclization, *via* either the sulphoxocarbene (110) or the pyrazoline (111). Cephalosporins (112) and (113), derived from these sulphoxides, exhibited weak antibacterial activity.

The action of methyl iodide and strong anhydrous base on 6β-tritylamino-penicillinates effects *S*-alkylation and 1,2- bond rupture, to form the 1,2-seco-penicillins (114).[75, 76] With *p*-methoxybenzyl esters, dialkylation may occur, giving (115;t) as a product.[75]

A novel photorearrangement of a cephalosporin in which rupture of the 1,2-bond occurs has been reported.[77] Photolysis of (2) in methanol gave the

[68] H. Tanida, R. Muneyuki, and T. Tsushima, *Tetrahedron Letters*, 1975, 3063.
[69] H. Tanida, R. Muneyuki, and T. Tsushima, *Bull. Chem. Soc. Japan*, 1975, **48**, 3429.
[70] R. D. Allan, D. H. R. Barton, M. Girijavallabhan, and P. G. Sammes, *J.C.S. Perkin I*, 1974, 1456.
[71] D. H. R. Barton, I. H. Coates, P. G. Sammes, and C. M. Cooper, *J.C.S. Perkin I*, 1974, 1459.
[72] R. Lattrell, *Annalen*, 1974, 1937.
[73] I. Ishimaru and T. Imamoto, *Bull. Chem. Soc. Japan*, 1975, **48**, 2989.
[74] S. R. Lammert and S. Kukolja, *J. Amer. Chem. Soc.*, 1975, **97**, 5583.
[75] E. G. Brain, I. McMillan, J. H. C. Nayler, R. Southgate, and P. Tolliday, *J.C.S. Perkin I* 1975, 562.
[76] J. P. Clayton, J. H. C. Nayler, M. J. Pearson, and R. Southgate, *J.C.S. Perkin I*, 1974, 22.
[77] Y. Maki and M. Sako, *J. Amer. Chem. Soc.*, 1975, **97**, 7168.

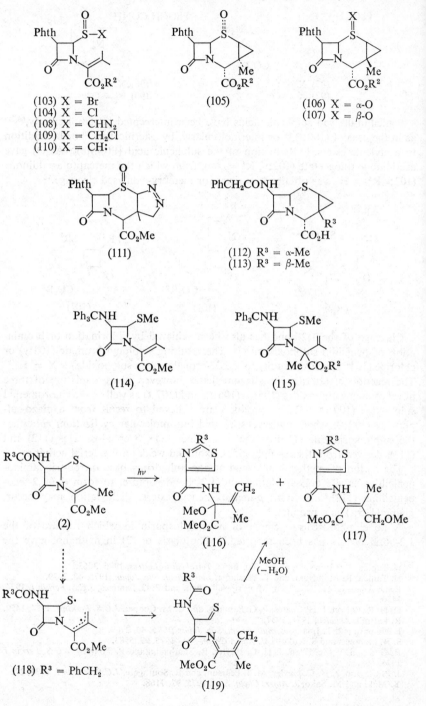

(103) X = Br
(104) X = Cl
(108) X = CHN$_2$
(109) X = CH$_2$Cl
(110) X = CH:

(105)

(106) X = α-O
(107) X = β-O

(111)

(112) R^3 = α-Me
(113) R^3 = β-Me

(114)

(115)

(2)

(116)

(117)

(118) R^3 = PhCH$_2$

(119)

thiazoles (116) and (117). A mechanism involving the diradical (118) has been proposed.

1,5-Bond Cleavage.—The second major type of penicillin rearrangement is 1,5-bond cleavage, usually a result of participation by the β-lactam nitrogen in the reaction of a sulphonium salt. For example, Chloramine-T reacts with the penicillinate (1;y) to give the β-lactam fused ylide (121), presumably by way of the iminium cation (120).[78-81] The structure of (121) was determined by *X*-ray crystallography.[78] Thermolysis of (121), in turn, effected ring-opening to give the azetidinone (122).[79, 80] A similar 1,5-cleavage of the seco-penicillin (114;x)

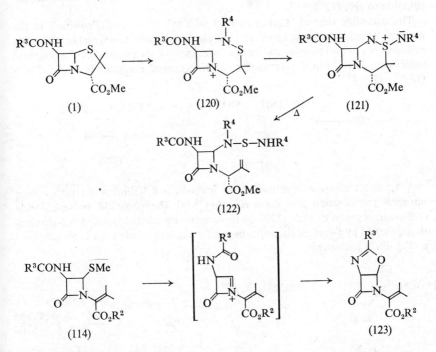

(1) (120) (121) (122) (114) (123)

with Chloramine-T gave a fused oxazoline-azetidinone (123;x),[82, 83] a product which is also obtained by the oxidation with Chloramine-T of certain trichloroethyl penicillinates.[81]

Oxazoline-azetidinones are also prepared by degradation of penicillinates with mercury(II) acetate.[84, 85] The mercuric salt (124) is converted into the propenyl azetidinone (126) when it reacts with DMSO, or the ester (127) with

[78] M. M. Campbell, G. Johnson, A. F. Cameron, and I. R. Cameron, *J.C.S. Chem. Comm.*, 1974, 868.
[79] M. M. Campbell and G. Johnson, *J.C.S. Chem. Comm.*, 1974, 974.
[80] M. M. Campbell and G. Johnson, *J.C.S. Perkin I*, 1975, 1212.
[81] M. M. Campbell, G. Johnson, F. A. Cameron, and I. R. Cameron, *J.C.S. Perkin I*, 1975, 1208.
[82] M. M. Campbell and G. Johnson, *J.C.S. Perkin I*, 1975, 1077.
[83] M. M. Campbell and G. Johnson, *J.C.S. Perkin I*, 1975, 1932.
[84] R. J. Stoodley and N. R. Whitehouse, *J.C.S. Perkin I*, 1974, 181.
[85] R. J. Stoodley and N. S. Watson, *J.C.S. Perkin I*, 1975, 883.

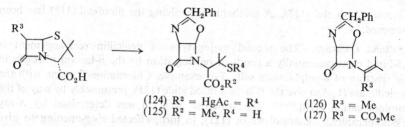

(124) R^2 = HgAc = R^4
(125) R^2 = Me, R^4 = H

(126) R^3 = Me
(127) R^3 = CO_2Me

diazomethane.[84] Methanolic sodium methoxide effects elimination of hydrogen sulphide to give (127).[85]

The oxazoline ring of (126) is cleaved with thiols [86, 87] and alcohols [88] in the presence of acid catalysts; however, ring-opening is non-stereospecific, affording mixtures of *cis*- and *trans*-azetidinones. Methods for the removal of the propenyl substituent have been developed for the ring-opened compounds [*e.g.* (126) → (128) → (129)].[89]

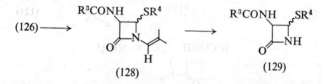

(126) ⟶ (128) ⟶ (129)

A 1,5-bond cleavage reaction which involves a 6,5-elimination rather than nitrogen participation has been reported.[89, 90] The penicillin imino-chloride (130) rearranges to oxazole (132) on treatment with triethylamine; a 6,5-elimination followed by β-lactam ring-opening and oxazole formation has been proposed as the likely pathway.[89]

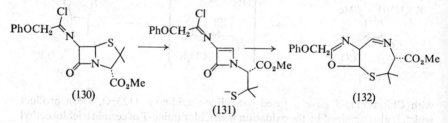

(130)   (131)   (132)

A combination of 1,2- and 1,5-bond cleavage rearrangements has permitted the synthesis of sulphur-free cepham derivatives.[91-93] Rearrangement of penicillin to anhydropenicillin (133) ($SOCl_2$, Et_3N) followed by chlorinolysis and esterification gave the sulphur-free azetidinone (134). Monobromination (NBS),

[86] D. F. Corbett and R. J. Stoodley, *J.C.S. Chem. Comm.*, 1974, 438.
[87] D. F. Corbett and R. J. Stoodley, *J.C.S. Perkin I*, 1975, 432.
[88] D. F. Corbett and R. J. Stoodley, *J.C.S. Perkin I*, 1974, 185.
[89] D. R. Carroll and L. M. Smith, *J. Heterocyclic Chem.*, 1975, **12**, 445.
[90] R. D. Carroll and L. L. Reed, *Tetrahedron Letters*, 1975, 3435.
[91] S. Wolfe, J. E. Ducep, K. C. Tin, and S. L. Lee, *Canad. J. Chem.*, 1974, **52**, 3996.
[92] Belg.P. 832 174.
[93] S. Wolfe, S. L. Lee, J. E. Ducep, G. Kannengiesser, and W. S. Lee, *Canad. J. Chem.*, 1975, **53**, 497.

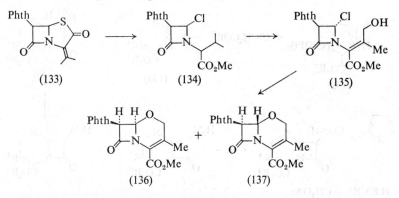

(133)   (134)   (135)

(136)   +   (137)

conversion into the formate ester (tetramethylguanidine–HCO_2H), and removal (HCl, MeOH–CH_2Cl_2) of the formate group afforded the alcohol (135), which cyclized ($SnCl_2$, dimethoxyethane) to a 1 : 1 mixture of *cis*- and *trans*-1-oxacephams (136) and (137).[91, 92] The 2-chloro-azetidinone has been converted into fused oxazoline-azetidinones which may prove to be valuable intermediates in the synthesis of new tricyclic β-lactam systems.[93]

6 Total Synthesis of Penicillins and Cephalosporins

Major advances have been achieved in the total syntheses of penicillins, cephalosporins, and related nuclear analogues in recent years. Although a number of synthetic approaches are under scrutiny, thus far, the Merck keten-imine approach and its many variants have produced the most promising results. However, the classical Sheehan synthesis, the Lowe diazo-amide photolysis approach, and a novel biogenetic-type synthesis have been utilized with success.

The well-known cycloaddition of a keten or keten precursors with an imine has been exploited by the Merck group to prepare a number of new antibacterial agents related to the cephalosporins.[94–98] Condensation of diethyl α-thioformamidophosphonoacetate (138) with α-chloromethyl ketones gave 6H-1,3-thiazine-4-carboxylates (139), which reacted with azidoacetyl chloride to give bicyclic β-lactams of the general structure (140) (Scheme 1). Reduction of the azide to the amine, epimerization *via* kinetic quenching of the anion (lithium salt) of the Schiff-base,[99] and acylation completed the synthesis. The generality of this approach is exemplified by the synthesis of cephalothin (141),[94] 10-methyl-cephalothin (142),[95] and the 3-arylcephalosporin (143).[96]

A similar route was employed by workers at the Syntex Laboratories to prepare the lactone (147).[100] Cycloaddition of azidoacetyl chloride to the furanothiazine

94   R. W. Ratcliffe and B. G. Christensen, *Tetrahedron Letters*, 1973, 4645, 4649.
95   N. G. Steinberg, R. W. Ratcliffe, and B. G. Christensen, *Tetrahedron Letters*, 1974, 3567.
96   R. A. Firestone, N. S. Maciejewicz, and B. G. Christensen, *J. Org. Chem.*, 1974, **39**, 3384.
97   L. D. Cama and B. G. Christensen, *J. Amer. Chem. Soc.*, 1974, **96**, 7582.
98   R. N. Guthikonda, L. D. Cama, and B. G. Christensen, *J. Amer. Chem. Soc.*, 1974, **96**, 7584.
99   R. A. Firestone, N. S. Maciejewicz, R. W. Ratcliffe, and B. G. Christensen, *J. Org. Chem.*, 1974, **39**, 437.
100   J. A. Edwards, A. Guzman, R. Johnson, P. J. Beeby, and J. H. Fried, *Tetrahedron Letters*, 1974, 2031.

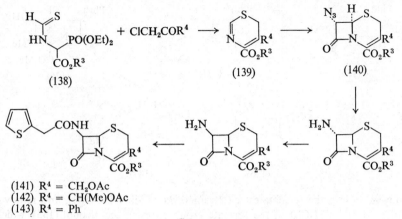

(141) R⁴ = CH₂OAc
(142) R⁴ = CH(Me)OAc
(143) R⁴ = Ph

Scheme 1

(144) gave the *trans* tricyclic β-lactam (145), which, on reduction, epimerization, and acylation, afforded the cephem (146;e). Bromination and subsequent hydrolysis gave the lactone (147;e).

1-Oxacephalothin (149) has been prepared[97] (Scheme 2) from the thioformamide (138). Alkylation of (138) gave a thioimino-ether, from which a monocyclic

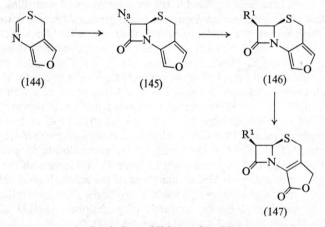

β-lactam (148) was obtained by addition of azidoketen. This monocyclic β-lactam was transformed into the oxacephalosporin (149) as depicted. Using a similar approach, the 1-carbocephalothin (150) has also been prepared.[98]

Penicillin derivatives are accessible *via* the keten-imine approach. Cyclo-addition of azidoacetyl chloride to the 2-phenyl-thiazoline (151;x) gave the 6α-azido-penam (152;x), which, after reduction and acylation, was epimerized to the 5-phenyl-penicillin V (153).[101] Cycloaddition reactions with oxazolines, however, are reported to give only non-β-lactam products.[102] Ketens derived

[101] H. Vanderhaeghe and J. Thomis, *J. Medicin. Chem.*, 1975, **18**, 486.
[102] B. T. Golding and D. R. Hall, *J.C.S. Perkin I*, 1975, 1302.

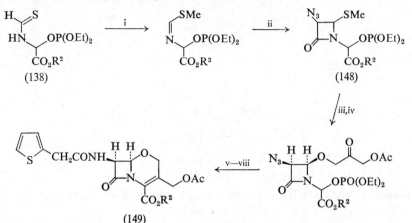

Reagents: i, MeI; ii, N₃CH₂COCl–Et₃N; iii, Cl₂; iv, HOCH₂COCH₂OAc–AgBF₄; v, NaH–DME; vi, separate the isomers; vii, H₂–Pd/C; viii, thienylacetyl chloride

Scheme 2

from acids other than azidoacetic acid have also been used. For example, addition of phthalimidoacetyl chloride to the thioformamidate (154) gave the *trans*-β-lactam (155), from which a *trans*-cepham (156) was prepared (Scheme 3).[103] An oxidative ring-expansion occurred on removal of the sulphur-protecting group of the corresponding unsaturated β-lactam (157).[104] A t-butoxy-carbonylglycine anhydride has been successfully employed as a keten precursor

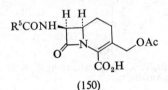

(150)

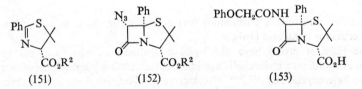

for addition to certain thioformamidates; however, the reaction is not general.[105] Keten precursors that do not contain nitrogen have been utilized for the synthesis of cephem derivatives. Addition of tosyloxyacetyl chloride to the thioimidate (158) gave the *trans*-azetidinone (159), which, on displacement with azide ion, reduction, and acylation, gave a *cis*-acylamino-azetidinone (Scheme 4).[106]

[103] M. D. Bachi and K. J. Ross-Petersen, *J.C.S. Perkin I*, 1975, 2525.
[104] M. D. Bachi and D. Goldberg, *J.C.S. Perkin I*, 1974, 1184.
[105] A. K. Bose, M. S. Manhas, H. P. S. Chawla, and B. Dayal, *J.C.S. Perkin I*, 1975, 1880.
[106] R. Lattrell and G. Lohaus, *Annalen*, 1974, 870, 901, 921.

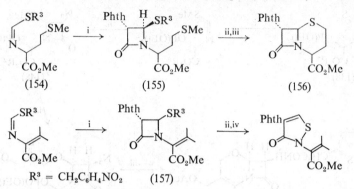

(154)   (155)   (156)

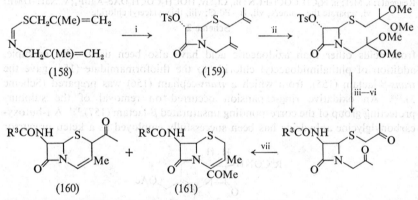

$R^3 = CH_2C_6H_4NO_2$ (157)

Reagents: i, PhthCH₂COCl; ii, sulphur deblocking; iii, MeI–NaI–DMF, HgCl₂; iv, DMSO

Scheme 3

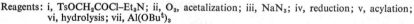

(158)   (159)

(160)   (161)

Reagents: i, TsOCH₂COCl–Et₃N; ii, O₃, acetalization; iii, NaN₃; iv, reduction; v, acylation; vi, hydrolysis; vii, Al(OBuᵗ)₃

Scheme 4

Cyclization with either aluminium tri(t-butoxide) or titanium tetra-t-butoxide gave cephems (160) and (161).

The Hoechst group have developed a synthesis of novel cephem systems which also utilizes a monocyclic azetidine as the starting point for the elaboration of the heterocyclic ring.[107–109] Displacement of acetate from azetidinone (162) by the complex thiol (163) gave the cepham carbinolamide (164); dehydration produced the cephem (165).[107, 108] Functionalization at C-7, *e.g.* (166) → (168), could be achieved by bromination followed by displacement by azide ion. Using this general approach, compounds (169) and (170) were prepared.[108]

An alternative route utilized a Wittig cyclization (Scheme 5); however, the method of functionalization at C-7 has not been disclosed.[109]

[107] H. W. Schnabel, D. Grimm, and H. Jensen, *Annalen*, 1974, 477.
[108] K. Kuhlein and H. Jensen, *Annalen*, 1974, 369.
[109] D. Borman, *Annalen*, 1974, 1391.

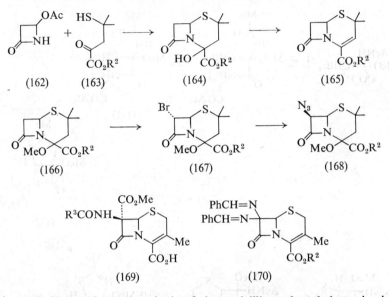

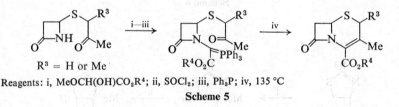

Reagents: i, MeOCH(OH)CO₂R⁴; ii, SOCl₂; iii, Ph₃P; iv, 135 °C

Scheme 5

A novel, biogenetic-type synthesis of the penicillin and cephalosporin ring-systems has been described.[110, 111] The key step in this approach is a stereo-controlled cyclization of the acyclic tripeptide (171) to the *cis*-β-lactam (172), an intermediate which may be transformed into both penicillin and cephalosporins (Scheme 6).[110] Alternatively, an oxidative cyclization of thiazoline (173) ulti-

mately leads to the same type of intermediate (Scheme 7).[111] β-Lactams have also been made from thiazolidine (174); conversion of (175) into the naturally occurring antibiotics has yet to be accomplished.[112]

An improved method for the transformation of the β-lactam thiazoline (176) into the penicillin nucleus has been reported.[113] Oxidation of (176) with *m*-chloroperoxybenzoic acid gave the sulphoxide (177), which underwent a radical-initiated rearrangement to the penicillin sulphoxide (1) in high yield.

Penicillin and cephalosporin synthesis by the classical intramolecular amide-coupling route continues to be employed successfully. The bisnorpenicillin V

[110] S. I. Nakatsuka, H. Tanino, and Y. Kishi, *J. Amer. Chem. Soc.*, 1975, **97**, 5008.
[111] S. I. Nakatsuka, H. Tanino, and Y. Kishi, *J. Amer. Chem. Soc.*, 1975, **97**, 5010.
[112] J. E. Baldwin, A. Au, M. Christie, S. B. Haber, and D. Hesson, *J. Amer. Chem. Soc.*, 1975, **97**, 5957.
[113] H. Tanimo, S. Kakatsuka, and Y. Kishi, *Tetrahedron Letters*, 1976, 571.

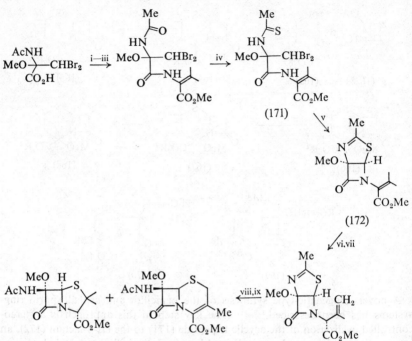

(171)

(172)

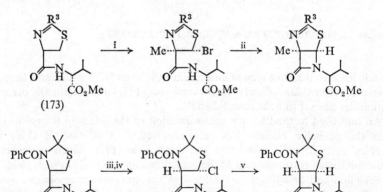

Reagents: i, penicillamine–DCC; ii, TsCl–C$_5$H$_5$N; iii, Et$_3$N; iv, P$_2$S$_5$–THF; v, NaH; vi, NBS–AIBN; vii, Zn–MeCO$_2$H; viii, MCPBA–TFA; ix, PCl$_3$

Scheme 6

(173)

(174)   (175)

Reagents: i, NBS; ii, KH–LiClO$_4$–THF; iii, (PhCO$_2$)$_2$–CCl$_4$; iv, HCl–CH$_2$Cl$_2$, 0 °C; v, NaH

Scheme 7

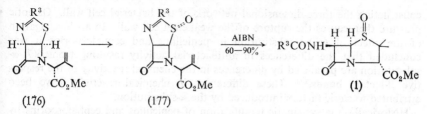

(176)   (177) $\xrightarrow[\text{60—90\%}]{\text{AIBN}}$ (1)

(179)[114] and the cephalosporin lactone (180)[115] were prepared by cyclization of the corresponding amino-acids. A photolytic Wolff rearrangement has been used to prepare the monocyclic β-lactam (181), which is structurally related to cephalosporins.[116] The oxapenam (183) was isolated from the photolysis of (182);

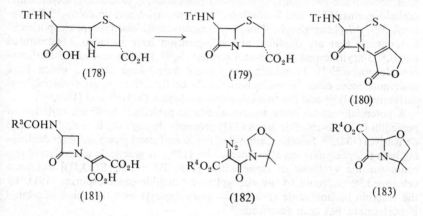

(178)   (179)   (180)

(181)   (182)   (183)

this β-lactam was very reactive towards nucleophilic reagents.[117] New syntheses of β-lactams by ring-expansion of cyclopropanone carbinolamides,[118] oxidative ring-contraction of α-keto-lactams,[119] photolysis of pyrazolidin-3-ones,[120] and cycloaddition of azlactones to imines [121] have been reported; all may provide new approaches to the synthesis of novel cephalosporins and penicillins.

7 Structure–Activity Considerations

Recent review articles [2, 7] and one monograph [1] provide an excellent survey of research on the mechanism of action and the effect of structural modifications on the antibiotic activity of the β-lactam antibiotics. The penicillins and cephalosporins are believed to exert their antibacterial activity by irreversible acylation of the bacterial transpeptidase responsible for the final cross-linking step in

[114] J. Hoogmartens, P. J. Claes, and H. Vanderhaeghe, *J. Medicin Chem.*, 1974, **17**, 389.
[115] R. Heymes, G. Amiard, and G. Nomine, *Bull. Soc. chim. France*, 1974, 563.
[116] G. Lowe and H. W. Yeung, *J.C.S. Perkin I*, 1973, 2907.
[117] B. T. Golding and D. R. Hall, *J.C.S. Perkin I*, 1975, 1517.
[118] H. H. Wasserman and E. Glazer, *J. Org. Chem.*, 1975, **40**, 1505.
[119] D. R. Bender, L. F. Bjeldaneo, D. R. Knapp, and H. Rapoport, *J. Org. Chem.*, 1975, **40**, 1264.
[120] C. E. Hatch and P. Y. Johnson, *Tetrahedron Letters*, 1974, 2719; P. Y. Johnson and C. E. Hatch, *J. Org. Chem.*, 1975, **40**, 3510; P. Y. Johnson and C. E. Hatch, *ibid.*, p. 3502.
[121] K. K. Prasad and T. Petrzilka, *Helv. Chim. Acta*, 1975, **58**, 2504.

constructing the three-dimensional network of the bacterial cell wall. Osmotic pressure then causes the rupture of the weakened cell wall. In an investigation of the effect of 6(7)-α-substitution in penicillins and cephalosporins it was concluded that the differences in antibacterial activity resulting from 6(7)-α-substitution are paralleled by differences in the chemical reactivity of the respective β-lactam bonds.[122] These differences in chemical reactivity have been attributed to steric factors introduced by the α-substitution.

Historically, the systematic modification of penicillins and cephalosporins by variation of the 6(7)-acyl group and, in the cephalosporins, by changes at the 3-position have been most successful in producing improved β-lactam antibiotics. Examples of such investigations are 6-quinoxalyl-di-*N*-oxide penicillins;[123] phenyl-substituted ampicillins;[124] 6-acylamino- and 6-(D-α-acylamino)-phenylacetamido-penicillins;[125] 6-aryloxy-penicillins;[126] α-sulpho-penicillins and cephalosporins;[127–129] and 7-amidino-,[130] 7-(1-methyl-4-pyridinothio)-,[131] 7-ferrocenyl-,[132, 133] 7-phenylglycyl-,[134] and 7-trifluoromethylthioacetyl-cephalosporins.[135] Compounds that are modified at the 3-position have included 3-(substituted vinyl-,[48] 3-acylthiomethyl-,[136] 3-acyloxymethyl-,[137] 3-halogeno-, and 3-methoxy-cephalosporins.[57, 58] Promising compounds from these studies, which have undergone more extensive evaluation, are the orally active *p*-hydroxyphenylglycyl derivative (184)[134] and the two injectable analogues (185)[135] and (186).[138]

A potentially useful route to semi-synthetic penicillins by direct acylation of penicillin G imino-chloride esters (187) proceeds through the intermediate diacyl derivatives (188).[139] Selective removal of the phenylacetyl group and ester deblocking with thiophenolate ion afford penicillins (189). A practical application of this route for the synthesis of carbenicillin [189; $R^3 = PhCH(CO_2H)$] has been developed.[140] A route to α-carboxyphenylacetamido-cephalosporins (191), by the reaction of isocyanate (190; R^2 = 4-nitrobenzyl) with the anion of t-butyl phenylacetate, has been described.[141]

[122] J. M. Indelicato and W. L. Wilkam, *J. Medicin. Chem.*, 1974, **17**, 528.
[123] M. L. Edwards, R. E. Bambury, and H. W. Ritter, *J. Medicin. Chem.*, 1976, **19**, 330.
[124] G. Schmidt and H. Rosenkranz, *Annalen*, 1976, 129.
[125] H. Ferres, M. J. Basker, and P. J. O'Hanlon, *J. Antibiotics*, 1974, **27**, 922.
[126] E. A. Ibrahim, S. M. Rida, Y. A. Beltogy, and M. M. Abdelkhalek, *Pharmazie*, 1974, **29**, 581.
[127] S. Morimoto, H. Nomura, T. Ishiguro, T. Fugono, and K. Maeda, *Heterocycles*, 1974, **2**, 61.
[128] H. Nomura, T. Fugono, T. Hitaka, I. Minami, T. Azuma, S. Morimoto, and T. Masuda, *J. Medicin. Chem.*, 1974, **17**, 1312.
[129] H. Nomura, T. Fugono, T. Hitaka, I. Minami, T. Azuma, S. Morimoto, and T. Masuda, *Heterocycles*, 1974, **2**, 67.
[130] J. Altman, E. Karoly, and N. Maoz, *J. Medicin. Chem.*, 1975, **18**, 627.
[131] F. Casey and G. P. Bodey, *J. Antibiotics*, 1974, **27**, 520.
[132] E. I. Edwards, R. Epton, and G. Marr, *J. Organometallic Chem.*, 1975, **85**, C23.
[133] E. I. Edwards, R. Epton, and G. Marr, *J. Organometallic Chem.*, 1976, **107**, 361.
[134] G. L. Dunn, J. R. E. Hoover, D. A. Berges, J. J. Taggart, L. D. Davis, E. M. Dietz, D. R. Jakas, N. Yim, P. Actor, J. V. Uri, and J. A. Weisbach, *J. Antibiotics*, 1976, **29**, 65.
[135] R. M. DeMarinis, J. R. E. Hoover, G. L. Dunn, P. Actor, J. V. Uri, and J. A. Weisbach, *J. Antibiotics*, 1975, **28**, 463.
[136] J. M. Essery, U. Corbin, V. Sprancmanis, L. B. Crast, R. G. Graham, P. F. Misco, D. Willner, D. N. McGregor, and L. C. Cheney, *J. Antibiotics*, 1974, **27**, 573.
[137] D. A. Berges, *J. Medicin. Chem.*, 1975, **18**, 1264.
[138] C. H. O'Callaghan, R. B. Sykes, D. M. Ryan, R. D. Foord, and P. W. Muggleton, *J. Antibiotics*, 1976, **29**, 29.
[139] I. Buskooszczapowicz, J. Kazimierczak, and J. Cieslak, *Roczniki Chem.*, 1974, **48**, 253.
[140] I. Buskooszczapowicz, J. Kazimierczak, and J. Cieslak, *Acta Polon. Pharm.*, 1975, **32**, 43.
[141] G. A. Koppel, *Tetrahedron Letters*, 1974, 2427.

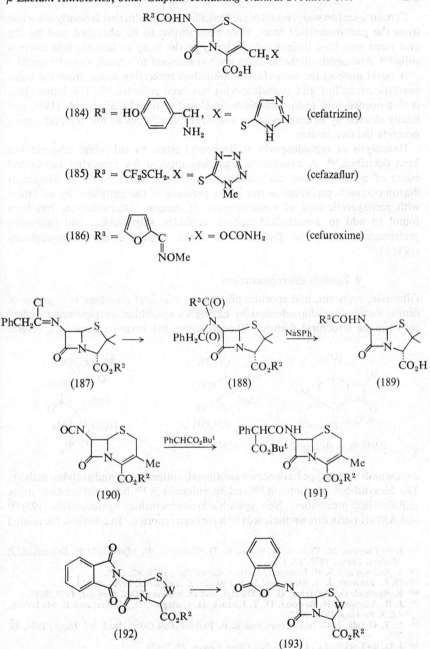

(184) R³ = HO—⟨ ⟩—CH—, X = (structure) (cefatrizine)
$\quad\quad\quad\quad\quad\quad\quad$ NH₂

(185) R³ = CF₃SCH₂, X = (structure) (cefazaflur)

(186) R³ = (structure), X = OCONH₂ (cefuroxime)

(187) → (188) → (189)

(190) → (191)

(192) → (193)

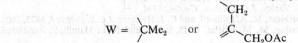

W = CMe₂ or C(structure)CH₂OAc

Certain α-carboxy-aryl esters of carbenicillin, a penicillin that is poorly absorbed from the gastrointestinal tract, have been shown to be absorbed well by the oral route and then hydrolysed efficiently in the body to liberate free carbenicillin.[142] Analogous aliphatic esters are too resistant to hydrolysis to be useful.

A novel method for removing the phthaloyl protecting group from the base-sensitive penicillins and cephalosporins has been reported.[143] The imide (192) is ring-opened with sodium sulphide, re-closed to the phthalisoimide (193), and finally cleaved by hydrazine under anhydrous conditions at low temperature to generate the free amine.

Hydrolysis of cephalosporin *p*-nitrobenzyl esters by microbial enzymes has been described.[144] A convenient high-yield method for preparing benzhydryl esters of penicillins and cephalosporins has been described.[145, 146] Diphenyl-diazomethane is generated *in situ* in the presence of the antibiotic by oxidation with peroxyacetic acid of benzophenone hydrazone. Diazomethane has been found to add to 3-methyl-Δ³-cephem sulphides, sulphoxides, and sulphones preferentially from the β-side to give the respective pyrazolino-cephams (194; b).[147]

8 Epidithiodioxopiperazines

Gliotoxin, aranotin, and sporidesmin are the principal members of a group of fungal metabolites characterized by having an epidithiodioxopiperazine moiety as a unique structural feature. Though somewhat toxic to mammals, several

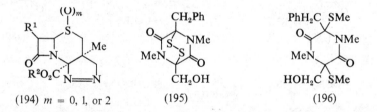

(194) m = 0, 1, or 2   (195)   (196)

compounds of this type have shown antifungal, antibacterial, and antiviral activity. The biosynthesis of gliotoxin[148] and sporidesmin A[149] has been studied, using radiolabelled precursors. New epidithiodioxopiperazines hyalodendrin (195)[150] and A30641 (structure undisclosed)[151] have been reported. In addition, the related

[142] J. P. Clayton, M. Cole, S. W. Elson, K. D. Hardy, L. W. Mizen, and R. Sutherland, *J. Medicin. Chem.*, 1975, **18**, 172.
[143] S. Kukolja and S. R. Lammert, *J. Amer. Chem. Soc.*, 1975, **97**, 5582.
[144] D. R. Brannon, J. A. Mabe, and D. S. Fukuda, *J. Antibiotics*, 1976, **29**, 121.
[145] R. Bywood, G. Gallagher, G. K. Sharma, and D. Walker, *J.C.S. Perkin I*, 1975, 2019.
[146] J. R. Adamson, R. Bywood, D. T. Eastlick, G. Gallagher, D. Walker, and E. M. Wilson, *J.C.S. Perkin I*, 1975, 2030.
[147] E. T. Gunda, J. C. Jaszberenyi, and E. R. Farkas, *Acta Chim. Acad. Sci. Hung.*, 1974, **83**, 205.
[148] J. D. Bu'Lock and C. Leigh, *J.C.S. Chem. Comm.*, 1975, 628.
[149] G. W. Kirby and M. J. Varley, *J.C.S. Chem. Comm.*, 1974, 833.
[150] G. M. Strunz, M. Kakushima, M. A. Stillwell, and C. J. Heissner, *J.C.S. Perkin I*, 1973, 2600.
[151] D. H. Berg, R. P. Massing, M. M. Hoehn, L. D. Boeck, and R. L. Hamill, *J. Antibiotics*, 1976, **29**, 394.

metabolite gliovictin (196),[152, 153] having the disulphide bridge ruptured, has been isolated from two fungal sources.

The synthesis of the natural epidithiodioxopiperazines and related structures has received considerable attention. The total syntheses of (±)-hyalodendrin (195)[154] and (±)-sporidesmin B (197)[155] have been described. Both syntheses

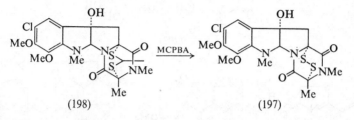

(198)   (197)

utilized an elegant oxidative ring-closure reaction in forming the disulphide bridge. For example, dithioketal (198) may be treated with *m*-chloroperoxybenzoic acid and then with boron trifluoride to give sporidesmin B (197).

Other synthetic studies have been directed at the introduction of sulphur-containing substituents into the dioxopiperazine ring *via* addition of thiols and H_2S to cyclodipeptides.[156-160] Syntheses of desthiomethylene- (199) and desthioseco- (200) analogues in the gliotoxin series have been described.[161-163]

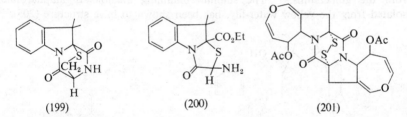

(199)   (200)   (201)

The reactivity of the disulphide linkage of acetylaranotin (201) towards a number of reagents (*e.g.* S, HCN, MeSH) has been investigated.[164]

9 Other Sulphur-containing Natural Products

The structure of griseoviridin (202), a broad-spectrum antibiotic isolated from *Streptomyces griseus*, has been established unequivocally, using *X*-ray

[152] F. Dorn and D. Arigoni, *Experientia*, 1974, **30**, 134.
[153] G. M. Strunz, C. J. Heissner, M. Kakushima, and M. A. Stillwell, *Canad. J. Chem.*, 1974, **52**, 325.
[154] G. M. Strunz and M. Kakushima, *Experientia*, 1974, **30**, 719.
[155] S. Nakatsuka, T. Fukuyama, and Y. Kishi, *Tetrahedron Letters*, 1974, 1549.
[156] P. J. Machin and P. G. Sammes, *J.C.S. Perkin I*, 1974, 698.
[157] J. Housler and U. Schmidt, *Chem. Ber.*, 1974, **107**, 2804.
[158] U. Schmidt, A. Perco, and E. Ohler, *Chem. Ber.*, 1974, **107**, 2816.
[159] E. Ohler and U. Schmidt, *Chem. Ber.*, 1975, **108**, 2907.
[160] H. Poisel and U. Schmidt, *Chem. Ber.*, 1975, **108**, 2917.
[161] H. E. Ottenheizin, J. A. Hulshof, and R. J. Nivard, *J. Org. Chem.*, 1975, **40**, 2147.
[162] H. C. J. Ottenheijm, A. D. Potman, and T. Vanvroonhoven, *Rec. Trav. chim.*, 1975, **94**, 135.
[163] H. C. J. Ottenheijm, J. J. M. L. Hoffman, P. T. M. Biessels, and A. D. Potman, *Rec. Trav. chim.*, 1975, **94**, 138.
[164] K. C. Murdock, *J. Medicin. Chem.*, 1974, **17**, 827.

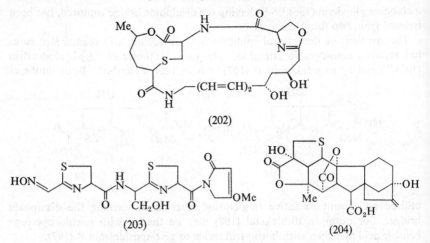

(202)

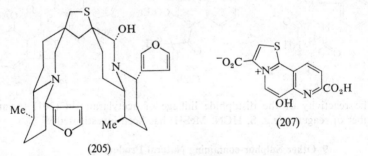

(203)   (204)

techniques.[165] Another broad-spectrum antibiotic, althiomycin, isolated from two species of *Streptomyces*, has been shown to have structure (203) by chemical and spectral methods.[166] Pharbitic acid (204), a diterpenoid isolated from seeds of the Japanese morning-glory, *Pharbitis nil*, is believed to arise biogenetically from the gibberellins.[167] The sulphur-containing alkaloid thionupharoline, isolated from the yellow water-lily, has been shown to have structure (205).[168]

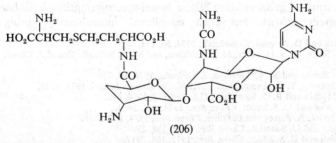

(205)   (207)

(206)

[165] G. I. Birnbaum and S. R. Hall, *J. Amer. Chem. Soc.*, 1976, **98**, 1926.
[166] B. W. Bycroft and R. Pinchin, *J.C.S. Chem. Comm.*, 1975, 121.
[167] T. Yokota, S. Yamazaki, N. Takahashi, and Y. Iitaka, *Tetrahedron Letters*, 1974, 2957.
[168] T. J. Martin, D. B. Maclean, J. T. Wrobel, A. Jwanow, and W. Starzec, *Canad. J. Chem.*, 1974, **52**, 2705.

The structure of an antifungal antibiotic, ezomycin A (206), isolated from a *Streptomyces* species, has been established, using degradative and spectral evidence.[169]

Berninamycinic acid, a degradation product of the antibiotic berninamycin, has been shown to have structure (207) by *X*-ray analysis.[170]

An elegant nineteen-step stereospecific total synthesis of *d*-biotin (208) from its biogenetic precursor L-(+)-cysteine has been described [171] (see Chapter 5, p. 221). Incorporation of [3-¹³C]propionate into the propionamide of the antibiotic ureothricin (209) has been demonstrated, using ¹³C n.m.r. techniques.[172]

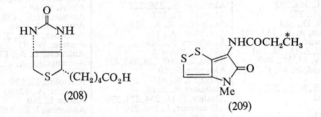

(208)   (209)

[169] K. Sakata, A. Sakurai, and S. Tamura, *Tetrahedron Letters*, 1974, 4327.
[170] J. M. Liesch, J. A. McMillan, R. C. Pandey, I. C. Paul, and K. L. Rinehart, jun., *J. Amer. Chem. Soc.*, 1976, **98**, 299.
[171] P. N. Confalone, G. Pizzolato, E. G. Baggiolini, D. Lollar, and M. R. Uskokovic, *J. Amer. Chem. Soc.*, 1975, **97**, 5936.
[172] M. Yamazaki, Y. Maebayashi, F. Katoh, and Y. Koyama, *J. Pharm. Soc. Japan*, 1975, **95**, 347.

Author Index

496

506   *Author Index*